Mutation Breeding, Genetic Diversity and Crop Adaptation to Climate Change

Mutation Breeding, Genetic Diversity and Crop Adaptation to Climate Change

Edited by Shoba Sivasankar[1], Noel Ellis[2], Ljupcho Jankuloski[1] and Ivan Ingelbrecht[1]

[1]Joint FAO/IAEA Centre, Department of Nuclear Sciences and Applications, Vienna, Austria
[2]School of Biological Sciences, University of Auckland, New Zealand

CABI

CABI is a trading name of CAB International

CABI
Nosworthy Way
Wallingford
Oxfordshire OX10 8DE
UK

Tel: +44 (0)1491 832111
Fax: +44 (0)1491 833508
E-mail: info@cabi.org
Website: www.cabi.org

CABI
WeWork
One Lincoln Street
24th Floor
Boston, MA 02111
USA

Tel: +1 (617)682-9015
E-mail: cabi-nao@cabi.org

A catalogue record for this book is available from the British Library, London, UK.

Open access provided with a grant from the International Atomic Energy Agency. The opinions expressed in this article are those of the authors and do not necessarily reflect the views of International Atomic Energy Agency nor its Member States. The IAEA has no responsibility for the persistence or accuracy or URLs for external or third-party Internet websites referred to in this publication and does not guarantee that any content on such website is or will remain accurate or appropriate.

Library of Congress Cataloging-in-Publication Data

Names: Sivasankar, Shoba, editor. | Ellis, Thomas Henry Noel, editor. | Jankuloski, Ljupcho, editor. | Ingelbrecht, Ivan, editor.
Title: Mutation breeding, genetic diversity and crop adaptation to climate change / edited by Shoba Sivasankar, Noel Ellis, Ljupcho Jankuloski, Ivan Ingelbrecht.
Description: Oxfordshire : CAB International, [2021] | Includes bibliographical references and index. | Summary: "International authors review achievements, new developments, trends and challenges in the field of plant mutation breeding, across the scientific community and the private sector. Chapters highlight specific challenges, such as emerging transboundary threats to crop production, and assess the overall importance of mutation breeding to food security"-- Provided by publisher.
Identifiers: LCCN 2021024084 (print) | LCCN 2021024085 (ebook) ISBN 9781789249095 (hardback) | ISBN 9781789249101 (ebook) | ISBN 9781789249118 (epub)
Subjects: LCSH: Plant mutation breeding. | Mutagenesis. | Plant biotechnology.
Classification: LCC SB123 .M88 2021 (print) | LCC SB123 (ebook) | DDC 572.8/38--dc23
LC record available at https://lccn.loc.gov/2021024084
LC ebook record available at https://lccn.loc.gov/2021024085

References to Internet websites (URLs) were accurate at the time of writing.

ISBN-13: 978 1 78924 909 5 (hardback)
978 1 78924 910 1 (ePDF)
978 1 78924 911 8 (ePub)

DOI: 10.1079/9781789249095.0000

Commissioning Editor: Rachael Russell/Rebecca Stubbs
Editorial Assistant: Emma McCann
Production Editor: Marta Patiño

Typeset by SPi, Pondicherry, India
Printed and bound in the UK by Severn, Gloucester

Contents

*Corresponding author (for contact details, see Contributors list, below)

Contributors (corresponding authors only)

Abdalla, Elgailani
El-Obeid Research Station, PO Box 429, El-Obeid, North Kordofan State, Sudan
elgailani_ers@hotmail.com
Abe, Tomoko
Ion Beam Breeding Team, Beam Mutagenesis Group, RIKEN Nishina Center for Accelerator-Based Science, Wako, Saitama 351-0198, Japan
tomoabe@riken.jp
Andrade, Alexander de
Empresa de Pesquisa Agropecuária e Extensão Rural de Santa Catarina, Estação Experimental de Itajaí (EEI), Rodovia Antônio Heil 6800, Itajaí, SC, Brazil, CEP: 88318-112
alexanderandrade@epagri.sc.gov.br
Azad, Md. Abul Kalam
Plant Breeding Division, Bangladesh Institute of Nuclear Agriculture, BAU Campus, Mymensingh-2202, Bangladesh
makazad.pbdbina@gmail.com
Bado, Souleymane
Phytoprise GmbH, Department of Plant Science, DI Rudolf Schober-Straße 4, A-7400 Oberwart, Austria.
bado_souleymane@hotmail.com
Bagheri Leila
Nuclear Agriculture Research School, Nuclear Science and Technology Research Institute, Atomic Energy Organization, Karaj, Iran
llbagheri@aeoi.org.ir
Barquero, Miguel
Investigation unit, Costa Rican Coffee Institute (ICAFE), Heredia, Costa Rica
mbarquero@icafe.cr
Begum, Shamsun Nahar
Plant Breeding Division, Bangladesh Institute of Nuclear Agriculture, BAU Campus, Mymensingh-2202, Bangladesh
sbluna98@yahoo.com
Choudhary, Shashi Bhushan
ICAR-Central Research Institute for Jute and Allied Fibres, Barrackpore, Kolkata-700121, India
shashigen@gmail.com

Costa de Oliveira, Antonio
Centro de Genômica e Fitomelhoramento, Universidade Federal de Pelotas, Capão do Leão, Brasila, Brazil
costol@cgfufpel.org
Engel, Karl-Heinz
Chair of General Food Technology, Technical University of Munich, Maximus-von-Imhof-Forum 2, D-85354 Freising-Weihenstephan, Germany
k.h.engel@wzw.tum.de
Gatica-Arias, Andrés
Laboratorio Biotecnología de Plantas, Escuela de Biología, Universidad de Costa Rica, 2060-11501, San Pedro de Montes de Oca, Costa Rica
andres.gatica@ucr.ac.cr
Geras'kin, Stanislav
Russian Institute of Radiology and Agroecology, Kievskoe shosse, 109 km, 249020, Obninsk, Russia
stgeraskin@gmail.com
Gomez-Pando, Luz
Cereals and Native Grains Research Program – Agronomy Faculty, National Agrarian University La Molina, La Molina, Lima Peru
luzgomez@lamolina.edu.pe
Guo, Huijun
Institute of Crop Sciences, Chinese Academy of Agricultural Sciences / National Center of Space Mutagenesis for Crop Improvement, 12 Zhongguancun South Street, Beijing 100081 China
guohuijun@caas.cn
Gustafson, J. Perry
University of Missouri, Columbia, Missouri 65211 USA
pgus@missouri.edu
Human, Soeranto
Pusat Aplikasi Isotop dan Radiasi (PAIR), Badan Tenaga Nuklir Nasional (BATAN) of Indonesia, Jl. Lebak Bulus Raya No. 49, Jakarta 12440
Soeranto@batan.go.id
Hussain, Manzoor
Cotton Breeding Group, Nuclear Institute for Agriculture & Biology (NIAB), Jhang Road, Faisalabad, Pakistan
mhmanj@gmail.com
Ibrahim, Rusli
Malaysian Nuclear Agency, Ministry of Science, Technology and Innovation (MOSTI), Bangi 43000 Kajang, Selangor, Malaysia
rusli.anm@gmail.com
Islam, Mirza Mofazzal
Plant Breeding Division, Bangladesh Institute of Nuclear Agriculture, BAU Campus, Mymensingh-2202, Bangladesh
mirza_islam@yahoo.com
Jambhulkar, Sanjay J.
Nuclear Agriculture & Biotechnology Division, Bhabha Atomic Research Centre, Mumbai-400085, India.
sjj@barc.gov.in
Jankowicz-Cieslak, Joanna
Joint FAO/IAEA Centre, Department of Nuclear Sciences and Applications, Vienna, Austria
j.jankowicz@iaea.org
Jankulovska, Mirjana
Faculty of Agricultural Sciences and Food-Skopje, Ss. Cyril and Methodius University in Skopje, 16-ta Mak. brigada no.3, 1000 Skopje, Republic of North Macedonia
mirjanajankulovska@yahoo.com

Jegadeesan, Souframanien
Nuclear Agriculture & Biotechnology Division, Bhabha Atomic Research Centre, Trombay, Mumbai-400085, India
souf@barc.gov.in
Karunananda, Dayani P.
Horticultural Crops Research and Development Institute, Gannoruwa, Peradeniya, Sri Lanka
dayani.karunananda@gmail.com
Kumar, Arun
Department of Genetics & Plant Breeding, Birsa Agricultural University, Ranchi, India
arun.bau4@yahoo.co.in
Kumar, Vikash
Nuclear Agriculture and Biotechnology Division, Bhabha Atomic Research Centre, Trombay, Mumbai 400085, India
kvikash@barc.gov.in, vikash007barc@gmail.com
Landau, Alejandra
Instituto de Genética "Ewald A. Favret", CICVyA (Centro de Investigación en Ciencias Veterinarias y Agronómicas), Instituto Nacional de Tecnología Agropecuaria, Nicolás Repetto y de los Reseros s/n (1686) Hurlingham, Buenos Aires, Argentina
landau.alejandra@inta.gob.ar
Latado, Rodrigo R.
Centro de Citricultura Sylvio Moreira / IAC, Rod. Anhanguera, km 158, Cx. Postal 4, Cordeirópolis, SP, CEP 13490-970, Brazil
rrlatado@gmail.com
Liu, Luxiang
Institute of Crop Sciences, Chinese Academy of Agricultural Sciences/National Center of Space Mutagenesis for Crop Improvement, 12 Zhongguancun South Street, Beijing 100081, China
liuluxiang@caas.cn
Lundqvist, Udda
Nordic Genetic Resource Centre (NordGen); Smedjevägen 3, SE-23053 Alnarp, Sweden
udda.lundqvist@nordgen.org
Malathy Parasuraman
Horticultural Crops Research and Development Institute, Department of Agriculture, Sri Lanka
mpmalathy@yahoo.com
Manova, Vasilissa
Genome Dynamics and Stability Laboratory, Institute of Plant Physiology and Genetics, Bulgarian Academy of Sciences, "Acad. G. Bonchev" Str., Bldg. 21, Sofia 1113, Bulgaria
vmanova@bas.bg
Mba, Chikelu
Plant Production and Protection Division, Food and Agriculture Organization of the United Nations (FAO), Viale delle Terme di Caracalla, 00153 Rome (RM), Italy
chikelu.mba@fao.org
Morales, Yonis
Genetic Improvement Program of the Honduran Coffee Institute (IHCAFE), Research and Development Department, Dr. Jesús Aguilar Paz Research and Training Center, La Fé, Ilama, Santa Bárbara, Honduras
ymorales71@gmail.com
Nakagawa, Hitoshi
Central Research Laboratory, Hamamatsu Photonics K.K., 5000 Hirakuchi, Hamakita-ku, Hamamatsu, Shizuoka, 434-8601 Japan
hitoshi.nakagawa@crl.hpk.co.jp
Oliveira, Antonio Costa de
Centro de Genômica e Fitomelhoramento, Universidade Federal de Pelotas, Capão do Leão, Brasila, Brazil.
acostol@terra.com.br

Pareek, Ashwani
Stress Physiology and Molecular Biology Laboratory, School of Life Sciences, Jawaharlal Nehru University, New Delhi 110067, India
ashwanip@mail.jnu.ac.in
Puripunyavanich, Vichai
Thailand Institute of Nuclear Technology, 9/9 Saimoon Ongkharak NakhonNayok, 26120, Thailand
vichaipuri@gmail.com
Rabefiraisana, Harimialimalala Jhonny
Plant Physiology Laboratory, Faculty of Sciences, University of Antananarivo, Madagascar
nyjahmalayah@mail.com
Roux, Nicolas
Bioversity International, Parc Scientifique Agropolis II, 34397 Montpellier Cedex 5, France
n.roux@cgiar.org
Saif, Abdulwahid
Department of Crops, Northern Highlands Research Station, Agricultural Research & Extension Authority (AREA), Yemen
Amozaid@yahoo.com
Schultheiss, Sebastian J.
Computomics GmbH, Machine Learning-based Data Analysis, Eisenbahnstr. 1, 72072 Tübingen, Germany
sebastian.schultheiss@computomics.com
Sevanthi, Amitha Mithra V.
ICAR-National Institute for Plant Biotechnology, Pusa Campus, New Delhi – 110112, India
amithamithra.nrcpb@gmail.com
Singh, Sarvjeet
Dept. of Plant Breeding & Genetics, Punjab Agricultural University, Ludhiana – 141004, Punjab, India
sarvjeetm@rediffmail.com
Sivasankar, Shoba
Joint FAO/IAEA Centre, Department of Nuclear Sciences and Applications, IAEA, 1400 Vienna, Austria
s.sivasankar@iaea.org
Tadele, Zerihun
University of Bern, Institute of Plant Sciences, Alternbergrain 21, 3013 Bern, Switzerland
zerihun.tadele@ips.unibe.ch
Thao Duc Le
Department of Mutation and Heterosis Breeding, Agricultural Genetics Institute, Pham Van Dong, Bac Tuliem, Hanoi, Vietnam
leducthao@agi.vaas.vn
Till, Bradley J.
Veterinary Genetics Laboratory, University of California, Davis, United States
till.brad@gmail.com
Tomlekova, Nasya
Molecular Biology Laboratory, Department of Breeding, Maritsa Vegetable Crops Research Institute, 32, Brezovsko shosse Str., 4003 Plovdiv, Bulgaria
nasia.tomlekova@gmail.com
Yusop, Mohd Rafii
Institute of Tropical Agriculture and Food Security, Universiti Putra Malaysia 43400 Serdang, Selangor, Malaysia
mrafii@upm.edu.my
Zhou, Libin
(1) Biophysics Group, Institute of Modern Physics, Chinese Academy of Sciences, 509 Nanchang Road, Lanzhou, Gansu Province 730000, People's Republic of China
(2) University of Chinese Academy of Sciences, 19(A) Yuquan Road, Shijingshan District, Beijing 100049, People's Republic of China
(3) Innovation Academy for Seed Design, Chinese Academy of Sciences, 1 West Beichen Road, Chaoyang District, Beijing 100101, People's Republic of China
libinzhoulz@gmail.com

Foreword

The successful application of radiation for the induction of plant mutations for the first time in the early 20th century prompted a swathe of research and development in its use to generate novel genetic diversity for crop improvement. Until then, spontaneous mutations were the only source of genetic diversity available for the domestication and breeding of plants and animals. The year 2018 marked the 90th anniversary of the first report on the induction of mutations in plants by L.J. Stadler, a plant geneticist, specifically in barley, oats, wheat and maize through irradiation, enabled by the discovery of X-rays and gamma-rays at the turn of the 20th century. Since then, induced mutations have been used to generate novel genetic variation in many food, feed and cash crops, leading to the release of large numbers of mutant varieties for cultivation across the globe, thus contributing to food and nutrition security and to farmers' income. More recently, increasingly cost- and time-efficient genomics technologies have been facilitating the identification of the DNA variants underlying mutations. The identity of the genes responsible for the spontaneous mutations in wheat and rice that drove the Green Revolution of the 1960s were thus identified in the 1999–2002 timeframe.

In 2018, the 90th anniversary year of that first report on plant mutation induction, the Joint FAO/IAEA Centre of Nuclear Techniques in Food and Agriculture held the International Symposium on Plant Mutation Breeding and Biotechnology in Vienna, Austria. The Symposium reviewed the current state of knowledge in mutation breeding and associated biotechnologies and the impact of mutation breeding in the release and wide cultivation of improved mutant varieties of crops across the globe.

This volume includes peer-reviewed articles of selected presentations from the Symposium reflecting its five main sessions and plenaries. It presents reviews on the application of the technology for crop improvement towards food and nutrition security, and research status on mutation breeding and associated biotechnologies in both seed crops and vegetatively propagated crops. It also presents perspectives on the significance of next-generation sequencing and bioinformatics in determining the molecular variants underlying mutations and on emerging biotechnologies such as gene editing. Reviews and articles are organized into five sections in the publication: (1) Contribution of Crop Mutant Varieties to Food Security; (2) Mutation Breeding in Crop Improvement and Climate-Change Adaptation; (3) Mutation Induction Techniques for Enhanced Genetic Variation; (4) Mutation Breeding in Vegetatively Propagated and Ornamental Crops; and (5) Induced Genetic Variation for Crop Improvement in the Genomic Era.

The increasing threat posed by climate change on biodiversity loss, crop productivity and, consequently, food and nutrition security and farmers' livelihoods demands the generation of novel genetic variation for the development of improved varieties that can adapt to adverse climate events. Global food and nutrition security in the coming years will depend heavily on the availability and cultivation of improved crop varieties that can perform well under the pressures of increasing frequencies and intensities of drought, flooding and coastal salinity, warming temperatures and the transboundary spread of intensifying plant pests and diseases triggered by warming temperatures. Recent years are witness to the increasing application of newer mutagen sources such as ion beams, electron beams and cosmic rays (space mutagenesis) for the improvement of plant species, and of functional genomics technologies that establish gene-to-phenotype relationships for the discovery of molecular variants underlying mutations to be used in marker-assisted breeding and gene editing.

The contents of this volume present excellent reference material for researchers, students and policy makers involved in the application of induced genetic variation in plants for the maintenance of biodiversity and the acceleration of crop adaptation to climate change to feed a growing global population in the coming years and decades.

Qu Liang
Director
Joint FAO/IAEA Centre of Nuclear Techniques in Food and Agriculture
IAEA

Introduction

Shoba Sivasankar[1] and Noel Ellis

Crop improvement is key to food security, as the global population is poised to surpass the 9 billion mark by 2050 and as pressures of climate change create devastating crop losses. Genetic diversity is central to crop improvement. Mutations occurring in the genome, either spontaneous or induced, constitute the basis for this genetic diversity. Crop domestication and improvement for agriculture over the past 10,000 years have been driven by spontaneous mutation combined with selection for desirable crop traits such as yield, plant and inflorescence architecture, non-shattering of seeds, etc. Induced mutations have been used to create novel genetic diversity in plants since as early as 1928 and systematic mutation breeding has been used for crop improvement for about 70 years. The Plant Breeding and Genetics team at the Joint FAO/IAEA Division of Nuclear Techniques in Food and Agriculture (NAFA) have been supporting Member States since 1964 to establish viable mutation breeding programmes to develop new and improved crop varieties through induced genetic variation that provides a wide genetic base for selection. This support is through technology development and transfer in the areas of mutation induction in seed and vegetatively propagated crops, tissue culture and single-cell regeneration methods, phenotypic/molecular selection, speed-breeding and associated biotechnologies. The Mutant Variety Database of the Joint FAO/IAEA Division currently holds more than 3300 voluntarily contributed records of mutant varieties from about 220 crop species across more than 70 countries. There is increasing emphasis on understanding the molecular variants underlying mutations for their use as markers in molecular breeding and as candidate genes in gene editing to accelerate the pace of crop improvement.

This volume is a compilation of articles from the FAO/IAEA International Symposium on Plant Mutation Breeding and Biotechnology held on 27–31 August 2018 in Vienna, Austria. The Symposium was attended by 350 delegates from 84 Member States, two non-Member States and four international organizations. It discussed current trends in mutation breeding and associated biotechnologies spanning five integral topics: (1) Contribution and Impact of Mutant Varieties on Food Security; (2) Mutation Breeding for Adaptation to Climate Change in Seed Propagated Crops; (3) Enhancing Agricultural Biodiversity through New Mutation Induction Techniques; (4) Mutation Breeding for Ornamental and Vegetatively Propagated Crops; and (5) New Challenges and Technologies in Plant Genomics and Breeding. While the book is organized slightly differently, it captures the above topics. As evident from the range of chapters

[1] Email: S.Sivasankar@iaea.org

presented and their contents, the two critical areas requiring continued research attention and technology development in mutation breeding are: (i) the induction of stable genetic variation in vegetatively propagated crops; and (ii) the development or fine-tuning of efficiency-enhancing techniques, including doubled haploidy, rapid cycling, marker-assisted selection and the requisite computational biology and bioinformatics.

The contents of this book bear witness to the contributions of mutation breeding and associated biotechnologies to the development of improved crop varieties, and thereby to food security, and to the application of new and emerging technologies in the breeding pipeline. A few cautionary notes remain important in experimental procedures and interpretations in mutation breeding for crop improvement. These are described below under two categories: (i) mutagenesis; and (ii) data analysis. While the latter has wide relevance for accurate selections in any type of breeding, the discussion here pertains specifically to the selection of mutants.

Mutagenesis

The articles in this volume discuss a wide range of mutagens, both physical (notably those articles describing the effects of directed electron beams) and chemical. Each mutagen acts in different ways at the levels of the genome. Therefore, for reproducibility, it is important that certain crucial factors are always reported in results from mutation breeding experimentation: the nature of the mutagen, the irradiation dose or the mutagen/chemical concentration, dose rate, duration of the treatment and moisture content of the irradiated tissue. These data, while critical for gauging reproducibility, are not always reported in the literature, but can be found, for example, in the articles in this volume by Nakagawa (Chapter 3), Abe *et al.* (Chapter 42) or Jankowicz-Cieslak *et al.* (Chapter 38).

The dose **rate** is significant because it can have an effect, for example, on the LD_{50}. Radiation doses are given in Grays (Gy), i.e. the absorbed energy per unit mass (J/kg). How this relates to the number of mutations induced is complex. Obviously, the energy (or energy spectrum) of the radiation is very important, because the ratio of the energy of the photon or particle to the dose determines how many individual interactions between the target material and the radiation can occur. Furthermore, the mass of the irradiated material has a complex relation to the number of mutations induced for a given dose. For example, large seeds such as *Vicia faba* are mostly cotyledon and the mass of the apical meristem (from which the reproductive cells are derived) is a very small proportion of the mass of the sample, while the converse is true for species such as *Dendrobium* (Ibrahim, Chapter 7). Furthermore, the structure and developmental morphology of the apical meristem in the seed determines how many different meristematic cells may experience an independent mutation. In chemical mutagenesis, seed architecture (or the structure of the mutagenized material) is again important, because the accessibility of the mutagen to the apex may not be a simple function of concentration.

The pathway from exposure to a mutagen and the observation of a mutant form is complex, with many independent stochastic processes. Thus, the statistical variation in the number of observed mutants of a given type is expected to be high as a consequence of multiple stochastic processes. This point is critical in drawing conclusions about the relative merit of different mutagenic treatments, as it is difficult to be sure that mean frequencies are actually different when the variances are high.

Types of mutation

Different mutagens induce different types of mutation. What is especially of interest is the difference between mutagens that preferentially promote base substitution, such as the chemical mutagen ethyl methanesulfonate (EMS), versus those that preferentially promote deletion, such as gamma irradiation (see discussion in Jankowicz-Cieslak *et al.*, Chapter 38). Genome architecture is also important in determining whether particular mutant forms are observed, for example whether genes are duplicated or not. This is not necessarily a fixed feature of a given species. For this reason, conclusions about differences between genotypes with respect to radiation sensitivity require care in their interpretation if based on the frequency of a single phenotype.

Base substitutions are often silent and those that are missense or nonsense mutations (or that affect structural elements of a gene) will most often correspond to loss-of-function mutations. Deletion mutations will usually correspond to loss-of-function mutations. These comments are important for understanding the frequency with which different mutant classes might be observed.

In clonally propagated material as in Jankowicz-Cieslak *et al.* (Chapter 38) and Bado *et al.* (Chapter 36), induced mutations will be expected to be in the heterozygous condition. As most mutations are recessive, there are relatively few opportunities for finding desirable mutations. Missense mutations are more likely to create rare dominant mutants than those that generate deletions. When the genetic background is highly heterozygous, a loss-of-function mutation may unmask some variation previously hidden by a dominant allele. This may be an explanation for the success of mutagenesis in generating variants in ornamental species (Ibrahim, Chapter 7).

Epigenetic changes and somaclonal variants are discussed. These are a broad spectrum of variants that are not yet well defined. For many polyploid species that are propagated clonally, the possibility of non-disjunction leading to aneuploidy or the substitution of a chromosome by a homologue may reveal variation previously hidden by dominance or epistasis. Similarly, mitotic crossing over can lead to the appearance of a hidden recessive phenotype in clonally propagated material of either a diploid or a polyploid.

Data Analyses

What the statistical approaches analyse

Among the papers collected in this volume, a variety of different experimental designs are used to determine whether any useful variation can be found among the progeny of a mutagenized line. Examples include the exploratory statistical approach of Zhou *et al.* (Chapter 44), the use of R by Jankulovska *et al.* (Chapter 21), or an Analysis of Variance as used in many of the articles. Analysis of Variance partitions variation into various categories, and that which is explained by the difference between lines in replicates is often taken to represent genetic variation. Indeed, genetic variation would contribute to such differences, but this is not necessarily the only systematic difference between such replicates; clearly differences between seed batches may contribute. The article by Geras'kin *et al.* (Chapter 43) addresses this nicely by analysing the performance of different seed lots in response to low hermetic levels of radiation. This study, examining the beneficial effects of low-level irradiation, points to important differences that may be a consequence of slight differences in seed physiology. Though priming of seeds is one way to overcome such physiological differences for practical benefit, further research is needed.

Identifying real differences

Typically, trial data involve some form of randomization. Often, a formal structure such as a randomized complete block design is used that is then subject to a standard statistical analysis such as Analysis of Variance coupled with some method of statistical inference such as multiple range tests (Begum *et al.*, Chapter 20), some type of discriminant analysis (Jankulovska *et al.*, Chapter 21) or multivariate analysis (Yusop *et al.*, Chapter 23). No statistical analysis is perfect and the need to find real differences in noisy data (avoiding type I error) has to be balanced against detecting differences, when in fact there are none (type II error). Some statistical treatments are more tolerant of type I errors at the risk of type II errors and vice versa. Statistical analysis therefore requires a validation step.

Genetics

All of the issues above concern the selection of biological material that is worth the effort to investigate further, i.e. to validate (or refute) what appears to be of interest. The underlying assumption, in a mutagenesis approach to crop improvement, is that allelic differences at single loci can be detected that are either informative or useful. If it is suspected that such variation has been uncovered, then there are two strong and testable predictions. The first is that the difference will segregate in crosses and follow Mendelian rules; and the second is that independently detected variants may be allelic. Mutation breeding lends itself well to exploiting these predictions to identify reliable sources of useful genetic variation.

Severity scores

It should be recognized that severity scores for disease susceptibility, lodging intensity, etc. are categorical scores. A scale of 1–9, for example, is sometimes used. In a scheme of this type, average scores need to be treated with extreme caution, because the difference between successive categories is not necessarily the same and it would be possible to arrive at the same mean from different sets of categories.

Nomenclature

In most cases the same nomenclature for mutant generations is used in these articles and the term M_1 is used to refer to the seed (or other tissue) immediately post mutagenesis. Exceptions to this scheme are noted. These are in fact the same biological generation as the non-mutagenized seed, but on selfing they generate progeny that segregate for the mutations that were generated by the mutagen and were in a heterozygous condition. Thus the next generation, the M_2, behaves like the F_2 of a classical cross, except for the fact that the M_1 is chimeric, with some cells carrying a mutation and others not.

Summary

The chapters included here cover: (i) the contribution of crop mutant varieties to food security; (ii) mutation breeding in crop improvement and climate-change adaptation; (iii) mutation induction techniques for enhancing genetic variation; (iv) mutation breeding in vegetatively propagated and ornamental crops; and (v) mutation breeding in the genomic era. There has been significant success in the development of new improved crop varieties through mutation breeding, especially in the Asia Pacific region. Progress is also reported in the testing or adaptation of next-generation technologies for efficiency enhancement in mutation breeding. Successful identification and use of novel genetics for crop improvement require careful attention to the factors influencing the intensity and frequency of mutagenesis and to both the factors affecting selection and the analytical methods used in selection.

Section 1.

Contribution of Crop Mutant Varieties to Food Security

1 World Food Supply: Problems and Prospects

J. Perry Gustafson[1] and Peter H. Raven

Abstract

The current world population of 7.6 billion is projected to reach 9.9 billion by 2050. The UN projects that agricultural output will need to increase by 70% simply to maintain current dietary standards, which does not include improving the diets of approximately 800 million malnourished people. Agricultural production increased at a rate insufficient to reach the goal set by the 2009 World Summit on Food Security to reduce by one half the number of malnourished people in the world by 2015. In spite of declining poverty rates, achieving this reduction in the number of malnourished people will be very difficult, as it is likely that the projected 2.3 billion additional people will be among the poorest of the poor. Food imports are expected to increase despite projected increased production, with many poor countries unable to afford those imports. Agriculture can improve sustainable world food production on the land currently under production and by doing so protect our fragile environment as much as possible.

Keywords: population growth • food supply • agriculture

1 Introduction

When our ancestors adopted crop agriculture, some 10,500 years ago, the entire global population amounted to about 1 million people, with only about 100,000 in all of Europe. It took until approximately 1810 for the world's population to reach 1 billion people for the first time, and today it has grown to more than 7.6 billion, increasing by about 200,000 per day, and is estimated to reach 9.9 billion, an increase of 29%, by 2050 (PRB, 2018) and then perhaps 11 billion people by the end of the century, thus placing tremendous pressures on increasing world food production of all kinds. The United Nations (UN) projects that by 2050 agriculture will need to increase production by 70% (Alexandratos and Bruinsma, 2010) in order to maintain the same dietary standards that we have today, but the world would still have more than 800 million severely undernourished people and more than 100 million living near starvation. In 2018, world cereal production was estimated to be 2.56 billion tonnes, thus the world's farmers will need to produce 3 billion tonnes by 2050 to meet the FAO projections – and according to Fischer *et al.* (2014) it is likely this can be achieved. As approximately 2.3 billion poor people could be added during that same period, it is just a rough estimate to give what increases

[1] Email: pgus@missouri.edu

DOI: 10.1079/9781789249095.0001

will be needed. Since 1990, the number of undernourished people has risen to 815 million on a global scale but did actually decline from 900 million in 2000. Over the whole period from 1990 to 2010, world food production per capita increased by 12% (Barrett, 2010). Unfortunately, agricultural production was projected to increase by only about 1.3% per year during the period to 2030 (Mann, 1999). Fischer *et al.* (2014) placed it higher, at around 1.5%, for cereals. Therefore, it did not seem possible for agricultural output to come close to reaching the 2009 World Summit on Food Security goal of reducing by 50% the number of hungry and malnourished people in the world by 2015 (FAOSTAT, 2010). The World Bank (2008) estimated that in 1990 approximately 1.13 billion people worldwide were living in poverty, but 10 years later this figure was thought to have dropped to about 800 million. In spite of all our technology and agricultural improvements, in 2010 roughly 1 billion people still faced hunger on a daily basis. However, one must look very closely at the numbers involved. Some claim that there is no longer a major world food problem, as currently evidenced by the improvements in many developing countries, where the middle classes grow and there is a decrease in the percentage of poor people. This is true if we look at the percentages, but we should not look at percentages; instead, we need to look at the numbers, since there are about 1 billion impoverished people today when the world population is over 7.6 billion (14%), but when the world population reaches over 9 billion people, all projections show a decline in percentage of poor people.

Even with the projected increase in food production, food imports, on a world scale, will continue to increase; for example, wheat imports were projected to increase from 30 to 75 million tonnes by 2020 (Pingali and Rosegrant, 1998). However, the US Department of Agriculture (USDA) projected that increases in world wheat imports for 2018 would be about 170 million tonnes. Imports cannot solve world food problems within a ceiling of inadequate production, especially since only countries that are wealthy enough will be able to afford them.

In the past, as the human population grew, crops and domesticated animals provided a stable source of food; villages, towns and cities were formed and within them was developed the essence of what we consider civilization today. People could specialize as political or religious leaders, music makers, storytellers, historians and gradually into all of the myriad professions that we practise today. By about 5000 years ago, writing had been invented and history became recorded much more formally than earlier. What were once small groups of people became larger and more powerful, warring with one another for land and wealth. More than 200 million people have died in wars over the past two centuries, while our overall population grew from 1 billion to 7.6 billion. Humans are certainly the dominant species now and we often seem to fail to recognize our limits as we drain the world's potentially sustainable resources and compete with one another for arable land and food supplies, wealth and ever-higher levels of consumption.

Plants, once domesticated, were continually improved by selecting seeds and individuals that produced the highest yields and at the same time were sustainable. In 2010, FAOSTAT data showed a phenomenal increase in world food production over the preceding 47 years to about 7.99 billion tonnes of total food production, including from both plant (6.90 billion tonnes) and animal (1.09 billion tonnes) sources, on basically the same amount of land (USDA-NASS, 2010). The dramatic increase in world crop production, which supplies feed for domestic animals and food for human consumption, has mainly come from improved crop cultivars, technology advances and improved management practices. Over the next 40 years, about 80% of the required increase in world food production was projected to result from yield increases (67%) and higher cropping intensities, i.e. better management (12%), with the remaining 21% coming from a minimal expansion of arable land under crop production (World Bank, 2008).

2 Requirements and Impediments

The main objectives in feeding the world's population include, first, the task of increasing world food production, and improving dietary standards of the chronically undernourished, for the expanding population. Secondly, even if agriculture can accomplish the daunting task of increasing production, we will have to deal with the overwhelming task of improving and expanding the world's infrastructure in order to distribute food equitably to all regions of the world to offset the current consequences of population growth and increasing world hunger. We will never see a lasting solution to the world food/hunger and poverty problem without a strong balance between food production and distribution – in other words, social justice. Thirdly, agriculture needs to accomplish these objectives with a minimum impact on the world's biodiversity and our fragile environment. Fourthly, the increased food production needs to be accomplished without significantly expanding existing levels of land under cultivation. Approximately 11% of the world's landmass is used for crop production, with an estimated 22% more used for pasture, most of which is natural grassland and manifestly unsustainable for crop production. No more than 10% of additional arable land is potentially available even for limited crop production (Bruinsma, 2003). The massive increase in food production between 1963 and 2005 was accomplished utilizing about the same amount of land as was under production in 2010 (USDA-NASS, 2010). For example, world grain yields more than doubled from 1.4 tonnes per hectare (t/ha) in 1961–1963 to 3.05 t/ha in 1997–1999, at the same time as the amount of land required for producing the increase in grain yield actually declined by approximately 56% (World Bank, 2008). About one-eighth of our necessary protein comes from the sea, but this supply does not appear elastic in relation to demand, and again, most of it is taken by wealthy countries.

There is clear scope for increases in world food production on existing agricultural land and these increases should be feasible utilizing existing and newly developed technology. It was estimated that in parts of South-east Asia, the average rice yields were only 60% of their average maximum climate-adjusted yields (Godfray *et al.*, 2010). In addition, 11 countries were producing 37% of the total world wheat tonnage on predominantly rainfed production conditions, well under their attainable yield potential (World Bank, 2008). If farmers around the world were able to produce closer to their potential yield, the world's production levels should significantly increase without any additional land being brought into cultivation. This increase in production should equal over 23% of the current world production. As has been shown in the past 50 years, any local or national improvement in cultivars and management will spill over into the rest of the world. The world needs to understand that agriculture is not a local or national industry and any changes that are made on a local level will have global impacts. On the other hand, the droughts in Australia and Argentina seem tied to global climate change, as does the irregularity of the monsoon season in India. Therefore, food production increases will surely be subject to a number of additional constraints yet to be defined.

Even though this discussion indicates the potential for major increases in world agricultural production on existing cultivated land, based on existing and newly developed technology, there are at least three impediments that could limit increased world food production.

1. The technology and management improvements required for advancing yield might not be accessible or applicable to all crops, regions and farmers of the world.
2. Advanced technology and management inputs in agriculture could spread into areas of the world where they could accelerate environmental problems and have an adverse impact on biodiversity.
3. The understanding of agriculture by the public and by farmers certainly needs to be vastly improved for productivity to be increased, because we need a cooperative environment where agriculture can grow as increased numbers of people move into cities and tend to place obstacles in the way of improving crop production or to pay insufficiently to support those who raise our food.

Intensifying agricultural technology on existing lands, therefore, has had and will continue to have a major role in preserving biodiversity and maintaining the sustainability of our fragile global environment overall.

3 Technology for the Creation and Deployment of Diversity

One vital need to eliminate hunger and poverty involves the preservation of sufficient genetic diversity in cultivated plants and their relatives to ensure that breeders have the capacity to create cultivars capable of resisting biotic and abiotic stresses and adapting to new environmental conditions. Existing, improved and newly developed biotechnological tools alongside traditional technology will play major roles in improving world food production, in the same manner as did the Green Revolution that occurred from the 1960s through the 1980s. This will amount to what Conway (1999) termed 'the doubly green revolution'. The main problem will be the degree to which individual countries, industrialized and developing, can manage existing and new technologies to adapt and improve their production with advantageous changes in agronomic practices and production costs without any adverse effects on the world's environment and biodiversity.

The domestication and development of crops over the past 10,500 years has involved progeny selection from the most productive individuals based on their phenotypes. Existing and new biotechnologies are taking plant improvement to new heights with the potential of greatly improving food quality and production. Existing and new agricultural technologies are composed of individual systems that involve both public and private breeders' plant selection programmes and involve many technologies, including the following.

1. Tissue culture, in which plants can be broken down into cell suspensions that are manipulated followed by regenerating plants, bypasses the traditional approaches to seed production.
2. The utilization of anther culture coupled with various chromosome-doubling techniques can successfully create double haploid populations, greatly reducing the time required to produce commercial cultivars.
3. Modern approaches to mutation technology have been and will continue to be successful in creating additional genetic variation necessary for crop improvement programmes.
4. More recently, the utilization of molecular marker-assisted selection, where various types of DNA marker systems are linked to traditionally difficult-to-screen value-added traits of interest, have already been successfully used in cultivar improvement programmes (for an excellent review see Tester and Langridge, 2010). Among the techniques most commonly being used today are: restriction fragment length polymorphism (RFLP) and amplified fragment length polymorphism (AFLP), which at the present time are not being widely used for marker-assisted selection; simple sequence repeat or microsatellite repeat (SSR); single nucleotide polymorphism (SNP); and diversity array technologies (DArT).
5. The application of genome-wide selection is mainly used for selection involving quantitative traits in both animals and plants (Goddard and Hayes, 2009; Mayor and Bernardo, 2009). However, genome-wide selection requires the availability of a massive number of very cheaply available markers (up to 20 markers per centimorgan (cM, the unit for measuring genetic linkage)) or a high-density chip specifically for the species being manipulated, which can be very expensive technology. Also, when selecting on a genome-wide scale one has to consider the presence of considerable linkage drag.
6. The application of genomic sequencing of individual plant genomes to expose the location and potential function of the entire genetic composition of an organism can be used in conjunction with other technologies to assist cultivar improvement programmes.
7. Plant transformation technology creates genetically modified organisms (GMOs). This involves bypassing the sexual process to transfer genes from one organism to another and has already been successful in several crops.

A brilliant new technique for improving the agronomic characteristics of plants has emerged in the past two decades: the application of clustered regularly interspaced short palindromic repeats (CRISPR) and CRISPR-associated (Cas) systems (reviewed by Bomgardner, 2017), which is currently showing considerable potential applications in biotechnology crop improvements. The identification of a target site within a target gene of a plant species is quite straightforward using CRISPR technology, which requires the editing components to be delivered to plant cells. Using the current technology, it is common to introduce the selected genes as transgenes, but CRISPR methods are more direct. Their use is spreading rapidly and will clearly bring great benefit to the characteristics in the future, a tremendously important advance given the major changes in climate that we are facing over the decades to come. The ability to produce targeted mutations in any plant gene is extremely valuable in manipulating specific genes within a crop which are likely to improve its most critical features.

Most of the traditional and newly developed technologies have been adapted to work efficiently in a more land- and labour-intensive form of agriculture improvement. It is clear that traditional organic farming applications are not capable of producing enough to feed or improving dietary standards of the existing world population, let alone the projected increase to 9.9 billion people by 2050. It has been estimated that a world population of only 4 billion people could be sustained if organic nitrogen farming systems were in place on a global scale (Buringh and van Heemst, 1977; Smil, 2001, 2004; Conner, 2008), though requiring significantly more land under production to generate the required organic nutrients. All things considered, we are not really sure how many people the world can support sustainably, but we clearly need to develop sustainable, productive agriculture to the extent that we can while working with other factors leading to sustainability at the same time.

4 Discussion

Existing data indicates that agriculture is capable of feeding the projected increased world population on approximately the same amount of land that is currently under production (World Bank, 2008). However, it will take all of the available technologies and skills of plant and animal breeders and agronomists, coupled with the actions of farmers, to achieve the desired goal of eliminating world hunger. In addition, integrated pest management (IPM), water management, precision farming, limiting chemical input and many other techniques must be wisely applied and improved simultaneously to maximize yield from existing farmland. Only the co-ordinated application of all these techniques will improve the productivity of the lands currently cultivated to the degree needed.

Significant progress has been made over the past several years in advancing our knowledge of biology, which is being applied to technology adaptable to improving agricultural production. We do not know the direction current research will take, but we can assume that any commercial application is going to be determined by world economic and social factors. The private sector and many funding agencies are increasingly addressing the agricultural needs of developing countries. That a 'back to nature' or the 'pure organic' approach can feed the world's people is a theory that does not take into account the current world population and the scale of human suffering from malnutrition and starvation. Embracing social justice is the only way that people can really survive and prosper and we need to utilize all of our resources to accomplish the goal of feeding humanity.

All crops will continue to be improved by traditional and biotechnological approaches to advance their yield potential. Building adaptable gene complexes into crop varieties for the future is something that we must do, especially in the face of global climate change and the world's increasing population. This process will require a much larger number of cultivars, with different genetic backgrounds, to be spread around the world than in the past. In future crop development, several factors seem especially important.

1. We need to understand the characterization of the genome structure, gene function and regulation, and evolution at macro- and micro-geographical scales of all crops and animals.
2. We need to combine single- and multi-locus value-added traits to produce a higher degree of cultivar development.
3. Crop genetic systems must be analysed to determine the genetic flexibility of various species in diverse ecological contexts, according to their breeding systems, mutation rates, genome recombination properties, genomic distribution and function of structural genes (primarily abiotic and biotic stress genes).
4. We need to characterize the interface between developing agricultural ecological dynamics and adaptive ecosystems in order to manipulate genome composition and limit the potential for gene contamination.

In the past, when modern agriculture competed with the traditional subsistence forms of agriculture, local landrace cultivars were often discarded in favour of the new high-yielding cultivars. Recently, massive efforts have been undertaken to preserve crop and animal diversity, which has resulted in the characterization of more old and new landrace diversity of all kinds, and these are being more widely used in agriculture and breeding programmes than 50 years ago. Extraordinary efforts are being made in this area, but much crop diversity is being lost at the same time. National and international seed banks are and will continue to be critically important to agriculture and the maintenance of the world's biodiversity. The continued long-term health of world food production is one of the foundations to world security. The stable future of humanity, our environment and our biodiversity are intimately tied to the improvement of crop production. Adequately feeding the world's population is clearly one of the most important challenges facing the world today and in the future.

At the same time, and as rapidly as we can, we need to achieve a global population level that is sustainable, an achievement that will clearly depend in a major way on empowering women throughout the world. Efforts to do this are increasing in some areas, but they need to be greatly strengthened so that women everywhere can gain the right to control their own destinies, make whatever contributions they can to our common good and determine their own reproductive decisions. We need to achieve just and sustainable levels of consumption worldwide; five planets would be necessary if everyone lived at the current level of consumption in the USA, while at the same time many of the poorest nations are entering what Mathis Wackernagel, founder of the Global Footprint Network, has termed an 'environmental poverty trap' as a condition that will deny them the resources to import food. There is little evidence that wealthy nations will step in to feed the countries of sub-Saharan Africa, for example, which makes the concept of producing enough food for the whole world somewhat of an illusion. We need to care enough to stop global climate change if agriculture is to stay close to its present levels of productivity, let alone improve them. Overall, a world in which national greed is dominating as a policy can never be a world where everyone is adequately fed, regardless of how much food might be produced. We need social justice and love for one another if we are ever going to achieve such a goal.

References

Alexandratos, N. and Bruinsma, J. (2010) *World Agriculture Towards 2030/50*. Agricultural Development Economics Division, FAO, Rome. Revised and updated 2012 (ESA Working Paper no. 12-03).

Barrett, C.B. (2010) Measuring food insecurity. *Science* 327, 825–828.

Bomgardner, M.M. (2017) CRISPR: A new toolbox for better crops. *Chemical & Engineering News* 95 (24), 30–34.

Bruinsma, J. (ed.) (2003) *World Agriculture: Towards 2015–2030 An FAO Perspective*. Earthscan Publications, London.

Buringh, P. and van Heemst, H.D.J. (1977) *An estimation of world food production based on labour-oriented agriculture*. Centre for World Food Market Research, Amsterdam.

Conner, D.J. (2008) Organic agriculture cannot feed the world. *Field Crops Research* 106, 187–190.

Conway, G. (1999) The doubly green revolution: food for all in the twenty-first century. Cornell University Press, Ithaca, New York.

FAOSTAT (2010). Food and Agriculture Data. Available at http://www.fao.org/faostat/en/ (accessed 2 September 2021).

Fischer, R.A., Byerlee, D. and Edmeades, G.O. (2014) *Crop yield and global food security: will yield increase continue to feed the world? ACIAR Monograph 158*. Australian Centre for International Research, Canberra. Available at: http://aciar.gov.au/publication/mn158 (accessed 2 September 2021).

Goddard, M.E. and Hayes, S.J. (2009) Mapping genes for complex traits in domestic animals and their use in breeding programmes. *Nature Reviews Genetics* 10, 381–391.

Godfray, H.C. *et al.* (2010) Food security: The challenge of feeding 9 billion people. *Science* 327, 812–818.

Mann, C.C. (1999) Future food: crop scientists seek a new revolution. *Science* 283, 310–314.

Mayor, P.J. and Bernardo, R. (2009) Genomewide selection and marker-assisted recurrent selection in doubled haploid versus F2 populations. *Crop Science* 49, 1719–1725.

Pingali, P.L. and Rosegrant, M.W. (1998) Supplying wheat for Asia's increasingly westernized diets. *American Journal of Agricultural Economics* 80, 954–959.

PRB (2018) *2018 World Population Data Sheet*. Population Reference Bureau, Washington, DC. Available at: www.worldpopdata.org (accessed 24 March 2021)

Smil, V. (2001) *Feeding the World: a Challenge for the Twenty-first Century*. MIT Press, Cambridge, Massachusetts.

Smil, V. (2004) *Enriching the Earth: Fritz Haber, Carl Bosch, and the Transformation of World Food Production*. MIT Press, Cambridge, Massachusetts.

Tester, M. and Langridge, P. (2010) Breeding technologies to increase crop production in a changing world. *Science* 327, 818–822.

USDA-NASS (2010) United States Department of Agriculture – National Agricultural Statistics Service. Available at https://www.nass.usda.gov (accessed 2 September 2021).

World Bank (2008) *World Development Report: Agriculture for Development*. The World Bank, Washington, DC.

2 Scandinavian Mutation Research During the Past 90 Years – a Historical Review

Udda Lundqvist[1]

Abstract

In 1928, the Swedish geneticists Herman Nilsson-Ehle and Åke Gustafsson started to act on their own ideas with the first experiments with induced mutations using diploid barley. They started with X-rays and UV irradiation. Very soon the first chlorophyll mutations were obtained and followed by the first 'vital' mutations *Erectoides* (*ert*) (Franckowiak and Lundqvist, 2001). Several other valuable mutations were identified as early maturity, high yielding, lodging resistant and characters with altered plant architecture. The experiments expanded to include other different types of irradiation, followed by chemical mutagenesis starting with mustard gas and concluding with sodium azide. The research brought a wealth of observations of general biological importance, such as the physiological effects of radiation as well as the difference in the mutation spectrum with respect to mutagens.

This research was non-commercial, even if some mutants have become of important agronomic value. It peaked in activity during the 1950s to 1980s and, throughout, barley was the main experimental crop. About 12,000 different morphological and physiological mutants with a very broad phenotypic diversity were brought together and are incorporated in the Nordic Genetic Resource Centre (NordGen), Sweden. Several important mutant groups have been analysed in more detail genetically, with regard to mutagen specificity and gene cloning. These are: (i) early maturity mutants (*Praematurum*); (ii) six-rowed and intermedium-spike mutants; (iii) mutants affecting surface wax coating (*Eceriferum*); and (iv) mutants affecting rachis spike density (*Erectoides*). Some of these groups are presented in more detail in this review.

Once work with induction of mutations began, it was evident that mutations should regularly be included in breeding programmes of crop plants. In Sweden, direct X-ray induced macro-mutants have been successfully released as cultivars, some of them having been used in combination breeding. Their importance for breeding is discussed in more detail.

Keywords: mutation • barley • genetics • breeding

1 Introduction

Swedish research on induced mutations in barley started in 1928 on a small scale at Svalöf, Sweden, initiated by the eminent Swedish geneticists Herman Nilsson-Ehle and Åke Gustafsson (Fig. 2.1). Although L.J. Stadler in the USA also published data on induced mutations in barley, he described them with much pessimism, stating that they would be of no use for plant breeding

[1] Email: udda.lundqvist@nordgen.org

This chapter is an updated version of a previously published article on the history of Scandinavian mutation research (*Hereditas* 151, 123–131). It is published here with the permission of the author and of *Hereditas*.

 Mutation Breeding, Genetic Diversity and Crop Adaptation to Climate Change (eds S. Sivasankar *et al.*)

DOI: 10.1079/9781789249095.0002

Herman Nilsson-Ehle (1873–1949) Åke Gustafsson (1908–1988)

Fig. 2.1. The eminent Swedish geneticists for mutation research and breeding.

(Stadler, 1928). Nilsson-Ehle and Gustafsson did not share this pessimism and at Gustafsson's suggestion experiments were initiated in barley (*Hordeum vulgare*) with induced mutations. The first treatments with X-rays and ultraviolet irradiation commenced using the Svalöf cultivar 'Gull', which was the most common barley cultivar grown in Sweden at that time. Different pre-treatments were also tested, since it was known that mutation frequency increased if seeds were soaked in water before irradiation (Stadler, 1928). The first chromosome aberrations were observed and the first phenotypic changes in the seedlings, chlorophyll mutations, occurred. At that time distinct categories *albina*, *viridis* and *xantha* (Fig. 2.2) were established, and the rare two-coloured striped and zoned phenotypes were recognized. These chlorophyll mutations were always the first indication of treatment success (Gustafsson, 1938, 1940) and their abundance served as the standard method for measuring the induced mutagenic effects.

In the mid-1930s the first viable mutations appeared and it was possible to distinguish two subgroups: 'Morphological' and 'Physiological' mutations. The most common group at that time was the *Erectoides* mutants characterized by a compact or dense spike. Morphologically they resemble the *erectum* barleys, in comparison with the normal *nutans* spikes in most of the barley cultivars. In the following years, several of the mutants that were produced were considered very valuable, with characteristics such as high yield, straw stiffness, straw length, early maturity, tillering capacity, changes in spikes, kernels and awns, changed pigmentation and so on (Gustafsson, 1941, 1947). These results from all the early experiments looked so promising, even for plant breeding, that in 1940 the Swedish Seed Association at Svalöf started to sponsor this new research with funding from the Swedish milling industry. This made it possible to extend the experiments considerably. It was possible to integrate theoretical and practical results. In 1948, the Wallenberg Foundation incorporated these mutation activities in their research programme and permitted Gustafsson to gather around himself a group of specialists to carry on the research work with a wider perspective. In 1953, at the instigation of the Swedish Government, the 'Group for Theoretical and Applied Mutation Research' (Fig. 2.3) was established with the aim of studying basic research problems in order to influence and improve the methods for breeding of cultivated plants. The Agricultural Research Council provided funding for most of the Mutation Group's scientific activities approved by the Swedish Parliament (Gustafsson, 1954). Its most intensive activities were during the 1950s, 1960s, 1970s and 1980s.

2 Materials and Methods

X-irradiation of dry seeds was the standard method for studying the mutation process, but

Albina *Xantha* *Viridis*

Fig. 2.2. The three distinct categories of chlorophyll mutations.

Fig. 2.3. The Swedish group for Theoretical and Applied Mutation Research (mid-1950s).

soon other types of irradiation such as γ-rays for sparsely ionizing radiation, neutrons, electrons, protons for densely ionizing radiation, then α-rays from radon, and β-rays from phosphorus 32 and sulfur 35 were included in the experiments. Applications of treatments with different soaking times of the seeds, both before and after irradiation, were also studied. Not only was the water content of the seeds an important trait in relation to radiation sensitivity but so also were different environmental conditions. When comparing the two irradiation types, sparsely versus densely irradiated, it can be summarized as follows: seeds are 20–30 times more sensitive to neutrons than to X-rays; germinating seeds are two to three times more sensitive to neutrons than to X-rays; neutrons are approximately 10 times as effective as X-rays in producing chromosome disturbances and 50–100 times more effective in increasing the mutation rate in the second generation; and neutrons produce relatively more chlorophyll mutations than X-rays (Ehrenberg *et al.*, 1952, 1959; Lundqvist, 1992).

Already in the mid-1940s, chemical mutagenesis had entered the scene and became included in experiments together with irradiation. The idea was to influence not only the mutation rate but also morphological and physiological types of mutations. The first to be applied were colchicine and mustard gas, followed by epoxides and epimines, most importantly ethylene oxide and ethylene imine, in addition to many different compounds such as purines and purine derivatives and various alkylating and oxidizing compounds. Then more complex compounds were applied, such as alkane sulfonic esters with monofunctional and di-and trifunctional sulfonates, N-alkyl-N-nitrosamides, organic sulfates and sulfonates, most commonly ethyl methanesulfonate (EMS), nitroso compounds, purine and acridine derivates and many other chemical mutagens. Finally, in the mid-1970s the inorganic chemical mutagen sodium azide was introduced and it became the most popular mutagen for the isolation of viable mutants for practical purposes in Swedish experiments. The mutation frequency in these experiments increased rapidly up to 80%; they were 20 times more effective than irradiation. Significant differences between the actions of ionizing radiation and chemical mutagens were demonstrated. It was shown that neutrons and sodium azide form two extremes: neutrons induce a relatively large number of chromosome changes, whereas sodium azide primarily causes gene mutations at the nucleotide level (Nilan *et al.*, 1976). But differences in the mutation spectrum were also noticed, first with regard to chlorophyll mutations and later in some morphological and physiological mutation groups. Some gene loci reacted in different ways. The aim was actually to control the direction of mutagenesis. During all the years of working with mutation research, the following properties were always studied and determined in the different M generations.

1. The numbers of chromosome disturbances in germinating seeds were studied.
2. The numbers of germinating plants were counted in the M_1 field generation.
3. The numbers of mature harvested plants in the M_1 generation were registered.
4. The determination rate of chlorophyll mutations was studied in the seedling stage of the M_2 generation. This part was very important, as it showed if the mutagenic treatment had been successful or not.
5. In the following generations all morphological and physiological mutants were studied and isolated.

This mutation research was non-commercial, even if some important mutants were used to develop commercial cultivars (Ehrenberg *et al.*, 1956, 1961; Lundqvist, 1992, 2014).

3 The Swedish Collection of Barley Mutants

Genetic diversity in barley has been of great importance and has long been studied in great detail. This is important not only for plant breeding but also for investigations and localization studies of the barley genome. The Russian geneticist N.I. Vavilov felt it necessary to explore the total genetic diversity of crop plants throughout the world as well as diversity of related wild species (Vavilov, 1992). There will always be a large demand for a broad diversity, including genetically characterized mutants. These mutants will serve as basic material for all kinds of barley research and methodical work will sooner or later lead to positive results (Lundqvist,1986).

Over the years, a large collection of morphological and physiological mutations (about 12,000 different mutant alleles) with a very broad range of phenotypes were brought together and were genetically and agronomically studied. The collection consists of five main categories (Table 2.1). Groups A and B are divided into 12 sub-collections (Tables 2.1 and 2.2) (Lundqvist, 2005, 2014) and they are incorporated into the Nordic Genetic Resource Centre (NordGen), Sweden. Germination tests are conducted regularly,

Table 2.1. The five main categories of mutants.

Group	Category name
A	Barley Morphological and Physiological characters
B	Barley Near Isogenic Lines in Bowman
C	Barley Near Isogenic Lines in Bonus
D	Barley Duplication Lines
E	Barley Translocation Lines

Table 2.2. The 12 morphological and physiological barley sub-collections.

Sub-group	Character
1.	Changes in spikes and spikelets
2.	Changes in awn length and formation
3.	Changes in days to maturity
4.	Changes in epicuticular wax composition
5.	Changes in leaf blades
6.	Changes in culm length and composition
7.	Changes in kernel development and formation
8.	Changes in growth habit
9.	Changed pigmentation
10.	Changes in chlorophyll pigmentation
11.	Intermedium double mutants
12.	Resistance to powdery mildew

regeneration is done if necessary and the accessions are checked for their homozygosity. Most passport data are included as far as possible in the NordGen's information system, SESTO. About 750 barley genetic stock descriptions (BGS) with all genes localized also with the Swedish mutants are published in the *Barley Genetics Newsletter* (BGN) and in the International Database for Barley Genes and Barley Genetic Stocks (Lundqvist *et al.*, 2016; NordGen BGS, 2021). The Swedish mutants have been induced in different cultivars. In the USA, Jerome D. Franckowiak undertook a tremendous effort and by skilful crossing work he transferred many of the genes into a common background cv. 'Bowman', a two-row high-malting barley cultivar. He established about 1000 'Barley Near Isogenic Lines in Bowman' (NIL); 60% of them include Swedish mutant genes. This NIL collection is very useful for linkage studies, assessment of specific marker genes and determination of linkage drag and has assisted gene transfer for all barley researchers worldwide. Due to this collection, several of the mutant genes have been cloned in recent decades (Druka *et al.*, 2011).

The Swedish collection is unique, since all alleles of investigated genes are conserved at the Nordic Genetic Resource Centre (NordGen), Sweden, and are available for all barley research and plant breeding. It is a major source for today's gene mapping and valuable for molecular genetic analyses of cloned mutant genes. It also forms outstanding material for investigations within radiobiology, genecology, gene physiology, ultrastructural research, plant biochemistry, gene localizations, genetic fine mapping and molecular marker research. It serves as an important gene pool. During the years of the peaks of mutation research, about half of all mutants have been analysed genetically in greater or less detail, but they form only a minor part of the range of mutant characters. The mutant groups shown in Table 2.3 were studied and isolated in more detail genetically and also with regard to mutagen specificity. These studies have increased the knowledge of the mutation process and the architecture of different characters (Lundqvist, 1986, 1992, 2005, 2008, 2014). Some of them are presented and described in more detail in this review.

3.1 *Praematurum* (early maturity) mutants

The demand for early-maturing cultivars has increased rapidly, because earliness has become an important goal for Swedish plant breeding and is an important feature under natural conditions. Farmers want an early crop to establish an effective crop rotation. Already in the 1940s, it was found that maturity in barley could easily be changed by using X-rays in either direction with both increased earliness and increased lateness. The time of heading was chosen as a safe character for screening induced early mutants, but early heading and early ripening are characters where environmental influences, especially climate conditions, may hamper a safe classification (Gustafsson, 1942; Gustafsson *et al.*, 1960).

Over the years about 1250 different *Praematurum* mutants have been isolated and studied, using various mutagenic treatments. Several cultivars ('Bonus', 'Foma', 'Kristina', 'Lotta', 'Semira', 'Frida', 'Golf' and 'Lina') were used. Very soon the mutants could be grouped into three categories according to their heading and maturity time. We established three different classes of earliness, all compared with the mother cultivars, as follows: (i) drastically altered earliness of at least 1 week; (ii) medium increase of earliness of 3–5 days; and (iii) slightly modified earliness of 1–2 days (Gustafsson and Lundqvist, 1976; Lundqvist, 1992). Long-term studies made it possible to localize 195 mutants and distribute them to nine *mat* loci (Table 2.4). The different loci show

Table 2.3. Survey of the genetically investigated Scandinavian mutant groups.

Mutant group	Number of alleles	Number of loci
Praematurum (Early maturity)	195	9
Erectoides (Dense spikes)	222	31
Breviaristatum (Short awns)	196	25
Eceriferum (Waxless, Glossy)	1580	79
Hexastichon (Six-rowed spike)	65	1
Intermedium-spike mutants	83	10
Elongated outer glume (*Macrolepis*)	54	1
Third outer glume (*Bracteatum*)	12	4
Calcaroides	22	5
Anthocyanin mutants	766	31
Liguleless (Auricleless and Eligulum)	25	3
Albino lemma (*Eburatum*)	5	1
Orange lemma (*Robiginosum*)	7	1
Powdery mildew resistant mutants	77	Several
Chlorophyll synthesis and chloroplast development	455	106

Table 2.4. Distribution of the early maturity mutants to the nine mat loci.

Locus	*mat-a*	*mat-b*	*mat-c*	*mat-d*	*mat-e*	*mat-f*	*mat-g*	*mat-h*	*mat-i*	Total
Mutants	85	49	31	2	9	7	4	2	6	195

in general quite distinct phenotypic characters. The mutations selected for earliness also change other properties of agricultural value. Significantly shorter straw with lower internode number is found in the extreme early mutant loci, *mat-a*, *mat-b* and *mat-c*. Mutants of locus *mat-a* are generally more resistant to lodging than mutants in locus *mat-b* and have generally a more reduced culm length. Thus, mutants of the *Praematurum* type may offer favourable materials for the selection of high-yielding and semi-dwarf types. Among these loci, the most dramatic is the *Praematurum-a.8* (*mat-a.8*) mutant, derived as a one-step X-ray mutant, a drastically early mutant which heads 8–10 days earlier than its progenitor cultivar 'Bonus'. It was approved and released as a commercial Swedish cultivar under the name 'Mari' in 1960 and was intended to replace early Swedish six-row cultivars. It was grown widely in Iceland and in the Mediterranean region and was also included in the International Maize and Wheat Improvement Centre (CYMMIT) barley breeding programme in Mexico (Hagberg, 1961; Sigurbjörnsson, 1976).

Not until the mid-1960s was it found that *mat-a.8* had a special property that definitely distinguished it from the 'Bonus' parent cultivar, namely, a profound change in the photo- and thermo-period reaction, making it both heading and seed fertile at 8 h of daylight (short-day tolerant). During the 1960s, large phytotron experiments were carried out at the Stockholm Forest University under different photoperiod conditions to compare different mutants and cultivars (Dormling *et al.*, 1966; Dormling and Gustafsson, 1969; Gustafsson *et al.*, 1982). Later, when labour costs became too expensive, a darkening arrangement, using a special plastic tissue, was used in an ordinary glasshouse with natural light lasting for 8 h. This type of arrangement was applied over many years for identifying daylength-neutral mutants (Fig. 2.4). It was possible to distinguish three genotypic categories under the extreme short-day conditions of 8 h of light: (i) genotypes with complete early heading and good seed set; (ii) genotypes with incomplete and late heading and seed set; and (iii) genotypes that never headed but remained in a purely vegetative and often luxuriant state (Fig. 2.4). Regarding the mutants in *mat-c* and *mat-e*, they were characterized by less pronounced daylength neutrality compared with *mat-a* mutants. The mutants in all other *mat* loci were long-day adapted like the parent cultivars, being rather

Fig. 2.4. Heading mutants. Left: arrangement using a special plastic sheet to keep plants in the dark. Right: plants at maturity time showing mature mutant plants that never head (i.e. remain vegetative).

productive at 16–24 h of light, more or less infertile and late heading at 12 h, and not heading at all at 8 h of light (Gustafsson and Lundqvist, 1976). The induced early mutants tested derived from the whole spectrum of mutagens. There were rather more than expected short-day adapted mutants under sulfonate treatments, whereas the long-day adapted cases seemed to be in excess when ethylene imine was applied. Other observations indicated that sodium azide was less efficient in producing daylength-neutral mutants (Lundqvist, 1991, 1992). At the same time, Japanese researchers isolated short-day neutral mutants (*ea8*, *eam8*) allelic to the *Praematurum-a* (*mat-a*) gene (Yasuda, 1977).

Due to the development of many molecular programmes, it was possible to clone two of the *Praematurum* genes: *mat-a* and *mat-c*. The *Praematurum-a* (*mat-a*) mutant has been identified as a homologue of the *Arabidopsis thaliana* circadian clock regulator *EARLY FLOWERING 3* (*ELF3*). From 85 induced *mat-a* and 2 *eam8* mutants, more than 20 different *mat-a* alleles were identified and these mutations predicted a defective ELF3 protein. Expression analysis of *HvElf3* and *HvGigantea* in mutant and wild-type plants further characterized the flowering pathway leading to the early phenotype of the *mat-a* mutants (Zakhrabekova *et al.*, 2012). The other early-maturity gene *Praematurum-c* (*mat-c*) was identified as a variant of the barley homologue of *Antirrhinum majus CENTRORADIALIS* (*HvCEN*) (Comadran *et al.*, 2012). Of the 31 examined *mat-c* mutants, 29 different alleles were found (Comadran *et al.*, 2012; Matyszczak, 2014) and led to the conclusion that variation in *HvCEN* was important in enabling geographical range extension. The *mat-c* mutants were also demonstrated to be alleles at the early maturity 6 (*Eam6*) locus (Comadran *et al.*, 2012; Matyszczak, 2014).

3.2 The six-row (*Hexastichon*) and *Intermedium-spike* mutant group

Barley is one of the oldest cultivated crops, and the number of rows of spikelets is a key character in inferring the origin of barley. After at least 100 years of discussion, it has been concluded that domesticated two-row barley is older than six-row barley, as is supported by archaeological specimens (Bothmer *et al.*, 1995). The six-row (*Hexastichon*) and *Intermedium-spike* mutants affect the development of the lateral spikelets, with genetic interaction leading to synergistic enhancements. This research has given an insight into the rather complex genetics of kernel rows in barley. Normal two-row barley carries two spikelets that are on opposite sides of the spike and central spikelets with two reduced sterile lateral spikelets. Two-row barley is able to produce six-row barley in a single mutational step. These mutants have well developed central and lateral spikelets, are fully fertile and have long awns. All the 65 isolated cases have been localized to only one locus *v1*, renamed *hex-v* or *vrs1*, located in the long arm of chromosome 2H (Gustafsson *et al.*, 1969; Lundqvist, 1992; NordGen BGS, 2021). But two-row barley may also produce mutants with spike development intermediate between the two-row and six-row states. These mutants have enlarged lateral spikelets that vary in characteristic ways with regard to awn and kernel development and fertility, not only among mutants, but also depending on environmental conditions. A total of 126 such *Intermedium-spike* mutants have been isolated and 83 of them have been localized to ten different *int* gene loci and studied in more detail (Table 2.5). All these mutations are recessive and show independent inheritance (Gustafsson and Lundqvist, 1980; Lundqvist and Lundqvist, 1988a). All *Intermedium-spike* mutants have been induced in the two-row barley cultivars 'Bonus', 'Foma', 'Kristina' and 'Nordal'. Both radiation types and most of the used chemical mutagens were applied but no specific type of mutation could be associated with a certain type of mutagen (Lundqvist, 1992).

During the 1970s Japanese researchers studied several such so-called 'six-rowed' mutants. All of them were recessive and they were assigned the gene symbols *v2–v5* (Fig. 2.5), and were localized on the barley genetic map. In further studies at Svalöf, allelism tests were carried out between the Swedish *int* loci and the Japanese *v* genes. The results are summarized in Table 2.5 (Lundqvist, 1992). Due to newly developed molecular methods, all of these described *vrs* loci have been cloned. *Six-rowed spike 1* encodes a homeodomain-leucine zipper I-class homeobox gene and its expression is strictly localized in the lateral spikelet primordial of immature spikes, suggesting that the Vrs1 protein suppresses development of lateral spikelets (Komatsuda *et al.*, 2007). *Vrs2* encodes a SHORT INTERNODES (SHI) transcriptional regulator which is expressed during barley inflorescence and shoot development (Youssef *et al.*, 2017). *Vrs3* encodes a putative histone Lys demethylase with a conserved zinc finger and Jumonji C and N domain and controls lateral spikelet development (Bull *et al.*, 2017; Esse *et al.*, 2017). *Vrs4* also affects lateral spikelets and their fertility. This gene was cloned as an orthologue of the maize inflorescence architecture gene *RAMOSA2* (*HvRA2*) which is associated with loss of spikelet determination (Koppolu *et al.*, 2013). *Vrs5* was cloned as an orthologue of the maize domestication gene *TEOSINTE BRANCHED 1 (HvTB1)*; this gene modifies lateral spikelet development (Ramsay *et al.*, 2011).

Table 2.5. Distribution of the *Intermedium-spike* mutants to the 10 gene loci.

Locus	*int-a*	*int-b*	*int-c*	*int-e*	*int-f*	*int-h*	*int-i*	*int-k*	*int-l*	*int-m*	Total
Alternative names	*vrs3*	*vrs2*	*vrs5*	*vrs4*							
Number	33	3	23	15	1	4	1	11	1	1	83

Fig. 2.5. Six-row and *Intermedium-spike* mutants.

3.3 Surface wax coating mutants: *Eceriferum*, *Glossy* (waxless)

Presence of a wax coating reduces evaporation of water from the plant and helps protect it against pathogens. Most surface wax mutants at the *Eceriferum* and *Glossy* loci affect the presence and type of epicuticular waxes on leaf blades, sheaths, culms and spikes. When the wax coating is completely absent, various organs appear as a bright, glossy green colour. Cooperation between Swedish and Danish researchers made this mutant type probably the best-known character complex of any cultivated plant. The mutants have been isolated, the genes localized, their influence on yield studied, loci have been mapped genetically and electronic microscopy and biochemical analyses have been done. Their reactions to various climates were studied in the phytotron. Phenotypically, three different organs of the barley plant were studied in regard to wax coating and composition. Five phenotypic categories were established: spike and leaf sheath; spike and leaf sheath partially; spike; leaf blade; spike, leaf sheath and leaf blade (Lundqvist and Wettstein, 1962; Lundqvist *et al.*, 1968). A total of 1580

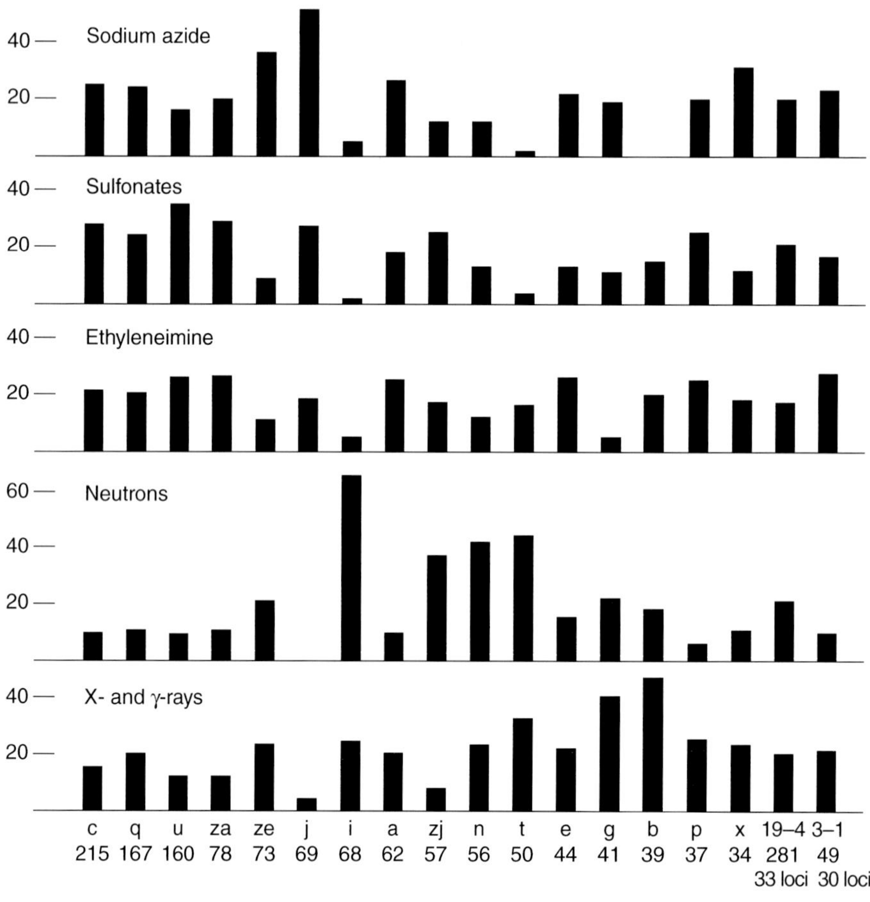

Fig. 2.6. Percentage of each *cer* mutation at each locus. Values are normalized for the five main kinds of mutagens (sodium azide, organic chemicals, ethyleneimine, neutrons and X-and γ-rays). The letters at the bottom of the diagram are the locus designations; underneath, the number of alleles at each locus is given. The numbers at the extreme right are the two classes of pooled loci with 19 down to 4 and 3 down to 1 mutation, respectively, for different mutagens. The numbers on the *y*-axis are the percentages (after Lundqvist and Lundqvist, 1988b; Lundqvist, 1992).

such *Eceriferum* mutants have been localized to 79 loci; 78 of them are recessive and one is dominant. Nearly all *Eceriferum* loci have been positioned on the barley genetic map. Seven types of mutagenic treatments have been applied and it is obvious that different loci show markedly different mutagen-specific reactions. Detailed analyses of mutation distribution led to the following results: (i) strong mutagenic differences between chemicals and ionizing radiation, especially neutrons; (ii) no significant differences among various kinds of organic chemicals; (iii) significant differences between organic chemicals and sodium azide; (iv) sodium azide differing strongly from X-rays and still more from neutrons; and (v) distinct differences between the two kinds of ionizing radiation (Lundqvist *et al.*, 1968; Lundqvist and Lundqvist, 1988b; Lundqvist, 1992). In summary, the wealth of alleles distributed on a large number of *cer* loci has provided important insights into the mutation process. These insights into the mutation process combined with knowledge of the localization of the different genes in the genome will add to our understanding of the mechanisms of mutagenesis and the organization of the eukaryotic genome (Lundqvist and Lundqvist, 1988b; Lundqvist, 1992).

The frequency distribution of the number of mutations observed at *cer* loci for different mutagen treatments is shown in Fig. 2.6. The most frequently observed mutations are at the *cer-c*, *cer-q* and *cer-u* loci, all affecting the wax coating on spike and leaf sheath. Most mutations in the *cer-c*, *cer-q* and *cer-u* loci were obtained after treatment with chemicals. They are all located in the short arm of chromosome 2H, in a very closely linked gene cluster *cer-cqu*. The locus *cer-b* affecting wax on the spike and leaf sheath was lacking mutations after treatments with sodium azide; for the locus *cer-i* affecting wax coating on the spike, the greatest number of mutations was caused by densely ionizing radiation; while *cer-j* (affecting wax coating on leaf blade) had a high mutation frequency for treatment with chemicals, but no mutant isolated with neutrons. These results emphasize the variability in the frequency of mutations with respect to the mutagen, the gene or the combination of these two (Lundqvist and Wettstein, 1962; Lundqvist *et al.*, 1968). Also 13 multiple *cer-cqu* mutants (seven triple and six double) were found, a rather high frequency of 2.5%; 11 were with neutrons, one with γ-rays and one with ethyl methanesulfonate (Lundqvist *et al.*, 1968; Wettstein-Knowles and Søgaard, 1980; Lundqvist, 1992; Lundqvist, unpublished).

Recently the *cer-cqu* gene cluster has been cloned (Schneider *et al.*, 2016) and the following results were found. The *Cer-c* gene is a chalcone synthase-like polyketide synthase designated diketone synthase with two exons. *Cer-q* encodes a lipase/carboxyl transferase with a single exon and the *Cer-u* (*AK373499*) gene is a P450 hydroxylase with five exons. It is now definitively shown that there are three different genes involved (Schneider *et al.*, 2016). The *Eceriferum* gene *glf1* (*cer-zh*) when investigated in the mutant *cer-zh.54* was identified as an elongase component, *β*-ketoacyl-CoA synthase (CER-ZH/*Hv*KCS1) (Li *et al.*, 2018).

3.4 *Erectoides* or dense spike mutants

The *Erectoides* (dense spike) mutants were the first viable mutants induced by irradiation and the most commonly induced morphological changes in ear density. They are characterized by compact, dense spikes, implying that the spike rachis internodes are shorter than in the mother strain (Fig. 2.7). They generally possess a very stiff and often short straw. The first uppermost internode of the culm is generally longer than in the mother cultivar and the basal ones are shorter. In all, about 1270 such *Erectoides* (*ert*) mutants have been isolated at Svalöf and studied intensively; all different cultivars as mentioned before were applied. Among 222 investigated mutants, 31 *ert* gene loci could be established. Most of the loci have distinct phenotypic characteristics, 30 are recessive and one is dominant; 21 loci have been localized and spread over the seven barley chromosomes. Differences in the mutation spectrum could be noticed: three of the *Erectoides* (*ert*) loci could be identified as mutagen specific, where more than 80%, 70% and 50%, respectively, of the alleles were induced by irradiation. The analysed mutant number is too small to state any gene preference. Most of the alleles of the *ert-b* locus have mutated in one special cultivar, 'Gull'. Many of the *Erectoides* mutants are fully viable and very promising from a practical point of view, and their productive capacity became tested continuously.

Fig. 2.7. Examples of some *Erectoides* mutants at different loci.

High-yielding *Erectoides* were formed by mutation in several gene loci. The most outstanding one is *Erectoides 32* in locus *ert-k*, which was released as the new very stiff-straw and high-yielding cultivar 'Pallas' in 1958 and was specially grown in Denmark and Great Britain (Gustafsson, 1941; Hagberg *et al.*, 1952, 1958; Hagberg, 1953; Fröjer *et al.*, 1958; Persson and Hagberg, 1968).

Already in the 1920s, cultivars with very short rachis internodes and a pyramid shape were found, and for these the name zeocriton (little barley) has been historically applied. This name was also applied to induced mutants having a similar phenotype (Hayes and Harlan, 1920). The dominant *Erectoides-r* (*Ert-r*) gene has been shown to be allelic to the *Zeocriton 1* (*Zeo1*) locus; the DNA sequencing showed that the *Zeo1* mutants occur in a *Hordeum vulgare APELATA2* (*AP2*)-like transcription factor, *HvAP2* (Nair *et al.*, 2010; Houston *et al.*, 2013); the dense spike phenotypes are a consequence of a perturbed interaction between microRNA 172 (*Hv-miR172*) and its corresponding binding site on the mRNA from the *HvAP2* gene, which acts early in spike development to regulate turnover of *HvAP2* mRNA. The *Zeo1* and *Ert-r* mutants occur in the last intron of *HvAP2*, the binding site of *Hv-miR172*, and prevent cleavage of the *HvAP2* mRNA (Nair *et al.*, 2010; Houston *et al.*, 2013). Another gene, *Erectoides-m* (*ert-m*), was identified as an orthologue of *ERECTA* in *Arabidopsis thaliana* (Zakhrabekova *et al.*, 2015). Several *ert-m* alleles also carry the *anthocyanin-less 1* (*ant1*) gene, which cannot be separated from the *ert-m* alleles. Sequencing of *HvERECTA* in the barley *ert-m* mutants revealed severe DNA changes in 15 mutants, including full gene deletions in *ert-m.40* and *ert-m.64*. Both deletions additionally cause anthocyanin deficiency associated with the closely linked *anthocyanin-less 1* (*ant1*) locus (Zakhrabekova *et al.*, 2015).

4 The Use of Mutants for Plant Breeding

Once the work with artificial induction of mutations began, it was evident that mutation programmes should be regularly included in breeding programmes of crop plants. The application of mutation research in plant breeding was the most important stimulus. It was shown already in the 1950s and 1960s that the work at Svalöf through the joint work with barley breeders and scientists can be used as an example of how mutation breeding can be employed in a crop improvement programme (Gustafsson, 1963). The main interest was focused on macro-mutations. Several characters such as earliness, straw stiffness, higher yields, semi-dwarfs, protein content and disease resistance are of major importance. Not only new direct mutants, but also the indirect use of induced mutations was applied. In the latter case, breeding work changed modifying systems by crossing mutants with various established cultivars and selecting the best recombinants homozygous for the mutations. In the Swedish programme, the use of macro-mutations has

Table 2.6. Induced barley mutants and their derivatives as released cultivars at Svalöf.

Primary mutant cultivars	Cultivars approved by mutant crosses	Cultivars approved by complex mutant crosses
44/3 extremely lodging resistant	'Gunilla' (1970)	
Pallas: *Erectoides-k.32*; lodging resistant	'Hellas' (1967)	'Visir' (1970) 'Senat' (1974) 'Jenny' (1980) 'Troja' (1981) 'Lina' (1982)
Mari: (*Praematurum-a.8*) Extreme early maturity and lodging resistant	'Kristina' (1969) 'Mona' (1970) 'Eva' (1973) 'Salve' (1974) 'Pernilla' (1979)	

proved to be more successful than recurrent mutagenic treatments (Gustafsson *et al.*, 1971).

A rather large number of mutant cultivars of two-row barley were registered as originals and 15 Swedish ones were commercially released. Two of these cultivars, 'Pallas', a stiff-straw, lodging-resistant and high-yielding *Erectoides* mutant, and 'Mari', an extremely early photo- and thermo-insensitive mutant, were produced directly by X-irradiation. All other cultivars derive from crosses and backcrosses with the X-ray induced mutants 'Pallas', 'Sv 44/3' and 'Mari' (Table 2.6). The series of cultivars obtained after crossing were found to be agriculturally suitable for different parts of Scandinavia and other parts of the world. The aim of this work was to demonstrate that original mutant materials can be used successfully in recombined breeding programmes and in the hands of skilful breeders. Additional methods ought to be used together, also today with many modern technologies, adding to the results of ordinary crossing and selection (Gustafsson, 1969, 1986).

In conclusion, the words of Åke Gustafsson, the father of mutations, from his last paper (Gustafsson, 1986) are summarized as follows.

Useful mutations in barley include a wide range of economically important characters that influence morphological as well as physiological and biochemical properties and will be an important tool in plant breeding, even more when the chemistry of the gene has been studied more intensively. Genetic instruments of artificial selection will increase the power and capacity of the plant breeder. It seems rather strange that also today there is a certain negative attitude towards the use of mutations in plant breeding or in most experiments concerning the general evolutionary theory. Such negative ideas are often associated with the view that mutationists ignore the natural sources of genetic variability and oppose the breeding value of primitive biotype collections.

References

Bothmer, R. von *et al.* (1995) An ecogeographical study of the genus *Hordeum*. Systematic and Ecogeographic Studies on Crop Genepools 7 (2nd edn). IBPGR, Rome.

Bull, H. *et al.* (2017) Barley *SIX-ROWED SPIKE3* encodes a putative Jumonji C-type H3K9me2/me3 demethylase that represses lateral spikelet fertility. *Nature Communications* 8, 936.

Comadran, J. *et al.* (2012) Natural variation in a homolog of *Antirrhinum CENTRORADIALIS* contributed to spring growth habit and environmental adaptation in cultivated barley. *Nature Genetics* 44, 1388–1392.

Dormling, I. and Gustafsson, Å. (1969) Phytotron cultivation of early barley mutants. *Theoretical and Applied Genetics* 39, 51–61.

Dormling, I. *et al.* (1966) Phytotron cultivation of Svalöf's Bonus barley and its mutant Svalöf's Mari. *Hereditas* 56, 221–237.

Druka, A. *et al.* (2011) Genetic dissection of barley morphology and development. *Plant Physiology* 155, 617–627.

Ehrenberg, L. *et al.* (1952) Effects of ionizing radiation in barley. *Arkiv Botanik*, Ser. 2.1 No. 17, 557–568.

Ehrenberg, L. *et al.* (1956) Chemically induced mutation and sterility in barley. *Acta Chemica Scandinavica* 10, 492–494.

Ehrenberg, L. *et al.* (1959) The mutagenic effects of ionizing radiation and reactive ethylene derivates in barley. *Hereditas* 45, 351–368.
Ehrenberg, L. *et al.* (1961) Viable mutants induced in barley by ionizing radiations and chemical mutagens. *Hereditas* 47, 243–282.
Esse, G.W. van *et al.* (2017) Six-rowed spike3 (VRS3) is a histone demethylase that controls lateral spikelet development in barley. *Plant Physiology* 174, 2397–2408.
Franckowiak, J.D. and Lundqvist, U. (2001) Rules for nomenclature and gene symbolisation in barley. *Barley Genetics Newsletter* 31, 76.
Fröjer, K. *et al.* (1958) Pallas barley, a variety produced by ionizing radiation: its significance for plant breeding and evolution. In: *United Nations Report 2nd Conf. on the peaceful uses of atomic energy in Geneva, August 1958.*
Gustafsson, Å. (1938) Studies on the genetic basis of chlorophyll formation and the mechanism of induced mutating. *Hereditas* 24, 33–93.
Gustafsson, Å. (1940) The mutation system of the chlorophyll apparatus. Lunds Univ. Årsskrift, N.F. Avd. 2. *Bd.* 36, 1–40.
Gustafsson, Å. (1941) Mutation experiments in barley. *Hereditas* 27, 225–242.
Gustafsson, Å. (1942) Mutationsforschung und Züchtung. *Der Züchter* 14, 57–64.
Gustafsson, Å. (1947) Mutations in agricultural plants. *Hereditas* 33, 1–100.
Gustafsson, Å. (1954) Swedish mutation work in plants: background and present organization. *Acta Agriculturae Scandinavica IV* 3, 361–364.
Gustafsson, Å. (1963) Productive mutations induced in barley by ionizing radiations and chemical mutagens. *Hereditas* 50, 211–263.
Gustafsson, Å. (1969) Positive Mutationen und ihre Verwendung in der Züchtung hochleistender Gerstensorten. *Bericht über die Arbeitstagung der Arbeitsgemeinschaft der Saatzuchtleiter in Gumpenstein, November* 1969, pp. 63–88.
Gustafsson, Å. (1986) Mutation and gene recombination – principal tools in plant breeding. In: Olsson, G. (ed.) *Svalöf 1886–1986. Research and Results in Plant Breeding*. LTs förlag, Stockholm, pp. 76–84.
Gustafsson, Å and Lundqvist, U. (1976) Controlled environment and short-day tolerance in barley mutants. In: *Induced Mutations in Cross Breeding*. Proceedings Panel Vienna, Vienna, 1975, pp. 45–53.
Gustafsson, Å. and Lundqvist, U. (1980) *Hexastichon* and *intermedium* mutants in barley. *Hereditas* 92, 229–236.
Gustafsson, Å. *et al.* (1960) The induction of early mutants in Bonus barley. *Hereditas* 46, 675–699.
Gustafsson, Å. *et al.* (1969) A proposed system of symbols for the collection of barley mutants at Svalöv. *Hereditas* 62, 409–414.
Gustafsson, Å. *et al.* (1971) Induced mutations and barley improvement. *Theoretical and Applied Genetics* 41, 239–248.
Gustafsson, Å. *et al.* (1982) Gene, genotype and barley climatology. *Biol. Zentbl.* 101, 763–782.
Hagberg, A. (1953) Heterozygosoty of *erectoides* mutants in barley. *Hereditas* 39, 161–178.
Hagberg, A. (1961) Svalöfs original Mari-korn. In: *Aktuellt från Svalöf. Allmänna Svenska Aktiebolaget*, pp. 13–16.
Hagberg, A. *et al.* (1952) Allelism of *erectoides* mutations in barley. *Hereditas* 38, 510–512.
Hagberg, A. *et al.* (1958) Sparsely *contra* densely ionizing radiations and the origin of *erectoid* mutations in barley. *Hereditas* 44, 523–530.
Hayes, H.K. and Harlan, H.V. (1920) *The inheritance of the length of internode in the rachis of the barley spike*. US Department of Agriculture, Bulletin 869. 26 pp.
Houston, K. *et al.* (2013) Variation in the interaction between alleles of *HvAPETALA2* and microRNA172 determines the density of grains on the barley inflorescence. *Proceedings National Academy of Sciences USA* 110, 16675–16680.
Komatsuda, T. *et al.* (2007) Six-rowed barley originated from a mutation in a homeodomain-leucine zipper I-class homeobox gene. *Proceedings National Academy of Sciences USA* 104, 1424–1429.
Koppolu, R. *et al.* (2013) Six-rowed spike4 (*Vrs4*) controls spikelet determinacy and row-type in barley. *Proceedings National Academy of Sciences USA* 110, 13198–13203.
Li, C. *et al.* (2018) The β-Ketoacyl-CoA Synthase *HvKCS1*, encoded by *Cer-zh*, plays a key role in synthesis of barley leaf wax and germination of barley powdery mildew. *Plant Cell Physiology* 59, 806–822.
Lundqvist, U. (1986) Barley mutants – diversity and genetics. In: Olsson, G. (ed.) *Svalöf 1886–1986. Research and Results in Plant Breeding*. LTs förlag, Stockholm, pp. 85–88.

Lundqvist, U. (1991) Swedish mutation research in barley with plant breeding aspects. A historical review. In: *Plant Mutation Breeding for Crop Improvement*. Proceedings International Symposium, Vienna 1990. IAEA, Vienna, pp. 135–148.

Lundqvist, U. (1992) Mutation research in barley. PhD thesis, Dept. of Plant Breeding Research, Swedish University of Agriculture Sciences. Available at: www.nordgen.org/ngdoc/plants/Udda_thesis.pdf (accessed April 2021)

Lundqvist, U. (2005) The Swedish collection of barley mutants held at the Nordic Genebank. *Barley Genetics Newsletter* 35, 150–154.

Lundqvist, U. (2008) Eighty years of Scandinavian barley mutation research and breeding. In: Shu, Q.Y. (ed.) *Proceedings FAO/IAEA International Symposium on Induced Mutations in Plants*, 12–15 August, 2008. Vienna, Austria, pp. 39–43.

Lundqvist, U. (2014) Scandinavian mutation research in barley – a historical review. *Hereditas* 15, 123–131.

Lundqvist, U. and Lundqvist, A. (1988a) Induced *intermedium* mutants in barley. Origin, morphology and inheritance. *Hereditas* 108, 13–26.

Lundqvist, U. and Lundqvist, A. (1988b) Mutagen specificity in barley for 1580 *eceriferum* mutants localized to 79 loci. *Hereditas* 108, 1–12.

Lundqvist, U. and Wettstein, D. von (1962) Induction of *eceriferum* mutants in barley by ionizing radiations and chemical mutagens. *Hereditas* 48, 342–362.

Lundqvist, U. *et al.* (1968) Induction of *eceriferum* mutants in barley by ionizing radiation and chemical mutagens. II. *Hereditas* 59, 473–504.

Lundqvist, U. *et al.* (2016) International Database for Barley Genes and Barley Genetic Stocks. Poster at the 12th International Barley Genetics Symposium, Minneapolis-St Paul, Minnesota, 26–30 June 2016. Available at: www.nordgen.org/bgs (accessed April 2021)

Matyszczak, I. (2014) Characterization of early maturity barley mutants *praematurum-a, -b* and *-c*. PhD Thesis, Aarhus University, Department of Molecular Biology and Genetics, Faculty of Science and Technology, Denmark.

Nair, S.K. *et al.* (2010) Cleistogamous flowering in barley arises from the suppression of microRNA-guided *HvAP2* mRNA cleavage. *Proceedings National Academy of Sciences USA* 107, 490–495.

Nilan, R.A. *et al.* (1976) Azide mutagenesis in barley. In: Gual, H. (ed.) *Barley Genetics III.,* Proceedings Third International Barley Genetic Symposium, Garching, 1975. Verlag G. Thiemig, München, pp. 113–122.

NordGen BGS (2021) International Database for Barley Genes and Barley Genetic Stocks. Nordic Genetic Resource Centre, Alnarp, Sweden. Available at: www.nordgen.org/bgs (accessed April 2021)

Persson, G. and Hagberg, A. (1968) Induced variation in a quantitative character in barley. Morphology and cytogenetics of *Erectoides* mutants. *Hereditas* 61, 115–178.

Ramsay, L. *et al.* (2011) *INTERMEDIUM.C*, a modifier of lateral spikelet fertility in barley, is an ortholog of the maize domestication gene *TEOSINTE BRANCHED 1*. *Nature Genetics* 43, 169–173.

Schneider, L.M. *et al.* (2016) The *Cer-cqu* gene cluster determines three key players in a β-diketone synthase polyketide pathway synthesizing aliphatics in epicuticular waxes. *Journal of Experimental Botany* 67, 2715–2730.

Sigurbjörnsson, B. (1976) The improvement of barley through induced mutation. In: Gaul, H. (ed.) *Barley Genetics III*, Proceedings Third International Barley Genetics Symposium, Garching, 1975. Verlag G. Thiemig München, pp. 84–95.

Stadler, L.J. (1928) The rate of induced mutation in relation to dormancy, temperature and dosage. *Anatomical Record* 41.

Vavilov, N.I. (1992) *Origin and Geography of Cultivated Plants*. Translated by Doris Love. Cambridge University Press, Cambridge, UK.

Wettstein-Knowles, P. von and Søgaard, B. (1980) The *cer-cqu* region in barley: gene cluster or multifunctional gene. *Carlsberg Research Communications* 45, 125–141.

Yasuda, S. (1977) Linkage of the earliness gene *eak* and its pleiotropic effects under different growth conditions. Ber. Ohara Inst. landw. Biol., Okayama Univ. 17, 15–28.

Youssef, H.M. *et al.* (2017) *VRS2* regulates hormone-mediated inflorescence patterning in barley. *Nature Genetics* 49, 157–161.

Zakhrabekova, S. *et al.* (2012) Induced mutations in circadian clock regulator *Mat-a* facilitated short-season adaptation and range extension in cultivated barley. *Proceedings National Academy of Sciences USA* 109, 4326–4331.

Zakhrabekova, S. *et al.* (2015) Genetic linkage facilitates cloning of *ERT-m* regulating plant architecture in barley and identified a strong candidate of *Ant1* involved in anthocyanin biosynthesis. *Plant Molecular Biology* 88, 609–626.

3 History of Mutation Breeding and Molecular Research Using Induced Mutations in Japan

Hitoshi Nakagawa[1]

Abstract

Following the construction of the Gamma Field at the Institute of Radiation Breeding in 1960, mutation breeding was accelerated in Japan. The facility is used, with a radiation dose up to 2 Gy/day (ca. 300,000 times that of natural background), to induce mutations at a higher frequency than occurs in nature. There have been 318 direct-use mutant cultivars representing 79 species generated through irradiation of gamma-rays, X-rays, ion beams and chemicals and somaclonal variation. Approximately 79% of these direct-use cultivars were induced by radiation. There have been 375 indirect-use mutant cultivars, including 332 rice, of which 162 cultivars (48.8%) were derived from the semi-dwarf mutant cv. 'Reimei'. The economic impact of these mutant cultivars, primarily of rice and soybean, is very large. Some useful mutations are discussed for rice, such as low digestible protein content, low amylose content, giant embryo and non-shattering. Useful mutations in soybean such as radiosensitivity, fatty acid composition and super-nodulation have been identified. Japanese pear and apple resistant to *Alternaria* disease have also been identified. The achievements of biological research such as characterization and determination of deletion size generated by gamma-rays, the effect of deletion size and the location, and a mechanism of dominant mutation induction are identified. Similarly, genetic studies on mutations generated through the use of gamma-ray induced mutations, such as phytochrome response, aluminium tolerance, stay-green (Mendel's gene) and epicuticular wax have also been conducted. Mutation breeding is a very useful technology for isolating genes and for elucidating gene functions and metabolic pathways in various crops.

Keywords: Gamma Field • deletion size • genetic study • mutant cultivar • molecular analysis

1 Introduction

After the construction of the Gamma Field, considered the world's largest radiation field (Fig. 3.1), at the Institute of Radiation Breeding (IRB) in Ibaraki, Japan, in 1960, mutation breeding was accelerated through cooperative research with national and prefectural breeding laboratories, private companies and universities in Japan (Yamaguchi, 2001).

In the *New York Times* (Broad, 2007), Dr P.J.L. Lagoda of the Joint FAO/IAEA was quoted as saying:

> Spontaneous mutations are the motor of evolution. We are mimicking nature in this. ... I'm not doing anything different from what

[1] Email: hitoshi.nakagawa@crl.hpk.co.jp

 Mutation Breeding, Genetic Diversity and Crop Adaptation to Climate Change (eds S. Sivasankar *et al.*)

DOI: 10.1079/9781789249095.0003

Fig. 3.1. Gamma Field, Institute of Radiation Breeding, NARO, Hitachi-Ohmiya, Ibaraki, Japan.

> nature does. I'm not using anything that was not in the genetic material itself. ... We're concentrating time and space for the breeder so he can do the job in his lifetime. We concentrate how often mutants appear – going through 10,000 to one million – to select just the right one.

The concept and objectives of the IRB's Gamma Field have the same goals for the plant breeder. The facility has an irradiation tower installed with an 88.8 TBq ^{60}Co at the centre of a circular field with a radius of 100 m (Nakagawa, 2010), to artificially induce mutations at a higher frequency than occurs in nature. The radiation dose at the nearest point in the field (10 m from the centre: ca. 2 Gy/day) is estimated to be about 300,000 times that of normal and natural background radiation when it is operated for 8 h per day. At the farthest point (100 m from the centre: ca. 0.01 Gy/day), the radiation dosage is about 2000 times the normal background radiation. Plants growing at the nearest point to the gamma-ray source are being treated to an equivalence of 1000 years of accumulated normal background rates of radiation per day and crop evolution is accelerated. Although we do not know all the genes or mechanisms of mutation, radiation breeding has produced many useful mutant cultivars and contributed greatly to the farms and industries of Japan. Following the genome sequencing of the 12 rice chromosomes completed in 2005 (International Rice Genome Sequence Project, 2005), molecular genetic studies based on the results of the genome sequencing project became the most powerful tool for selecting mutants of certain characteristics in rice.

In this report, the mutant cultivars are mainly developed by gamma-ray irradiation. In addition, descriptions of the molecular studies performed on the mutation at the DNA level and the cultivars' economic impacts in Japan are discussed.

Part of this report was presented at the 53rd Gamma Field Symposium held in Mito, Ibaraki, Japan, in 2014 and a review (Nakagawa and Kato, 2017).

2 Number of Mutant Cultivars

In a 2018 search regarding the number of induced-mutation varieties in the IAEA database (http://mvgs.iaea.org/AboutMutantVarities.aspx, accessed 2019), China had the most described induced-mutation varieties with 810, Japan was second with 479 and India was third with 335.

2.1 Number of cultivars developed by mutation breeding

The numbers of direct-use and indirect-use mutant cultivars registered in Japan in each 5-year period from 1961 to 2018 are shown in Fig. 3.2. The number of registered direct-use cultivars rapidly increased until 2000, when 65 cultivars were registered in 5 years. This number has since fallen but 33 cultivars were registered from 2011 to 2015. The number of indirect-use cultivars, primarily generated in rice, steadily increased to 85 during 2001–2005 and 81 during 2006–2010 but decreased to 61 during 2011–2015. The utilization of agronomically useful direct-use mutant cultivars, such as cv. 'Reimei', possessing the *sd1* dwarf gene (Ashikari *et al.*, 2002) can increase this number if further utilized by rice breeders.

There have been 318 direct-use mutant cultivars comprising 79 species generated through irradiation utilizing gamma-rays, X-rays and ion beams, chemical mutagenesis and *in vitro* culture (somaclonal variation) registered and released in Japan (Fig. 3.3); approximately 79% of these were induced by radiation. Those induced by somaclonal variation and chemical mutagens (not including those with doubled chromosome numbers through colchicine treatment) are 14.5% and 6.6%, respectively. Recently, the development of mutant cultivars generated by ion beam irradiation has been a growing area of mutation induction in Japan.

Table 3.1 shows the number of registered direct-use mutant cultivars of some crops developed by radiation, gamma-rays and those irradiated at the IRB. These include 64 mutant cultivars of chrysanthemum (*Chrysanthemum*), 45 of rice (*Oryza sativa*), 18 of soybean (*Glycine max*), 15 of carnation (*Dianthus caryophyllus*), ten of rose (*Rosa*), four of wheat (*Triticum aestivum*) and four of barley (*Hordeum vulgare*). Among them, 136 cultivars (ca. 42.8%) have been generated through gamma-ray irradiation at the IRB. This high percentage of gamma-ray

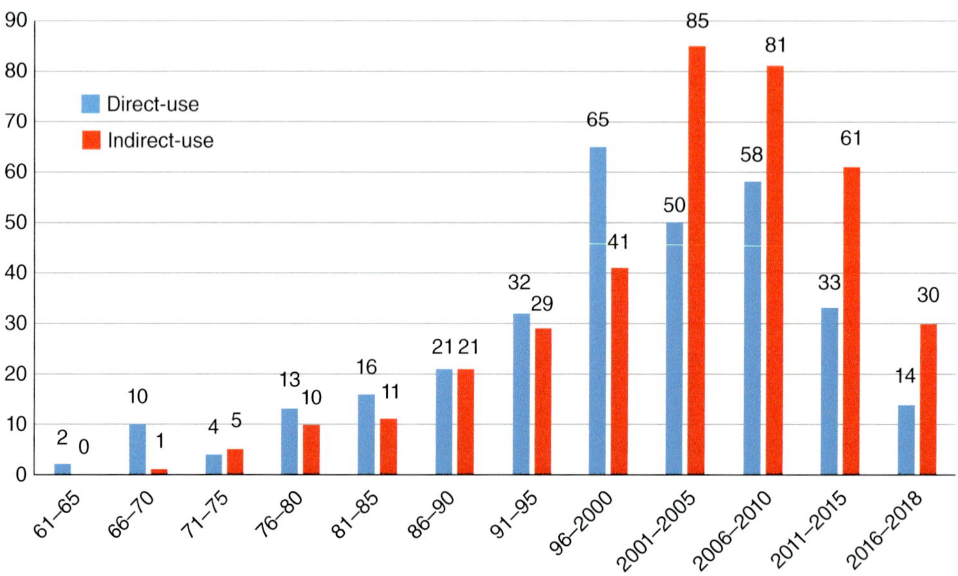

Fig. 3.2. Number of direct-use (318) and indirect-use (375) mutant cultivars registered and released in Japan in each 5-year period between 1961 and 2018.

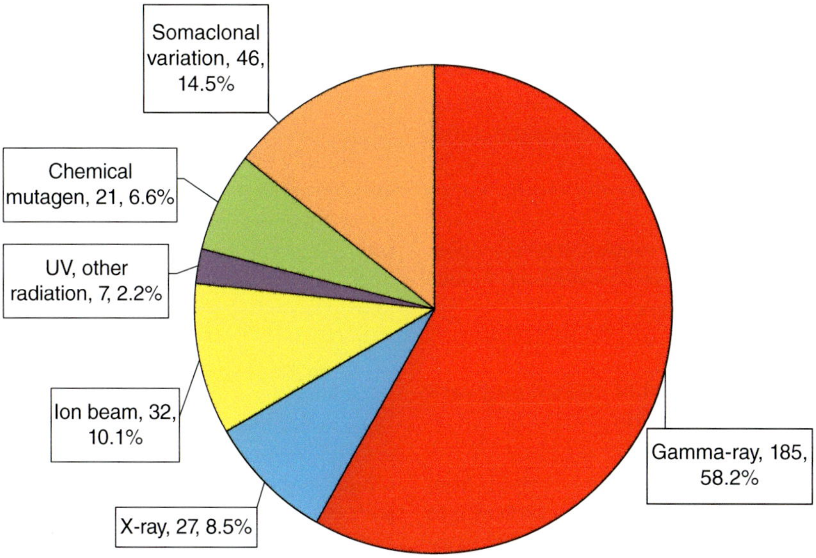

Fig. 3.3. Numbers and percentages of all 318 direct-use cultivars developed by mutation breeding using various methods in Japan (up to 2018).

Table 3.1. Number of direct-use mutant cultivars registered in Japan (2018).

	Number registered			
		Induced by		
	Mutant cultivars	Radiation	Gamma-rays	At IRB
79 Crops	**318**	**251**	**185**	**136**
Chrysanthemum	64	61	43	43
Rice	45	23	22	21
Soybean	18	17	16	9
Carnation	15	11	3	2
Rose	10	7	7	6
Wheat	4	2	2	0
Barley	4	4	3	0
Others	158	126	89	55

irradiated mutants indicates that mutation breeding via gamma-ray irradiation is an effective and highly successful approach for generation of commercial cultivars.

The first mutant rice cultivar developed using gamma-ray irradiation was cv. 'Reimei'. The cultivar, registered in 1966 (Futsuhara, 1968), was a successful case of an irradiation-induced semi-dwarf mutant. This cultivar exhibits a mutation of the *sd-1* locus (Ashikari *et al.*, 2002), which is the same as the mutation of a miracle rice, cv. 'IR8', through cv. 'Dee-geo-woo-gen' which later contributed to the 'Green Revolution' of rice and shows a culm 15 cm shorter than the original cv. 'Fujiminori'. The semi-dwarf character is associated with high-yielding ability and recorded the highest yield in Japan in 1967 (Futsuhara, 1968).

In Japan, the total number of indirect-use mutant cultivars in 2018 was 375, which included 332 rice, 16 soybean, eight barley, six wheat, three tomato, four lettuce (*Lactuca sativa*), one eggplant (*Solanum melongena*), two Japanese lawngrass (*Zoysia japonica*), two mat rush (*Juncus effusus*) and one Job's tear (*Coix lacryma-jobi*). Interestingly, among the 332 indirect-use mutant rice cultivars, 162 cultivars (48.8%) were derived from the direct-use mutant cultivar 'Reimei', or its offspring. This demonstrates that agronomically useful mutations can be efficiently and intensely utilized as parental lines to develop new cultivars with the same characteristics and over time the particular mutant gene multiplies itself in farmers' fields and provides agronomic advantages to the growers.

2.2 Economic impact of mutant cultivars in Japan

The number of rice cultivars derived from mutants generated by gamma-rays and sown in farmers' fields has increased in Japan since 1960. The cv. 'Reimei' was first cultivated on 61,598 ha in 1968 (http://ineweb.narcc.affrc.go.jp/, accessed 2019). The number of mutant cultivars derived from cv. 'Reimei' has been increasing and 99 mutant cultivars (two direct-use and 97 indirect-use cultivars) were in cultivation in 2005 (Nakagawa and Kato, 2017). The total cultivated area of mutant cultivars, mostly derived from gamma-ray irradiation, increased after cv. 'Reimei' was released for cultivation in 1968. The peak use of induced-mutation cultivars was 250,000 ha in 1986 and slightly exceeded 200,000 ha during 1994–2005. In 2005, the total cultivated area of mutant cultivars was 210,692 ha, which was 12.4% of the 1,702,000 ha cultivated for paddy rice in Japan (Nakagawa and Kato, 2017). The amount of total crude income of farmers selling the brown rice of mutant cultivars was estimated to be approximately 250 billion Yen (US$2.34 billion) in 2005 (Nakagawa and Kato, 2017).

The latest data for the top three paddy rice cultivars, in 47 prefectures of Japan in 2013, showed that 14 indirect-use mutation cultivars were listed in 180,233 ha of paddy field (11.03% of total paddy field) (Nakagawa and Kato, 2017). Among them, cv. 'Kinuhikari', which is descended from a gamma-ray induced semi-dwarf and lodging-tolerant mutant line, cv. 'Hokuriku 100 Gou' (Samoto and Kanai, 1975; Koga *et al.*, 1989), was the most popular indirect-use mutation cultivar with a good taste and shorter stems. The cv. 'Kinuhikari' covered 48,187 ha in 11 prefectures in 2013 and the total cultivated area of cv. 'Kinuhikari' and its descendants was 95,103 ha, or 5.8% of all the paddy field in Japan. The real cultivated areas of mutant cultivars exceeded these values, because the data included only the top three cultivars of prefectures and did not include all mutant cultivars ranked lower than the top three. This means that the economic impacts of mutant cultivars of rice are huge and the roles of cvs 'Reimei', 'Kinuhikari' and 'Mineasahi' are very important in Japan.

There have been 16 direct-use mutant cultivars of soybean registered in Japan since cvs 'Raiden' and 'Raikou' were developed by gamma-ray irradiation in 1960. The improved characteristics were early and late maturity, yellow hilum, seed coat colour, short stems, number of pods per stem, lipoxygenase-free and low allergens (Nakagawa *et al.*, 2011a). Among them, cv. 'Mura-yutaka' with a yellow hilum colour mutation was induced by X-rays at Saga University and is preferred in Japan (Nakamura *et al.*, 1991). There are 15 indirect-use mutant soybean cultivars, of which four, including cv. 'Ryuhou', are descended from a mutant cultivar 'Raiden' with induced early-maturity characteristics. Among the direct-use mutant cultivars, the cultivated area of cv. 'Mura-yutaka' was 1173 ha (1403 ha in 2011), cv. 'Kosuzu' 134 ha (194 ha in 2011) and cv. 'Akita-midori' 40 ha (0 ha in 2011) in 2014. Among the indirect-use mutant cultivars, the cultivated area of cv. 'Ryuhou' represented the third of all soybean cultivars planted in Japan with 10,548 ha in 2011 and fourth with 9600 ha in 2014. The total cultivated area of mutant cultivars in farmers' fields was 14,399 ha (10.5% of total cultivated area of 136,700 ha of soybean in Japan) in 2011 and 12,614 ha (9.5% of total area of 131,900 ha) in 2014. The total farmers' crude income in 2011 was estimated as 11.6 billion Yen (ca. US$116 million) (Nakagawa and Kato, 2017).

3 Achievements from Biological Research on Mutations Induced by Gamma-ray Irradiation

After the genome sequencing of rice chromosomes was completed, molecular genetic studies became the most powerful tool for identifying mutations in rice. There are two interesting questions:

1. What kind of base changes occurred in the genome through gamma-ray irradiations?
2. Can we induce dominant mutations?

Today, we have the answers. In addition, molecular studies can elucidate the mutation that generates interesting phenotypes at the DNA level.

3.1 Molecular changes generated by gamma-ray irradiation

Naito *et al.* (2005) studied the deletion sizes of transmissible and non-transmissible mutations induced through gamma-ray and carbon-ion beam irradiation by utilizing pollen-irradiation methods in Arabidopsis. Many mutants induced through these ionizing irradiations possess non-transmissible, extremely large deletions of more than 6 Mbp, or normally transmissible small deletions of 1 or 4 bp. Morita *et al.* (2009) researched the deletion sizes of transmissible mutations through gamma-ray irradiation in rice. Among 24 gamma-ray induced mutations that are transmittable to the next generation, four groups can be identified: (i) those with a base substitution; (ii) those with small deletions (1–16 bp); (iii) those with extremely large deletions; and (iv) those with a large inversion of over 1 Mbp including small deletions (Table 3.2). However, it is not known how difficult it may be to generate mutants with medium-sized deletions (1.0–5.0 kbp) through gamma-ray irradiation, although some reports mention that neutrons can induce deletions ranging between 300 bp and 12 kbp (Li *et al.*, 2001; Nagano *et al.*, 2008). However, it is interesting that inversions are not considered rare events following gamma-ray irradiation, as is the case for sorghum (Mizuno *et al.*, 2013). A possible explanation for this is provided in the following section.

Table 3.2. Gamma-ray irradiation-induced deletions, base substitutions and inversions and the size of the lesion (Morita *et al.*, 2009)

Mutation type	Gene	Size (bp)
Small Deletion	*CAO* (*cao-g1*)	1
	CAO (*cao-g2*)	3
	CPS (*cps-g1*)	1
	GA3ox (*ga3ox-g1*)	1
	GA3ox (*ga3ox-g2*)	3
	GID1 (*gid1-g1*)	1
	GluA1 (*gluA1-g1*)	1
	GluA2 (*gluA2-g1*)	1
	KAO (*kao-g1*)	4
	KAO (*kao-g2*)	16
	PLA1 (*pla1-g1*)	5
	PLA2 (*pla2-g1*)	5
	Wx (*wx-g1*)	2
	Wx (*wx-g2*)	5
	Wx (*wx-g3*)	6
Large Deletion	*GID2* (*gid2-g1*)	42,200
	Glb (*glb1*)	62,800
	GluB4/B5 (*glu1*)	129,700
	Wx (*wx-g4*)	9,400
Base Substitution	*GluA2* (*gluA2-g2*)	1
	PLA1 (*pla1-g2*)	1
	Wx (*wx-g5*)	1
Inversion	*Wx*	1,284,800
	PLA2	3,208,500

CAO: chlorophyll b deficiency;
CPS, *KAO* and *Ga3ox2*: gibberellin deficiency;
GID: gibberellin insensitivity;
GluA and *GluB*: glutelin deficiency;
Glb: alpha-globulin deficiency;
PLA: shortened plastochron;
Wx: glutinous endosperm

3.2 Different size and location of deletion generates different kinds of phenotypes

In the course of plant evolution, genes are often duplicated in tandem, resulting in functional redundancy. Glutelin is a major digestible seed storage protein of grasses and is encoded by a multigene family. Mutants of tandem-duplicated glutelin genes were investigated for their genotypes and phenotypes. They represent an inverted repeated two-locus event (Fig. 3.4), with both repeated units encoding glutelin mRNA. Various mutants with low glutelin content have been isolated using SDS-PAGE (Iida *et al.*, 1993, 1997). The mechanisms by which low glutelin content arises in the mutants that have been studied suggest that the size and position of the deletion determines the characteristics of the mutant. *Low glutelin content* (*Lgc-1*) is a dominant mutation that reduces glutelin content in the rice grain (Iida *et al.*, 1993). Kusaba *et al.* (2003) reported that in *Lgc-1* homozygotes contain a 3.5 kb deletion

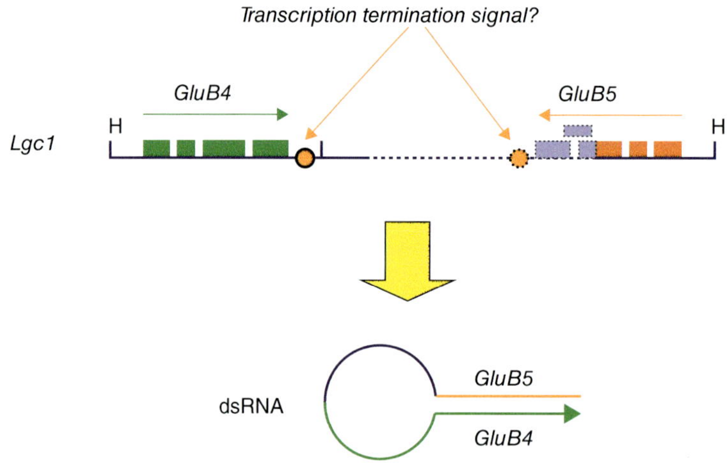

Fig. 3.4. Molecular characterization of *Lgc1* (Low glutelin rice). A deletion of 3500 bp, including a transcription terminal signal, generates a long RNA with homologous GluB4 and GluB5 related sequences. As this would be a double-stranded RNA, which could play a role in the RNA inhibition of the two genes, these two genes would be silenced (Kusaba *et al.*, 2003). (Courtesy of Prof. Kusaba of Hiroshima University.)

between two highly similar glutelin genes that forms a tail-to-tail inverted repeat, from which a double-stranded RNA molecule could be produced, which would be a potent inducer of RNA silencing (Fig. 3.4). As a result of this inverted repeat, glutelin synthesis would be suppressed and the glutelin content lowered. This was the first report that showed that the mechanism of a mutation was RNA interference (RNAi) in plants. The *Lgc-1* provides an interesting example of RNA silencing among genes that exhibit various levels of similarity to a gene-induced RNA silencing.

The gamma-ray induced 'glu1' mutant of rice lacks an acidic subunit of glutelin. Morita *et al.* (2007) showed that the *glu1* gene of the 'glu1' mutant harbours a 129.7 kb deletion involving two highly similar, tandem-repeated glutelin genes, *GluB5* and *GluB4*. The deletion eliminates the entirety of the *GluB5* and *GluB4* genes except for half of the first exon of *GluB5*. As a result, the phenotype of the *glu1* gene completely lacks the acidic subunit of glutelin and acts as a recessive gene for low glutelin content in rice grains.

The above examples demonstrate that the position and size of deletions at the same locus can dramatically change the phenotype of mutant gene expression through the processes of transcription and translation.

Furthermore, *GluB5* and *GluB4* have the same amino acid sequence in their acidic subunit, suggesting that only the mutation involving both *GluB5* and *GluB4* generates the resultant phenotype.

Sequenced plant genomes showed that more than 14% of their genes formed tandem arrays (Arabidopsis Genome Initiative, 2000; International Rice Genome Sequence Project, 2005; Morita *et al.*, 2007). This finding suggests that gamma-rays can be an effective mutagen to generate knockout mutants of both loci and to be used for analysis of tandem repeat and functionally redundant genes.

4 Useful Mutations from Various Screening Methods

4.1 Low digestible protein content

The protein content of cooked white rice is ca. 7%. A mutant line with a low glutelin content was obtained from the ethyleneimine (EI) treatment of cv. 'Nihon-masari'. The cv. 'LGC-1' with lower glutelin content was developed from a mutant line induced through EI treatment of cv. 'Nihon-masari' and backcrossing of this mutant with the original cultivar (Iida *et al.*, 1993). The seed protein composition of cv. 'LGC-1' is altered by a dominant mutation which causes a reduction in the amount of digestible glutelin, although it remains the main component, and this reduction is compensated for by an increase in the amount of indigestible prolamine. The characterization of this mutation and the mechanism were mentioned in the previous section. This protein composition is disadvantageous for human digestion of rice grains, although the total amount of protein is similar to the original cultivar. As a result, cv. 'LGC-1' is useful as a low-protein rice, and some clinical trials indicate that the cooked rice of this variety is a useful and effective daily food for patients with kidney disease (Mochizuki and Hara, 2000). The defects of the cv. 'LGC-1' are its inferior taste and the presence of other loci that control the biosynthesis of digestible globulin. Nishimura *et al.* (2005) induced a mutant line that exhibited a deficiency in globulin through gamma-ray irradiation from the leading Japanese cv. 'Koshihikari' with a good taste. The mutant line was hybridized with cv. 'LGC-1', and cvs 'LGC-Katsu' and 'LGC-Jun' were selected from the hybrids. The glutelin content of these two cultivars is as low as that of cv. 'LGC-1' and the globulin content is zero. The total digestible protein content is about 30% of ordinary rice. As the eating quality is improved and digestible protein content is lower than for cv. 'LGC-1', these two cultivars greatly help in the dietary management of proteins in cases of chronic renal failure.

4.2 Glutinous rice (low amylose content)

In general, people in Japan as well as Korea, northern Thailand, Myanmar and southern China prefer sticky rice. Amylose contents are closely related to this character and range from 0 (waxy: glutinous) to higher than 20% w/w, in indica-type rice. In Japan, glutinous rice has a special utilization for festivals and celebrations, while non-glutinous popular cultivars are used for daily cooking, and these exhibit ca. 17% amylose content. The waxy locus (*Wx*) was mapped to chromosome 6 of rice (Iwata and Omura, 1971) and knockout of *Wx* makes non-glutinous ordinary rice completely glutinous (*wx*). The *waxy* genes were identified to encode granule-bound starch synthesis (Nelson and Pan, 1995). In Japan, several glutinous rice cultivars have been induced from non-glutinous cultivars through gamma-ray irradiation in the Gamma Field (Toda, 1982; Imbe *et al.*, 2004).

There is another type of endosperm starch mutation, 'dull', whose amylose content is ca. 10%, which is lower than the non-waxy (*Wx*) rice, and this exhibits partial stickiness when cooked. Genetic analysis of dull mutants showed that most of these mutations were controlled by a single recessive gene which is non-allelic to the *wx* alleles (Okuno *et al.*, 1983). One of the most popular dull cultivars is cv. 'Milky Queen' induced by chemical mutagen MNU (N-methyl-N-nitrosourea) treatment of cv. 'Koshihikari', and this mutant cultivar has an amylose content of 9–12% (Ise *et al.*, 2001). However, the dull mutation in cv. 'Milky Queen' is not like other dull mutants, because it is caused by a mutation at *wx* allele (*Wx-mq*) (Sato *et al.*, 2002).

4.3 Giant embryo

When the embryo grows after soaking rice in water, gamma-aminobutyric acid (GABA) accumulates following transformation from glutamic acid (Saikusa *et al.*, 1994). Defatted rice embryos enriched with GABA are useful as a functional food. GABA-accumulated brown rice is already on the market as a health food based on an ordinary rice variety. The giant embryo mutant lines were induced by MNU treatment (Satoh, 1981) for this purpose. The embryo volume of these lines is 3–4 times that of ordinary rice cultivars. The giant embryo trait of the mutant line 'EM40' is controlled by one recessive gene (*ge*) located on chromosome 7 (Satoh and Iwata, 1990). The first giant embryo mutant cv. 'Haiminori' was developed by the hybridization between the 'EM40' and cv. 'Akenohoshi' (Nemoto *et al.*, 2001). The embryo volume of cv. 'Haiminori' is 3–4 times that of ordinary rice. After soaking in water for 4 h, the amount of accumulated GABA in cv. 'Haiminori' is about 4 times that of the traditional cv. 'Nipponbare'. The *GIANT EMBRYO* (*GE*) gene was identified as essential for controlling the size balance in rice. The function of *GE*, which encodes CYP78A13 and is predominantly expressed in the interfacing tissues of both the embryo and endosperm, is in controlling cell size in the embryo and cell death in the endosperm (Nagasawa *et al.*, 2013). Development of giant embryo lines with good seedling establishment will be a continuing objective of the breeding, because a common defect of giant embryo cultivars is that they have a lower germination rate (Nemoto *et al.*, 2001).

4.4 Non-shattering

Much attention has been paid recently to indica-type high-yielding genetic resources for improving biomass productivity of forage rice. Seed shattering is one of the undesirable characteristics of indica-type rice. Mutation in the dominant *qSH1* gene in domesticated rice eliminates the abscission layer and results in non-shattering seeds (Konishi *et al.*, 2006) and the knockout of this dominant gene generates a non-shattering rice. Several non-shattering mutant rice cultivars have been developed by the knockout of this gene (Kato *et al.*, 2006; Sakai *et al.*, 2013). Through genetic analyses of rice cultivars, its wild relatives, chromosomal segment substitution lines and an induced shattering mutant line derived from gamma-ray irradiation, several quantitative trait loci associated with seed shattering, such as *SH4* (Li *et al.*, 2006), *OsCPL1* (*Oryza sativa CTD phosphatase-like 1*) (Ji *et al.*, 2010) and *SHAT1* (*Shattering Abortion1*) (Zhou *et al.*, 2012), have been identified.

4.5 Radiosensitivity

Takagi (1969) identified two major genes that control radiosensitivity in some soybean cultivars by evaluating the radiosensitivity of root length and concluding that this trait was associated with acute irradiation of seeds or chronic irradiation of plants for the entire growth period. The cv. 'Lexington' ($rs_1rs_1rs_2rs_2$) and cv. 'Shinmejiro' ($rs_1rs_1Rs_2Rs_2$) are 2–5 times as sensitive to radiation as the resistant cv. 'Tachisuzunari' ($Rs_1Rs_1Rs_2Rs_2$). The differences in radiosensitivity between the cultivars to chronic irradiation in the Gamma Field are controlled by a single recessive rs_1 gene. A second recessive gene, rs_2, was discovered in cv. 'Goishi-shirobana' ($Rs_1Rs_1rs_2rs_2$), whose activity is only expressed following acute seed radiation.

4.6 Fatty acid composition

Soybean is the most widely used source of edible oil for human consumption. There is an extremely high diversity of fatty acid compositions across the oil crops. The biosynthesis of fatty acids common to all oil crops is as follows: a carbon elongation process (palmitic acid (16:0) → stearic acid (18:0)) and unsaturated reaction process (stearic acid (18:0) → oleic acid (18:1) → linoleic acid (18:2) → linolenic acid (18:3)). Another biosynthesis route is a carbon elongation reaction unique to traditional rapeseed of oleic acid (18:1) → eicosenoic acid (20:1) → erucic acid (22:1). The particular fatty acid contents of cultivars result from differences in activity or reactivity of their enzymes and the number of genes encoding each enzyme involved in the various steps related to carbon elongation and unsaturated reaction.

Takagi and his colleagues identified 46 mutant lines generated by X-ray application to cv. 'Bay' (Takagi and Rahman, 1995; Nakagawa *et al.*, 2011a). This induction of fatty acid variability confirms that artificial mutation is useful for enhancing fatty acid diversity of soybean. Following hybridization and examination of the subsequent generations, the inheritance of genes conditioning low and high palmitic, high stearic, high oleic and low and high linolenic acids was elucidated.

Allele-specific genotypic selection, through use of molecular markers related to these gene families, will be superior to phenotypic selection using gas–liquid chromatography. Multiple sources of alleles for each candidate gene isoform can provide further benefits in the breeding of germplasm with superior fatty acid composition and minimize fixation of alleles linked to target genes (Takagi *et al.*, 1998; Takagi and Anai, 2006; Nakagawa *et al.*, 2011a).

4.7 Super-nodulation

Super-nodulation is a character that generates a large number of root nodules in leguminous plants. Super-nodulating mutants of soybean may provide new genetic resources for improving soybean productivity by higher nitrogen fixation as well as for elucidating mechanisms of rhizobium–plant interactions. A super-nodulating mutant En6500 was isolated from an M_2 population generated by treating cv. 'Enrei' with 0.5% of ethyl methanesulfonate (EMS) (Akao and Kouchi, 1992). The En6500 had an increased number of nodules, inherited as a Mendelian recessive trait (Kokubun and Akao, 1994). However, its growth and yield performance were less than that of cv. 'Enrei'. This reduced yield performance is presumably caused by the high consumption of carbohydrates by nodules and a decrease of total nitrogen assimilation through the reduction of nitrate absorption (Takahashi *et al.*, 1995). A second attempt utilizing traditional backcross breeding to cv. 'Enrei' was initiated and the super-nodulating cv. 'Kanto 100' was selected. This cultivar exhibits super-nodulation and the yield is similar to that of cv. 'Enrei' (Takahashi *et al.*, 2003). This super-nodulating mutation has potential impact on the development of high-efficiency, nitrogen-fixing soybeans as well as for elucidation of rhizobium–plant interaction.

4.8 Japanese pear and apple resistant to *Alternaria* disease

A popular cultivar of Japanese pear (*Pyrus serotina* Rehd. var. *culta* Rehd.), cv. 'Nijisseiki', was a leading variety that occupied 28% of the total cultivated area of Japanese pear in Japan in 1990. However, the cultivar is highly susceptible to the black spot disease *Alternaria kikuchiana* Tanaka (Nishimura *et al.*, 1978). Growers spray fungicides several times during the growing season to counter the disease. To induce mutations resistant to the disease, small plants of the cv. 'Nijisseiki' were planted within 37–63 m from the ^{60}Co source in 1962 and chronic gamma-ray irradiation was applied (30×10^{-2} to 4×10^{-2} Gy/day) in the Gamma Field (Sanada *et al.*, 1993). In 1981, a twig bearing green leaves exhibited no symptoms of the disease, while all the other twigs were infected by the fungus and most leaves had many black spots. The resistant twig was found on a plant planted 53 m from the radiation source. Cuttings from this twig were vegetatively multiplied, registered and released in 1991 with the name cv. 'Gold Nijisseiki' (Sanada *et al.*, 1993). It was registered under the same name in Australia in 2004 (Certificate Number 2533).

At the same time, Nakashima *et al.* (1982, 1985) isolated and identified the chemical structure of the toxin that is named as 'AK-toxin' produced by the fungus. The IRB breeding group entered into a cooperative research programme with the chemistry group in Kyoto University and established a unique bioassay method. When punched leaf discs are placed on filter paper in a Petri dish soaked with AK-toxin obtained either from the extracts of the fungal body or artificially synthesized, and kept for 2 days at 25°C, leaves of susceptible varieties turn black and leaves of resistant varieties stay green (Sanada, 1988).

Following its development, two new mutant cvs 'Osa-Gold' (Masuda *et al.*, 1997) and 'Kotobuki Shinsui' (Kitagawa *et al.*, 1999) were developed in 3–4 years using this unique screening method, although it took 19 years to develop cv. 'Gold Nijisseiki'. The economic effect of this research has been extraordinary.

Furthermore, the same bioassay method was utilized for the selection of plants resistant to *Alternaria* blotch disease using the AM-toxin (Ueno *et al.*, 1977) produced by the fungus *Alternaria mali* that attacks apple. A new mutant apple cv. 'Houiku Indo' was selected from gamma-ray irradiated micro-propagated shoots of susceptible cv. 'Indo' and found to be resistant to this disease (Yoshioka *et al.*, 2001).

This research suggests that breeding of fruit trees requires patience and that development of simple, efficient and precise screening methods is very important for mutation breeding.

4.9 Other mutations

Many other interesting mutations have been identified using various screening methods in Japan. Additional information about unique mutations, screening methods and released cultivars is given in the Appendices of a publication of Nakagawa and Kato (2017). *Gamma Field Symposia* (a series of proceedings, available at http://www.nias.affrc.go.jp/eng/public/index2.html, accessed 2019) and *Radiation Breeding Technologies* (ibid.) also provide a wide range of information.

5 Genetic Studies of Useful Mutations Induced by Acute or Chronic Gamma-ray Irradiation

Spontaneous and induced mutation resources have played important roles not only for mutation breeding but also in genetic studies and the elucidation of gene functions.

5.1 Phytochrome

Takano *et al.* (2005) isolated *phytochrome B* (*phyB*) and *phyC* mutants from rice in a population irradiated in the Gamma Field and produced all combinations of double mutants. Seedlings of *phyB* and *phyB phyC* mutants exhibited a partial loss of sensitivity to continuous red light but still exhibited significant de-etiolation responses. The response to red light was completely lost in *phyA phyB* double mutants. These results indicate that *phyA* and *phyB* act in a highly redundant manner to control de-etiolation under red light. They also found that mutations in either the *phyB* or *phyC* locus caused moderate early flowering under a long-day photoperiod, but monogenic *phyA* mutations had little effect on flowering time. However, the *phyA* mutation in combination with the *phyB* or *phyC* mutation caused dramatically early flowering. The *phyB* mutants were generated by chronic gamma-ray irradiation with dose rates of 3–6 Gy/day (Takano *et al.*, 2005).

5.2 Aluminium tolerance

Ma *et al.* (2005) isolated a mutant with high sensitivity to aluminium (Al) from cv. 'Koshihikari', an Al-resistant japonica rice (Wu *et al.*, 1997). The mutant was induced through chronic gamma-ray irradiation in the Gamma Field and exhibited the same phenotype as the wild-type, but without Al tolerance. The root elongation of the mutant was highly inhibited in the presence of 10 μM Al. The mutant also exhibited poorer root growth in acid soil. Genetic analysis showed that the high sensitivity to Al is controlled by a single recessive gene, *als1*, that is mapped to the long arm of chromosome 6 (Ma *et al.*, 2005).

5.3 Mendel's gene: stay-green character

Gregor Mendel determined the law of heredity by using seven pairs of characteristics in garden pea. Among them were shape (dominant: smooth; recessive: wrinkled) (Bhattachayya *et al.*, 1990), stem length (dominant: tall; recessive: short) (Lester *et al.*, 1997) and flower (and seed) colour character (Hellens *et al.*, 2010). These traits have been genetically identified as natural mutations. Sato *et al.* (2007) identified molecular background of cotyledon colour (dominant: yellow; recessive: green) through the use of a mutant induced through chronic gamma-ray irradiation in the Gamma Field. The results present evidence that Mendel's green cotyledon gene is the *SGR* in pea (*PsSGR*). Furthermore, they reported that the *OsSGR* gene, present in a stay-green rice, was strongly expressed in senescent leaves and involved in activation of the chlorophyll degradation pathway through translational or post-translational regulation of chlorophyll-degrading enzymes. Mutants induced in the Gamma Field that retain greenness of leaves during senescence are known as 'stay-green' mutants of rice. That is to say, the most famous stay-green mutant is Mendel's green cotyledon pea. Pea plants with a homozygous recessive mutation (known as *i*) maintain green colour of the cotyledon during seed maturation and leaf senescence. They found tight linkage between the *I* locus and stay-green gene originally found in rice, SGR. Molecular analysis confirmed that the *I* gene encodes SGR in pea and functional analysis of *sgr* mutants in pea and rice revealed that SGR is primarily involved in chlorophyll degradation (Fig. 3.5).

5.4 Wax-free mutation of sorghum

Sorghum (*Sorghum bicolor* L. Moench) deposits a bloom or epicuticular wax on the surface of stems and leaves. The deposition of this wax is a dominant trait (*Bm*) with variation (Jordan *et al.*, 1983), and bloomless (*bm*) mutants have no bloom on the surface of plant parts. This wax is thought to contribute to tolerance to drought through reduced cuticular transpiration (Blum, 1975) and protection from

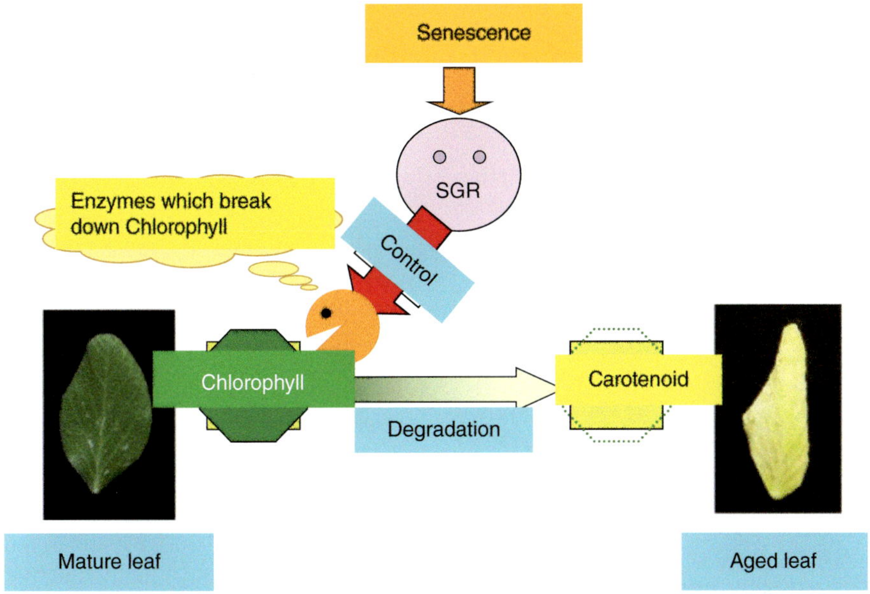

Fig. 3.5. Mechanism of stay-green phenotype. The *SGR* involved in activation of the chlorophyll degradation and carotenoid generation pathway through translational or post-translational regulation of chlorophyll degrading enzymes. (Courtesy of Dr R. Morita, RIKEN.)

ultraviolet (UV) light (Jordan *et al.*, 1983). Reducing cuticular wax is also known to increase susceptibility to the fungal disease of Northern corn leaf blight by *Exserohilum turcicum* (Jenks *et al.*, 1994). However, *bm* sorghum tends to have increased resistance to biotic stresses such as resistance to greenbug attacks (Peiretti *et al.*, 1980; Nakagawa *et al.*, 2011b) and sheath blight (Kasuga *et al.*, 2001).

In Japan, sweet sorghum has recently become a candidate crop for bioethanol production by utilizing the sugars and lignocellulose of stems and leaves through fermentation. The development of sorghum varieties resistant to biotic stresses will have advantages for low input without pesticides and fungicide and sustainable biofuel production. In 2008, Nakagawa *et al.* (2011c) induced two *bm* mutant plants of cv. 'Italian' without visible epicuticular wax (bloom) on the stems. The two self-pollinated lines were planted in the field and identified as pure lines with no variation. At the same time, the F_2 generation of the hybridization of these mutants and the original cv. 'Italian' plant segregated for individuals with and without epicuticular wax at a frequency of approximately 3:1 segregation ratio, suggesting that the *bm* phenotype was controlled by a single recessive nuclear gene (Nakagawa *et al.*, 2011c; Mizuno *et al.*, 2013).

The leaf sheath of F_2 plants with the wild-type phenotype and those with the *bm* phenotype were subjected to RNA-seq analysis (Mizuno *et al.*, 2013).

Total RNA of each plant was extracted from leaves and converted to cDNA for massive parallel sequencing in an Illumina Genome Analyzer and differentially expressed genes were identified. Of the 31 downregulated genes, one gene was similar to the ABC transporter responsible for wax secretion in Arabidopsis (Bird *et al.*, 2007; Panikashvili *et al.*, 2007; Ukitsu *et al.*, 2007). This induced *bm* mutant of sorghum was identified to carry a 1.4 Mb genomic inversion proximal to the promoter region of *Sb06g023280*, which is the candidate gene of the *bm* mutant, with small deletions at both ends. The analysis proved that the inversion involving the *Sb06g023280* gene inhibited wax secretion in the bloomless sorghum, although the epicuticular wax was synthesized inside the cells.

6 Activities Contributing to Mutation Breeding in Japan

6.1 Gamma Field Symposium

The 1st Gamma Field symposium was inaugurated in 1962 and these symposia terminated in 2014. The 1st Gamma Field Symposium was held at the Institute of Radiation Breeding for exchanging information and discussions among scientists in national agricultural experiment stations and institutes of Ministry of Agriculture and Forestry, national universities and institutes of Ministry of Education, and seed companies in this new research area of mutation breeding, and for providing a seminar for students of the universities. During its 53-year history, the symposium committee has selected various themes related to mutation and breeding and has invited leading scientists with expertise in these areas as lecturers to provide results of their research on a wide variety of related topics. After the symposium a series of publications, *Gamma Field Symposia* (Nos 1–50), were published (available at http://www.naro.affrc.go.jp/archive/nias/newsletter/#symposia, accessed 2019). The publication includes contributed papers from the invited lecturers in English and the discussions in Japanese. This publication will help plant breeders, researchers and students to realize the contributions of mutation breeding to plant sciences.

6.2 Forum for Nuclear Cooperation in Asia (FNCA)

Food production policy is among the most important subjects confronting nations exhibiting rapid population growth and a diminishing capacity to produce food. FNCA has been conducting a Mutation Breeding Project to utilize mutation breeding technology with irradiation to crops that are highly required by the people of Asia. The aim is to develop varieties that are more resistant to disease, insects and drought, or give higher yields and offer higher quality. Mutation breeding is achieving wider acceptance as a method of breeding superior germplasm following the development of rice and other agricultural crops in the ten member countries (Bangladesh, China, Indonesia, Japan, Korea, Malaysia, Mongolia, Philippines, Thailand and Vietnam) representing the FNCA.

At the beginning of the project, the main activity had been information exchange, but it was shifted to sub-project activities focused on target crops that are highly needed in Asian countries, such as: (i) disease resistance in banana (2004–2010); (ii) insect resistance in orchid (2003–2009); (iii) drought tolerance in sorghum and soybean (2002–2006); (iv) composition or quality in rice (2007–2012); and (v) mutation breeding for sustainable agriculture in Asia (2013–present).

The FNCA website homepage (https://www.fnca.mext.go.jp/english/mb/e_introduction.html, accessed 2019) includes: (i) a mutation breeding database; (ii) mutation breeding manual; and (iii) achievement of the sub-projects mentioned above.

7 Conclusion

In the Preface of Mutation Breeding – Theory and Practical Application, A. M. van Harten (1998) stated:

> An explanation for the decreasing interest in mutation breeding, at least in most 'developed' countries, may be that during the past two decades attention has become more and more directed towards studying the possibilities offered to plant breeding by various new molecular technologies. ... As a result of these developments mutation breeding seems to have lost part of its previous attraction for young researchers. It is even not inconceivable that mutation breeding, as a discrete branch of plant breeding, may sink into oblivion and that, as a consequence, much valuable knowledge on this topic built up throughout the years, will be lost.

The record has shown that mutation induction is a very useful conventional breeding tool for developing crop cultivars (Nakagawa and Kato, 2017: Appendix 1 and 2). Today, site-directed mutagenesis can be envisioned and many researchers are conducting programmes in this direction. The author anticipates that new fields of science and technologies will be developed with continued achievements through the application of traditional or classical methods.

Acknowledgements

The author would like to express sincere thanks to Dr Bryan Kindiger, USDA, ARS-SR, Grazinglands Research Laboratory, for his critical reading of the manuscript. The author thanks Prof. Makoto Kusaba of Hiroshima University and Dr Ryouhei Morita of Riken Nishina Center for Accelerator-based Science for permission to use figures in this report.

References

Akao, S. and Kouchi, H. (1992) A supernodulating mutant isolated from soybean cultivar Enrei. *Soil Science and Plant Nutrition* 38, 183–187.

Arabidopsis Genome Initiative (2000) Analysis of the genome sequencing of the flowering plant *Arabidopsis thaliana. Nature* 408, 796–815.

Ashikari, M. *et al.* (2002) Loss-of-function of a rice gibberellin biosynthetic gene, *GA20 oxidase* (*GA20ox-2*), led to the rice 'Green Revolution'. *Breeding Science* 52, 143–150.

Bhattachayya, M.K. *et al.* (1990) The wrinkled-seed character of pea described by Mendel is caused by a transposon-like insertion in a gene encoding starch-branching enzyme. *Cell* 60, 115–122.

Bird, D. *et al.* (2007) Characterization of Arabidopsis ABCG11/WBC11, an ATP binding cassette (ABC) transporter that is required for cuticular lipid secretion. *Plant Journal* 52, 485–498.

Blum, A. (1975) Effect of the *Bm* gene on epicuticular wax and water relations of *Sorghum bicolor* L. (Moench). *Israel Journal of Botany* 24, 50–51.

Broad, W.J. (2007) Useful mutants, bread with radiation. *New York Times*, August 27, 2007.

Futsuhara, Y. (1968) Breeding of a new rice variety Reimei by gamma-ray irradiation. *Gamma Field Symposia* 7, 87–109.

Hellens, R.P. *et al.* (2010) Identification of Mendel's white flower character. *PLoS ONE* 5(10), e13230.

Iida, S. *et al.* (1993) A rice (*Oryza sativa* L.) mutant having a low content of glutelin and a high content of prolamine. *Theoretical and Applied Genetics* 87, 374–378.

Iida, S. *et al.* (1997) Mutants lacking glutelin subunits in rice: mapping and combination of mutated glutelin genes. *Theoretical and Applied Genetics* 94, 177–183.

Imbe, T. *et al.* (2004) Odorokimochi – a multipurpose high-yield glutinous rice variety. *Bulletin of the National Institute of Crop Sciences* 6, 19–35 [in Japanese with English summary].

International Rice Genome Sequence Project (2005) The map-based sequence of the rice genome. *Nature* 436, 793–800.

Ise, K. *et al.* (2001) 'Milky Queen', a new high-quality rice cultivar with low amylose content in endosperm. *Bulletin of the National Institute of Crop Sciences* 2, 39–61 [in Japanese with English summary].

Iwata, M. and Omura, T. (1971) Linkage analysis of reciprocal translocation method in rice plant (*Oryza sativa* L.). II. Linkage groups corresponding to the chromosome 5, 6, 8, 9, 10 and 11. *Science Bulletin of the Faculty of Agriculture, Kyushu University* 30, 137–153 [in Japanese with English summary].

Jenks, M.A. *et al.* (1994) Chemically-induced cuticle mutation affecting epidermal conductance to water vapor and disease susceptibility in *Sorghum bicolor* (L.) Moench. *Plant Physiology* 105, 1239–1245.

Ji, H. *et al.* (2010) Inactivation of the CTD phosphatase-like gene *OsCPL1* enhances the development of the abscission layer and seed shattering in rice. *The Plant Journal* 61, 98–106.

Jordan, W.R. *et al.* (1983) Environmental physiology of sorghum: I. Environmental and genetic control of epicuticular wax load. *Crop Science* 23, 552–558.

Kasuga, S. *et al.* (2001) Effects of brown midrib and bloomless genes on the resistance to sheath blight (*Rhizoctonia solani* Kuehn) in sorghum. *Grassland Science* 47, 256–261.

Kato, H. *et al.* (2006) A newly bred rice variety 'Minamiyutaka' for whole-crop silage. *Bulletin of Miyazaki Agricultural Experiment Station* 41, 51–60 [in Japanese with English summary].

Kitagawa, K. *et al.* (1999) A new Japanese pear cultivar, 'Kotobuki Shinsui'. *Bulletin of Tottori Horticultural Experimental Station* 3, 1–13 [in Japanese].

Koga, Y. *et al.* (1989) Breeding a new rice cultivar 'Kinuhikari'. *Bulletin of Hokuriku National Agricultural Experimental Station* 30, 1–24 [in Japanese with English summary].

Kokubun, M. and Akao, S. (1994) Inheritance of supernodulation in soybean mutant En6500. *Soil Science and Plant Nutrition* 40, 715–718.

Konishi, S. *et al.* (2006) An SNP caused loss of seed shattering during rice domestication. *Science* 312, 1392–1396.

Kusaba, M. *et al.* (2003) *Low glutelin content1*: a dominant mutation that suppresses the glutelin multigene family via RNA silencing in rice. *Plant Cell* 15, 1455–1467.

Lester, D.R. *et al.* (1997) Mendel's stem length gene (*Le*) encodes a gibberellin 3P-hydroxylase. *Plant Cell* 9, 1435–1443.

Li, C. *et al.* (2006) Genetic analysis of rice domestication syndrome with the wild annual species, *Oryza nivara*. *New Phytologist* 170, 185–194.

Li, X. *et al.* (2001) A fast neutron deletion mutagenesis-based reverse genetic system for plants. *Plant Journal* 27, 235–242.

Ma, J.F. *et al.* (2005) Isolation and characterization of a rice mutant hypersensitive to Al. *Plant Cell Physiology* 46, 1054–1061.

Masuda, T. *et al.* (1997) Selection of mutants resistant to black spot disease by chronic irradiation of gamma-rays in Japanese pear 'Osanijisseiki'. *Journal of the Japanese Society for Horticultural Science* 66, 85–92.

Mizuno, H. *et al.* (2013) Genome inversion caused by gamma irradiation contributes to downregulation of a WBC11 homolog in bloomless sorghum. *Theoretical and Applied Genetics* 126, 1513–1520.

Mochizuki, T. and Hara, S. (2000) Usefulness of low protein rice in diet therapy in patients with chronic renal failure. *Japanese Journal of Nephrology* 42, 24–29 [in Japanese with English summary].

Morita, R. *et al.* (2007) Knockout of glutelin genes which form a tandem array with a high level of homology in rice by gamma irradiation. *Genes & Genetic Systems* 82, 321–327.

Morita, R. *et al.* (2009) Molecular characterization of mutations induced by gamma irradiation in rice. *Genes & Genetic Systems* 84, 361–370.

Nagano, A.J. *et al.* (2008) AtMap1: a DNA microarray for genome deletion mapping in *Arabidopsis thaliana*. *Plant Journal* 46, 1058–1065.

Nagasawa, N. *et al.* (2013) *GIANT EMBRYO* encodes CYP78A13, required for proper size balance between embryo and endosperm in rice. *Plant Journal* 75, 592–605.

Naito, K. *et al.* (2005) Transmissible and nontransmissible mutations induced by irradiating *Arabidopsis thaliana* pollen with γ-rays and carbon ions. *Genetics* 169, 881–889.

Nakagawa, H. (2010) Gamma radiation facilities at the Institute of Radiation Breeding. *Plant Mutation Reports* 2, 55–56.

Nakagawa, H. and Kato, H. (2017) Induced mutations for food and energy security: challenge of inducing unique mutants for new cultivars and molecular research. *Bulletin of the NARO Institute of Crop Science* 1, 33–124 Available at: http://www.naro.affrc.go.jp/publicity_report/pub2016_or_later/files/nics_report01_4.pdf (accessed 24 August 2021).

Nakagawa, H. *et al.* (2011a) Mutation breeding of soybean in Japan. In: Khan, S. and Kozgar, M.I. (eds) *Breeding of Pulse Crops*. Kalyani Publishers, New Delhi, India, pp. 55–84.

Nakagawa, H. *et al.* (2011b) Agronomic characteristics and aphid resistance of *bm* mutant lines of sweet sorghum developed through gamma-ray irradiation. *Japanese Journal of Grassland Science* 57 (Suppl.), 103 [in Japanese].

Nakagawa, H. *et al.* (2011c) Characteristics and inheritance of *bm* mutant induced by gamma-ray irradiation. *Breeding Research* 13 (Suppl. 1), 111 [in Japanese].

Nakamura, D. *et al.* (1991) A new soybean cultivar 'Murayutaka'. *Bulletin of Saga Agricultural Experiment Station* 27, 21–42 [in Japanese with English summary].

Nakashima, T. *et al.* (1982) Structure elucidation of AK-toxins, host specific phyto-toxic metabolites produced by *Alternaria kikuchiana* Tanaka. *Tetrahedron Letters* 23, 4469–4472.

Nakashima, T. *et al.* (1985) Isolation and structure of AK-toxin I and II. Host-specific phytotoxic metabolites produced by *Alternaria alternata* Japanese phytotype. *Agricultural and Biological Chemistry* 49, 807–815.

Nelson, O. and Pan, D. (1995) Starch synthesis in maize endosperms. *Annual Review of Plant Physiology and Plant Molecular Biology* 46, 475–496.

Nemoto, H. *et al.* (2001) A new rice cultivar with giant embryo 'Haiminori'. *Bulletin of Chugoku National Agricultural Experiment Station* 22, 25–40 [in Japanese with English summary].

Nishimura, M. *et al.* (2005) New rice varieties with low levels of easy-to-digest protein, 'LGC-Katsu' and 'LGC-Jun'. *Breeding Science* 55, 103–105.

Nishimura, S. *et al.* (1978) Two different phases in pathogenicity of the *Alternaria* pathogen causing black spot disease of Japanese pear. *Journal of Faculty of Agriculture, Tottori University* 13, 1–10.

Okuno, K. *et al.* (1983) A new mutant gene lowering amylose content in endosperm starch of rice, *Oryza sativa* L. *Japanese Journal of Breeding* 33, 387–394.

Panikashvili, D. (2007) The *Arabidopsis* DESPERADO/AtWBC11 transporter is required for cutin and wax secretion. *Plant Physiology* 145, 1345–1360.

Peiretti, R.A. *et al.* (1980) Relationship of bloomless (*bm bm*) sorghum to greenbug resistance. *Crop Science* 20, 173–176.

Saikusa, T. *et al.* (1994) Accumulation of γ-aminobutyric acid (GABA) in the rice germ during water soaking. *Bioscience, Biotechnology, and Biochemistry* 58, 2291–2292.

Sakai, M. *et al.* (2013) 'Ruriaoba', a new rice forage variety for double harvesting. *Bulletin of National Agricultural Research for Kyushu Okinawa Regions.* 60, 1–12 [in Japanese with English summary].

Samoto, S. and Kanai, D. (1975) Studies on the mutation breeding in rice. I. Short stiff mutations induced by gamma-ray irradiation to the rice variety Koshihikari. *Japanese Journal of Breeding* 25, 1–7 [in Japanese with English summary].

Sanada, S. (1988) Selection of resistant mutants to black spot disease of Japanese pear by using host specific toxin. *Japanese Journal of Breeding* 38, 198–204.

Sanada, T. *et al.* (1993) A new Japanese pear cultivar 'Gold Nijisseiki', resistant to black spot disease of Japanese pear. *Japanese Journal of Breeding* 43, 455–461 [in Japanese with English summary].

Sato, H. *et al.* (2002) Molecular characterization of *Wx-mq*, a novel mutant gene for low amylose content in endosperm of rice (*Oryza sativa* L.). *Breeding Science* 52, 131–135.

Sato, Y. *et al.* (2007) Mendel's green cotyledon gene encodes a positive regulator of the chlorophyll-degrading pathway. *Proceedings National Academy of Sciences USA* 104, 14169–14174.

Satoh, H. (1981) New endosperm mutations induced by chemical mutagens in rice, *Oryza sativa* L. *Japanese Journal of Breeding* 31, 316–326.

Satoh, H. and Iwata, N. (1990) Linkage analysis in rice. On three mutant loci for endosperm properties, *ge* (giant embryo), *du-4* (dull endosperm-4) and *flo-1*(floury endsperm-1). *Japanese Journal of Breeding* 40 (Suppl. 2), 268 [in Japanese].

Takagi, Y. (1969) The second type of gamma-ray sensitive gene RS_2 in soybean *Glycine max* (L.) Merrill. *Gamma Field Symposia* 8, 83–94.

Takagi, Y. and Anai, T. (2006) Development of novel fatty acid composition in soybean oil by induced mutation. *Oleoscience* 6, 195–203 [in Japanese with English abstract].

Takagi, Y. and Rahman, S.M. (1995) Variation of different fatty acids in mutants in comparison with natural soybean varieties. *Bulletin of Faculty of Agriculture, Saga University* 79, 23–67.

Takagi, Y. *et al.* (1998) Construction of novel fatty acid composition in soybean oil by induced mutation. *Gamma Field Symposia* 37, 17–28.

Takahashi, M. *et al.* (1995) Characterization of nitrogen assimilation in a supernodulating soybean mutant En6500. *Soil Science and Plant Nutrition* 41, 567–575.

Takahashi, M. *et al.* (2003) Breeding of supernodulating soybean cultivar 'Sakukei 4'. *Bulletin of National Institute of Crop Science* 4, 17–28.

Takano, M. *et al.* (2005) Distinct and cooperative function of phytochromes A, B, and C in the control of deetiolation and flowering in rice. *Plant Cell* 17, 3311–3325.

Toda, M. (1982) The breeding of four new mutant varieties by gamma-rays in rice. *Gamma Field Symposia* 21, 7–18.

Ueno, T. *et al.* (1977) Mass spectrometry of *Alternaria mali* toxins and related cyclodepsipeptides. *Biomedical Mass Spectrometry* 4, 134–142.

Ukitsu, H. *et al.* (2007) Cytological and biochemical analysis of COF1, an *Arabidopsis* mutant of an ABC transporter gene. *Plant and Cell Physiology* 48, 1524–1533.

van Harten, A.M. (1998) *Mutation Breeding – Theory and Practical Applications*. Cambridge University Press, Cambridge, UK.

Wu, P. *et al.* (1997) Genetic control of seedling tolerance to aluminum toxicity in rice. *Euphytica* 97, 289–293.

Yamaguchi, I. (2001) Forty years of mutation breeding in Japan – research and fruits. *Gamma Field Symposia* 40, 1–14.

Yoshioka, T. *et al.* (2001) Induction of mutant resistant to *Alternaria* blotch of apple by gamma-ray irradiation. *Gamma Field Symposia* 39, 69–79.

Zhou, Y. *et al.* (2012) Genetic control of seed shattering in rice by the APETALA2 transcription factor *SHATTERING ABORTION1*. *Plant Cell* 24, 1034–1048.

4 Soybean Breeding Through Induced Mutation in Vietnam

Thao Duc Le[1] and Chung Thi Bao Pham

Abstract

In Vietnam, soybean is one of the traditional crops and plays an important role in crop rotation, soil improvement and meeting the nutritional needs of humans and livestock. With the aim of generating genetic variability in soybean and creating new soybean varieties to meet the needs of production, induced mutation research has been carried out since the 1980s and has gained outstanding achievements. Induction of modified traits and their incorporation into an ideal genotype was achieved by judicious use of the induced mutation technique. So far, outstanding soybean varieties such as DT84, DT90, DT99, DT2008 and several promising lines have been developed in Vietnam by incorporating desirable traits like high and stable yield (2.0–3.5 t/ha), good quality, drought tolerance, disease resistance (rust, powdery mildew, downy mildew), short growth duration (70–100 days), wide adaptability and suitability for cropping systems and ecological regions in the whole country. The most outstanding variety, DT84, occupies over 50% of the total production area and 80% in Central and North Vietnam (about 70,000–80,000 ha/year). These varieties have also been used as materials for developing several additional improved soybean varieties. Thus, induced mutation research has played an important role in improving soybean varieties in Vietnam.

Keywords: DT84 • DT90 • DT2008 • mutation • soybean

1 Introduction

Soybean is one of the traditional food crops and plays an important role in crop rotation, soil improvement and providing food for humans and livestock in Vietnam (Ngo *et al.*, 1999). In 1980, according to the General Statistics Office (GSO) of Vietnam (www.gso.gov.vn, accessed 2019), the soybean production area reached 49,800 ha, but with a low yield of 0.68 t/ha. The existing soybean varieties had low yield with narrow adaptability, unstable productivity, and suitability for either hot season (summer) or cold season (spring and winter) (Mai *et al.*, 2000a,b).

With the aim of generating genetic variability in soybean and creating new soybean varieties to meet the needs of production, induced mutation research has been carried out since the late 1980s and has gained outstanding achievements. Ten mutant varieties have been released and adopted by farmers, and in addition many promising lines have been generated.

[1] Email: leducthao@agi.vaas.vn

 Mutation Breeding, Genetic Diversity and Crop Adaptation to Climate Change (eds S. Sivasankar *et al.*)
DOI: 10.1079/9781789249095.0004

These varieties have stability with three crops per year, good quality with a high protein content of 38–47%, wide adaptability for many ecological regions and good resistance to disease. The most outstanding variety, DT84, occupies over 50% of the total production area and 80% in Central and North Vietnam (about 70,000–80,000 ha/year). After 20 years of release, no other soybean variety can overcome DT84 in its wide adaptability, resistance and quality. Varieties DT90 and DT2008 occupy about 5000 ha/year, and DT99 about 10,000 ha/year. They contributed to increasing soybean productivity from 0.66 t/ha in 1980 to 1.44 t/ha in 2013.

The demand for soybean is increasing, especially for livestock. After corn, soybean is now considered as the country's second strategic crop contributing to sustainable livestock development. In 2014–2016, soybean production in Vietnam reached 103,000 ha with an average yield of 1.5 t/ha. The average annual import of soybean was 1.6 million tonnes with the value of US$780 million (USDA GAIN, 2019). The production of soybean is now meeting only 8–10% of domestic demand. It was predicted that Vietnam would have to import about 5 million tonnes of soybean grain by 2020. There are many problems restricting soybean production in Vietnam, including the following.

1. Price competition with genetically modified (GM) soybean imported (mainly for animal feed) has reduced the attractiveness of the crop for farmers to expand the production area.
2. Cultivated land is generally in small parcels and is scattered in area. This terrain makes it difficult to mechanize production, resulting in high production costs.
3. Climate changes such as drought and extremes of hot and cold are presenting challenges.
4. High yield and quality varieties are lacking and there is development of pests and diseases.
5. Farming techniques are poor, especially in postharvest management. The limitation of mechanization in soybean production in Vietnam due to difficult weather conditions and uneven terrain has increased production costs. The high price of domestically produced soybean puts it at a disadvantage in the market; buyers prefer the cheaper imports, so the demand for domestically produced soybean is low and consequently farmers are reluctant to increase the production area.

2 Materials and Methods

2.1 Materials

Materials included local and introduced soybean varieties ('Coc chum', ĐH4, AK04, V74), hybrid lines (D.3/033, D.98-099, 2001HC) and created varieties (DT2008, DT96, ĐT26).

2.2 Methods

Radiation method

Dry seeds were irradiated with ^{60}Co gamma-rays at doses of 50, 100, 150, 180, 200, 220 and 250 Gy.

Chemical treatments

Dry seeds were treated with chemical mutagens of DES, EMS, NMU and NDMU at concentrations of 0.02, 0.04, 0.06 and 0.08% w/v for 2, 4, 6 and 8 h. Ethyleneimine concentration was 0.008, 0.02 and 0.05%.

Selection methods

Single-seed descent and pedigree methods were applied from the M_2 to M_5 generations.

Evaluation of promising mutant lines

Evaluations were carried out according to Test Guidelines No. 10TCN 339-2006 and QCVN 01-58: 2011/BNNPTN at experimental stations of the Agricultural Genetics Institute, Legumes Research and Development Centre and Vietnam National University of Agriculture in Hanoi.

Release of a new variety

The elite mutant soybean lines were evaluated based on the National Technical Regulation on Testing for Value of Cultivation and Use of Soybean varieties No. 10TCN-339/2006. The promising soybean lines were assayed and introduced into trial production in different ecological regions and in different seasons. National Testing and release were as for a new variety.

3 Results

Over nearly 30 years (1990–2018), 11 mutant soybean varieties have been released, including six national varieties (DT84, M103, ĐT22, DT55 (AK06), DT90 and DT2008) and three regional varieties (S-31, DT99 and DT95) together with promising mutant lines (DT2008ĐB, DT96ĐB, ĐT26ĐB ...) with many significant improved characteristics. The radiation doses that have been successfully applied in soybean breeding are 50, 100, 150, 180 and 200 Gy (Table 4.1).

Breakthrough traits of these varieties are wide adaptation, suitability of productivity for three crops per year (spring, summer, winter) in the eco-regions from north to south (DT84, DT2008, DT95, DT55, DT90, DT99, DT83), drought tolerance (DT2008), resistance to rust (DT95, DT2008, DT96ĐB, DT84, DT2008ĐB, DT99), resistance to powdery mildew (DT90, DT2008, DT2008ĐB), resistance to lodging (ĐT22, DT90, M103, DT96ĐB), good quality (large seeds with high protein content of 40–47%) (DT90, DT2008, DT2008ĐB) and high productivity of 2.0–4.0 t/ha (DT2008, DT2008ĐB) (Table 4.2).

4 Discussion

The technique of induced mutation has been applied in Vietnam since the 1980s with physical and chemical mutagens. It has been used to create more variation and plays an important role in creating new soybean varieties in Vietnam. Up to 2018, there were ten soybean mutant varieties and promising mutant lines. The main mutagenic treatment of soybean breeding in Vietnam has been irradiation of dry seed with gamma-rays from a ^{60}Co source. Ten (77%) of the 13 soybean mutants (DT84, M103, DT90, DT95, DT99, ĐT22, DT2008, DT2008ĐB, DT96ĐB and ĐT26ĐB) were created by ^{60}Co gamma irradiation. The other three were created by chemical mutagenesis or a combination of chemical mutagenesis and gamma irradiation.

In particular, DT84 was created by ^{60}Co gamma irradiation of dry seeds of D.3/033 line (DT80 × ĐH4) at a dose of 180 Gy. DT84 has resistance to rust, wide adaptability in different ecological regions from north to south (Northern Mountainous area, Red River Delta, Central Highlands, Mekong Delta) and is cultivated as three crops/year (spring, summer and winter). In 2003, DT84 ranked first in area with 47,576 ha (Pham *et al.*, 2003). So far it has been the main variety in the Northern provinces. In 2005, DT84 was awarded a national prize of Science & Technology VIFOTEC – 2005. From 2007 to 2011, the production area of DT84 reached 81% of the winter crop and 60% of the summer crop in the Northern provinces (Department of Horticulture, 2012). In 2017, the production area of DT84 reached 28,900 ha in the Hanoi soybean production area (accounting for 90%) and 40,500 ha of the whole country (accounting for 40%). Today DT84 has been developing in Lao and Cambodia.

DT90 was created by ^{60}Co gamma irradiation at 180 Gy on dry seeds of the hybrid line (K7002 × 'Coc chum'). DT90 has good quality (large seeds, nice colour, yellow hilum, high protein content of 47%), is resistant to powdery mildew and has good resistance to lodging. It is suitable for intercropping, thanks to its canopy. Furthermore, it is tolerant to cold (therefore suitable for winter and spring crop in the Red River Delta and for spring and summer (August) crop in the Northern Mountainous provinces) (Mai *et al.*, 2000a,b, 2007a, 2009; Duong, 2005). During 2000–2005, the production of DT90 reached an average area of 3000 ha per year.

DT99 was created by ^{60}Co gamma irradiation at 150 Gy on dry seeds of the hybrid line D.98-099 (IS011 × Cuc). It has wide adaptation with three crops/year and is suitable for intercropping and rotation with other crops thanks to its short growth duration of 70–80 days. During 2000–2003, the average production area of DT99 reached 5000 ha per year.

DT2008 was created by ^{60}Co gamma irradiation at 180 Gy on dry seeds of the hybrid line 2001HC (DT2001 × HC100). It has a high yield of 2.0–4.0 t/ha (23–45% higher than that of DT84) (Mai *et al.*, 2010, 2012; Pham, 2015), better tolerance to drought than W82 (Ha *et al.*, 2012; Sulieman *et al.*, 2015) and good resistance to rust, downy mildew and bacterial blight (Mai *et al.*, 2008, 2010, 2012; Pham, 2015). During 2008–2017, DT2008 was tested and introduced into production at Ha Noi, Thai Nguyen, Tuyen Quang, Cao Bang, Phu Tho, Yen Bai, Ha Giang, Dak Lak and Dak Nong (Pham, 2015).

Table 4.1. Mutant soybean varieties released in Vietnam (1990–2018).

No.	Mutant line/ variety name	Registration year and level	Origin	Mutagen	Traits improved	Reference
1	S-31	1995 Regional var.		Gamma-ray ^{60}Co-180 Gy + EI-0.04%	High yield and tolerance to low temperature	Mai *et al.*, 2000a,b
2	M103	1994 National var.	ĐH4	Gamma-ray ^{60}Co-50 Gy + EI-0.01%	High grain yield, resistance to lodging and high grain quality	Duong, 2005; Mai *et al.*, 2009
3	DT83	1990 National var.	Coc chum	EI-0.04%	Seed coat colour (green to yellow), 60% higher 100-seed weight, 70% higher yield, 50% higher stem height	Duong, 2005; Mai *et al.*, 2000a,b, 2007a,b, 2009
4	DT84	1995 National var.	D.3/033-F3 (ĐH4 × ĐT80)	Gamma-ray ^{60}Co-180 Gy	Yield 30–40% higher, heat and cold tolerance	Duong, 2005; Mai *et al.*, 2000a,b, 2007a,b, 2009
5	DT95	1998 Regional var.	AK04	Gamma-ray ^{60}Co-180 Gy	Seed coat colour (green to yellow), 15–20% higher yield, early maturity	Duong, 2005; Nguyen *et al.*, 2006; Mai *et al.*, 2000a,b, 2007a,b, 2009
6	DT99	2000 Regional var.	D.98-099-F4 (IS01 × Cuc)	Gamma-ray ^{60}Co-150 Gy	Early maturity, good adaptability, resistance to rust, tolerance to high and low temperature	Duong, 2005; Mai *et al.*, 2000a,b, 2007a,b, 2009
7	AK06 (DT55)	2000 National var.	V74	Gamma-ray ^{60}Co-100 Gy	Early maturity and high yield	Duong, 2005; Mai *et al.*, 2000a,b, 2007a,b, 2009
8	DT90	2002 National var.	Hybrid line-F4 (K7002 × Coc chum)	Gamma-ray ^{60}Co-180 Gy	20–30% higher 100-seed weight, high yield and high protein content	Duong, 2005; Mai *et al.*, 2000a,b, 2007a,b, 2009
9	ĐT22	2006 National var.	Hybrid line (ĐT12 × DT95)	Gamma-ray ^{60}Co-150 Gy	Early maturity	Duong, 2005; Tran *et al.*, 2007
10	DT2008	2010 National var.	2001HC-F4 (DT2001 × HC100)	Gamma-ray ^{60}Co-180 Gy	High yield, resistance to rust, downy mildew, bacterial posture and drought tolerance	Mai *et al.*, 2008, 2010, 2012; Ha *et al.*, 2012; Pham, 2015
11	DT2008ĐB	2015 National Test	DT2008	Gamma-ray ^{60}Co-200 Gy	Seed coat colour (yellow to black), higher content of carotenoid, omega 3 and omega 6 of 58, 30 and 12% respectively, earliness of 5–8 days	Le *et al.*, 2015a; Nguyen *et al.*, 2016a,b; Nguyen, 2018
12	DT96ĐB	2015 National Test	DT96	Gamma-ray ^{60}Co-200 Gy	Resistance to lodging, hair colour of main stem (brown to grey)	Le *et al.*, 2015a,b, 2017b; Nguyen, 2018
13	ĐT26ĐB	2015 Promising line	ĐT26	Gamma ray ^{60}Co-150 Gy	Seed coat colour (yellow to black)	Le *et al.*, 2015a, 2017a; Nguyen, 2018

Table 4.2. Characteristics of mutant soybean varieties created by induced mutation in Vietnam (1990–2018).

Mutant line/variety name	Flower colour	Seed coat colour	Plant height (cm)	Growth duration (days)	Productivity (t/ha)	1000-seed weight (g)	Protein content (%)	Tolerance and resistance	Adaptability in crop seasons
S-31	Violet	Yellow	45–60	85–90	1.7–2.5	150–170	38	Tolerance to low temperature	Spring and winter
M103	Violet	Yellow	55–70	85–90	1.7–2.5	160–180	40	Heat tolerance and lodging resistance	Summer
DT83	Violet	Yellow	40–50	85–94	1.2–2.8	138–145	38	Resistance to powdery mildew	Spring, summer, winter
DT84	Violet	Yellow	45–50	85–90	1.7–3.0	160–180	41	Resistance to rust, tolerance to high and low temperature	Spring, summer, winter
DT95	Violet	Yellow	50–80	90–103	1.8–3.6	160–180	41	Resistance to rust	Spring, summer, winter
DT99	White	Yellow	35–45	70–80	1.4–2.3	150–170	41	Resistance to rust, tolerance to high and low temperature	Spring, summer, winter
DT55 (AK06)	White	Yellow	45–60	85–95	1.7–3.0	160–180	41		Spring, summer, winter
DT90	White	Yellow	45–50	90–100	1.8–3.0	180–200	47	Tolerance to canopy, resistance to powdery mildew, lodging	Spring, summer, winter
ĐT22	White	Yellow	45–50	85–90	1.8–2.7	155–160	–	Resistance to lodging	Spring, summer, winter
DT2008	Violet	Yellow	50–70	95–110	2.0–4.0	200–220	40	Drought tolerance and resistance to rust, downy mildew, powdery mildew, bacterial posture	Spring, summer, winter
DT2008ĐB	Violet	Black	50–70	90–100	2.0–3.5	200–220	42	Drought tolerance and resistance to rust, downy mildew, powdery mildew, bacterial posture	Spring, summer, winter
DT96ĐB	Violet	Yellow	45–55	90–95	2.0–3.0	180–200	40	Resistance to rust, powdery mildew and lodging	Spring, summer, winter
ĐT26ĐB	White	Black	45–60	90–95	1.8–2.7	165–180	41		Spring, winter

DT2008ĐB was created by ^{60}Co gamma irradiation at 200 Gy on dry seeds of DT2008. It has a black seed coat. Compared with its parent DT2008, it has the same good growth and development, with only mild infection by some diseases (e.g. rust, downy mildew), and has a high yield (2.46–3.18 t/ha) but it has a higher nutrient content and is 5–8 days shorter in growth duration (90–100 days) (Le *et al.*, 2015a; Nguyen *et al.*, 2016a,b; Nguyen, 2018).

In general, the technique of induced mutation has been an effective tool for improving soybean varieties in Vietnam. The following significant characteristics have been obtained from the application of irradiation in soybean breeding in Vietnam (Mai *et al.*, 2009):

1. **Improving yield:** Increases in yield compared with the original were recorded in DT83, DT84 and DT95 at 70%, 30–40% and 15–20%, respectively.
2. **Improving seed quality and colour:** A change of seed coat colour from blue to yellow was observed in DT83 and DT95 varieties. Increases in 1000-seed weight were recorded in DT83 and DT90 at 60% and 20–30% higher, respectively, compared with the originals.
3. **Improving adaptability:** The varieties DT84, DT90, DT95, DT99, AK06 (DT55) and DT2008 can be cultivated as three crops/year in the Northern provinces of Vietnam by combining the heat and cold tolerance of their parent. AK06 can be planted as three crops in North Vietnam after improving the character of non-tolerance to heat in the original variety V74. Mutant variety DT95 showed its resistance to 7/10 strains of rust (*Phakopsora pachirhizi* Sydow) (Nguyen *et al.*, 2006). The mutant variety DT2008 showed high resistance to three kinds of diseases (rust, downy mildew and bacterial blight) and in addition exhibits drought tolerance.
4. **Improving growth duration:** Early maturity was recorded in DT95, DT99, AK06, ĐT22 and DT2008ĐB.

5 Conclusion

Through ^{60}Co gamma irradiation at doses of 50, 100, 150, 180 and 200 Gy, the Agricultural Genetics Institute has successfully created 13 soybean varieties/lines such as DT84, DT90, DT99, DT2008, etc. These have wide adaptation with three crops/year, high and stable yield, good quality and good resistance to diseases, and are suitable for different ecological regions. The mutant varieties cover about 70% of the production area in the whole country.

References

Department of Horticulture (2012) Soybean. In: *Investigation on the System of Rice, Peanut and Soybean in 2011 Crop*. Ministry of Agriculture and Rural Development, Hanoi.

Duong, H.D. (2005) 450 high-yielding new crop varieties, Vol. II: 281 varieties of food and cereal crops. Labour and Society, May 2005. Ministry of Labour, Hanoi.

Ha, V.C. *et al.* (2012) Characterization of the newly developed soybean cultivar DT2008 in relation to the model variety W82 reveals a new genetic resource for comparative and functional genomics for improved drought tolerance. *Journal of Biomedicine and Biotechnology* 2012.

Le, D.T. *et al.* (2015a) The effect of ^{60}Co gamma irradiation doses on creating mutants beneficial for soybean breeding. *Journal of Vietnam Agricultural Science and Technology* 2(9), 5–9.

Le, D.T. *et al.* (2015b) Research for improving soybean variety DT96 through ^{60}Co gamma irradiation on dry seeds. *Journal of Vietnam Agricultural Science and Technology* 4(57), 18–21.

Le, D.T. *et al.* (2017a) Research for improving soybean variety ĐT26 through ^{60}Co gamma irradiation on dry seeds. *Journal of Agriculture and Rural Development (Special Subject of Crop varieties and Animals)* 1, 65–68.

Le, D.T. *et al.* (2017b) Results of evaluating promising mutant lines derived from DT96 through Co^{60} gamma irradiation. *Journal of Agriculture and Rural Development (Special Subject of Crop varieties and Animals)* 1, 69–71.

Mai, Q.V. *et al.* (2000a) Twenty-year achievements (1984–2004) for researching and developing soybean varieties with high yield, wide adaptability, good quality (DT84, DT90, DT96, DT55-AK06, DT99,

DT94, DT95, DT83) created by Agricultural Genetics Institute. In: Award for creating science and technology in Vietnam and WIPO award in 2005, pp.137–140.

Mai, Q.V. *et al.* (2000b) Twenty-year results on application of induced mutation in soybean (Glycine max (L.) Merr.) breeding at Agricultural Genetics Institute (AGI), Hanoi, Vietnam. *Proceeding of the 8th Workshop on Plant Mutation Breeding – Effective Use of Physical/Chemical Mutagens* October 9–13, 2000, Hanoi, Vietnam.

Mai, Q.V. *et al.* (2007a) Results of breeding and development of soybean mutant varieties DT83, DT84, DT90, DT95, DT96, DT99, DT2001, DT55(AK06), S31 created by AGI in period from 1990–2007. Report in Scientific Council of MARD. Ministry of Agriculture and Rural Development, Hanoi.

Mai, Q.V. *et al.* (2007b) Some results of ^{60}Co gamma irradiation application on creating new soybean varieties of Agricultural Genetics Institute. In: *Proceedings of 7th National Science Conference of Nuclear Technology*, Da Nang 8/2007, pp. 111–112.

Mai, Q.V. *et al.* (2008) Some application results of ^{60}Co gamma irradiation for improving soybean varieties of Agricultural Genetics Institute 2001–2007. In: *Proceedings of the 7th National Conference on Nuclear Science and Technology*, pp. 442–448.

Mai, Q.V. (2008) Creation and development of drought soybean varieties with wide adaptation and tolerance to difficult conditions due to climate change in Northern mountains. In: *Proceedings of the Conference 'Scientific and Technological Report for Agriculture and Rural Development in Northern Mountainous Provinces*. Scientific and Technological Report for Agriculture and Rural Development, Phu Tho 12/2008, pp. 72–77.

Mai, Q.V. *et al.* (2009) Current status and research directions of induced mutation application to seed crops improvement in Vietnam. In: *Induced Plant Mutations in the Genomics Era*. Food and Agriculture Organization of the United Nations, Rome, pp. 341–345.

Mai, Q.V. *et al.* (2010) Results of researching and creating drought soybean variety DT2008. *Journal of Vietnam Agricultural Technology and Science* 2(15), 46–50.

Mai, Q.V. *et al.* (2012) Results of evaluating and selecting drought and high-yielding soybean varieties suitable for difficult ecological condition due to climate change in Central Highlands. *Journal of Vietnam Agricultural Science and Technology* 2(15), 29–35.

Ngo, T.D. *et al.* (1999) *Soybean*. Agricultural Publishing, Hanoi, 366 pp.

Nguyen, T.B. *et al.* (2006) Characterizing resistance of soybean accessions to soybean rust (Phakopsora pachyrhizi). In: *National Soybean Rush Symposium*. Nov. 29– Dec. 1, 2006. St. Louis, Missouri, USA.

Nguyen, V.M. (2018) Research on improving soybean varieties DT2008, DT96 and ĐT26 through Co^{60} gamma irradiation. PhD Thesis in Agriculture, Vietnam Academy of Agricultural Science, 130 pp.

Nguyen, V.M. *et al.* (2016a) Results of researching and creating black soybean variety DT2008ĐB. In: *Proceedings of 2nd National Crop Science Conference*, Can Tho 11–12/8/2016, pp. 488–493.

Nguyen, V.M. *et al.* (2016b) Results of evaluating promising mutant lines derived from DT2008 through ^{60}Co gamma irradiation. *Journal of Agriculture and Rural Development (Special Subject of Crop Varieties and Animals)* 2, 162–165

Pham, D.Q. *et al.* (2003) Report of investigating crop varieties in the whole country in 2003. Results of assaying and testing crop varieties in 2003, pp. 189–193.

Pham, T.B.C. (2015) Research on creating soybean varieties suitable for some Northern mountainous provinces in Vietnam. PhD Thesis in Agriculture, Vietnam Academy of Agricultural Science, 128 pp.

Sulieman, S. *et al.* (2015) DT2008: a promising new genetic resource for improved drought tolerance in soybean when solely dependent on symbiotic N_2 fixation. *BioMed Research International* (Special Issue, 2015), article 687213. doi: 10.1155/2015/687213

Tran, D.L. *et al.* (2007) Results of creating soybean variety ĐT22 (ĐT22.2) by mutation method. In: *Conference of Agricultural Advanced Techniques*, 2007, Hanoi.

USDA GAIN (2019) REPORT 2019: Vietnam Oilseeds and Products Annual. US Department of Agriculture Global Agricultural Information Network. Report VM9013.

5 New Mutation Techniques for Crop Improvement in China

Luxiang Liu[1], Yongdun Xie, Huijun Guo, Linshu Zhao, Hongchun Xiong, Jiayu Gu and Shirong Zhao

Abstract

There are at least 1 billion hungry people worldwide and the Asia and Pacific region harbours the biggest estimated regional distribution of hunger. Lifting a billion people out of poverty and feeding more than 9 billion by 2050 will require increasing cereal production by 70%. Accelerating the development of agriculture to continually increase productivity should be the final approach to end poverty. Mutation techniques have played very significant roles in ensuring food security by developing new mutant germplasm and mutant varieties in China, which have generated a tremendous socio-economic impact. New mutagenesis approaches were initiated in the late 1980s by Chinese scientists, including spaceflight and heavy-ion beam irradiation used as new effective and alternative ways for crop genetic improvement. Protocols for crop mutation induction by space radiation with high-energy heavy-ion beams have been established and applied for crop breeding. More than 1030 mutant varieties with high-yielding, fine-quality and multi-resistant traits have been developed and officially released mainly in cereals, oil and vegetable crops. They have been playing an important role in agricultural production. Hundreds of rare mutant germplasm accessions with a possible breakthrough effect on main economic traits such as grain yield and quality were also identified and applied in conventional breeding programmes. The development of new mutation techniques will be heavily based on, and associated with, not only effective use of nuclear and aerospace research platforms, but also advanced plant omics and molecular biology.

Keywords: induced mutation • crop breeding • gamma • heavy-ion beam • aerospace

1 Introduction

Food production must increase to meet the demand of expanding populations in the next 30 years and developing elite crop varieties could serve as an effective way to secure future food supplies. Their genomes provide the genetic basis for cultivating elite varieties with high and stable yield potential, good quality, and high nutrient use efficiency. Although genetic diversity in wild populations of crop precursor species may remain high, the process of domestication and breeding has selected a restricted subpopulation of this diversity, especially for important agronomic traits such as seed shattering, flowering time adaptation and disease resistance. (Bevan *et al.*, 2017). This is particularly the case for polyploid wheat, which supplies more than 20% of the protein and caloric intake of the human diet. Cultivated hexaploid and tetraploid wheats have suffered a dramatic loss of genetic diversity compared with their wild relatives. The

[1] Email: liuluxiang@caas.cn

DOI: 10.1079/9781789249095.0005

genetic variation of wheat is less than 30% of that in wild *Triticum* species, which is much lower than that in maize and millet (Borrill *et al.*, 2018). Maize and millet have retained over 60% of the genetic diversity of their wild relatives (Wright *et al.*, 2005), yet this is an insufficient proportion to support the development of breakthrough varieties. Broadening crop genetic diversity has become an urgent need for developing climate-smart crop varieties with characteristics of heat tolerance, drought tolerance, resistance to new diseases and high nutritional value.

Induced mutation has been playing a significant role worldwide in enriching genetic diversity, agriculturally important gene identification and elite crop variety development (Bado *et al.*, 2015). Induced mutation breeding in crops has served as an effective strategy in China for more than 60 years. The mutant varieties have also generated a tremendous socio-economic impact. To report and display new results on mutation breeding, the Chinese Society of Nuclear Agricultural Sciences hosted by the Chinese Academy of Agricultural Sciences (CAAS) publishes the monthly *Journal of Nuclear Agricultural Sciences* (available at http://www.hnxb.org.cn/CN/volumn/current.shtml, accessed April 2021).

2 Developmental Process of Mutation Breeding in China

Plant mutation induction in China experienced four different developmental stages between 1957 and 1969. In 1957, the first research laboratory for the use and application of atomic energy in agriculture was established in CAAS. Since then, institutes and laboratories have been consecutively set up in many provinces for the application of mutation induction in breeding. During this period, national training courses on mutation induction were held and more than 300 scientists were trained. Mutation induction facilities were built gradually and applied in mutation breeding of crops including rice, wheat and soybean.

The second stage was a rapid development stage that involved the speedy application of mutation induction, mutant screening and mutant dissemination from 1970 to 1989. In this period, various crops were subjected to mutation induction and breeding; and both the number of mutant varieties and their cultivated area increased rapidly. The breeders who developed the ten most famous mutant varieties in rice, wheat, soybean, maize, cotton and mulberry won National Invention Awards. In 1981, the second edition of the *Manual on Mutation Breeding*, which was published in 1977 by the Joint FAO/IAEA Division, was translated into Chinese, helping local breeders enrich their fundamental knowledge of induced mutations in breeding.

The third stage, from 1990 to 1999, was of steady development. During this period, mutation techniques were integrated with biotechnologies, offering highlighted new opportunities for the application of mutagenesis. Space mutagenesis was initially used for mutation induction and crop improvement (Liu and Zheng, 1997). Mutation induction mechanisms by irradiation methods such as the use of ion beams have been a focus for local scientists (Lin *et al.*, 1996). At this stage, mutant libraries covering important agronomic traits for different crops started to be built. Mutagenesis using plant tissue culture, acting as an important discipline and research area, was quickly spread in mutation research programmes. The combination of conventional crosses, mutation induction and cell engineering (doubled haploids) became a new breeding system for crop improvement.

The fourth stage has been of molecular mutagenesis from 2000 to date. During this period, mutation induction approaches have been deeply associated with biotechnologies for mutation induction, mutant analysis and development of good germplasm (Wang *et al.*, 2015a,b; Zhao *et al.*, 2015). New technical systems for mutation induction, such as space mutagenesis and ion beam irradiation, coupled with efficient mutation identification, such as Targeting Induced Local Lesions in Genomes (TILLING) and proteomic platforms, were successfully established and applied in mutation research and breeding programmes. Some mutants with desired agronomic traits were analysed in detail for dissection of the underlying developmental and mutational mechanism (Zhang *et al.*, 2016; Li *et al.*, 2017; Xiong *et al.*, 2017, 2018).

2.1 Mutation induction techniques developed and applied in China

As is the case throughout the world, gamma-ray irradiation is the commonest and most widely used approach for mutation induction and breeding in China. More than 50 gamma-ray facilities have been established and applied in agricultural research institutions. To date, more than 80% of the released crop mutant varieties were developed based on, or combined with, gamma-ray irradiation. Due to its relatively easy sample preparation, gamma-ray irradiation has been widely applied in almost all crops. Gamma-ray irradiation still plays a great part in mutation breeding. In order to obtain a deep understanding of the factors affecting radiosensitivity of hexaploid wheat to gamma irradiation, we studied the biological effects by using the highly radiosensitive wheat variety HY1 and insensitive control J411, which were screened from a natural population. Our results showed that both free-radical content and total antioxidant capacity exhibited significant differences between the two varieties (Han *et al.*, 2016).

Ion beam irradiation has been developed in the past 30 years. Among the mutation induction methods, heavy-ion beam irradiation is a novel and efficient approach to induce genetic variation. Lithium, carbon and protons are the most commonly used ion sources for mutation induction. With collaboration between the Chinese Institute of Atomic Energy and the Institute of Modern Physics of the Chinese Academy of Sciences, heavy-ion beam mutagenesis pipelines working for national and international researchers have been successfully developed. Different from traditional gamma-rays or X-rays, heavy-ion beam irradiation can deposit both energy and mass locally. It was also found that heavy-ion beam irradiation could cause higher relative biological effectiveness (RBE) (Guo *et al.*, 2007). Due to the advantages in induction of both new phenotypic features of mutants and their broad mutation spectrum of phenotypes, heavy-ion beam irradiation was rapidly adopted for mutation research (Tanaka *et al.*, 2010). By using heavy-ion beam irradiation, we identified a mutant displaying a novel stem development pattern in winter wheat (Zhang *et al.*, 2016). During the jointing stage, the stems of the mutant elongated much more quickly than that of wild-type counterparts. Consequently, the mutant plants were taller than the wild-type plants in this stage. However, the final plant height of both mutant and wild-type plants was the same after anthesis. This stem development mutant supplies a new gibberellin-sensitive semi-dwarf germplasm and is a potential gene resource for wheat improvement.

A simulated cosmic ray (CR) irradiation method for plant mutagenesis was developed as a result of the collaboration between the Institute of High Energy Physics (IHEP) and the Chinese Academy of Sciences. By using the facilities of the Beijing Electron Positron Collider (BEPC) in IHEP, we mimicked cosmic ray irradiations, which are mainly composed of mesons, photons, protons, electrons, positrons, etc. Irradiated by CR, wheat plants showed inhibited mitosis and various chromosome aberrations. Semi-dwarf genotypes have played an important role in developing high-yielding varieties of wheat and rice, which triggered the first Green Revolution. We applied this approach to wheat mutagenesis aiming to obtain ideal mutant resources with agronomic significance. Taking D6-3 as the initial genotype, a gibberellin-sensitive semi-dwarf mutant with shortened lifetime was successfully identified and used in breeding programmes (Xiang *et al.*, 2014).

Spaceflight-induced mutation, also called space breeding, is a technique using an aerospace environment to generate genetic variation. The plant seeds or tissues are taken to aerospace by recoverable spacecraft or high-altitude balloons and kept there for one or several weeks. The space environment is characterized by its mixture of cosmic ray irradiation, microgravity, weak geomagnetic field and super vacuum, and is super clean. We launched and recovered a specific breeding satellite, Shijian 8, in 2006 to conduct basic research and for the application of space breeding. This satellite, with a total mass of 208.8 kg, carried 2020 sets of crop seeds and microorganism strains in nine categories, covering 152 species, together with a set of instruments for space environment exploration. Sensitivity to aerospace environment differs obviously among plant species and crop varieties. Different from traditional gamma-ray irradiation and other mutagenic treatments, the first generation (SP1) produced by spaceflight did not show strong damage effects. For some crop species, such as wheat, barley, maize, cotton and sunflower, spaceflight could even enhance the growth of SP1 seedlings. Remarkably, although the genetic variation in the SP2 generation of wheat was lower than that of gamma-rays at

200 Gy in the M_2 generation, a much wider phenotypic variation rate was observed in the SP2 generation, ranging from 2.2% to 11.1% (Liu *et al.*, 1996).

2.2 Biotechnologies for mutant characterization and mutant variety development

TILLING, a powerful and efficient reverse genetics technique for identifying genetic variation, was introduced in 2000. For the identification of mutants, this method relies on the formation of DNA heteroduplexes that are formed by wild-type and mutated genomic fragments. A 'bubble' structure will form at the mismatch of two DNA fragments, which is then cleaved by CEL1 nuclease. The products can be separated and analysed by our Li-COR 4300 system, which was supported by IAEA. We have successfully screened novel mutant alleles of the wheat starch synthesis gene *TaSSIVb-D* from our mutant library (Guo *et al.*, 2017). We identified 54 mutations in the *TaSSIVb-D*, with a mutation density of 1/165 Kb using the TILLING method. The expression level of *TaSSIVb-D* in these mutants is significantly reduced, which subsequently affected both granule number in chloroplast and photosynthetic activity.

Transcriptome and proteome analysis could provide an overall understanding of gene and protein expression. In a combination of transcriptome and proteome analyses and mutant lines with agronomic significance, we will get clear and beneficial understanding of the mechanisms underlying the mutated phenotypes. The specific nucleic acid fragments and proteins can be also used as molecular markers in subsequent breeding programmes to accelerate development of new varieties. From a wheat segregating population induced by spaceflight, we obtained a mutant characterized by its enhanced salinity tolerance, named *salinity tolerance 1* (*st1*). Results of the comparison of transcriptome information between wild-type and *st1* showed that both the homeostasis of oxidation–reduction process and the 'butanoate metabolism' pathway are responsible for the mutant's salt-tolerant activity (Xiong *et al.*, 2017). For another spaceflight-induced wheat albino mutant *mta*, a combined analysis of transcriptome and proteome was conducted (Shi *et al.*, 2017). Using transcriptome and chloroplast proteome profiling of *mta* and wild-type, we identified differentially expressed genes and proteins. Our results showed that transcription factors including members of the PIF3, GLK and MYB families might participate in chloroplast development and chlorophyll biosynthesis pathways.

Cell engineering, mainly doubled haploids and *in vitro* mutagenesis, plays an important role in accelerating phenotype stabilization and development of new varieties. By using a combination of gamma-ray irradiation and anther culture, we developed the new wheat germplasm H307, with high anther culture ability and with the highest callus induction frequency of 94% and green plantlet production rate of 31% (Zhao *et al.*, 2015). H307 could be used as a worthwhile germplasm for tangible improvement of agronomic traits in winter wheat.

3 Elite Crop Mutant Varieties Released in Recent Years

Mutation induction techniques have been proven to be a powerful and fruitful method widely applied for crop improvement in China. Mutant varieties are playing very important roles in securing our food supplies. By the end of 2018, the total number of officially released mutant varieties had reached more than 1030, covering 46 crop and ornamental species. The maximum annual dissemination area of mutant varieties was more than 9 million hectares. The mutant varieties yielded more than 1.5 million tonnes of crop production annually, which was valued at US$500 million.

The Wheat Mutation Breeding Group (WMBG) led by Professor Luxiang Liu in the Institute of Crop Sciences (ICS) of CAAS has developed and officially released 14 wheat mutant varieties with improved salt tolerance, higher yielding potential, better quality, or enhanced drought resistance. The wheat mutant variety H6756, characterized by its salt tolerance, produces 17% higher yield than the control variety. Hangmai247, which was released nationally in 2016, has high and stable yield potential in a large area. The national variety Luyuan502, which was developed by collaboration with Institute of Agro-food Science and Technology, Shandong Academy of Agricultural Sciences, is a widely adaptable, sprouting-resistant, lodging-resistant and high-yielding variety. The average yield of this variety is 8.2 t/ha and the maximum yield is 12.2 t/ha. The dissemination area of Luyuan502 has been increasing quickly since its release. Its yield is 11% higher than the national control variety and it is also more tolerant to drought and the main diseases. By 2017, the variety had a cumulative cultivation area of over 3.6 million hectares – almost as large as Switzerland. Hangmai2566, Hangmai287 and Hangmai501 are new mutant varieties, developed by WMBG and released in 2018, that are starting to contribute their benefits for higher grain yield, higher farmers' income and a better environment. Due to their contribution to mutation breeding, WMBG was awarded the 'Achievement Award for Plant Mutation Breeding' by the Joint FAO/IAEA Division in 2014.

4 Mutation Breeding Research Network in China and in Asia & Pacific Region

Since the 1980s, the National Centre of Space Mutagenesis for Crop Improvement of the Institute of Crop Sciences (ICS) of CAAS in Beijing has been the coordinator of national research projects on crop mutation breeding and genetics. We are leading two research networks, National Network of Crop Nuclear Radiation Breeding and National Network of Crop Space Mutagenesis Research, and two academic networks of the Chinese Society of Nuclear Agricultural Sciences and Crop Biotechnology Branch of Chinese Society of Agriculture Biotechnology. We have received continuous support for these projects from both the National Government and Provincial Governments through regular 5-year plans. The newest national project under the National Key R&D Program is Induced Mutations for Crop Improvement (2016–2020), which involves more than 40 national research groups working together.

The Regional Cooperative Agreement (RCA) for Research, Development and Training is an intergovernmental agreement established in 1972 under the auspices of the International Atomic Energy Agency (IAEA), which is open to the participating Member States of the IAEA from South Asia, South-east Asia and the Pacific or the Far East. We established a website (www.plantmutagenesis.net, accessed 2019) through IAEA TC project (RCA) on plant mutation breeding as an effective information exchange platform for global mutation research community through the IAEA platform.

5 Conclusions and Perspectives

Mutants are widely used in plant research, such as plant physiology, genetics and plant breeding. Mutation breeding has played in the past, and will also play in the future, very important roles in elite crop variety development to secure the world food supply. The role of nuclear techniques in enhancing genetic diversity will become even more important in the future.

In the future, mutation induction approaches will be closely associated with biotechnology platforms, such as high-throughput genotyping and phenotyping for the introduction of new genetic alleles for higher water and nutrient efficiency. The target of mutation breeding will transfer from yield traits to quality traits and stress tolerance for adapting to climate change. Collaboration among national, regional and worldwide partners will be further strengthened to achieve the goal of achieving food and nutrition security.

Acknowledgement

This work was financially supported by The National Key Research and Development Program of China (Grant No. 2016YFD0102100), China Agriculture Research System (Grant No. CARS-03), and IAEA TC Projects RAS5077 and CPR5024.

References

Bado, S. *et al.* (2015) Plant mutation breeding: current progress and future assessment. In: Janick, J. (ed.) *Plant Breeding Reviews*: *Vol. 39*. Wiley, Hoboken, New Jersey, pp. 23–88.

Bevan, M.W. *et al.* (2017) Genomic innovation for crop improvement. *Nature* 543(7645), 346–354.

Borrill, P., Harrington, S.A. and Uauy, C. (2018) Applying the latest advances in genomics and phenomics for trait discovery in polyploid wheat. *Plant Journal* 97(1), 56–72.

Guo, H.J. *et al.* (2007) Biological effects of high energy 7Li ion beams implantation on wheat. *Plant Mutation Rep* 1, 31–35.

Guo, H.J. *et al.* (2017) Novel mutant alleles of the starch synthesis gene TaSSIVb-D result in the reduction of starch granule number per chloroplast in wheat. *BMC Genomics* 18, 10.

Han, B. *et al.* (2016) Factors affecting the radiosensitivity of hexaploid wheat to irradiation: radiosensitivity of hexaploid wheat (*Triticum aestivum* L.). *PLoS ONE* 11(8), 15.

Li, W.J. *et al.* (2017) Identification of novel alleles induced by EMS-mutagenesis in key genes of kernel hardness and starch biosynthesis in wheat by TILLING. *Genes & Genomics* 39(4), 387–395.

Lin, M. *et al.* (1996) The damaging effects to hydrogen bond in DNA molecules caused by 7Li ion beam. *Chinese Journal Of Agricultural Biotechnology* 1, 93–94.

Liu, L. and Zheng, Q. (1997) Space mutagenesis and crop improvement. *China Nuclear Science and Technology Report* 475–485.

Liu, S. *et al.* (1996) The biological effects of the near-ground space treatment on M1 of wheat varieties. *Acta Agriculturae Nucleatae Sinica* 10(4), 205–210.

Shi, K. *et al.* (2017) Transcriptome and proteomic analyses reveal multiple differences associated with chloroplast development in the spaceflight-induced wheat albino mutant mta. *PLoS ONE* 12(5), 24.

Tanaka, A., Shikazono, N. and Hase, Y. (2010) Studies on biological effects of ion beams on lethality, molecular nature of mutation, mutation rate, and spectrum of mutation phenotype for mutation breeding in higher plants. *Journal of Radiation Research* 51(3), 223–233.

Wang, J-S. *et al.* (2015a) Generation of peanut mutants by fast neutron irradiation combined with in vitro culture. *Journal of Radiation Research* 56(3), 437–445.

Wang, J.S., *et al.* (2015b) Performance of peanut mutants and their offspring generated from mixed high-energy particle field radiation and tissue culture. *Genetic Molecular Research* 14(3), 10837–10848.

Wright, S.I. *et al.* (2005) The effects of artificial selection on the maize genome. *Science* 308(5726), 1310–1314.

Xiang, X. *et al.* (2014) Expression analysis for key genes of GA signaling pathway in wheat dwarf mutant DC20. *Acta Agriculturae Nucleatae Sinica* 28(2), 208–216.

Xiong, H.C. *et al.* (2017) RNAseq analysis reveals pathways and candidate genes associated with salinity tolerance in a spaceflight-induced wheat mutant. *Scientific Reports* 7, 13.

Xiong, H.C. *et al.* (2018) Enhancement of dwarf wheat germplasm with high-yield potential derived from induced mutagenesis. *Plant Genetic Resources-Characterization and Utilization* 16(1), 74–81.

Zhang, N. *et al.* (2016) Gibberellins regulate the stem elongation rate without affecting the mature plant height of a quick development mutant of winter wheat (*Triticum aestivum* L.). *Plant Physiology and Biochemistry* 107, 228–236.

Zhao, L. *et al.* (2015) Development of a new wheat germplasm with high anther culture ability by using a combination of gamma-ray irradiation and anther culture. *Journal of the Science of Food and Agriculture* 95(1), 120–125.

6 High-yielding NERICA Mutant Rice for Upland Areas and Hope for Bangladeshi Farmers

Mirza Mofazzal Islam[1], Shamsun Nahar Begum and Rigyan Gupta

Abstract

Drought is an important stress phenomenon in Bangladesh that greatly hampers crop production. So, it is imperative to develop drought-tolerant rice varieties. Low-yielding, non-uniform flowering and late-maturing Africa rice – New Rice for Africa (NERICA), viz. NERICA-1, NERICA-4 and NERICA-10 varieties – were irradiated with different doses of gamma-rays (250, 300 and 350 Gy) in 2010. M_1 plants were grown and M_2 plants were selected based on earliness and higher grain yield. The desired mutants along with other mutants were grown as the M_3 generation during 2011. A total of 37 mutants from NERICA-1, NERICA-4 and NERICA-10 were selected on the basis of plant height, short duration, drought tolerance and high yield in the M_4 generation. In the M_5 generation, six mutants were selected for drought tolerance, earliness, grain quality and higher yield. With respect to days to maturity and grain yield (t/ha), the mutant N_1/250/P-2-6-1 of NERICA-1 matured earlier (108 days) and had higher grain yield (5.1 t/ha) than the parent. The mutant N_4/350/P-4(5) of NERICA-4 also showed a higher grain yield (6.2 t/ha) than its parent and other mutants. On the other hand, NERICA-10 mutant N_{10}/350/P-5-4 matured earlier and had a higher yield (4.5 t/ha) than its parent. Finally, based on agronomic performance and drought tolerance, the two mutants N_4/350/P-4(5) and N_{10}/350/P-5-4 were selected and were evaluated in drought-prone and upland areas during 2016 and 2017. These two mutants performed well with higher grain yield than the released upland rice varieties. They will be released soon for commercial cultivation and are anticipated to play a vital role in food security in Bangladesh.

Keywords: rice • NERICA • New Rice for Africa • mutant • upland

1 Introduction

Bangladesh is a developing country where the main staple food is rice. Without rice, no meal will be acceptable for the people of this country. The total land area of Bangladesh is 147,570 km^2 and the total cultivable land is 85,052 km^2 (BBS, 2016). Currently, the population of Bangladesh is increasing and the land area is decreasing. So, Bangladesh needs to produce more rice with this limited cultivable land. There are many approaches to developing a rice variety and mutation breeding is one of them. From 1930 to 2014 more than 3200 crop varieties derived from mutants were released (Schouten and Jacobsen, 2007; PBGR, 2014) and these have been derived either as direct mutants (70%), or indirectly from their progeny (30%) (Maluszynski *et al.*, 2000). Crop plants account for 75% of species for which direct or indirect

[1] Email: mirza_islam@yahoo.com

 Mutation Breeding, Genetic Diversity and Crop Adaptation to Climate Change (eds S. Sivasankar *et al.*)
DOI: 10.1079/9781789249095.0006

mutant varieties have been released, with the remaining 25% being ornamentals or decorative plants (Ahloowali, 2004).

Bangladesh needs stress-tolerant crop varieties because about 40% of cultivable land is saline or prone to drought, submergence or excesses of cold or heat. Moreover, infestations of insects and other pests are also major problems for rice production. Under drought conditions rice does not grow properly, due to the water-stress condition; rice could be considered a semi-aquatic plant. In practice, in Bangladesh, the two most critical stages of rice that are susceptible to drought are tillering and flowering. In these two stages, if water is not supplied, growth will be retarded and yield will be lower.

Between 1960 and 1991, drought events occurred 19 times in Bangladesh. Very strong droughts hit the country in 1961, 1975, 1981, 1982, 1984, 1989, 1994 and 2000. Approximately every 5 years, Bangladesh is affected by major country-wide droughts. However, local droughts occur regularly and affect crop production. Agricultural drought, corresponding to soil moisture scarcity, occurs at different stages of crop growth, development and reproduction. Monsoon failure reduces crop production drastically and often brings famine to the affected region. North-western regions of Bangladesh are particularly exposed to droughts. In Bangladesh, the yields of Aus (upland) varieties of local rice cultivars are 1.5–2.0 t/ha, while for high-yielding varieties yields are 3.5–4.0 t/ha. Aus rice is sown in summer and harvested in the autumn and the Aus season is the water-deficit season in Bangladesh. A strong drought can cause greater than 40% damage to broadcast Aus rice. During the months of July–September, drought causes significant destruction of the Aman (rainfed) crop in approximately 2.32 million hectares every year (Aman rice is sown in the rainy season in July and August and harvested in winter). In the October–March crop-growing season known as rabi (from the Arabic word for 'spring', when the winter-sown crops are harvested), about 1.2 million hectares of agricultural land face droughts of different magnitudes. Apart from the agricultural loss, droughts have important effects on livestock population, land degradation, health and employment.

NERICA (New Rice for Africa) are the varieties that are developed by the Africa Rice Center. NERICA originated from the progeny of *O. sativa* × *O. glaberrima* crosses. NERICA is upland rice grown in drought conditions, with growth duration ranging from 110 to 125 days. The crop is photo-insensitive, which would be suitable for cultivation in the Aus season. The adaptability of NERICA lines has been tested over the past several years, but these varieties are not adapted or accepted in Bangladesh. The short duration and photo-insensitivity of NERICA presented some opportunities for farmers in Bangladesh, so the challenge to take advantage of them meant that these barriers to adoption needed to be overcome. With this in view, an irradiation programme was initiated to develop early, drought-tolerant and high-yielding rice varieties.

2 Materials and Methods

Dry seeds of NERICA-1, NERICA-4 and NERICA-10 were irradiated using the ^{60}Co source at the Bangladesh Institute of Nuclear Agriculture (BINA) in 2010 with gamma radiation doses of 250, 300 and 350 Gy with the aim to develop drought-tolerant rice varieties. Two thousand seeds from each variety were irradiated and 12,000–13,000 M_1 plants were grown in the M_2 generation. M_2 plant families were grown in a panicle-to-row design for each variety; three NERICA varieties were irradiated and, from these, three mutant M_2 populations were derived. In the M_2 generation, 500–600 plants were selected for desirable characteristics from each variety. From 556 plants selected in the M_3 generation, 37 desirable mutants were then selected in the M_4 generation from NERICA-1, NERICA-4 and NERICA-10. On the basis of plant height, earliness, drought tolerance (reduction in dry matter, number of seeds per panicle and grain yield per plant) and higher yield, six mutants from the M_5 generation were selected.

Yield trials were done under rainfed and irrigated conditions at BINA HQ (Mymensingh), Godagari (Rajshahi), Nachole (Chapainawabgonj) and Kaliganj (Jhenaidah). Data were collected on plant height, number of effective tillers per plant, number of filled grains, days to maturity and grain yield (t/ha). The plot size was 5 m × 2 m and spacing was 20 cm × 20 cm. The design used was a replicated complete block design (RCBD) with three replications; recommended doses of fertilizer were applied. Selected populations were further screened for earliness, grain quality (grain quality is considered as grain shape, size, and length and breadth ratio), drought tolerance and yield.

Field evaluation with the two selected mutants, N_4/350/P-4(5) and N_{10}/350/P-5-4, was carried out in farmers' fields and on-station at Mymensingh, Magura, Godagari, Nokla (Sherpur), and Khagrachari for N_4/350/P-4(5) along with high-yielding upland rice variety BRRI dhan48, and N_{10}/350/P-5-4 along with high-yielding upland rice variety HYV BR26 during the Aus growing season in 2016 and 2017, respectively. The field evaluations were laid out in RCBD with three replications. Unit plot size was 5 m × 6 m and spacing between hills and rows was 15 cm × 20 cm. Intercultural practices were done as and when necessary. Grain yield (kg) per plot was recorded from each plot and was converted to t/ha.

Ten hills were collected from each plot and data were recorded on plant height, number of effective tillers per plant, number of panicles per plant, panicle length (cm), number of filled grains per panicle, number of unfilled grains per panicle, 1000-seed weight (g), days to maturity and grain yield (t/ha).

Analysis of variance of the data was performed using MSTAT-C (1988) and the significance of difference of means was adjudged by Duncan's Multiple Range Test (Gomez and Gomez, 1984).

3 Results

The relevant phenotypes of the plants discussed below are presented in Figs 6.1 to 6.3.

Parent (NERICA-1)

N_1/250/P-6-2-6-1

N_1/250/P-7-3-12-2-1

Fig. 6.1. Pictorial view of NERICA-1 (Wild) and its mutants.

Fig. 6.2. Pictorial view of NERICA-4 (Wild-type) and its mutants.

3.1 Performance of two NERICA-1 mutants compared with their parent

Two mutants were selected from NERICA-1, viz. N_1/250/P-6-2-6-1 and N_1/250/P-7-3-12-2-1. These mutants were shorter than NERICA-1 (Table 6.1).

Days to maturity is an important character because short-duration varieties may fit better with cropping patterns. Table 6.1 shows that the maturity period of the mutants is 8–12 days earlier than that of their parent NERICA-1. The average number of effective tillers per hill was found to be highest in the mutant N_1/250/P-6-2-6-1, followed by N_1/250/P-7-3-12-2-1, and then the parent NERICA-1. Panicle length of the mutant N_1/250/P-7-3-12-2-1 was highest (33.9 cm), followed by N_1/250/P-6-2-6-1 (25.6 cm) and the parent NERICA-1 (24.2 cm). The number of filled grains per panicle was highest in N_1/250/P-6-2-6-1 (180) and lowest in NERICA-1 (104). The number of unfilled grains per panicle was lowest (27) in N_1/250/P-6-2-6-1 and highest (49) in NERICA-1. Thousand-seed

Fig. 6.3. Pictorial view of NERICA-10 (Wild-type) and its mutant.

weight was lower in N_1/250/P-7-3-12-2-1 (22.9 g) than in NERICA-1 (26.9 g). Grain yield was higher (5.1 t/ha) in N_1/250/P-6-2-6-1 than in NERICA-1 (2.9 t/ha).

3.2 Performance of three NERICA-4 mutants compared with their parent

Three mutants were selected from NERICA-4: N_4/250/P-1(2), N_4/350/P-4(5) and N_4/250/P-2(6)-26. It was observed that plant height of N_4/250/P-1(2) was higher (107 cm) than its parent NERICA-4 (99 cm) (Table 6.1). It was found that the period of days to maturity for N_4/250/P-1(2) was lower (111 days) than its parent NERICA-4 (118 days). The number of effective tillers per plant was higher (14) in the mutant N_4/250/P-2(6)-26 than its parent NERICA-4 (7). Panicle lengths of the mutants and their parent were almost identical. The number of filled grains per panicle was found to be higher (212) in the N_4/350/P-4(5) mutant than in the parent NERICA-4. The number of unfilled grains per panicle was found to be lower (16) in N_4/350/P-4(5) compared with NERICA-4. Thousand-seed weight was lower (22.2 g) in N_4/250/P-2(6)-26 compared with NERICA-4. Grain yield was higher (6.2 t/ha) in N_4/350/P-4(5) compared with its parent NERICA-4 (3.9 t/ha) (Table 6.1).

3.3 Performance of one mutant of NERICA-10

There was one mutant, N_{10}/350/P-5-4, selected from NERICA-10 for which plant height was different; NERICA-10 was higher (98 cm) than the mutant N_{10}/350/P-5-4 (91 cm) (Table 6.1).

Table 6.1. Mean values of various plant parameters at M_6 generation for ten plants of parents and selected mutants of NERICA-1, NERICA-4 and NERICA-10.

Mutant	Days to maturity	Plant height (cm)	Effective tiller (no.)	Panicle length (cm)	Filled grains/ panicle (no.)	Unfilled grains/ panicle (no.)	1000-seed wt (g)	Grain yield (t/ha)
N_1/250/P-6-2-6-1	108	102	13	25.6	180	27	23.8	5.1
N_1/250/P-7-3-12-2-1	112	120	10	33.9	145	35	22.9	4.6
N_4/250/P-1(2)	111	107	13	24.5	195	25	22.6	5.8
N_4/350/P-4(5)	113	106	12	24.4	212	16	25.4	6.2
N_4/250/P-2(6)-26	112	104	14	24.2	145	26	22.2	5.7
N_{10}/350/P-5-4	106	91	14	13	125	14	23.3	4.5
NERICA-1(Parent)	120	121	7	24.2	104	49	26.9	2.9
NERICA-4 (Parent)	118	99	7	23.8	105	37	28.1	3.9
NERICA-10 (Parent)	123	98	23.5	7	98	27	28.0	3.2

N_1 = NERICA-1; N_4 = NERICA-4; N_{10} = NERICA-10

Considering days to maturity, it was found that the maturity period of N_{10}/350/P-5-4 was earlier (106 days) than the parent NERICA-10. The number of effective tillers per plant was observed to be higher (14) in N_{10}/350/P-5-4 mutant than in the parent NERICA-10. Panicle lengths of the N_{10}/350/P-5-4 mutant and the parent NERICA-10 were almost identical. The number of filled grains per panicle was higher (125) in N_{10}/350/P-5-4 than in the parent NERICA-10. The number of unfilled grains per panicle was lower (14) in N_{10}/350/P-5-4 than in the parent NERICA-10. Thousand-seed weight was lower (23.3 g) in N_{10}/350/P-5-4 mutant than in the parent NERICA-10. In terms of grain yield, N_{10}/350/P-5-4 performed better (4.5 t/ha) than its parent NERICA-10 (Table 6.1).

3.4 Field evaluation of N_4/350/P-4(5)

For field evaluation of drought-tolerant rice mutants, drought stress was imposed at 16 days before flowering and continued up to 10 days after flowering in a rainout shelter. In this stress situation, the tolerant mutant had a yield loss to a maximum of 30% while susceptible mutants had yield losses in the range of 70–100% (Table 6.2). For field evaluation of N_4/350/P-4(5), high-yielding BRRI dhan48 was considered as check variety for the Aus (upland) season. It was observed that plant height of N_4/350/P-4(5) was higher (106 cm) than that of BRRI dhan48, days to maturity were earlier (100) in N_4/350/P-4(5) mutant than in BRRI dhan48 and the number of effective tillers per plant was higher (12) in N_4/350/P-4(5) than in BRRI dhan48. It was found that the thousand-grain weight of the N_4/350/P-4(5) mutant was lower (24.0 g) than in the check variety BRRI dhan48 (Table 6.3).

It was observed that the grain yield of N_4/350/P-4(5) was higher (4.99 t/ha) than in the check variety BRRI dhan48 (4.48 t/ha) in all the locations (Table 6.4).

The comparisons between N_4/350/P-4(5) and the check variety BRRI dhan48 for the grain characteristics, viz. the milling yield (%), head rice yield (%), chalkiness, whole grain length (mm), dehulled grain per kernel and amylose (%), are given in Table 6.5.

3.5 Field evaluation of the drought-tolerant mutant N_{10}/350/P-5-4 in Aus (upland) season

Field evaluation of N_{10}/350/P-5-4 was performed with HYV BR26 as a check variety for the Aus (upland) season. It was shown that N_{10}/350/P-5-4 was shorter (98 cm) than HYV BR26. N_{10}/350/P-5-4 had fewer days to maturity (105) than HYV BR26 and the number of filled grains per panicle was higher (140) than for HYV BR26. The number of effective tillers per plant was higher (15) in N_{10}/350/P-5-4 than in BR26 and the thousand-grain weight of N_{10}/350/P-5-4 was higher (23.5 g) than in the check, HYV BR26 (Table 6.6).

Table 6.7 shows that the grain yield of N_{10}/350/P-5-4 was higher (4.40 t/ha) than in the check, BR26 (4.10 t/ha), in all the locations.

With respect to grain characteristics, the comparisons between N_{10}/350/P-5-4 and BR26 are given in Table 6.8 for the milling yield (%), head rice yield (%), chalkiness, whole grain length (mm), dehulled grain per kernel and amylose (%).

The other four mutants are also being tested in farmers' fields and the best performing mutant will be selected and evaluated for variety release.

4 Discussion

Rice is the staple food in Bangladesh but yields of traditional varieties have fallen in recent years due to changing weather patterns, including higher temperatures, severe droughts, floods, salinity and diseases, coupled with more erratic rainfall. In upland conditions (Aus season – mid-March to June), growing of rice is very difficult due to the scarcity of water and the farmer mostly depends on rainfall. Irradiation induces changes in the DNA and speeds up the natural process of mutation in plants, increasing the diversity in crop varieties available to farmers. These rice mutant lines possess favourable traits, with some including higher yield, long slender grain and stronger climate resilience.

Nuruzzaman *et al.* (2016) reported that the average range of plant height among the NERICA-1 mutants was 73.43–114.77 cm. Chowhan *et al.* (2018) found that the NERICA mutant

Table 6.2. Performance of rice mutants under drought stress at the reproductive stage.

Name of the genotype	Number of effective tillers per plant	Number of non-effective tillers per plant	Number of filled grains per plant	Panicle length (cm)	Grain yield (t/ha)
N_1/250/P-2-6-1	10.27 ± 0.45	5.87 ± 0.11	331.19 ± 5.61	21.59 ± 0.46	2.68 ± 0.10
N_1/300/P-1-8-1	2.84 ± 0.34	6.77 ± 0.81	41.51 ± 8.43	18.74 ± 0.46	0.35 ± 0.11
N_{10}/350/P-5-4	9.88 ± 0.15	7.57 ± 0.49	315.58 ± 6.74	12.78 ± 0.16	2.50 ± 0.12
N_4/350/P-4(5)	10.50 ± 0.66	6.67 ± 0.29	364.75 ± 7.38	20.56 ± 0.45	3.15 ± 0.10
N_4/250/P-2-3-6	2.24 ± 0.43	5.48 ± 0.58	65.00 ± 4.78	17.13 ± 0.47	0.61 ± 0.10
NERICA-1	8.63 ± 0.43	7.23 ± 0.60	135.14 ± 5.27	18.74 ± 0.33	1.24 ± 0.11
NERICA-4	7.43 ± 0.48	6.54 ± 0.45	207.24 ± 9.81	19.20 ± 0.35	1.98 ± 0.14
NERICA-10	7.64 ± 0.48	7.33 ± 0.48	173.11 ± 7.60	18.50 ± 0.52	1.64 ± 0.10

Standard errors, *n* = 3

Table 6.3. Morphological, agronomic and grain characters of the drought-tolerant NERICA-4 mutant during Aus (upland) growing season of 2016.

Mutant/variety	Plant height (cm)	Days to maturity	No. of effective tillers per plant	No. of filled grains per panicle	1000-grain wt (g)
N_4/350/P-4(5)	106 ± 0.11	100 ± 1.15	12 ± 0.58	108 ± 6.43	24.0 ± 0.06
BRRI dhan48 (check)	103 ± 0.58	104 ± 0.58	10 ± 0.50	93 ± 2.65	25.0 ± 0.06

Standard errors, *n* = 3

Table 6.4. Yield performance of the NERICA-4 mutant at multi-location trials grown in Aus (upland) season of 2016.

	Grain yield (t/ha) ± Standard error (*n* = 3)					
Mutant/variety	L1	L2	L3	L4	L5	Mean
N_4/350/P-4(5)	6.11 ± 0.12	4.73 ± 0.20	6.22 ± 0.19	3.94 ± 0.29	3.95 ± 0.18	4.97 ± 0.15
BRRI dhan48 (check)	5.74 ± 0.12	3.90 ± 0.15	5.83 ± 0.12	3.43 ± 0.13	3.49 ± 0.13	4.48 ± 0.09

L1, Magura; L2, Mymensingh; L3, Khagrachari; L4, Godagari (Rajshahi); L5, Nokla (Sherpur)

N_4/350/P-4(5) had a lower percentage of unfilled grains per hill (17.62), earlier maturity (101 DAD) compared with BRRI dhan48 and higher harvest index (47.77%). Straw (11.70 t/ha) and biological yield (13.80 t/ha) were greatest for NERICA-4 and least for N_4/350/P-4(5) (straw 4.80 t/ha and biological yield 9.10 t/ha). Grain yield (4.3 t/ha) was higher in N_4/350/P-4(5) followed by BRRI dhan48 (4.0 t/ha) and NERICA-4 (2.30 t/ha).

Afrin *et al.* (2017) studied days to flowering, plant height, tiller number per plant, effective tiller number per plant, 100-seed weight, harvest index, etc. to measure the nature of association between these traits and yield. Plant height, number of tillers per hill, number of filled grains per panicle, total dry matter per hill, 1000-grain weight, grain yield and harvest index decreased with increasing water stress levels among the susceptible lines. Responses of the rice genotypes to different water stresses varied significantly. There had been different degrees of reduction (at different stages of plant growth) to the yield, contributing characters for the stress (Zubaer *et al.*, 2007). Grain yield per plant had a significant positive correlation with biological yield per plant, harvest index, panicles per plant, plant height, spikelets per panicle, panicle

Table 6.5. Grain characteristics of the NERICA-4 mutant (N_4/350/P-4(5)).

Mutant/ variety	Milling yield (%)	Head rice yield (%)	Chalkiness[a]	Whole grain length (mm)	Dehulled grain per kernel: Length (mm)	Breadth (mm)	L/B ratio	Size and shape	Amylose (%)
N_4/350/P-4(5)	72 ± 0.11	93 ± 0.57	wb1	8.1 ± 0.01	6.2 ± 0.01	3.0 ± 0.01	2.06 ± 0.01	Medium medium	23.5 ± 0.30
BRRI dhan48 (check)	71 ± 0.08	90 ± 0.72	wb1	8.2 ± 0.01	6.3 ± 0.01	2.9 ± 0.01	2.17 ± 0.01	Medium medium	24.0 ± 0.11

[a]wb1 = less than 10% chalkiness. Standard errors, *n* = 3

Table 6.6. Yield and yield-contributing characters of the proposed drought-tolerant mutant rice and check variety grown in Aus (upland) season 2016–2017.

Mutant/variety	Plant height (cm)	Days to maturity	No. of filled grains per panicle	No. of effective tillers per plant	1000-grain wt (g)
N_{10}/350/P-5-4	98 ± 0.26	105 ± 0.58	140 ± 2.08	15 ± 0.58	23.5 ± 0.03
BR26 (check)	110 ± 0.56	110 ± 1.15	127 ± 6.24	11 ± 0.58	21.5 ± 0.15

Standard errors, *n* = 3

Table 6.7. Yield performance of the NERICA mutant at multi-locational trials grown in Aus (upland) season 2017.

Designation	Grain yield (t/ha) ± Standard error, *n* = 3					Mean
	L1	L2	L3	L4	L5	
N_{10}/350/P-5-4	4.48 ± 0.78	4.73 ± 0.06	4.92 ± 0.08	3.95 ± 0.26	3.80 ± 0.11	4.38 ± 0.01
BR26 (check)	3.86 ± 0.06	4.56 ± 0.12	4.82 ± 0.07	3.70 ± 0.07	3.60 ± 0.08	4.10 ± 0.05
Yield advantage	0.62	0.19	0.10	0.25	0.20	0.30

L1, Fulbaria (Mymensingh); L2, BINA sub-station, Nalitabari; L3, BINA sub-station, Magura; L4, Tanore (Rajshahi); L5, Khagrachari

length, test weight, spikelet fertility and flag leaf length (Sravan *et al.*, 2012).

The mutant lines N_1/250/P-2-6-1, N_4/350/P-4(5) and N_{10}/350/P-5-4 from different NERICA accessions performed well with respect to number of effective tillers per plant, number of non-effective tillers per plant, number of filled grains per plant and panicle length, which influenced their overall higher yield. It was also noted that these selected mutants exhibited higher grain yield than their respective parents under drought stress (Table 6.2).

5 Conclusion

In Bangladesh, in upland conditions there is a water deficit from mid-March to June. There are no good varieties in the country that cope well with these constraints. Farmers generally use local cultivars which are very low yielding (1.5–2.0 t/ha). African rice NERICA cultivars have some problems in Bangladesh, such as non-uniform flowering, late in maturity and low yield. With this in mind, we irradiated the varieties NERICA-1, NERICA-4 and NERICA-10 in an attempt to overcome and eliminate these defects. Six promising mutants that are well adapted to Bangladesh conditions have been identified for uniform flowering, early maturity and higher yield. Two mutants, N_4/350/P-4(5) and N_{10}/350/P-5-4, have been selected for such improved traits in Aus (upland) conditions. Days to maturity of these mutants are 100–105 days and grain yield 4.5–5.0 t/ha. These two mutants will be released for commercial cultivation in Bangladesh. We have another four mutants that are also being tested in farmers' fields in Aus (upland) conditions. These upcoming mutant varieties will be grown in fragile environments in upland conditions and should help ensure sustainable food security in Bangladesh.

Acknowledgement

The authors would like to thank and acknowledge the great assistance of AfricaRice (Africa Rice Center) for providing seed materials of NERICA.

Table 6.8. Grain characteristics of the proposed variety.

Proposed variety	Milling yield (%)	Head rice yield (%)	Chalkiness[a]	Whole grain length (mm)	Dehulled grains per kernel				Amylose content (%)
					Length (mm)	Breadth (mm)	L/B ratio	Size and shape	
N_{10}/350/P-5-4	74 ± 0.02	93 ± 0.10	wb1	9.3 ± 0.06	7.2 ± 0.03	2.0 ± 0.01	3.6 ± 0.01	Long slender	25.1 ± 0.10
BR26 (check)	71 ± 0.62	91 ± 0.12	wb1	9.1 ± 0.10	7.0 ± 0.01	2.0 ± 0.01	3.5 ± 0.01	Long slender	22.7 ± 0.12

[a]wb1 = less than 10% chalkiness. Standard errors, *n* = 3

References

Afrin, W., Hossain, M.A. and Islam, M.M. (2017) Evaluation of yield and yield contributing traits for submergence tolerance of rice. *Progressive Agriculture* 28, 7–11.

Ahloowali, B.S. (2004) Global impact of mutation-derived varieties. *Euphytica* 135, 187–204.

BBS (2016) *Yearbook of Agricultural Statistics – 2016*. 28th Series. Planning Division, Ministry of Planning, Dhaka, Bangladesh, pp. 17–66.

Chowhan, S. *et al.* (2018) Evaluation of NERICA rice mutant in Jhum cultivation. *International Journal of Agronomy and Agricultural Research* 12(2), 24–31.

Gomez, K.A. and Gomez, A.A. (1984) *Statistical Procedures for Agricultural Research*, 2nd edn. Wiley, New York.

Maluszynski, M.K. *et al.* (2000) Officially released mutant varieties – the FAO/IAEA Database. *Mutation Breeding Review* 12, 1–84.

MSTAT-C (1988) MSTAT-C, a Microcomputer Program for the Design, Arrangement, and Analysis of Agronomic Research. Michigan State University, East Lansing.

Nuruzzaman, M. *et al.* (2016) Morphological characterization and assessment of genetic parameters of NERICA mutant rice lines under rainfed condition. *Asian Journal of Medical and Biological Research* 2, 532–540.

PBGR (2014) *Plant Breeding and Genetics Report*. Joint FAO/IAEA Division of Nuclear Techniques in Food and Agriculture, Vienna.

Schouten, H.J. and Jacobsen, E. (2007) Are mutations in genetically modified plants dangerous? *Journal of Biomedicine and Biotechnology* 1, article 82612.

Sravan, T. *et al.* (2012) Genetic variability and character association in rainfed upland rice (*Oryza sativa* L.). *Journal of Rice Research* 5 (1 & 2), 24–29.

Zubaer, M.A. *et al.* (2007) Effects of water stress on growth and yield attributes of Aman rice genotypes. *International Journal of Sustainable Crop Production* 2, 25–30.

7 Impact of Mutant Varieties in Malaysia: Challenges and Future Perspectives for Mutation Breeding

Rusli Ibrahim[1]

Abstract

Malaysia has made substantial progress in plant mutation breeding with the use of nuclear techniques and related biotechnologies, not only in the development of new mutant varieties but also in the establishment of an excellent nuclear research centre. A total of 53 mutant varieties have been developed, including rice *Oryza sativa* (19), banana *Musa acuminata* (one), groundnut *Arachis hypogaea* (two), orchid *Dendrobium* 'Sonia' (six), chrysanthemum *Chrysanthemum morifolium* (seven), hibiscus *Hibiscus rosa-sinensis* (three), roselles *Hibiscus sabdariffa* L. (three) and other ornamental and landscaping plants (12). Most of the new ornamental varieties have been developed by both acute and chronic gamma-ray irradiation of seeds, rooted cuttings, bulbs and tissue cultures. Food crops that have an economic impact on sustainable agricultural production are mutant varieties of banana ('Novaria') and rice (MRQ74, MR219-9 and MR219-4). 'Novaria' is a selection made from a mutant, 'GN-60A', of 'Grande Naine' (AAA *Musa*) identified from gamma-ray treated populations of the Biotechnology Laboratory in Seibersdorf, Austria. 'Novaria' was the first mutant variety, officially released in 1995 by the Malaysian Nuclear Agency as a new variety for its improved characteristics such as early flowering, short stature and high yield. MRQ74 is a type of high-quality fragrant rice with newly induced traits such as resistance to blast, long and slender grain shape, non-sticky and with the elongation properties of cooked rice similar to those of Basmati-type rice. It is an indirect mutant variety released in 2003 and one of its parental lines for cross-breeding was the mutant 'Mahsuri', which was developed through mutation breeding using gamma-rays. In 2014, two new mutant varieties of rice, 'MR219-9' and 'MR219-4', which are drought tolerant, high yielding and resistant to blast, were selected from gamma irradiated material. Despite these achievements, applications of induced mutation have decreased during the past 10 years due to reduced funding. Mutation breeding is still a promising technique for the development of novel varieties which in combination with advanced molecular genetics can bring plant mutation breeding into a new era.

Keywords: plant mutation breeding • nuclear techniques • mutant varieties • irradiation • Novaria

1 Introduction

Rice is the staple food crop in Malaysia and its production depends largely on the irrigated lowland production system. Malaysia managed to achieve 72% self-sufficiency in rice with an average rice yield of 3.7 t/ha (Siwar *et al.*, 2014). The first application of nuclear technology in mutation breeding for the improvement of rice was conducted in 1984 for a Coordinated Research Project entitled 'Semi-Dwarf Mutants for Rice Improvement in Asia and Pacific' under

[1] Email: rusli.anm@gmail.com

DOI: 10.1079/9781789249095.0007

the Joint FAO/IAEA Division. Within 5 years, 101 semi-dwarf mutant lines were identified. Twenty-nine of the semi-dwarf mutant lines had grain yields between 6000 and 7300 kg/ha; these were higher than the parent variety 'Manik', which yielded 5700 kg/ha. These were classified as potentially good yielding mutants (Othman *et al.*, 1989). One of the mutant lines, designated as MA03, showed a drastic change in its characteristics and performed better than the parent and other mutant lines. This mutant line was popularly known as mutant 'Tongkat Ali' because of its outstanding agronomic traits, with very erect panicles even after grain-filling, very strong culm and resistance to lodging (Othman *et al.*, 1989). In 2012, under the 10th Malaysia Plan, the Malaysian Nuclear Agency initiated a Mutation Breeding Program for Varietal Improvement of Irrigated Rice under Minimal Water Conditions, which focused on developing new rice varieties with high yield under minimal water requirement. This research programme was supported by the National IAEA Technical Cooperation Project on Applying Mutation Breeding and Optimized Soil, Water and Nutrient Management for Enhanced and Sustainable Rice Production (MAL/5/029). A total of 38 potential mutant families from 'MR219' were evaluated in the field with minimal water requirement. Two mutant lines, designated as MR219-4 and MR219-9, were selected for further field trials under saturated soil and flooded conditions, based on their high yields.

Banana is one of the important fruit crops cultivated in Malaysia. It is ranked second in terms of production area and fourth in export revenue based on balance-of-trade figures. In Malaysia, banana covers about 26,000 ha with a total production of 530,000 t and with more than 15% of the total acreage under fruits; it contributes more than RM30 million (US$8 million) to the balance of trade. About 50% of the banana-growing land is cultivated with 'Pisang Berangan' and the Cavendish type. Most of the bananas produced are consumed locally and about 10% are exported, mainly to Singapore, Brunei, Hong Kong and the Middle East. 'Novaria' is a selection made from 'GN-60A', a mutant of 'Grande Naine', identified in gamma-ray treated populations of the FAO/IAEA Agriculture and Biotechnology Laboratories, Seibersdorf, Austria. Released in 1995, 'Novaria' was the first mutant variety officially released by the Malaysian Nuclear Agency as a new variety for its improved characteristics such as early flowering, short stature, high yield, strong fruit pedicel and good flavour and pulp texture (Maluszynski, 2001; Mak *et al.*, 1996).

A mutation breeding programme for the improvement of groundnut was started in 1985 under an FAO/IAEA Coordinated Research Project and later funded by the Ministry of Science, Technology & Innovation (MOSTI) under Intensified Research Priority Areas (IRPAs) in 1987. The main objectives were to induce new mutant varieties resistant to *Cercospora* leaf spot disease with high yield and high nitrogen fixation. In Malaysia, groundnuts and peanuts are considered as underutilized crops, like many tuber crops such as cassava or tapioca, sweet potato, yams or taro, which are rich in starch. The mutant groundnut varieties 'KARISMA Sweet' and 'KARISMA Serene' were developed through collaboration between the Malaysian Nuclear Agency and the Rubber Research Institute of Malaysia (RRIM) under an FAO/IAEA Coordinated Research Project that started in 1985. 'KARISMA Sweet' and 'KARISMA Serene' were released with improved traits such as high yield, resistance to *Cercospora* leaf spot disease, high nitrogen fixation, sweet taste and uniform seed pod (Rusli *et al.*, 1998).

The Malaysian Nuclear Agency also started mutation breeding of ornamental plants in 1990. The main objective was to create new varieties with attractive colour and long-lasting flowers for cut flowers as well as landscaping plants. Most of the ornamental plants in Malaysia are vegetatively propagated and some are sterile (have no seed or extremely low seed set). To widen the genetic variation of these plants, *in vitro* culture was used in combination with mutation induction. *In vitro* cultures of selected flowering plants such as *Alpinia*, *Petunia*, *Amaryllis*, chrysanthemums and orchids were established and irradiated with gamma-rays. Irradiated seedlings were then hardened and planted for morphological screening.

At present, a total of 53 mutant varieties have been developed, including rice *Oryza sativa* (19), banana *Musa acuminata* (one), groundnut *Arachis hypogaea* (two), orchid *Dendrobium* 'Sonia' (six), chrysanthemums *Chrysanthemum morifolium* (seven), hibiscus *Hibiscus rosa-sinensis* (three), roselles *Hibiscus sabdariffa* (three) and other ornamental and landscaping plants (12). This chapter describes techniques in mutation induction that produced some of the mutant varieties which have been officially released and successfully distributed to the end users, especially local farmers and growers, and which have contributed to greater food security and have had a positive economic impact development.

2 Materials and Methods

2.1 Mutation induction in rice

The three projects conducted in rice were: (i) development of 'Semi-dwarf mutants for rice improvement in Asia and Pacific'; (ii) development of new varieties of rice adaptable to climate change or drought tolerance using advanced mutation breeding and optimized soil, nutrient and water management practices; and (iii) cross-breeding programme for the development of specialty rice. Mutation induction for 'Semi-Dwarf Mutants for Rice Improvement' was initiated in 1984 under a Coordinated Research Project of the Joint FAO/IAEA Division. Preliminary experiments were conducted by irradiating seeds of parental variety 'Manik' with a series of gamma-ray doses ranging from 0 Gy to 1000 Gy to determine the LD_{50}. Screening for targeted traits with improved characteristics started in the M_2 generation and continued until the M_7 and M_8 generations. For the identification of drought-tolerant mutants, seeds of the popular variety 'MR219' were first irradiated with gamma-rays using a gamma cell from ^{60}Co source at doses from 0 Gy to 1000 Gy. Once LD_{50} had been identified, 300 Gy was selected for the main experiment. Screening for drought tolerance was first conducted under glasshouse and field conditions starting in the M_2 generation (Fig. 7.1A) and later in the M_7 and M_8 generations for replicated yield trials in farmers' plots with the application of biofertilizer and oligochitosan to identify superior genotypes for use with minimal water requirement (Fig. 7.1B). For the development of specialty rice, a cross-breeding programme was initiated by crossing the long-grain rice variety 'Q34' with the original mutant variety 'Mahsuri', which is aromatic and has resistance to blast and had been developed through mutation breeding using gamma-rays.

2.2 Mutation induction in banana

Shoot tip meristems from *in vitro* shoots of the 'Grande Naine' variety of Cavendish-type banana were first irradiated with gamma-rays from ^{60}Co source at 15, 30, 45 and 60 Gy at a dose rate of 8 Gy/min at the FAO/IAEA Agriculture and Biotechnology Laboratories, Seibersdorf, Austria (Novak *et al.*, 1990). An early-flowering putative mutant designated as 'GN-60A' was identified among the regenerated plants obtained from gamma-ray treated populations in the glasshouses. *In vitro* shoots of the 'GN-60A' clone were sent to the Malaysian Nuclear Agency for field testing. With the collaboration of the University of Malaya and United Plantations Sdn Bhd, 'GN-60A' clones were micropropagated and subcultured until M_1V_5 for multiplication and to minimize chimerism. A total of 2000 plants were planted at United Plantations Sdn Bhd to evaluate their performance in terms of earliness and other agronomic traits.

2.3 Mutation induction in groundnut

Two popular local varieties, 'Mat Jan' and 'V13', were chosen as the starting materials to induce mutation with the objectives of producing new mutant varieties of groundnut resistant to *Cercospora* leaf spot disease and with high yield and high nitrogen fixation. A total of 3000 seeds per dose per variety were irradiated with gamma-rays at 200, 300 and 400 Gy using a gamma cell with ^{60}Co source. Irradiated seeds were inoculated with *Bradyrhizobium* obtained from the RRIM to induce nodulation and then planted in the field. After initial selection in the M_2 generation, two promising mutant lines resistant to *Cercospora* leaf spot were selected for further evaluation in multi-location yield trials and compared with their parent variety.

2.4 Mutation induction in ornamentals

For vegetatively propagated ornamental plants such as *Hibiscus rosa-sinensis*, a radiosensitivity

Fig. 7.1. Screening and selection for drought-tolerant mutants of gamma-ray irradiated plants. **(A)** Early screening in the greenhouse (left) and field screening under aerobic conditions (right) in the M_2 generation. **(B)** Evaluation of yield performance of advanced mutant lines MR219-4 (left) and MR219-9 (right) under low-water condition in M_7 and M_8 generations.

test was carried out to determine the growth rate or growth reduction (GR_{50}) and optimal irradiation doses by irradiating stem cuttings with gamma-rays, using a gamma cell with a ^{60}Co source at 10, 20, 30 and 60 Gy at a dose rate of 1.66 Gy/s. Irradiated stem cuttings were then planted in sand beds and data on the number of growing shoots were recorded. Increasing the gamma-ray dose resulted in a reduction in the number of growing shoots of irradiated stem cuttings. The GR_{50} (50% growth reduction) for stem cuttings was 36.2 Gy and the GR_{75} (75% growth reduction) was 17.15 Gy. Based on these results, 20–30 Gy was selected as an optimal dose range to induce mutation in *Hibiscus rosa-sinensis*. The GR_{50} can no longer be used as a reference. The tendency is now to use relatively low doses, because they produce less chromosomal damage and other negative side effects than stronger treatments. In gamma irradiation, M_1 seedling growth reduction of 30–50% or a survival rate of 40–60% in control plants has often been considered the criterion for promising gamma-ray irradiation treatment (Yamaguchi, 2013). Based on these results, 20–30 Gy (60–70% growing shoots) was selected as an optimal dose to induce mutation in *Hibiscus rosa-sinensis*. Irradiated stem cuttings were planted in sand beds for rooting. After 1 month, rooted stem cuttings were transferred to polybags and allowed to grow under plastic netting with 70% shade. New shoots produced in M_1V_1 were cut back and propagated as M_1V_2 and the variants that appeared were subsequently observed. This technique of cutting back was repeated at least 4–5 times (M_1V_4–M_1V_5). A similar technique was applied using rhizomes in *Canna hybrida* and shoots in *Tradescantia spathacea* as source explants.

For ornamental orchids, chrysanthemums and petunia, tissue culture materials such as protocorm-like bodies (PLBs), callus and *in vitro* shoots, respectively, were used for *in vitro* mutagenesis. A radiosensitivity test was first carried out to determine GR_{50} and optimal doses for the main experiment. Irradiated explants were multiplied until the fourth subculture cycle (M_1V_4) before they were transferred to the nursery. Observations on morphological changes in irradiated plants were carried out both *in vitro* and *in vivo*. Selected mutant(s) were then conventionally propagated or micropropagated to obtain clonal mutant lines.

3 Results

3.1 Development of drought-tolerant mutants and specialty rice

For the project on 'Development of Semi-Dwarf Mutants for Rice Improvement in Asia and Pacific', which was initiated in 1984 (Othman *et al.*, 1989), within 5 years 101 semi-dwarf mutant lines were identified which had the potential to be used as varieties or as parents in cross-breeding programmes. One of the mutant lines, MA03, popular among farmers and known as 'Tongkat Ali', has a strong stem with an erect panicle, even after grain filling, and is resistant to lodging (Othman *et al.*, 1989). Even though it had not been officially released, it had been planted commercially by a few farmers, especially in the northern part of Malaysia. During the first season, the yield was recorded at 6589 kg/ha and the second season at 5715–7504 kg/ha. In MA03, genetic analysis of semi-dwarfism was shown to be determined by a single recessive gene (Mohamad Farazi, 1994). This recessive gene was non-allelic to sd_1 and was named as sd_8 (Mohamad Farazi, 1994).

For drought-tolerant mutants, 55 mutant lines were evaluated in the M_4 generation, of which 38 were selected for having high percentage of grain filling. However, only two lines, MR219-4 and MR219-9, were selected in the M_7 and M_8 generations. When evaluated under saturated soil and flooded conditions at the MADA Experimental Station, the yield recorded for MR219-4 was 5.9 t/ha under saturated soil conditions and 7.2 t/ha under flooded conditions. For MR219-9, the yield was 6.8 t/ha and 6.1 t/ha under saturated and flooded conditions, respectively. These mutant lines also performed satisfactorily when grown under aerobic soil condition at the Seberang Perai Experimental Station of MARDI (Malaysian Agricultural Research and Development Institute). Grain yields as high as 6.3 t/ha for MR219-4 and 3.4 t/ha for MR219-9 were achieved under aerobic conditions. Since MR219-4 performed better than MR219-9 in terms of high yield under aerobic condition, it was selected by the local farmers for yield trials. When tested in farmers' plots with the use of normal fertilizer practices, within 1 year with two planting seasons the average yield recorded was 6.3 t/ha. However, with the application of biofertilizer and oligochitosan produced by the Malaysian Nuclear Agency, the yield recorded was 8.05 t/ha, an increase of about 22%.

MRQ74, commonly known as 'Mas Wangi', is a type of high-quality fragrant rice which is resistant to blast and is an indirect mutant variety released in 2003. One of its parental lines for cross-breeding was the mutant 'Mahsuri', which was developed through mutation breeding using gamma-rays. Since it is a specialty rice, large-scale cultivation was done mainly by commercial plantations and at present there is a total acreage of about 1000 ha under cultivation with MRQ74.

3.2 Development of the mutant banana, 'Novaria'

Early-flowering plants obtained in the field were multiplied again by tissue culture for further screening. In September 1993, 2000 plants were field planted at United Plantations Sdn Bhd to evaluate their performance in terms of earliness and other induced traits (Fig. 7.2A). From this population, a selection was made for mutated plants. These flowered about 10 weeks earlier than the original parental clone, 'Grande Naine', and had other improved traits such as short stature, high yield, strong fruit pedicel and good flavour and pulp texture. The yield increase was 30%, from 20 kg/bunch to 30 kg/bunch. This mutant selection was launched under the name of 'Novaria' in 1995 and released as a new variety (Mak *et al.*, 1996). It is still widely planted by farmers in Peninsula Malaysia, especially in Perak (Fig. 7.2B) and also in East Malaysia in the state of Sarawak.

3.3 Development of new mutant varieties of groundnut resistant to *Cercospora* leaf spot disease

After a few generations of field screening and multi-locational yield trials, two advanced mutant lines, MJ40/42 and MJ20/165-5 (Rusli *et al.*, 1998), were selected from the variety 'Mat Jan', with improved agronomic traits such as resistance to *Cercospora* leaf spot disease, high yield and early flowering. These two mutant lines, MJ40/42 and MJ20/165-5, were officially released in 2005 as new varieties, 'KARISMA Sweet' and 'KARISMA Serene', respectively (Fig. 7.3). 'KARISMA Sweet' is high yielding (4.6 t/ha), resistant to *Cercospora* leaf spot disease and with high nitrogen fixation (60% of its N from N_2-fixation). It flowers 25 days after planting and its maturity period is 90–100 days. It has two seeds per pod which are uniform in size; it is sweeter than the parent variety 'Mat Jan' and preferred by the snack industry. 'KARISMA Serene' is also high yielding (4.8 t/ha), resistant to *Cercospora* leaf spot disease and shows high nitrogen fixation (> 70% of its N from N_2-fixation). It flowers 25 days after planting and its maturity period is 85–90 days.

3.4 Development of new mutant varieties of ornamental plants

In orchids, irradiation of protocorm-like bodies (PLBs) using gamma-rays on 'Sonia', a commercial hybrid *Dendrobium*, at a dose of 35 Gy resulted in variability of phenotypic characteristics of the flowers, inflorescences, shelf life and growth habit. Five new varieties of orchids have been successfully generated with commercial potential, namely *Dendrobium* 'Sonia Keena Oval', *Dendrobium* 'Sonia Keena Radiant', *Dendrobium* 'Sonia Keena Hieng Ding', *Dendrobium* 'Sonia Keena Ahmad

Fig. 7.2. **(A)** Field screening of mutant 'Novaria' at United Plantations Sdn Bhd and **(B)** commercial plantations in farmers' plots.

Fig. 7.3. New mutant varieties of groundnut induced by gamma irradiation. **(A)** Mutant 'KARISMA Sweet' (left) and mutant 'KARISMA Serene' (right). **(B)** Seeds and pods of 'KARISMA Sweet' (left) and seeds and pods of 'KARISMA Serene' (right).

Sobri' and *Dendrobium* 'Sonia Keena Pearl' (Fig. 7.4). Similarly, irradiation of stem cuttings of *Hibiscus rosa-sinensis* with acute gamma-rays at 20 Gy generated four new varieties with unique and interesting flower colours and shape which have been registered as new mutant varieties: 'Siti Hasmah Red Shine', 'Siti Hasmah Pink Beauty', 'Nori' and a mutant with compact curly petals (Fig. 7.5). Irradiation of other landscaping and pot plants using both acute and chronic gamma radiation has also generated interesting and unique flower colour mutants, especially in *Alpinia, Petunia, Hippeastrum, Turnera* and *Canna* (Zaiton *et al.*, 2012) (Fig. 7.6).

4 Discussion

As a nuclear research centre with excellent research facilities in plant mutation breeding and biotechnology, the Malaysian Nuclear Agency had generated 53 mutant varieties, as outlined in the Introduction above, including rice, banana, groundnut, orchids, chrysanthemums, hibiscus and other ornamental and landscaping plants. The actual number is probably higher, since some of the mutants generated were not recorded in the national mutant database and were developed by other national research institutions and universities, especially MARDI and the Universiti Kebangsaan Malaysia (UKM) and Universiti Pertanian Malaysia (UPM). The release of some of the mutant varieties highlights the important application of nuclear technology that contributes to a greater economic impact in agriculture and sustainable food production.

Farmers now grow mutant and mutant-derived rice varieties that have increased yields and improved end-use traits such as quality and fragrance. Two mutant lines, MR219-4 and MR219-9, were found to be adaptable to different soil water conditions (flooded, saturated, field capacity) and are tolerant to blast. They also have the potential to be developed as drought-tolerant varieties or 'aerobic

Fig. 7.4. New mutant varieties of orchid '*Dendrobium* Sonia' induced by acute gamma irradiation. Top row on left: the original parent, reddish purple in colour with new mutant varieties, 'Keena Hieng Ding' with broad petals (middle) and 'Keena Pearl' with white and purple edges (right). Bottom row, left to right: mutant varieties, 'Keena Ahmad Sobri' with uniform diamond-shape purple petals, 'Keena Pastel' with soft purple-shade petals and 'Keena Radiant' with pale purple flowers.

(A)

(B)

Fig. 7.5. Flower colour mutants of *Hibiscus rosa-sinensis* induced by acute gamma irradiation. Top row **(A)** on left is original parent variety, pink in colour. Mutant varieties 'Siti Hasmah Red Shine' (middle) and 'Siti Hasmah Pink Beauty' (right). Bottom row **(B)** on the left is original parent variety with striking red colour. New mutant varieties, 'Nori', red in colour with multiple petal layers (middle) and mutant plant with compact and curly bright red petals (right).

rice'. With the use of an efficient technology package for new rice production or new mutant varieties, such as the application of biofertilizer and oligochitosan as growth promoter and elicitor (for pest and disease control), these varieties resulted in yield improvement in experimental plots and farmers' fields. Specialty rice MRQ74, commonly known as 'Mas Wangi', is a type of high-quality fragrant rice which is resistant to blast and was an indirect mutant variety. Large-scale cultivation mainly by commercial plantations at present has a total acreage of about 1000 ha.

The release of the banana mutant 'Novaria' was made possible with collaboration between IAEA, the University of Malaya and the commercial tissue culture company, United Plantations Sdn Bhd. It was the first mutant variety released by the Malaysian Nuclear Agency in 1995 (Mak *et al.*, 1996; Maluszynski, 2001) and is also the most popular mutant variety from Malaysia. It was successfully developed through an IAEA Technical Cooperation Project with technical support from the Joint FAO/IAEA Division. Mutant 'Novaria' seedlings were propagated using tissue culture techniques by the United Plantations Sdn Bhd, which produced about 200,000 plantlets per year for sale to growers in Malaysia, corresponding to a plantation area of about 1400 acres (566 ha) from 2000 to 2005 (Mak *et al.*, 1998). However, 'Novaria' is susceptible to Fusarium wilt, a soil-borne disease caused by *Fusarium oxysporum* f. sp. *cubense* Race 4, the most serious disease in banana, which affects many important varieties of banana and plantain. Evaluation of banana germplasm for resistance to Fusarium wilt using *in vitro* mutation and selection, along with somaclonal variation and somatic hybridization, could improve banana breeding efficiency for resistance against the disease (Siamak and Zheng, 2018).

The new mutant varieties of groundnut, 'KARISMA Sweet' and 'KARISMA Serene', with improved traits such as resistance to *Cercospora* leaf spot disease, high yield and high N-fixation, had initially been grown on a large scale by local growers. However, at present, due to the high

Fig. 7.6. New mutant varieties of landscaping and pot plants induced by acute and chronic gamma irradiation. **(A)** *Alpinia purpurata* red control (left) and pink mutant (right). **(B)** *Petunia hybrid* purple control (left) and pink mutant, 'NK Tropicana' (right). **(C)** *Hippeastrum puniceum* red control (left) and orange mutant 'Orange BioGamma' (right). **(D)** *Turnera subulata* non-flowering control (left) and mutant with blooming flowers at 1 pm (right). **(E)** *Canna generalis* pink control (left) and pink-orange mutant (right). **(F)** *Canna generalis* red-orange control (left) and orange-yellow mutant (right).

cost of production, not many growers plant groundnut and so the acreage is currently low. Under a seed exchange programme of the Forum for Nuclear Cooperation in Asia (FNCA), other Member States, such as Sri Lanka, used the mutant variety 'KARISMA Serene' for their own mutation breeding programmes and have produced promising mutant lines with improved characteristics and adaptation to local conditions (Luxiang, 2013).

Mutation breeding has been more successful in ornamental plants, because changes in phenotypic characteristics like flower colour, shape and size, chlorophyll variegation in leaves and growth habit can be easily detected as early as the M_2 generation (Datta, 2009, 2012). In addition, the heterozygous nature of many ornamental plants offers high frequency of mutant phenotypes (Broertjes and Van Harten, 1988). The use of heterozygous starting material (*Aa*) is more practical for mutation work than homozygous material (*AA* or *aa*), because induced mutations are mostly recessive, hence they can only be expressed in mutants derived from *Aa* genotypes where the dominant allele is mutated. Mutations from recessive to dominant occur at a very low frequency (Suprasanna and Nakagawa, 2012). Some ornamental plants take longer to produce mutants, like orchids. Others, like hibiscus, chrysanthemums or petunias, may take a shorter time, depending on their breeding cycle. New mutant varieties of orchids were mass-propagated by a local commercial tissue culture collaborator, Hexagon Green Sdn Bhd, for a project entitled 'Pre-commercialization of Mutant Orchids for Cut Flower Industry'. This project focuses on observation and determination of stalk length, floral architecture, flower quality and quantity, pest and disease susceptibility/damages, postharvest evaluation and quality control compliance for the cut-flower markets in Asia, Australia, Japan and Europe. Other mutant varieties, like those of hibiscus, have been successfully propagated in collaboration with growers for use as landscaping plants.

The shortage of funding for continuous research programmes and the lack of trained and qualified human resources, particularly in mutation breeding, poses a challenge to the country. With continuous support from IAEA this gap can be addressed by developing training programmes in new techniques in mutation induction, conducting regional and national training courses and workshops, delivering expert missions, facilitating scientific visits, and placing national candidates at relevant institutions in other regions. Impacts of climatic change such as drought, soil acidity and depletion of soil organic matter leading to reduced soil productivity continue to be a constraint and will become more serious under climate change and variability.

Mutation breeding has its own advantages and disadvantages. The advantages include the creation of new alleles that do not exist in germplasm pools. In vegetatively propagated crops, mutation breeding has the ability to change one or a few characters of an otherwise outstanding variety without altering the remaining and often unique part of the genotype (Broertjes and Van Harten, 1978). Breeders are able to develop new varieties in a short breeding cycle since homozygous lines can be obtained in the M_4 generation (four crop cycles) at the latest, whereas it would take at least eight to ten generations to produce a similar line through cross-breeding (Shu *et al.*, 2012). The disadvantage of mutation breeding is its limited power in generating dominant alleles that might be desired (Shu, 2009). Even though there are so many challenges, it can be foreseen that mutation screening technology will become more high-throughput, powerful and affordable with the rapid development of DNA technologies, including high-throughput DNA sequencing techniques.

5 Conclusion

In total, 53 new mutant varieties have been developed by the Malaysian Nuclear Agency. Mutation breeding has been especially successful in ornamentals as indicated by new varieties of orchids, hibiscus, chrysanthemums and pot and landscaping plants such as petunias, cannas, etc. In the case of rice as the major food crop, two drought-tolerant and high-yielding mutants (MR219-4 and MR219-9) and fragrant rice (MRQ74) have been generated and are grown commercially by local farmers. Groundnut is a legume categorized as an underutilized crop which has commercial potential and is well adapted to marginal soil and climate conditions. Two potential mutant varieties have been generated ('KARISMA Sweet' and 'KARISMA Serene'), which are resistant to *Cercospora* leaf spot disease and have a high yield and high N-fixation. The most popular mutant variety, banana 'Novaria', was produced in collaboration with the IAEA, the University of Malaya and United Plantations Sdn Bhd; this variety has improved agronomic traits, some derived from mutant lines, such as early flowering, high yield and a strong fruit pedicel. It also has good flavour and pulp texture. Mutation breeding is still an attractive method for creating genetic variability and it has become a routine technique in many seed and vegetatively propagated crops.

References

Broertjes, C. and Van Harten, A.M. (1978) *Application of Mutation Breeding Methods in the Improvement of Vegetatively Propagated Crops*. Elsevier Scientific Publishing Co., Amsterdam.

Broertjes, C. and Van Harten, A.M. (1988) *Applied Mutation Breeding for Vegetatively Propagated Crops*. Elsevier Scientific Publishing Co., Amsterdam.

Datta, S.K. (2009) Role of classical mutagenesis for development of new ornamental varieties. In: Shu, Q.Y. (ed.) *Induced Plant Mutations in the Genomics Era*. Proceedings of an International Joint FAO/IAEA Symposium. International Atomic Energy Agency, Vienna, Austria. pp. 300–302.

Datta, S.K. (2012) Success story of induced mutagenesis for development of new ornamental varieties. *Bioremediation, Biodiversity and Bioavailability* 6 (Special Issue 1), 15–26.

Luxiang, L. (2013) Towards sustainable development of food & agriculture in Asia and the Pacific – significant roles of IAEA / RCA projects. In: *Proceedings of Open Seminar for FNCA Mutation Breeding Project on Radiation and Nuclear Technologies for Crop Improvement and Productivity in Sustainable Agriculture*. FNCA/MEXT, pp. 68–72.

Mak, C. *et al.* (1996) 'Novaria' – a new banana mutant induced by gamma irradiation. *Infomusa* 5, 35–36.

Mak, C. *et al.* (1998) Micropropagation and mutation breeding techniques for the improvement of bananas. *Acta Horticulturae* 461, 219–224.

Maluszynski, M. (2001) Officially released mutant varieties – the FAO/IAEA Database. *Plant Cell, Tissue and Organ Culture* 65, 175–177.

Mohamad Farazi, J. (1994) Inheritance of semi dwarfism and panicle erectness in mutant line of rice. MSc Thesis, Universiti Kebangsaan, Malaysia.

Novak, F.J. *et al.* (1990) Mutation induction by gamma irradiation of in vitro cultured shoot-tips of banana and plantain (*Musa* cvs). *Tropical Agriculture (Trinidad)* 67, 21–28.

Othman, R. *et al.* (1989) Semi-dwarf mutants for rice improvement. In: *Proceedings of the 6th International Congress of SABRAO*, pp. 401–404

Rusli, I. *et al.* (1998) Evaluation of yield and N_2 fixation of mutant lines of groundnut in Malaysia. IAEA-TEC-DOC-1027, 87–94.

Shu, Q.Y. (2009) Turning plant mutation breeding into a new era: molecular mutation breeding. In: Shu, Q.Y. (ed.) *Induced Plant Mutations in the Genomic Era*. Food and Agriculture Organization of the United Nations (FAO), Rome, pp. 425–427.

Shu, Q.Y. *et al.* (2012) Principles and applications of plant mutation breeding. In: Shu, Q.Y., Forster, B.P. and Nakagawa, H. (eds) *Plant Mutation Breeding and Biotechnology*. Plant Breeding and Genetics Section, Joint FAO/IAEA Division of Nuclear Techniques in Food and Agriculture International Atomic Energy Agency, Vienna, Austria.

Siamak, S.B. and Zheng, S. (2018) Banana Fusarium wilt (*Fusarium oxysporum* f. sp. *cubense*) control and resistance, in the context of developing wilt-resistant bananas within sustainable production systems. *Horticultural Plant Journal* 4(5), 208–218.

Siwar, C. *et al.* (2014) Issues and challenges facing rice production and food security in the granary areas in the East Coast Economic Region (ECER), Malaysia. *Research Journal of Applied Sciences, Engineering and Technology* 7(4), 711–722.

Suprasanna, P. and Nakagawa, H. (2012) Mutation breeding of vegetatively propagated crops. In: Shu, Q.Y., Forster, B. and Nakagawa, H. (eds) *Plant Mutation Breeding and Biotechnology*. Joint FAO/IAEA Programme, Nuclear Techniques in Food and Agriculture, Vienna, Austria, pp. 347–369.

Yamaguchi, H. (2013) Characteristics of ion beams as mutagens for mutation breeding in rice and chrysanthemums. *Japan Agriculture Research Quarterly* 47(4), 339–346.

Zaiton, A. *et al.* (2012) Improvement of Malaysian ornamental plants through induced mutation. *Pertanika Journal of Tropical Agricultural Science* 35(3), 631–636.

8 Application of Mutation Breeding Techniques in the Development of Green Crop Varieties in Sri Lanka: the Way Forward

Malathy Parasuraman[1] and Priyantha Weerasinghe

Abstract

The Department of Agriculture (DOA) in Sri Lanka initiated mutation breeding in the 1960s with the introduction of a cobalt-60 source. The first rice mutant variety, MI 273, was released for general cultivation in 1971. M1 273, derived from irradiation of the H-4 variety, was identified as a drought-tolerant variety. An indirect rice mutant variety, developed by crossing the short mutant line BW267-3 with a highly adaptable variety, was released as BW 372 in 2013. It is moderately tolerant to blast, bacterial leaf blight, brown plant hopper, gall midge and iron toxicity, and thus increases productivity to 3–4 t/ha on lands prone to iron toxicity. The most popular groundnut variety cultivated in the country, 'Tissa', is a mutant developed by irradiation with gamma-rays at 200 Gy. It showed attributes of high yield, medium maturity (90–100 days) and high oil content (42%). 'Tissa' presently covers 80% of the groundnut cultivated area in Sri Lanka. A sesame mutant line, derived from the variety MI-3 irradiated at 200 Gy with ^{60}Co gamma-rays, was released as 'Malee' (ANK-S2) in 1993. It is a high-yielding variety (1.1–1.8 t/ha) resistant to Phytophthora blight. A cherry-type mutant tomato variety, developed by irradiation of seeds with gamma-rays (320 Gy), was released as 'Lanka Cherry' in 2010. Improved attributes are pear-shaped fruits and bacterial wilt resistance. Narrow genetic variability in many crops is a constraint to the development of new varieties adapted to the changing climate. Hence, the DOA is emphasizing integration of induced mutagenesis in conventional breeding programmes to develop resistant/tolerant varieties having high yield, quality and health-promoting functional properties in field and horticultural crops. The newly installed gamma irradiation chamber facilitates the creation of genetic variability in food crops, thus paving the way for the development of greener varieties.

Keywords: groundnut • mutation techniques • rice • sesame • Sri Lanka

1 Introduction

Sri Lanka is an island located to the south-east of the Indian subcontinent (5° 55' to 9° 51' N and 79° 42' to 81° 53' E) and to the south-west of the Bay of Bengal. The country is characterized by a variety of land forms, central highlands (1060–2420 m asl), uplands (270–1060 m asl) and low flat lands (0–270 m asl). The climate is tropical and warm, often moderated by ocean winds. The rainfall varies between 800 mm and more than 2500 mm, while the mean annual temperature ranges from 27°C in the coastal lowlands to 16°C in the central highlands. There are two cultivation seasons, named *maha* and *yala*, which are synonymous with two monsoons.

[1] Email: mpmalathy@yahoo.com

DOI: 10.1079/9781789249095.0008

The *maha* season falls during the 'North-east monsoon' from September to March. The *yala* season is effective during the period from May to end of August. Sri Lanka has three major climatic zones, the Wet, Intermediate and Dry Zones, that further divide into 46 Agro-Ecological Zones.

Paddy is the major food crop cultivated in the country, along with vegetables, fruits and other field crops (OFC), such as pulses, grains, condiments and oil crops. Tropical vegetables are grown in most of the country, while temperate vegetable types are generally cultivated in the up-country wet and intermediate zones. Fruit production is primarily of tropical crops, with a few temperate types. Cultivation of OFC is mostly restricted to dry zone areas. Except for temperate-type crops, most of the varieties cultivated in this country are developed and released by the Department of Agriculture (DOA). Some major food crops cultivated in Sri Lanka are shown in Table 8.1.

Breeding of crops for agricultural purposes has been conducted with new methods being introduced continuously. Conventional breeding approaches resulted in many useful outcomes for cereal crops, especially rice. However, narrow genetic variability in food crops limits the development of new varieties to achieve food security under a changing climate. Thus, germplasm introduction and techniques that could generate desired genetic variability could play a pivotal role in variety development in the Sri Lankan context. Mutation breeding is a commonly practised method used in plant breeding (Dubinin, 1961). In recent times, mutation breeding has increased in popularity as a successful tool for crop improvement in breeding programmes and a potent method of enhancing crop variability and crop productivity (Dubinin, 1961; Acharya *et al.*, 2007; Bhosale and Hallale, 2011). Mutation breeding creates genetic variation for developing drought or disease resistance and other desirable traits, enabling increasing crop productivity for sustainable food security as the erratic weather events and climate change impact on agriculture, reducing the productivity of existing local varieties (Acharya *et al.*, 2007; Bhosale and Hallale, 2011).

Mutation breeding is conducted by exposing genetic materials to mutagens, which cause changes in the genome and produce new traits in the organism (Bhosale and Hallale, 2011). Mutagens or mutating agents can induce mutations. A commonly used mutagen is gamma-ray irradiation. In recent times, gamma-rays from radioactive sources such as ^{60}Co and ^{137}Cs have become popular and are available to many countries through the IAEA. A large number of desirable mutant crop varieties have been developed and are being cultivated globally (Pathirana, 2011).

Table 8.1. Extent, production, imports and exports of selected major food crops in Sri Lanka (source: Agstat, 2018).

Crop	Extent (ha)	Production (t)	Imports (t)	Exports (t)
Paddy	791,679	2,383,153	747,800	4,968
Maize	52,544	195,744	179,589	–
Green gram	7,371	9,392	15,541	476
Black gram	8,089	7,329	12,767	446
Sesame	9,065	7,754	223	1,704
Groundnut	12,639	22,475	3,876	–
Chilli	10,937	51,827 (green chilli)	51,692 (dry chilli)	–
Big onion	3,026	53,603	232,318	–
Potato	4,457	73,358	151,438	–
Manioc	22,087	306,347	–	–
Bean	7,723	87,385	–	–
Tomato	5,329	80,839	–	–
Brinjal	9,665	108,856	–	–
Pumpkin	6,159	82,394	–	–
Banana	49,307	750,588	–	15,018
Mango	28,272	151,733	–	222
Pineapple	4,783	52,786	–	1,002

The DOA in Sri Lanka initiated mutation breeding in the late 1960s with the introduction of a cobalt-60 (^{60}Co) source to the country. It released the first mutant rice variety, MI 273, in 1971. This success led to the production of direct and indirect mutant varieties of rice, sesame, groundnut and tomato (IAEA, 2018). Improved mutant lines were either used to develop new varieties or conserved at the Plant Genetic Resources Centre (PGRC) of the DOA for future use.

The DOA has participated in many of the IAEA's coordinated research projects and national and regional technical cooperation projects to build necessary infrastructure for genetic improvements of crops through mutation breeding. Furthermore, the IAEA has assisted in establishing the Sri Lanka National Centre for Nuclear Agriculture by providing a new cobalt-60 source, through the IAEA's Technical Cooperation programme, to expand the country's mutation breeding activities. Additionally, the IAEA has been offering fellowships and training since the 1970s to assist the development of mutant crop varieties (IAEA, 2018). Currently, this facility caters to the research needs of all stakeholders by providing a service of mutation breeding for the country.

Being the organization responsible for the food crops sector in Sri Lanka, the DOA is emphasizing the integration of induced mutagenesis with conventional breeding to develop resistant/tolerant varieties with high yield, quality and health-promoting functional properties in rice, vegetables, fruits and other field crops. Improved varieties can be used either directly for commercial cultivation or to develop desirable genetic material as donor parents in conventional breeding programmes or as a parent in hybrid breeding programmes.

2 Crops

2.1 Rice

Rice (*Oryza sativa*) is the staple food of Sri Lanka and is the livelihood of more than 1.8 million farmers. It is cultivated on 0.79 million hectares, nearly 34% of the total cultivated land area, with the production of 2.38 million tonnes per annum (Agstat, 2018). The average productivity is around 4.29 t/ha. The per capita consumption of rice in the country is above 100 kg.

MI 273, the first mutant rice variety of the DOA, was released in 1971 for general cultivation. This variety originated after irradiating the H-4 variety with 350 Gy gamma-rays. Because of its shorter plant stature it was renamed as 'H-4 Dwarf mutant' in 1982. This rice variety became popular, as it can withstand occasional drought conditions. However, a lower degree of shattering was observed during late harvesting. This variety is rarely grown at present, owing to the introduction of new improved varieties that have a greater yield potential.

Since then, conventional breeding programmes have taken the leading role in rice breeding to improve the genetic potential of new cultivars and several varieties were released subsequently. From late 2000 onwards, mutation breeding was again in practice to incorporate a non-lodging trait into the improved rice varieties. This technique is currently used to develop quality rice varieties and varieties tolerant to emerging biotic and abiotic stresses under climate change.

In 2013, an indirect mutant variety, BW 372, was released for general cultivation. It is derived from the cross of the mutant line BW 267-3 with the popular variety, BG 359. This short mutant line is moderately resistant to many biotic stresses, such as blast, bacterial leaf blight, brown plant hopper and gall midge. This variety is also tolerant to iron toxicity. The average productivity of BW 372 in lands prone to iron toxicity is nearly 3–4 t/ha. Thus, it is cultivated in 0.03% of the total paddy area, which is approximately 275 ha, and produces about 1210 t with a value of Rs.42 million (US$0.26 million).

Apart from these two released rice mutant varieties, there are several lines identified as suitable to be released in the near future. BW 03-1198, an indirect mutant line, was produced by crossing Kahata wee/IRLON 98-217 with the mutant line BW 267-3. It matures in 3½ months and has a yield potential of about 5.4 t/ha. BW 03-1198 has good grain quality parameters, such as high gelatinization temperature, high amylose content and intermediate translucency. Furthermore, it is resistant to blast and gall midge and moderately resistant to brown plant hopper. It also shows tolerance to iron toxicity. This line is being evaluated for its adaptability in diverse farmers' fields.

BW 12-574 is another indirect mutant line, developed from the cross of BW 361 with the indirect mutant BW 372, suitable to be released in the near future. This line matures in 3 months and gives an average yield of about 3.8 t/ha. It has high contents of zinc (29.7 mg Zn/kg) and iron (27.92 mg Fe/kg) and shows high gelatinization temperatures and a chalkiness index of WC 2. It is resistant to blast and moderately resistant to brown plant hopper and gall midge. In addition, this indirect mutant line shows tolerance to iron toxicity.

There is a growing demand for Sri Lankan rice in the export market; hence the DOA emphasizes the development of quality rice varieties. For this, mutational breeding has been considered as one of the breeding options. The traditional variety 'Suwandel' and improved varieties, BG 94-1 and BG 1165-6, were irradiated by gamma radiation with specified doses to improve grain quality. This has resulted in nine promising mutant populations that have good grain quality traits preferred by farmers and consumers. These lines are being evaluated at the M_6 generation.

2.2 Other field crops

Field crops other than rice fall into this category. Since rice is the staple food of the nation, it has been considered as a special crop in Sri Lanka. Hence, field crops other than rice are considered as 'other field crops' (OFCs), which include grains, pulses, condiments and oil crops. OFCs are mainly grown in the non-paddy cultivating lands in the dry zone but are also grown in paddy lands during the water-scarce *yala* season. These crops are cultivated on 0.17 million hectares with annual production of 0.57 million tonnes. The per capita consumption is around 70 g per person and approximately 0.48 million tonnes are imported into Sri Lanka due to insufficient supply for different purposes (e.g. soybean meal, red lentils, chickpea, maize).

Groundnut (*Arachis hypogaea*) is a popular oilseed crop in Sri Lanka and about 5% of farm families earn their income by cultivating groundnut. 'Tissa' is the most popular groundnut variety cultivated in Sri Lanka. This is a mutant variety, developed by the DOA in 1993 by irradiating a Vietnamese groundnut variety with gamma-rays (200 Gy). The improved attributes of the mutant are disease resistance, drought tolerance and high yield. It shows an average potential pod yield of 2.0–2.5 t/ha (IAEA, 2018). Moreover, variety 'Tissa' has a high oil content of 42% and a low shelling percentage. It is a short-duration crop which matures in 90–100 days. 'Tissa' is cultivated on 12,933 ha, which is around 78% of total groundnut cultivation in this country. The annual production is approximately 20,822 t, which is valued at nearly Rs.3165 million (US$20 million). With the introduction of this mutant groundnut variety, groundnut importation to the island has reduced by 50% and it has been selected by many farmers (80%) as their preferred type of groundnut variety (IAEA, 2018).

Sesame (*Sesamum indicum* L.) is another oilseed crop that is popular in Sri Lanka, mainly in the Dry Zone. A mutant sesame variety 'Malee' (ANK – S2) was released by the DOA in 1993 to expand sesame cultivation. This variety was developed by irradiating MI-3 with ^{60}Co gamma-rays at 200 Gy (Pathirana, 1992; Pathirana *et al.*, 2000). The key attributes of this mutant are high yield (1.15–1.89 t/ha) and resistance to the *Phytophthora* blight disease. It is a short-duration crop and matures in 78–80 days. 'Malee' is suitable for southern dry regions of Sri Lanka where *Phytophthora* blight disease is prominent. It is cultivated in around 1.5% of the total extent of the sesame crop area (approximately 268 ha) in Sri Lanka with an annual production of 198 t, which is valued at nearly Rs.31 million (US$0.2 million). Sesame oil is valued for its use in industry as well as in food.

Mung bean (*Vigna radiata*) is another popular field crop cultivated especially in the Dry Zone areas. It is also cultivated as a third seasonal crop after the paddy harvest, using residual moisture in the paddy land with additional irrigation water. Introduction of short-duration varieties is required for successful third-season cultivation. Hence ongoing mutation breeding programmes are aiming to develop new varieties having drought tolerance, short duration (< 60 days) and synchronized maturity. Some promising results have been obtained so far, such as shorter maturity period and plant stature. Among these, a mutant line MIMB (M) – 1029 having short maturity (around 55 days) was selected as a parental line for the conventional breeding programme. In addition, dwarf mung bean lines were produced through irradiation. At present, evaluation of 14 different progenies (M_3) that were created through irradiation at 600 Gy (^{60}Co) is underway.

Soybean (*Glycine max*) is cultivated both in the Dry and Intermediate Zones of Sri Lanka. This crop is cultivated on a limited scale in paddy fields as well as in uplands in both the *yala* (dry) and *maha* (wet) seasons. Sri Lanka needs 12,000 t of soybean annually for the production of *triposha*, which is a protein-rich food supplement containing soybean meal together with rice and maize flour. Hence, it is vital to introduce new desirable varieties for promoting soybean cultivation on the island. The main attributes considered in soybean improvement are yield (6 t/ha under irrigated conditions; 3 t/ha under rainfed conditions), nutritional composition such as high protein (> 40%) and oil (> 20%) content and resistance to pod shattering and purple seed stain. The mutant lines (M_7) developed so far are presently being tested for desirable traits.

Chilli (*Capsicum annuum* L.) mutation breeding activities were carried out with Chilli to develop inbred lines having resistance/tolerance to Chilli leaf curl virus complex (CLCV). Currently, seven mutant progenies at M_4 generation that showed low incidences of CLCV symptoms in the field have been selected for further studies.

Finger millet (*Eleusine coracana*) is now being promoted as one of the major crops for the Dry Zone. The aim is to develop mutant lines having resistance/tolerance to blast disease and short to medium maturity. Seeds of recommended varieties were irradiated with different dosages of gamma-rays. Screening of early mutant generations is underway.

2.3 Vegetables

Vegetable production is carried out throughout the year in Sri Lanka and the entire local demand is met through local production. Vegetables are cultivated in about 0.9 million hectares with an annual production of 1.1 million tonnes. The majority of the vegetable cultivation in the country is in the Central and Uva provinces, which account for more than 60% of the total production (Agstat, 2018). The productivity of vegetables is around 12.8 t/ha, and the per capita consumption is around 136 g, excluding root and tuber crops. Vegetable crops are vulnerable to climate change; hence, productivity improvement and continuous cultivation are targeted through developing climate-resilient varieties by adopting various breeding techniques.

Tomato (*Solanum lycopersicum*) is one of the major vegetables, widely grown, and has high consumer demand. 'Lanka Cherry' is a cherry-type mutant variety released in 2010 by irradiation of seeds with 320 Gy gamma-rays. The key attributes of this mutant variety are its easily identifiable pear shape (mutant trait) and resistance to bacterial wilt. This mutant variety is preferred by the hotel industry, due to its distinct shape. However, owing to the very specific demands of this variety, the present extent of cultivation has been limited to a few hectares. The aim is to develop quality tomato varieties tolerant to bacterial wilt, heat and drought. Screening of mutant populations (M_2) is ongoing.

Common bean (*Phaseolus vulgaris*) is another major vegetable crop having high farmer and consumer demand in Sri Lanka. The available varieties are highly susceptible to virus yellowing and also sensitive to moisture stress. Hence, mutation breeding activities were initiated to develop high-yielding, quality bean varieties having tolerance to virus yellowing. Screening is continuing with some promising mutant lines identified at the M_4 generation with the aim of developing bean varieties tolerant to moisture stress. Seeds of two released varieties were irradiated with different dosages, viz. 150, 200 and 250 Gy. *In vivo* and *in vitro* studies revealed that seeds irradiated with 200 Gy increased the root volume in variety 'Gannoruwa Green'. Screening is continuing with the mutated population. It is expected that this characteristic could successfully be employed in generating climate-resilient varieties for future use.

2.4 Fruits

The area under fruit cultivation is around 0.11 million hectares with the production of 1 million tonnes of different fruits per annum. Banana, pineapple and papaya are the most popularly grown fruit crops in Sri Lanka and banana is the most extensively cultivated fruit crop, with nearly 40,000 ha of cultivation. Fruit productivity in Sri Lanka is around 8.6 t/ha, and the per capita consumption is extremely low at 87 g. Therefore, attempts are being made to increase the productivity of fruit crops by adopting various technologies and also to increase production.

Banana (*Musa* spp.) plays an important role among the fruit crops in Sri Lanka in terms of extent of production and consumption. From the cultivars and varieties available the silk banana variety 'Agra' is widely cultivated in the island. However, panama is one of the most important diseases affecting the banana industry in Sri Lanka and identifying tolerant varieties for this disease remains unresolved. The recommended silk banana (Kolikuttu) variety 'Agra' is also susceptible to panama disease caused by the soil-borne fungus, *Fusarium oxysporum* f. sp. *cubense* (Foc). Banana improvement through conventional techniques is extremely difficult, due to sterility and polyploidy. Somaclonal variation and induced mutation are hence considered as effective ways to improve banana and several varieties have been developed elsewhere using these techniques (Roux, 2004; Smith *et al.*, 2005).

Mutation breeding of banana in the island was initiated with chemical mutation. Shoot tip explants of banana variety 'Agra' were treated with 1% EMS + 2% DMSO for an incubation period of 3 h. Mass treatment was done batch-wise over 3 years (2013–2016). Further, about 1100 shoot tip explants were subjected to 40 Gy gamma irradiation using ^{60}Co source. The plants derived from treated shoot tips were screened using different screening methods (*in vitro*, early screening in protected houses, sick plot and Fusarium 'hotspots'). More than 8000 plants were screened under field and screen house conditions while 1100 explants were screened under *in vitro* conditions. Variants were observed at the nursery stage and at field level, but most of these did not perform well. One mutant plant that survived in the sick plot for more than 21 months and another mutant having short stature (plant height < 2.3 m) were selected for further screening and agronomic studies. It is expected that mutation breeding activities of the crop could provide a breakthrough in searching for a variety that shows tolerance to this extremely challenging disease in the industry.

3 Conclusion

Narrow genetic variability in many crops is a constraint to the development of new varieties adapted to climate change. Hence, the DOA is emphasizing the integration of induced mutagenesis into conventional breeding programmes to develop resistant/tolerant varieties with high yield, quality and health-promoting functional properties in field and horticultural crops. The fellowships and trainings offered by the IAEA for the capacity building of scientists assist in the successful implementation of the programme. The newly installed gamma irradiation chamber facilitated through the IAEA/TCP/SRL 5045 project helps to create genetic variability in food crops, thus paving the way for the development of greener varieties.

References

Acharya, S.N. *et al.* (2007) Improvement in the medicinal and nutritional properties of fenugreek (*Trigonella foenum-graecum* L.). In: Acharya, S.J. and Thomas, J.E. (eds) *Advances in Medicinal Plant Research*. Research Signpost, Kerala, India, pp. 491–512.

Agstat (2018) *Agricultural Statistics*. Agriculture, Peradeniya. Volume XV. Socio Economics and Planning Center, Department of Agriculture, Peradeniya.

Bhosale, U.P. and Hallale, B.V. (2011) Gamma radiation induced mutations in black gram. *Asian Journal of Plant Science and Research* 1(2), 96–100.

Dubinin, N. (1961) *Problems of Radiation Genetics.* Oliver and Boyd, London.

IAEA (2018) *Farmers benefit from mutation bred groundnut for 25 years.* IAEA NewsCentre, Vienna, 27 August 2018. Available at: https://www.iaea.org/newscenter/news/sri-lankan-farmers-benefit-from-mutation-bred-groundnut-for-25-years (accessed 4 October 2018)

Pathirana, R. (1992) Gamma ray-induced field tolerance to Phytophthora blight in sesame. *Plant Breeding* 108(4), 314–319.

Pathirana, R. (2011) Plant mutation breeding in agriculture. *CAB Reviews Perspectives in Agriculture Veterinary Science Nutrition and Natural Resources* 6(032), 1–20.

Pathirana, R. *et al.* (2000) Development and release of gamma ray induced sesame mutant ANK-S2 in Sri Lanka. *Tropical Agricultural Research and Extension* 3(1), 19–24.

Roux, N.S. (2004) Mutation induction in Musa – review. In: Jain, S.M. and Swennen, R. (eds) *Banana Improvement: Cellular, Molecular Biology and Induced Mutations*. Science Publishers, Enfield, New Hampshire, pp. 23–32.

Smith, M.K. *et al.* (2005) *Musa* spp. Banana and plantain. In: Litz, R.E. (ed.) *Biotechnology of Fruit and Nut Crops*. CAB International, Wallingford, UK, pp. 366–392.

9 Mutation Breeding in Rice for Sustainable Crop Production and Food Security in India

Vikash Kumar[1], Anjali Chauhan, Avinash Kumar Shinde, Ramesh L. Kunkerkar, Deepak Sharma and Bikram Kishore Das

Abstract

With the inevitable risk posed by global climate change affecting crop yield and the ever-increasing demands of agricultural produce, crop improvement techniques need to be more precise in developing smart crop varieties. The rice crop, a staple food for the majority of the world population, has a significant role to play in alleviating the global hunger problem. With the world population burgeoning at an unprecedented rate, limited fertile land resources, climate change, emerging new races of pests and diseases and consumer preferences for quality attributes, it is imperative to increase crop diversity, and this requires better selection efficiency addressing the challenges of future rice production. Mutation breeding is a fundamental and very successful tool helping to increase crop diversity and allowing plant breeders to exercise their skill in developing desirable crop varieties. The induction of mutations has been used to enhance yield, improve nutritional quality and widen the adaptability of the world's most important crops such as wheat, rice, pulses, millets and oilseeds. India is considered to be one of the primary centres of origin of crop species with the concomitant very high genetic diversity in traditional landraces for different agronomic traits of economic importance. Plant architecture, such as plant height, branching habit (tiller number), leaf shape and patterns, floral and grain traits and quality traits such as aroma, amylose content and cooking quality are of tremendous importance for rice improvement programmes. Traditional landraces of rice have premium grain quality, fetching a premium price, but their cultivation is being marginalized due to their tall stature, proneness to lodging, late maturity and poor yield. Mutation breeding technology has been successfully implemented in rice improvement programmes, which have resulted in the improvement of aromatic rice varieties, such as 'Pusa Basmati 1', 'Dubraj and Jawaphool'. Two high-yielding mutant rice varieties, TCDM-1 ('Trombay Chhattisgarh Dubraj Mutant-1') and TKR Kolam ('Trombay Karjat Rice Kolam'), have been released for cultivation in Chhattisgarh and the Konkan region of Maharashtra. Both these varieties possess dwarf plant stature (110 cm), medium maturity (130 days), premium grain quality and resistance to major pests and diseases. Improvement of other traditional rice varieties is underway which will bring these varieties back into cultivation and help in improving the tribal and marginal farmers' economy.

Keywords: mutation • rice • plant architecture • grain quality • landraces

1 Introduction

Rice (*Oryza sativa* L.) is the most important cereal crop which is grown under a wide range of climatic and geographical conditions on all five major continents. More than 60% of the global population and more than 75% of the Asian population is dependent on the rice-based

[1] Email: kvikash@barc.gov.in

DOI: 10.1079/9781789249095.0009

cropping system for food and economy. The importance of this crop lies in the fact that it has shaped the culture, diet and economy of millions of people, particularly in Asia. Considering the role rice plays in providing food and nutritional security, and eradicating poverty, the United Nations designated the year 2004 as the International Year of Rice. In order to achieve stable growth in rice production, stronger efforts are required to boost productivity, break yield barriers and provide safety against fluctuations in climatic conditions.

Global rice production statistics from the Statista website (available at https://www.statista.com, accessed April 2021) showed production levels of 486.2 million tonnes of milled rice in 2016–2017 with an estimated world acreage of 161.1 million hectares. India, with a rice production volume of approximately 111.3 million tonnes, ranked second in global rice production in 2016–2017, with an estimated harvest area of about 44.5 million hectares of rice. With the world population increasing at an unprecedented rate, climate change, limited land resources, diseases and consumer preferences for quality attributes and emerging new races of pests, it is necessary to increase crop diversity with better selection efficiency addressing the challenges of future rice production. In order to meet the dietary requirements of our increasing population and the challenges emerging due to the adverse impacts of climatic changes, genetic gain in the grain yield per hectare of major crops like rice needs to rise faster than the current rate (Ray *et al.*, 2012; Challinor *et al.*, 2014).

1.1 Food security in India: significance of rice production and associated challenges

Food security is the backbone of national prosperity and well-being. The progress and growth of any nation is directly linked to food security. Being the most important carbohydrate source in India, an increase in production and productivity of rice, especially for small and marginal farmers, is the most important dimension of food security. Indian food and nutritional security are being challenged by many social, economic and environmental factors, such as increase in the population, increasing urbanization and increasing demand for food due to raised income levels. In addition, dietary preferences, including a high demand for specialized rice in different parts of India (such as short or medium bold in southern India, aromatic and long slender in northern India, small slender and fine-grain rice in western, central and eastern India) are creating a need for diverse types of rice in cultivation. Rice rich in micronutrients to address malnutrition problems for poor people, rice having speciality uses and consumption of more processed rice products are also building up more pressure on the food supply system. In the current scenario, rice production in India is challenged by various factors as outlined below.

Climate change

This is one of the most impactful challenges to long-term rice production and ultimately national food security, as it could lead to dramatic scarcity of fresh water in the northern and peninsular regions of the country. Various estimates suggest that India will experience an increase of 2.2–2.9°C in average annual temperature by 2050, affecting overall rice production. Rainfed rice farming, which covers 60% of the cultivated land in the country, will feel the heat drastically. Declining and degrading land resources also pose a serious threat to food security, as the availability of land per capita is declining sharply due to increases in population. Decline in soil biodiversity, salinization, waterlogging, heavy metal and pesticide contamination, organic carbon deterioration and soil erosion, ultimately leading to nutrient depletion, are some of the major threats to soil quality (Issaka *et al.*, 2017).

Biotic and abiotic stresses

In rice production, the introduction of semi-dwarf varieties and the intensive use of inputs like fertilizers and insecticides have changed the incidence of pest occurrence from low to high for several pests, such as brown plant hoppers, stem borers and leaf folders, and diseases such as blast, bacterial blight and sheath rot (Hongxing *et al.*, 2017). This is basically due to increased cropping intensity, from single-crop cultivation to multiple crops per year, thereby inducing a favourable environment for insect pests and pathogen multiplication. To help minimize these problems, the development of tolerant varieties and sustainable agronomic management strategies are the challenges for new-generation scientists. In the context of global warming, it is crucial to address the outcomes of increased CO_2 and temperature, deficient and excess rainfall and expected sea-level rise, through the development of climate-resilient rice varieties and farming practices. The increase in temperature may not be the real constraint for Indian rice, as the reproductive stage of the *kharif* rice, which is the dominant rice system in India, coincides with the onset of winter.

Rice grain quality

The combination of depreciation in price of rice in the world market and the increase in per capita income led to increased domestic and international demand for quality rice. However, the preferred rice grain quality is highly variable in different regions and countries and among different consumers. Important physical traits determining the popularity and marketability of the rice grain include grain shape, size, chalkiness, degree of milling and head rice recovery; major decisive chemical parameters are amylose content, gelatinization temperature and gel consistency. The amylose:amylopectin ratio, gelatinization temperature, viscosity, texture of cooked rice, flavour and aroma determine the cooking and eating qualities of rice, while the nutritional quality of the grain is largely governed by its chemical composition. Most of these rice grain quality characteristics are likely to be adversely affected by climate change and global warming (Thornton *et al.*, 2014).

There is a perception that China has, by and large, solved its 'food problem', whereas India has not (Timmer, 2014). The core of India's food security problem today pertains not so much to increasing food availability or production, but to the distribution of food. This is not to suggest that the challenges associated with ensuring food availability in sustainable ways is not a policy concern, but rather, in terms of the immediacy of challenges, ensuring food access appears to score above concerns over food availability. Despite lagging growth rates in the agricultural sector relative to targets, India has seen impressive growth in food grain production in recent years. The National Food Security Mission has played a key role in augmenting production in cereals and pulses. Most of this has come from yield increases in the eastern parts of India where the Green Revolution did not take place. At the same time, there has also been a strong and continuing trend for diversification into non-cereal and high-value commodities such as dairy, fruits and vegetables, which implies the possibility of higher-quality diets. Investments in the agricultural sector shot up after 2004–2005, in both public and private domains, with private gross capital formation accounting for an increasing share in all endeavours.

Despite the large increase in production, access to food continues to be a serious issue, especially in the context of extraordinarily high inflation rates in food commodities in recent years and limited access to high-quality diets. The seriousness of the challenge of food security derives from recent evidence, from India and elsewhere, suggesting that income growth might not always translate fully or quickly enough to improve the health and nutritional status of children, implying that this issue needs attention (Haddad *et al.*, 2003; Lee *et al.*, 2012; Coffey *et al.*, 2014). This weak link between income growth and nutritional outcomes implies that food security, in the sense defined earlier, would require special attention of policy makers and cannot be presumed to follow as a consequence of growth. This is quite apart from a parallel discourse that argues for a rights-based approach to food security, so that primary responsibility rests with the state. In general, there is a broad agreement on the imperatives of food security in India, but there are deep disagreements on how to achieve this.

2 Mutation Breeding: a Novel Tool for Crop Improvement

Genetic improvement work without the existence of wide variability in the population is not possible. Intraspecific variation, which forms the basis for selection and improvement, is a prerequisite for initiating any plant improvement programme (Datta, 2005). The number of genes expressed during the lifetime of a particular plant is estimated to be between 16,000 and 33,000 (Gibson *et al.*, 1993). The genetic architecture of living organisms is influenced by mutations, in a complex manner. Mutations form the foundation for selection, which ultimately is the driving force for progressive breeding among crop plants. The mutants may have a natural (spontaneous) origin or may be artificially induced. Spontaneous mutations, the naturally occurring heritable changes to genetic material, are rather rare and random events. Not only can this infrequent process be greatly accelerated through artificial induction of mutation, but it also augments biodiversity. Germplasm resources act as a reservoir and can include genotypes with tolerance to various emerging stresses which can be used in breeding programmes to help introgression of traits to elite crop varieties. Conventional breeding is limited by tight linkages of undesirable traits and genetic differentiation at the expense of genetic diversity (Rauf *et al.*, 2010). For rice breeders, a continuous task is to find the new genomic variations that can be directly used in rice breeding programmes (Lee *et al.*, 2015). Mutation breeding is an efficient technology in the hands of plant breeders which helps in breaking undesirable allelic combinations and also results in genetic changes desirable for breeders in the background of an elite variety. In rice, according to the FAO/IAEA Mutant Variety Database (MVD, 2018), more than 820 mutant varieties have been developed through mutation breeding across the globe.

Mutagenesis is an important tool in crop improvement and is free of the regulatory restrictions imposed on genetically modified organisms. The forward genetic approach enables the identification of improved or novel phenotypes that can be exploited in conventional breeding programmes. Powerful reverse genetic strategies that allow the detection of induced point mutations in individuals of the mutagenized populations can address the major challenge of linking sequence information to the biological function of genes and can also identify novel variation for plant breeding (Parry *et al.*, 2009). These strategies also allow accessing cryptic variations, revealing allelic combinations that produce a desirable phenotype, which is of particular significance for traits governed by multiple genes.

2.1 Role of mutation breeding in crop improvement

The role of mutation breeding in increasing food production and providing sustainable nutrition is well established (Goyal *et al.*, 2009; Wani *et al.*, 2011). Food security has been variously defined in an economic context, but the most widely accepted definition is the one by the World Bank: 'access by all people at all times to enough food for an active, healthy life'. Likewise, on food plan action, the World Food Summit at Rome in 1996, also known as the Rome Declaration on World Food Security (FAO, 1996), observed that:

> Food security at the individual, household, national and global level exists where all people at all times have physical and economic access to sufficient, safe and nutritious food to meet their dietary needs and food preferences for an active and healthy life.

In both definitions, the emphasis has been on the physical availability and economic accessibility of food to the people. Mutant varieties have been grown on a large scale by farmers in their fields, and any increase in food production achieved through the cultivation of the mutant varieties could be translated into increased food security through availability to a greater proportion of the population. In a little less than a century, induced mutagenesis has bagged the credit for the development of several superior crop varieties that are being grown all over the world. A total of 3281 crop mutants have been released by different countries in the world and registered in the Mutant Variety Database of the IAEA in Vienna (MVD, 2018). China is the leading country in mutation breeding activities and has developed the highest number of crop mutant varieties (810), followed by Japan, which has developed just over 480 crop mutants. In this list, India ranks third (335 crop mutant varieties developed so far). Globally, the highest number of crop mutants (821) was developed for rice, followed by barley (304) and chrysanthemum (281) (Figs 9.1 and 9.2).

In the approximately 80-year history of induced mutations, there are many examples of the development of new and valuable alterations in plant traits significantly contributing to increased yield potential of specific crops. The primary motive of mutation breeding is to enlarge the frequency and spectrum of mutations (Khan *et al.*, 1992)

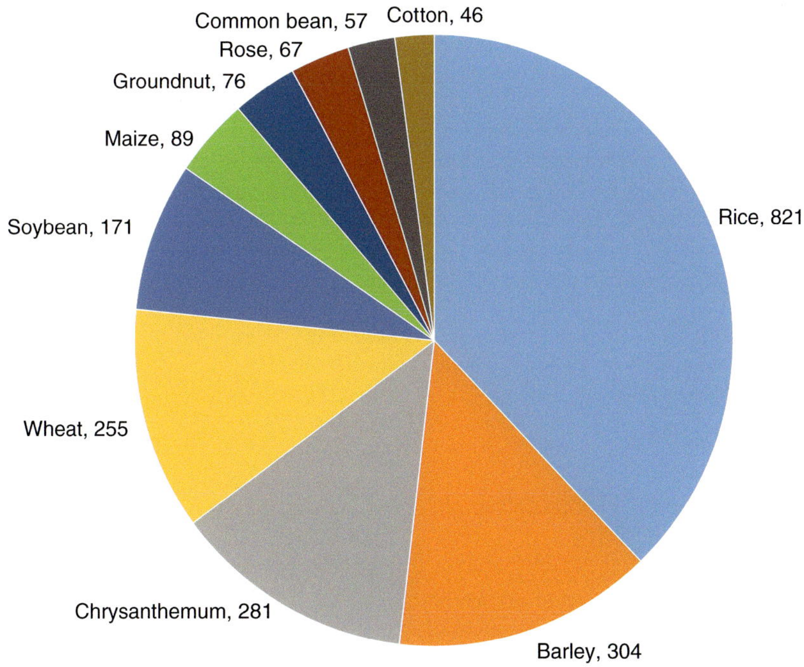

Fig. 9.1. Number of mutant varieties released for a selection of crop species including both direct mutant and mutant derivatives.

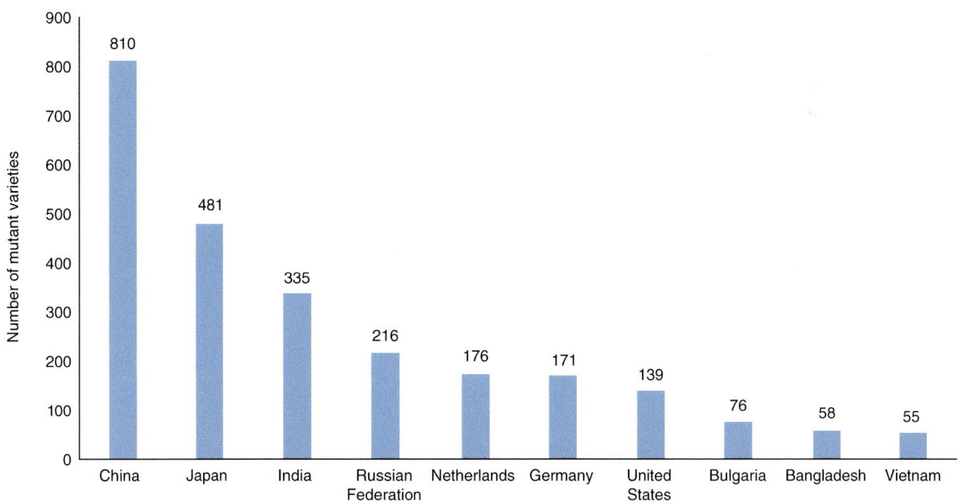

Fig. 9.2. Number of mutant varieties of crop plants released in different countries.

and also to increase the incidence of viable mutations. The main focus has been on upgrading well-adapted varieties by altering traits like maturity, seed size and disease resistance which play vital roles in increasing yield and yield-attributing characters. The attributes that have been improved through mutation breeding include a wide range of characters, such as tolerance to abiotic and biotic stresses, duration of maturity and flowering and other yield-contributing characters (Micke, 1985). Cereals and legumes represent important food crops. Improvement in these food crops has been a major concern for plant breeders over the years. In the post-Mendelian era, these crops have been improved through introduction, selection and hybridization using either natural or induced genetic variability. In the present era, induced mutagenesis provides an opportunity to create hitherto unknown alleles, leading to wide genetic variability. This possibility has been exploited in both cereals and legumes, as is evident from the list of mutant cultivars developed in legumes and cereals in India (Fig. 9.3).

2.2 Global scenario for mutation breeding in rice

The impact of induced mutations in applied research is best exemplified by the development of improved rice varieties through mutation breeding. During the past five decades, more than 800 varieties of rice have been developed across the globe, either directly from induced mutations or as a result of crossing such mutants with other breeding lines (Kharkwal *et al.*, 2009). In 1957, China released the first rice varieties 'KT 20-74' and 'SH 30-21' through induced mutation and a cross-breeding programme with a mutant resulted in the first rice variety 'Yenhsing-1' (Rutger, 1992). 'Reimei' was a semi-dwarf mutant released in Japan (Futsuhara, 1968), which had significantly increased yield because of its lodging resistance. 'Calrose 76' and 'Basmati 370', semi-dwarf varieties of rice with short stiff straw, brought a revolution in rice production in the USA and Pakistan, respectively. 'Kashmir Basmati', developed in Pakistan, matures early, has cold tolerance and retains the aroma and cooking quality of the parent; it was derived from an induced mutation in 'Basmati 370' (Awan, 1991). The PNR series of rice mutants, with high yield, were released in India and some of these were early in maturity and had short stature (Chakrabarti, 1995). Among these, two early-ripening and aromatic mutation-derived rice varieties,

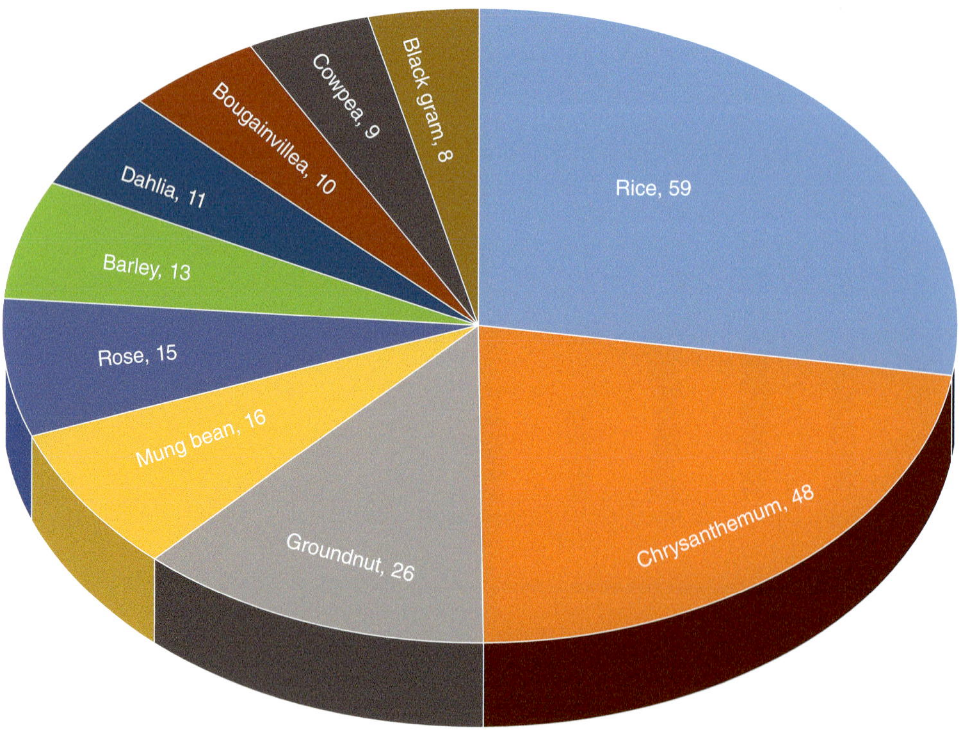

Fig. 9.3. Number of mutant cultivars of cereals, legumes and other crops released in India.

'PNR-381' and 'PNR-102', are popular for cultivation in Haryana and Utter Pradesh. The rice mutant 'Zhefu 802' was cultivated on more than 10.6 million hectares of land area in China over a span of 10 years. The RD series of rice mutants was very popular in Thailand and one such aromatic indica variety of rice, 'RD6' released in 1977, became very popular. It was extensively grown on 2.4 million hectares during the year 1994–1995. According to the Bureau of Economic and Agricultural Statistics in Bangkok in 1995, a similar mutant, 'RD15' released in 1978, was grown on 0.2 million hectares, equivalent to 3.2% of the area under rice. In Australia, nine rice mutant varieties – 'Amaroo' (1987), 'Bogan' (1987), 'Echua' (1989), 'Harra' (1991), 'Illabong' (1993), 'Jarrah' (1993), 'Langi' (1994), 'Millin' (1995) and 'Namaga' (1997) – have been developed. The induction of a thermosensitive genic male-sterile (TGMS) mutant in japonica rice mutant 'PL-12', which is controlled by a single recessive gene, has had an immense contribution in designing the strategies for the production of hybrid rice varieties (Maruyama *et al.*, 1991). In China '26 Zhaizao' was developed by gamma-ray irradiation of indica rice (Shen *et al.*, 1993). These mutants play an important role in two-line heterosis breeding. In India, out of 335 mutant varieties released, rice, with 60 mutant varieties, stands ahead of other plant species (Table 9.1).

Table 9.1. List of rice varieties developed through mutation breeding in India.

Year	Varieties registered
1967	K84
1968	Vellayani
1972	IIT 48, IIT 60
1973	Hybrid Mutant 95
1975	Jagannath (BSS-873), PL-56
1976	Au-1, Rasmi
1979	CNM 25, CNM 31
1980	CNM 6 (Lakshmi), CNM 20, Indira
1982	Biraj
1983	Mohan (CSR4), Sattari, Savitri
1984	HPU 8020, Prabhavati
1985	Keshari
1987	Hari (TR-RNR-21)
1988	Intan mutant, Padmini, Pusa-NR-162
1989	Dharitri (IET-6272), Gayatri (IET-8020), Pusa-NR-166, Pusa-NR-381
1990	HUR 36, Pusa-NR-519, Pusa-NR-555-5, Pusa-NR-570-17, Pusa-NR-571
1992	Lunisree
1993	MDU-4 (ACM-15)
1994	ADT 41
1996	Gautam
1997	Malviya Dhan-36, Pusa-NR-550-1-2 (JD-8), Pusa-NR-551-4-20 (JD-6), Pusa-NR-555-28 (JD-10), Radhi, Tapaswini
1998	Pusa-NR-555-5 (JD-3)
1999	CRM 49, CRM 51, CRM 53, MO 15 (Remanika), Padmanth (IET-11876), Ramchandi (IET-13354)
2000	Dhanu
2002	Kaum-20-19-4
2003	Pusa-NR-546
2005	CRM 2007-1 (Geetanjali)
2006	Anashwara, Shivam
2019	Trombay Chhattisgarh Dubraj Mutant-1 (TCDM-1)

3 Traditional Landraces of Rice

India has a diverse genetic wealth of rice and it has been estimated from various surveys that the country is endowed with more than 2 lakh (i.e. 200,000) rice varieties, a rich biodiversity that no other country on earth possesses (Arumugasamy *et al.*, 2006; Sathya, 2007; Ashraf *et al.*, 2017). The switch to high-yielding varieties with the spread of modern agriculture has posed a great threat to the security of the age-old practice of growing traditional varieties and landraces, which may serve as valuable resource for important traits (Richharia, 1979; Roy, 1985; Sharma *et al.*, 1987; Patra, 2000). The diversity hotspots existing in the North-Eastern region, Central India and Southern India are enriched with unique scented and aromatic, medicinal and quality rice. Some important rice landraces of India are listed in Table 9.2.

Traditional and aromatic rice varieties are unique in terms of their pleasant aroma, fine grain size, better cooking quality and suitability to consumer preference for special culinary preparations (Fig. 9.4). Farmers get a premium price from these traditional varieties due to their premium quality and lower production owing to their poor yield, lodging losses, cultivation by tribal and marginal farmers, etc. Improvement of these agronomic attributes will help to increase the production potential of these varieties and bring them back into cultivation.

Table 9.2. Important rice landraces of India with their quality attributes and undesirable traits.

Landrace name	Special quality attributes	Undesirable character
Ilayachi	Short grain, good grain quality, good for mouth ulcers, improves digestion	Late maturity (140–150 days), tall (140–145 cm), poor yield
Bangalaya	Aromatic short grain rice (ASG)	Late maturity and tall stature
Kalanamak	ASG, better grain quality	Very late maturity and tall stature
Tulsi Manjari	ASG, good grain quality, suitable for *Kheer,* head rice recovery (HRR) (> 65%)	Late maturity and poor grain yield
Gopal Bhog	ASG, good grain quality, strong aroma, head rice recovery (> 65%)	Late maturity, lodging, poor grain yield
Tilkasturi	ASG, head rice recovery (> 65%)	Late maturity, lodging, poor grain yield
Kali Kamod	ASG, better grain quality, head rice recovery (> 65%)	Late maturity
Kumhda Phool	Bold, good for scented Poha/Chiwda[a], good grain quality	Late maturity and lodging
Alsakar	Bold, good for aromatic Poha/Chiwda[a]	Late maturity, lodging, HRR (45%)
Danighoda	Bold, good for Poha/Chiwda[a]	Late maturity, HRR (45%)
Adanga Dhan	High biomass, bold, good for Poha[a]	Poor grain yield and HRR (48%)
Rudra Dhan	Good grain quality, short bold, good Idli making	Poor grain yield and HRR (49%)
Roti Dhan	Suitable for roti and Poha[a] making, coarse grain	Chalky grain, late maturity, poor yield
Makdo Dhan	Bold grain, good for Poha[a]	Lodging plant type, poor grain yield
Dubraj	Aromatic, premium grain quality, very fine grain	Late maturity and lodging plant type
Jawaphool	Short slender and aromatic grain, better grain quality	Lodging, poor grain yield

[a]Poha and Chiwda are local names for flattened/beaten rice, used for preparation of a very popular Indian breakfast dish

3.1 Improvement of aromatic and traditional landraces in rice

India is presently facing a food crisis mainly due to the erosion of its biodiversity, poor productivity of landraces and increase of monocropping mega varieties in agriculture. Climate change has led to frequent floods and prolonged droughts and as modern high-yielding rice varieties and hybrids suffer most, this has led to partial or total loss of crops. The major reason for the disappearance of thousands of local rice varieties is their steady replacement with the high-yielding varieties (HYVs) introduced in the 1960s coinciding with the Green Revolution (GR). This has slowly led to the gradual extinction of traditional landraces and aromatic rice varieties, due to their poor yield, undesirable plant architecture, lodging problems resulting in yield losses and late to very late maturity making them unsuitable for fitting into their cropping pattern. In India, every state has some distinctive rice landraces and premium aromatic varieties which are being cultivated by the marginal farmers of that region only. Improvement of agronomic traits in these traditional landraces will help in expanding the area under cultivation of these improved rice varieties.

3.2 Involvement of BARC in rice improvement using radiation technology

Development of TCDM-1 ('Trombay Chhattisgarh Dubraj Mutant-1') rice variety

TCDM-1 is an improved mutant of the highly priced and aromatic local rice variety 'Dubraj' which is known for its premium grain quality and aroma. This variety was released by the state variety release committee of Chhattisgarh state in 2018 under the joint collaboration of Bhabha Atomic Research Centre (BARC), Trombay, and Indira Gandhi Krishi Vishwa Vidyalaya

Fig. 9.4. Diverse rice landraces of India varying in grain size, colour and their quality attributes. First row (left to right): Jaigundi, Javaphool, Jeeradhan, Shyamjeera, Srikamal, and T. Manjari. Second row (left to right): Jeeraphool, Jouphool, Kalikamod, T. Prasad, Tilkasturi and Tulsimanjari. Third row (left to right): Kubarimohar, Lallu, Londhi, Anterved, Atmasheetal and Badshshbhog. Fourth row (left to right): Mai Dubraj, Samudrafen, Shuldaphool, Bisni, Chhatri and Chinnor.

(IGKV), Raipur. The salient features of TCDM-1 include dwarf plant type (90–95 cm), longer and denser panicles (235–240 grains per panicle), more tillers per plant (10–12 tillers per plant), mid-late maturity (135–140 days), better aroma, better grain quality, resistance to major pests and diseases and higher yield (4.7 t/ha) (Fig. 9.5). This variety has become very popular among the farmers of Chhattisgarh state during adaptive trials because of its lodging tolerance (preventing yield losses during late monsoon rains), improved aroma and better post-cooking grain quality.

Development of TKR Kolam rice ('Trombay Karjat Kolam') variety

The Konkan region of Maharashtra is the major rice-producing area of Maharashtra state. An area of nearly 0.4 million hectares of the Konkan is under rice crop with production of 1.6 million tonnes of rice. Consumers in Central and South-West India prefer small and fine-grain (Short Slender) rice varieties, colloquially named as Kolam type rice. Traditional Kolam rice varieties have specific ecological niches, required for their specific grain quality. 'Wada Kolam', 'Surti Kolam', 'Silk Kolam', 'Lachkari Kolam' and 'HMT Kolam' are popular Kolam rice varieties predominant in Western and Southern India. These rice landraces are very tall (> 150 cm) and very late (> 160 days) in maturity. Their yield potential is also very poor. These varieties are highly susceptible to lodging during late monsoon rains, amounting to complete economic loss to the farmers. BARC has embarked upon a strong mutation breeding programme in rice. BARC has developed an improved version of Kolam rice which is dwarf (110 cm), mid-late in maturity (130–135 days), lodging resistant and has a high grain yield (4.5–5.0 t/ha) as compared with traditional

Fig. 9.5. Varietal characteristics of TCDM-1. **(a)** Field view of local Dubraj (parent) and TCDM-1 (dwarf mutant of Dubraj). **(b)** Panicle of TCDM-1. **(c)** Paddy and rice grains of TCDM-1. The most important feature of TCDM-1 is its milled rice recovery (71.6%) and head rice recovery (69.2%), resulting in higher grain yield with less percentage of broken grains, a characteristic feature of premium rice. This rice variety fetches premium price (Rs.100–150/kg) in the local market, which will help to double farmers' income.

Kolam rice, which yields < 2.5 t/ha. TKR Kolam rice ('Trombay Karjat Rice Kolam') variety (Fig. 9.6) was released for the farmers in the Konkan region of Maharashtra in 2018 and notified for commercial cultivation in 2019. This variety has been developed through mutation breeding of rice variety 'PB1' for over 6 years. It was followed by a Maharashtra state and national All India Coordinated Rice Improvement Programme (AICRIP) yield trial.

3.3 Improvement of other traditional landraces of rice

India is known for its richness in diversity for rice, especially for aromatic and quality rice. BARC has a target-oriented mutation breeding programme for improvement of traditional landraces of Chhattisgarh, Maharashtra, West Bengal, Haryana, Tamil Nadu, Andhra Pradesh and other states of India in collaboration with the IGKV, Dr Bala Sahab Konkan Krishi Vidyapeeth (DBSKKV), Dapoli, and other State Agricultural Universities (SAUs). Gamma-ray and proton beam induced mutagenesis have been employed to improve traditional rice landraces such as 'Luchai', 'Hundar', 'Lokti Machi', 'Jawaphool', 'Sonagathi', 'Badsahbhog', 'Safri', 'Tilkormel', 'Swarna', 'Chinoor', 'Ambe Mohar', etc. We have been able to develop dwarf and mid-early high-yielding mutants of these varieties which are in state and national yield trials (Fig. 9.7). 'Jhilli dhan Wada Kolam', 'Zinia', 'Bangalaya' and 'Palghar' series of fine quality rice are very popular in different pockets of Maharashtra state. These traditional landraces are also being improved by radiation-induced mutation breeding.

3.4 Improvement of other important agronomic traits in rice using radiation technology

BARC is a pioneer institute in developing radiation technology for crop improvement. The rice crop suffers from challenges of abiotic and biotic

Fig. 9.6. 'Trombay Karjat Kolam' rice (TKR Kolam) variety. **(a)** Field view. **(b)** Husked rice. **(c)** Polished rice.

Fig. 9.7. (a) Field view of Safri-17 parent (tall and late) and its mutants (dwarf/dwarf and early). **(b)** Field view of Luchai parent (tall and very late) and its mutant (dwarf, early and high yielding).

stresses. Among abiotic stresses, salinity stress is a major challenge in coastal and inland saline areas where the productivity is very low (< 1.5 t/ha). Gamma-ray induced mutagenesis has been used to develop salt-tolerant lines for coastal saline areas of India, in particular Maharashtra state. A mutant selection 'BARCKKV 16' (mutant of 'CSR30') has shown great promise in performance for coastal saline areas with average yields of more than 3.5 t/ha in salinity-affected areas and is likely to be released (Fig. 9.8a). Improvement of other traditional salt-tolerant rice varieties, viz. 'CSR27', 'CSR36', 'Bhura rata', 'Pokkali', 'White Ponni' and 'Trichi 2', is underway. Drought is another major stress, especially at terminal stages. Mutation breeding has been initiated to improve the drought tolerance and other agronomic traits of rice varieties 'N-22', 'Dagad Deshi', 'IBD-1', 'Sahbhagi Dhan' and 'CR Dhan 201', using gamma and proton beam induced mutagenesis. In India, rice is not only consumed as cooked rice as a main source of carbohydrate, but it is also used in making different culinary preparations. *Poha* (flattened rice) made out of scented rice 'Barhasal' is very popular, but its cultivation is constrained by poor agronomic traits. A high-yielding and non-lodging Barhasal mutant has been developed and is being tested in yield trials (Fig. 9.8b). *Modak*, a sweet preparation, is popularly used by people in South India during festivals, for which rice having moderate to low amylose and bold grains is preferred. 'Botwel' is a very popular rice variety used to make *modak* and is very popular among farmers but poor in agronomic traits and yield. BM4 is a direct mutant of 'Botwel' and has been recently released for cultivation. It has excellent *modak*-making quality, with higher yield (4.5 t/ha) and non-lodging characters (Fig. 9.8c).

Traditional rice varieties are highly photoperiod sensitive and cannot be grown in the season. 'Tulai Panji' is a highly scented, photoperiod-sensitive and extremely lodging-susceptible rice variety which is used for making *paysam* (sweet preparation). Photoperiod-insensitive mutants obtained using gamma-ray irradiation will help farmers to grow this variety throughout the year (Fig. 9.9a). To address the food

Fig. 9.8. **(a)** BARCKKV 16 salt-tolerant mutant rice. **(b)** Improved Barhasal (popular for *poha* making) mutant with higher yield (inset: poha). **(c)** BM-4, improved Botwel mutant, for making *modak* (a sweet Indian dish made on festivals and special occasions) (inset: *modak*).

Fig. 9.9. **(a)** Tulaipanji parent and photoperiod-insensitive mutant. **(b)** TGMS and extruded stigma type mutant in ADT43 rice variety for hybrid rice development.

security problem, hybrid rice technology will be the most important tool for higher rice production and better productivity. Diversification of the cytoplasmic male sterility (CMS) system is mandatory, since most of the hybrid rice programme in India is based on the IR 58025A CMS system. Seed setting is another constraint in hybrid rice production technology. Gamma-ray induced mutation breeding helped us to develop temperature-sensitive genetic male sterile (TGMS) mutants with an extruded stigma which will not only help us to improve hybrid rice grain quality, but also solve the hybrid rice seed production problems (Fig. 9.9b). This will help to increase rice production in India and address the hunger problem.

In recent years, rice breeding efforts have shifted towards developing better-quality grain rice to address hidden hunger. Red rice is a special variety of rice that looks red in colour due to its high anthocyanin content and is a good source of iron, manganese and zinc that helps accelerate wound healing and maintain immune-system function. It has a high fibre content and low glycaemic index. BARC has several red rice accessions collected from different parts of the country, viz. 'Lalkada', 'Foxtail', 'RTN 73', 'Kolhapur Sunil', 'Halvi Sal', 'RP 4-14', 'RTN 711', 'Phule Radha', 'Barmail', 'Surak', 'Pawana', 'Mahadi', 'Kachari Local', 'Munga', 'RKM 21', 'RKM 1', etc. 'Surak' is a local red rice landrace with short bold grain and a spreading type of plant suffering yield losses. Dwarf, early and slender-grain type mutants have been developed using gamma-ray induced mutagenesis with higher yields and higher Fe and Zn contents (Fig. 9.10).

4 Utilization of Novel Radiation Technology for Crop Improvement

Crop mutagenesis has gained significant momentum in recent years, driven by the surging need for continuous improvement in crop production and the availability of more efficient techniques. So far, a plethora of chemical and physical agents have been used to generate the

Fig. 9.10. **(a)** Surak plant with spreading panicle and red grain (indicated in the inset). **(b)** Semi-dwarf mutants of Surak with higher yield. **(c)** Mutants with long and compact panicles. **(d)** Red grain mutants with improved grain type in the periphery and control (Surak) in the centre. Control grain type is medium bold with red pericarp, whereas mutant grains include short slender, long slender, medium slender, short bold types, with some of them possessing red endosperm.

desirable genetic diversity required to achieve targeted improvements that do not exist or are rare in the natural population. In this direction, ion beam technology is being focused upon to exploit the high mutagenic efficiency bestowed by higher LET values, as compared with X-rays and gamma-rays which have been used since the 1930s (Kharkwal, 2012). The particulate nature, mass and charge possessed by ions result in their interaction being different and consequentially dense, providing local deposition of energy (Mei *et al.*, 1994) that may lead to production of spectra of variations that are distinct from those produced through exposure to other mutagens (Mba, 2013). Larger local damage to DNA will possibly break linkage drag between undesirable and desirable traits. The effects are subject to penetration of the plant propagule to the desired depth without much energy loss in the medium. A proton beam, accelerated to 14 MeV using a BARC-TIFR pelletron accelerator facility, achieved penetration of the embryo of both wheat and rice seeds. Proton beam irradiation technology was optimized in rice for increasing the efficiency of mutation and increasing the spectrum of mutants. Proton beam irradiation has been initiated to develop tolerance to the glyphosate herbicide in rice and initial success has been achieved in rice varieties 'CSR27' and 'IBD-1' (Fig. 9.11). A study estimated that at the highest weed density and highest labour cost in upland rice, herbicide application is approximately 80% (about US$200/ha) more profitable than hand weeding (Beltran *et al.*, 2012). These mutants will help to develop herbicide-tolerant rice in the future, which will be a boon for upland and aerobic rice cultivation.

5 Future Challenges

Climate change and unpredictable weather conditions are impacting agricultural production and productivity across the globe. The challenges in rice production, such as enhancing yield and quality, preventing or combating pest, diseases and weeds, and generating crops adaptive to ensuing ecological conditions for future environments, are issues that require immediate attention and action. In the short term, rainfed agriculture is set to be highly insecure, based on the predictions of meteorological departments that the weather in general, and rainfall in particular, is going to become more erratic in the coming years. Nutrient depletion, soil health deterioration, uncertainty about monsoon and unpredictable environmental conditions, floods and pest and disease outbreaks have made rice production highly unstable. With the organized cash crop sector being lucrative, a sizeable area of rice lands might be diverted to cash crops. However, efforts to diversify the rainfed uplands with other crops have not received much success so far. Because of several socio-economic concerns, these lands will continue to be cultivated with rice.

5.1 Frontier areas of rice mutation breeding research

Hybrid rice

In the coming decades, to give a quantum jump to overall rice production in India, hybrid rice is expected to be the most appropriate technology through which increased heterosis can be achieved

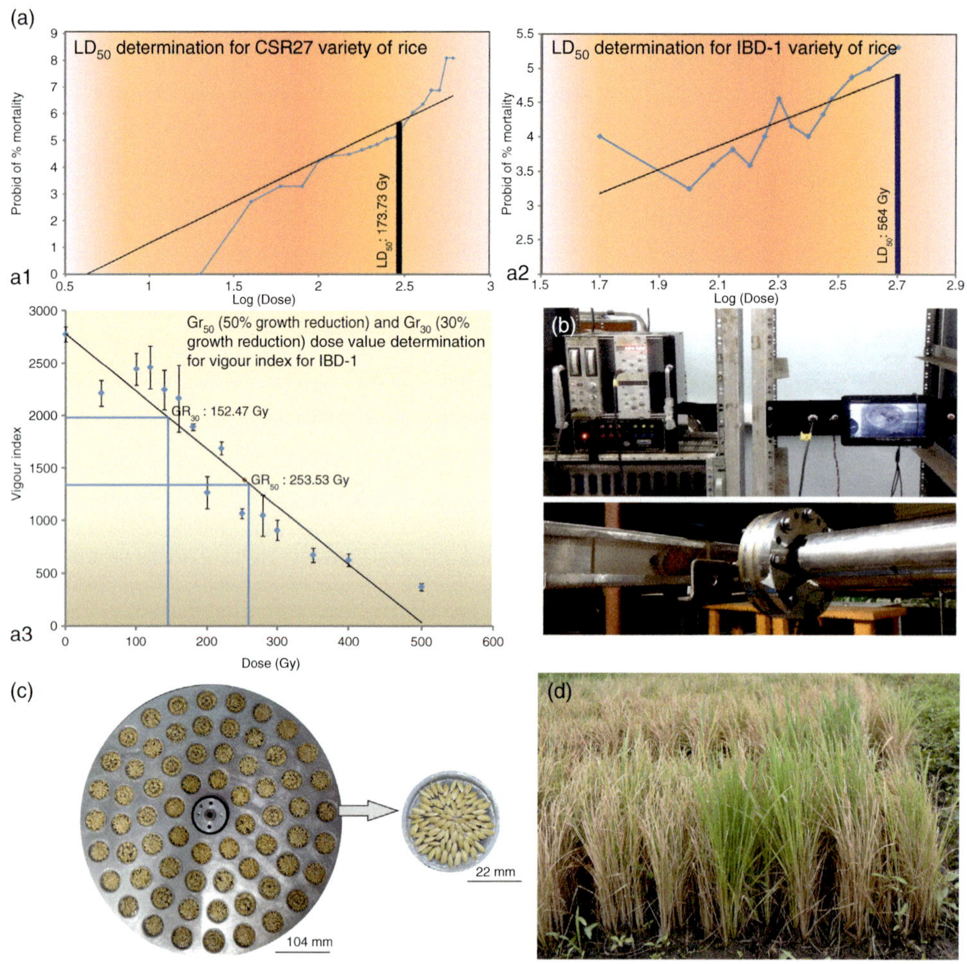

Fig. 9.11. (a) LD_{50} determination for two different rice varieties, viz. CSR27 (a1) and IBD1 (a2), and GR_{50} and GR_{30} determination for IBD-1 (a3). **(b)** Set-up for proton beam irradiation of rice at BARC-TIFR Pelletron Facility. The accelerated ion beam is brought into air through a titanium window (top panel); to irradiate seeds arranged on plates (bottom panel). **(c)** Arrangement of plates with rice seeds for hassle-free irradiation. (Inset) Concentric arrangement of seeds on single plate. **(d)** Glyphosate tolerant mutant in proton beam irradiated CSR27 rice variety.

along with the introgression of known pest and disease tolerance genes in the parents. Mutation breeding technology will help in diversification of cytoplasmic male sterility, having drought and salt tolerance, improvement of hybrid seed production, improvement of restorers, etc. so that the yield potential of rainfed and irrigated lowland areas can be increased considerably.

Bio-fortified rice

Micronutrient malnutrition among the poor is a major nutritional challenge to eliminate. Bio-fortification of staple food crops like rice will serve as a weapon to combat micronutrient malnutrition. Mutation breeding of traditional landraces rich in micronutrients for agronomic traits or improvement of high-yielding varieties for high Fe/Zn with less phytate will help to boost bio-fortification and eradicate malnutrition in developing countries.

Tolerance towards multiple abiotic stress

Major factors that prevent crops from realizing their full yield potential are drought, heat, cold, salinity and submergence. Some of these stresses often occur concurrently, such as heat and drought stress, submergence and salinity. Inducing mutations using gamma-ray or other mutagenic tools in popular rice varieties and screening for resistance to multiple stresses under field conditions will help to develop widely adapted mutants with higher yield and tolerance to multiple stresses.

Apomictic rice and doubled haploid breeding

Apomixis in rice can potentially be introduced using mutagenic and genomic approaches, through diverse species germplasm as well as interspecific hybrids, followed by the use of efficient screening techniques. Mutagenesis is being used as a potent tool to identify apomictic plants, which would be a tool to fix heterosis in the future and so to increase food grain production.

Less resource-intensive rice (green rice)

Judicious utilization of locally available resources, minimum tillage, residue management and crop diversification are some of the techniques to sustain productivity and form a part of conservation agriculture, which may help to face the challenges to produce more food at less cost and to improve water and nutrient use efficiency and also to mitigate the effects of climate change with regard to the emission of greenhouse gases in rice–rice systems. Rice requiring fewer agronomic inputs and management or rice competitive to weeds will be helpful to large numbers of small and marginal farmers.

Nitrogen-fixing rice

Modern cultivars are fertilizer responsive but high input of fertilizers has become a major concern for farmers due to the rising prices of fertilizers. Also, indiscriminate use of nitrogenous fertilizers has an adverse effect on environmental sustainability and contributes more than half of the carbon footprint of agriculture. Keeping environmental safety in mind, mutation breeding for low-input rice can be screened under low- or no-nutrient conditions. We may look for 'stay-green' mutants in M_2 and M_3 and finally confirm them using protein content, chlorophyll content, activity of nitrogen fixing genes, etc.

C4-type rice

For the same amount of light energy, C4 plants produce higher yields with double the water-use efficiency of C3 plants and about 40% less nitrogen requirement to achieve 50% higher yields. In C4 plants, biochemical, cellular and anatomical changes result in a mechanism that first concentrates CO_2 and then supplies it to RuBisCO in the C3 pathway. Overcoming the inefficiency of the C3 mechanism could be the key to achieving a leap in the amount of food or energy a plant can produce from the same amount of sunlight and revitalizing the Green Revolution, which has been slowing as the yields of elite cultivars are approaching a plateau. This may be achieved by looking for mutations in RuBisCO subunits using TILLING or other genomic technology so as to have high photosynthetic efficiency under high-temperature and low-CO_2 environments, low or no oxygenase activity resulting in low photo-respiratory losses (screening under high O_2 environment), high biomasses under low CO_2, high O_2 and high temperature environments, etc. This way we may achieve the partial goal towards increasing the photosynthetic efficiency like C4 crops under high temperature and high O_2 environments.

6 Conclusion

The genetic diversity of the majority of cultivated rice is lower than in the past, due to the use of more popular and hybrid rice varieties in India over a long period of time. Induced mutagenesis is one of the most important and effective approaches for broadening the genetic diversity in rice to circumvent the bottleneck. This becomes more evident as future rice production will have immense pressure from consumers' preferences, changing climatic conditions, resurgence of new races of pests and diseases, demand for bio-fortified rice, etc. Induced mutagenesis has demonstrated the potential of broadening the genetic base of crops and thereby availing plant breeders with the raw materials required to address various problems related to rice production and consumer preferences. Crop varieties generated through the exploitation of mutation breeding are significantly contributing to global food and nutritional security and improved livelihoods.

Acknowledgement

The authors express thanks to Dr V.P. Venugopalan (Associate Director, Bioscience Group) and Dr P. Suprasanna, (Head, NABTD, BARC) for their continuous support and guidance. The BARC-TIFR Pelletron Accelerator facility, Mumbai, is duly acknowledged for providing proton beam facility for seed irradiation.

References

Arumugasamy, S.K. *et al.* (2006) Parambaria Nel rakangal – Tamil (traditional paddy varieties). Centre for Indigenous Knowledge Systems, Chennai.

Ashraf, A.M. *et al.* (2017) A review of rice landraces in india and its inherent medicinal values – the nutritive food values for future. *International Journal of Current Microbiology and Applied Sciences* 6(12), 348–354.

Awan, M.A. (1991) Use of induced mutations for crop improvement in Pakistan. *Plant Mutation Breeding for Crop Improvement* 1, 67–72.

Beltran, J.C. *et al.* (2012) Economic implications of herbicide resistance and high labour costs for management of annual barnyard grass (*Echinochloa* crus-galli) in Philippine rice farming systems. *Crop Protection* 31, 31–39.

Chakrabarti, S.N. (1995) Mutation breeding in India with particular reference to PNR rice varieties. *Journal of Nuclear Agriculture and Biology* 24, 73–82.

Challinor, A. *et al.* (2014) A meta-analysis of crop yield under climate change and adaptation. *Nature Climate Change* 4(4), 287–291.

Coffey, D. *et al.* (2014) Wealth and health of children in India: a state-level analysis. *Economic & Political Weekly* 49(15), 64–70.

Datta, S.K. (2005) Mutation Breeding for Crop Improvement: a review. In: Datta, S.K. (ed.) *Role of Classical Mutation Breeding in Crop Improvement*. Daya Publishing House, Delhi, pp. 20–35.

FAO (1996) *Declaration on World Food Security*. FAO, Rome.

Futsuhara, Y. (1968) Breeding of a new rice variety Reimei by gamma ray irradiation. *Japanese Journal of Breeding* 17, 13–18.

Gibson, S. *et al.* (1993) Isolating plant genes. *Trends in Biotechnology* 11, 306–313.

Goyal, S. *et al.* (2009) Comparative study of chromosomal aberrations in Vigna mungo induced by ethylmethane sulphonate and hydrazine hydrate. *Thai Journal of Agricultural Science* 42(3), 177–182.

Haddad, L. *et al.* (2003) Reducing child undernutrition: how far does income growth take us? *World Bank Economic Review* 17(1), 107–131.

Hong-xing, X. *et al.* (2017) Sustainable management of rice insect pests by non-chemical-insecticide technologies in China. *Rice Science* 24(2), 61–72.

Issaka, S. *et al.* (2017) Impact of soil erosion and degradation on water quality: a review. *Geology, Ecology and Landscapes* 1(1), 1–11.

Khan, S. *et al.* (1992) Effect of selection for improvement of quantitative characters in mutated population of moong bean (*Vigna radiata* L. Wilczek). *Journal of the Indian Botanical Society* 71, 69–71.

Kharkwal, M.C. (2012) A brief history of plant mutagenesis. In: Shu, Q.Y. (ed.) *Plant Mutation Breeding and Biotechnology*. CAB International, Wallingford, UK, pp. 21–30.

Kharkwal, M.C. *et al.* (2009) The role of induced mutations in world food security. In: Shu, Q.Y. (ed.) *Induced Plant Mutations in the Genomics Era*. FAO, Rome, pp. 33–38.

Lee, J.S. *et al.* (2012) Food insecurity and health across the lifespan. *Advances in Nutrition* 3(5), 744–745.

Lee, K.J. *et al.* (2015) Genetic diversity among Korean rice landraces (*Oryza sativa* L.) based on characters and SSR markers. *Plant Breeding and Biotechnology* 3(3), 216–225.

Maruyama, K. *et al.* (1991) Thermosensitive genetic male sterility induced by irradiation. *Rice Genetics II*, 227–235.

Mba, C. (2013) Induced mutations unleash the potentials of plant genetic resources for food and agriculture. *Agronomy* 3(1), 200–231.

Mei, M. *et al.* (1994) Mutagenic effects of heavy ion radiation in plants. *Advances in Space Research* 14(10), 363–372.

Micke, A. (1985) Plant cultivars derived from mutation induction or the use of induced mutants in cross breeding (INIS-XA--389). International Atomic Energy Agency (IAEA), Vienna.
MVD (2018) Joint FAO/IAEA Mutant Variety Database. International Atomic Energy Agency, Vienna. Available at: http://mvd.IAEA.org (accessed April 2021)
Parry, M.A.J. *et al.* (2009) Mutation discovery for crop improvement. *Journal of Experimental Botany* 60(10), 2817–2825.
Patra, B.C. (2000) Collection and characterization of rice genetic resources from Keonjhar district of Orissa. *Oryza* 34, 324–326.
Rauf, S. *et al.* (2010) Consequences of plant breeding on genetic diversity. *International Journal of Plant Breeding* 41, 1–21.
Ray, D. *et al.* (2012) Recent patterns of crop yield growth and stagnation. *Nature Communications* 3(1), 1–7.
Richharia, R.H. (1979) An aspect of genetic diversity in rice. *Oryza* 16, 1–31.
Roy, J.K. (1985) Collection and evaluation of genetic resources of rice in India. *Phytobreedon* 1, 1–9.
Rutger, J.N. (1992) Searching for apomixis in rice. *Proceedings of Apomixis Workshop*. US Department of Agriculture, Beltsville, Maryland, pp. 36–39.
Sathya, A. (2007) Evaluation of biodiversity in traditional paddy varieties of Tamil Nadu. PhD Thesis, Tamil University, Thanjavur, Tamil Nadu (India).
Sharma, D.K. *et al.* (1987) Present status of rice germplasm in Madhya Pradesh (India). *Oryza* 24, 127–131.
Shen, Y.W. *et al.* (1993) A new thermosensitive radiation induced male sterile rice line. *Rice Genetics Newsletter* 10, 97–98.
Thornton, P.K. *et al.* (2014) Climate variability and vulnerability to climate change: a review. *Global Change Biology* 20(11), 3313–3328.
Timmer, C.P. (2014) Food security in Asia and the Pacific: the rapidly changing role of rice. *Asia and the Pacific Policy Studies* 1(1), 73–90.
Wani, M.R. *et al.* (2011) Induced chlorophyll mutations. I. Mutagenic effectiveness and efficiency of EMS, HZ and SA in mungbean. *Frontiers of Agriculture in China* 5(4), 514–518.

Section 2.

Mutation Breeding in Crop Improvement and Climate-Change Adaptation

10 Isolation and Characterization of Yellow Rust Resistant Mutants in Wheat

Suman Bakshi, Johar Singh and Sanjay J. Jambhulkar[1]

Abstract

Stripe rust, also known as yellow rust, caused by *Puccinia striiformis* f. sp. *tritici* (*Pst*), is a major threat to wheat production leading to yield losses up to 84%. Due to climate change, new races of the yellow rust pathogen are appearing for which no durable source of resistance has been observed in the present high-yielding varieties. A mutation breeding programme was initiated in two popular varieties, namely PBW343 and HD2967, using gamma-ray and electron beam irradiation. Gamma-ray doses of 250, 300 and 350 Gy and electron beam doses of 150, 200 and 250 Gy were used for seed irradiation. The M_2 population was screened in the field from seedling to adult plant stage by spraying a mixture of urediniospores of *Pst* pathotypes. Disease severity was recorded as the percentage of leaf area covered by the rust pathogen following a modified Cobb's scale. A total of 52 putative yellow rust resistant mutants in HD2967 and 63 in PBW343 were isolated. The number of mutants was higher in the electron beam irradiated population compared with gamma-rays. The absence of sporulation and spore production of the rust pathogen on the mutants indicated resistance. Mutant plants showing seedling resistance also showed resistance at adult plant stage. Seed yield and its contributing characters were better in the mutants compared with the parents. These rust resistant mutants could be novel sources of stripe rust or yellow rust resistance. The plant-to-row progenies of these mutants were confirmed and characterized in the M_3 generation.

Keywords: stripe rust/yellow rust • mutation breeding • gamma rays • electron beam • wheat

1 Introduction

Stripe rust (or yellow rust) is caused by the biotrophic fungal pathogen *Puccinia striiformis* f. sp. *tritici* (*Pst* pathotype) and is one of the most damaging foliar diseases of the world's wheat-growing regions (Wellings, 2011). It affects the productivity and quality of wheat throughout the world. In India, the disease is usually more widespread in cooler areas and prevails in the plains of Jammu and Kashmir, foothills of Punjab, Tarai region of Uttrakhand and parts of Haryana (Sharma and Saharan, 2011). The yellow rust pathogen is fast mutating and may adapt to higher temperatures, resulting in the susceptibility of hitherto resistant cultivars (Solh *et al.*, 2012). It is airborne and can traverse long distances under favourable conditions of proliferation, leading to epidemics (Brown and Hovmoller, 2002; Chen, 2005). The evolution of more virulent pathotypes has led to high disease severity due to the breakdown of widely used sources of resistance in wheat (Nayar *et al.*, 2001; Prashar *et al.*, 2007). Resistance

[1] Email: sjj@barc.gov.in

DOI: 10.1079/9781789249095.0010

governed by a single major gene is likely to break down, resulting in the appearance of new rust races and the further spread of disease over large areas. For instance, the gene *Yr27* present in popular cultivars PB343 and PBW 373 was overcome by the virulence of pathotype 78S84. All commercially grown cultivars in India became susceptible on emergence of the new pathotype (78S84). Cultivation of resistant varieties is the most effective, economically safe and environmentally friendly approach to mitigate the effect of this disease. A large number of stripe rust resistance genes (> 78) have been identified and mapped (McIntosh *et al.*, 2017). Sustainable wheat production predominately depends on new sources of resistance to maintain resistance against newly emerging pathotypes. Induced mutagenesis is a proven technique to create genetic variation for a trait of interest in crop plants (Micke *et al.*, 1990; Ahloowalia *et al.*, 2004; Maluszynski *et al.*, 2004). PBW343, carrying gene(s) for resistance to 46S119, was released in 1995. It is being cultivated on around 10 million hectares but is now susceptible for stripe rust race 78S84. It was replaced by the popular variety, HD2967, which was released in 2011 and is currently cultivated on 12 million hectares in India. This variety was also found susceptible to stripe rust disease. Therefore, a mutation breeding programme was initiated in both varieties (PBW343 and HD2967) to isolate yellow rust resistant mutants with high yield potential.

2 Materials and Methods

2.1 Parental material

Seeds of yellow rust susceptible varieties PBW343 and HD2967 were used for irradiation.

2.2 Irradiation of seeds

Five thousand seeds per variety per dose were treated with 250, 300 and 350 Gy of gamma-rays and 150, 200 and 250 Gy of electron beam using cobalt-60 source for gamma-rays at the Nuclear Agriculture & Biotechnology Division of the Bhabha Atomic Research Centre, Mumbai, and Linac reactor source for electron beam at Raja Ramanna Centre for Advanced Technology, Indore.

2.3 Development of mutant populations

Six M_1 populations each of PBW343 and HD2967 were space planted in the *rabi* season of 2015–2016 at Punjab Agricultural University, Ludhiana. The dose-wise number of plants that survived in the M_1 population was recorded and is presented in Table 10.1. Seeds of the M_1 individual plants were harvested separately. The M_2 population was raised as plant-to-progeny row along with the susceptible parents PBW343 and HD2967, which were planted after every ten M_2 progeny rows. Putative mutants were isolated in the M_2 generation and their breeding behaviour was confirmed in the M_3 generation.

2.4 Screening of yellow/stripe rust disease

Stripe rust infection data were recorded at three stages: flag leaf initiation; ear emergence; and anthesis. Data on disease severity was recorded on putative mutants in the M_2 generation and progeny rows in the M_3 generation, using modified Cobb scale (Peterson *et al.*, 1948). Disease severity was recorded as percentage of rust reaction on the plant and the intervals used were 0 (immune), 5% (5S), 10% (10S), 20% (20S) and 40% (40S). Field response was categorized qualitatively. Zero (0) means no visible infection on the plant and is called immune. Resistant (R) means visible necrotic tissue with 5S and 10S reaction on leaf with no urediniospores. Moderately resistant (MR) has 20S reaction with urediniospore-producing pustules surrounded by necrotic tissue. Moderately susceptible (MS) indicates a 21S to 40S reaction which shows urediniospore-producing pustules surrounded by chlorotic tissue. Susceptible (S) indicates more than 40S reaction with urediniospore-producing pustules surrounded by green tissue. Disease severity and field response were combined to record stripe rust data. The mutation frequency was calculated as number of yellow rust resistant plants to total number of M_2 plants in respective mutagen and dose.

Table 10.1. Plant survival in the M_1 generation and frequency of putative yellow rust resistant mutants in M_2 generation of varieties HD2967 and PBW343.

Variety	Treatment	Number of seeds treated	Number of M_1 plants at harvest	Number of M_2 plant at harvest	Number of Putative mutants for yellow rust resistance	Observed mutant frequency
HD2967	Control (0 Gy)	1,000	970	–	–	
HD2967	Gamma-rays					
	250 Gy	5,000	2,200	110,000	5	0.0045
	300 Gy	5,000	1,410	64,000	8	0.0125
	350 Gy	5,000	1,105	55,000	3	0.0054
	Total	15,000	4,715	229,000	16	0.0069
	Electron beam					
	150 Gy	5,000	2,520	125,000	8	0.0064
	200 Gy	5,000	2,250	90,000	20	0.0222
	250 Gy	5,000	2,205	78,000	8	0.0102
	Total	15,000	6,975	293,000	36	0.0122
PBW343	Control (0 Gy)	1,000	962			
PBW343	Gamma-ray					
	250 Gy	5,000	1,913	95,900	7	0.0072
	300 Gy	5,000	1,752	86,000	13	0.0151
	350 Gy	5,000	845	38,000	8	0.0210
	Total	15,000	4,510	219,900	26	0.0118
	Electron beam					
	150 Gy	5,000	2,400	108,000	14	0.0130
	200 Gy	5,000	2,150	81,200	11	0.0135
	250 Gy	5,000	1,850	59,400	12	0.0202
		150,000	6,400	248,600	37	0.0148

2.5 Yield and yield components

Data on grain yield per plant, plant height and yield components which include spike length, tiller number, spikelets per spike and grain yield per spike were recorded for five plants of the variety HD2967 only.

3 Results

3.1 Genotype sensitivity and occurrence of yellow rust resistant mutants

The results showed that genotypic sensitivity does exist in response to doses of gamma-ray and electron beam, as indicated by the number of plants that survived to maturity in the M_1 generation. The frequency of putative yellow rust resistant mutants was higher for gamma-ray dose of 300 Gy and electron beam dose of 200 Gy of HD2967. In variety PBW343, the gamma-ray dose of 350 Gy and electron beam dose of 250 Gy were found most effective for inducing yellow rust resistant mutants (Table 10.1). For both varieties, the frequency of yellow rust resistant mutants was higher in electron beam irradiated seed than in those irradiated by gamma-rays (Table 10.1). A total of 52 putative mutants were isolated from 522,000 M_2 plants of HD2967 and 63 mutants from 468,000 M_2 plants of PBW343. The resistance scores ranged from 0 to 20S (Fig. 10.1).

3.2 M_3 generation

Individual plant progenies of 52 putative yellow rust resistant mutants in HD2967 and 63 in PBW343 were grown in the field at Punjab Agricultural University, Ludhiana, during 2018–2019. In the M_3 generation of PBW343, three progenies exhibited an immune response, six M_3 progenies scored 5S and three scored 10S, while the parent line had a yellow rust reaction score in the range 60S–80S (Table 10.2). In the case of HD2967 (Fig. 10.2 A–E), two plant progenies

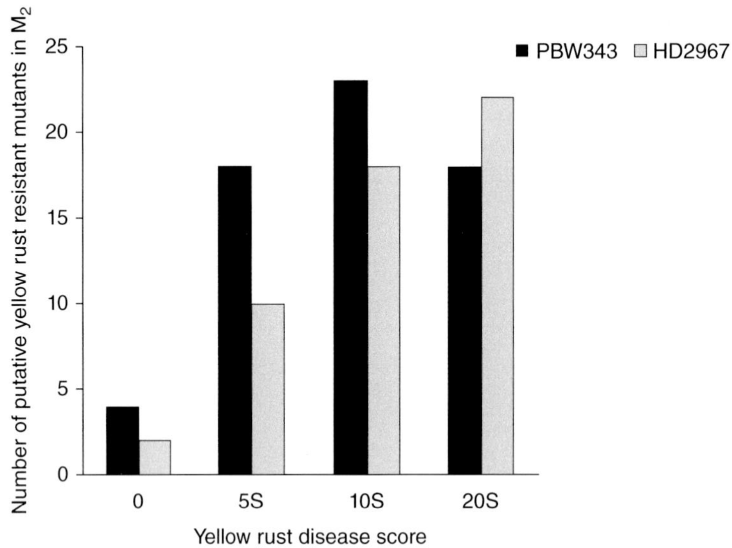

Fig. 10.1. Frequency of putative yellow rust resistant mutants in M_2 generation of PBW343 and HD2967.

Table 10.2. Yellow rust reaction of resistant mutants in M_3 progenies of PBW343.

	Yellow rust resistance score			
Mutant	25/01/2017	10/02/2017	25/02/2017	Final score
PBW343Mutant-1	0	0	0	Immune
PBW343Mutant-2	0	0	0	Immune
PBW343Mutant-3	0	0	0	Immune
PBW343Mutant-4	0	5S	5S	Resistant
PBW343Mutant-5	0	5S	5S	Resistant
PBW343Mutant-6	0	5S	5S	Resistant
PBW343Mutant-7	5S	5S	5S	Resistant
PBW343Mutant-8	5S	5S	5S	Resistant
PBW343Mutant-9	5S	5S	5S	Resistant
PBW343Mutant-10	5S	10S	10S	Resistant
PBW343Mutant-11	5S	10S	10S	Resistant
PBW343Mutant-12	10S	10S	10S	Resistant
PBW343 (Parent)	40S	60S	80S	Susceptible

were immune (Fig. 10.2 C–D), two were resistant with a 5S score, seven were resistant with a 10S score and three were moderately resistant (20S), while parent HD2967 was scored as 60S (Table 10.3). The yellow rust reaction was uniform in all the plants of each progeny in the M_3 generation. A notable feature was that mutants showing resistance at seedling stage were also found resistant as adult plants.

3.3 Yield and yield components

Data on yield and yield components were recorded for five plants of parent (HD2967) and resistant mutants (Table 10.4). Resistant mutants had a greater seed yield (13.6–20.4 g per plant) compared with the parent (12.2 ± 0.22 g). This could be due to the increased spike length (9.9–12.6 cm) and grain number (64–76) of the

Fig. 10.2. (A) HD2967-80S plant row. **(B)** Single leaf of a susceptible plant. **(C, D)** Immune mutants lacking stripe rust symptoms. **(E)** Single leaf blade of immune mutant.

Table 10.3. Yellow rust reaction of resistant mutants in M_3 progenies of HD2967.

	Yellow rust resistance score			
Mutant	30/01/2017	12/02/2017	28/02/2017	Final score
HD2967Mutant-1	0	0	0	Immune
HD2967Mutant-2	0	0	0	Immune
HD2967Mutant-3	5S	5S	5S	Resistant
HD2967Mutant-4	5S	5S	5S	Resistant
HD2967Mutant-5	10S	10S	10S	Resistant
HD2967Mutant-6	10S	10S	10S	Resistant
HD2967Mutant-7	10S	10S	10S	Resistant
HD2967Mutant-8	5S	10S	10S	Resistant
HD2967Mutant-9	10S	20S	10S	Resistant
HD2967Mutant-10	5S	10S	10S	Resistant
HD2967Mutant-11	5S	10S	10S	Resistant
HD2967Mutant-12	5S	20S	20S	Moderately Resistant
HD2967Mutant-13	10S	10S	20S	Moderately Resistant
HD2967Mutant-14	5S	20S	20S	Moderately Resistant
HD2967 (Parent)	20S	60S	60S	Susceptible

mutants compared with the spike length of 9.4 cm and grain number of 64 in the parental line. These data serve as an indicator for performance of a single plant in the M_3 generation which needs confirmation in subsequent generations.

4 Discussion

The variety PBW343 was released in 1996 and adopted over 7 million hectares in the fertile Indo-Gangetic plains (Singh *et al.*, 2017). This variety

Table 10.4. Grain yield per plant and yield components of 14 mutants and parent (HD2967) in M_3 generation.

Mutant	Plant height (cm)	Tiller number	Spike length (cm)	Spikelets per spike	Grain number per spike	Seed weight per spike (g)	Seed weight per plant (g)
Mutant-1	96.0 ± 1.3	11.6 ± 0.51	12.1 ± 0.25	24.2 ± 0.5	76.0 ± 1.14	3.2 ± 0.05	20.2 ± 0.96
Mutant-2	99.5 ± 1.6	10.3 ± 0.40	10.0 ± 0.31	22.2 ± 1.0	68.4 ± 0.93	2.3 ± 0.08	18.2 ± 0.49
Mutant-3	93.5 ± 0.7	9.8 ± 0.58	9.9 ± 0.50	24.6 ± 0.4	74.6 ± 1.32	3.0 ± 0.14	20.4 ± 0.70
Mutant-4	92.5 ± 0.8	10.0 ± 0.55	10.4 ± 0.40	22.3 ± 0.7	72.2 ± 1.12	2.6 ± 0.05	16.4 ± 0.39
Mutant-5	87.0 ± 1.3	10.4 ± 0.51	10.6 ± 0.21	22.8 ± 0.8	70.0 ± 0.55	3.2 ± 0.07	18.6 ± 0.53
Mutant-6	90.4 ± 0.8	10.0 ± 0.55	10.2 ± 0.40	22.2 ± 0.5	68.2 ± 0.66	2.8 ± 0.14	16.8 ± 0.36
Mutant-7	94.5 ± 1.0	10.8 ± 0.66	10.0 ± 0.70	23.0 ± 0.6	70.0 ± 0.71	2.8 ± 0.17	18.1 ± 0.73
Mutant-8	96.0 ± 1.2	11.0 ± 0.55	11.0 ± 0.30	22.4 ± 0.6	64.0 ± 0.72	2.4 ± 0.09	14.2 ± 0.18
Mutant-9	92.0 ± 1.3	12.4 ± 0.51	12.6 ± 0.34	25.2 ± 0.6	72.0 ± 0.55	3.4 ± 0.07	17.8 ± 0.45
Mutant-10	92.0 ± 1.2	11.8 ± 0.37	12.4 ± 0.29	24.0 ± 0.4	73.2 ± 0.66	3.6 ± 0.09	15.4 ± 0.30
Mutant-11	87.5 ± 0.9	10.8 ± 0.58	11.2 ± 0.35	23.2 ± 0.9	68.0 ± 0.84	2.6 ± 0.09	13.6 ± 0.24
Mutant-12	93.0 ± 1.8	10.4 ± 0.68	10.4 ± 0.37	22.8 ± 0.6	64.2 ± 0.49	2.8 ± 0.08	14.8 ± 0.27
Mutant-13	88.0 ± 0.7	10.8 ± 0.66	12.2 ± 0.26	24.6 ± 0.4	70.4 ± 0.81	3.0 ± 0.06	16.2 ± 0.34
Mutant-14	95.0 ± 0.6	11.0 ± 0.71	11.2 ± 0.30	23.0 ± 0.9	68.0 ± 0.71	2.8 ± 0.07	14.8 ± 0.34
HD2967 (Parent)	86.0 ± 1.1	8.0 ± 0.55	9.4 ± 0.38	22.4 ± 0.8	64.0 ± 0.71	2.4 ± 0.06	12.2 ± 0.22

is known to carry resistance to yellow rust disease along with wider adaptability and high productivity which could be due to the presence of 1B/1R wheat–rye translocation (Rajaram *et al.*, 1983). Another high-yielding, yellow rust resistant variety, HD2967, was released in 2011 but showed loss of resistance over time, presumably due to genetic changes in the yellow rust pathogen resulting in new pathogenic races. Severe epidemics of yellow rust started occurring in these varieties due to the appearance of new pathotypes in the north-western plain zone and the situation demands continuous efforts for identification of new sources of resistance (Virdi *et al.*, 2016). The widely cultivated species of wheat, *Triticum aestivum*, has a narrow genetic base and creation of genetic variability for any trait using conventional approaches, such as hybridization with primary and secondary gene pool, is time consuming. The pace at which the yellow rust fungus mutates and spreads is extremely high compared with the rate of developing new resistant cultivars through hybridization. Mutation breeding is one of the best approaches to combat such a situation and has been proved to induce disease resistance (Boyd *et al.*, 2001; Chong *et al.*, 2004; Goddard *et al.*, 2014). A mutation breeding programme to develop yellow rust resistant mutants in the popular wheat varieties HD2967 and PBW343 was undertaken using gamma-rays and electron beam irradiation at Bhabha Atomic Research Centre, Mumbai, India. Gamma-rays as well as the electron beam were found to be effective for the isolation of yellow rust resistant mutants. A total of 52 putative rust resistant mutants in HD2967 and 63 in PBW343 were isolated. Success of mutation breeding was realized in the past for eradicating the deadly wheat stem rust race *Ug99* caused by *Puccinia graminis* f. sp. *Tritici* (*Pgt*) which was virulent in most wheat varieties cultivated around the world. However, a global initiative on mutation breeding initiated by FAO/IAEA through a multi-country project led to the development of 13 resistant mutants. Field evaluation of these resistant mutants in Kenya led to the release of the world's first Ug99-resistant wheat mutant variety in February 2014 and was named 'Eldo Ngano 1' (Forster, 2014).

The new mutagen, electron beam, was employed for the first time for inducing mutation in wheat. The number of yellow rust resistant mutants was higher in electron beam irradiation than in gamma-ray. Guo *et al.* (1982) compared the effect of electron beam (5MeV) and gamma (^{60}Co) irradiation in rice and showed that electron beam induced lower damage and a higher frequency of occurrence of mutants than gamma radiation. Mutants isolated in this study showed different degrees of resistance, including immune response to the rust pathogen, which could be due to mutations occurring at many rust resistance loci. This is a novel source of resistance for yellow rust in India. The superior yield potential of these mutations over the parent could be helpful for developing high-yielding varieties with yellow rust resistance.

5 Conclusion

The initiative taken to develop yellow rust resistant mutants of landmark high-yielding wheat cultivars through induced mutagenesis proved that mutations, though occurring randomly and in low frequency, are the solution to evolve disease-resistant mutations. Gamma-rays as well as electron beam were found to be effective for the isolation of yellow rust resistant mutants, which could be a novel source of durable resistance. These mutants could be exploited in wheat breeding programmes to develop yellow rust resistant high-yielding varieties.

Acknowledgements

The authors thankfully acknowledge Dr V.P. Venugopalan, AD (A) Bioscience Group, for his kind support.

References

Ahloowalia *et al.* (2004) Global impact of mutation-derived varieties. *Euphytica* 135, 187–204.

Boyd, L.A. *et al.* (2001) Wheat mutants showing altered adult plant disease resistance. *Euphytica* 122, 361–368.

Brown, J.K.M. and Hovmoller, M.S. (2002) Aerial dispersal of pathogens on the global and continental scales and its impact on plant disease. *Science* 297, 537–541.

Chen, X.M. (2005) Epidemiology and control of stripe rust [*Puccinia striiformis* f. sp. *tritici*] on wheat. *Canadian Journal of Plant Pathology* 27, 314–337.

Chong, U. *et al.* (2004) Identification of rice genes induced in a rice blast resistant mutant. *Molecular Cell* 17, 462–468.

Forster, B.P. (2014) The development of 'Eldo Ngano I': the world's first Ug99 resistant mutant wheat variety. *IAEA Bulletin* 55, 18–19.

Goddard, R. *et al.* (2014) Enhanced disease resistance caused by BRI1 mutation is conserved between *Brachypodium distachyon* and barley (*Hordeum vulgare*). *Molecular Plant–Microbe Interactions* 27, 1095–1106.

Guo, J. *et al.* (1982) Studies on the mutagenic effect of 5MeV electron irradiation on rice. *Acta Genetica Sinica* 9, 461–467.

Maluszynski, M. *et al.* (2004) Mutation techniques. *Encyclopedia of Applied Plant Sciences* 1–3, 186–201.

McIntosh, R.A. *et al.* (2017) *Catalogue of Gene Symbols for Wheat: 2017 Supplement*. In: KOMUGI–Integrated Wheat Science Database. Available at: http://shigen.nig.ac.jp/wheat/komugi/genes/macgene/supplement2017.pdf (accessed 2019)

Micke, A.M. *et al.* (1990) Induced mutations for crop improvement. *Mutation Breeding Review* 7, 1–41.

Nayar, S.K. *et al.* (2001) *Revised Catalogue of genes that accord resistance to* Puccinia *species*. *Wheat Research Bulletin* No. 3, 48 pp.

Peterson, R.F. *et al.* (1948) A diagrammatic scale for rust intensity on leaves and stems of cereals. *Canadian Journal of Research* 26, 496–500.

Prashar, M. *et al.* (2007) Pathogenic evolution in *Puccinia striiformis* in India during 1995–2004. *Australian Journal of Agricultural Research* 58, 602–604.

Rajaram, S. *et al.* (1983) Adaptation, stability and high yield potential of certain 1B/1R CIMMYT wheats. In: Sakamoto, S. (ed.) *Proceedings 6th International Wheat Genetics Symposium*, Kyoto, Japan, pp. 613–621.

Sharma, I. and Saharan, M.S. (2011) Status of wheat disease in India with a special reference to stripe rust. *Plant Disease Research* 26, 156.

Singh, R.P. *et al.* (2017) Bread wheat cultivar PBW 343 carries residual additive resistance against virulent stripe rust pathotype. *Journal of Crop Improvement* 31, 183–191.

Solh, M. *et al.* (2012) The growing threat of stripe rust worldwide. *Borlaug Global Rust Initiative, Technical Workshop Proceedings, China*, pp. 1–10.
Virdi, R. *et al.* (2016) Genetic characterization of resistance to stripe rust, leaf rust, karnal bunt and cereal cyst nematode in a multiple disease resistant wheat stock W8627. *Indian Journal of Genetics* 76, 40–46.
Wellings, C.R. (2011) Global status of stripe rust: a review of historical and current threats. *Euphytica* 179, 129–141.

11 Identification of Rice Mutants Tolerant to Cold Stress at the Germination Stage by TILLING

Viviane Kopp da Luz, Vívian Ebeling Viana, Gabriela Magalhães da Fonseca, Camila Pegoraro, Luciano Carlos da Maia and Antonio Costa de Oliveira[1]

Abstract

Cold stress is a common factor affecting rice culture in temperate regions, which impairs seed germination, crop establishment and grain yield. This work aimed to identify, through a TILLING assay, rice mutant families displaying cold tolerance during the germination stage. The mutant analyses were performed in 4000 M_3 plants obtained through chemical mutagenesis with ethyl methanesulfonate. We screened for mutations in the *Os03g0103300* (qLTG3–1) gene, which is responsible for cold tolerance during germination. The TILLING assay identified a mutant (516 A3) which was tested for germination efficiency in cold stress (13°C). The mutant genotype showed a higher relative performance in germination and germination velocity index, which was more than 50% higher compared with wild-type. The mutation induction was efficient in creating genetic variability for cold stress tolerance during germination. Gene expression analyses demonstrate that *Os03g0103300* was downregulated in stage S3 in the mutant and wild-type plants germinated under cold stress. However, downregulation in the *Os03g0103300* gene was less severe in the mutant, which suggests that the expression related to germination ability under cold stress may be detected in the previous stages, embryo activation and weakening of the tissues that cover the embryo. Overall, the mutant 516 A3 presents a new genetic variant for cold tolerance during germination.

Keywords: germination • mutagenesis • cold tolerance

1 Introduction

Cold stress includes chilling (0–15°C) and freezing (< 0°C), which directly affect plant growth and grain yield, impairing diverse plant cell process. Cold firstly affects cell membranes, causing stiffening which activates a range of stress responses, followed by protein stability disruption and reduction of enzyme activity leading to photo-inhibition and impairing photosynthesis, gene expression and protein synthesis (Ding *et al.*, 2019). Temperature plays an important role in seed germination during embryo activation and post-germination stages which directly affects plant growth. In this sense, temperatures below 15°C cause a delay in rice plant emergence, since the optimum temperature range for rice germination is between 20°C and 35°C. At temperatures below 10°C, rice seeds are unable to germinate. Thus, chilling causes slow and non-uniform germination, resulting in irregular emergence and small plant population (Cruz and Milach, 2004).

[1] Email: acostol@terra.com.br

 Mutation Breeding, Genetic Diversity and Crop Adaptation to Climate Change (eds S. Sivasankar *et al.*)
DOI: 10.1079/9781789249095.0011

The successive selection cycles toward the improvement of desirable agronomic traits led to a reduction of the genetic variability in crops, decreasing their resilience to abiotic stresses. In this sense, hybridizations to unadapted genotypes lead to a change in the adjustment of genotypes to modern cultivation management techniques. The spontaneous mutation rate in higher plants is very low (Jiang and Ramachandran, 2010), making mutation induction an effective alternative strategy to raise the frequency of mutations (Luz *et al.*, 2016). Mutation induction can be applied to obtain new tolerant variants, with the advantage of causing single or few changes in adapted genotypes. Targeting Induced Local Lesions In Genomes (TILLING) is an efficient tool for point mutation detection; a DNA pool of mutated plants is amplified through polymerase chain reaction (PCR) using specific oligonucleotides for the desired region (McCallum *et al.*, 2000a,b). The heteroduplex formed due to the mismatching between mutant and wild-type DNA during PCR is recognized and cleaved by CEL I (mismatch-specific celery endonuclease). The presence of a mutation in the DNA pool is identified through separation of the digestion products in a denaturing gel. Later, individual DNAs are screened, identifying the mutant plant and the approximate mutant position along the sequence (Till *et al.*, 2003, 2006). The TILLING assay has been applied in many crops, such as maize, wheat, rice, barley, soybean and canola (Tadele, 2016). Thus, TILLING can be used as a tool for detection of variants in specific regions which may be related to resistance responses against abiotic stress, such as cold.

Gene responses for cold tolerance are complex, as they depend on the intensity, period of occurrence, stage of plant development and occurrence of other stresses affecting the cell molecular and biochemical dynamics, which in turn affect the physiological and morphological responses to stress (Wang *et al.*, 2003; Chinnusamy *et al.*, 2004). Many quantitative trait loci (QTL) associated with cold during germination have been reported (Fujino and Matsuda, 2010). However, the molecular basis of this trait is not completely elucidated. The qTLG3-1 is an important QTL identified on chromosome 3 of rice; it is associated with germination under cold stress and is responsible for more than 30% of the phenotypic variation (Fujino *et al.*, 2008). Through unknown elicitors, qTLG3-1 acts to upregulate genes involved in phytoalexin biosynthesis and the pathogen-related protein PBZ1, which plays a role in programmed cell death, promoting the weakening of the tissues that cover the embryo, leading to seed germination (Fujino and Matsuda, 2010). The *Os03g0103300* gene is localized within qTLG3-1 and is responsible for rice seed germination potential under cold (Fujino *et al.*, 2008). The expression of *Os03g0103300* was detected in embryos during germination and showed an increase in expression during the germination process in cold. Overall, qTLG3-1 is associated with the germination process as well as with tolerance to abiotic stresses (Fujino and Matsuda, 2010).

In this work, mutations were induced in the rice variety 'BRS Querência', with the chemical mutagen ethyl methanesulfonate (EMS). 'BRS Querência' is a short-cycling rice with high tillering capacity and strong stems. It stands out for presenting a long panicle with many fertile spikelets and for moderate cold tolerance during germination and emergence (SOSBAI, 2018). Through TILLING targeting the *Os03g0103300* gene (localized on qTLG3-1), the 516 A3 mutant was identified. The 516 A3 mutant displayed a germination and emergence rate higher than the wild-type. The time of exposure and concentration of the mutagenic EMS was efficient in promoting the desired genetic variability and the 516 A3 mutant is promising for inclusion in breeding programmes aiming to introduce cold resistance at germination.

2 Materials and Methods

2.1 Obtaining the mutant families and collecting plant material

Mutant families were obtained by treating rice seeds of 'BRS Querência' with the chemical mutagen, EMS, at a concentration of 1.5% (v/v) (0.15 M). The seeds were soaked in distilled water for 6 h and then were subjected to the mutagen for 2 h. After chemical treatment, the seeds were sown in an experimental field for generation advancement (Luz *et al.*, 2016). Leaf samples (200 mg) were collected, according to Fulton *et al.* (1995), from 4000 M_3 plants, fixed in liquid nitrogen and stored at –80°C.

2.2 Identification of mutants through TILLING

Identification of mutants within the families (pools)

Mutations in *Os03g0103300* were sought by screening 4000 M_3 plants in 500 pools of eight plants. The DNA pools of eight plants were extracted with the 2% CTAB method, as described by Saghai-Maroof *et al.* (1984). DNA concentration and quality were accessed through agarose gel electrophoresis, comparing the Low DNA Mass Ladder (Invitrogen). PCR with the DNA pools and wild-type DNA was performed using 50 ng of DNA and amplified with Platinum™ Taq DNA Polymerase High Fidelity (Invitrogen) using oligonucleotides targeting the *Os03g0103300* gene (Table 11.1). For mutation detection, the SURVEYOR® Mutation Detection Kits for Standard Gel Electrophoresis (Transgenomic) kit was used according to manufacturer's recommendations. The digestion products were separated by agarose gel electrophoresis and stained with ethidium bromide.

Identification of mutants within pools

Pools which showed a product profile different from the wild-type were analysed through PCR. Seeds from the M_4 mutant family and wild-type 'BRS Querência' were sown in wet paper rolls and germinated in a germination chamber (BOD) for 7 days, after which leaf samples were collected. DNA extraction, PCR conditions and digestion were performed as described above.

2.3 Germination tests

After mutant identification, phenotypic characterization was performed under cold conditions. Seeds from the 516 A3 mutant (M_7), identified by TILLING, and from 'BRS Querência' (wild-type) were sown in a germination box containing germination paper. The paper was wetted with water in a proportion equivalent to 2.5 times the substrate weight, following the criteria established by the Rules for Seed Analysis (MARA Brazil, 2009). For the germination tests (% of seeds germinated), seeds were germinated in the germination chamber in two different conditions: 13°C for 28 days and 25°C for 7 days (Cruz and Milach, 2004). To determine the germination velocity index (GVI), daily counts were performed for 28 days. The GVI was calculated following Maguire (1962). The experiments were performed in randomized blocks with three replicates.

A variance analysis ($p \le 0.05$) was performed and means were compared by Dunnett's test ($p \le 0.05$). Both analyses were performed with the software Genes (Cruz, 2001). For the comparison of means, the relative performance (RP) was calculated comparing the temperature of 13°C with that of 25°C, according to the equation

$$RP = 100(X13°C / X25°C)$$

Table 11.1. Primer sequences used for the identification of mutants by TILLING and for the quantification of transcripts by qRT-PCR.

Gene	F/R	Sequence	Annealing temperature °C
Os03g0103300 (T)[a]	F	GCTAGGTAGAGGCCAGGCCATAGATCG	63
	R	GGCATGCAGAAAAGACGAGATGCAG	
Os03g0103300 (GE)[b]	F	TGAATGGGCTGATAAACGTG	60
	R	AGGTTGAGGTTGATGCCAAG	
NAPB - LOC_Os06g11170 (RG)[c]	F	GGAATGTGGACGGTGACACT	62
	R	TCAAAATAGAGTCCAGTAGATTTGTCA	
EF1-α - LOC_Os03g08020 (RG)[c]	F	TGGTATGGTGGTGACCTTTG	61
	R	GTACCCACGCTTCAGATCCT	
AK059783 (RG[3])	F	CTACGTCCCTGCCCTTTGTACA	63
	R	ACACTTCACCGGACCATTCAA	

[a] T = TILLING; [b] GE = Gene Expression; [c] RG = Reference Gene

with *X* representing the observed value. Percentage data of the relative performance were transformed into the square root of *X*.

2.4 Real-time quantitative reverse transcription-PCR (qRT-PCR) analysis

To understand the effect of the mutations on *Os03g0103300* regulation, gene expression analysis by qRT-PCR was performed. Seeds from the 'BRS Querência' wild-type and from the 516 A3 mutant were sown as described above. Later, they were placed in a germinating chamber (BOD) under control (25°C, 16 h light) and cold (13°C, 16 h light) conditions. Shoot samples were collected at the S3 seedling state following the scale described by Counce *et al.* (2000) and fixed in liquid nitrogen. Total RNA was extracted from 2 g of fresh tissues using Trizol® reagent (Invitrogen, Carlsbad, California). The quantity of the RNA was assessed by spectrophotometry and the quality by agarose gel electrophoresis. Samples were treated with DNAse I (Invitrogen) and control reactions were performed with qRT-PCR reactions using RNA samples after DNAse treatment to confirm the absence of genomic DNA. Each sample (2 μg) was reverse transcribed into cDNA using oligo(dT) with commercial kit SuperScript® III first-strand system for RT-PCR (Invitrogen).

The qRT-PCR experiment was performed according to MIQE guidelines (Bustin *et al.*, 2009) using oligonucleotide pairs for *Os03g0103300* gene and three reference genes, *AK059783*, *Nucleic acid binding protein* (*NAPB*) and *Eukaryotic elongation factor 1-alpha* (*EF1-α*) (Table 11.1). Oligonucleotides for *Os03g0103300* gene were designed from sequences deposited in the MSU Rice Genome Annotation Project (Ouyang *et al.*, 2007) using Primer 3 plus (Untergasser *et al.*, 2007). Oligonucleotides for the reference genes were obtained from Jain *et al.* (2006) and Narsai *et al.* (2010). Validation experiments were performed using four concentrations of the cDNA pool using a dilution factor of 10× the target gene and the reference genes to determine if the amplification efficiencies were equal. The gene expression assay was conducted in an Applied Biosystems 7500 fast real-time PCR system using SYBR™ Green PCR Master Mix (Invitrogen). Three independent biological replicates of each sample and three technical replicates of each biological replicate were used for real-time PCR analysis. Negative control reactions were also run to confirm the absence of genomic DNA. The quantification was performed according to the ΔΔCt method (Livak and Schmittgen, 2001). The expression data of reference genes were analysed in DataAssist™ v3.0 Software (Applied Biosystems) and the *Os03g0103300* expression data were normalized with reference genes showing scores of < 1.0.

3 Results

3.1 Identification of the mutant through TILLING

The TILLING assay was used to identify the presence of mutations in the *Os03g0103300* gene caused by EMS treatment. We observed a pool containing putative mutants for the *Os03g0103300* gene in the M_3 families derived from the EMS treatment (Fig. 11.1A).

Subsequently, identification of the mutant plant within the pool was performed by amplification of the gene of interest in individual plants. The TILLING assay suggested the presence of a mutation, in an individual designated 516 A3, for the *Os03g0103300* gene (Fig. 11.1B).

3.2 Germination and germination velocity index

To verify the effects of the mutation in 561 A3 on germination performance under cold stress, germination and germination velocity index (GVI) analyses were performed. The results presented by analysis of variance showed significant differences at 5% significance by an F-test, between the 516 A3 mutant and the 'BRS Querência' wild-type for the relative performance of the germination and GVI (Table 11.2).

The results indicated that both variables demonstrated a performance average superior to the wild-type, suggesting that the mutation promoted a beneficial variation associated with cold tolerance at the germination stage (Fig. 11.2).

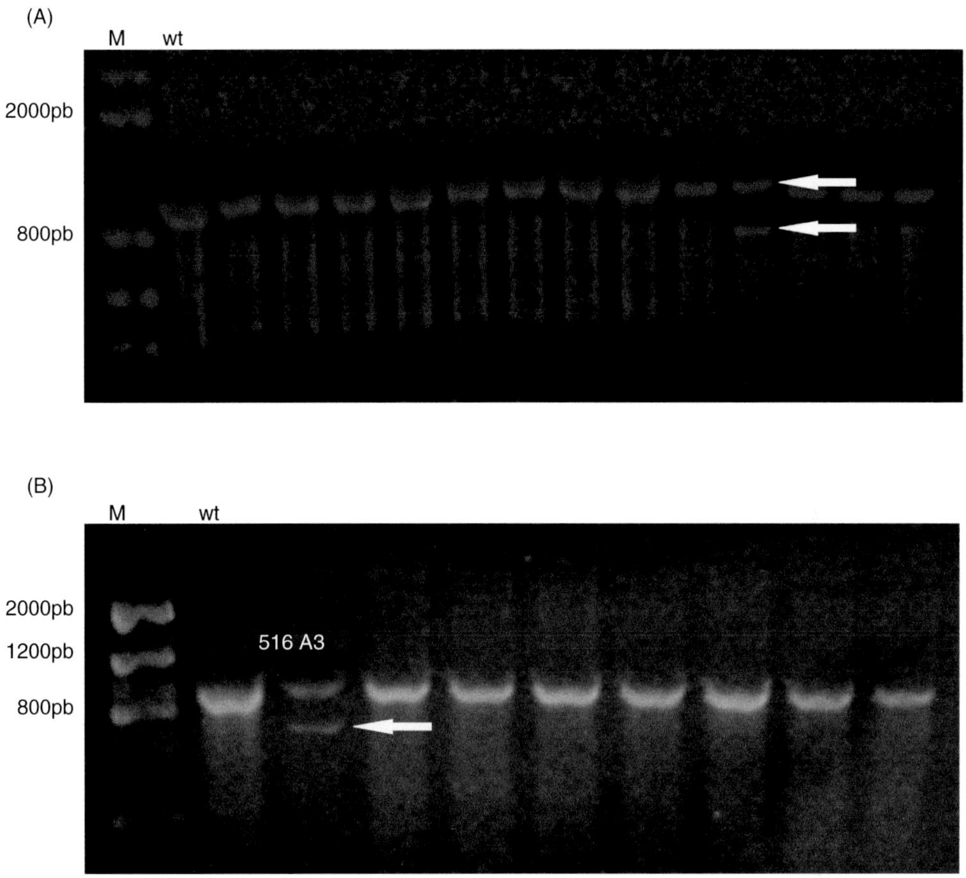

Fig. 11.1. Analysis of mutant families derived from 'BRS Querência' seeds treated with EMS for the presence of mutations targeting *Os03g0103300* gene through TILLING. **(A)** The slots represent the DNA pools of 8 plants (plants in M_3). **(B)** Each slot represents the DNA of each plant in the pool where the mutation was identified (plants in M_4). M, Low DNA Mass Ladder (Invitrogen); wt, amplification pattern of the wild-type sequence; 516 A3, amplification pattern of the mutant sequence. White arrow indicates the pool containing putative mutants for the gene of interest.

Table 11.2. Summary of variance analysis for relative performance (RP) of germination (RP G) and germination velocity index (RP GVI), for the mutant 516 A3 and 'BRS Querência' (wild-type).

Source of variation	DF	Mean square	
		RP G	RP GVI
Genotype	1	15.1801*	29.0102*
Blocks	2	0.0003	0.1783
Residue	2	0.4071	0.1478
Mean	–	8.3752	7.568
CV (%)	–	7.6186	5.0799

*Significant values at the 5% probability by the F-test.
DF = degrees of freedom.

The mutant genotype displayed superior performance compared with the wild-type for both variables, germination and GVI, reaching more than 50% for germination and 60% for GVI.

These results are consistent with those found in the molecular analysis for the *Os03g0103300* gene (which detected changes in the gene sequence), indicating a sequence change related to the seed germination process. The mutant 516 A3 presented a higher germination performance as well as in the GVI, demonstrating: (i) that the mutagenic agent was efficient in creating genetic variability; and (ii) genotype 516 A3 shows tolerance in germination under cold conditions and

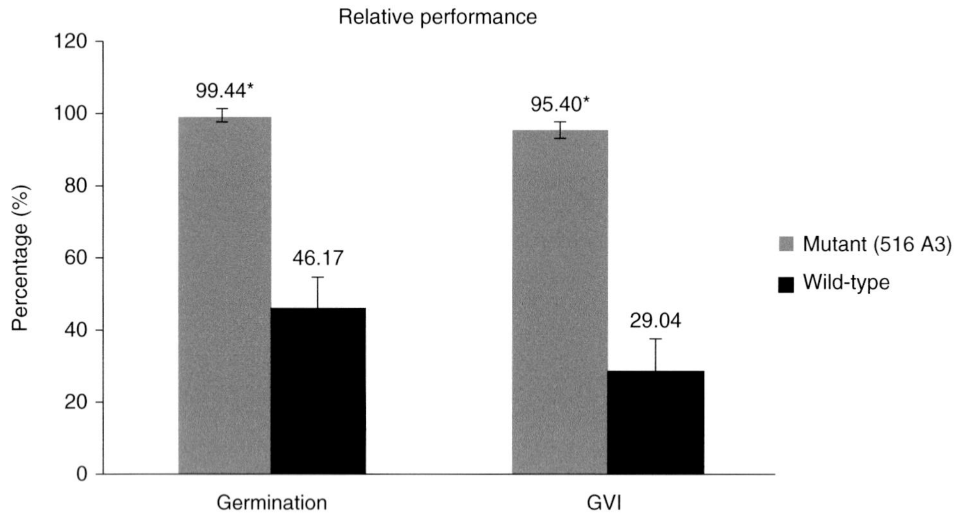

Fig. 11.2. Relative performance of germination and germination velocity index (GVI) of the 516 A3 mutant and of the 'BRS Querência' wild-type. *Statistically different from the wild-type 'BRS Querência' (Dunnett $p \leq 0.05$).

can be used as a source for direct or indirect use in breeding programmes aiming for cold tolerance.

3.3 Expression of the *Os03g0103300* gene in qRT-PCR

The three reference genes used presented a score < 1.0 when analysed in the DataAssist software, indicating that they are stable and may be used to normalize the *Os03g0103300* gene expression (Table 11.3).

The expression assay demonstrates that the *Os03g0103300* gene was downregulated in both the wild-type and in the mutant genotype in the S3 stage seedlings when submitted to a temperature of 13°C (Fig. 11.3). However, the results demonstrated that the *Os03g0103300* gene was more severely downregulated in the wild-type.

The *Os03g0103300* gene is not as highly expressed in cold conditions as at normal temperature that may be associated with the tissue and stage analysed in this study, or processes prior to the S3 stage, such as embryo activation and weakening of the tissues that cover the embryo, which are determinants for germination performance. These results, coupled with knowledge of the function of this gene in seed germination under cold stress, together with mutant performance during germination, suggest that the *Os03g0103300* gene is more expressed in the initial, rather than later, stages of metabolic activation of the embryo after water uptake and radicle protrusion.

Table 11.3. Ranking of the three reference genes according to the transcription stability obtained through DataAssist Software.

Reference gene		Score	Reference
NAPB	*Nucleic acid binding protein*	0.9992	Narsai *et al.*, 2010
EF1-α	*Eukaryotic elongation factor 1-alpha*	0.6039	Jain *et al.*, 2006
AK059783	*AK059783*	0.7815	Jain *et al.*, 2006

4 Discussion

Temperature has a determining role in seed germination, affecting the activation stage and post-germination growth. Low temperatures impair plant development and photosynthesis.

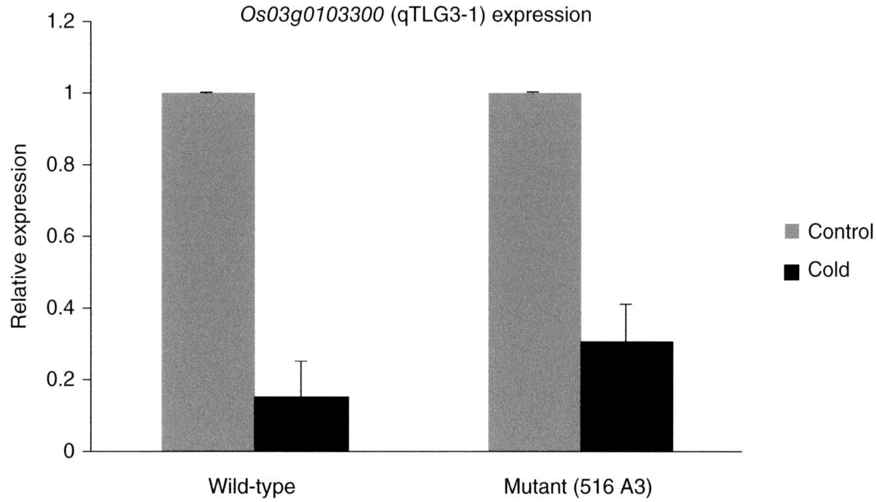

Fig. 11.3. Relative expression of the *Os03g0103300* gene in the wild-type ('BRS Querência') and in the mutant (516 A3) under control conditions (25°C) and cold conditions (13°C). Error bars represent three biological replicates containing three technical replicates each.

Also, cold may affect rice yield, due to plant population reduction as a result of a delay and/or decrease in seed germination (Cruz and Milach, 2004). Many methods have been developed to evaluate cold tolerance in rice. At germination, these evaluations have been performed by subjecting seeds to temperatures ranging from 10°C to 25°C for periods of 3–35 days. The percentage and velocity of germination and the length of coleoptile and radicle are the most commonly evaluated traits (Bertin *et al.*, 1996; Sthapit and Witcombe, 1998; Cruz and Milach, 2004; Wang *et al.*, 2009; Sharifi, 2010; Ñanculao *et al.*, 2013; Pouramir-Dashtmian *et al.*, 2013; Lone *et al.*, 2018). In this work, germination and GVI were evaluated. The results demonstrated that the mutant 516 A3 (Fig. 11.1) displayed higher performance in GVI and germination when subjected to cold stress (Fig. 11.2). The germination rates presented by the 516 A3 mutant were greater than 50% with respect to the 'BRS Querência' wild-type. Thus, the 516 A3 mutant presented tolerance to cold stress during germination and can be used as a source for increases in genetic variability in breeding programmes aimed at cold tolerance in the germination stage. The molecular identification of qTLG3-1 in the control of germination under cold temperatures was important for the elucidation of the complex genetic regulation of rice germination under stress conditions. Transcriptome analysis demonstrated that the expression of the gene *Os03g0103300* (located in qLTG3-1) coincides with the upregulation of genes involved in processes related to weakening of the tissues that cover the embryo during the germination of seeds, resulting in a reduction in mechanical resistance to coleoptile growth (Fujino *et al.*, 2008; Fujino and Matsuda, 2010). In addition, this QTL is responsible for more than 30% of the phenotypic variation for cold tolerance at germination stage (Fujino *et al.*, 2004). Mutation in the *Os03g0103300* gene, identified in this study through TILLING, can be associated to the effect observed in GVI and the germination of the 516 A3 mutant seeds under cold stress.

In previous studies, Fujino *et al.* (2008) identified the expression profile of the *Os03g0103300* gene in several tissues and during germination. During seed germination the *Os03g0103300* gene was expressed in the embryo. Expression in the aerial part of seedlings (4–7 days of age) and in the young panicle was also reported. However, the expression of this gene was not detected in the endosperm during germination and 2-month-old leaves. Despite the significant difference in germination and GVI observed in this study, the qRT-PCR analysis revealed that the *Os03g0103300* gene was downregulated in the wild-type and mutant genotypes at the S3 stage (Fig. 11.3). This indicates that differential expression may be associated with earlier stages during germination. Even though qLTG3-1 is responsible for more than 30% of the phenotype variation, the induction of mutation performed may have affected other genes, or the response is associated with modification in the molecular network and in other genes related to the traits evaluated. New studies should be performed at stages prior to S3 to elucidate the effects of the mutation identified by TILLING in the *Os03g0103300* gene.

5 Conclusion

The induction of a mutation using 1.5% (v/v = 0.15 M) EMS is efficient in increasing the genetic variability in rice. The analysis of mutant families by TILLING allowed the identification of a genotype (516 A3) with mutations in the *Os03g0103300* gene (located in qLTG3-1), responsible for a higher rice germination potential at cold stress. The *Os03g0103300* transcript abundance was lower in the mutant than in the wild-type at the S3 stage, and further analyses should be performed in earlier stages of germination.

References

Bertin P. *et al.* (1996) Evaluation of chilling sensitivity in different rice varieties. Relationship between screening procedures applied during germination and vegetative growth. *Euphytica* 89, 201–210.

Bustin S.A. *et al.* (2009) The MIQE guidelines: minimum information for publication of quantitative real-time PCR experiments. *Clinical Chemistry* 55, 611–622.

Chinnusamy, V. *et al.* (2004) Molecular genetic perspectives on cross-talk and specificity in abiotic stress signaling in plants. *Journal of Experimental Botany* 55, 225–236.

Cruz, C.D. (2001) *Programa genes: aplicativo computacional em genética e estatística*. UFV, Viçosa, 648 pp.

Cruz, R.P. da and Milach, S.C.K. (2004) Cold tolerance at the germination stage of rice: methods of evaluation and characterization of genotypes. *Scientia Agrícola* 61, 1–8.

Counce, P.A. *et al.* (2000) A uniform, objective, and adaptive system for expressing rice development. *Crop Science* 40, 436–443.

Ding, Y. *et al.* (2019) Advances and challenges in uncovering cold tolerance regulatory mechanisms in plants. *New Phytologist* 222, 1690–1704.

Fujino, K. and Matsuda, Y. (2010) Genome-wide analysis of genes targeted by qLTG3-1 controlling low-temperature germinability in rice. *Plant Molecular Biology* 72, 137–152.

Fujino, K. *et al.* (2004) Mapping of quantitative trait loci controlling low-temperature germinability in rice (*Oryza sativa* L.). *Theoretical and Applied Genetics* 108, 794–799.

Fujino, K. *et al.* (2008) Molecular identification of a major quantitative trait locus, qLTG3-1, controlling low-temperature germinability in rice. *Proceedings National Academy of Sciences USA* 105, 12623–12628.

Fulton, T.M. *et al.* (1995) Microprep protocol for extraction of DNA from tomato and other herbaceous plants. *Plant Molecular Biology Reporter* 13, 207–209.

Jain, M. *et al.* (2006) Validation of housekeeping genes as internal control for studying gene expression in rice by quantitative real-time PCR. *Biochemical and Biophysical Research Communications* 345, 646–651.

Jiang, S.-Y. and Ramachandran, S. (2010) Assigning biological functions to rice genes by genome annotation, expression analysis and mutagenesis. *Biotechnology Letters* 32, 1753–1763.

Livak, K.J. and Schmittgen, T.D. (2001) Analysis of relative gene expression data using real-time quantitative PCR and the 2 -ΔΔCT Method. *Methods* 25, 402–408.

Lone, J.A. *et al.* (2018) Cold tolerance at germination and seedling stages of rice: methods of evaluation and characterization of thirty rice genotypes under stress conditions. *International Journal of Current Microbiology and Applied Sciences* 7, 1103–1109.

Luz, V.K. da *et al.* (2016) Identification of variability for agronomically important traits in rice mutant families. *Bragantia* 75, 41–50.

Maguire, J.D. (1962) Speed of germination-aid selection and evaluation for seedling emergence and vigor. *Crop Science* 2, 176–177.

MARA Brazil (2009) *Regras para análise de sementes*. Ministério da Agricultura e da Reforma Agrária. Brasília, DF: SNDA/DNDV/CLAV, 365 pp.

McCallum, C.M. *et al.* (2000a) Targeted screening for induced mutations. *Nature Biotechnology* 18, 455–457.

McCallum, C.M. *et al.* (2000b) Targeting Induced Local Lesions In Genomes (TILLING) for plant functional genomics. *Plant Physiology* 123, 439–442.

Ñanculao G.D. *et al.* (2013) Cold tolerance evaluation in Chilean rice genotypes at the germination stage. *Chilean Journal of Agricultural Research* 73, 3–8.

Narsai, R. *et al.* (2010) Defining reference genes in *Oryza sativa* using organ, development, biotic and abiotic transcriptome datasets. *BMC Plant Biology* 10, 56.

Ouyang, S. *et al.* (2007) The TIGR Rice Genome Annotation Resource: improvements and new features. NAR 35 Database Issue, D846–851. Available at http://rice.plantbiology.msu.edu/home_overview.shtml (accessed April 2021)

Pouramir-Dashtmian, F. *et al.* (2013) Methods for rice genotypes cold tolerance evaluation at germination stage. *International Journal of Agriculture and Crop Sciences* 5, 2111–2116.

Saghai-Maroof, M.A. *et al.* (1984) Ribosomal DNA spacer-length polymorphisms in barley: Mendelian inheritance, chromosomal location and population dynamics. *Proceedings of the National Academy of Sciences of USA* 81, 8014–8018.

Sharifi, P. (2010) Evaluation on sixty-eight rice germplasms in cold tolerance at germination stage. *Rice Science* 17, 77–81.

SOSBAI (2018) *Arroz irrigado: recomendações técnicas da pesquisa para o Sul do Brasil*. Sociedade Sul Brasileira de Arroz Irrigado, Itajaí.

Sthapit, B.R. and Witcombe, J.R. (1998) Inheritance of tolerance to chilling stress in rice during germination and plumule greening. *Crop Science* 38, 660–665.

Tadele, Z. (2016) Mutagenesis and TILLING to dissect gene function in plants. *Current Genomics* 17, 499–508.

Till, B.J. *et al.* (2003) Large-scale discovery of induced point mutations with high-throughput TILLING. *Genome Research* 13, 524–530.

Till, B.J. *et al.* (2006) High-throughput TILLING for Arabidopsis. *Methods in Molecular Biology* 323, 127–135.

Untergasser, A. *et al.* (2007) Primer3Plus, an enhanced web interface to Primer 3. *Nucleic Acids Research* 35, W71–W74.

Wang, X. *et al.* (2003) Plant responses to drought, salinity and extreme temperatures: towards genetic engineering for stress tolerance. *Planta* 218, 1–14.

Wang, Z.F. *et al.* (2009) Genetic control of germination ability under cold stress in rice. *Rice Science* 16, 173–180.

12 Mutation Breeding of Sorghum to Support Climate-Smart Agriculture

Soeranto Human[1], Sihono and Wijaya Murti Indriatama

Abstract

Global climate change effects in agricultural fields often increase plant stress. For mitigating the negative effects of climate change, climate-smart agricultural policies should be developed, for example through the improvement of crop adaptability, productivity and quality in environments impacted by climate change. Attempts to increase crop genetic variability must be sought to aid in mitigating adverse consequences of climate change. For that purpose, mutation breeding plays an important role since it can increase genetic variation of important crops. By selecting desired mutant genotypes, the plant breeder can advance their germplasm by progressing lines with good adaptability, high productivity and quality under adverse conditions. For Indonesia, significant adverse impacts of climate change have appeared in some agricultural regions, such as prolonged drought problems in the east. To face the worsening conditions brought about by climate change and variability, a crop was sought that would require less agricultural input, being drought tolerant, having good adaptability and with high economic value. The choice fell on sorghum (*Sorghum bicolor*). In certain areas sorghum is recognized as a source of food, feed and fuel. Mutation breeding of sorghum has been conducted at the Centre for Isotopes and Radiation Application (CIRA) of the National Nuclear Energy Agency of Indonesia (BATAN). Sorghum mutation breeding is relevant to the national programme on food and energy diversification to support food and energy security in the country. The breeding objectives are to improve sorghum genotypes for improved yield and quality, and with tolerance to adverse conditions brought about by climate change, especially prolonged drought. Three sorghum mutant varieties have now been obtained and are being developed further by stakeholders. Sorghum cultivation in Indonesia has made significant impacts on mitigating the effects of climate change and supporting the food and energy diversification programme for maintaining food and energy security in the country. It has also promoted economic growth in rural areas impacted by climate change.

Keywords: sorghum • mutation breeding • mutant variety • climate change

1 Introduction

For Indonesia, the negative effects of climate change include increased adverse conditions for agricultural development, mostly in drought-prone areas, especially in the eastern part of the country. To face the worsening conditions brought about by climate change and variability, a crop was sought that would require less agricultural input, was drought tolerant and had good adaptability, productivity, quality and high economic value. The choice fell on sorghum,

[1] Email: Soeranto@batan.go.id

DOI: 10.1079/9781789249095.0012

since sorghum has become a potential crop for Indonesia. In certain areas sorghum had been recognized as a source of food, feed and fuel. However, sorghum was still regarded as a minor crop and its cultivation was limited. Sorghum is not of Indonesian origin and so the readily available plant genetic variability was low. Hence, a sorghum breeding programme is needed for Indonesia in order to increase genetic variation for further improved varieties (Human, 2015a).

Sorghum (*Sorghum bicolor* L.) is a cereal crop commonly grown in the hot and dry (arid or semi-arid) areas and is regarded as a multipurpose crop owing to its use for food, feed or fuel. From an agronomic point of view, sorghum has been known as a low-input crop and is very efficient in utilizing fertilizer and sunshine (Aznur and Suwarto dan Purnawati, 2017; Suminar and Suwarto dan Purnawati, 2017). Sorghum grain contains carbohydrate that can be used as a food source and its stems and leaves (stover) can be used for animal feed. One kind of sorghum (known as sweet sorghum) has stems containing a high concentration of sugar; it is usually used as raw material for making solutions of sugar and its products have been used by many people suffering from diabetes (Human, 2015a). The glycaemic index indicates the food's effect on a person's blood glucose (also called blood sugar level). In many countries, sweet sorghum is commonly used as raw material for making bioethanol (as renewable bioenergy) through a process of fermentation of its sugar juice.

Sorghum can grow and adapt well in Indonesia, especially in the drought-prone areas of the eastern part of the country, but it has low productivity and quality. Farmers in the Nusa Tenggara Barat (NTB) and Nusa Tenggara Timur (NTT) provinces have cultivated local sorghum varieties for a long time and have used it as their main food because other food crops such as rice and maize cannot be grown well there during the dry season. Local sorghum landraces mostly have undesirable traits such as low productivity, tall plant stature (making them susceptible to lodging), low-quality grains of dark colour indicating a high tannin content, and late maturity. The dark colour of sorghum grains has been studied and it is controlled not only by additive dominant genes but also by epistatic genes with medium heritability, which means that improvement of grain colour can be achieved in a plant breeding programme (Trikoesoemaningtyas *et al.*, 2017). The breeding objectives in this work are to improve the agronomic character of sorghum, including plant stature, growth duration, yield, quality and plant adaptation for tolerance to adverse conditions brought about by climate change, such as prolonged drought.

2 Materials and Methods

In Indonesia, improvement of sorghum varieties has been achieved through a plant breeding programme using mutation techniques conducted at the Center for Isotopes and Radiation Application (CIRA) of the National Nuclear Energy Agency (BATAN). The main facility available in this Center is a gamma irradiator with a cobalt-60 source which can be used for irradiating breeding materials in order to increase plant genetic variation (Fig. 12.1). Parental plant material used was the variety 'Zhengzu', which was introduced from China. Seeds were treated with gamma radiation with doses of 0, 100, 200, 300, 400, 500, 600, 700, 800, 900 and 1000 Gy and then grown in trays of sandy soil in the greenhouse for radiosensitivity analysis in the M_1 generation. Seedling growth was measured for each irradiation treatment and the irradiation doses to be used were determined by using lethal dose (LD) values giving the highest plant genetic variation in the M_2 generation, as has been practised in wheat mutation breeding (Nur *et al.*, 2014). LD values were determined by using best-fitting curve software, which is based on the function of measured lethality (Y) for given irradiation doses (X).

Plant screening for drought tolerance was started in the M_2 generation using indirect screening in the greenhouse followed with direct screening in the field (Fig. 12.2). The indirect screening was done at the seeding stage by growing seedlings in a medium containing polyethylene glycol (PEG 6000). PEG 6000 is a chemical that can influence the amount of water absorbed by plant roots (Van den Berg and Zeng, 2006; Human and Sihono, 2010). The higher the concentration of PEG in the growth medium, the less water that can be absorbed by the root. The concentration of 25% w/v PEG is equivalent to critical drought exposure in the

Fig. 12.1. Two gamma irradiators **(A, B)** Serial Type GC-4000 made in India and **(C)** model GC-220 made in Canada with an installed cobalt-60 source available at CIRA, BATAN for irradiating breeding materials in mutation breeding programmes.

Fig. 12.2. Screening methods for drought tolerance. **(A)** Indirect screening in the greenhouse using 25% w/v PEG 600 showing plants that are resistant (r) or susceptible (s) to this induced water deficit. **(B)** Resistant seedlings emerging from the soil in direct screening in the field.

field causing permanent wilting of leaves in cereals. The use of PEG for screening for drought tolerance has been reported in sorghum mutation breeding programmes previously (Human and Sihono, 2010; Human, 2015b). The selected tolerant seedlings were transplanted to the field for seed multiplication and further direct screening was conducted in the field during the dry season. In subsequent generations, yield trials were performed in the field with putatively drought-tolerant plants. Sorghum mutation breeding follows that of a self-pollinated crop which is directed to the formation of pure or homozygous lines of the selected mutants (Richards, 1997; van Harten, 1998). Homogeneity and stability of the mutant lines were studied in the M_5 generation (Human *et al.*, 2011) and homogenous lines with the desired properties were denoted as 'promising mutant lines'. Multi-location yield trials and grain quality analyses were conducted for the promising mutant lines before they were proposed for submission for official release to the Varietal Release Team, which is under the Ministry of Agriculture. Sorghum quality analysis was conducted at the Center for Chemistry at the Indonesian Academy of Science (LIPI) at Serpong, Banten, Indonesia.

3 Results

From a radiosensitivity study in the M_1 generation it was found that LD_{20} and LD_{50} values for sorghum were around 350 and 500 Gy, respectively (Fig. 12.3). From this dose range, the genetic variation (measured as the statistical variance of the population) was found to be highest in the M_2 plant population (Fig. 12.4), where plant selection was started. Selection was continued in subsequent generations until homozygous plants were obtained.

In the M_5 generation, 17 drought-tolerant mutants were identified and together with their parental and check varieties were evaluated in multi-location trials. Agronomic data were collected, analysed and reported to the Varietal Release Team, and finally only three mutant lines, ZH-30, PATIR-1 and PATIR-4, were accepted for release as new sorghum mutant varieties. These were given the names 'Pahat', 'Samurai-1' and 'Samurai-2', respectively. 'Pahat' and 'Samurai-2' were recommended as grain sorghum to be used as foodstuff, while 'Samurai-1' was recommended as a sweet sorghum variety for sugar syrup production or further processing for bioethanol (for bioenergy). The panicles of these mutant varieties are presented in Fig. 12.5.

Data on grain productivity and quality are presented in Table 12.1. All mutant varieties had higher grain productivity than the parental variety 'Zhengzu' (38–52% increase) and the national check variety 'Mandau'. The grain of mutant varieties contained food-quality carbohydrate. However, based on the tannin content, only 'Pahat' and 'Samurai-2' met requirements for food. According to food regulation in Indonesia, tannin content should be less than 0.025% in food. 'Pahat' is the most promising variety to be used as food, since the varietal release proposal described it as having good carbohydrate, high protein and fibre but low tannin content. From an agronomic point of view, 'Pahat' is early maturing (89 days) and a semi-dwarf (148 cm) that is easy to harvest and resistant to lodging in strong winds. As one of the negative impacts of climate change is strong winds, the semi-dwarf variety 'Pahat' can be used for mitigating the adverse effects of climate change.

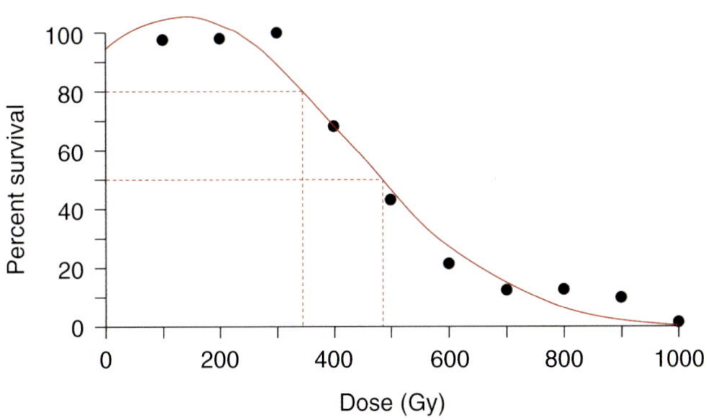

Fig. 12.3. Estimation of LD_{20} and LD_{50} from a Gaussian curve fitted to survival data (using the software CurveExpert 1.3). LD_{20} = 351 ± 33 Gy, LD_{50} = 487 ± 25 Gy.

Fig. 12.4. Plant variation in the M_2 generation.

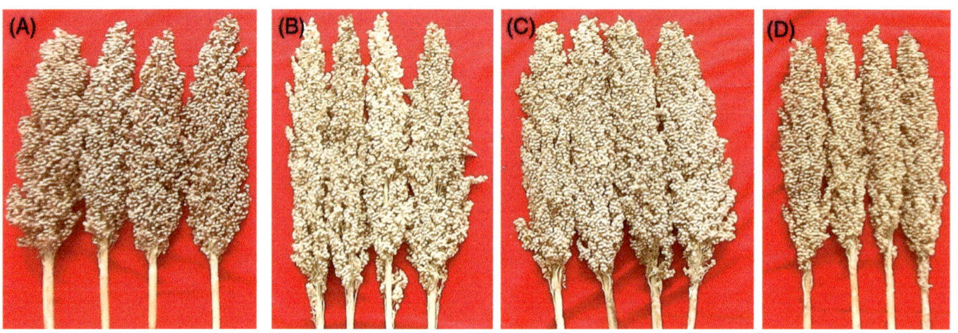

Fig. 12.5. Mature panicles of the mutant varieties of sorghum. **(A)** Patir-1 ('Samurai-1'). **(B)** Patir-9. **(C)** Patir-4 ('Samurai-2'). **(D)** ZH-30 ('Pahat'). The scale bar to the right is 10 cm.

'Pahat' is now widely grown in Indonesia, especially to support food security in drought-prone areas in the eastern part of the country, which is very much impacted by climate change. Its short growth duration (only 3 months) also gives an advantage of avoiding drought during the prolonged dry season.

'Samurai-1' is the mutant variety with the highest sugar content or brix (12%) in its stem juice and this variety can be classified as a sweet sorghum type. Even though it has high productivity, 'Samurai-1' grain is not recommended for use as food, owing to its high content of tannin (0.030%). Therefore, 'Samurai-1' is recommended mainly for use as a source of making syrup or further processing through fermentation for making bioethanol. The Government of Indonesia has recently promoted the use of bioenergy as a renewable energy and many companies are now dealing with the business of bioenergy.

Based on breeder seed of mutant varieties produced by BATAN and its distribution, in 2017–2018 the growing areas of sorghum mutant varieties in Indonesia were estimated to be about 800,000 ha. The growers or stakeholders included farmers, private companies, local government offices, universities, etc. For example, 'Pahat' was grown widely in the Banyuwangi District of East Java Province by the private company PTPN XII (Fig. 12.6). Cultivation of 'Pahat' was intended to increase land productivity impacted by climate change (drought-prone areas) and to promote food security in the Province, especially during a prolonged dry season. Meanwhile, the Ministry of Energy and Mineral Resources of Indonesia (ESDM) is currently developing the sweet sorghum mutant variety 'Samurai-1' in Merauke, Irian Province, for making bioethanol where the industry has a capacity of 40 kilo litres per day (40 KLPD). Irian Province

Table 12.1. Grain productivity and quality of sorghum mutant varieties (source: Indonesian Ministry of Agriculture).

	Analysed components						
Sorghum variety	Productivity (t/ha)	Carbohydrate (%)	Protein (%)	Fat (%)	Fibre (%)	Tannin (%)	Brix (%)
'Samurai-1'	6.1	73.59	12.55	2.56	2.20	0.030	12
'Samurai-2'	6.4	75.53	12.07	2.80	1.53	0.016	8
'Pahat'	5.8	72.86	12.80	2.42	2.21	0.012	5
'Zhengzu' (Parent)	4.2	70.63	10.71	2.59	1.39	0.016	5
'Mandau' (National Variety)	5.1	75.40	12.73	2.97	1.22	0.013	6

Fig. 12.6. A field of 'Pahat' in Banyuwangi, East Java, Indonesia, grown by the private company PTPN XII. The variety is widely grown by this company and this image shows the condition of the variety during the dry season.

is located in Eastern Indonesia and is one of the provinces adversely impacted by climate change in Indonesia.

4 Discussion

The optimal gamma irradiation doses for sorghum mutation breeding purposes were about 200–450 Gy. Mutant selection started in the M_2 generation and screening for drought tolerance continued in subsequent generations. Seventeen mutant lines were obtained and tested for drought tolerance and these were then evaluated in multi-location trials. Finally, only three mutant lines, Zh-30, PATIR-1 and PATIR-4, fulfilled the requirements (relatively high yield and quality better than the controls) for release by the Varietal Release Team under the Ministry of Agriculture. The three mutant varieties of sorghum were given the names 'Pahat', 'Samurai-1' and 'Samurai-2', respectively.

5 Conclusion

The mutant varieties 'Pahat', 'Samurai-1' and 'Samurai-2' are now grown widely in Indonesia, with an estimated area of about 800,000 ha in 2017–2018. Stakeholders include farmers, private companies and local government officials. Sorghum cultivation in Indonesia provides significant impact in mitigating the negative impacts of climate change while supporting the food and energy diversification programme for maintaining food and energy security in the country.

References

Aznur, F. and Suwarto dan Purnawati, H. (2017) Efisiensi penggunaan cahaya matahari dan partisi karbohihrat tanaman sorgum pada berbagai tingkat pemupukan nitrogen. *Jurnal Agronomi Indonesia* 45(3), 278–284. ISSN 2085-2916 [in Indonesian].

Harten, A.M. van (1998) *Mutation Breeding: Theory and Practical Applications*. Cambridge University Press, Cambridge, UK.

Human, S. (2015a) Sorghum mutant varieties: from research to market. *Plant Breeding and Genetics Newsletter*, No. 34. Joint FAO/IAEA Programme.

Human, S. (2015b) Sorghum mutation breeding for improving tolerance to abiotic stresses brought about by climate change. *Joint FAO/IAEA Programme*. Available at: http://www-naweb.iaea.org/nafa/news/pbg-sorghum-mutation.html (accessed 2019)

Human, S. and Sihono, S. (2010) Sorghum breeding for improved drought tolerance using induced mutation with gamma irradiation. *Jurnal Agronomi Indonesia* 38(2).

Human, S. *et al.* (2011) Stability test for sorghum mutant lines derived from induced mutations with gamma-ray irradiation. *Atom Indonesia Journal* 37(3).

Nur, A. *et al.* (2014) Keragaman genetik gandum populasi mutan M3 di agroekosistem tropis. *Jurnal Aplikasi Isotop dan Radiasi* 10(1) [in Indonesian].

Richards, A.J. (1997) *Plant Breeding Systems*, 2nd edn. Chapman & Hall, London.

Suminar, R. and Suwarto dan Purnawati, H. (2017) Pertumbuhan dan hasil sorgum di tanah latosol dengan aplikasi dosis pupuk nitrogen dan fosfor yang berbeda. *Jurnal Agronomi Indonesia* 45(3), 271–277 [in Indonesian].

Trikoesoemaningtyas *et al.* (2017) Kendali genetik karakter morfologi dan agronomi pada tiga populasi sorgum (*Sorghum bicolor* (L.) Mounch). *Jurnal Agronomi Indonesia* 45(3), 271–277 [in Indonesian].

Van den Berg, L. and Zeng, Y.J. (2006) Response of South African indigenous grass species to drought stress induced by polyethylene glycol (PEG). *South African Journal of Botany* 72(2).

13 Production of Haploid Embryos and Plants in Iranian Melon (*Cucumis melo* L.) Through Irradiated Pollen-induced Parthenogenesis

Leila Bagheri[1], Mahmoud Lotfi and Mansour Nori

Abstract

The irradiated pollen technique (IPT) is the most successful haploidization technique within Cucurbitaceae. The influence of gamma-ray doses (250, 350, 450 and 550 Gy), genotypes and stage of development of embryos obtained by IPT on the induction of haploid embryos were studied in several Iranian melon cultivars as well as their hybrids with alien cultivars. Female flowers were pollinated using pollen that had been irradiated with gamma rays. Different shapes and stages of embryos were excised 21–25 days after pollination and cultured on E20A medium. Direct culture, liquid culture and integrated culture methods were used; integrated culture and liquid culture methods showed advantages in increasing the efficiency of haploid plant production in melon breeding programmes. Results revealed that 550 Gy of gamma irradiation was successful in inducing parthenogenesis and fruit development, whereas lower irradiation doses were not effective in inducing haploid embryos. The percentages of embryos per seed were the highest in 'Samsoori' (1.2%) and 'Saveh' (1.1%) cultivars. Some of the heart-shaped and cotyledon-shaped embryos developed into haploid plants. In total, 52 parthenogenic melon plantlets were recovered from 274 embryos via IPT. Production of haploid embryos and haploid plants was strongly influenced by gamma-ray dose, embryo stage and genotype. Indirect methods and chromosome counting performed on the root cells of regenerated plants showed that these plants were haploid ($n = x = 12$).

Keywords: parthenogenesis • gamma irradiation • irradiated pollen • Iranian melon

1 Introduction

Melon, *Cucumis melo* (L.), is one of the most important horticultural species within the Cucurbitaceae family. Melon and cantaloupe genotypes are cultivated for nutritional value and human health benefits in Iran (Paris *et al.*, 2012). These plants are characterized by a high degree of heterozygosity, due to open pollination. For this reason, progress in breeding programmes through conventional methods (i.e. generations of crossing and selection) is time consuming. Biotechnological techniques such as production of haploids offer new opportunities for genetic research in breeding programmes (Hofer *et al.*, 2003). Haploidization methods facilitate the production of pure lines and represent significant advantages for breeders and geneticists (Pochard *et al.*, 1971). Induction of haploidy followed by doubling helps to fix traits in the homozygous state in a single step. These homozygous individuals are useful for genome mapping and the identification of the location of major genes and quantitative trait loci for economically important traits

[1] Email: llbagheri@aeoi.org.ir

 Mutation Breeding, Genetic Diversity and Crop Adaptation to Climate Change (eds S. Sivasankar *et al.*)
DOI: 10.1079/9781789249095.0013

(Kush and Virmani, 1996). Early attempts to produce haploids of melon by anther or ovule culture were not successful (Dryanovska *et al.*, 1983). The first reported success in obtaining haploid or doubled haploid (DH) melons was achieved by rescue of parthenogenetic embryos induced by pollination with irradiated pollen (Sauton *et al.*, 1987). The selection of an efficient radiation dose, the optimization of the pollination method, seed collection times, the developmental stage, culture media and culture conditions are all important factors affecting the success of this technique and the number of haploid embryos rescued (Germanà, 2009). The purpose of this research was to evaluate the use of gamma-irradiated pollen for production of parthenogenetic haploid embryos and plants in Iranian melon.

2 Materials and Methods

2.1 Plant materials

Six cultivars of Iranian melon and cantaloupe ('Khatooni Mashhad', 'Sooski Zard Ivanake', 'Sooski Sabz Ivanake', 'Samsoori', 'Saveh' and 'Garmak Esfahan'), two alien cultivars ('Ananasi' and 'Yellow Canary') and six hybrids between native and alien cultivars ('Ananasi' × 'Garmak', 'Ananasi' × 'Samsoori', 'Ananasi' × 'Saveh', 'Ananasi' × 'Sooski Zard Ivanake', 'Honeydew' × 'Sooski Zard Ivanake' and 'Khatooni Mashhad' × 'Honeydew') were used as maternal plants.

2.2 Irradiation and pollination

The plants were grown in a greenhouse. Male flowers were collected in the afternoon on the day before anthesis. At the same time, hermaphrodite flowers were emasculated and covered with a capsule to prevent uncontrolled pollination. After removal of petals from the male flower, pollen grains were irradiated with gamma-ray doses (250, 350, 450 and 550 Gy) using a ^{60}Co source (Gammacell PX-30, Russia) and then stored overnight at room temperature. On the following day, female flowers were pollinated using pollen that had been irradiated with gamma-rays. After 21–25 days, induced parthenocarpic fruits were harvested (Fig. 13.1).

Fig. 13.1. **(A)** Collection of male flowers. **(B)** Emasculation of hermaphrodite flowers. **(C)** Covering of female flowers with a capsule to prevent uncontrolled pollination. **(D)** Irradiation of pollen grains with 550 Gy dose of gamma-rays. **(E)** Gammacell with ^{60}Co source. **(F)** Pollination of female flowers with irradiated male flowers.

2.3 Recovery of haploid plants from parthenogenetic embryos

Seeds were removed from the fruits and washed under running water in the laboratory, sterilized for 10 min with a 1% solution of sodium hypochlorite and rinsed with sterile water. For detecting and rescue of induced embryos, three methods were used and the efficiency of all methods was evaluated. In the direct method, seeds were excised individually in a laminar flow hood and opened to remove any visible embryo. These embryos were cultured (two to three embryos per dish) in 100 × 20 mm Petri dishes containing 20 ml E20A medium (Sauton *et al.*, 1987) solidified with 0.8% Phytagar. In the liquid method, the rest of the seeds were sown in liquid E20A medium (about 40 seeds per dish). This medium was used as a nutrient medium; haploid embryos were not identified in these seeds. Sown seeds were observed for about 10–15 days over a light box containing a small white fluorescent lamp, and checked for whether they contained an embryo. When embryos were seen, the seeds were opened aseptically. Embryos were excised and transferred to Petri dishes of solid E20A medium by an integrated method (liquid and direct method). All cultures were placed at 25°C with a photoperiod of 16 h light to 8 h dark, with occasional shaking by hand, and were cultured for approximately 2–3 weeks. Embryos that grew well on plates and had 1–1.5 cm shoots were moved to large boxes containing 40 ml of solid E20A medium. Subsequently, each plantlet was propagated by putting the shoot tip and single stem nodes into fresh medium. Rooted plants were transplanted to soil, acclimatized and then transferred to the greenhouse (Fig. 13.2).

2.4 Determination of ploidy level

The ploidy levels of the plants were determined by direct and indirect methods. In the indirect method, based on comparison of morphology of flower and leaf size, the growth power and fertility of the plants were judged by maternal plants. Due to these apparent features, it is possible to find the level of ploidy, but it was necessary to verify with a more reliable method. Therefore, in the direct method, chromosome counting was performed on the plants' root tips (Darlington *et al.*, 1976).

3 Results

Pollen grains that were irradiated with the dose of 250 Gy led to the highest number of fruits, but 81% of seeds were full. By increasing the

Fig. 13.2. Parthenocarpic fruits after pollination by irradiated pollen in **(A)** 'Khatooni Mashhad' and **(B)** 'Saveh' melons. **(C)** Detection of parthenocarpic melon embryos by liquid method. **(D)** Heart-shaped embryo dissected from the parthenocarpic seed. **(E)** Micropropagation of parthenocarpic plantlets. **(F)** Rooted plants for transplanting to soil.

dose to 350 and then 450 Gy, the percentage of hollow seeds increased. At 550 Gy, all seeds were hollow, therefore this dose was selected as the effective dose for the production of haploid embryos (Table 13.1).

The findings in Table 13.1 showed that a small percentage of hollow seeds had an embryo. These embryos were in different developmental stages, such as globular heart-shaped or with cotyledons, which were classified for each maternal parent. Overall, 274 embryos were rescued. Most of them were globular (43.8%) and heart-shaped (27.38%) from all maternal parents. With the direct method, abortion and getting brown were especially observed. Not all the embryos that were in the cotyledon stage had the ability to regenerate plantlets. During the various cultures, 134 fruits of 14 types of maternal parents were obtained. In total, 52 parthenogenetic melon plantlets were recovered. The highest percentage of regeneration was in 'Samsoori' and 'Saveh' genotypes, at 1.2% and 1.1%, respectively. The lowest percentage of regeneration was in the 'Ananasi' × 'Samsoori' hybrid, with 0.32% (Table 13.2).

In the liquid method, 15,455 seeds (obtained from two fertilization periods and 59 fruits) were cultured in 283 Petri dishes containing liquid E20A medium. After the examination of the seeds, 137 embryos were obtained (Tables 13.3 and 13.4). The number of rescued embryos was higher than from the direct method (almost fourfold) and a high percentage of these embryos

Table 13.1. Effect of irradiation of pollen with different doses of gamma-rays on fruit set and seed set in Iranian melons.

Dose (Gy)	Number of pollinated flowers	Fruit set (%)	Fruit containing seed (%)	Seedless fruit (%)
250	86	68 a	81 a	19 c
350	29	66 a	58 ab	42 bc
450	17	59 a	30 b	70 b
550	123	58 a	0 c	100 a
Total	255	61	39	61

Values followed by the same letter are not significantly different as determined by protected Duncan ($p \leq 0.05$).

Table 13.2 Induction of parthenogentic embryos and the number of haploid plants obtained by pollination with irradiated pollen in Iranian melons.

					Embryo stage				
Maternal plants	Fruit number	Seed number	Embryo number	Embryos per 100 seeds	Globular	Heart-shaped	Cotyledon	Plantlets obtained	Embryos/ fruit
Khatooni Mashhad	19	4,913	39	0.79	15	10	14	8	2.05
Sooski Zard Ivanake	9	2,437	21	0.86	13	4	4	3	2.33
Sooski Sabz Ivanake	12	3,175	27	0.85	15	7	8	5	2.25
Samsoori	13	3,095	37	1.20	16	9	11	7	2.85
Saveh	16	3,892	43	1.10	14	14	13	10	2.69
Garmak Esfahan	17	4,691	45	0.96	20	14	12	7	2.65
Ananasi	4	1,218	9	0.74	2	2	5	3	2.25
Yellow canary	2	678	3	0.44	0	1	2	1	1.50
Ananasi × garmak	9	2,054	16	0.78	4	7	4	5	1.78
Ananasi × Samsoori	11	2,200	7	0.32	3	2	2	1	0.64
Ananasi × Saveh	6	1,200	7	0.58	5	1	1	1	1.16
Sooski Zard Ivanake × Ananasi	7	1,400	8	0.57	6	2	0	0	1.14
Sooski Zard Ivanake × Honeydew	5	1,000	8	0.80	4	2	2	1	1.60
Khatooni Mashhad x Honeydew	4	800	4	0.50	2	1	1	0	1
Total/mean	134	32,753	274	0.75	119	75	79	52	2.04

Table 13.3. Comparison of different methods for the isolation of haploid embryos induced by irradiated pollen in Iranian melons.

	Type of embryo			
Method of embryo rescue	Cotyledonary	Heart-shaped	Globular	Total
Direct culture	7	18	15	40
Liquid medium	54	40	43	137
Integrated culture	18	17	62	97
Total	79	75	120	274

Table 13.4. Comparison of the efficiency of two methods of extraction of the embryo: liquid method and integrated method.

Genotype	Extraction method	Final seed number	Number of embryos obtained	Percentage of embryos obtained
All	Integrated	8,771	98	1.1%
All	Liquid	15,455	137	0.88%

were in the developmental stage of cotyledon, which increased the probability of regeneration of the plant (Table 13.3).

The efficiency of integrated and liquid methods was compared (Table 13.4).

In addition to the two methods mentioned above, the integrated method was also evaluated. In this method, after direct observation for detection of embryos, liquid medium was used to improve the chance of development of embryos. In the integrated method, from 8771 seeds (obtained from one fertilization and 34 fruits) in different cultivars, 97 embryos were obtained (1.1%) (Table 13.4).

3.1 Confirmation of haploidy

In the indirect method it was observed that vegetative organs like leaves and flowers were smaller in the haploid plants than in the diploid plants. Also, during the transition to the pot in greenhouse conditions, a large number of male flowers and gradually a number of female flowers appeared that were distinguished by the separation of the petals. The results of chromosome counting in root tips of plants (in the direct method) confirmed the haploidy of these plants, i.e. with $n = 12$ chromosomes, compared with the control with $2n = 24$ chromosomes (Fig. 13.3).

4 Discussion

The plants obtained from pollen irradiated at 250, 350 and 450 Gy suggested that these doses are insufficient for pollen sterility (Froelicher *et al.*, 2007). Low levels of irradiation may damage only part of the generative nucleus while maintaining its capacity to fertilize the egg cell and lead to hybridization (Sestili and Ficcadenti, 1996). All haploids were obtained by irradiation at 550 Gy. Gamma-irradiated pollen with 550 Gy can germinate on the stigma, grow through the style and reach the embryo sac. Despite being unable to fertilize the egg cell and polar nuclei, it stimulates the development of haploid embryos (Musial *et al.*, 1998).

The rate of embryo conversion into a plant changed according to the type and developmental stage of the embryo. In general, the rate of haploid embryo conversion into plant increased with more advanced stages of embryo development. Globular-type embryos gave the lowest percentage conversion. Cotyledon and heart-type embryos gave the best results (Sari *et al.*, 1999).

The seeds that contained embryos were easily detected over a light source in the liquid method. Some of these embryos had turned green within the seeds and germinated like normal seeds. In addition, excision of embryos from seeds grown in this medium caused less

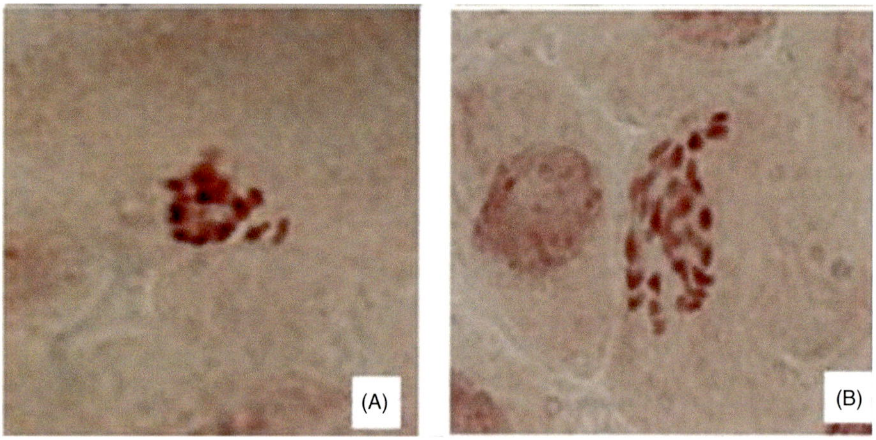

Fig. 13.3. Chromosome number in root tip cells of haploid plant (n = 12) **(A)** compared with the diploid control ($2n$ = 24) **(B)** in 'Samsoori' melon.

injury to the embryos. In contrast, opening all the seeds within a fruit took several hours of careful and tedious work, as in the direct method (Sauton, 1989).

Some Petri dishes of seeds cultured in liquid medium were contaminated. This problem was due to a single contaminated seed that could result in loss of the entire Petri dish. In comparison with the advantages of liquid medium detailed above, the contamination problem was not severe and could be minimized by the culture of fewer seeds per plate. It would also be advisable to put the Petri dishes on a shaker at low speed (30 rpm) for better aeration.

5 Conclusion

The irradiated pollen technique is the most successful haploidization technique within Cucurbitaceae. Results revealed that a dose of 550 Gy of gamma irradiation was successful in inducing parthenogenesis and fruit development, whereas lower irradiation doses were not effective in inducing haploid embryos. Regarding the results, it was observed that the combination of integrated and liquid methods was more effective than the liquid method alone in melon and more embryos were regenerated into seedlings. Thus, although the role of liquid culture was confirmed for increasing the efficiency in production of haploid plants, the combination of this method with the direct method for dissection of embryos showed higher efficiency. So the combination can be readily applied for the production of haploid plants in breeding programmes. Also, the rate of haploid embryo conversion into plants increased with more advanced stages of embryo development. Cotyledon and heart-type embryos gave the best results. The efficiency of embryo induction in this species is determined by several factors, such as radiation dose, embryo stage and genotype. Chromosome counting in root tips of 52 regenerated plantlets confirmed that they were at the haploid level (n = 12).

References

Darlington, C.D. *et al.* (1976) *The Handling of Chromosomes*, 6th edn. Allen and Unwin, London.

Dryanovska, O.A. *et al.* (1983) In vitro anther and ovule cultures in muskmelon (*Cucumis melo* L.). *Proceedings of the Bulgarian Academy of Sciences (C R Acad Bulg Sci)* 36, 1107–1110.

Froelicher, Y. *et al.* (2007) Induced parthenogenesis in mandarin for haploid production: induction procedures and genetic analysis of plantlets. *Plant Cell Reports* 26, 937–944.

Germanà, M.A. (2009) Haploids and doubled haploids in fruit trees. In: Touraev, A., Forster, B.P. and Jain, S.M. (eds) *Advances in Haploid Production in Higher Plants*. Springer, Dordrecht, pp. 241–263.

Hofer, M. *et al.* (2003) Induction of doubled haploids in sweet cherry (*Prunus avium*). *Euphytica* 130, 191–197.

Kush, G.S. and Virmani, S.S. (1996) Haploids in plant breeding. In: Jain, S.M. *et al.* (eds) *In Vitro Haploid Production in Higher Plants*. Kluwer, Dordrecht, pp. 11–34.

Musial, K. *et al.* (1998) Influence of irradiated pollen on embryo and endosperm development in Kiwifruit. *Annals of Botany* 82, 747–756.

Paris, H.S. *et al.* (2012) Medieval emergence of sweet melons, *Cucumis melo* (Cucurbitaceae). *Annals of Botany* 110, 23–33.

Pochard, E. *et al.* (1971) La monoploidie chez le piment (*Capsicum annuum* L.). *Zeitschrift fur Pflanzenzuchtung [Journal of Plant Breeding]* 65, 23–46.

Sari, N. *et al.* (1999) Investigation of new protected cultivation melon lines resistant to *Fusarium oxysporum* f. sp. *melonis* using dihaploidization. Turkey IIIrd National Horticultural Congress, Ankara, pp. 498–506.

Sauton, A. (1989) Haploid gynogenesis in *Cucumis sativus* by irradiated pollen. *Cucurbit Genetics Cooperative Report* 12, 22–23.

Sauton, A. *et al.* (1987) Obtention de plantes haploids chez le melon (*Cucumis melo* L.) par gynogenes induite par du pollen irradie. *Agronomie* 7, 141–148.

Sestili, S. and Ficcadenti, N. (1996) Irradiated pollen for haploid production. In: Jain, S.M. *et al.* (eds) *In Vitro Haploid Production in Higher Plants*. Springer, Dordrecht, pp. 263–247.

14 Application of Mutation Breeding to the Improvement of the Under-studied Crop Tef (*Eragrostis tef* (Zucc.) Trotter)

Zerihun Tadele[1], Kebebew Assefa, Solomon Chanyalew, Abate Bekele, Annett Weichert, Mirjam Schnell, Nora Röckel, Negussu Hussein and Gina Cannarozzi

Abstract

Induced mutation has been playing a significant role in the improvement of diverse crop types. This led to the release of over 3200 crop varieties in over 70 countries. We implemented induced mutation on tef (*Eragrostis tef* (Zucc.) Trotter), one of the most important cereal crops in the Horn of Africa, especially in Ethiopia, where it is annually cultivated on over 3 million hectares of land, equivalent to 30% of the total area allocated to cereals. Although tef is extensively cultivated in Ethiopia due to its resilience to diverse environmental stresses, the productivity of the crop is very low. The Tef Improvement Project based at the University of Bern in Switzerland employs mutation breeding to tackle major constraints in tef in order to enhance crop productivity. About 12,000 EMS (ethyl methanesulfonate) mutagenized M_2 families were generated from four improved tef varieties, namely 'Tsedey', 'Dukem', 'Kora' and 'Dagim'. Screening for major traits of importance helped us to obtain several candidate lines, including semi-dwarf and lodging-tolerant, drought-tolerant and acid-soil-tolerant lines. Among these, the most promising ones were introgressed to locally adapted improved varieties followed by several years of testing at representative locations for traits of interest. As a result, a new variety called 'Tesfa' with a novel and desirable combination of traits was approved for release to the farming community. This shows that the project has been actively involved in all three phases of induced mutation: mutation induction, mutation detection and mutation breeding.

Keywords: EMS • *Eragrostis tef* • mutation breeding • mutation detection • mutation induction

1 Role of Induced Mutation in Crop Improvement

Crop production is under continuous challenge from diverse environmental constraints, which include a variety of biotic and abiotic stresses, as well as policy-related constraints. The urgency of boosting crop productivity at the present time is due to: (i) the higher rate of population growth compared with the increase in food production; (ii) the widespread problem of biotic and abiotic stresses; (iii) weakness in the inherent properties of the plant (e.g. susceptibility of the plant to lodging); (iv) the cultivation of plants for biofuel production at the expense of food crops; (v) shortcomings in land and investment policies; and (vi) the negative impact of climate change on crop production. Although multiple challenges exist, boosting productivity of crops per unit area is important as it narrows

[1] Email: zerihun.tadele@ips.unibe.ch

 Mutation Breeding, Genetic Diversity and Crop Adaptation to Climate Change (eds S. Sivasankar *et al.*)
DOI: 10.1079/9781789249095.0014

down the wide yield gap that exists between the potential and current yield for most crops, especially in the developing world (Tadele, 2017).

Crop improvement has long been dependent on harnessing the naturally existing huge diversity for the trait of interest. However, the required level of diversity might not exist in landraces for some key agronomic traits. This might be due to the tight genetic linkage between the trait of interest and other undesirable traits. It is extremely difficult, if not impossible, to separate closely linked traits using conventional crossing and selection methods of plant breeding.

Induced mutation is, however, considered powerful as it can randomly mutate any trait of interest. In addition to creating variability from which breeders can select for any trait of their choice, induced mutation can also precisely knock out a single gene from a pair of tightly linked ones. Hence, the problem of linkage drag can be minimized.

Induced mutation began about seven decades ago to improve the productivity and/or quality of plants, as it creates stable and heritable alterations in the genetic material of the organism. Since then, mutation breeding has contributed significantly to the improvement of many economically important crops. Crops descended from this technique were superior to the original cultivars in productivity and/or tolerance to biotic and abiotic stresses. The list of officially released and commercially available crop varieties that originated from induced mutation is available in the Mutant Variety Database (MVD) of the Joint FAO/IAEA Division of Nuclear Techniques in Food and Agriculture (IAEA, 2018). The majority of the 3281 varieties recorded so far are from Asia (60%) and Europe (30%) (Table 14.1). Asian countries such as China, Japan and India together with European countries such as Russia and Germany have invested significantly in mutation breeding and have released a large number of varieties of food crops. An exception to this is The Netherlands, where the majority of mutant varieties released have been ornamental rather than food crops. In Africa and Latin America, on the contrary, not enough emphasis has been given to mutation breeding as yet and only a few varieties have been released using this technique. Of all the globally released crop varieties, about 50% are cereals, including rice, wheat and barley. Some of these varieties have been playing a significant role in the economy of some countries in providing substantial contributions to food, feed, fibre and brewing.

2 Tef as a Crop of Choice for Induced Mutation

Tef (*Eragrostis tef* (Zucc.) Trotter) is the most important cereal crop in the Horn of Africa, especially in Ethiopia, where it is annually cultivated on over 3 million hectares of land, which is equivalent to 30% of the total area allocated to cereals (CSA, 2014). The crop is preferred both by farmers and by consumers. Farmers prefer cultivating tef to other cereals since tef is more resilient to environmental stresses such as poor soil drainage during the rainy season and also to moisture scarcity. In addition, as a cash crop, both the grain and the straw of tef fetch a higher price than respective products from other cereals. Consumers prefer tef not only because it makes good quality *injera* (a pancake-like soft bread) but also because it is nutritious, due to its high protein and mineral content (Bultosa *et al.*, 2002; Abebe *et al.*, 2007), and the absence of gluten (Spaenij-Dekking *et al.*, 2005) which makes it an alternative food for people suffering from celiac disease. In general, tef plays a vital role in food security, nutrition and income generation to smallholder farmers.

Despite its versatility in adapting to extreme environmental conditions, the productivity of tef is very low in Ethiopia at 1.5 t/ha, compared with 3.2 t/ha for maize (CSA, 2014). The major yield-limiting factors are lack of cultivars tolerant to lodging and drought (Assefa *et al.*, 2011), as well as small seed size. This is also related to the widespread use of landraces and cultivars lacking desirable agronomic traits. Lodging (permanent displacement of the stem from the upright position) is the major bottleneck in tef production. Tef possesses tall and weak stems that easily succumb to lodging caused by wind or rain. In addition, lodging hinders the use of high-input crop husbandry, as the application of increased amounts of nitrogen fertilizer to boost the yield results in severe lodging. Consequently, both the yield and the quality of the grain and the straw are severely reduced. The lodged plant also poses difficulties in harvesting.

Table 14.1. The number and type of crop varieties released from induced mutation in different continents and countries. Major food crops that benefited from these programs are indicated. Adapted from IAEA (2018).

Continent	Varieties released	Top countries	Varieties released	Major food crops
Asia	1993	China	810	Rice, wheat, soybean, maize
		Japan	479	Rice
		India	335	Rice, barley, peanut
		Bangladesh	70	Lentil, chickpea, peanut, rice
		Pakistan	59	Mung bean, rice, wheat, chickpea
		Vietnam	58	Rice, soybean
Europe	995	Russia	216	Barley, buckwheat, wheat
		Netherlands	176	No food crop
		Germany	171	Barley
		France	39	Barley, rice
		Italy	35	Durum wheat
		UK	34	Barley
North America	200	USA	139	Barley, bean, rice
		Canada	40	Beans
Africa	72	Côte d'Ivoire	25	Rice
		Mali	15	Rice, sorghum
Latin America	51	Guyana	26	Rice
		Brazil	14	Bean
Australia & Pacific	10	Australia	9	Oats

The National Tef Improvement Program at Debre Zeit Agricultural Research Centre of the Ethiopian Institute of Agricultural Research began using induced mutation techniques in the early 1970s, specifically in 1972 with technical support from the IAEA. The method was then considered as the only option for creating variability in tef, as it was believed that tef flowers were entirely cleistogamous and not amenable to cross-breeding. This belief was later proven to be wrong after it was discovered that the cleistogamous nature of tef florets is incomplete and that they open early in the morning. This discovery eventually enabled the development of the artificial tef hybridization technique by Tareke Berhe in 1974 (Berhe, 1975). In the early efforts, through the technical support of the IAEA, a gamma-ray irradiation facility (Gamma Cell 220) was established at Debre Zeit. In addition, there is an ongoing project on induced mutation at the centre. In all the efforts, the major traits of interest were improvement of lodging tolerance, resistance to leaf rust disease (*Uromyces eragrostidis* Tracy) and resistance to shattering. In spite of repeated attempts made over the years, however, no tef varieties have been released through the conventional induced mutation techniques employed.

The Tef Improvement Project at the University of Bern in Switzerland was established a decade ago to boost the productivity of tef by tackling major production constraints through developing cultivars with desirable agronomic and nutritional traits. This project focuses on problem-oriented and demand-driven research. Priority has been given to developing cultivars with resistance to lodging and drought tolerance, since these two constraints contribute towards significant yield losses in tef (Assefa *et al.*, 2011).

The strategy and pipeline to develop new cultivars with valuable agronomic traits have been presented previously (Cannarozzi *et al.*, 2018). The three phases of the project are technology generation, technology transfer and technology delivery. Under the *Technology Generation* phase, modern genetic, molecular and genomic tools are applied to obtain candidate tef lines for traits of interest. In the *Technology*

Transfer phase, promising tef lines harbouring traits of choice are sent to the Ethiopian Institute of Agricultural Research, where they are introgressed into high-yielding and widely adapted cultivars and evaluated for several generations at the on-station and on-farm sites across Ethiopia before release to the farming community. In the *Technology Delivery* phase, seeds of newly released varieties are multiplied and disseminated through private and public institutions.

There are over 5000 tef landraces collected and conserved at the Ethiopian Biodiversity Institute. Although the entire germplasm has not yet been thoroughly evaluated, huge diversity was reported and some of these accessions were very distinct as assessed by morphological and agronomic traits (Chanyalew *et al.*, 2013; Plaza-Wüthrich *et al.*, 2013; Assefa *et al.*, 2015; Jifar *et al.*, 2015). However, this substantial phenotypic diversity had not enabled us to identify lodging-tolerant tef lines.

As a result, the Tef Improvement Project embarked on a large-scale mutation breeding project.

2.1 Mutation induction in tef

Mutation induction, which refers to the creation of genetic diversity, can be investigated based on at least four aspects: (i) source of mutation (natural or induced); (ii) type of mutagen (physical or chemical); (iii) patterns of DNA cleavage intensity of mutation (point mutation, indels or rearrangements) and spectrum of mutation (nonsense, missense, silent or splice junction); and (iv) precision of the mutation (random or targeted) (Tadele, 2016). In order to establish mutagenized populations, seeds of four improved tef varieties, namely, 'Tsedey' (DZ-Cr-37), 'Dukem' (DZ-01-974), 'Kora' (DZ-Cr-438 RIL-133B) and 'Dagim' (DZ-Cr-438 RIL91A), were used for mutagenesis. A widely applied chemical mutagen, ethyl methanesulfonate (EMS), was used to treat tef seeds (Sikora *et al.*, 2011; Mba, 2013). EMS mainly creates point mutations (G to A transition) in which a single nucleotide is altered. Before embarking on large-scale mutagenesis, the optimum level of EMS was determined (Fig. 14.1). Based on the germination and growth of treated plants, treatment with 0.2%

Fig. 14.1. Effect of different concentrations of ethyl methanesulfonate (EMS) on germination and seedlings of two tef varieties ('Tsedey' and 'Dukem'). Seeds were treated for 8 h each with four concentrations of EMS shown as %v/v at the top of the figure.

(v/v) EMS for 8 h was selected for large-scale mutagenesis of tef seeds.

Approximately 20,000 seeds from the four elite cultivars treated with EMS as above were grown as individual M_1 plants to set seeds. About 12,000 M_2 families derived from these M_1 plants have been used in the screening for diverse trait(s) of interest (Table 14.2) (Fig. 14.2).

2.2 Mutation detection in tef

Mutation detection refers to the discovery of either the altered gene or the mutant line. In forward genetics, the genes altered in the candidate lines are elucidated, whereas in reverse genetics, mutant lines with defects in the genes of interest are identified. We applied both approaches for the discovery of either the gene or the mutant line.

In the forward genetics approach, the mutagenized tef population was used for phenotypic screening to obtain candidate mutant lines for the traits of interest. Moreover, the same populations were used in reverse genetics approaches: TILLING (Targeting Induced Local Lesions in Genomes) to screen for mutant lines that harbour DNA lesions in genes of interest. TILLING is a non-transgenic method and it has been applied to several crops, including major and orphan crops (Tadele *et al.*, 2010; Esfeld *et al.*, 2013). Some benefits of TILLING are as follows.

1. It produces a spectrum of allelic mutations that are useful for genetic analysis.
2. It reveals mutations that are difficult to identify by forward genetics, since TILLING can focus on a particular gene of interest.
3. It applies to any organism regardless of genome size and ploidy level.
4. It produces stable mutations.

Table 14.2. The number of mutagenized tef populations from diverse background genotypes and their current status.

Background genotype				
Common name	Variety name	No. of M_2 populations	Traits screened for	Current status
Tsedey	DZ-Cr-37	5000	Diverse agronomic traits	Different stages
Dukem	DZ-01-974	2000	Drought tolerance	Inbred line testing at multi-locations
Kora	DZ-Cr-438 RIL133B	2500	Diverse agronomic traits	Different stages
Dagim	DZ-Cr-438 RIL91A	2400	To be determined	Planning phase

Fig. 14.2. Partial view of large-scale cultivation of mutagenized tef plants in the greenhouse. **(A)** M_1 individual lines. **(B)** M_2 populations derived from individual M_1 lines.

5. Since no exogenous DNA is introduced into the plant by the chemical mutagenesis, the product is considered non-transgenic and is exempted from regulatory procedures imposed on transgenic products (Tadele *et al.*, 2010; Alonso and Ecker, 2006).

The technique of TILLING comprises the following main steps: (i) mutagenesis; (ii) development of a non-chimeric population; (iii) preparation of a germplasm stock; (iv) DNA extraction and sample pooling; (v) screening M_2 families to detect mutations in the desired gene; and (vi) identification of mutant lines and sequencing of target gene (Till *et al.*, 2006). As indicated earlier (Tadele *et al.*, 2010; Alonso and Ecker, 2006), TILLING has been applied in M_2 mutagenized tef families. Focus has been given to identifying mutant lines altered in plant height and starch type or content. The TILLING project has benefited immensely from the Tef Genome Sequencing Project, especially in designing unique primers for the allotetraploid tef with two very close genomes.

Phenotypic screenings for the traits of interest were done mainly at the University of Bern. A partial list of promising candidate mutant lines obtained from diverse screenings is shown in Table 14.3. Emphasis was given to screening for semi-dwarf lines, since semi-dwarfism during the Green Revolution in wheat and rice has been associated with lodging tolerance (Peng *et al.*, 1999; Spielmeyer *et al.*, 2002; Hedden, 2003).

Kegne was the first semi-dwarf candidate line obtained through phenotypic screening of over 5000 mutagenized M_2 families. In addition to semi-dwarfism and lodging tolerance, *kegne* plants have unique properties. Among these, one is the twisting of the leaves and stem of the plant

Table 14.3. Selected candidate lines from EMS mutagenized tef populations screened for traits of interest.

Candidate line	Background genotype	Desirable trait	Screening institution	References
Kegne	*Tsedey*	Semi-dwarf	University of Bern	Jost *et al.* (2015)
Kinde	*Tsedey*	Semi-dwarf	University of Bern	Cannarozzi *et al.* (2018)
dtt2	*Tsedey*	Early drought tolerance	University of Bern	Schneider (2011)
dtt13	*Tsedey*	Early drought tolerance	University of Bern	Schneider (2011)
tdt4-5	*Dukem*	Terminal drought tolerance	University of Bern	Rindisbacher (2015)
tdt4-15	*Dukem*	Terminal drought tolerance	University of Bern	Rindisbacher (2015)
tdt4-19	*Dukem*	Terminal drought tolerance	University of Bern	Rindisbacher (2015)
ml-209	*Tsedey*	Acid soil tolerance	Kwazu Natal – ARARI	Desta *et al.* (2017)
ml-153	*Tsedey*	Acid soil tolerance	Kwazu Natal – ARARI	Desta *et al.* (2017)
meten 1	*Kora*	Semi-dwarf	University of Bern	This work
meten 2	*Kora*	Semi-dwarf	University of Bern	This work
meten 3	*Kora*	Semi-dwarf	University of Bern	This work
Several candidates	*Tsedey*	Starch altered	ETH Zurich	W. Wang (personal communication)
sde (semi-dwarf & early)	*Kora*	Semi-dwarf and early maturing	University of Bern	This work
Several	*Kora*	Salinity tolerance	University of Bern	This work
Several	*Kora*	Acid soil tolerance	University of Bern	This work

to the right side (Jost *et al.*, 2015). This twisting in *kegne* plants (about 12°) is similar to *twisted dwarf 1* (*tid1*) in rice (Sunohara *et al.*, 2009) and *spr1* or *spr2* in Arabidopsis (Furutani *et al.*, 2000). The candidate gene approach allowed us to identify the microtubule-associated gene affected in the *kegne* mutant and also to design the Cleaved Amplified Polymorphic Sequences (CAPS) marker to trace the mutant after crossing to other tef genotypes (Jost *et al.*, 2015).

The other promising semi-dwarf line obtained from phenotypic screening was *kinde* (Fig. 14.3). This line possesses a number of desirable properties, including higher tillering number and increased lodging tolerance.

Screening for drought tolerance targeted both early drought, which occurs during the onset of the growing season, and terminal drought, which occurs during the flowering or maturity period of tef plants. Mutations in two candidate genes for early drought tolerance named *drought tolerant tef* (*dtt2* and *dtt13*) tolerate about 3 weeks of moisture scarcity, while the original tef lines could not withstand this level of drought (Schneider, 2011). Both *dtt* lines have fewer and smaller stomata compared with the original line and these might limit the loss of water through transpiration.

Screening for terminal drought-tolerant lines also resulted in three candidate lines: *tdt4-5*, *tdt4-15* and *tdt4-19* (Rindisbacher, 2015).

3 Mutation Breeding

Candidate tef lines from diverse screening programmes were sent to the National Tef Research Program in Ethiopia where they were hybridized to locally adapted and high-yielding varieties. Due to their desirable traits, candidate mutant lines were used as parental lines in over 50 crosses. A partial list of the crossings and current status is shown in Table 14.4. Each cross generated about 500 F_2 lines from which selection has been made for the trait of interest. While some of these crosses are at an early breeding stage, others are at an advanced stage where they are undergoing multi-location testing at representative sites across the country (Table 14.5). In general, these breeding programmes correspond to 16 experiments under early and late maturity groups. While germplasm in the early-maturity groups is tested at semi-arid locations with moisture deficit, those with late maturity are evaluated in areas with adequate precipitation.

After years of multi-location testing, the new variety called ‘Tesfa’ was approved for release by the National Variety Release Committee in Ethiopia. The variety ‘Tesfa’ has the following desirable traits: compact panicle, tolerance to lodging, non-shattering and thick culm making it suitable under irrigated conditions (Kebede *et al.*, 2018) (Fig. 14.4). Currently, this newly released ‘Tesfa’ is being

Fig. 14.3. The semi-dwarf and lodging tolerant *kinde* line (left) obtained from screening a mutagenized population in the background of ‘Tsedey’ cultivar (right). Bar = 10 cm.

Table 14.4. A selection of crosses made to candidate mutant tef lines developed by the Tef Improvement Project and their current status.

Mutant line		Crossed to		
Name	Desirable trait	Name	Desirable trait	Current status[a]
Kegne	Lodging-tolerant	*Key Murri*	High culm strength	VVT
		Magna	White-seeded variety	NVT
		Quncho	Popular variety	NVT
		Tsedey	Drought tolerant	F_5 seeds
Kinde	Lodging-tolerant	*Alba*	Early maturing	VVT
		Boset	Drought tolerant	F_4 seeds
		Dukem	High yielding	F_4 seeds
		Key Murri	High culm strength	NVT
		Kora	High yielding	F_4 seeds
		Magna	White-seeded	F_4 seeds
		Quncho	Popular variety	NVT
		Tsedey	Drought tolerant	F_5 seeds
dtt2	Drought-tolerant	*dtt13*	Drought tolerant	OBN
		Kegne	Semi-dwarf	F_5 seeds
		Key Murri	High culm strength	PVT
		Magna	White-seeded	F_4 seeds
		Quncho	Popular variety	OBN
dtt13	Drought-tolerant	*Boset*	Drought tolerant	F_4 seeds
		dtt2	Drought tolerant	OBN
		Key Murri	High culm strength	F_4 seeds
		Magna	White-seeded	F_4 seeds
tdt4-15	Terminal drought tolerance	*Dukem*	High yielding	F_3 seeds
		Kora	High yielding	F_3 seeds
		Magna	White-seeded	F_4 seeds
tdt4-19	Terminal drought tolerance	*Boset*	Drought tolerant	F_3 seeds
		Dukem	High yielding	F_3 seeds
		Kora	High yielding	F_3 seeds
		RIL44	Semi-dwarf	F_3 seeds

[a]NVT, National Variety Trial; OBN, Observation Nursery; PVT, Preliminary Variety Trial; VVT, Variety Verification Trial

outscaled in four districts in central Ethiopia together with other improved technologies (Bekele *et al.*, 2017).

4 Discussion

The goal of this manuscript is to show the highlights of our initiative and the progress made in mutation breeding on tef.

5 Conclusion

Mutation breeding is the cornerstone of the Tef Improvement Project based in Bern, Switzerland, which closely collaborates with the National Tef Improvement Program in Ethiopia. Since the project focuses on problem-oriented research by tackling major constraints affecting tef productivity, emphasis has been given to developing lodging and drought-tolerant cultivars. Candidate mutant lines identified in Bern were sent to Ethiopia, where they were incorporated into the national breeding programme. 'Tesfa', the first tef variety released from the mutation background, has received high acceptance by smallholder farmers. In general, our value-chain approach which starts from the basic research of identifying candidate mutant lines to the breeding and dissemination of improved technology relied on mutagenized populations.

Table 14.5. Locations in Ethiopia where germplasm from the Tef Improvement Project have been evaluated after crossing to improved tef varieties.

Centre/site	Geographical coordinates	Distance and direction from Addis Ababa	Altitude (m asl)	Climate
EIAR (Ethiopian Institute of Agricultural Research)				
Adadi Mariam	8°37′N, 38°30′E	55 km S	1900	Sub-humid
Akaki	8°53′N, 38°47′E	10 km S	2300	Cool-wet
Alem Tena	8°18′N, 38°57′E	110 km S	1650	Semi-arid
Ambo	8°59′N, 37°51′E	115 km W	2185	Temperate
Asosa	10°04′N, 34°31′E	655 km W	1590	Warm to sub-humid
Chefe Donsa	7°32′N, 40°38′E	40 km SE	2400	Cool-wet
Debre Zeit-black soil	8°44′N, 39°00′E	45 km S	1800	Temperate
Debre Zeit-light soil	8°44′N, 39°00′E	45 km S	1800	Temperate
Dhera	7°44′N, 39°29′E	122 km SE	1680	Semi-arid
Ginchi	8°54′N, 38°09′E	90 km W	2200	Tepid-moist
Holetta	9°03′N, 38°30′E	35 km W	2390	Cool-wet
Jimma	7°40′N, 36°50′E	365 km SW	1760	Sub-humid
Melkassa	8°24′N, 39°21′E	115 km SE	1550	Semi-arid
Minjar	9°09′N, 39°19′E	110 km SE	1800	Semi-arid to moist
Pawe	11°19′N, 36°19′E	575 km NW	1000	Sub-humid
Werer	9°16′N, 40°09′E	280 km NE	750	Warm-arid
Wolenchiti	8°40′N, 39°26′E	120 km SE	1400	Semi-arid
ARARI (Amhara Regional Agricultural Research Institute)				
Adet	11°16′N, 37°29′E	445 km NW	2240	Moist-cool
Bichena	10°27′N, 38°12′E	270 km NW	2500	Cool-wet
Kobo	12°09′N 39°38′E	590 km N	1470	Warm-moist
Koga	11°24′N, 37°09′E	665 km NW	1980	Tepid-moist
Metema	12°58′N, 36°12′E	900 km NW	690	Sub-humid
Shewa Robit	10°00′N, 39°54′E	210 km N	1280	Semi-arid
Simada	11°29′N, 38°14′E	770 km NW	2470	Semi-arid
Sirinka	11°45′N, 39°36′E	510 km N	1850	Semi-arid
Oromia (Oromia Agricultural Research Institute)				
Bako	9°06′N, 37°09′E	250 km W	1590	Sub-humid
Shambu	9°40′N, 36°59′E	305 km W	2500	Cool-wet
TARI (Tigray Agricultural Research Institute)				
Axsum	14°16′N, 39°09′E	955 km N	2100	Semi-arid
Humera	14°00′N, 37°00E	975 km NW	600	Warm-moist lowland
Mehoni	12°39′N, 39°44′E	635 km N	2400	Arid
SARI (South Agricultural Research Institute)				
Jinka	5°46′N, 36°33′E	580 km SW	1500	Sub-humid lowland
GARI (Gambella Agricultural Research Institute)				
Abobo	7°49′N, 34°29′E	750 km W	500	Sub-humid
Universities				
Haramaya (Hirna)	9°24′N, 42°01′E	365 km E	1775	Cool-wet
Jigjiga	9°21′N, 42°48′E	600 km E	1600	Semi-arid
Wolkite	8°17′N, 37°47′E	190 km SW	1920	Sub-humid

Fig. 14.4. Mutation breeding allowed us to release the first variety with improved properties. A new tef variety (left) is compared to the standard variety (right) near Debre Zeit. (Photo: Z. Tadele, 5 October 2018.)

References

Abebe, Y. *et al.* (2007) Phytate, zinc, iron and calcium content of selected raw and prepared foods consumed in rural Sidama, Southern Ethiopia, and implications for bioavailability. *Journal of Food Composition and Analysis* 20, 161–168.

Alonso, J.M. & Ecker, J.R. (2006) Moving forward in reverse: genetic technologies to enable genome-wide phenomic screens in Arabidopsis. *Nature Reviews Genetics* 7, 524–536.

Assefa, K. *et al.* (2011) Breeding tef [*Eragrostis tef* (Zucc.) Trotter]: conventional and molecular approaches. *Plant Breeding* 130, 1–9.

Assefa, K. *et al.* (2015) Genetic diversity in tef [*Eragrostis tef* (Zucc.) Trotter]. *Frontiers in Plant Science* 26, 177.

Bekele, A. *et al.* (2017) Lead farmers' approach in disseminating improved tef production technologies. *Ethiopian Journal of Agricultural Sciences* 27, 25–36.

Berhe, T. (1975) *A Break-through in Tef-breeding Techniques*. FAO, Rome.

Bultosa, G. *et al.* (2002) Physico-chemical characterization of grain tef [*Eragrostis tef* (Zucc.) Trotter] starch. *Starch-Starke* 54, 461–468.

Cannarozzi, G. *et al.* (2018) Technology generation to dissemination: lessons learned from the tef improvement project. *Euphytica* 214, 31.

Chanyalew, S. *et al.* (2013) Phenotypic and molecular diversity in tef. In: Assefa, K., Chanyalew, S. and Tadele Z. (eds) *Achievements and Prospects of Tef Improvement*. EIAR-University of Bern, Bern, Switzerland, pp. 21–31.

CSA (2014) Agricultural Sample Survey for 2013/14. In: *Statistical Bulletin 532*. Addis Ababa, Ethiopia.

Desta, E.A. *et al.* (2017) Screening of ethyl methane sulphonate mutagenized tef [*Eragrostis tef* (Zucc.) Trotter] population identifies Al-tolerant lines. *Journal of Plant Interactions* 12, 70–176.

Esfeld, K. *et al.* (2013) Application of TILLING for orphan crop improvement. In: Jain, S.M. and Gupta, S.D. (eds) *Biotechnology of Neglected and Underutilized Crops*. Springer, Dordrecht, pp. 83–113.

Furutani, I. *et al.* (2000) The SPIRAL genes are required for directional control of cell elongation in *Arabidopsis thaliana*. *Development* 127, 4443–4453.

Hedden, P. (2003) The genes of the Green Revolution. *Trends in Genetics* 19, 5–9.

IAEA (2018) Joint FAO/IAEA Mutant Variety Database. Available at: https://mvd.iaea.org/ (accessed 9 November 2018)

Jifar, H. *et al.* (2015) Grain yield variation and association of major traits in brown-seeded genotypes of tef [*Eragrostis tef* (Zucc.) Trotter]. *Agriculture & Food Security* 4, 7.

Jost, M. *et al.* (2015) Semi-dwarfism and lodging tolerance in tef (*Eragrostis tef*) is linked to a mutation in the *alpha-Tubulin 1* gene. *Journal of Experimental Botany* 66, 933–944.

Kebede, W. *et al.* (2018) Tef (*Eragrostis tef*) variety Tesfa. *Ethiopian Journal of Agricultural Sciences* 28, 107–112.

Mba, C. (2013) Induced mutations unleash the potentials of plant genetic resources for food and agriculture. *Agronomy* 3, 200–231.

Peng, J.R. *et al.* (1999) 'Green revolution' genes encode mutant gibberellin response modulators. *Nature* 400, 256–261.

Plaza-Wüthrich, S. *et al.* (2013) Genetic and phenotypic diversity in selected genotypes of tef [*Eragrostis tef* (Zucc.) Trotter]. *African Journal of Agricultural Research* 8, 1041–1049.

Rindisbacher, A. (2015) Studying drought tolerance in tef [*Eragrostis tef* (Zucc.) Trotter]. MSc thesis, University of Bern, Institute of Plant Sciences, Switzerland.

Schneider, R. (2011) Investigation of drought tolerance in tef [*Eragrostis tef (*Zucc.) Trotter]. MSc thesis, University of Bern, Institute of Plant Sciences, Switzerland.

Sikora, P. *et al.* (2011) Mutagenesis as a tool in plant genetics, functional genomics, and breeding. *International Journal of Plant Genomics* 2011, 314829.

Spaenij-Dekking, L. *et al.* (2005) The Ethiopian cereal tef in celiac disease. *New England Journal of Medicine* 353, 1748–1749.

Spielmeyer, W. *et al.* (2002) *Semidwarf* (*sd-1*), 'green revolution' rice, contains a defective gibberellin 20-oxidase gene. *Proceedings of the National Academy of Sciences of the United States of America* 99, 9043–9048.

Sunohara, H. *et al.* (2009) A dominant mutation of TWISTED DWARF 1 encoding an alpha-tubulin protein causes severe dwarfism and right helical growth in rice. *Genes & Genetic Systems* 84, 209–218.

Tadele, Z. (2016) Mutagenesis and TILLING to dissect gene function in plants. *Current Genomics* 17, 499–508.

Tadele, Z. (2017) Raising crop productivity in Africa through intensification. *Agronomy Journal* 7, 22.

Tadele, Z. *et al.* (2010) TILLING for mutations in model plants and crops. In: Jain, S.M. and Brar, D.S. (eds) *Molecular Techniques in Crop Improvement*, 2nd edn. Springer, Dordrecht, pp. 307–332.

Till, B.J. *et al.* (2006) A protocol for TILLING and ecotilling in plants and animals. *Nature Protocols* 1, 2465–2477.

15 Improving Sustainable Cotton Production Through Enhanced Resilience to Climate Change Using Mutation Breeding

Manzoor Hussain[1], Ljupcho Jankuloski, M. Habib-ur-Rahman, Massoud Malek, Md. Kamrul Islam, M. Reza Raheemi, Jawdat Dana, Khin Myat Lwin, Fiaz Ahmad, Muhammad Rizwan, Ghulam Mohyuddin Talha, M. Asif and S. Ali

Abstract

Cotton, being a leading commercial fibre crop, is grown on 20.5 million hectares in three major cotton-producing countries: China, India and Pakistan. Wide differences in yield per hectare exist among these countries and these are being aggravated by changing climate conditions, i.e. higher temperatures and significant seasonal and regional fluctuation in rainfall. Pakistan is one of the countries most affected by climate change. The disastrous effects of extreme periods of heat stress in cotton were very prominent in Pakistan during the growing seasons 2013–2014 (40–50% fruit abortion) and 2016–2017 (33% shortfall), which posed an alarming threat to the cotton-based economy of Pakistan. Poor resilience of the most commonly grown cotton varieties against extreme periods of heat stress are considered to be major factors for this drastic downfall in cotton production in Pakistan. Using the approach of induced mutation breeding, the Nuclear Institute for Agriculture and Biology (NIAB), Faisalabad, Pakistan, has demonstrated its capabilities in developing cotton mutants that can tolerate the changed climatic conditions and sustain high yields under contrasting environments. The results of studies on the phenological and physiological traits conferring heat tolerance are presented here for thermo-tolerant cotton mutants (NIAB-878, NIAB-545, NIAB-1048, NIAB-444, NIAB-1089, NIAB-1064, NIAB-1042) relative to FH-142 and FH-Lalazar. NIAB-878 excelled in heat tolerance by maintaining the highest anther dehiscence (82%) and minimum cell injury percentage (39%) along with maximum stomatal conductance (27.7 mmol $CO_2/m^2/s$), transpiration rate (6.89 µmol $H_2O/m^2/s$), net photosynthetic rate (44.6 mmol $CO_2/m^2/s$) and physiological water use efficiency (6.81 mmol CO_2/µmol H_2O) under the prevailing high temperatures.

Keywords: sustainable • climate change • cotton productivity • phenological and physiological traits

1 Introduction

Cotton has a special significance and plays an important role in the economies of Australia, China, India, Iran, Myanmar, Pakistan, Vietnam and Bangladesh. This leading fibre crop is grown on 20.5 million hectares in the three main cotton-producing countries of Asia and its Pacific region, i.e. China, India and Pakistan, with their annual contribution of about 60–65% of total world cotton production. Bangladesh, Myanmar, Vietnam and Iran have a very nominal role in total

[1] Email: mhmanj@gmail.com

 Mutation Breeding, Genetic Diversity and Crop Adaptation to Climate Change (eds S. Sivasankar *et al.*)
DOI: 10.1079/9781789249095.0015

world cotton production. However, emerging demands from Vietnam and Bangladesh for their cotton mill use signifies the increased role of cotton production in the economy of regional countries. Huge yield gap differences exist amongst the top three cotton-producing countries of the region (China 1484 kg/ha; India 529 kg/ha; and Pakistan 700 kg/ha) and also amongst other countries of the region (Vietnam 453 kg/ha; Bangladesh 608 kg/ha; Iran 594 kg/ha; and Myanmar 653 kg/ha). These yield differences are being further aggravated due to changing climate associated with higher temperature stresses (Rahman *et al.*, 2018). Pakistan is likely to be the country that is most affected by climate change as far as agriculture and cotton production are concerned (Ahmad *et al.*, 2015). Year-to-year variation in cotton crop yields due to climatic variability is not only impacting the farming industry negatively, but is also straining the positive development of cotton-based industries in the region. The cotton belts in countries like Bangladesh, China, Iran and Pakistan are located in a high-temperature zone, where the maximum temperature often exceeds 40°C during the cotton-growing season. Increasing temperatures lead to the inhibition of plant growth, increased photorespiration and poor control of insects and pests, enhancing the requirements for inputs such as irrigation and fertilizer, which in turn lead to higher production costs. Heat stress has negative effects on crop growth development and ultimately on yield, a fact that was most exaggerated during the cotton-growing season of 2013–2014 when there was early termination of the crop with 40–50% fruit abortion. The quality of lint and cotton seed was affected in general, while cotton seed yield was particularly affected due to the high temperatures and heavy rainfalls. The sensitivity of commercial cotton varieties to periods of extreme heat stress coupled with enhanced requirements for inputs like water and fertilizer and with poor seed germination are considered to be major contributory factors. The impact of global warming on cotton production (Anonymous, 2013a,b), especially the effects of unexpected periodic episodes of extreme heat stress on cotton in China (Zhou *et al.*, 1996; Liu *et al.*, 2006), India (Anonymous, 2013b) and Pakistan, is gaining significance as an emerging challenge for researchers and cotton producers. Reduced pollen viability (Burke *et al.*, 2004), the physiological response of cotton to high temperatures (Bibi *et al.*, 2003), lower fertilization efficiency (Snider *et al.*, 2009), reduced boll size and seed number per locule (Pettigrew, 2008) and increased fruit shedding (Hodges *et al.*, 1993; Reddy *et al.*, 1992, 1999) are the main effects associated with reduction of cotton yield during periods of heat stress. Different techniques have been used previously for the screening of cotton genotypes for high temperature (Liu *et al.*, 2006), such as cellular membrane thermo-stability for heat tolerance (Rahman *et al.*, 2004), genetic diversity for stomatal conductance (Radin *et al.*,1994), multi-level determination of heat tolerance under field conditions (Cottee *et al.*, 2007, 2010) and screening of upland cotton under field conditions (Karademir *et al.*, 2012). There are challenges for the development of heat-tolerant cotton genotypes as high temperature stress affects floral development and cotton crop yield (Oosterhuis *et al.*, 2008; Oosterhuis and Snider, 2011). This effect is further exaggerated due to climate change, as there is uncertainty in climate change projections as depicted by a multi-model study for cotton production in Pakistan (Rahman *et al.*, 2018). Using the approach of induced mutation breeding, the Nuclear Institute for Agriculture and Biology (NIAB), Faisalabad, Pakistan, has demonstrated its capabilities in developing cotton mutants that can withstand heat stress. So far, NIAB has developed 14 cotton cultivars through the use of induced mutation, including the famous cultivars NIAB-78, NIAB-Krishma, NIAB-111 and, most recently, the high-yielding and fine-fibre cotton varieties NIAB-KIRAN and NIAB-878, NIAB-545 and NIAB-1048, respectively. The cotton mutants developed through the use of induced mutation showed enhanced resilience against high temperatures under field conditions and also showed significant variation in their root length (30, 60, 90, 120 and 150 days after sowing). Our results suggest that these cotton mutants have the ability to sustain their yield under the high temperature conditions projected from current climate change models.

2 Materials and Methods

Seven advanced cotton mutant lines, NIAB-878, NIAB-545, NIAB-1011/48, NIAB-444, NIAB-1011/89, NIAB-1011/64, NIAB-1011/42, along with two unrelated controls, FH-142 and FH-Lalazar, were evaluated for their response to high temperature under field conditions at NIAB using a randomized complete block design (RCBD). Data regarding morphological and phenological parameters conferring heat tolerance were recorded. The same lines were also evaluated for their heat tolerance under field conditions at the Central Cotton Research Institute, Multan. Sowing at NIAB was in the first week of May 2016, whilst sowing at the Central Cotton Research Institute in Multan was in mid-April, 2016, so that the fruiting phase of the crop would coincide with the hottest period of the season during the cotton-growing year of 2016. Data regarding other phenological traits and plant height, number of squares, number of flowers and total bolls formed (opened or unopened) were recorded on five guarded plants of each line, in four replicates, starting 75 days after planting and thereafter every 15 days up to 150 days.

Data were recorded regarding physiological parameters that potentially contribute towards heat tolerance in cotton, such as relative cell injury, electrical conductivity, anther dehiscence, pollen viability and gas exchange characteristics, including stomatal conductance, transpiration rate, net photosynthesis rate and physiological water use efficiency. Also recorded were some morphological characteristics, namely: first sympodial node number and its height; sympodial node number bearing the first effective boll; height of the first sympodial node bearing an effective boll; percentage boll set on first position along sympodia; percentage boll set on second position along sympodia; total fruiting positions per intact points (total fruiting positions); and number of locules and seeds per boll up to tenth sympodium. A porometer (Porometer LI-1600; Steady State Porometer, Germany) was also used to measure leaf transpiration rate, diffusive resistance, relative humidity and leaf temperature. Gas exchange characteristics (stomatal conductance, transpiration rate, net photosynthetic rate and physiological water use efficiency) were also computed from the data generated by the porometer for three leaves on three plants of each genotype in four replications after 75 days of planting. Data were collected during the midday period, 10.00–13.00 h, and in the absence of cloud cover (radiation levels of > 1200 μmol/m^2/s). The porometer measures stomatal diffusive resistance from which stomatal conductance g_s is calculated (in mmol CO_2/m^2/s). Further, given the transpiration rate E (in μmol H_2O/m^2/s) and the net photosynthetic rate P_N (in mmol CO_2/m^2/s), the physiological water use efficiency (P_N/E) (in mmol CO_2/μmol H_2O) can be calculated.

Also, at maturity, data were recorded on number of plants, plant height, number of bolls/m^2 and average yield per plant. An analysis of seed parameters was also performed, which showed the effect of heat stress on seed features and it may contribute towards heat tolerance. Seed protein content was measured by using the micro-Kjeldahl method as reported by AOAC (1990). Seed coat percentage, seed embryo percentage and their indices were calculated as follows.

- Seed coat (%) = (coat dry weight/seed dry weight) × 100
- Seed embryo (%) = (embryo dry weight/seed dry weight) × 100
- Seed coat index (SCI) = coat dry weight/seed dry weight
- Embryo index (EI) = embryo dry weight/seed dry weight

Agronomic practices were kept uniform throughout the experiment period.

3 Results

The measured values of various phenological attributes thought to contribute to heat tolerance are given in Tables 15.1, 15.2 and 15.3. An analysis of seed traits (seed protein, seed coat and embryo percentages) is given in Fig. 15.1. Data regarding various physiological parameters such as relative cell injury, electrical conductivity, anther dehiscence, pollen viability, gas exchange characteristics (i.e. stomatal conductance, transpiration rate, net photosynthesis rate and physiological water use efficiency) are given in Figs 15.2 and 15.3. It is clearly evident that there are different responses of these mutant cotton lines, compared with the controls, to heat stress under field conditions.

Generally, NIAB genotypes performed better in studied parameters against the standard varieties (Table 15.1). Highest fruit retention capacity up to tenth sympodium was found for NIAB-1089 (56%) while all other genotypes belonging to NIAB retained more fruit than the standard varieties (i.e. FH-142 and FH-Lalazar).

Table 15.1. Morphological, yield and fruit retention traits related to heat tolerance in cotton genotypes studied at NIAB, Faisalabad, 2016–2017.

Trait name[a]	NIAB-545	NIAB-878	NIAB-1089	NIAB-1011/64	NIAB-444	NIAB-1042	NIAB-1011/48	FH-142[a]	FH-Lalazar[a]
Node of first sympodium	13 ± 0.65	12 ± 0.41	12 ± 0.63	12 ± 0.41	11 ± 0.48	14 ± 0.41	13 ± 0.41	10 ± 0.48	11 ± 0.25
1st syp-node height (cm)	23 ± 0.82	23 ± 0.82	24 ± 0.50	25 ± 1.03	25 ± 1.32	27 ± 0.58	26 ± 0.96	20 ± 0.85	20 ± 0.63
Syp-height bearing 1st intact boll (cm)	30 ± 1.85	36 ± 1.47	29 ± 1049	37 ± 1.80	31 ± 0.48	41 ± 2.18	37 ± 1.65	37 ± 2.48	32 ± 1.85
TFP – up to 10th sympodium of the plant	35 ± 0.63	29 ± 2.75	35 ± 2.90	32 ± 2.33	47 ± 1.55	29 ± 1.85	26 ± 2.06	28 ± 1.55	27 ± 0.25
TIP – up to 10th sympodium of the plant	17 ± 1.04	13 ± 1.44	19 ± 1.89	10 ± 0.29	18 ± 1.25	11 ± 1.60	11 ± 1.03	7 ± 1.47	7 ± 0.85
SPs – up to 10th sympodium of the plant	18 ± 1.44	16 ± 1.89	15 ± 1.65	22 ± 2.75	29 ± 2.21	18 ± 1.03	14 ± 2.25	21 ± 2.22	20 ± 0.85
Fruit retention capacity (%)	47 ± 3.41	46 ± 3.74	56 ± 0.89	30 ± 3.24	38 ± 3.25	38 ± 4.32	42 ± 5.45	25 ± 5.47	25 ± 3.11
TNB – up to 10th sympodium	16 ± 3.10	16 ± 1.61	21 ± 0.87	12 ± 0.33	17 ± 3.38	12 ± 1.34	11 ± 0.98	9 ± 2.44	7 ± 0.80
TNL – up to 10th sympodium	65 ± 11.05	60 ± 8.52	79 ± 5.34	47 ± 1.66	69 ± 13.48	46 ± 6.91	46 ± 3.46	36 ± 9.69	29 ± 4.25
TNS – up to 10th sympodium	316 ± 63.94	282 ± 39.01	357 ± 23.24	225 ± 2.26	336 ± 62.24	230 ± 33.30	223 ± 21.25	189 ± 47.11	141 ± 28.85
Yield (g) up to 10th sympodium of the plant	35 ± 6.86	32 ± 4.15	40 ± 3.10	26 ± 0.70	40 ± 6.95	26 ± 3.96	24 ± 3.20	22 ± 5.64	17 ± 3.66
Plant height (cm)	160 ± 4.01	168 ± 6.38	149 ± 3.28	160 ± 5.57	147 ± 11.60	171 ± 4.24	162 ± 4.39	164 ± 6.18	168 ± 10.78
Number of bolls m^{-2}	916 ± 72.51	853 ± 69.55	891 ± 30.36	892 ± 33.59	871 ± 85.64	926 ± 65.74	948 ± 87.02	782 ± 55.17	723 ± 58.63
Yield (kg ha^{-1})	5004 ± 201	5384 ± 192	4997 ± 222	5475 ± 183	4745 ± 158	5121±184	5477±125	4465±241	3328±215

Abbreviations: Syp, Sympodium; TFP, Total fruiting points; TIP, Total intact points; SPs, Shedding points; TNB, Total number of bolls; TNL, Total number of locules; TNS, Total number of seeds; 10th sympodium – means sympodia branch at 10th node of the plant

[a]For control/standards

Table 15.2. Values of different phenological parameters studied in cotton genotypes at NIAB, Faisalabad, 2016–2017.

	Days after planting: 75				Days after planting: 90				Days after planting: 105			
Genotypes	PH	NB/P	NF/P	NS/P	PH	NB/P	NF/P	NS/P	PH	NB/P	NF/P	NS/P
NIAB-545	93 ± 1.44	14 ± 3.38	6 ± 1.38	38 ± 3.34	106 ± 2.25	32 ± 3.35	6 ± 1.27	47 ± 2.05	128 ± 3.77	56 ± 3.51	6 ± 1.07	35 ± 1.10
NIAB-878	93 ± 2.24	7 ± 1.89	4 ± 0.71	42 ± 1.47	108 ± 2.62	23 ± 1.93	6 ± 0.13	39 ± 3.05	146 ± 7.02	48 ± 4.99	7 ± 1.54	36 ± 2.63
NIAB-1089	89 ± 2.43	14 ± 2.58	5 ± 1.61	39 ± 3.37	99 ± 3.48	32 ± 3.51	6 ± 0.44	45 ± 4.33	127 ± 5.11	57 ± 3.12	6 ± 1.35	31 ± 1.09
NIAB-1011/64	92 ± 2.61	13 ± 2.87	5 ± 1.06	43 ± 6.30	102 ± 4.94	29 ± 2.63	7 ± 1.44	49 ± 5.66	129 ± 8.33	56 ± 7.47	6 ± 0.99	36 ± 4.66
NIAB-444	92 ± 2.68	12 ± 3.19	5 ± 1.18	42 ± 4.71	101 ± 5.86	26 ± 3.78	6 ± 0.34	40 ± 2.81	120 ± 11.52	50 ± 3.15	5 ± 1.54	32 ± 3.76
NIAB-1042	99 ± 4.35	9 ± 2.53	4 ± 0.44	38 ± 0.87	116 ± 7.44	28 ± 2.01	6 ± 1.01	46 ± 3.34	150 ± 7.24	48 ± 3.76	8 ± 1.03	39 ± 4.10
NIAB-1011/48	95 ± 2.66	10 ± 2.82	4 ± 0.97	38 ± 1.74	108 ± 3.38	26 ± 2.42	6 ± 1.26	41 ± 1.72	130 ± 1.83	52 ± 3.61	7 ± 0.76	29 ± 1.30
FH-142[a]	95 ± 3.49	9 ± 4.86	5 ± 1.65	43 ± 4.83	112 ± 3.02	26 ± 5.15	7 ± 0.60	43 ± 3.84	140 ± 9.31	47 ± 3.50	9 ± 1.54	30 ± 3.07
FH-Lalazar[a]	86 ± 4.80	9 ± 2.24	3 ± 0.52	40 ± 2.68	101 ± 5.91	21 ± 3.31	6 ± 1.14	45 ± 1.51	128 ± 8.19	40 ± 4.00	5 ± 1.51	34 ± 4.17
	Days after planting (120)				Days after planting (135)				Days after planting (150)			
Genotypes	PH	NB/P	NF/P	NS/P	PH	NB/P	NF/P	NS/P	PH	NB/P	NF/P	NS/P
NIAB-545	149 ± 2.87	63 ± 3.14	3 ± 0.32	16 ± 3.29	152 ± 4.01	83 ± 2.69	1 ± 0.25	3 ± 0.29	160 ± 4.01	88 ± 7.56	1 ± 0.50	10 ± 5.33
NIAB-878	161 ± 8.11	56 ± 1.97	2 ± 0.43	13 ± 1.31	162 ± 6.42	70 ± 4.39	1 ± 0.25	3 ± 0.75	168 ± 6.38	74 ± 8.61	1 ± 0.29	7 ± 2.04
NIAB-1089	139 ± 5.36	70 ± 3.57	1 ± 0.24	14 ± 2.22	145 ± 4.10	86 ± 2.48	2 ± 0.85	5 ± 2.35	149 ± 3.28	89 ± 7.24	0 ± 0.25	7 ± 3.82
NIAB-1011/64	149 ± 6.39	64 ± 9.47	3 ± 0.52	18 ± 2.74	155 ± 6.06	86 ± 11.64	2 ± 0.85	5 ± 2.52	160 ± 5.57	92 ± 8.82	1 ± 0.29	6 ± 1.47
NIAB-444	141 ± 12.83	57 ± 4.48	3 ± 0.39	20 ± 3.55	145 ± 12.89	72 ± 4.78	2 ± 0.58	4 ± 0.85	147 ± 11.60	78 ± 8.06	1 ± 0.29	7 ± 3.47
NIAB-1042	162 ± 5.32	65 ± 2.63	2 ± 0.24	15 ± 1.42	162 ± 3.90	79 ± 2.68	1 ± 0.25	2 ± 0.63	171 ± 4.24	85 ± 3.45	1 ± 0.50	10 ± 3.18
NIAB-1011/48	150 ± 6.57	50 ± 1.41	2 ± 0.67	15 ± 1.64	157 ± 4.48	73 ± 9.71	1 ± 0.48	4 ± 0.82	162 ± 4.39	74 ± 5.71	1 ± 0.48	7 ± 1.41
FH-142[a]	153 ± 7.59	57 ± 2.39	3 ± 0.33	16 ± 3.51	156 ± 5.79	75 ± 6.57	1 ± 0.29	3 ± 0.65	164 ± 6.18	84 ± 9.59	1 ± 0.75	10 ± 2.75
FH-Lalazar[a]	151 ± 8.45	50 ± 4.56	4 ± 0.24	17 ± 2.62	152 ± 8.38	64 ± 3.11	0 ± 0.25	3 ± 1.11	168 ± 10.78	65 ± 2.02	1 ± 0.48	10 ± 5.88

Abbreviations: PH, Plant height; NB/P, Number of bolls/plant; NF/P, Number of flowers/plant; NS/P, Number of squares/plant

[a]For control/standards

Table 15.3. Phenological data on number of unopened and opened bolls per plant with a 15-day interval at NIAB, Faisalabad, 2016–2017.

Genotypes	Days after planting: 105		Days after planting: 120		Days after planting: 135		Days after planting: 150	
	No. of unopened bolls/plant	No. of opened bolls/plant	No. of unopened bolls/plant	No. of opened bolls/plant	No. of unopened bolls/plant	No. of opened bolls/plant	No. of unopened bolls/plant	No. of opened bolls/plant
NIAB-545	43 ± 1.85	12 ± 2.58	35 ± 2.84	29 ± 1.65	20 ± 2.52	63 ± 3.04	7 ± 1.89	81 ± 6.22
NIAB-878	41 ± 4.39	6 ± 1.49	37 ± 0.65	20 ± 1.93	18 ± 2.48	53 ± 2.40	6 ± 0.65	69 ± 8.57
NIAB-1089	46 ± 3.07	12 ± 0.88	37 ± 4.51	33 ± 2.53	24 ± 3.59	63 ± 4.09	11 ± 2.29	78 ± 7.82
NIAB-1011/64	46 ± 4.80	10 ± 3.01	39 ± 9.84	25 ± 3.35	27 ± 2.95	58 ± 8.63	10 ± 2.69	82 ± 6.36
NIAB-444	38 ± 2.06	12 ± 1.76	31 ± 7.72	27 ± 4.63	23 ± 2.53	49 ± 3.84	7 ± 0.75	71 ± 7.92
NIAB-1042	42 ± 4.18	6 ± 0.76	43 ± 1.65	22 ± 2.56	17 ± 0.96	63 ± 3.28	5 ± 1.75	80 ± 3.07
NIAB-1011/48	42 ± 2.97	10 ± 2.87	29 ± 2.86	22 ± 3.12	25 ± 6.89	49 ± 3.35	9 ± 2.42	65 ± 4.41
FH-142[a]	36 ± 0.85	11 ± 4.15	37 ± 6.96	21 ± 6.05	21 ± 5.34	56 ± 2.87	7 ± 2.35	75 ± 8.59
FH-Lalazar[a]	30 ± 2.61	10 ± 1.92	33 ± 4.97	17 ± 2.50	21 ± 2.84	43 ± 1.49	5 ± 0.50	60 ± 1.58

[a]For control/standards

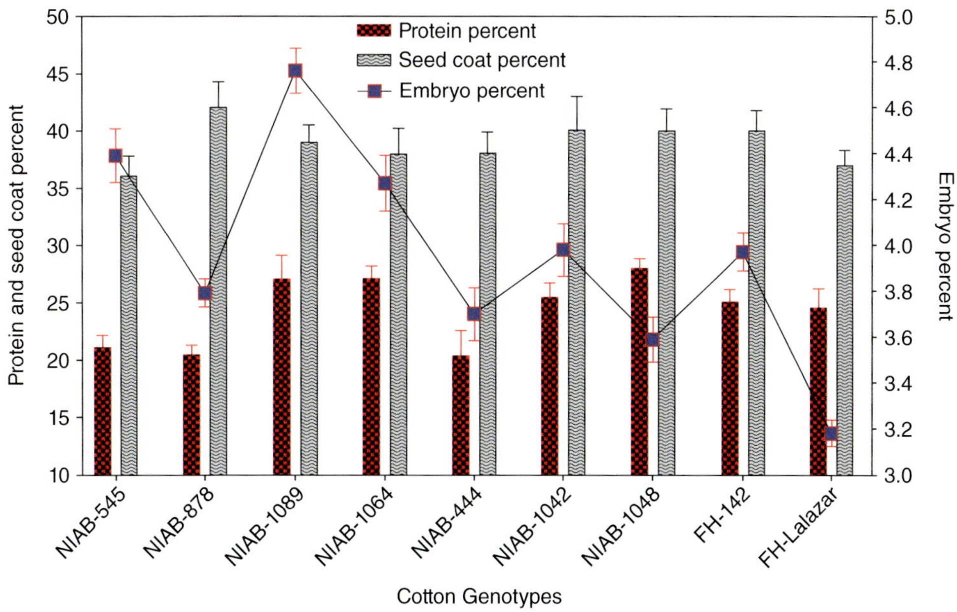

Fig. 15.1. Seed trait analysis for determining heat tolerance in cotton genotypes (seed cotton collected up to 10th sympodial of the plant). The experiment was conducted at NIAB, Faisalabad, during growing year of 2016–2017. These mean values were derived from four replicates and values of standard errors are mentioned against each parameter. Experiments were conducted under field conditions and the climate of the study was arid to semi-arid.

Similarly, results of all other studied parameters revealed that NIAB-1089 produced the highest number of bolls (21), locules (79) and seeds (357) as compared with other genotypes. The same trend was followed by NIAB-444, NIAB-545 while lower numbers of bolls (9 and 7), locules (36 and 29) and seeds (189 and 141) were recorded for the standard varieties FH-142 and FH-Lalazar, respectively. A higher yield of 45–57% was recorded for genotypes belonging to NIAB (NIAB-1089, NIAB-545 and NIAB-878) as compared with standard varieties (FH-142 and FH-Lalazar).

Data recorded at different days after planting (DAP) with an interval of 15 days for plant phenological traits are presented in Table 15.2. NIAB-1042 showed the highest plant height (171 cm) against the minimum plant height of 147 cm for NIAB-444. Plant height of genotypes ranged from 149 to 168 cm. The maximum number of bolls (92) was produced by NIAB-1011/64 at 150 DAP, followed by NIAB-1089 (89) and NIAB-545 (88) against FH-Lalazar and FH-142 (65 and 84). Data on flowers per plant were consistent for all genotypes except FH-Lalazar, which showed three flowers at interval of 75 DAP. Flowering ceased in FH-Lalazar at 135 DAP. Data on number of squares per plant of studied genotypes is presented in Table 15.2, which shows the difference among genotypes. An almost similar trend was found as discussed above. There was an increase in number of squares for all genotypes between 75 DAP and 90 DAP. After this interval the number of squares started to decline until 135 DAP. After this interval all genotypes again showed an increase up to 150 DAP which ranged from 6 to 10 squares per plant. NIAB-1011/64 surpassed all genotypes with 82 opened bolls per plant against FH-Lalazar (60) and FH-142 (75) after 150 DAP, also as given in Table 15.3. From the given data in Table 15.3 it is evident that NIAB-1089 showed the largest number of unopened bolls (11) followed by NIAB-1011/64 (10) against FH-Lalazar (5) and FH-142 (7) after 150 DAP. From the yield results data as given in Table 15.1 it is evident that NIAB-1011/48 surpassed all the genotypes in yielding highest (5477 kg/ha) followed by NIAB

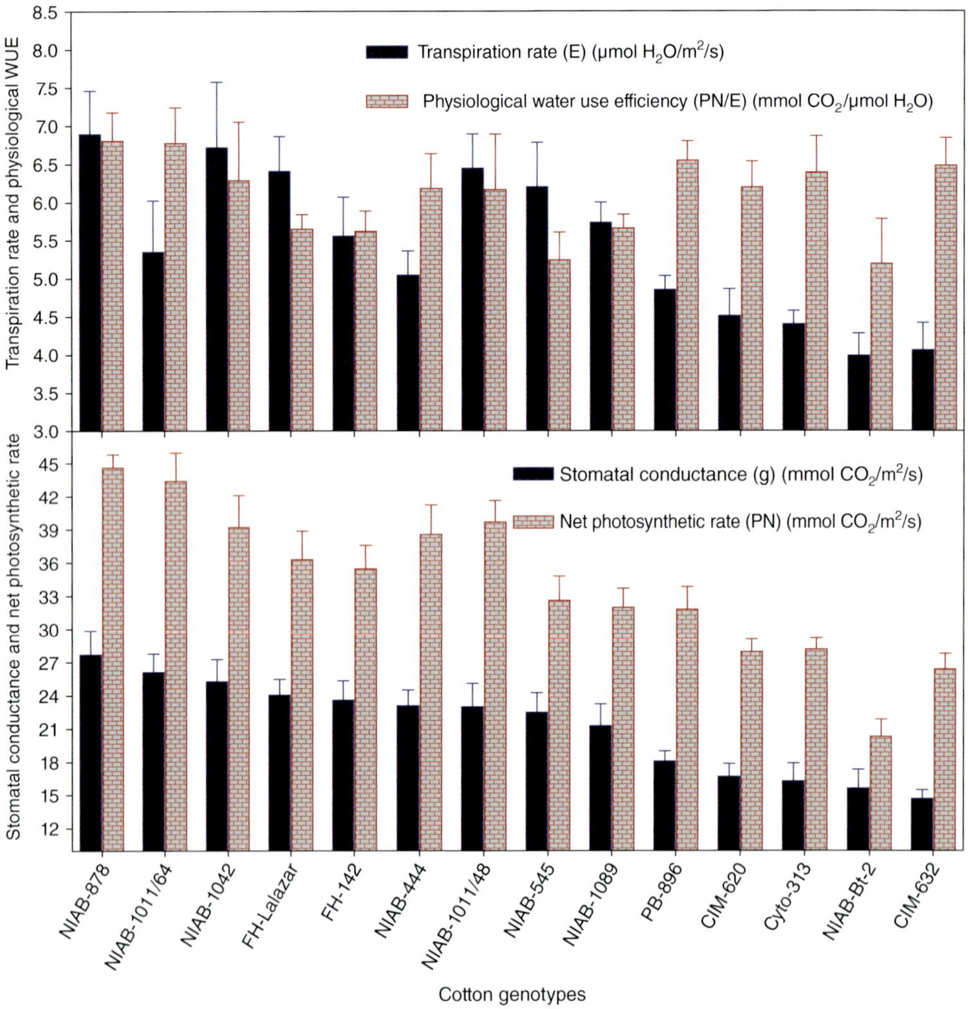

Fig. 15.2. Physiological characteristics and attributes of cotton genotypes related to heat tolerance for the field experiments conducted during cotton growing year of 2016–2017 under arid climatic conditions of CCRI, Multan. These mean values were derived from four replicates and values of standard errors are mentioned against each parameter. Ten randomly selected plants from each plot were assessed and examined for said parameters.

1011/64 (5475 kg/ha) and NIAB-878 (5384 kg/ha) against the yields of FH-142 (4465 kg/ha) and FH-Lalazar (3328 kg/ha).

From the data values shown in Fig. 15.1, it is evident that the largest seed coat percentage (42%) was shown by NIAB-878, followed by NIAB-1011/48 (40%) against the lowest seed coat percentage value of FH-Lalazar (37%). For embryo percentage, NIAB-1089 showed largest portion of embryo (4.76%) followed by NIAB-545 (4.39%) and NIAB-1011/64 (4.27%) as compared with standard FH-Lalazar, which showed 3.18% embryo. Regarding seed protein percentage, NIAB-1011/48 had the highest content (28%) followed by NIAB-1011/64 (27.13%) and NIAB-1089 (27.04%) against FH-Lalazar (24.59%).

Data regarding various physiological parameters of cotton genotypes studied under open field conditions are given in Figs 15.2 and 15.3.

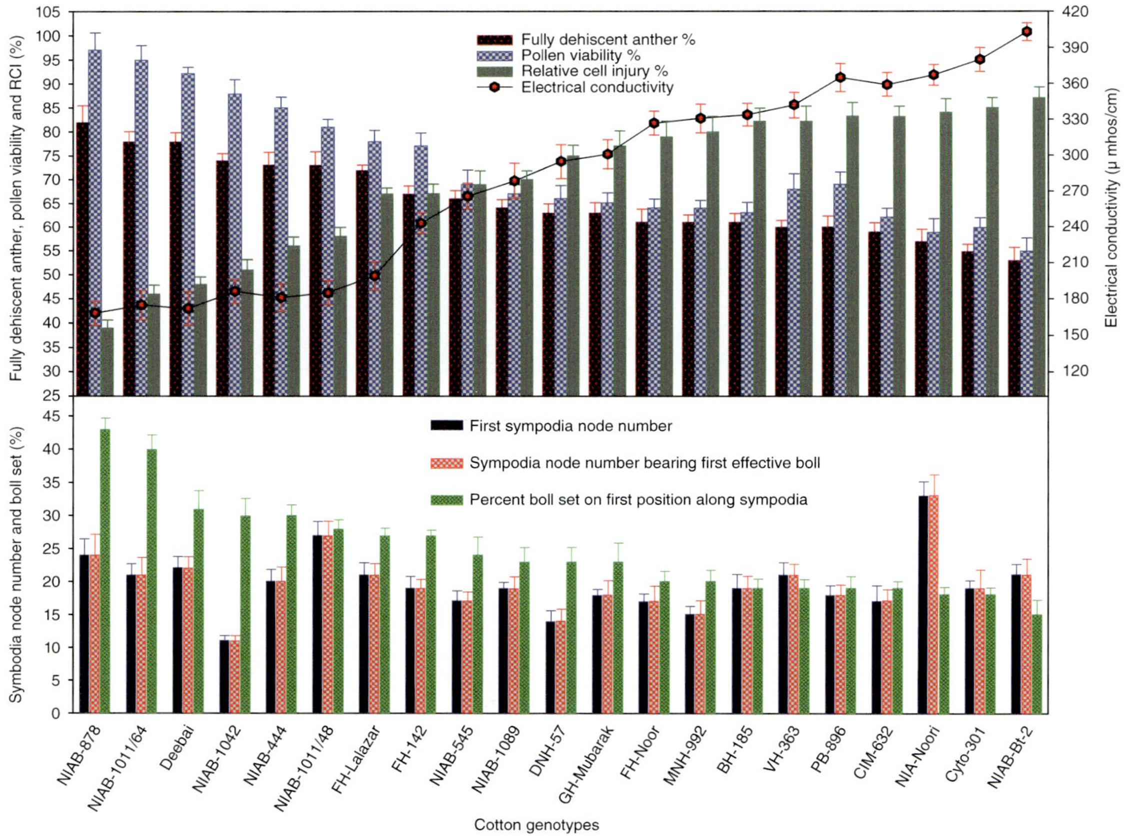

Fig. 15.3. Measuring of various physiological and morphological traits of cotton genotypes related to heat tolerance under field experiments conducted during cotton growing year of 2016–2017 under arid climatic conditions of CCRI, Multan. These mean values were derived from four replicates and values of standard errors are mentioned against each parameter. Ten randomly selected plants from each plot were assessed and examined for said parameters.

Data given in Fig. 15.3 illustrate that NIAB-878 maintained the highest anther dehiscence (82%), pollen viability (97%) and lowest cell injury percentage (39%) as compared with other genotypes and standard varieties. Lowest anther dehiscence (53%), pollen viability (55%) and maximum cell injury percentage (87%) were shown by NIAB-Bt-2, which was found to be the genotype most susceptible to heat stress. Data regarding gas exchange characteristics such as stomatal conductance (*g*), transpiration rate (*E*) and net photosynthetic rate (*PN*) varied among the genotypes as given in Fig. 15.2. The stomatal conductance (*g*) varied from 21.3 to 27.7 mmol $CO_2/m^2/s$), transpiration rate (*E*) from 5.04 to 6.89 µmol $H_2O/m^2/s$ and net photosynthetic rate (*PN*) from 32.0 to 44.6 mmol $CO_2/m^2/s$. The physiological water use efficiency (*PN*/*E*) varied from 5.25 to 6.81 mmol CO_2/µmol H_2O in different genotypes. Amongst the evaluated genotypes, NIAB-878 maintained the highest values of net photosynthetic rate and physiological water use efficiency under the prevailing high temperature conditions. NIAB-878 also performed better than the other genotypes by manifesting the maximum value of stomatal conductance (27.7 mmol $CO_2/m^2/s$), transpiration rate (6.89 µmol $H_2O/m^2/s$), net photosynthetic rate (44.6 mmol $CO_2/m^2/s$) and physiological water use efficiency (6.81 mmol CO_2/µmol H_2O) under prevailing high temperature conditions. NIAB-Bt-2 being the most susceptible genotype to heat tolerance showed minimum stomatal conductance (15.6 mmol $CO_2/m^2/s$), transpiration rate (3.99 µmol $H_2O/m^2/s$), net photosynthetic rate (20.3 mmol $CO_2/m^2/s$) and physiological water use efficiency (5.20 mmol CO_2/µmol H_2O) among all studied genotypes.

4 Discussion

These studies were conducted to identify thermotolerant and sensitive cotton genotypes, under the prevailing high temperatures during the months of June, July and August under field conditions in Pakistan using various morphological and physiological attributes of cotton genotypes. Amongst seven advanced cotton genotypes, namely, NIAB-878, NIAB-545, NIAB-1011/48, NIAB-444, NIAB-1011/89, NIAB-1011/64, NIAB-1011/42 (F_8 generation) and two controls, namely, FH-142 and FH-Lalazar, evaluated for their responses to high temperature under controlled conditions, it was revealed that the greatest fruit retention capacity, highest number of bolls/locules and number of seeds, and maximum seed cotton yield per plant up to tenth sympodium of the plant were shown by NIAB-1011/89 compared with standard sensitive cotton varieties (i.e. FH-142 and FH-Lalazar). The standard cotton varieties had lower values of fruit retention, lowest number of bolls/locules and number of seeds, and yield per plant up to tenth sympodium of the plant with ultimate loss/shedding of bolls showing their inability to survive under the prevailing high temperature. Fruit retention ability up to tenth sympodium of the genotypes studied was used as a marker for developing heat-tolerant cotton mutants, because the reproductive organs of the plant are strongly affected during the peak of high temperature (July–August) up to this part of the plant. In the months of July–August, day (40°C) and night temperatures (27°C) often exceeded the optimal limit. The seeds of NIAB-878 had the largest proportion by weight of seed coat, followed by NIAB-1011/48, while the lowest seed coat percentage value was for FH-Lalazar. The data on seed-coat characteristic variation suggested that this might be exploited in cotton to aid improvement of its environmental protectant characteristics, particularly against high temperatures, protecting the development of the cotyledons within the developing bolls of the cotton plant. Embryo percentage studies of genotypes showed that higher embryo percentage (largest embryo per unit seed weight) was shown by NIAB-1011/89, followed by NIAB-545 in comparison with standard FH-Lalazar.

High values of anther dehiscence and pollen viability coupled with low values for cell injury percentage reflect the resilience of cotton plants against high temperatures. NIAB-878 excelled in heat tolerance by maintaining the highest values for anther dehiscence, pollen viability and minimum cell injury percentage as compared with other genotypes and standards, i.e. FH-142 and FH-Lalazar. NIAB-878 also surpassed all the genotypes and standards by manifesting maximum stomatal conductance, transpiration rate, net photosynthetic rate and physiological water use efficiency under prevailing high-temperature conditions, indicating its ability to survive under water-deficit conditions under high-temperature stress. Based on the overall performance of cotton genotypes in these trials on heat tolerance under controlled conditions, NIAB-878, NIAB-1011/89 and NIAB-1011/48 were found to be tolerant and better able to withstand heat stress as compared with the susceptible genotypes, FH-142 and FH-Lalazar. Further, based on the results of cotton genotypes screened for physiological, gas exchange characteristics, phenological traits and seed-related parameters under field conditions, the advanced line NIAB-878 was found to be more resilient to high temperatures compared with the standards. This material showing high temperature stress tolerance is being evaluated in independent government varietal evaluations, where it is reported as proving of extreme worth in enabling successful cultivation of cotton in heat-prone cotton-growing areas of Pakistan, with maximum return to end users.

5 Conclusion

Studies were conducted during the cotton-growing season of 2016–2017 under field environmental conditions to assess the effects of high temperature and response of cotton genotypes, to identify the heat-tolerant and sensitive cotton genotypes using data of various morphological and physiological traits. Amongst evaluated cotton mutants, NIAB-1011/89 revealed the highest fruit retention capacity, highest number of bolls/locules and number of seeds, and maximum seed cotton yield per plant up to tenth sympodium of the plant against the standard sensitive cotton varieties (FH-142 and FH-Lalazar). Since standard cotton varieties produced lower yield, boll, poor fruit retention capacity and generally showed poor response in all studied parameters, they may be inefficient under prevailing conditions of high temperature. The greatest seed coat percentage was found in NIAB-878, followed by NIAB-1011/48, with the smallest seed coat percentage value in FH-Lalazar. The studied parameters are consistent with the fact that seed coat in cotton acts as an environment protectant against high temperatures, particularly during the development of cotyledons at cotton boll development stage. The developed vigorous cotyledons ultimately lead to higher weight of seed cotton per boll, with vigorous germination and optimum crop plant stand for next year's crop. Further, the higher portion by weight of the embryo in NIAB-545 compared with FH-Lalazar reflected the minimal effect of high temperature in the development of the embryo during the peak period of high temperatures (July–August) in Pakistan.

The results of physiological studies showed that NIAB-878 excelled in heat tolerance, maintaining the highest values for anther dehiscence, pollen viability and minimum cell injury percentage as compared with other genotypes and the standards FH-142 and FH-Lalazar. Higher values of anther dehiscence and pollen viability together with lower values for relative cell injury correspond to better tolerance to high temperatures. NIAB-878 surpassed all the genotypes and standards in the measures of maximum stomatal conductance, transpiration rate, net photosynthetic rate, physiological water-use efficiency under prevailing high temperatures of field conditions, and the ability to survive under water-deficit and high-temperature stress. Based on the overall performance of cotton genotypes regarding heat tolerance under field conditions, NIAB-878, NIAB-1011/89 and NIAB-1011/48 were found to be tolerant and could better withstand the heat stress as compared with susceptible genotypes, i.e. FH-142 and FH-Lalazar. These tolerant lines are also proving to be valuable for the successful cultivation of cotton in heat-prone cotton-growing areas of Pakistan and have the potential to maximize returns to cotton growers and farmers.

Acknowledgements

The lead author is extremely thankful to the management of IAEA at Vienna, Austria, for providing the financial support to attend and present this paper in the FAO/IAEA International Symposium on Plant Mutation Breeding & Biotechnology, 27–31 August 2018 at Vienna, Austria, under the Technical Cooperation Project RAS 5/075 'Improving Sustainable Cotton Production through Enhanced Resilience to Climate Change'. The author is also thankful to the management of PARC, ALP, Planning Division, Pakistan, for the award of the national project on cotton ALP (CS-197) under which current investigations were carried out.

Last but not least, a lot of thanks to the management of PAEC, HQ Islamabad and my team members at NIAB for their encouraging support for conducting these studies.

References

Ahmad, A. *et al.* (2015) Impact of climate change on the rice–wheat cropping system of Pakistan. In: Rosenzweig, C. and Hillel, D. (eds) Chapter 7 in *Handbook of Climate Change and Agroecosystems: The Agricultural Model Inter-comparison and Improvement Project (AgMIP) Integrated Crop and Economic Assessments, Part 2*. Joint Publication with American Society of Agronomy, Crop Science Society of America and Soil Science Society of America, pp. 219–258.

Anonymous (2013a) Global warming and cotton production. ICAC press release, 3 October 2013. International Cotton Advisory Committee, Washington, DC.

Anonymous (2013b) Impact of heat and other stresses on boll development in cotton plant. Available at: www.cicr/org.in (accessed 2019)

AOAC (1990*) Official Methods of Analysis*, 15th edn. Association of Official Analytical Chemists, Washington, DC.

Bibi, A.C. *et al.* (2003) The physiological response of cotton to high temperatures for germplasm screening. In: Oosterhuis, D.M. (ed.) *Summaries of Arkansas Cotton Research*. Arkansas Agricultural Experiment Station, Fayetteville, Arkansas, 521, pp. 87–93.

Burke, J.J. *et al.* (2004) In vitro analysis of cotton pollen germination. *Agronomy Journal* 96, 359–368.

Cottee, N.S. *et al.* (2010) Multi-level determination of heat tolerance in cotton (*Gossypium hirsutum* L.) under field conditions. *Crop Science* 50, 2553–2564.

Cottee, N.S. *et al.* (2007) Screening cotton cultivars for thermo tolerance under field conditions. In: *Proceedings 4th World Cotton Research Conference*, Lubbock, Texas, 10–14 September.

Hodges, H.F. *et al.* (1993) *Temperature effects on cotton*. Mississippi Agricultural & Forestry Experiment Station, Mississippi State University.

Karademir E. *et al.* (2012) Screening cotton varieties (*Gossypium hirsutum* L.) for heat tolerance under field conditions. *African Journal of Agricultural Research* 7(47), 6335–6342.

Liu, Z. *et al.* (2006) Screening for high-temperature tolerant cotton cultivars by testing in vitro pollen germination, pollen tube growth and boll retention. *Journal of Integrative Plant Biology* 48, 706–714.

Oosterhuis, D.M. and Snider, J.L. (2011) High temperature stress on floral development and yield of cotton: stress physiology in cotton. In: Oosterhuis, D.M. (ed.) *Stress Physiology in Cotton*. The Cotton Foundation, Cordova, Tennessee, pp. 1–25.

Oosterhuis D.M. *et al.* (2008) Screening for temperature tolerance in cotton. In: Oosterhuis, D.-M. (ed.) *Summaries of Arkansas Cotton Research*. Arkansas Agricultural Experiment Station, Fayetteville, Arkansas, 573, pp. 37–41.

Pettigrew, W.T. (2008) The effect of higher temperatures on cotton lint yield production and fiber quality. *Crop Science* 48, 278–285.

Radin J.W. *et al.* (1994) Genetic variability for stomatal conductance in Pima cotton and its relation to improvements of heat adaptation. *Proceedings National Academy of Sciences USA* 91, 7217–7221.

Rahman, H. *et al.* (2004) Heat tolerance of upland cotton during the fruiting stage evaluated using cellular membrane thermo stability. *Field Crops Research* 85, 149–158.

Rahman H.M. *et al.* (2018) Multi-model projections of future climate and climate change impacts uncertainty assessment for cotton production in Pakistan. *Agricultural and Forest Meteorology* 253–254, 94–113.

Reddy K.R. *et al.* (1992) Temperature effects on cotton fruit retention. *Agronomy Journal* 84, 26–30.

Reddy K.R. *et al.* (1999) Temperature regime and carbon dioxide enrichment alter cotton boll development and fiber properties. *Agronomy Journal* 91, 851–858.

Reddy V.R. *et al.* (1991) Temperature effects on growth and development of cotton during the fruiting period. *Agronomy Journal* 83, 211–217.

Snider, J.L. *et al.* (2009) Heat stress-induced limitations to reproductive success in *Gossypium hirsutum*. *Physiologia Plantarum* 137, 125–138.

Zhou G.Y. *et al.* (1996) The influence of climatic differences in the Yangtze River valley on cotton yield and fiber quality. *China Cottons* 10, 15–18.

16 Development of Climate-Adaptable/Resilient Crop Varieties Through Induced Mutation

Md. Abul Kalam Azad[1], Fahmina Yasmine, Md. Kamruzzaman, Md. Hasanuzzaman Rani and Hosne Ara Begum

Abstract

For crop production to cope with problems driven by climate change, such as salinity, drought and extreme temperatures, the Bangladesh Institute of Nuclear Agriculture (BINA) released a late Boro rice variety, 'Binadhan-14' in 2013 which is tolerant to high temperature, has short duration (105–115 days) and gives average yield of 6.9 t/ha. This variety was developed by irradiating the seeds of 'Ashfal', a local salt-tolerant landrace of rice, with 200 Gy of carbon-ion beams. The late-transplanting potential of this variety also helps in avoiding seedling injury due to severe cold. Another variety, 'Binadhan-19', was developed by irradiating the seeds of 'NERICA-10' rice with 40 Gy of carbon-ion beams. This was released by the National Seed Board of Bangladesh (NSB) in 2017 as a drought-tolerant, short-duration (95–105 days) and high-yielding (average 4.0 t/ha) variety for the *Aus* growing season. BINA developed a salt-tolerant wheat variety, 'Binagom-1', by selecting from a segregating population, obtained from NIAB, Pakistan. This variety was released in 2016; it can tolerate salinity (up to 12 dS/m) and produces an average yield of 2.8 t/ha. Apart from these, BINA developed four salt-tolerant groundnut varieties ('Binachinabadam-5', 'Binachinabadam-6', 'Binachinabadam-7' and 'Binachinabadam-9') by irradiation with gamma-rays. All these four varieties can tolerate salinity (up to 8 dS/m) from flowering to maturity and produce pods at 1.8–3.4 t/ha under saline soil conditions. These climate-resilient varieties are playing a significant role in food security and enhancing the nutritional status of the people of Bangladesh.

Keywords: mutation breeding • carbon-ion beam irradiation • drought tolerant rice • salt tolerant wheat • salt tolerant groundnut

1 Introduction

Bangladesh is a small country with a big population. It has to feed nearly 160 million people but has only 8.6 million hectares of cultivable land. The area of cultivable land is decreasing by 0.73% per year yet needs to meet the demand of an increasing population (Imamul Huq and Hassan, 2015). Moreover, due to climate change, salinity, drought, high/low temperatures, river and flash floods, tidal surges, cyclones, hailstorms, etc. are recurrent phenomena in Bangladesh. To cope with these natural disasters, agriculture requires the development of climate-resilient crop varieties. Genetic improvement of traits, either qualitative or quantitative in nature, has been successful with mutation breeding. Shamsuzzaman *et al.* (1998), Azam

[1] Email: makazad.pbdbina@gmail.com

DOI: 10.1079/9781789249095.0016

and Uddin (1999), Hamid *et al.* (2006), Azad *et al.* (2010, 2012, 2013a,b,c, 2014), Lagoda and Foster (2013), Chakraborty (2014), Hoque (2014) and Ferdoush (2017) collated data that documented the release of 3218 crop varieties worldwide that were developed using this technique.

2 Materials and Methods

2.1 Development of a high-temperature tolerant *Boro* rice variety using the carbon-ion beam irradiation technique

Seed irradiation, growing M_1 generation, screening M_2–M_3 populations

Dehusked seeds of 'Ashfal', a salt-tolerant local landrace of the *Aman* growing-season (July–November) rice, were exposed to 26.7 MeV/n carbon ions with doses of 0, 10, 20, 40, 60, 80, 100, 120, 160 and 200 Gy in January 2009 at the Japan Atomic Energy Agency. The details of seed irradiation, growing the M_1 generation and screening M_2–M_3 populations were described by Azad *et al.* (2013a). A schematic diagram of variety development starting from seed irradiation to release of the variety is given in Fig. 16.1. For shortening the breeding cycle, two generations were grown per year, one in the *Boro* growing season (December–May) and the other in the *Aman* growing season. Evaluation during the M_2 generation to zonal yield trial identified two short-duration high-yielding mutant lines.

On-farm, on-station and field evaluation trials

These trials were conducted in the *Boro* growing season with two selected short-duration mutant lines. On-farm and on-station trials were conducted in 2012 at nine locations and the field evaluation trial in 2013 at 12 locations in Bangladesh. Delayed transplantation of seedlings between 8 February and 8 March was followed as a measure of high-temperature tolerance test, which was also followed in wheat by Islam *et al.* (2013) and Hossain *et al.* (2018). In Bangladesh conditions, irrigated *Boro* rice is transplanted in December–January when minimum and maximum air temperatures remain low (Figs 16.2 and 16.3). From February, the temperature rises sharply and reaches the maximum in April (Fig. 16.3). The unit plot sizes were 5.1 m × 4.0 m and 5.0 m × 6.0 m, respectively. 'BRRI dhan28' was used as a standard check variety. At a seedling age between 35 and 50 days, seedlings were transplanted to 15 cm spacing within rows which were 20 cm apart.

2.2 Development of a drought-tolerant *Aus* rice variety using carbon-ion beam irradiation

Irradiation of 'NERICA-10' seeds and growing the M_1 generation

With a view to developing high-yielding, stable varieties that can be grown under rainfed condition, dehusked seeds of the 'NERICA-10' rice variety introduced from the West African Rice Development Association (WARDA) were irradiated with 0, 40, 60, 80, 120, 160, 180 and 200 Gy of carbon ions (see 2.1.1 above). Two hundred seeds were used for each dose. After receiving the irradiated seeds from Japan, they were washed with 3.7 mM $MgCl_2$ solution for 10 min and then placed on separate Petri dishes according to dose and variety. The sprouted seeds were then sown in earthen pots on 4 March 2013. Finally, on 2 April 2013, seedlings were transplanted to the field, 15 cm apart along rows that were 20 cm apart, separately according to dose and variety. A plant was found with conspicuous changes in seed colour, seed size and plant stature compared with the parent variety which was confirmed not to be a contaminant of another variety by Nafis (2016). At maturity, M_1 seeds were harvested and kept separately according to dose and variety and the M_2 generation was grown in the next *Aus* growing season (April–July). A schematic diagram of variety development and shortening of breeding cycle are shown in Fig. 16.1. Evaluations during the M_2 generation to zonal yield trial identified the mutant N_{10}-40(C)-1-5 as the best.

Field evaluation trial

A field evaluation trial of the mutant line N10-40(C)-1-5 in the *Aus* growing season

Fig. 16.1. Schematic diagram of the development of 'Binadhan-14' and 'Binadhan-19'.

was carried out to assess yield performance under direct seeding and rainfed conditions at different locations. Direct seeding (dibbling) under rainfed conditions was used for testing drought tolerance. 'BRRI dhan43' and 'NERICA-10' were used as check varieties. 'BRRI dhan43' is a direct seeded rice variety for the *Aus* growing season released by Bangladesh

Rice Research Institute (BRRI). Seeds were sown between 7 April and 12 May 2016 at 12 locations. The names of the locations and location-wise total rainfall during the crop growing period are given in Table 16.1. Three to four sprouted seeds were sown together at 15 cm apart along rows which were 20 cm apart. No supplemental irrigation was applied except a life-saving irrigation in case of severe drought. A unit plot size was 6.0 m × 5.0 m.

2.3 Development of a salt-tolerant wheat variety

Purification and comparative performance evaluation of salt-tolerant wheat mutants

Four segregating populations, L-885, L-61, L-879 and L-880, developed by gamma irradiation, were collected from the National Institute of Agriculture and Biology (NIAB), Pakistan. A total of 156

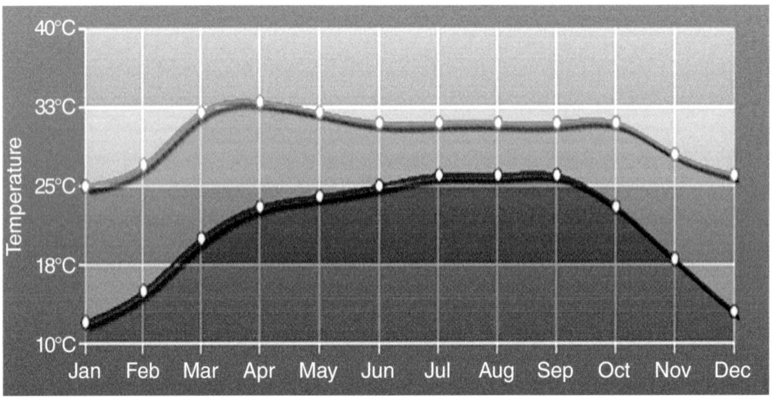

Fig. 16.2. Trends of monthly maximum and minimum temperatures in Bangladesh.

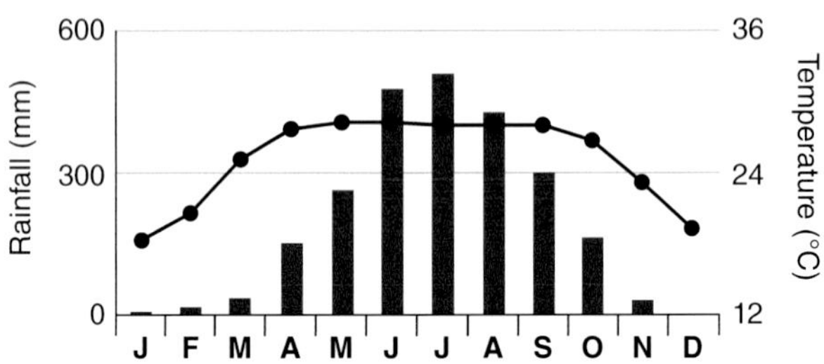

Fig. 16.3. Trends of average monthly temperature and rainfall in Bangladesh.

Table 16.1. Locations with total rainfall during the growing season of the mutant, parent and the check variety.

Location	L1	L2	L3	L4	L5	L6	L7	L8	L9	L10	L11	L12
Total rainfall (mm)	1264	1921	1921	1552	782	762	1127	1127	2194	2194	1854	1854

L1 = Mymensingh HQ; L2 = Nalitabari sub-station; L3 = Nalitabari farmer's field; L4 = Rangpur farmer's field; L5 = Nachole, Chapainawabgonj; L6 = Godagari, Rajshahi; L7 = Magura sub-station; L8 = Magura farmer's field; L9 = Raicha, Bandarban; L10 = Mrulongpara, Bandarban; L11 = Alutila, Khagrachari; L12 = Matiranga, Khagrachari spikes were obtained by selecting the longest spike with the highest number of spikelets from each hill of each population; 20 spikes were obtained from L-885, 37 from L-61, 40 from L-879 and 50 from L-880. These populations were grown in a head-to-row design together with the check 'Prodip'. Seeds were sown in November 2010 at BINA farm, Mymensingh. Thirty-nine lines were selected based on shorter plant height and maturity and higher yield than the check variety, 'Prodip'. Further evaluations in preliminary yield trial (PYT) to advanced yield trial (AYT) identified the mutant L-880-43 as the best.

Zonal yield and field evaluation trials

These were carried out with L-880-43 along with the only released salt-tolerant wheat variety in Bangladesh, 'BARI Gom-25', in 2013 and 2014 at four and seven locations, respectively. In zonal yield trials (ZYT), the locations were Satkhira, Khulna and Patuakhali (saline area) and Ishurdi (non-saline area). In the field evaluation trial, five locations were in the saline area and the other two (Ishurdi and Rangpur) were in the non-saline area. The unit plot sizes were 5.0 m × 4.0 m and 6.0 m × 5.0 m, respectively.

In both trials, soil salinity records were gathered from the experimental plots in the saline areas at 0–6 cm depth during sowing, vegetative and flowering stages. The soil salinity of the top soil in ZYT ranged between 0.82 and 3.94 dS/m, which was comparatively low, but that in the field evaluation trial was high (Fig. 16.4). As NaCl is the most common salt in solution in ground or river water, a 'rule of thumb' is to multiply the electrical conductivity (EC) by ten to equate it to salt molarity, i.e. 1 dS/m = 10 mM NaCl.

2.4 Development of salt-tolerant groundnut varieties using gamma irradiation

Development of salt-tolerant groundnut varieties 'Binachinabadam-5' and 'Binachinabadam-6'

The details of the methodology for development of the salt tolerant groundnut varieties were described by Azad *et al.* (2014).

Development of salt-tolerant groundnut varieties 'Binachinabadam-7' and 'Binachinabadam-9'

Seed irradiation, growing of M_1 generation and evaluation of M_2 and M_3 populations. Two hundred seeds of each of 'Dacca-1' and PK-1 were irradiated with 200 and 250 Gy of gamma-rays from a ^{60}Co source. The M_1 seeds were sown immediately at BINA farm, Mymensingh, on 1 February 2006. In the M_2 generation, 54 progenies of 'Dacca-1' and 59 progenies of PK-1 were sown on 28 November in the same year following a plant-progeny-row protocol. Selection and evaluation during the M_3 generation to AYT was made in non-saline soils based on higher pod yield than the respective parents and only two mutants were identified.

Zonal, on-farm and on-station trials. ZYT was carried out to assess performance of two M_8 mutants in saline and non-saline areas of Bangladesh. The parents of these mutants, 'Dacca-1' and PK-1, were also included in this experiment. Seeds were sown on 18 December 2012 and 17 January 2013 at BINA sub-station farms at Rangpur and Ishurdi, respectively, in the non-saline area, on 8 January 2013 at Subarnachar, Noakhali, and on 6 January 2013 at Regional Research Station of Bangladesh Jute Research Institute (BJRI), Pakhimara, Patuakhali. The salinity of topsoil (0–6 cm depth) at Patuakhali ranged between 2.40 and 11.1 dS/m during the growing period. The highest salinity was developed during flowering and pegging stages in March. At Noakhali, it ranged between 0.05 and 2.75 dS/m. On-farm and on-station trials were performed to assess the performance of the same mutants in saline and non-saline areas of Bangladesh. The popular variety 'Dacca-1' was included in this experiment. Seeds were sown in non-saline soil on 18 and 24 December 2013 at BINA sub-station farms, Rangpur and Ishurdi, respectively, and in farmers' fields at Jhenaidah, Lalmonirhat and Natore on 8 February 2013 and 2 March 2014, respectively. In saline areas seeds were sown on 29 December 2013 and 11 January 2014, at Patuakhali, Bhola and Noakhali, respectively. Soil salinity during the growing period ranged between 0.41 and 2.38 dS/m at Patuakhali, 1.08–7.16 dS/m at Bhola and 0.53–7.7 dS/m at Noakhali.

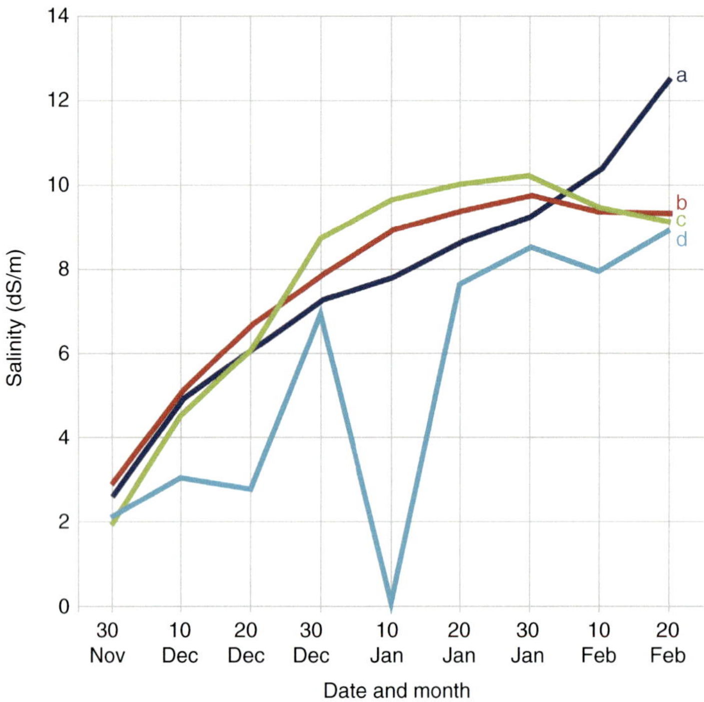

Fig. 16.4. Salinity records of the experimental plot at 0–6 cm depth during the growing season. **(a)** Batiaghata, Khulna; **(b)** BJRI sub-station, Patuakhali; **(c)** farmer's field, Patuakhali; **(d)** Ghatkhali, Borguna.

2.5 Statistical design

In all the above yield trials like PYT, AYT, ZYT, on-farm and on-station and field evaluation trials, randomized complete block design (RCBD) with three replications was used.

2.6 Fertilizer dose and method of application

Rice

Fertilizers were applied at the rate of N 90 kg, P 10 kg, K 35 kg, S 10 kg and Zn 2.0 kg/ha in the *Boro* growing season and N 75 kg, P 9kg, K 25 kg and S 9 kg/ha in the *Aus* growing season in the forms of urea, triple super phosphate (TSP, $Ca(H_2PO_4)$), muriate of potash (MoP, KCl), gypsum (hydrated $CaSO_4$) and $ZnSO_4$, respectively. In the *Aman* and *Boro* growing seasons, all the above-mentioned fertilizers were applied during the final land preparation as basal dose except for urea, which was applied in three equal splits. The first split was applied immediately after seedling establishment, the second at maximum tillering stage and the third 5–7 days before panicle initiation (PI). In the *Aus* growing season, of the three equal-split applications of urea, the first split was applied during final land preparation as basal dose.

Wheat

Fertilizers were applied at the rate of N 110 kg, P 28 kg, K 65 kg, S 14 kg, Zn 2.0 kg and B 1.0 kg/ha in the forms mentioned already in rice except boron, which was applied in the form of boric acid. Two-thirds of the nitrogen and all of the phosphorus, potassium, sulfur, zinc and boron were applied as a basal dose during final land preparation. The remaining one-third of nitrogen was applied 17–21 days after sowing.

Groundnut

Fertilizers were applied at the rate of N 20 kg, P 35 kg, K 68 kg and S 20 kg/ha in the forms mentioned above. All the fertilizers were applied as a basal dose during the final land preparation.

2.7 Data recording

Rice

Data on plant height, days to maturity, number of effective tillers per hill, panicle length, filled and unfilled grains per panicle and yield per plot were recorded. Maturity was recorded on a plot basis while plant height, effective tiller number, panicle length, filled and unfilled grains per panicle were recorded from five randomly selected competitive plants at harvest. Grain yield (at 14% moisture) was recorded from areas of 1.0 m^2 in the PYT, AYT and ZYT but from 10 m^2 in on-farm, on-station and field evaluation trials (later converted to t/ha). Moisture data were recorded with a grain moisture meter.

Wheat

Data on plant height, effective tiller number, spike length and grains per spike were recorded at harvest from five randomly selected competitive plants and grain yield was recorded following the same procedure as mentioned in the case of rice.

Groundnut

Data on plant height, pod number, pod yield per plant and 100-pod weight were recorded from ten randomly selected competitive plants at maturity. Like rice and wheat, pod yield per plot was recorded from 1.0 m^2 in earlier yield trials but from 10 m^2 in the later yield trials such as ZYT, on-farm and on-station trials. Pod yield per plant, 100-pod weight, 100-kernel weight and pod yield per plot were recorded after sun drying. Pod yield per plot was later converted to pod yield per hectare. Shelling percentage was calculated as

$$\frac{\text{Kernel weight}}{\text{Pod weight}} \times 100$$

2.8 Statistical analysis

Means and standard errors were calculated for the yield attributes and yield of the mutants and varieties in all the trials using the location-wise mean data. Means were separated by least significant difference (LSD) at $p < 0.05$ level.

3 Results

3.1 Development of a high-temperature tolerant *Boro* rice variety using the carbon-ion beam irradiation technique

In on-farm and on-station trials, the two mutants differed significantly ($p < 0.05$) from the check variety, 'BRRI dhan28', for plant height, panicle length and filled and unfilled grains (Table 16.2a). Compared with the check variety, the two mutants were shorter at maturity but had higher 1000-grain weights. On the other hand, the check variety had a longer panicle and a higher number of filled and unfilled grains. Grain yields of the two mutants were also significantly ($p < 0.05$) higher than the check variety (Fig. 16.5). Between the two mutants, there were no significant differences for plant height, panicle length, filled and unfilled grains and maturity period (Table 16.2a) but grain yield differed at BINA headquarters farm, Mymensingh, Magura, Rajshahi and a farmer's field at Rangpur (Fig. 16.5). Thousand-grain weight was a little higher in the mutant RM(1)-200(C)-1-17 than the other mutant. The effect of locations on plant height, effective tiller number, panicle length, unfilled grain number and maturity of the two mutants was comparatively less dispersed than the other mutant and the check variety (Table 16.2a) but the filled grain number of the mutant RM(1)-200(C)-1-17 was more dispersed than the other mutant and the check variety.

In the field evaluation trial, plant heights of the two mutants were shorter than the check variety, 'BRRI dhan28', despite having no remarkable differences in number of effective tiller and filled grains per panicle (Table 16.2b). The two mutants matured a little earlier than the check variety. Grain yield of the two mutants was significantly higher at all locations than the check variety (Fig. 16.6) despite having shorter

Table 16.2. Means of yield and yield attributes of the rice mutants versus the check variety 'BRRI dhan28' under late transplanting condition in (a) on-farm and on-station, averaged over nine locations, and (b) field evaluation trial, averaged over 11 locations.

Mutants/check variety	Plant height (cm)	Effective tiller (no.)	Panicle length (cm)	Filled grains/ panicle (no.)	Unfilled grains/ panicle (no.)	1000-grain weight (g)	Maturity (days)
(a) On-farm and on-station trial, 2012							
RM(1)-200-(C)-1-10	87.61 ± 2.01	10.47 ± 0.77	21.81 ± 0.44	92.14 ± 3.67	16.10 ± 2.06	23.18	127 ± 1.39
RM(1)-200-(C)-1-17	87.79 ± 1.95	10.34 ± 0.71	22.49 ± 0.38	96.55 ± 5.85	14.32 ± 1.43	23.64	127 ± 1.38
'BRRI dhan28'	98.27 ± 2.48	10.36 ± 0.78	23.22 ± 0.60	103.81 ± 5.43	19.61 ± 4.59	22.90	130 ± 1.41
LSD (0.05)	1.61	NS	0.43	4.91	2.33		–
(b) Field evaluation trial, 2013							
RM(1)-200-(C)-1-10	89.22 ± 1.65	10.93 ± 0.61	21.12 ± 0.37	97.68 ± 3.94	23.11 ± 1.65	23.09	125 ± 1.10
RM(1)-200-(C)-1-17	89.60 ± 1.41	10.68 ± 0.63	21.15 ± 0.36	100.16 ± 5.27	21.16 ± 2.43	23.33	125 ± 1.10
'BRRI dhan28'	109.96 ± 3.08	10.63 ± 0.36	23.35 ± 0.34	101.79 ± 5.09	30.14 ± 2.03	22.06	128 ± 1.20
LSD (0.05)	1.02	NS	0.36	NS	2.57	–	–

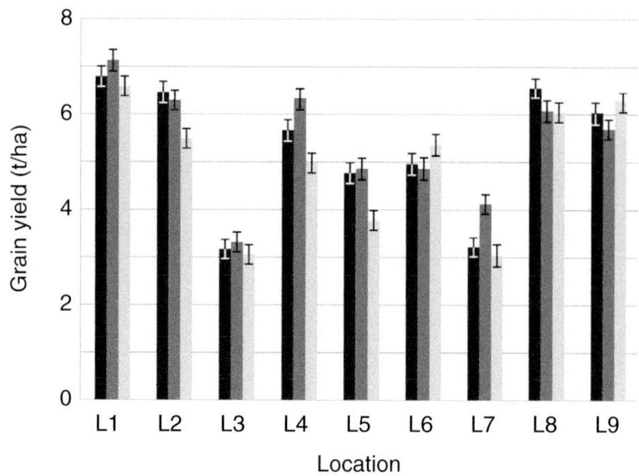

Fig. 16.5. Grain yield of the two mutants RM(1)-200(C)-1-10 (black) and RM(1)-200(C)-1-17 (grey) versus the check variety 'BRRI dhan28' (light grey) in on-farm and on-station trial at nine locations in Bangladesh under late transplanting condition. Error bars represent LSD values at $p < 0.05$. L1 = BINA Farm, Mymensingh; L2 = Ghunti, Mymensingh; L3 = Satkhira; L4 = Magura; L5 = Barishal; L6 = Natore; L7 = Rajshahi; L8 = farmer's field, Rangpur; L9 = BINA sub-station farm, Rangpur.

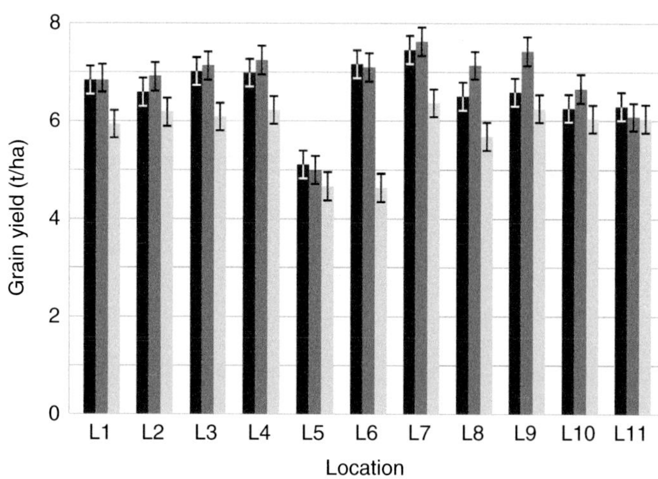

Fig. 16.6. Grain yield of the two mutants RM(1)-200(C)-1-10 (black) and RM(1)-200(C)-1-17 (grey) versus the check variety 'BRRI dhan28' (light grey) in field evaluation trial at 11 locations of Bangladesh under late transplanting condition. Error bars represent LSD values at $p < 0.05$. L1 = BINA Farm, Mymensingh; L2 = Maijbari, Mymensingh; L3 = BINA sub-station farm, Jamalpur; L4 = Komorpur, Jamalpur; L5 = BINA sub-station farm, Magura; L6 = Shibrampur, Magura; L7 = BINA sub-station farm, Ishurdi; L8 = Kalikapur, Ishurdi; L9 = BINA sub-station farm, Rangpur; L10 = Paskhifanda, Rangpur; L11 = BINA sub-station farm, Barishal.

panicle length (Table 16.2b). This was mostly attributed to their 1000-grain weight (Table 16.2b). As with the on-farm and on-station trials, the effect of locations in this trial on plant height and maturity period of the two mutants was comparatively less dispersed than on the check variety (Table 16.2a), but the other attributes were more dispersed in the two mutants.

3.2 Development of a drought-tolerant Aus rice variety using carbon-ion beam irradiation

Zonal yield trials with M_6 mutants of NERICA-10 under direct seeding and rainfed conditions in the Aus growing season

In the zonal yield trial, plant height of the three mutants was significantly ($p < 0.05$) shorter than the check variety, 'BRRI dhan43', but the mutants did not differ from the check variety for effective tiller number and filled grains per panicle (Table 16.3a). Maturity period of the mutants was a little shorter than that of the check variety. All three mutants produced significantly higher grain yields at all locations except Ishurdi (Fig. 16.7). At Ishurdi, grain yields of the three mutants did not differ significantly from the check variety. The effect of location on plant height and panicle length of the check variety was more dispersed than the three mutants (Table 16.3a), but the three mutants had more dispersion for effective tiller number, unfilled grains per panicle and maturity period.

In the field evaluation trial, plant height of the mutant was significantly ($p < 0.05$) shorter than the check varieties, 'BRRI dhan43' and 'NERICA-10' (Table 16.3b), but the mutants did not differ from the check variety for effective tiller number and filled grains per panicle (Table 16.3a). Maturity periods of the mutants were a little shorter than both the check varieties. All three mutants produced significantly higher grain yields at all locations except L12 (Matiranga, Khagrachari) (Fig. 16.8). At L12, grain yield of the check variety 'BRRI dhan43' was significantly higher. The effect of location on plant height of 'NERICA-10' was most dispersed followed by the check variety 'BRRI dhan43' (Table 16.3b), but filled and unfilled grains were more dispersed in the mutant RM-N10-40(C)-1-5, while maturity period was more dispersed in 'NERICA-10' followed by the mutant.

3.3 Development of a salt-tolerant wheat variety

In the zonal yield trial, plant height of the tested mutant L-880-43 was significantly ($p < 0.05$) greater than the check variety 'BARI Gom-25' but effective tiller number, grains per panicle and even grain yield/ha did not differ (Table 16.4a). In the field evaluation trial, plant height of the tested mutant once again appeared significantly greater than that of the check variety along with grains per spike and grain yield/ha (Table 16.4b). Higher grain yield/ha was mostly attributed to its significantly higher grain yield in all the locations (Fig. 16.9).

3.4 Development of salt-tolerant groundnut varieties

The details of the development of 'Binachinabadam-5' and 'Binachinabadam-6' are described in Azad *et al.* (2014).

Development of salt-tolerant groundnut varieties 'Binachinabadam-7' and 'Binachinabadam-9'

In zonal, on-farm and on-station trials, the mutant RS/25/3-1 had the shortest plant height, the highest number of mature pods, the highest pod weight per plant and 100-pod weight (Table 16.5a and b), but pod yield/ha of this mutant in zonal yield trial was a little lower than the check PK-1 although higher than the other mutant and the check variety 'Dacca-1'. In on-farm and on-station trials, the same mutant RS/25/3-1 had the highest pod yield/ha while the other mutant had significantly lower yield than it but higher than the check variety 'Dacca-1'. The location-wise pod yield of the mutants in on-farm and on-station trials showed the superior performance of both the mutants in both saline and non-saline areas (Fig. 16.10).

4 Discussion

Based on their performance in the field evaluation trial at 11 locations and the opinions of farmers, extension agents and the evaluation team of the Seed Certification Agency, it can be concluded that the mutant RM(1)-200(C)-1-17 is more suitable than the variety 'BRRI dhan28' for late transplanting after harvest of most of the

Table 16.3. Yield and yield attributes of some NERICA rice mutants under direct seeding (dibbling) and rainfed conditions compared with the check variety 'BRRI dhan43' in (a) zonal yield trial, averaged over seven locations, and (b) field evaluation trial, averaged over 12 locations.

Mutant/check	Plant height (cm)	Effective tillers (no.)	Panicle length (cm)	Filled grains/ panicle (no.)	Unfilled grains/ panicle (no.)	Days to maturity	1000-grain weight (g)
(a) Zonal yield trial, 2015							
RM-N_{10}-40(C)-1-1	85.93 ± 5.10	11.47 ± 0.69	20.61 ± 0.64	61.16 ± 11.17	23.76 ± 9.93	100.5 ± 0.86	–
RM-N_{10}-40(C)-1-5	85.46 ± 4.94	11.91 ± 0.70	20.61 ± 0.67	61.85 ± 9.85	21.41 ± 7.51	100.5 ± 0.86	–
RM-N_{10}-40(C)-1-7	86.17 ± 4.58	10.93 ± 0.39	20.5 ± 0.51	60.63 ± 7.64	25.27 ± 10.53	100.5 ± 0.86	–
'BRRI dhan43'	107.64 ± 5.45	12.49 ± 0.14	22.25 ± 0.83	56.15 ± 8.17	27.94 ± 7.21	97.5 ± 0.58	–
LSD (0.05)	4.6	NS	1.6	NS	4.2	–	–
(b) Field evaluation trial, 2016							
RM-N_{10}-40(C)-1-5	94.08 ± 0.85	10.56 ± 0.40	21.73 ± 0.21	64.02±2.92	15.18 ± 1.31	102.83 ± 2.44	23.0
'BRRI dhan43'	111.16 ± 1.29	10.45 ± 0.38	22.19 ± 0.16	58.31 ± 2.37	26.85 ± 0.85	103.08 ± 2.37	21.0
NERICA-10	107.66 ± 2.97	6.25 ± 0.24	22.94 ± 0.19	54.96 ± 2.37	27.16 ± 1.22	105.67 ± 2.81	22.0
LSD (0.05)	3.13	NS	1.33	NS	6.85	–	–

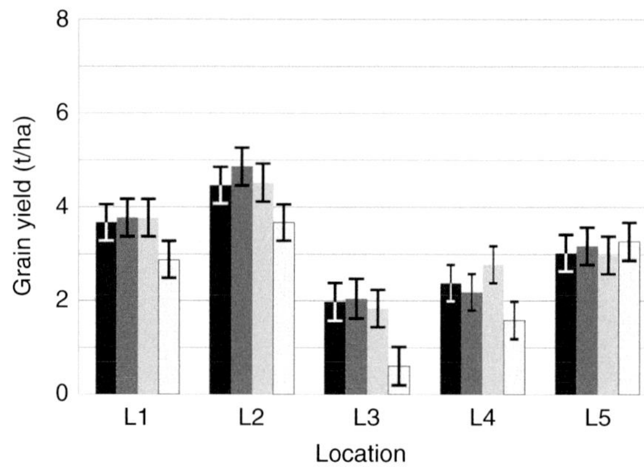

Fig. 16.7. Grain yields of the three mutants RM-N10-40(C)-1-1 (black), RM-N10-40(C)-1-5 (grey) and RM-N10-40(C)-1-7 (light grey) versus the check variety 'BRRI dhan43' (stippled) under direct seeding and rainfed conditions in zonal yield trial at five locations in Bangladesh. Error bars represent LSD value at $p < 0.05$. L1 = Mymensingh; L2 = Rangpur; L3 = Chapai Nawabganj; L4 = Magura; L5 = Ishurdi.

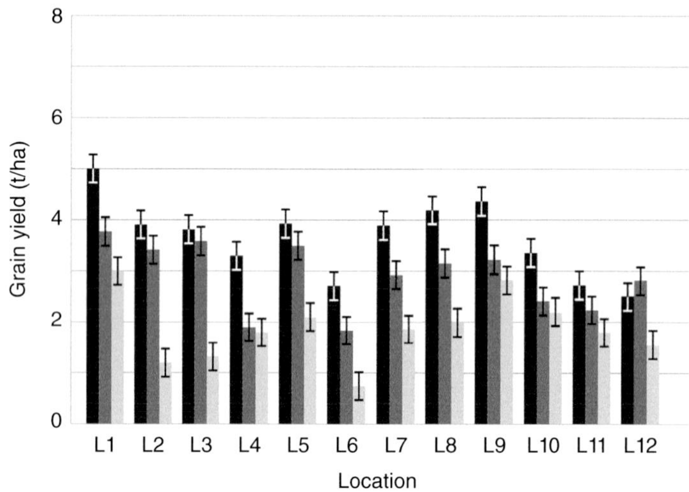

Fig. 16.8. Grain yields of the mutant RM-N10-40(C)-1-5 (black), a check variety 'BRRI dhan43' (grey) and the parent *NERICA-10* (light grey) at 12 different locations in Bangladesh, including drought-prone and hilly areas under direct seeding (dibbling) and rainfed conditions. Error bars represent LSD value at $p < 0.05$. L1 = BINA Farm, Mymensingh; L2 = BINA sub-station farm, Nalitabari; L3 = farmer's field, Nalitabari; L4 = farmer's field, Rangpur; L5 = Nachole, Chapai Nawabganj; L6 = Godagari, Rajshahi; L7 = BINA sub-station farm, Magura; L8 = farmer's field, Magura; L9 = Raicha, Bandarban; L10 = Mrulongpara, Bandarban; L11 = Alutila, Khagrachari; L12 = Matiranga, Khagrachari.

winter crops (such as mustard/rapeseed, wheat, potato, pulses and vegetables) as it had an average yield of 6.85 t/ha (ranging from 5.02 to 7.66 t/ha), which was 17% higher than the check variety 'BRRI dhan28'. Moreover, this mutant was shorter, lodging resistant with erect leaves, shorter duration, matured in 119–130 days and had a similar grain quality as the check variety. Considering all these factors, the National Seed Board (NSB) released this mutant as 'Binadhan-14' in 2014 for commercial cultivation in the late Boro season (from mid-February to mid-March).

In contrast, the mutant N10-40(C)-1-5 had an average yield of 3.46 t/ha (ranging

Table 16.4. Yield and yield attributes of a salt-tolerant line of wheat compared with the salt-tolerant released variety in Bangladesh 'BARI Gom-25' in (a) zonal yield trial, averaged over three saline and one non-saline location, and (b) field evaluation trial, averaged over five saline and two non-saline locations.

Mutant/check	Plant height (cm)	Effective tiller (no.)	Spike length (cm)	Grains/spike (no.)	1000-grain weight (g)	Grain yield (t/ha)
(a) Zonal yield trial, 2013						
L-880-43	88.25 ± 7.76	5.70 ± 1.06	8.74 ± 0.70	45.47 ± 5.09	36.55 ± 0.55	3.13 ± 0.37
'BARI Gom-25'	84.50 ± 7.49	5.23 ±1.03	10.07 ± 0.45	39.47 ± 3.39	50.89 ± 0.86	2.99 ± 0.38
LSD (0.05)	1.02	NS	0.38	NS	1.52	NS
(b) Field evaluation trial, 2015						
L-880-43	93.57 ± 2.4	5.0 ± 0.43	9.28 ± 0.29	46.57 ± 0.53	36.57 ± 0.87	3.20 ± 0.23
'BARI Gom-25'	89.14 ± 2.41	5.0 ± 0.48	10.28 ± 0.18	38.71 ± 1.27	48.14 ± 0.83	2.85 ± 0.23
LSD (0.05)	1.56	NS	0.27	1.97	1.17	0.11

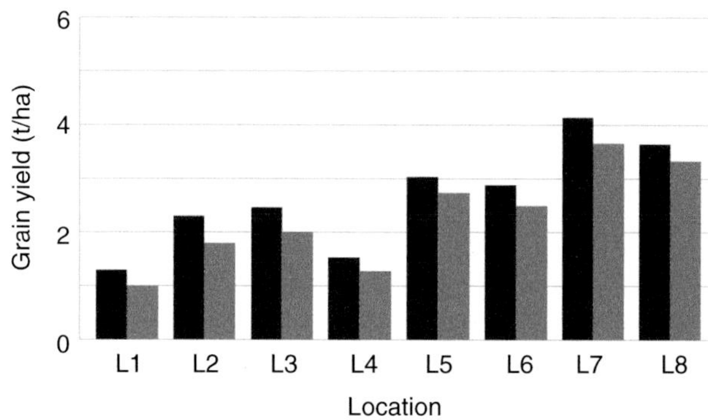

Fig. 16.9. Comparison of grain yields of the mutant L-880-43 (black) versus the salt-tolerant wheat variety 'BARI Gom-25' (grey) in four saline and four non-saline areas during 2015–2016 growing season. L1 = Batiaghata, Khulna; L2 = farmer's field, Patuakhali; L3 = BJRI sub-station, Patuakhali; L4 = Ghatkhali, Borguna; L5 = Patirajpur, Ishurdi; L6 = BINA sub-station farm, Ishurdi; L7 = BINA sub-station farm, Rangpur; L8 = Lahirirhat, Rangpur. Error bars are omitted because the grain yield of this mutant and variety did not differ at $p < 0.05$ from location to location.

from 2.5 to 5.0 t/ha) over 11 locations, including drought-prone Barind and hilly areas under rainfed dibbling conditions. This mutant produced 20% and 86% higher yield than the check variety, 'BRRI dhan43' and the parent NERICA-10, respectively. N10-40(C)-1-5 matured in 91–117 days, had long slender grains, was shorter and had lodging resistance. Considering all these factors, NSB released this mutant in 2017 as a drought-tolerant direct-seeded (dibbling) variety for cultivation in the *Aus* growing season with the name of 'Binadhan-19'.

The wheat mutant line L-880-43 had an average yield of 3.20 t/ha (ranging from 2.31 to 4.08 t/ha) under 2.0–2 dS/m salinity at five saline and two non-saline locations. This mutant had a 12% higher yield than the check variety, 'BARI Gom-25'. Considering grain yield under saline and non-saline conditions compared with 'BARI Gom-25', the National Seed Board released it for commercial cultivation in both saline and non-saline areas of Bangladesh in 2016 using the name of 'Binagom-1'.

Considering the performance of the two groundnut mutants both in saline and non-saline areas of Bangladesh, NSB registered the two mutants, D1/20/17-1 and RS/25/3-1, for commercial cultivation in both saline and non-saline areas of Bangladesh in 2014. They were given the names 'Binachinabadam-7' and 'Binachinabadam-9', respectively. In the same way, based on evaluation in saline and non-saline areas of Bangladesh, two superior mutants were

Table 16.5. Yield and yield attributes of two M_8 mutant lines of groundnut compared with 'Dacca-1' and PK-1, in (a) zonal yield trial, averaged over two saline and two non-saline locations, and (b) on-farm and on-station trials, averaged over three saline and five non-saline locations.

Mutants/ varieties	Plant height (cm)	Mature pods/ plant (no.)	Pod weight/ plant (g)	100-pod weight (g)	Pod yield (t/ha)
(a) Zonal yield trial, 2012–2013					
D1/20/17-1	43.57 ± 14.40	23.64 ± 5.49	12.26 ± 2.37	53.0 ± 3.57	2.46 ± 0.32
RS/25/3-1	38.01 ± 10.95	25.01 ± 8.33	14.86 ± 4.80	59.49 ± 3.19	2.61 ± 0.56
'Dacca-1'	40.04 ± 13.90	22.00 ± 6.56	12.66 ± 3.41	59.12 ± 4.05	2.24 ± 0.29
PK-1	41.02 ± 12.74	21.58 ± 4.84	13.62 ± 3.10	63.13 ± 4.35	2.70 ± 0.51
LSD (0.05)	1.41	1.74	1.03	4.44	0.24

Mutants/ varieties	Plant height (cm)	Mature pods/plant (no.)	Pod weight/ plant(g)	100-pod weight (g)	100-kernel weight (g)	Shelling (%)	Pod yield (t/ha)
(b) On-farm and on-station trial, 2013–2014							
D1/20/17-1	36.39 ± 5.37	23.85 ± 1.69	12.71 ± 1.63	58.63 ± 3.30	25.72 ± 1.12	71.62 ± 2.42	2.66 ± 0.30
RS/25/3-1	34.97 ± 4.86	24.31 ± 3.14	14.55 ± 1.90	66.90 ± 3.50	29.95 ± 1.53	70.57 ± 2.40	3.11 ± 0.42
'Dacca-1'	41.53 ± 5.79	21.15 ± 2.16	12.01 ± 1.65	61.71 ± 4.91	28.10 ± 1.94	69.84 ± 2.68	2.62 ± 0.34
LSD (0.05)	0.23	NS	0.24	2.30	0.38	NS	0.19

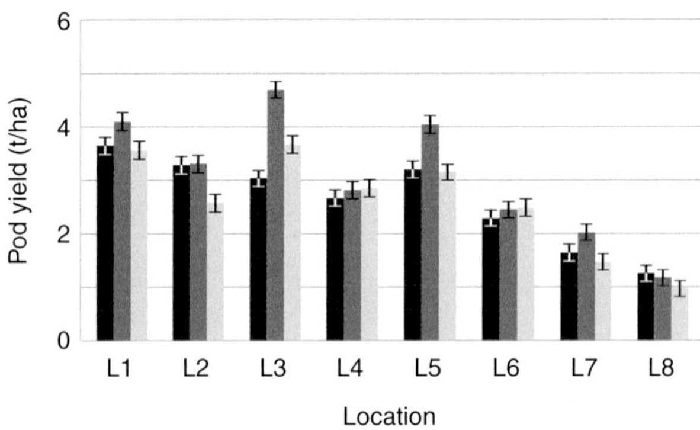

Fig. 16.10. Pod yield of the mutants D1/20/17-1 (black) and RS/25/3-1 (grey) versus the check variety 'Dacca-1' (light grey) in saline and non-saline locations of Bangladesh. The vertical bars represent LSD at $p < 0.05$. L1 = Rangpur; L2 = Lalmonirhat; L3 = Ishurdi; L4 = Natore; L5 = Jhenaidah; L6 = Noakhali; L7 = Patuakhali; L8 = Bhola.

released as 'Binachinabadam-5' and 'Binachinabadam-6' in 2011.

5 Conclusion

All these climate-resilient mutant varieties will play a significant role in food and nutritional security of the ever-growing population of Bangladesh, and will do so from a decreasing area of cultivable land. Moreover, it is important to note that the carbon-ion beam irradiation technique can help in shortening the breeding cycle of photoperiod-insensitive or even a photoperiod-sensitive rice breeding programme by at least 3–4 years.

Acknowledgements

The authors acknowledge the cooperation of Dr A.N.K. Mamun of Bangladesh Atomic Energy Agency (BAEC), Dr A. Tanaka, Dr Y. Hase and other team members of the Japan Atomic Energy Agency for rice seed irradiation with carbon ion beams.

References

Azad, M.A.K. *et al.* (2010) Binachinabadam-4: a high-yielding mutant variety of groundnut with medium pod size. *Plant Mutation Report* 2(2), 45–47.

Azad, M.A.K. *et al.* (2012) Tolerance to salinity stress in peanut (*Arachis hypogaea* L.) through osmotic adjustment and undamaged chloroplast. *Legume Research* 35(4), 271–284.

Azad, M.A.K. *et al.* (2013a) Photoperiod insensitive mutants with shorter plant height identified in M1 generation of rice irradiated with carbon ion beams. *SABRAO Journal of Breeding and Genetics* 45(2), 179–186.

Azad, M.A.K. *et al.* (2013b) Modification of salt tolerance level in groundnut (*Arachis hypogaea* L.) through induced mutation. *Legume Research* 36(3), 224–233.

Azad, M.A.K. *et al.* (2013c) Low $Ca2^{+}/Na^{+}$ ratio and efficiency of mobilization of Ca^{2+} from shoot tissues to kernel determines salinity tolerance in peanut (*Arachis hypogaea* L.). *Legume Research* 36(5), 396–405.

Azad, M.A.K. *et al.* (2014) Enhancing abiotic stress tolerance in groundnut through induced mutation. In: Tomlekova, N.B., Kozgar, M.I. and Wani, M.R. (eds) *Mutagenesis: Exploring Genetic Diversity of Crops*. Academic Publishers, Wageningen, Netherlands, pp. 329–344.

Azam, M.A. and Uddin, I. (1999) Binadhan-4, an improved rice variety bred through induced mutation. *Bangladesh Journal of Nuclear Agriculture* 15, 59–65.

Chakraborty, M. (2014) Cytological investigation and molecular characterization of selected mutant lines of onion against purple blotch using SSR markers. MS thesis, Department of Genetics and Plant Breeding, Bangladesh Agricultural University, Mymensingh, Bangladesh.

Ferdoush, J. (2017) Cytological and molecular marker aided selection of indigenous onion (*Allium cepa* L.) genotypes. MS thesis, Department of Genetics and Plant Breeding, Bangladesh Agricultural University, Mymensingh, Bangladesh.

Hamid, M.A. *et al.* (2006) Development of three groundnut varieties with improved quantitative and qualitative traits through induced mutation. *Plant Mutation Report* 1(2), 14–16.

Hoque, A. (2014) *In vitro* regeneration and cytological studies of selected mutant lines of onion (*Allium cepa* L.). MS thesis, Department of Genetics and Plant Breeding, Bangladesh Agricultural University, Mymensingh, Bangladesh.

Hossain, M. *et al.* (2018) Evaluation of fifty irrigated spring wheat genotypes grown under late sowed heat stress condition in multiple environments of Bangladesh. *Fresenius Environmental Bulletin* 27(9), 5993–6004.

Imamul Huq, S.M. and Hassan, M.N. (2015) *Sustainable Management and Protection of Soil Resources*. Bangladesh country Paper presented at the Workshop on Sustainable management and Protection of Soil Resources, held at Bangkok, Thailand, 13–15 May 2015 organized by Global Soil Partnership of FAO.

Islam, R. *et al.* (2013) Screening of heat tolerant wheat genotypes in Bangladesh. *Journal of Natural Products* 6, 132–140.

Lagoda, P.J.L. and Foster, B.P. (2013) Success in plant mutation breeding. Lecture delivered in the Regional Training Course on Mutation Induction, Mutation Detection and Pre-breeding held 3–14 June 2013 at Joint FAO/IAEA Laboratories, Seibersdorf, Austria.

Nafis, M.H. (2016) Confirming mutant, assessing genetic variability and identifying markers for drought tolerance in NERICA rice mutant. MS thesis, Bangladesh Agricultural University, Mymensingh, Bangladesh.

Shamsuzzaman, K.M. *et al.* (1998) Development of early maturing and high-yielding cotton genotypes through induced mutations. *Bangladesh Journal of Nuclear Agriculture* 14, 1–8.

17 Anthracnose Resistance Induction in Chilli by Electron Beam Irradiation

Vichai Puripunyavanich[1], Penjan Sutthanukul, Rakchai Kurubunjerdjit and Prartana Kewsuwan

Abstract

Chilli seeds were irradiated with 0.3 kGy at 8 MeV from the electron beam source at the Thailand Institute of Nuclear Technology. M_1 seeds were planted in Sukhothai Horticultural Research Centre and from these the line CA1131 was selected as suitable for growing in this area. Thirty anthracnose-resistant M_2 chilli plants were selected after the appearance of anthracnose disease, caused by *Colletotrichum gloeosporioides*, in Sukhothai province from an initial M_1 mutant population of 123 individuals. However, chilli fruits from 17 plants showed resistance after laboratory inoculation experiments. These chilli plants were crossed with the 'Hoarue Huaisai', which has large fruit. The F_2 progenies were selected for anthracnose resistance and large fruits. Two hybrids with anthracnose resistance (derived from the cross CA1131 × 'Hoarue Huaisai') were identified and used for field anthracnose resistance tests in 2015. Resistant plants with large fruits were discovered in the F_3 inbred line no. 6-1-4 grown during the dry season, but this line did not show strong disease resistance in the rainy season. A further 63 F_5 inbred lines showed anthracnose resistance in the field experiment. Five samples per line of each of the 63 inbred lines were inoculated in the laboratory at Thailand Institute of Nuclear Technology. The fruits of inbred line no. 32-2-8 showed complete anthracnose resistance and seven lines were segregating as resistant. All eight of these lines are being used in the ongoing chilli project aimed at developing chilli varieties with broad resistance to anthracnose caused by three *Colletotrichum* species that are prevalent in Thailand.

Keywords: chilli • electron beam • anthracnose

1 Introduction

Chilli is considered as one of the most important commercial spice crops and is a widely used universal spice, sometimes called 'wonder spice'. Different varieties are cultivated for varied uses such as vegetable, pickles, spice and condiments. Chilli (botanically known as *Capsicum annuum* L.; *Capsicum frutescens* L.), also called red pepper, belongs to the genus *Capsicum* in the family Solanaceae; it is believed to have originated in Latin America (Kraft *et al.*, 2013). Chilli is referred to as chilli (or chili), hot pepper, bell pepper, red pepper and capsicum in different parts of the world. Chilli is integral to and the most important ingredient in many different cuisines around the world as it adds pungency, taste, flavour and colour to the dishes (Ministry of Agriculture of India, 2009). World production was reported by FAOSTAT in 2017 to be approximately 36.1 and 4.6 million tonnes of fresh and dried chilli fruit. Anthracnose is a serious problem for chilli in

[1] Email: vichaipuri@gmail.com

 Mutation Breeding, Genetic Diversity and Crop Adaptation to Climate Change (eds S. Sivasankar *et al.*)

DOI: 10.1079/9781789249095.0017

Asia and is also a major constraint to chilli production in the tropics and subtropics worldwide, for example in the USA and Latin America (Mahasuk *et al.*, 2013). Thailand chilli is world famous for two important commercial qualities: its red colour and hot taste levels. The famous red colour is due to the pigment capsanthin and the hot taste is due to a high concentration of capsaicin. The other quality parameters in chilli are length, width and skin thickness.

Some species of the fungal genus *Colletotrichum* are important plant pathogens that cause the economically important disease anthracnose in a wide range of hosts, including cereals, legumes and vegetables. Among these hosts, chilli (*Capsicum* spp.), an important economic crop worldwide, is severely affected by anthracnose, which may cause yield losses of up to 50%. Typical anthracnose symptoms on chilli fruit include sunken necrotic tissues, with concentric rings of acervuli (Than *et al.*, 2008). Anthracnose in chilli fruit is one of the most destructive diseases of chilli-growing areas in Thailand, tropical Asia and worldwide (Ratanacherdchai *et al.*, 2007). Three major *Colletotrichum* species, *C. acutatum*, *C. capsici* and *C. gloeosporioides*, have been reported as the causal agents of chilli anthracnose in Thailand (Mongkolporn *et al.*, 2010). There is little information concerning the interactions between the complexes of species involved in chilli anthracnose. This information is necessary for plant breeding purposes and disease management. Some *Colletotrichum* species respond differently to various control measures; for example, *C. acutatum* was found to be moderately susceptible to the fungicide benzimidazole, while *C. gloeosporioides* was highly susceptible. Correct and accurate identification will ultimately lead to more effective disease control and management, such as selecting appropriate fungicides, or the breeding of resistant cultivars (Than *et al.*, 2008).

Although resistance to anthracnose caused by *C. gloeosporioides* was identified in chilli varieties by the Asian Vegetable Research and Development Centre (AVRDC) at Kasetsart University, Kamphaengsaen campus, Nakhon Pathom, none of them were stable. Six anthracnose-resistant varieties donated by AVRDC were provided to this project for translational research in Sukhothai Horticulture Research Centre and the nearby area in which anthracnose disease was widespread and serious. The objective of this project was to investigate new chilli genotypes, generated by electron beam irradiation, that are stably resistant to *C. gloeosporioides* infection.

Study of the molecular mechanism of a high mutation frequency induced by high-energy pulsed electron beam radiation elucidated the effects of electron beam radiation on yeast cells (Zhu *et al.*, 2008). The results showed that the viability of yeast cells declined with increasing absorbed dose. The beam-induced DNA damage including single-strand breaks and double-strand breaks could lead to gene mutations. Electron beams may therefore be a new effective method for induced mutation breeding and deserve further research (Luo *et al.*, 2012).

2 Materials and Methods

2.1 Dose optimization for seed irradiation

Preliminary seed irradiation by electron beam with 0, 0.1, 0.2, 0.3, 0.4, 0.5 and 0.6 kGy at the energy level 8 MeV was used to determine the most suitable dose rate for the research project. One hundred irradiated seeds per treatment were soaked and planted on soft paper in a plastic box. After 12 days the number of germinated seedlings was counted and compared with the non-irradiated control.

2.2 Seeds irradiation in M_1 and bulk seed selection in M_2

One hundred and eighty-four seeds of CA1131 and 100 seeds of ‘Huarue Huaisai’ chilli varieties were irradiated at the energy level of 8 MeV (suitable dose rate from a preliminary experiment) and 123 survivors from CA1131 were planted in Sukhothai Horticulture Research Centre (SHRC) and observed for variety compatibility with the climate and environment in Sukhothai province.

The progeny of these M_1 plants were grown to maturity as the M_2 generation and a variety of characters were recorded. The best performing 1–2 M_2 plants were retained for anthracnose resistance screening.

2.3 Selection for anthracnose resistance in the M_2 generation

M_2 seed was collected and sown and 1922 M_2 plants were screened for anthracnose resistance in an experimental field.

2.4 *C. gloeosporioides* inoculation of selected M_2 fruits

A *C. gloeosporioides* cell suspension (10^6 cells/ml) was prepared for inoculation of selected fruit (those that had not shown disease symptoms in the field). Inoculated fruits were put in parafilm-sealed Petri dishes and incubated at 37°C for 7 days.

2.5 'Huarue Huaisai' crosses

Crosses between two selected CA1131 lines and 'Huarue Huaisai' were performed in order to combine the large fruit size of 'Huarue Huaisai' with the putative anthracnose resistance trait from the mutant lines.

Selected CA1131 M_2 individuals and the variety 'Huarue Huaisai' were planted at SHRC where F_1 seed was collected from reciprocal crosses between selected field-grown CA1131 M_2 individuals and the variety 'Huarue Huaisai'.

2.6 Analysis of the F_2 of 'Huarue Huaisai' crosses

Anthracnose resistance was investigated in the F_2 and later generations of the 'Huarue Huaisai' crosses at SHRC. F_3–F_5 plants derived from healthy individuals in the previous generation were planted at the Sukhothai Horticulture Research Centre.

2.7 Anthracnose resistance screening

Anthracnose resistance was assessed for large-fruited individuals by the response to inoculation of fruit collected from healthy F_3–F_5 field-grown plants. This *C. gloeosporioides* inoculation and screening were performed in the laboratory as described in 2.4 above.

3 Results

3.1 Dose–response experiments of chilli seed irradiation by electron beam

Seeds of CA1131 were irradiated with electron beams at a range of doses (0, 0.1, 0.2, 0.3, 0.4, 0.5 and 0.6 kGy) at an energy level of 8 MeV in order to determine the most suitable dose for these experiments. The LD_{50} could be determined to be around 0.27 kGy. The results showed that the 0.2–0.4 kGy dose was most appropriate for use with chilli seeds, with survival rates of 64% and 40% (Table 17.1). The 0.3 kGy dose rate was applied for this project because of the larger number of healthy plants available for selection.

3.2 Selection of chilli varieties for mutagenesis

The 184 seeds of CA1131 were irradiated again with 0.3 kGy from an 8 MeV electron beam (suitable dose rate from a preliminary experiment) and

Table 17.1. Electron beam dose optimization for CA1131 chilli seed irradiation.

EB doses (kGy)	Number of seeds	Survival after 12 days (%)	Survival (control adjusted) (%)
Control	100	84	100
0.1	100	73	86.90
0.2	100	54	64.28
0.3	100	54	64.28
0.4	100	34	40.47
0.5	100	0	0
0.6	100	0	0

showed 123 survivals. The M_1 irradiated seeds of this variety were planted at SHRC and it was found that CA1131 was able to adapt to the Sukhothai climate and environment.

3.3 Selection in M_2 plants for anthracnose resistance and laboratory inoculation experiment

M_1 and M_2 plants were derived from irradiated seeds of CA1131 and were planted at SHRC for the investigation of anthracnose resistance. Thirty anthracnose-resistant plants were selected after the appearance of the disease in Sukhothai province at the time of growing the M_2 generation. The excised fruits from 17 of these plants showed complete resistance in an inoculation experiment performed in the laboratory of the Thailand Institute of Nuclear Technology.

3.4 Crossing between selected CA1131 lines with 'Huarue Huaisai'

The 17 candidate anthracnose-resistant CA1131 M_2 plants were crossed with the variety 'Huarue Huaisai'. Their F_2 progenies were selected for anthracnose resistance, presumably derived from a CA1131 mutant gene. These plants were large and healthy. The two F_2 plants that had anthracnose-resistant fruits ('Huarue Huaisai' 5-34 × CA 1131-2-27 (Fig. 17.1) and CA 1131(1) × 'Huarue Huaisai' 5-34) were identified and their selfed progeny were used for field anthracnose resistance screening.

3.5 Investigation of anthracnose-resistant hybrid chilli

F_3-F_5 individuals were planted and investigated for anthracnose-resistant hybrid chilli at SHRC and were screened for fruit size and quality. In the 2015–2016 field experiment, 63 inbred lines showed anthracnose resistance; five samples per line of each were inoculated in the laboratory. The inbred line fruits of no. 32-2-8 showed complete anthracnose resistance (Fig. 17.2) and seven lines showed segregating resistance (Table 17.2). Seed production in the F_5 showed that these plants were healthy with a high yield (1.55 kg per plant).

4 Discussion

From a preliminary experiment, the LD_{50} was determined to be around 0.27 kGy which has a correlation coefficient of 0.96 for electron-beam treated chilli. The germination rate of chilli seeds showed a linear or log-linear dose–response curve against varying doses of electron beam. Each chilli cultivar had its own LD_{50} dose. If the

Fig. 17.1. Anthracnose-resistant fruits of 'Huarue Huaisai' 5-34 × CA 1131-2-27 in 60 × 15 mm Petri dishes.

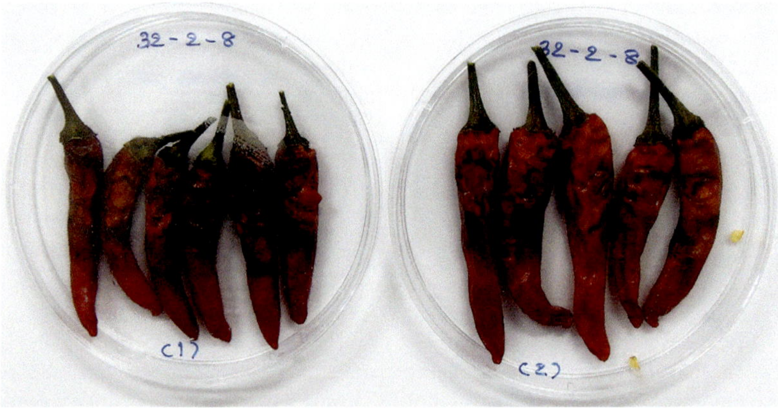

Fig. 17.2. Anthracnose-resistant fruits of inbred line no. 32-2-8 in 60 × 15 mm Petri dishes.

Table 17.2. Resistance to anthracnose assessed on fruit.

Plant number	Observed phenotype
32-2-8	All 5 fruits are anthracnose resistant
1-2-8	1 fruit is anthracnose resistant
7-2-7	1 fruit is anthracnose resistant
7-3-8	1 fruit is anthracnose resistant
19-1-8	1 fruit is anthracnose resistant
34-1-8	1 fruit is anthracnose resistant
35-2-5	2 fruits are anthracnose resistant
39-1-2	1 fruit is anthracnose resistant

seeds were exposed over their LD_{50} dose, their germination rates would be reduced. The LD_{50} electron beam dose for chilli irradiation, defined as the appropriate dose, ranged from 0.2 kGy to 0.4 kGy (Table 17.1).

Anthracnose infection in Sukhothai province and the nearby area was caused by *C. gloeosporioides* and no resistant variety was available. Therefore, mutation induction to create a new source was undertaken. However, recessive resistance is usually more durable than dominant resistance (IAEA, 2018). This information can benefit chilli breeding programmes for the production of anthracnose-resistant varieties (Kim *et al.*, 2007). Our results showed that anthracnose resistance could be induced by electron beam irradiation. The dose of 0.3 kGy at 8 MeV is suggested for mutation breeding in chilli varieties. Although there were differences in the observed LD_{50} for the different varieties, these were not statistically significant.

Eight F_5 lines showing anthracnose resistance were obtained in this project. Only line no. 32-2-8 showed uniform anthracnose resistance and seven lines showed segregating resistance after inoculation in the laboratory. However, all eight lines can be used to study resistance to anthracnose caused by *C. gloeosporioides* and the other two strains, *C. capsici* or *C. acutatum*, which are present in other areas (Mongkolporn *et al.*, 2010). The project of mutation induction for anthracnose resistance in chilli by electron beam irradiation was carried out based on this research because Sukhothai province in northern Thailand has only one strain of the fungus that causes anthracnose disease (Pongpisutta *et al.*, 2010). So, the resistant lines from SHRC will be used as parent lines in the next step of the breeding programme. We will investigate the resistance of anthracnose disease caused by *C. capsici* and *C. acutatum* in Nan Agricultural Research Centre, Thailand. This research centre is located in an isolated mountainous area and always encounters anthracnose disease and infections of multiple strains of the fungus every year.

5 Conclusion

1. The LD_{50} electron beam dose for chilli irradiation ranged from 0.2 kGy to 0.4 kGy.

2. Seventeen plants showed complete anthracnose resistance in an inoculation experiment at the M_2 generation.

3. Two anthracnose-resistant chilli fruits derived from crosses between 'Huarue Huaisai' and CA1131 were discovered, but these two hybrids produce rather small fruits.

4. After a 3-year field experiment, five samples per line of each of 63 inbred lines were inoculated in the laboratory. Fruits of the inbred line no. 32-2-8 were all anthracnose resistant and seven other lines showed segregating resistance.

Only *C. gloeosporioides* has been recorded in Sukhothai province and nearby (Pongpisutta *et al.*, 2010) but all three species are found in some parts of Thailand, such as Nan province. The aim of the ongoing chilli project is to select anthracnose resistance in chilli with the ability to resist all three fungi species.

Acknowledgements

We would like to thank AVRDC (Asian Vegetable Research and Development Centre, Kasetsart University, Kamphaengsaen Campus, Nakhon Pathom) for its materials support.

References

IAEA (2018) *Manual on Mutation Breeding*, 3rd edn. Joint FAO/IAEA Programme. Nuclear Techniques in Food and Agriculture.

Kim, S.H. *et al.* (2007) Resistance to anthracnose caused by *Colletotrichum acutatum* in chilli pepper (*Capsicum annuum* L.). *Journal of Crop Science and Biotechnology* 10, 277–280.

Kraft, K.H. *et al.* (2013) Multiple lines of evidence for the origin of domesticated chilli pepper, *Capsicum annuum*, in Mexico. *Proceedings of the National Academy of Sciences* 111(17), 6165–6170.

Luo, W.X. *et al.* (2012) Effect of electron beam radiation on trait mutation on azuki bean (*Vigna angularis*). *African Journal of Biotechnology* 11, 12939–12950.

Mahasuk, P. *et al.* (2013) Differential resistances to anthracnose in *Capsicum baccatum* as responding to two *Colletotrichum* pathotypes and inoculation methods. *Breed Science* 63, 333–338.

Ministry of Agriculture of India (2009) Post-harvest profile of chilli. Available at: http://agris.fao.org/agris-search/search.do?recordID= US201300009374 (accessed 2019)

Mongkolporn, O. *et al.* (2010) Differential reactions on mature green and ripe chilli fruit infected by three *Colletotrichum* spp. *Plant Disease Journal* 94, 306–310.

Pongpisutta, R. *et al.* (2010) Morphological variation of *Colletotrichum* spp, causal agents of chill anthracnose. *Agricultural Science Journal* 41 (Suppl. 1), 318–321.

Ratanacherdchai, K. *et al.* (2007) RAPD analysis of *Colletotrichum* species causing chilli anthracnose disease in Thailand. *Journal of Agricultural Technology* 3, 211–219.

Than, P.P. *et al.* (2008) Chilli anthracnose disease caused by *Colletotrichum* species. *Journal of Zhejiang University Science B* 9, 764–778.

Zhu, H. *et al.* (2008) Effects of high-energy pulse electron beam radiation on biomacromolecules. *Science in China Series B: Chemistry* 51, 86–91.

18 Induced Mutagenesis for Improvement of Bean (*Phaseolus vulgaris* L.) Production in Bulgaria

Svetla Sofkova-Bobcheva, Ivelin Pantchev, Ivan Kiryakov, Petar Chavdarov, Yordan Muhovski, Fatma Sarsu and Nasya Tomlekova[1]

Abstract

Although historically a surplus food producer, Bulgarian agriculture has faced a downturn in recent decades. Local legume cultivars have lost favour with farmers and the canning industry, due to their low productivity in comparison with imported ones. Diseases and abiotic stresses are the most important factors limiting the production of edible legumes, costing farmers hundreds of euros in lost revenue each year. The overall objective of our on-going bean mutation breeding programme was to enrich the gene pool of *Phaseolus vulgaris* L. and to develop genotypes resistant to *Xanthomonas axonopodis* pv. *phaseoli* (Smith) (*Xap*) and *Pseudomonas savastanoi* pv. *phaseolicola* (Burkh.) (*Psp*) using EMS. An elite line and common cultivar (an heirloom and a snap bean type) in Bulgaria, were selected as parents and the chemical mutagen EMS was used for generating mutations. In total, 1000 seeds were treated and the two generated M_1 populations were grown in the field. All M_2 mutant plants (1650 from initial line IP564 and 2420 from initial cultivar 'Mastilen 11b') were grown in field conditions and a number of phenotypic changes were observed on these mutated plants. They were also screened for *Xap* disease resistance via leaf artificial inoculation under greenhouse conditions. Individual plant selection was performed for the putatively resistant M_2 plants. In the M_3 generation these lines were screened using artificial inoculation with *Xap* and *Psp* pathogens (leaves and pods) under field conditions. Selected M_3–M_4 lines with confirmed disease resistance were tested for fresh pod quality. Yield tests were started in M_4 and M_5 generations and, according to their productivity performance, mutants were advanced to the M_6/M_7 generation for validation. The expression patterns of genes putatively involved in the resistance reactions towards two races of *Psp* were determined using qRT-PCR for the specific and reference genes. In conclusion, 50 plants with visible morphological changes and/or increased tolerance to the two targeted bacterial diseases were selected. A total of 20 advanced mutant bean lines are currently being evaluated for their competitiveness in multiple sites.

Keywords: common bean • induced mutagenesis • disease resistance • halo blight • common bacterial blight

1 Introduction

Common bean is a very important component of the diets of people in several continents, representing an important source of minerals and proteins (Gepts *et al.*, 2008). Consumption of dry and vegetable (snap) beans in Bulgaria has its traditions long back in the country's history and it continues to rise in response to consumer demand and scientific recognition of beans as a major health food. In addition to being high in fibre and protein, beans serve as an important natural

[1] Email: nasia.tomlekova@gmail.com

DOI: 10.1079/9781789249095.0018

source of folate and other B-vitamins, minerals and antioxidants. Snap beans are produced in almost every region of Bulgaria chiefly on small and medium-size farms, and dry beans are produced mainly in Southern Dobrudja, a hilly region with an average altitude of about 200–300 m and a continental climate; these beans are part of crop rotations and intercropping systems.

Common bean is exposed to a large number of diverse yield constraints during its growth. Diseases and abiotic stresses (drought, soil compaction, low soil fertility, high temperatures, etc.) are the most important factors limiting the production of beans and thus contributing to the large gap between actual and potential yield, costing farmers hundreds of euros in lost revenue each year. Sustainable agriculture requires reduced pesticide use, so cultivars with better pathogen tolerance are not only environmentally friendly but they also lower costs and raise product quality. Genetic resistance provides the best means of controlling common bacterial blight, a devastating seed-borne disease caused by *Xanthomonas axonopodis* pv. *phaseoli*, which plagues bean production in South-east Europe (Sofkova *et al.*, 2009; Singh and Miklas, 2015). Resistant cultivars can be selected from naturally occurring variation or induced mutants.

Common bean possesses a high level of natural mutations (Rukmanski, 2005). Induced mutagenesis was applied as a method for generating novel genetic diversity and obtaining new bean lines or cultivars in Bulgaria (Svetleva *et al.*, 1999a,b). Chemical mutagenesis in common bean is efficient and reliable. Although different agents have been used, mutagenesis with ethyl methanesulfonate (EMS) has been the most successful for different breeding traits, including increased yield and nodulation (Park and Buttery, 1988; Svetleva, 2004; Ramandeep *et al.*, 2018). EMS alkylates primarily guanine (G), leading to mispairing; alkylated G pairs to thymine (T) instead of cytosine (C). EMS mutations can be useful in obtaining genotypes with higher yield potential, or with altered plant architecture so that they are suitable for mechanical harvesting or resistant to lodging, or those that are early ripening, have increased levels of protein and beta-carotene, disease-, pest- and drought-resistance, or that have ecological plasticity (Tomlekova *et al.*, 2014a,b).

For the past few years our breeding efforts have been directed towards developing bean cultivars with improved disease resistance. The overall objective of our common bean programme is to enrich the gene pool of *P. vulgaris* L. through developing new genotypes for breeding by using induced mutagenesis. One of the sub-objectives is to develop and identify mutant bean genotypes resistant to *Pseudomonas savastanoi* pv. *phaseolicola* (*Psp*) and *X. axonopodis* pv. *phaseoli* (*Xap*) bacterial diseases. Moreover, we aimed to study the expression patterns of genes putatively involved in the resistance reaction to *Psp* by using quantitative reverse transcription polymerase chain reaction (qRT-PCR). The qRT-PCR has allowed for the calculation of differential gene expression in different organs or tissues under several treatments using cDNA molecules synthesized from mRNA (Heid *et al.*, 1996; Bustin, 2002). qRT-PCR is noted for its accuracy, precision and relative ease of use due to its speed, sensitivity and specificity (Reece, 2004; Bustin *et al.*, 2009). Given these features, we decided to use this method in our study.

2 Materials and Methods

2.1 Plant material

The line IP564 was developed in the Maritsa Vegetable Crops Research Institute (VCRI), Plovdiv, by a phased sexual hybridization between multiple parental components. Plants are characterized by a determinate type of growth with height of 50–55 cm, with a strong and erect stem and medium-size shrub. It has cylindrical, long (13–15 cm), straight green pods. The vegetation period is on average 48–50 days from germination to the occurrence of technological maturity of the pods. The seeds are white, kidney-shaped and of medium size (30.5 g/100 seeds). Plants of line IP564 have resistance to anthracnose (*Colletotrichum lindemuthianum* races 6 and 81), rust (*Uromyces appendiculatus* races 20-0, 20-2 and 20-3) and bean common mosaic virus (BCMV), but are moderately susceptible to halo blight (*P. savastanoi* pv. *phaseolicola* (Burkh.) Gardan *et al.* races 1 and 6) (*Psp*) and sensitive to the agent of bacterial blight (*X. axanopodis* pv. *phaseoli*) (*Xap*).

Cultivar 'Mastilen 11b' was developed in the VCRI from the local land race 'Mastilen'. The plant has a determinate growth habit and average height of about 40–45 cm, with good foliage distribution. The stem is erect with 6–8 branches, each typically approximately 18–20 cm in length. Pods are set in the lower part of the plant and some of them rest on the ground. The leaves are dark green and large. Lamina is ovate, with a rather pointed tip, 10–12 cm long and 7–10 cm wide. The flowers are purple and of medium size. Pods are flat, broad, slightly curved; the main colour is green with anthocyanin patterns. Dimensions of pods are: length 11–14 cm, width 13–16 mm. One peduncle bears on average two to three pods. At harvest pods are fragile, juicy, with small cavities without fibres and without a sclerenchyma layer and they have an average of four to five seeds per pod. Seeds are of medium size, ellipsoidal, convex, with a main colour of beige-violet and anthocyanin patterns, and weigh about 35.5 g/100 seeds. The hilum is white with a light brown halo. Days to maturity are 45–50 on average. 'Mastilen 11b' is sensitive to bacterial blight *Xap* and halo blight *Psp*.

2.2 Mutation breeding towards tolerance to bacterial pathogens

Initially, 100 seeds (M_0) from each parent genotype (line IP564 and cv. 'Mastilen 11b') were treated with five different concentrations of EMS. All plants from the M_1 generation were grown in the field, and then half of the M_2 seeds obtained from each M_1 plant were used for disease screening in the greenhouse; the other half were grown for seed reproduction.

In the M_2 generation, mutant lines with differential disease reaction were selected and further screened for disease resistance in the field. Identified disease-resistant M_3–M_4 lines were tested for grain and pod quality. Yield tests were started in the M_4 and M_5 generations, according to quality results.

Our molecular breeding approach was based on the expression analysis of plant pathogenesis-related proteins (PRP) in M_6 advanced mutant lines.

EMS-treatment assay

The mutagenesis protocol was first established in the molecular biology laboratory of the Maritsa VCRI with the assistance and guidance of Professors Christov and Svetleva. One hundred seeds for every treatment from two snap bean genotypes were treated with five different EMS concentrations (1.55, 3.1, 6.2, 12.4 and 24.8 mM) for 6 h at room temperature with slow shaking, adding 1% methanol as an intermediary solvent. As a result of an assay conducted before treatment, the time of absorption of water by the seeds was assessed and the protocol was modified; the treatment with EMS was conducted for 6 h. Before the present study, the reported results for the number of grown seedlings, survived plants and plants with formed seeds helped to establish protocols for the treatment of bean seeds with EMS (Svetleva, 2004; Christov *et al.*, 2014). A 30-min washing with running tap water was performed after treatment. The untreated seeds were pre-soaked in distilled water and were used as control. Seeds were rinsed and sown in the field in a complete randomized-block design (CRBD) with two replicates, to the first mutant (M_1) generation. Each plot had two rows 3 m long with a row-to-row and plant-to-plant distance of 70 cm and 10 cm, respectively.

Inoculum preparation

Bacterial cultures of *X. axanopodis* pv. *phaseoli* (*Xap*) and races 1 and 6 of *P. syringae* pv. *phaseolicola* (*Psp*) were provided from the collection of Dobrudja Agricultural Institute General Toshevo, and propagation potato dextrose agar medium at the Laboratory of Phytopathology of Institute of Plant Genetic Resources 'Konstatin Malkov', Sadovo. For the inoculation treatment, a 2-day-old bacterial suspension of 10^8 cfu/ml, further adjusted on optical spectrophotometer on wavelength 600 nm to OD = 0.3, was used to inoculate the fully expanded first leaf of each plant using multiple needle technique as described by Aggour *et al.* (1989).

Mutants in M_2–M_4 progenies were screened for *Xap* and *Psp* disease resistance by field inoculation of fully developed trifoliate leaves and green immature pods by pricking with a syringe according to the method of Aggour *et al.* (1989).

Bioassays

M_6 plants from four mutant lines were inoculated with *Psp* bacteria under greenhouse conditions using the same methods as described above. Parental line IP564 was used as a positive control due to its resistance to race 1 and moderate suitability to race 2 of *Psp* (unpublished data). Ten plants were planted in potting mix under controlled conditions at 22/16°C (day/night) and ca. 80% humidity.

Symptoms were recorded in the inoculated area 10 days after inoculation by using a 1–9 severity scale where 1 = no visible symptoms and 9 = very severe symptoms (Aggour *et al.*, 1989). Leaves from five plants from untreated control and inoculated treatments were separately collected at different periods: 6, 24, 48 and 74 h post inoculation (hpi). Finally, each sample represented a bulk from five individual plants. In addition, the experiment was repeated twice and another total RNA bulk was obtained, resulting in two biologically replicated experimental bulks. Plant materials were immediately frozen in liquid nitrogen and maintained at –80°C until RNA isolation.

2.3 Gene selection and primer design

The following target PRP genes were chosen for the experiment: *FLS2*, *FLS2.1*, *FLS2.3*, *RPG1-B* (AAR19097.1), *PTO* (AAK52036.1) and *PTO1* (Table 18.1). The gene sequences used in this study were obtained through bibliographical reviews of studies involving biotic stresses in common bean (Trabanco *et al.*, 2014). All of the target genes contained typical protein domains involved in resistance responses, such as leucine-rich repeat (LRR) domains. Genes that encode typical R proteins co-localized with the major resistance genes *Pse-3* and *Pse-1* (at the end of LGs Pv02 and Pv10).

Reference genes actin (*act*11) and ß-tubulin (*tub*8) were selected according to Borges *et al.* (2011).

The gene-specific primers were designed by Primer3 software (http://bioinfo.ut.ee/primer 3-0.4.0/, accessed 2019) using default parameters and were ordered from EUROGENTEC (Belgium). Primer specificity was checked with end-point polymerase chain reaction using bean cDNA or gDNA as template (Table 18.2).

2.4 Quantitative RT-PCR

Total RNA extraction

For each sample, 100 mg of plant material was crushed in 1 ml of TRI REAGENT® (MRC) and the total RNA was extracted following the manufacturer's recommendations. DNAse I (Life Technologies) treatment was applied to remove any residual DNA contamination. RNA quantity was evaluated spectrophotometrically (IMPLEN) at 260 nm and purity was determined by evaluating the absorbance ratio at 260/280 nm. RNA quality was verified on 1.0% agarose gel for the presence of the two predominant intact bands corresponding to 28S and 18S rRNA.

First strand cDNA synthesis

For each sample, 2.5 μl (5 μg) total RNA was added to: 4 μl of 5 × Reaction Buffer (Thermo Scientific®); 1 μl of RiboLock RNase Inhibitor (20 U/μl); 2 μl of 10 mM dNTP Mix and 1 μl (200 U/μl) of MuLV Reverse Transcriptase (Thermo Scientific® RevertAid H Minus First Strand cDNA Synthesis Kit), following procedures recommended by the manufacturer. The reverse transcription reaction was performed at 42°C for 1 h. The reverse transcription reaction was either used directly in further PCR applications or stored at –20°C.

qRT-PCR

Quantitative reverse transcription polymerase chain reactions for the specific and reference genes were carried out on a CFX96 cycler (Bio-Rad) in one step using 2.5 μl of cDNA (100 ng), 12.5 μl 2X SYBR Go Taq® qPCR Mastermix (Promega) and 300 nM of each primer in final volume of 25 μl. Two technical replicates were performed. Melting curve analysis was applied in order to verify primer specificity. A primer efficiency test was performed using tenfold serial dilution.

2.5 Data analyses

Raw data of fluorescence levels were processed and C_T values were provided from Bio-Rad Manager software. This programme performs baseline correction and linear regression analysis on each amplification curve. Further on, this data was used to calculate the relative gene expression of the target and reference genes using the equation described by Livak and Schmittgen (2001) for the $2^{-\Delta\Delta C_T}$ method:

$$\Delta\Delta C_T = (C_{T,\ \text{Target gene}} - C_{T,\ \text{reference gene}})_{\text{Time X}} - (C_{T,\ \text{Target gene}} - C_{T,\ \text{reference gene}})_{\text{Time 0}}$$

Table 18.1. Genes involved in the resistance to *Pseudomonas savastanoi* pv. *phaseolicola*.

Psp resistance genes	Species of origin	Gene bank Accession no.	Identified candidate gene	Annotated function in *P. vulgaris* genome	Identity %	E value	Physical location	LG	QTL mapped
FLS2	*A. thaliana*	NP_199445.1[a]	Phvul.004G136500.1	LRR receptor serine/threonine protein kinase	31.1	$5.4e^{-102}$	Chr04: 41556561–41560351	4	*Psp4*[812XC]
FLS2.1	*A. thaliana*	NP_199445.1[a]	Phvul.004G175700.1	LRR receptor serine/threonine protein kinase	30.0	$3.4e^{-103}$	Chr04: 45622442–45626397	4	*Psp4*[812XC]
FLS2.3	*A. thaliana*	NP_199445.1[a]	Phvul.004G175900.1	LRR receptor serine/threonine protein kinase	30.5	$3.6e^{-96}$	Chr04: 45637544–45641336	4	*Psp4*[812XC]
Pto	*G. max*	NP_001241285.1[a]	Phvul.004G164000.1	serine/threonine protein kinase	45.9	$8.5e^{-65}$	Chr04: 44601151–44605163	4	*Psp4*[812XC]
Pto1	*P. vulgaris*	AAK52036.1	Phvul.006G080500.1	serine/threonine protein kinase	46.3	$4.7e^{-50}$	Chr06: 19953718–19967156	6	*Psp6*[812XC]
RPG1-B	*G. max*	AAR19097.1	Phvul.006G066800.1	LRR-containing protein ADP binding	57.7	0	Chr06: 18542599–18546221	6	*Psp6*[812XC]

[a] Sequences obtained from RefSeq database

Table 18.2. Primers designed for qRT-PCR analyses.

Gene	Expected size (bp)	Specific primer sequence
PvFLS2 Phvul.004G136500.1 LRR receptor serine/threonine protein kinase	102	*PvFLS2*-D CAACCTCATCCCTGGTGACT *PvFLS2*-R TCAAGAGGAACCTTGCCATC
PvFLS2.1 Phvul.004G175700.1 LRR receptor serine/threonine protein kinase	119	*PvFLS2.1*-D GCATGCTTTCTCAAACCACA *PvFLS2.1*-R TTTCTGGGATTTCCCCTACC
PvFLS2.3 Phvul.004G175900.1 LRR receptor serine/threonine protein kinase	120	*PvFLS2.3*-D TCCTTCCAAGTTCCCATCAG *PvFLS2.3*-R TTCTCCAAAGCCACCATTTC
PvPto Phvul.004G164000.1 Serine/threonine protein kinase	115	*PvPTO*-D TGGTGAACCACTTCCTCCTC *PvPTO*-R AAGGAGGTTCGCATCAGAGA
PvPto.1 Phvul.006G080500.1 Serine/threonine protein kinase	108	*PvPTO.1*-D GATCGGCTTCAATCTTTTGG *PvPTO.1*-R CAAGCAATTGCCCCTACAAT
RPG1-B Phvul.006G066800.1 LRR -containing protein ADP binding	109	*PvRPG1-B*-D CAAGCCAAAGGGGTGATCTA *PvRPG1-B*-R TCAGATCCAACAACCAACGA
Actin-1 (reference gene 1)	190	*Act11*-D TGCATACGTTGGTGATGAGG *Act11*-R AGCCTTGGGGTTAAGAGGAG
Tubulin beta-8 (reference gene 2)	163	*Tub8*-D AATGTGAAGTCCAGCGTGTG *Tub8*-R CTTCCCCAGTGTACCAATGC

3 Results and discussion

3.1 Mutagenesis, plant maintenance and phenotype screening

The highest number of changes in the morphology of the plant was recorded at 6.2 mM mutagen concentration.

Initial treatment by EMS of M_0 seeds was followed by propagation from M_1 to M_6 generation.

In the M_2 generation, a total of 218 mutants derived from the line IP564 and cultivar 'Mastilen 11b' were selected for their morphological changes and/or better tolerance to bacterial blight (Fig. 18.1). After artificial infection with isolate *Xap*, 124 mutagenized plants of IP564 and 22 plants of 'Mastilen 11b' were selected in which morphological changes were observed in combination with a better response of tolerance to bacterial blight compared with the initial line.

Morphological description of elite plants was performed and seeds were collected from all of them. EMS treatment resulted in a number of phenotypic changes in the progeny (Fig. 18.2).

The following phenotypic changes compared with the initial genotype were observed:

- type of growth: from determinate to indeterminate and transformation of the apical bud from an inflorescence meristem to a vegetative meristem in mutant plants originating from IP564;
- length and number of internodes, fruit stems and branches;
- colour of flowers from magenta to white in mutants from the initial cultivar 'Mastilen 11b';
- colour and shape of the pods (from green to dark green, green with purple stripes and/or spots on the branches of the initial IP564 and vice versa) in mutants of the initial cultivar 'Mastilen 11b'; and
- colour, size and shape of the seeds.

As a result of the field screening conducted with *Xap* inoculation, 1000 M_3 plants from 216 M_2

Fig. 18.1. (a) Descriptive assessment scale for a bean plant infected with *Xanthomonas axonopodis* pv. *phaseoli (Xap)*. **(b, c, d)** Differential reaction on leaves and pods of the mutant bean lines 14 days after inoculation with *Xap*.

Fig. 18.2. Observed phenotypic changes in mutant bean plants of M_2 and M_3 generations obtained by treatment with EMS. **(a)** Leaf chlorophyll mutation. **(b)** Albino mutation. **(c)** Unifoliate seedling. **(d)** Flower colouration. **(e), (f)** Plant habit mutation. **(g)** Multi-pinnate leaf mutation. **(h)** Pod colouration.

mutant lines with phenotypic changes, confirmed in the M_3 generation, had in addition increased tolerance to bacterial blight compared with the initial lines (Table 18.3).

The results from the extended screens for pathogen resistance at Dobrudja Agricultural Institute's field facilities for disease screening are presented in Table 18.4. Seventeen lines have shown an increase in their degree of resistance against both pathogens. These results demonstrate that EMS mutagenesis is an efficient approach to generate a significant number of

Table 18.3. List of the bean (*P. vulgaris* L.) mutant lines identified with multiple resistance of leaves and pods to *Pseudomonas savastanoi* pv. *phaseolicola* (*Psp*) races 1 and 6, and two isolates of *Xanthomonas axonopodis* pv. *phaseoli* (*Xap*) according to the inoculation assays.

	Pseudomonas savastanoi pv. *phaseolicola*				*Xanthomonas axonopodis* pv. *phaseoli*			
	Race 1		Race 6		XB96221		XB99132	
Ref # 2014	leaf	pods	leaf	pods	leaf	pods	leaf	pods
Artificial inoculation under field conditions								
34-1-1	R	R	R	R	R	MR	MR	MS
34-1-2	R	R	R	MR	R	MS	MR	S
43-1-1	R	R	R	MR	R	MR	MR	MS
46-1-1	R	MR	R	R	R	I	MR	R
175-3-1	R	R	R	R	R	MS	R	MS
175-3-3	R	I	R	R	R	R	MR	R
175-3-4	R	I	R	I	R	MS	R	MR
190-1-1	R	R	R	R	MR	MR	R	MR
190-3-6	R	R	R	MR	R	S	R	S
190-4-1	R	R	R	R	R	S	R	S
191-1-1	R	MR	R	R	R	MS	MR	MS
193-9-2	R	R	R	MS	R	S	MR	S
Artificial inoculation under greenhouse conditions								
564-5-1	R	R	R	R	R	MR	MR	MS
564-22-1	R	R	R	MR	R	MS	MR	S
564-32-1	R	R	R	MR	R	MR	MR	MS
564-42-2	R	MR	R	R	R	I	MR	R
564-60-1	R	R	R	R	R	MS	R	MS
564-69-1	R	MR	R	R	MR	MR	MR	MR
564-69-3	R	MR	R	R	R	MR	R	MR
564-69-4	R	MR	R	R	R	MR	R	MR
564-69-5	R		R	R	R	MR	R	MR
564-69-6	R	MR	R	R	R	MR	R	MR
564-69-7	R	MR	R	R	R	MR	R	MR
564-69-8	R	MR	R	R	R	MR	R	MR
564-69-12	R	MR	R	R	R	MR	R	MR
564-69-32	R	R	R	R	MR	MS	MS	MS
564-69-35	R	MR	R	R	MR	MS	MS	MS
564-71-1	R	I	R	R	R	R	MR	R
564-74-4	R	I	R	I	R	MS	R	MR
564-75-1	R	R	R	R	MR	MR	R	MR
564-85-1	R	R	R	MR	R	S	R	S
564-89-1	R	R	R	R	R	S	R	S
564-110-1	R	MR	R	R	R	MS	MR	MS
564-161-15	R	R	R	R	R	S	R	S
564-12-1	R	MR	R	R	R	MS	MR	MS

Abbreviations: I, immune; R, resistant; MR, moderately resistant; MS, moderately sensitive; S, sensitive

mutants with multiple resistances. Seeds from selected mutants were collected to study their agronomic and yield characteristics in a field trial. Screening for plant morphology, grain quality and yield was performed in the M_4–M_5 generations.

3.2 Molecular assays

Halo blight is an economically significant disease of leguminous agricultural crops caused by the Gram-negative bacterial pathogen *Pseudomonas*

Table 18.4. List of bean (*P. vulgaris* L.) mutant lines identified with multiple resistance of leaves and seeds to *Pseudomonas savastanoi* pv. *phaseolicola* (*Psp*) races 1 and 6, and two isolates of *Xanthomonas axonopodis* pv. *phaseoli* (*Xap*).

		Pseudomonas savastanoi pv. *phaseolicola*				*Xanthomonas axonopodis* pv. *phaseoli*			
Ref №	Ref №	Race 1		Race 6		XB96221		XB99132	
2014	2013	leaf	seed	leaf	seed	leaf	seed	leaf	seed
16-1	34-1-1	R	R	R	R	R	MR	MR	MS
16-2	34-1-2	R	R	R	MR	R	MS	MR	S
18	43-1-1	R	R	R	MR	R	MR	MR	MS
20	46-1-1	R	MR	R	R	R	I	MR	R
31-1	56-4-1	R	R	R	MS	R	HS	R	HS
31-2	56-4-2	R	R	R	R	R	HS	R	HS
31-3	56-4-3	R	R	R	MS	R	MR	R	MR
32	56-5-1	R	R	R	MR	R	HS	MR	S
33	56-12-1	R	R	R	MR	R	HS	MR	S
38	64-1-1	R	R	R	R	R	S	MR	S
49	70-9-1	R	R	R	MR	MR	HS	MR	HS
53	73-1-1	R	MS	R	S	R	HS	MR	HS
54-1	73-2-1	R	MS	R	HS	R	HS	MR	HS
54-2	73-2-2	R	MS	R	S	R	MS	MR	S
55-1	73-3-1	R	MS	R	MS	R	S	MR	S
55-2	73-3-2	R	MR	R	MS	R	HS	MR	HS
55-3	73-3-3	R	I	R	MR	R	MR	MR	MS
55-4	73-3-4	R	MS	R	MS	R	HS	MR	HS
55-5	73-3-5	R	R	R	MR	R	HS	MR	S
55-6	73-3-6	R	MR	R	MS	R	HS	MR	HS
55-7	73-3-7	R	R	R	S	R	HS	MR	HS
57	73-5-1	R	R	R	MS	R	HS	MR	HS
58-1	73-6-1	R	R	R	MR	R	HS	MR	S
58-2	73-6-2	R	R	R	MS	R	HS	MR	HS
58-3	73-6-3	R	MR	R	MS	R	HS	MR	S
61	74-3-1	R	R	R	MS	R	HS	MR	S
63	74-6-1	R	MR	R	MR	R	HS	MR	S
93	172-2-1	R	R	R	R	R	MS	MR	MS
98-1	175-3-1	R	R	R	R	R	MS	R	MS
98-2	175-3-2	R	I	R	MR	R	S	MR	S
98-3	175-3-3	R	I	R	R	R	R	MR	R
98-4	175-3-4	R	I	R	I	R	MS	R	MR
118	190-1-1	R	R	R	R	MR	MR	R	MR
120-1	190-3-1	R	R	R	R	R	MR	R	MS
120-2	190-3-2	R	R	R	R	R	MR	R	MR
120-3	190-3-3	R	I	R	R	R	MR	R	MS
120-4	190-3-4	R	R	R	MR	R	MR	R	MS
120-5	190-3-5	R	MR	R	R	R	MR	R	MS
120-6	190-3-6	R	R	R	MR	R	S	R	S
120-7	190-3-7	R	R	R	R	R	R	R	R
121	190-4-1	R	R	R	R	R	S	R	S
122-1	191-1-1	R	MR	R	R	R	MS	MR	MS
122-2	191-1-2	R	MR	R	MR	R	S	R	S
125-1	192-2-1	R	S	R	S	MR	HS	R	HS
125-2	192-2-2	R	MS	R	MS	R	S	R	S
126	192-3-1	R	R	R	R	R	MR	MR	MR
132-1	193-9-1	R	R	R	MS	R	MS	MR	MS
132-2	193-9-2	R	R	R	MS	R	S	MR	S

Abbreviations: I, immune; R, resistant; MR, moderately resistant; MS, moderately sensitive; S, sensitive; HS, highly sensitive *syringae* pv. *phaseolicola* (*Psp*) (Romantschuk and Bamford, 1986).

An integral part of the research was to study: (i) the variation in the expression of selected common bean mutants and its relationship with resistance to *Psp*; (ii) the expression levels of six genes involved in the plant-pathogen resistance at different time points after inoculation with *Psp* (races 1 and 6); and (iii) the differences in gene expression between races 1 and 6 of *Psp* with different levels of aggressiveness.

Five advanced mutant lines derived from IP564 initial genotype were selected for this analysis based on their resistance to *Psp*. They were inoculated with races 1 and 6 of *Psp* and served as an mRNA source together with non-inoculated plants from each mutant line used as control. The susceptible initial genotype showed halo blight reaction and the mutant lines varied in their reaction to the pathogen races from resistance to moderate resistance, as expected.

A number of selected genes (LRR receptor serine/threonine protein kinase; serine/threonine protein kinase/LRR-containing protein ADP binding) potentially involved in *Psp* resistance were used (Table 18.1).

Relative quantification of gene expression was analysed using the comparative Ct method. The Ct value of one target gene was compared with another reference gene *Tub*8 (e.g. housekeeping gene) using the formula $2^{-\Delta\Delta CT}$ in a single sample.

Quantitative PCR analysis of reference genes

Selection of reference genes is an essential consideration to increase the precision and quality of relative expression analysis by the qRT-PCR method. For validating the comparative Ct method, the efficiency (E) of the target amplification (gene of interest) and the efficiency of the reference amplification must be approximately equal. The efficiency was then calculated based on the curve slope: $E\ (\%) = [(10^{-1/\text{slope}}) - 1] \times 100$. The desired amplification efficiencies range from 90% to 110%.

It was found that both reference genes showed high qRT-PCR efficiency rates; for *Act11* = 97.1%; and *Tub8* = 103.8% with correlation ($R \geq 0.999$) (Fig. 18.3). However, *Tub8* performed with better stability under both inoculated and controlled (not inoculated) conditions. Thus, *Tub8* was selected as the preferred reference gene for the expression study in the common bean/*Psp* pathosystem.

By studying gene expression with real-time polymerase chain reaction (qRT-PCR), we aimed to investigate changes (increases or decreases) in the expression of the set of genes by measuring the level of gene-specific transcripts. The investigation monitored the response of each gene to the inoculation with two races of *Psp*, after 24, 48 and 72 h post inoculation.

The relative gene expression levels in the M_6 lines showed differential regulation of all selected transcripts upon *Psp* pathogen inoculation (Figs 18.4 and 18.5). The expression levels of the target genes were dependent on the genotype, pathogen race level of aggressiveness and time post inoculation.

Plant interaction with race 1 of Psp

The transcripts of the selected genes responding to different time points were detected in resistant parental genotype and susceptible mutants but at different levels (Fig. 18.4). Overall, we observed greater induction of gene transcripts in 48 hpi and 72 hpi of interaction with race 1 of the pathogen compared with the 24 hpi. Having said that, *FSL2*, *FSL2.3*, *PTO1* and *RBG1-B* showed enhanced expression in inoculated resistant mutants compared with inoculated susceptible parental genotype at every time point after inoculation; the expression was significantly upregulated in all three time points for one mutant only (M192).

In contrast, *FLS2.2* and *PTO* showed inconsistent and low expression in challenged mutants until 72 h after infection, whereas infected susceptible parental genotype showed enhanced expression after 24 h post infection.

The result of expression analysis also showed that the mutant genotypes responded to the pathogen infection at a later time point than the susceptible genotype, thus proving their increased level of resistance. *Psp*-race 1 upregulated the expression of all the *P. vulgaris* defence-related genes analysed, with *FSL2.3* reaching the highest relative value. In the parental susceptible genotype compared with resistant mutants, only *FSL2*, *FSL2.3*, *PTO*, *PTO1* and *RBG1-B* were significantly upregulated in all resistant mutants, raising comparative expression values. Finally, M192 showed consistent and significant upregulation

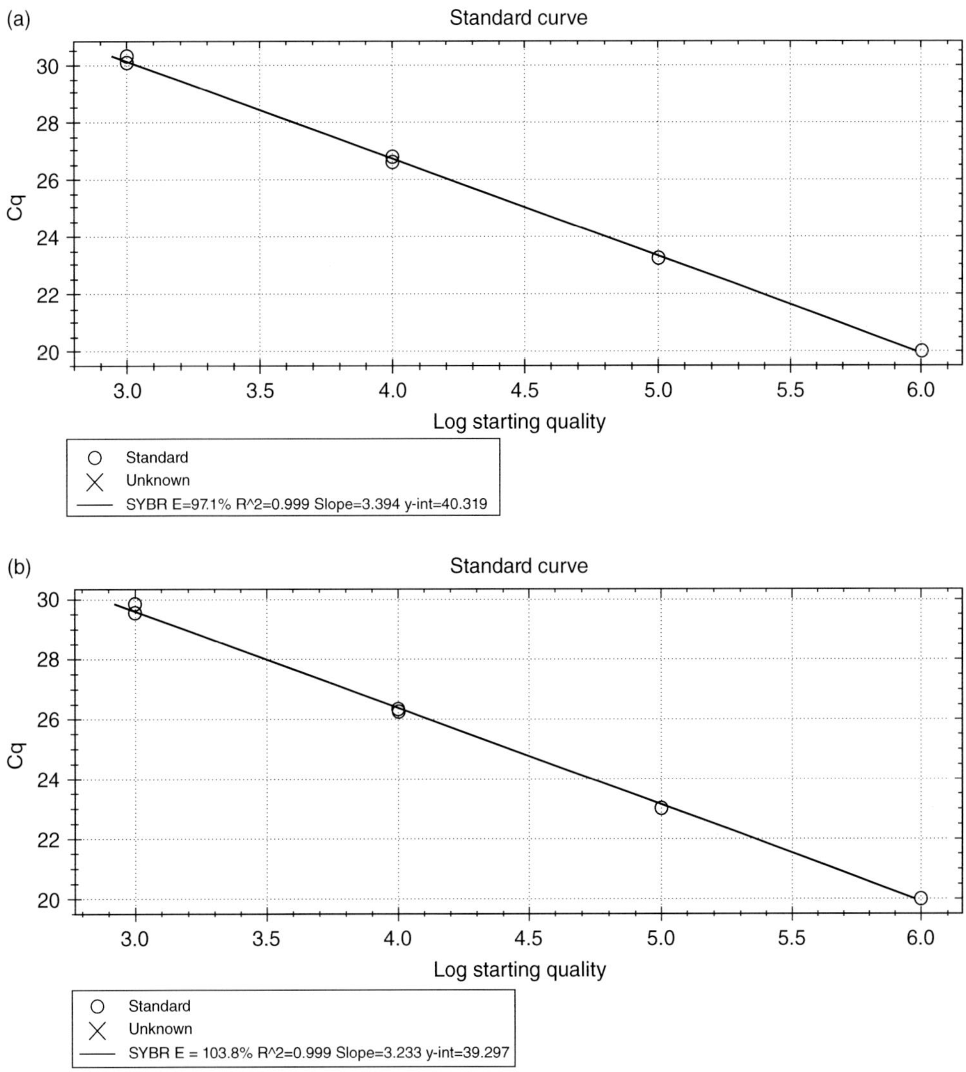

Fig. 18.3. Determination of qRT-PCR efficiencies of reference genes *Act11* **(a)** and *Tub8* **(b).** Cycle number of crossing point (Cq) versus cDNA (reverse transcribed total RNA) were plotted to calculate the slope. The corresponding real-time PCR efficiency was calculated according to equation: E (%) = $[(10^{-1/\text{slope}}) - 1] \times 100$.

of the expression ratio of all the analysed genes, with *FSL2*, *FSL2.3* and *PTO1* being the highest.

Plant interaction with race 6 of Psp

For the *Psp*-race 6 interaction, we observed greater inconsistency of gene expression among the mutant lines compared with race 1 (Fig. 18.5). The results included in Fig. 18.5 showed the following.

1. Like *Psp*-race 1, race 6 upregulated the expression of all the *P. vulgaris* defence-related genes analysed, with *FSL2.3* reaching the highest relative value again.

2. In parental susceptible genotype compared with resistant mutants, only *FSL2.3*, *PTO1* and *RBG1-B* were significantly upregulated in all time points, raising comparative expression values.

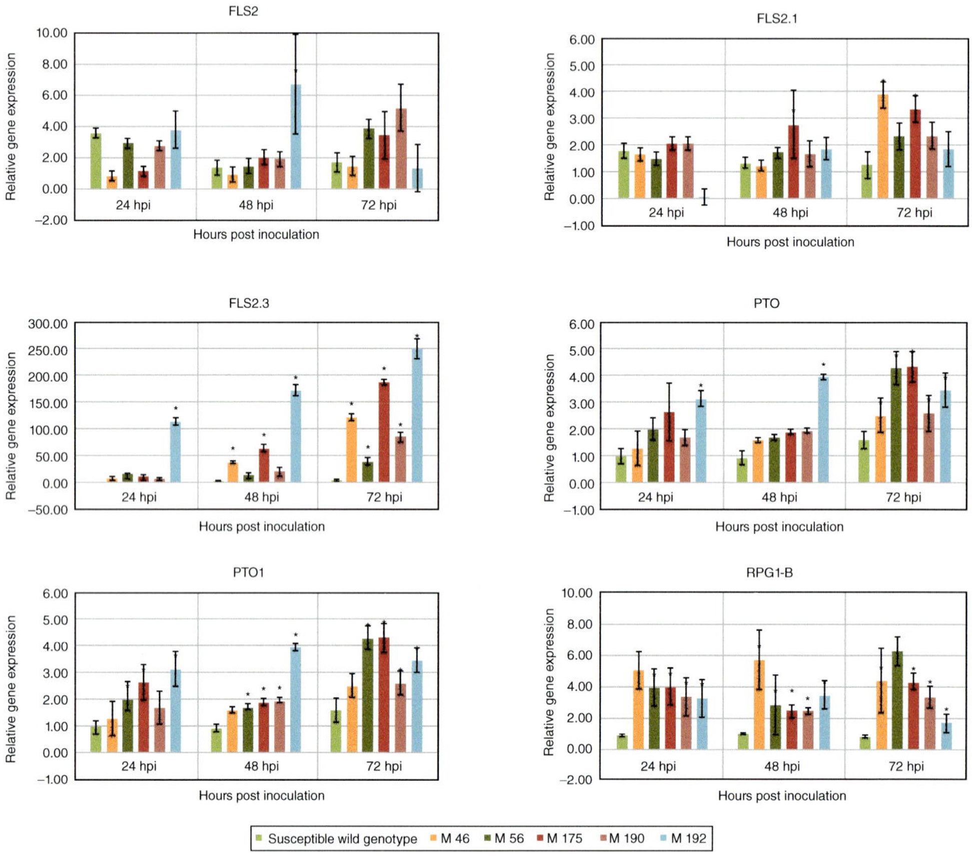

Fig. 18.4. Quantitative real-time PCR (qRT-PCR) analyses of six PR genes in response to *Psp*-race 1 infection in susceptible (wild) and resistant mutant common bean genotypes. The data were analysed by the $2^{-\Delta\Delta Ct}$ method. Probability values between infected wild-type and mutant plants were estimated by the Student's *t*-test, and statistically significant differences ($p < 0.05$) are indicated with an asterisk.

3. Finally, M192 showed consistent and significant upregulation of the expression ratio of all the analysed genes, except for *FLS2.1* and *RBG1-B*.

Although the avirulence functions of some effector proteins (e.g. *PTO*) have been described in detail (Kim *et al.*, 2002; Zhao *et al.*, 2003), the relatively constant regulation under both races of *Psp* indicates that these genes may be related to basic cell functions suppressed on infection.

The highest upregulation of *FLS2.3* may be related to the plant recognition mechanisms or to hypersensitive response (HR). It is now known that genes belonging to the family of PR proteins have been strongly linked to the resistance mechanisms of the plants against pathogens (Edreva, 2005).

4 Conclusion

Plant breeding achievements in Bulgaria confirm that experimental mutagenesis is a useful method capable of enhancing crop improvement. We have tested the chemical mutagen EMS in several concentrations on seeds from Bulgarian snap bean genotypes and developed advanced mutant lines. In these mutants, increased tolerance to the targeted halo blight and common bacterial blight diseases was observed under field conditions as well as in artificial inoculations under controlled (glasshouse) conditions. Individual plants with morphological and phenological mutations and increased resistance have been identified in the M_2 generation,

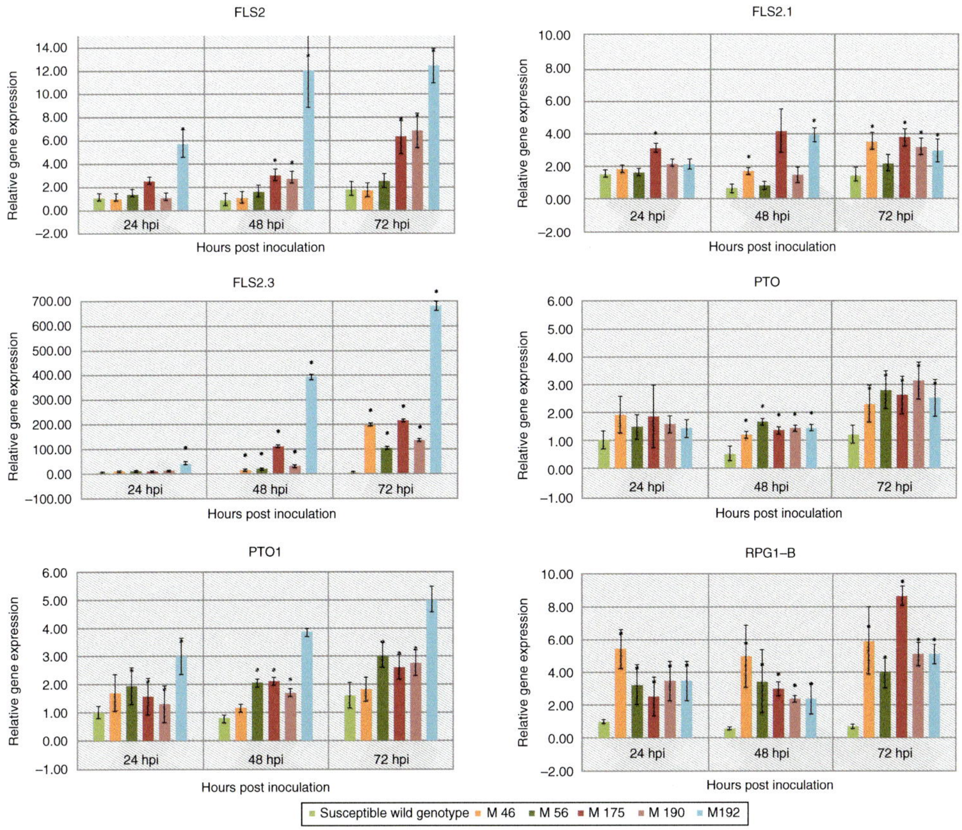

Fig. 18.5. Quantitative real-time PCR (qRT-PCR) analyses of six PR genes in response to *Psp*-race 6 infection in susceptible (parental) and resistant (mutant) common bean genotypes. The data were analysed by the $2^{-\Delta\Delta Ct}$ method. Probability values between infected wild-type and mutant plants were estimated by the Student's *t*-test, and statistically significant differences *($p < 0.05$)* are indicated with an asterisk.

further selected and tracked in the M_3, M_4 and M_5 generations. Agronomic and morphological traits in mutant lines different from the initial genotypes have been recorded for three successive mutant generations. Adding mutant lines with numerous beneficial agronomic traits to the pipeline of the bean breeding programme will boost competitiveness in international markets. However, there is still the need to overcome some undesirable side effects that have occurred from the mutagenic treatment by means of a conventional crossing and further selection programme.

In addition, this study has provided a valuable new insight into the molecular mechanisms of resistance to halo blight (*Psp*) in common bean by finding six genes to be differentially expressed in a time-dependent manner in bean leaves during the interaction with *Psp*. Resistance of the mutant plants to *Psp* was correlated with a high expression of PR genes.

The comparison of susceptible parental line IP564 and mutant lines was used to investigate the profile of gene expression in the interaction between two races of *Psp* and common bean at different stages of the infection process, and provided additional evidence for an active resistance response in the resistant mutant genotypes. The result of expression analysis shows that the mutant genotypes responded to the pathogen infection at a later time point than the susceptible genotype, thus proving their increased level of resistance. Furthermore, expression of most defence-related genes analysed was not significantly different in the mutants than in the parent in the earlier stages of infection, but higher and significantly different in later stages. Data generated from this study will contribute to the understanding of the molecular mechanisms associated with plant defence against halo blight in common bean. Pyramiding of major resistance genes and QTLs associated with resistance may contribute to more durable resistance against this important pathogen. Moreover, our results also revealed a differential reaction against races of *Psp*.

Acknowledgements

This work was conducted under the IAEA Technical Cooperation Project BUL/5/015 and the National Science Program (NSP) 'Healthy Foods for a Strong Bioeconomy and Quality of Life' by the Bulgarian Ministry of Education and Science. We are thankful for the contribution of Prof. Kiril Hristov and Prof. Diana Svetleva in the establishment of the protocol for treatment with chemical mutagenic agent.

References

Aggour, A.R. *et al.* (1989) Comparison of leaf and pod disease reactions of beans (*Phaseolus vulgaris* L.) inoculated by different methods with strains of *Xanthomonas campestris* pv. *phaseoli* (Smith) dye. *Euphytica* 43, 143–152.

Borges, A. *et al.* (2011) Validation of reference genes for RT-qPCR normalization in common bean during biotic and abiotic stresses. *Plant Cell Reports* 31, 827–838.

Bustin, S.A. (2002) Quantification of mRNA using real-time reverse transcription PCR (RT-PCR): trends and problems. *Journal of Molecular Endocrinology* 29, 23–39.

Bustin, S.A. *et al.* (2009) The MIQE guidelines: minimum information for publication of quantitative real-time PCR experiments. *Clinical Chemistry* 55, 611–622.

Christov, N.K. *et al.* (2014) Chemical mutagenesis, mutation breeding and quantitative genetic analyses of maize mutants: from theory to practice. In: Tomlekova, N.B., Kozgar, M.I. and Wani, M.R. (eds) *Mutagenesis: Exploring Genetic Diversity of Crops*. Wageningen Academic Publishers, The Netherlands, pp. 169–196.

Edreva, A. (2005) Pathogenesis-related proteins: research progress in the last 15 years. *General and Applied Plant Physiology* 31, 105–124.

Gepts, P. *et al.* (2008) Genomics of Phaseolus beans, a major source of dietary protein and micronutrients in the tropics. In: Moore, P.H. and Ming, R. (eds) (2008) *Genomics of Tropical Crops*. Springer, New York, pp. 113–143.

Heid, C.A. *et al.* (1996) Real time quantitative PCR. *Genome Research* 6, 986–994.

Kim, Y.J., Lin, N. and Martin, G.B. (2002) Two distinct Pseudomonas effector proteins interact with the *Pto* kinase and activate plant immunity. *Cell* 109, 589–598.

Livak, K.J. and Schmittgen, T. (2001) Analysis of relative gene expression data using real-time quantitative PCR and the 2–ΔΔCT method. *Methods* 25, 402–408.

Park, S.J. and Buttery, B.R. (1988) Nodulation mutants of white bean (*Phaseolus vulgaris* L.) induced by ethyl methane sulphonate. *Canadian Journal of Plant Science* 68, 199–202.

Ramandeep, Dhillon, *et al.* (2018) Effect of mutagen ethyl methane sulphonate on yield increasing parameters of French bean (*Phaseolus vulgaris* L.). *Genetika-Belgrade* 50(1), 199–207.

Reece, R.J. (2004) *Analysis of Genes and Genomes*. Wiley, Chichester, UK.

Romantschuk, M. and Bamford, D.H. (1986) The causal agent of halo blight in bean, *Pseudomonas syringae* pv. *phaseolicola*, attaches to stomata via its pili. *Microbial Pathogenesis* 1, 139–148.

Rukmanski, Y. (2005) Role of the mutations in the living nature, pp. 128

Singh, S.P. and Miklas, P.N. (2015) Breeding common bean for resistance to common blight: a review. *Crop Science* 55, 971–984.

Sofkova, S. *et al.* (2009) Breeding green beans (*Phaseolus vulgaris* L.) for complex disease resistance. *Genetics and Breeding* 38 (3–4), 79–89.

Svetleva, D. (2004) Induction of chlorophyll mutants in common bean under the action of chemical mutagens ENU and EMS. *Journal of Central European Agriculture*, 5(2), 85–89.

Svetleva, D. *et al.* (1999a) Plovdiv 11M-the first mutation bean cultivar in Bulgaria. *Plant Science (Bg)* 36, 303–309.

Svetleva, D. *et al.* (1999b) Characterization of mutant bean lines. *Rasteniev^dni nauki, (Bg)* 36, 456–459.

Tomlekova, N.B. *et al.* (2014a) *Mutagenesis: Exploring Genetic Diversity of Crops*. Wageningen Academic Publishers, The Netherlands.
Tomlekova, N.B. *et al.* (2014b) *Mutagenesis: Exploring Novel Genes and Pathways*. Wageningen Academic Publishers, The Netherlands.
Trabanco, N. *et al.* (2014) Identification of quantitative trait loci involved in the response of common bean to *Pseudomonas syringae* pv. *phaseolicola*. *Molecular Breeding* 33, 577–588.
Zhao, Y. *et al.* (2003) Virulence systems of *Pseudomonas syringae* pv. *tomato* promote bacterial speck disease in tomato by targeting the jasmonate signalling pathway. *The Plant Journal* 36, 485–499.

19 Mutation Induction to Improve Quinoa (*Chenopodium quinoa*) Resistance to Downy Mildew (*Peronospora variabilis*)

Luz Gomez-Pando[1], Jesus Bernardo-Rojas, Denisse Deza-Montoya, Martha Ibañez-Tremolada and Enrique Aguilar-Castellanos

Abstract
Quinoa is an important crop due to its nutritional characteristics (better than cereals) and its tolerance to abiotic stresses. However, various factors such as high susceptibility to diseases, especially downy mildew caused by *Peronospora variabilis*, limit its agricultural performance. Genetic improvement of quinoa could reduce the need to use fungicides for this crop and maintain the organic quality of Peruvian production in small-scale farms. Seeds of var. 'Amarilla Marangani', irradiated with 150 and 250 Gy of gamma-rays (^{60}Co), were evaluated in two experimental locations in Peru: coastland at La Molina and highland at Huancayo. Resistance to downy mildew and other agricultural traits in the M_3 and M_4 generations was studied. In both locations, downy mildew was observed in susceptible plants under natural infection, from the seedling stage to plant maturity. At the coastland site, six mutants with 30% leaf infection were obtained in the progeny of plants exposed to 150 Gy. Five additional mutants with 40% leaf infection were found in the progeny of plants exposed to 250 Gy. In the highland trial, only seven lines were identified with 30% severity (foliar area with symptoms) among the plants from the 150 Gy experiment. The parent materials showed 70–80% disease severity. Mutant lines with quantitative resistance and tolerance to downy mildew, high yield potential, reduced duration, shorter plant height, altered inflorescence shape and grain colour mutations were selected from both doses. This study showed that quantitative resistance and tolerance to downy mildew could be obtained in quinoa and this resulted in increased grain yields.

Keywords: quinoa • downy mildew • mutation induction • resistance

1 Introduction

Quinoa (*Chenopodium quinoa* Willd.) is an important crop in the Andean Region due to the nutritional value of its high-quality proteins with a balanced composition of essential amino acids (Repo-Carrasco *et al.*, 2003). Moreover, this plant has been described as drought and salinity tolerant (Jacobsen *et al.*, 2003, 2005; Razzaghi *et al.*, 2012, 2015; Shabala *et al.*, 2013). The increase of its area of cultivation in different environments has exposed negative factors of diverse nature that reduce yields and grain quality. Among the adverse biotic factors, the downy mildew disease caused by *Peronospora variabilis* Gäum (formerly *P. farinosa* f. sp. *chenopodii* Byford) (Choi *et al.*, 2010) stands out.

[1] Email: luzgomez@lamolina.edu.pe

DOI: 10.1079/9781789249095.0019

Downy mildew is considered the most important endemic disease of quinoa (Aragón and Gutierrez, 1992; Danielsen *et al.*, 2001, 2003; Testen *et al.*, 2014). It is an omnipresent limiting factor in quinoa-producing areas when conditions are favourable. Damage is reported throughout the Andean Region from Colombia to Chile and in Canada (Tewari and Boyetchko, 1990), Europe (Danielsen *et al.*, 2002), India (Kumar *et al.*, 2006), the USA (Testen *et al.*, 2012) and Korea (Choi *et al.*, 2014).

According to Danielsen *et al.* (2003), little is known about the variation in the incidence of mildew in different geographical areas and whether this is due to variation in the environment or to variation within the population of the pathogen. Downy mildew manifests with different intensities in the Andean Region, occurring with greater intensity in the areas with higher humidity and precipitation, especially in the areas surrounding Lake Titicaca and the inter-Andean valleys that have more moisture than the highlands. According to Danielsen and Munk (2004), yield losses vary from 33% to 99% in conditions of high severity. They found that the level of severity was tenfold higher in Huancayo (var. 'Utusaya', 'LP-4B' and 'Blanca de Juli') where losses due to infection ranged from 75% to 99%. On the other hand, var. 'La Molina 89' and var. 'Ingapirca' had yields reduced by 58% and 46%, respectively.

Fully resistant varieties are not yet available but responses of different severity have been observed in quinoa genetic material. In studies conducted in Peru, Otazú *et al.* (1976) reported 11 genotypes and identified a number of promising accessions but not the mechanisms of resistance. They pointed out that susceptibility is correlated with the life cycle, with the most precocious plants being the most susceptible. Danielsen *et al.* (2003) indicated that the greatest resistance is found in the valley quinoa ecotypes of Peru, Bolivia and Ecuador, and the most susceptible varieties are those of the southern highlands or altiplano. Kumar *et al.* (2006) evaluated 34 accessions of *Chenopodium* species, including *Chenopodium quinoa* (27), *C. berlandieri* ssp. *nuttalliae* (two) and one each of *C. bushianum*, *C. ugandae*, *C. strictum*, *C. ficifolium* and *C. opulifolium*, for their response to downy mildew (*P. farinosa* f. sp. *chenopodii* Byford) and found four accessions of quinoa that were immune or resistant to downy mildew (PI 510532, CHEN 67/78, Ames 22158, CHEN 7/81), suggesting physiological specialization of the pathogen. Gabriel *et al.* (2012) found that accessions resistant to mildew were late, tall, with thick stems and with a low 100-seed weight, which is typical in valley accessions compared with altiplano materials.

Several breeding programmes have been initiated for quinoa improvement in morphological and physiological characteristics such as heat tolerance, downy mildew resistance and efficiency in the use of agricultural inputs (Bazile *et al.*, 2015; Bonifacio and Saravia, 1999; Bonifacio *et al.*, 2015). New sources of heat-tolerance or disease-resistance genes from wild relatives are being introgressed into commercial varieties by plant breeders; however, undesirable characteristics from the donor species such as seed dormancy or shattering can decelerate a quinoa breeding programme (Zurita-Silva *et al.*, 2014).

Mutagenesis is a valuable tool used in the improvement of crops and is free from the regulations imposed on transgenics. Mutations change few characteristics in the parent or parental material (among them, for example, the susceptibility to resistance), while preserving the good combination of genes for agronomic characters, adaptation and quality of the germplasm. There is evidence that mutants with resistance to diseases in other crops can be obtained. According to FAO/IAEA data, 320 crop varieties with improved resistance to diseases have been obtained by mutagenesis in crops such as rice, barley, corn, wheat, beans and peas (Kozjak and Meglič, 2012). The induction of mutations is a conventional technique, used in various countries such as China, Japan, India, Russian Federation, Holland, Germany and the USA (MVGS, 2021) where more than 3230 varieties of 214 plant species have been obtained using mutagenesis. On the other hand, the application of transgenesis would limit the commercialization of quinoa, because one of the attributes of organic production is the use of varieties obtained through conventional methods.

M_3 and M_4 generations are used to identify mutants with different characters, since they are the generations where progeny tests of M_2 candidate mutants are carried out. These tests allow the verification of a mutant trait change observed in M_2 if it is transmitted to the offspring generation (Maluszynski *et al.*, 2001; Chen *et al.*, 2006).

The experiments shown here were developed to verify that the induction of mutations in quinoa using gamma radiation allows the generation of new genetic variation expressed in the appearance of mutants and the identification of new sources of resistance to downy mildew and other important morphological and agronomic characteristics.

2 Materials and Methods

2.1 Locations

The experiments were conducted at La Molina (240 m asl), in Lima province, and at the highland site of Huancayo (3200 m asl) in the central highlands of the Junin region. At the first location, the research was carried out in a mesh house and in field conditions. In the highlands, the experiments were established under field conditions. In both locations, natural infection occurred due to favourable environments for the disease, with moisture levels above 70% and temperatures between 18 and 25°C. The experiments were conducted during two growing seasons: 2015–2016 and 2016–2017.

2.2 Genetic material

Seeds from individual plant selections in the M_2 population or 'candidate mutants' were used in the experiments evaluated in the highland site, and seeds from bulk harvest of the M_2 population were used in the experiments at the La Molina location.

The mutant populations were obtained through the irradiation of seeds of the commercial variety 'Amarilla Marangani' with doses of 150 Gy and 250 Gy of gamma-rays (^{60}Co). Additionally, non-irradiated seeds were used as control.

2.3 Methodology

At La Molina, M_3 bulk seeds were used and planted in the mesh house using plastic trays with 104 cells per tray. A mixture of sterilized compost and garden soil was used as a substrate (2:1). Ten seeds were placed per tray cell to ensure germination. In total, 420,000 seedlings were evaluated regarding resistance to downy mildew derived from 14,095 plants from the 150 Gy dose and 20,965 plants from the 250 Gy dose of bulk seed harvested in the M_2 generation. From these materials, resistant plantlets were transplanted to seedbeds in the field at 30 days of growth and were evaluated by their response and level of resistance to downy mildew in the rest of the growth period. In the following M_4 generation the plants were evaluated by their response to downy mildew and agronomic performance.

In the highland location, seeds from 2441 individual plants selected at M_2 were sown in field conditions and evaluated for their response to downy mildew in two growing seasons (M_3 and M_4).

The three-leaf method (Danielsen and Munk, 2004) was used for the evaluation of the disease. It is based on the evaluation of the disease severity or percentage of leaf surface with disease symptoms. Three leaves per plant were selected randomly from the lower, middle and upper part of the plant. The symptoms of the disease are chlorotic lesions on the upper leaf surface and greyish spore masses on the lower side of the leaf.

3 Results

3.1 La Molina

In the growing season 2015–2016, in mesh-house conditions, 420,000 M_3 seedlings were evaluated under high natural infection (Fig. 19.1). The percentage leaf area infected by downy mildew had values from 10% to 100%. The values observed in parent material seedlings were between 80% and 90%. In this population, 735 and 701 seedlings with 10–20% of severity were selected from the doses of 150 Gy and 250 Gy, respectively. The selected seedlings were transplanted to beds for evaluation of response to downy mildew during the life cycle of the crop and only 116 resistant plants from 150 Gy and 172 resistant plants from 250 Gy were harvested

Fig. 19.1. Management of plants in the experiments at the different locations. **(A)** M_3 plantlets at the mesh house, La Molina. **(B)** Quinoa M_3 plants at the bed field, La Molina. **(C)** M_4 quinoa plants at field condition, La Molina. **(D)** M_3 quinoa plants at the highland site in Huancayo, Junin. **(E)** Downy mildew symptoms at field condition at Huancayo. **(F)** Selected quinoa plants at Huancayo.

that had 20–30% leaf area infected. These 288 resistant M_4 'plant candidates' were sown in the 2015–2016 agricultural campaign in field conditions. Again, parental material showed about 85% damage. At the end of the agricultural campaign, only six mutants were selected from the 150 Gy treatment and five mutants from 250 Gy. These showed between 30% and 40% leaf area infected.

Morphological leaf mutations were observed at the seedling stage and they were classified as fasciation, multiple leaves, deformed leaf, different leaf shape, different colour, bifurcated leaf and doubled leaf blade. During the life cycle, mutations of inflorescence morphology, life-cycle duration, grain colour and plant colour were observed. Table 19.1 and Fig. 19.2 present the frequency and the spectrum of leaf mutations found in generation M_3.

3.2 Highland of Huancayo, Junin

This experiment was conducted in field conditions (Fig. 19.1) in the 2015–2016 season. A high incidence of the disease was observed from the cotyledon leaf stage to the phase of grain development, reaching levels of severity of 100% in this location. In total, there were 2441 M_2 candidate mutant plants, selected individually based on morphological and physiological modifications. Within the progeny of M_3 plants, 101 mutant plants were selected from the 150 Gy treatment and 50 plants from the dose of 250 Gy; all of these had 30–40% leaf infection. In total, 313 mutant lines were selected and sown in the season 2016–2017 as M_4. Only seven lines were identified with 30% of foliar area covered by symptoms of the disease.

In this M_4 generation, five early mutants, four white grain mutants, one short plant and 83 lines with higher agronomic performance were selected from the 150 Gy material. In the same way, among the 50 mutant lines with low downy mildew severity, ten early mutants and 59 lines with higher yield potential were selected from the dose of 250 Gy treatment (Fig. 19.2). Some mutant lines were selected for their higher yield potential compared with the parental materials but with 60–70% of foliar damage; they showed severe defoliation of diseased leaves and also a fast development of new clean leaves that may have built up the photosynthetic area and contributed directly to the grain filling.

Table 19.1. Frequency and spectrum of mutations of quinoa (*Chenopodium quinoa*) observed in the M_3 generation at La Molina (Lima, Peru, 2016).

Mutation		Dose 150 Gy			Dose 250 Gy		
		No. plants	No. mutant plants	Frequency (%)	No. plants	No. mutant plants	Frequency (%)
Leaf morphology	Fasciation leaf	14,095	30	0.21	20,965	99	0.47
	Multiple leaves	14,095	23	0.16	20,965	60	0.29
	Deformed leaf	14,095	34	0.24	20,965	27	0.13
	Leaf shape	14,095	18	0.13	20,965	30	0.14
	Leaf colour	14,095	65	0.46	20,965	111	0.53
	Bifurcated leaf	14,095	7	0.05	20,965	13	0.06
	Two-leaf blade	14,095	6	0.04	20,965	18	0.09
Inflorescence	Shape	14,095	2	0.01	20,965	4	0.02
	Colour	14,095	1	0.01	20,965	2	0.01
Precocity		14,095	24	0.17	20,965	51	0.24
Plant height reduction		14,095	5	0.04	20,965	9	0.04
Quantitative resistance to mildew[a]		14,095	116	0.82	20,965	172	0.82

[a]In the M_4 generation six mutants were selected from the 150 Gy treatment and five mutants from 250 Gy with resistance to downy mildew.

Fig. 19.2. Mutation spectrum at M_3 generation. **(A), (B)** Bifurcated leaf and cotyledon blade. **(C), (D)** deformed leaves. **(E)** Two-blade leaf. **(F)** Leaf fasciation. **(G)** Multiple leaves. **(H), (I)** Leaf colours. **(J)** Inflorescence shape. **(K)** Inflorescence colour orange to green. **(L)** Early plants (precocity).

4 Discussion

The study and the selection of resistant mutants were carried out at two locations of Peru: the La Molina coastland, a new cultivation area; and the highland location in Huancayo, a traditional cultivation area. Some studies have shown geographical differences in the population of downy mildew. According to Danielsen *et al.* (2003), *P. variabilis* populations in the Andean countries show diversity; mildew is detected in various climatic and geographical areas, indicating its high degree of adaptability, and follows a similar distribution to the quinoa crop host. There is frequent sexual reproduction in downy mildew (an oomycete), which generates new pathogen genetic diversity and probably new pathotype development (Danielsen, 2001; Peterson *et al.*, 2015). Choi *et al.* (2010), based on phylogeny of the internal transcribed spacer (ITS) region of the ribosomal RNA genes (rDNA), showed differences between the populations from Europe and South America. Using *COX2* phylogeny, Danielsen and Lubeck (2010) observed differences between US and South American samples of downy mildew.

The locations in the highland and coastland regions are ideal areas for evaluation because their environmental conditions favour the development of the pathogen; the disease is always present. Danielsen and Munk (2004) made a similar evaluation of quinoa cultivars' response to downy mildew at the Huancayo highland site and reported 90% severity in susceptible quinoa varieties and a greater incidence of the pathogen than in La Molina. Favourable conditions for the development of the oomycete are relative humidity higher than 80% and temperatures of 15–25°C (Otazú *et al.*, 1976; Alandia *et al.*, 1979; Danielsen and Ames, 2000).

The evaluation, identification and selection of resistant and tolerant mutants were made in fields with natural infection of the pathogen during several generations in order to verify if the quantitative resistance to the disease was maintained in time and in different localities. The oospores of *P. variabilis* can be seed-borne and remain viable (Danielsen, 2004; Testen *et al.*, 2014) and could be viable in the crop residues in the field between crop seasons or in related weed species. The evaluations were made during two growing seasons in both locations to confirm the level of resistance of the genotypes. Evaluations of the severity of the disease or the size of foliar area covered by the disease symptoms could be under- or overestimated, because several variables could influence the response, including the level of inoculum, the distribution of focal points of infection and the composition of the population of the pathogen.

In total, 18 mutant lines with infection severity between 30% and 40% were selected among more than 420,000 plants and 2441 plant progenies of M_3 initially observed at the La Molina and highland locations. Different studies have suggested that downy mildew resistance in quinoa is a complex trait regulated by multiple resistance genes (Kumar *et al.*, 2006; Kitz *et al.*, 2009). Mhada *et al.* (2014) reported this type of resistance of quinoa to downy mildew in the evaluation of 79 accessions and also observed different levels of quantitative resistance. Gandarillas *et al.* (2015) recognized that horizontal (broad-spectrum) resistance is the most common resistance reported for downy mildew in quinoa and also that there are different degrees of resistance depending on the number of resistance genes. They indicated that varieties with a long life cycle have better mildew resistance than early-maturing varieties. Contrastingly, susceptible varieties have larger grains. Horizontal resistance can be useful in breeding programmes, since it would be more difficult for downy mildew to overcome it, but it is also difficult to transfer to other genotypes (Kitz, 2008).

One effect of mildew disease is defoliation. It is observed, however, that the degree of defoliation depends on the variety and its response to the pathogen. Danielsen and Munk (2004) pointed out that those varieties such as 'La Molina 89' and 'Amarilla Marangani' with similar levels of response to mildew and high yield potential showed different degrees of defoliation. In the present experiments, mutants with higher yield potential than the parental material and with an equal response to the pathogen (70–80%) were identified and selected as tolerant. They showed a high capacity for growth and speed of renewing the leaves and forming a new photosynthetic apparatus in the phase of grain growth and filling. According to Fortes *et al.* (2004), the tolerance mechanisms that promote growth after defoliation could be based on morphological factors (number and source of meristems) and/or physiological mechanisms (compensatory photosynthesis, carbon distribution or diverse carbohydrate reserves).

Some morphological and physiological mutations identified in this study (Table 19.1), such as reduction of plant height and life cycle, are valuable for the improvement of agronomic management and harvesting processes on large-scale farms. We previously reported these types of mutations in quinoa (var. 'Pasankalla'. and LM 89 varieties) (Gomez-Pando, 2014).

5 Conclusion

Quinoa seed irradiation with gamma rays at doses of 150 GY and 250 Gy induced morphological and physiological mutations conferring resistance to the mildew disease in var. 'Amarilla Marangani'. The selection in different localities and years allowed us to identify 18 mutant lines with quantitative resistance to *Peronospora variabilis* and to identify lines with tolerance to this pathogen.

Acknowledgement

Proyecto de investigación Contrato 014 - 2015 - INIA - PNIA - UPMSI/IE: 'Identificación de líneas con resistencia genética al Mildiú en poblaciones obtenidas por inducción de mutaciones y en el germoplasma colectado en Perú para una producción con enfoque orgánico y sostenible del cultivo de Quinua (*Chenopodium quinoa* Willd)'.

Project 2: Development of Value Chains for Biodiversity Conservation and Improvement of Rural Livelihoods. VLIRUOS – UNALM.

References

Alandia, S. *et al.* (1979) Enfermedades. In: Tapia, M. *et al.* (eds) *Quinua y Kañwa*. Editorial IICA, Bogotá, pp.137–148.

Aragón, L. and Gutierrez, W. (1992) El mildiú en cuatro especies de Chenopodium. *Fitopatologia* 27, 104–109.

Bazile, D. *et al.* (eds) (2015) *State of the Art Report on Quinoa around the World in 2013.* FAO and CIRAD, Rome. Available at: http://www.fao.org/3/a-i4042e.pdf (accessed April 2021)

Bonifacio, A. and Saravia, R. (1999) Evaluacion de la resistencia al mildiu en quinua. In: *Proceedings del Tercer Taller de Preduza en Resistencia Duradera en Cultivos Altoandinoss en la Zona Andina*, Cochabamba, Bolivia, pp. 49–59.

Bonifacio, A. *et al.* (2015) Quinoa breeding and modern variety development. In: Bazile, D. *et al.* (eds) *State of the Art Report on Quinoa around the World in 2013.* FAO and CIRAD, Rome, pp. 172–191.

Chen, Z. *et al.* (2006) Mutations in ABO1/ELO2, a subunit of holo-Elongator, increase abscisic acid sensitivity and drought tolerance in *Arabidopsis thaliana. Molecular and Cellular Biology* 26, 6902–6912.

Choi, Y.J. *et al.* (2010) Morphological and molecular characterization of the causal agent of downy mildew on quinoa (*Chenopodium quinoa*). *Mycopathologia* 169, 403–412.

Choi, Y.J. *et al.* (2014) First report of quinoa downy mildew caused by *Peronospora variabilis* in Republic of Korea. *Plant Disease* 98(7), 1003.

Danielsen, S. (2001) Heterothallism in *Peronospora farinosa* f. sp. *chenopodii*, the causal agent of downy mildew in quinoa (*Chenopodium quinoa*). *Journal of Basic Microbiology* 41(5), 305–308.

Danielsen, S. (2004) Seed transmission of downy mildew (*Peronospora farinosa* f. sp. *chenopodii*) in quinoa and effect of relative humidity on seedling infection. *Seed Science and Technology* 32, 91–98.

Danielsen, S. and Ames, T. (2000) *El mildiu (Peronospora farinosa) de la quinua (Chenopodium quinoa) en la zona andina. Manual practico para el estudio de la enfermedad y del patogeno.* International Potato Center, Lima, Peru.

Danielsen, S. and Lubeck, M. (2010) Universally primed-PCR indicates geographical variation of *Peronospora farinosa* ex. *Chenopodium quinoa. Journal of Basic Microbiology* 50, 104–109.

Danielsen, S. and Munk, L. (2004) Evaluation of disease assessment methods in quinoa for their ability to predict yield loss caused by downy mildew. *Crop Protection* 23, 219–228.

Danielsen, S. *et al.* (2001) Impact of downy mildew on the yield of quinoa. In: *CIP Program Report 1999–2000.* International Potato Center, Lima, Peru, pp. 397–401.

Danielsen, S. *et al.* (2002) First report of downy mildew of quinoa caused by *Peronospora farinosa* f. sp. *chenopodii* in Denmark. *Plant Disease* 86, 1175.

Danielsen, S. *et al.* (2003) Diseases of quinoa (*Chenopodium quinoa*). *Food Reviews International* 19(1), 43–59.

Fortes, D. *et al.* (2004) Estrategias para la resistencia de las plantas a la defoliación. *Revista Cubana de Ciencia Agricola* 38(2), 111–119.

Gabriel, J. *et al.* (2012) Quinua de valle (*Chenopodium quinoa* Willd) fuente valiosa de resistencia genética al mildiú (*Perenospora farinosa* Willd). *Journal of the Selva Andina Research Society* 3(2), 27–44. Selva Andina Research Society, La Paz, Bolivia.

Gandarillas, A. *et al.* (2015) Principle quinoa pests and diseases. In: Bazile, D. *et al.* (eds) (2015) *State of the Art Report on Quinoa around the World in 2013.* FAO and CIRAD, Rome, pp. 192–215.

Gomez-Pando, L. (2014) Development of improved varieties of native grains through radiation-induced mutagenesis. In: Tomlekova, N.B., Kozgar, M.I. and Wani, M.R. (eds) *Mutagenesis: Exploring Novel Genes and Pathways.* Wageningen Academic Publishers, Wageningen, Netherlands, pp. 105–123.

Jacobsen, S.E. *et al.* (2003) The resistance of quinoa (*Chenopodium quinoa* Willd.) to adverse abiotic factors. *Food Reviews International* 19, 99–109.

Jacobsen, S.E. *et al.* (2005). Plant responses of quinoa (*Chenopodium quinoa* Willd.) to frost at various phenological stages. *European Journal of Agronomy*, 22, 131–139.

Kitz, L. (2008) Evaluation of downy mildew (*Peronospora farinosa* f. sp. *Chenopodii*) resistance among quinoa genotypes and investigation of *P. farinosa* growth using Scanning Electron Microscopy. Thesis, *Brigham Young University, Provo, Utah.* Available at: https://scholarsarchive.byu.edu/etd/1512 (accessed April 2021)

Kitz, L. *et al.* (2009) Downy mildew resistance in four breeding lines of quinoa. *Phytopathology* 99, S184.

Kozjak, P. and Meglič, V. (2012) Mutagenesis in plant breeding for disease and pest resistance. In: Mishra, R. (ed.) *Mutagenesis.* InTech, Rijeka, pp. 195–220.

Kumar, A. *et al.* (2006) Screening of exotic *Chenopodium quinoa* accessions for downy mildew resistance under mid-eastern conditions of India. *Crop Protection* 25, 879–889.

Maluszynski, M. *et al.* (2001) Induced mutations – a new paradigm in plant breeding. *Euphytica* 118, 167–173.

Mhada, M. *et al.* (2014) Assessment of downy mildew resistance (*Peronospora farinosa*) in a quinoa (*Chenopodium quinoa* Willd.) germplasm. *International Journal of Biological, Biomolecular, Agricultural, Food and Biotechnological Engineering* 8(3), 277–280.

MVGS (2021) FAO/IAEA Database of mutant varieties and genetic stock. Available at: http://mvgs.iaea.org (accessed April 2021)

Otazú, V. *et al.* (1976) Resistencia en quinua (*Chenopodium quinoa*) al mildiu (*Peronospora effusa*). *Fitopatologia* 11, 47–49.
Peterson, A.J. *et al.* (2015) Quinoa in the United States of America and Canada. In: Bazile, D. *et al.* (eds) (2015) *State of the Art Report on Quinoa around the World in 2013*. FAO and CIRAD, Rome, pp. 549–562.
Razzaghi, F. *et al.* (2012) Effects of salinity and soil-drying on radiation use efficiency, water productivity and yield of quinoa (*Chenopodium quinoa* Willd.). *Journal of Agronomy and Crop Science* 198, 173–184.
Razzaghi, F. *et al.* (2015) Ionic and photosynthetic homeostasis in quinoa challenged by salinity and drought – mechanisms of tolerance. *Functional Plant Biology* 42, 136–148.
Repo-Carrasco, R. *et al.* (2003) Nutritional value and use of the Andean crops quinoa (*Chenopodium quinoa*) and kañiwa (*Chenopodium pallidicaule*). *Food Reviews International* 19, 179–189.
Shabala, S. *et al.* (2013) Genotypic difference in salinity tolerance in quinoa is determined by differential control of xylem Na+ loading and stomatal density. *Journal of Plant Physiology* 170, 906–914.
Testen, A.L. *et al.* (2012) First report of quinoa downy mildew caused by *Peronospora variabilis* in the United States. *Plant Disease* 96, 146.
Testen, A.L. *et al.* (2014) Molecular detection of *Peronospora variabilis* in quinoa seed and phylogeny of the quinoa downy mildew pathogen in South America and the United States. *Phytopathology* 104, 379–386.
Tewari, J.P. and Boyetchko, S.M. (1990) Occurrence of *Peronospora farinosa* f.sp. *chenopodii* on quinoa in Canada. *Canadian Plant Disease Survey* 70, 127–128.
Zurita-Silva, A. *et al.* (2014) Breeding quinoa (*Chenopodium quinoa* Willd.): potential and perspectives. *Molecular Breeding* 34,13–30.

20 Development of the First Kabuli Type Chickpea Mutant Variety in Bangladesh

Shamsun Nahar Begum[1], Mirza Mofazzal Islam and Rigyan Gupta

Abstract

Chickpea has a high yield potential, nutritional importance and diversity of use. A mutation breeding programme was undertaken at Bangladesh Institute of Nuclear Agriculture (BINA) with a view to developing early-maturing, large-seeded and high-yielding varieties of chickpea. Seeds of the popular chickpea variety 'Desi Binasola-2' were treated with different doses of gamma-rays (200, 300 and 400 Gy). The treated seeds were grown in batches according to dose for raising the M_1 generation. M_2 seeds were collected from individual M_1 plants and subsequently grown in plant-progeny rows in the M_2 generation and selections were made from the M_2 families. Only 85 plants were selected from the M_2 population and these were grown in the M_3 generation. The mutant 'CPM-kabuli' and 28 other mutants were selected from M_3 and were grown in the M_4 generation. Only five mutants, including 'CPM-kabuli', were selected from M_4 and were grown in M_5. The selected mutant 'CPM-kabuli' along with check varieties were put into preliminary yield trials. Finally, the mutant lines were evaluated, with respect to two check varieties, in advanced, zonal-yield, on-farm and on-station trials in successive generations. All the selected mutant lines were grown at different locations in Bangladesh to observe the yield and other characteristics. The performance of the mutants was evaluated under two management practices: research management and farmers' management. Contrary to its parent, 'CPM-kabuli' was found to be tolerant to root rot and Botrytis grey mould, and also showed greater tolerance to pod borer insect-pest infestation than other mutants and check varieties. The main improved attributes are a cream seed coat colour, which reflects kabuli type, larger seed size and higher seed yield. The mutant 'CPM-kabuli' matures in the range of 115–125 days and is high yielding (1.7 t/ha). Considering all these, the bold Kabuli type chickpea mutant 'CPM-kabuli' was registered as the variety 'Binasola-9' for commercial cultivation during 2017 and is suitable for farmers in drought-prone areas in Bangladesh.

Keywords: chickpea • mutant • kabuli type

1 Introduction

Chickpea, an important pulse legume, can be improved by harnessing the diversity of wild-crop relatives. They provide evidence of ancestral adaptations for seed coat colour crypsis, the impact of environment on genetic structure and trait values and they demonstrate variation between wild and cultivated accessions for agronomic properties. Chickpea is a member of the Fabaceae family and subfamily Faboidae; it was domesticated in southern Turkey (von Wettberg *et al.*, 2018). Chickpea is widely used as a fodder crop or as green manure. It is an important source of protein, carbohydrate, B-group vitamins and certain minerals (Williams *et al.*, 1988),

[1] Email: sbluna98@yahoo,.com

DOI: 10.1079/9781789249095.0020

particularly for the populations of developing nations (Chavan *et al.*, 1987). It is an important pulse crop in Bangladesh, especially in the north-western part of the country, and ranks third among the pulses in terms of preference. However, compared with other chickpea-producing countries of the region, the production and acreage of chickpea in Bangladesh are comparatively low. The average yield of chickpea in the major growing regions is around 0.5–0.7 t/ha (BBS, 2004). Bangladesh has to import a large quantity of chickpea from abroad and recent information indicates that Bangladesh is the second largest importer of chickpea (FAOSTAT, 2015; Muehlbauer *et al.*, 2017).

Seed size and number of seeds are important factors in determining yield in chickpea, as is well recognized by chickpea breeding programmes throughout the world (Akhtar *et al.*, 2011). Upadhyaya *et al.* (2006) reported large seed size as a profitable trade-related trait and a component of yield and adaptation in chickpea. Many other researchers have reported the importance of large seed size in yield improvement (Mehla *et al.*, 2000; Bicer, 2009). Induced mutation has been reported as an efficient technique to breed chickpea genotypes with large seed size (Khattak *et al.*, 2007; Barshile *et al.*, 2009).

Based upon seed size and colour, chickpea is classified as kabuli type (or macrosperma) and desi type (or microsperma). Kabuli type chickpea seed is large and has a thin seed coat ranging from white to pale cream in colour. Desi type chickpea seed is smaller than the kabuli type and has a thicker seed coat, ranging in colour from brown to yellow. The kabuli type chickpea is believed to have developed from the desi type through natural mutation and selection (Moreno *et al.*, 1978; Hawtin *et al.*, 1980; Salimath *et al.*, 1984; Gil *et al.*, 1993). Toker (2009) reported that kabuli chickpea could have originated from spontaneous mutants of *C. reticulatum*. However, in the market, both are simply designated as chickpeas. Kabuli chickpeas are the type most commonly found in American supermarkets. Kabuli chickpeas are usually sold whole and also often as split cotyledons or flour, so seed size and appearance are critically important. They flower at a similar time as the desi types.

Global warming as well as unfavourable climate, coupled with the ever-increasing human population, has led to growing hunger and increased malnutrition. With limited arable land, attaining sustainable food production is a challenge for farmers and appropriate technology development is the challenge for the scientific community. Although past breeding efforts to enhance production led to development of improved cultivars, they also created genetic bottlenecks where the founder effects resulted in a narrow genetic base, especially in crops like chickpea (Abbo *et al.*, 2003). As a result, susceptibility of the crop to several biotic and abiotic stresses is severe and production potential is seriously hampered. To overcome these constraints, genomic approaches have been deployed in recent years to understand the genetic basis of such complex quantitative traits and for trait improvement in chickpea (Thudi *et al.*, 2014).

Several superior lines with enhanced drought tolerance and resistance to Fusarium wilt (FW) (caused by *Fusarium oxysporum* f. sp. *ciceris*) and Ascochyta blight (AB) (caused by *Ascochyta rabiei*) have been developed (Varshney *et al.*, 2013, 2014).

The objectives of this study were to develop higher seed yield, larger seed size and resistance/tolerance to major diseases (such as Botrytis grey mould) and infestation by insects (such as pod borer).

2 Materials and Methods

Seeds of 'Binasola-2' were irradiated with gamma-rays from a ^{60}Co source at doses of 200, 300 and 400 Gy. In the first year, treated seeds were grown in batches according to dose and the M_1 generation was grown in the 2008–2009 growing season. M_2 seeds were collected from 10,425 individual M_1 plants and were grown in plant-progeny rows as the M_2 generation during 2009–2010. Only 85 individual plants were selected from the M_2 population and grown in the M_3 generation in 2010–2011. From the M_3 population, the selected mutant 'CPM-kabuli' and 28 other mutants were grown in the M_4 generation in 2011–2012 along with check varieties. Five selected M_5 mutants, including 'CPM-kabuli', along with check varieties were put into a preliminary yield trial in two locations (Rajshahi and Magura districts) during 2012–2013. The mutant 'CPM-kabuli' was selected based on cream seed colour (genetic marker), larger seed size and higher seed yield. It was derived from the 300 Gy gamma-ray treated seed. Finally, the mutants were evaluated by comparison with two check varieties in advanced zonal yield trials as well as on-farm and on-station trials in the successive generations from 2013–2014 to 2015–2016. All the selected mutants were grown at three different locations (Ishurdi, Magura and Rajshahi districts) in Bangladesh to observe their yield and other characteristics. The performance of the mutants was evaluated under two management practices: research management and farmers' management. In the research management practices, trials had a randomized complete block design with three replications, and recommended doses of fertilizer were used. Cultivation practices such as weeding and irrigation were done to maximize growth and Mancozeb-group fungicide was sprayed to control diseases. The mutant 'CPM-kabuli' was found to be tolerant to root rot and Botrytis grey mould and also showed greater tolerance to pod borer insect-pest infestation than other mutants and check varieties. Data on days to maturity, plant height, pods per plant, seeds per pod and seed yield per plot were recorded. Plot seed yield was converted to kg/ha. Mean values were used for statistical analyses following Gomez *et al.* (1984).

3 Results and Discussion

3.1 Preliminary yield trial

On the basis of seed size, attractive seed coat colour, earliness and high seed yield, five mutants viz. CPM-7-400, CPM-7-300, CPM-4-300, 'CPM-kabuli' and CPM-2-400 were selected from the M_4 generation. These five selected mutants, along with two check varieties, were grown in a preliminary yield trial at Rajshahi and Magura during 2012–2013. On average (mean over two locations) it was observed that CPM-2-400 had the greatest plant height (45.3 cm) among other mutants and check varieties (Table 20.1). Maturity period varied from 118 to 122 days. The highest number of primary branches was produced by CPM-7-300, while the lowest number was found in 'Binasola-6' (check). The highest number of pods per plant and seeds per pod was found in the mutant 'CPM-kabuli'. The 100-seed weight of 'CPM-kabuli' was the highest (21.3 g) followed by CPM-2-400 (21.0 g). Among the five mutants and check varieties, 'CPM-kabuli' had the highest seed yield (1709 kg/ha) and CPM-2-400 produced the second highest (1698 kg/ha). 'CPM-kabuli', for its cream colour and larger seed size, and three other mutants were selected for a further trial in the next growing season in 2013–2014.

3.2 Advanced yield trial

In the M_6 generation four chickpea mutants were selected on the basis of seed yield and grain size and tested along with two check varieties ('BARI Sola-7' and 'Binasola-7') at Ishurdi, Rajshahi and Magura during 2013–2014. The

Table 20.1. Mean performance of five chickpea mutants combined over two locations for yield and yield contributing characters during 2012–2013.

Variety/mutants	Days to maturity	Plant height (cm)	Primary branches per plant	Pods per plant (no.)	Seeds per pod (no.)	100-seed weight (g)	Seed yield (kg/ha)
CPM-7-400	122a	39.8b	5.3ab	60.7a	1.6b	14.5bc	1595b
CPM-7-300	118b	36.4c	5.9a	56.6ab	1.5b	14.1bc	1415c
CPM-4-300	118b	39.9b	5.7a	64.8a	1.7b	17.1ab	1688a
CPM-kabuli	118b	36.2c	5.0ab	67.7a	2.1a	21.3a	1709a
CPM-2-400	115bc	45.3a	5.0ab	66.6a	2.0a	21.0a	1698a
BARI Sola-7 (check)	116bc	37.3bc	5.1ab	54.2b	1.4bc	15.6b	1459c
Binasola-6 (check)	120a	38.2b	4.8b	56.3ab	1.5b	16.1b	1547b
SD	2.34	3.14	0.40	5.45	0.27	2.94	119.05
SE (±)	0.96	1.28	0.16	2.23	0.11	1.20	48.60

In a column, values with same letter(s) do not differ significantly at 5% level by Duncan's Multiple Range test (DMRT).

Table 20.2. Mean performance of four chickpea mutants along with two check varieties combined over three locations for different characters during 2013–2014.

Variety/mutants	Days to maturity	Plant height (cm)	Primary branches per plant	Pods per plant (no.)	Seeds per pod (no.)	100-seed weight (g)	Seed yield (kg/ha)
CPM-kabuli	116ab	56.6ab	4.5^{ns}	63.8a	1.9b	21.3a	1732a
CPM-4-300	116ab	63.9a	4.1	57.2ab	1.6a	17.2c	1642bc
CPM-7-400	116ab	59.8ab	4.0	58.9ab	1.6a	14.6c	1552c
CPM-2-400	116ab	61.2a	4.1	62.1a	1.7a	21.1a	1693b
BARI Sola-7 (check)	119a	61.4a	4.2	49.5b	1.5ab	15.6c	1483d
Binasola-7 (check)	118a	61.8a	4.0	49.8b	1.4b	19.2b	1544c
SD	1.33	2.44	0.19	6.06	0.17	2.82	96.55
SE (±)	0.52	0.70	0.19	1.10	0.19	0.75	4.39

^{ns}Not significant. In a column, values with same letter(s) do not differ significantly at 5% level by DMRT.

methodology was the same as followed in the previous experiment. Mean performance values for the mutants and check varieties for individual locations and combined over locations are presented in Table 20.2.

From the mean of two locations, 'CPM-kabuli' and CPM-2-400 mutants were found to be promising candidates for high seed yield, with 1732 and 1693 kg/ha, respectively. The mutant 'CPM-kabuli' produced the highest number of pods per plant. The 100-seed weight of 'CPM-kabuli' was the highest (21.3 g), followed by CPM-2-400 (21.1 g) and 'Binasola-7' (19.2 g). Three mutants were selected for further trials in the next growing season.

3.3 Zonal yield trial

In the M_7 generation, three chickpea mutants were selected on the basis of higher seed yield and large grain size. These three mutants, along with two check varieties, were put into zonal yield trials at Rajshahi, Magura and Ishurdi during 2014–2015. The methodology was the same as followed in the previous experiment. Mean performance of the mutants and check varieties for individual locations and mean over locations are presented in Table 20.3.

Over the locations, 'Binasola-8' had the greatest plant height. The mutant 'CPM-kabuli' produced the highest number of pods per plant, followed by 'Binasola-8' and CPM-2-400. The 100-seed weight of 'CPM-kabuli' was the highest (21.3 g) followed by CPM-2-400 (21.1 g). The mutant 'CPM-kabuli' produced the highest seed yield (1739 kg/ha) among the mutants and check varieties. With respect to days to maturity, there was no significant difference among the genotypes. Significant differences were found among the genotypes for the characteristics of plant height, number of primary branches per plant, number of pods per plant, number of seeds per pod, 100-seed weight and seed yield. Two mutants were selected for on-farm and on-station trials in the next growing season (a parental check was used up to the M_4 generation, but for the different yield trials national standard check varieties were used).

3.4 On-station and farmers' field trials

On-station and farmers' field trials were conducted with two mutants that had performed well in previous trials ('CPM-kabuli' and CPM-2-400) and two check varieties ('Binasola-8' and 'BARI Sola-7') at Ishurdi, Magura and Rajshahi during 2015–2016. On-station trials were replicated and farmers' field trials were non-replicated. Only seed yield per plot was recorded, which was then converted into kg/ha. On average, the mutant 'CPM-kabuli' produced the highest seed yield (1746 kg/ha), followed by CPM-2-400 (1691 kg/ha) in the research management practice. A similar trend of seed yield produced by the entries was found in the farmers' management trial (Table 20.4). 'CPM-kabuli' has a cream seed

Table 20.3. Performance of four chickpea mutants along with two check varieties for different characteristics during 2014–2015.

Variety/mutants	Days to maturity	Plant height (cm)	Primary branches per plant	Pods per plant (no.)	Seeds per pod (no.)	100-seed weight (g)	Seed yield (kg/ha)
CPM-kabuli	120	52.4b	4.3a	62.9a	2.1a	21.3a	1739a
CPM-7-400	119	55.6b	4.1a	50.9b	1.9ab	14.5b	1583b
CPM-2-400	120	60.8a	4.1a	60.9a	1.9ab	21.1a	1719a
BARI Sola-7 (check)	119	60.6a	3.8ab	57.4ab	1.5bc	15.6b	1567b
Binasola-8 (check)	121	62.3a	4.1a	61.1a	1.7b	20.0a	1709a
SD	1.67	4.17	0.18	4.76	0.23	3.21	81.61
SE (±)	0.33	1.02	0.21	1.09	0.24	0.90	4.52

In a column, values with same letter(s) do not differ significantly at 5% level by DMRT.

Table 20.4. Seed yield of the selected mutants grown at research station and farmer's field during 2015–2016.

	Seed yield (kg/ha) ± Standard Error[a], *n* = 3			
Genotypes/varieties	Ishurdi	Magura	Rajshahi	Average
Research station				
CPM-2-400	1695 ± 8.14	1705 ± 8.54	1673 ± 9.64	1691 ± 9.45
CPM-kabuli	1745 ± 6.81	1760 ± 9.29	1735 ± 6.81	1747 ± 7.26
Binasola-8 (check)	1682 ± 9.16	1710 ± 8.72	1655 ± 6.08	1682 ± 15.88
BARI Sola-7 (check)	1596 ± 8.32	1585 ± 6.03	1550 ± 9.54	1577 ± 13.87
Farmer's field				
CPM-2-400	1661 ± 5.03	1685 ± 8.54	1656 ± 9.16	1667 ± 8.95
CPM-kabuli	1711 ± 6.24	1737 ± 5.29	1690 ± 6.03	1713 ± 13.59
Binasola-8 (check)	1655 ± 9.86	1685 ± 6.92	1610 ± 8.96	1650 ± 21.79
BARI Sola-7 (check)	1576 ± 6.81	1559 ± 6.11	1532 ± 7.51	1556 ± 12.81

[a]At the three sites the Standard Errors are the errors from replicated measurements of the same sample. The Average is the average of the three means and the Standard Error is the error associated with these three estimates of the yield. There is no statistical significance between the values from the three locations.

coat colour and large seed type like kabuli chickpea (the kabuli type is characterized by white cream colour and smooth seed coat) which is not available on the market in Bangladesh.

Seed size is stable, highly heritable and easy to select in segregating populations and is an important yield component for the improvement of seed yield in chickpea (Gan *et al.*, 2003; Upadhyaya *et al.*, 2006; Khattak *et al.*, 2007; Bicer, 2009).

In pulses the number of pods per plant, seed size and number of branches per plant are key yield-contributing factors (Haq *et al.*, 2003; Khattak *et al.*, 2007).

4 Conclusion

On the basis of seed size, seed coat colour and seed yield, BINA has registered 'CPM-kabuli' as 'Binasola-9', which is now an important chickpea variety for the National Seed Board of Bangladesh.

References

Abbo, S. *et al.* (2003) Viewpoint: evolution of cultivated chickpea: four bottlenecks limit diversity and constrain adaptation. *Functional Plant Biology* 30, 1081–1087.

Akhtar, L.H. *et al.* (2011) Genetic divergence and inter-relationship studies in chickpea (*Cicer arietinum* L.). *Pakistan Journal of AgriScience* 48, 35–39.

Barshile, J.D. *et al.* (2009) Genetic enhancement of chickpea through induced mutagenesis. *Journal of Food Legumes* 22(1), 26–29.

BBS (2004) *Year Book of Agricultural Statistics of Bangladesh*. Bangladesh Bureau of Statistics, Ministry of Planning, Govt of the People's Republic of Bangladesh, Dhaka.

Bicer, B.T. (2009) The effect of seed size on yield and yield components of chickpea and lentil. *African Journal of Biotechnology* 8(8), 1482–1487.

Chavan, J.K. *et al.* (1987) Biochemistry and technology of chickpea (*Cicer arietinum* L.) seeds. *Critical Reviews in Food Science and Nutrition* 25(2), 107–158.

FAOSTAT (2015) Statistical Database of the United Nations Food and Agriculture Organization, Statistical Division. Rome. Available at: http://faostat.fao.org/site/ 339/default.aspx. (accessed 3 September 2021).

Gan, Y.T. *et al.* (2003) Response of kabuli chickpea to seed size and planting depth. *Canadian Journal of Plant Science* 83, 39–46.

Gil, J. *et al.* (1993) Inheritance of seed coat thickness in chickpea (*Cicer arietinum* L.) and its evolutionary implications. *Plant Breeding* 111, 257–260.

Gomez, K.A. *et al.* (1984) *Statistical Procedures for Agricultural Research*, 2nd edn. Wiley, New York.

Haq, M.A. *et al.* (2003) Induction of genetic variability for plant type and disease resistance in chickpea and its utilization in breeding. In: Anwar, R., Bhatti, M.S., Takahshi, J. and Masood, S. (eds) *Sustainable Utilization of Plant Genetic Resources for Agricultural Production*. Proceedings of Seminar, 17–19 December 2002, NARC, Islamabad, Pakistan. Pakistan Agricultural Research Council, Islamabad, Pakistan, pp. 28–37.

Hawtin, G.C. *et al.* (1980) Kabuli-desi introgression: problems and *Cicer arietinum* prospects. In: *Proceedings of International Workshop on Chickpea Improvement*, ICRISAT, Hyderabad, AP, India, 28 February–2 March 1979, pp. 51–60.

Khattak, G.S.S. *et al.* (2007) High yielding desi chickpea (L.) variety NIFA 2005. *Pakistan Journal of Botany* 39(1), 93–102.

Mehla, I.S. *et al.* (2000) Association of seed mass groups and seed yield in kabuli chickpea. *International Chickpea Newsletter* 7, 7–8.

Moreno, M.T. *et al.* (1978) Variation in *Cicer arietinum*. *Euphytica* 27, 465–485.

Muehlbauer, F.J. *et al.* (2017) Economic importance of chickpea: production, value, and world trade. In: Varshney, R.K., Thudi, M. and Muehlbauer, F.J. (eds) *The Chickpea Genome*. Springer International Publishing, Cham, Switzerland, pp. 5–12.

Salimath, P.M. *et al.* (1984) Genetic diversity in chickpea (*Cicer arietinum*.). *Zeitschrift für Pflanzenzuchtung* [*Journal of Plant Breeding*] 92, 52–60.

Thudi, M. *et al.* (2014) Genomics-assisted breeding for drought tolerance: a dream comes true in chickpea! *Functional Plant Biology* 41, 1178–1190.

Toker, C. (2009) A note on the evolution of kabuli chickpeas as shown by induced mutations in *Cicer arietinum* Ladizinsky. *Genetic Resources and Crop Evolution* 56, 7–12.

Upadhyaya, H. *et al.* (2006) Two major genes for seed size in chickpea (*Cicer arietinum* L.). *Euphytica* 147(3), 311–315.

Varshney, R.K. *et al.* (2013) Fast-track introgression of 'QTL-hotspot' for root traits and other drought tolerance traits in JG 11, an elite and leading variety of chickpea. *The Plant Genome* 6, 3.

Varshney, R.K. *et al.* (2014) Marker-assisted backcrossing to introgress resistance to Fusarium wilt race 1 and Ascochyta blight in C 214, an elite cultivar of chickpea. *The Plant Genome* 7, 1.

von Wettberg, E.J.B. *et al.* (2018) Ecology and genomics of an important crop wild relative as a prelude to agricultural innovation. *Nature Communications* 9, 649.

Williams, P.C. *et al.* (1988) Quality screening and evaluation in pulse breeding. In: Summerfield, RJ. (ed.) *World Crops, Cool Season Food Legumes*. Kluwer Academic Publishers, Dordrecht, pp. 445–457.

21 Evaluation of Advanced Wheat Mutant Lines for Food and Feed Quality

Mirjana Jankulovska[1], Sonja Ivanovska, Ljupcho Jankuloski, Mile Markoski, Biljana Kuzmanovska and Dane Boshev

Abstract

The main goals of this study were to evaluate the agronomic performance of wheat mutant lines; to detect the effect of genotype, location and different fertilizer levels on analysed traits; to assess seed and feed quality; and to select best performing mutant lines for dual-purpose growing. Ten wheat mutant lines were sown on two locations in Macedonia, for evaluation of their agronomic performance. At both locations, grain yield, straw mass, harvest index, nitrogen use efficiency, nitrogen and protein content in seed and straw, neutral detergent fibre and acid detergent fibre in the straw were determined. In order to classify the genotypes based on all analysed traits, two-way cluster analysis was applied. According to their overall performance, at both locations and with the three different fertilization treatments, the mutant lines were classified in two main groups. The first cluster consisted of mutants 5/1-8, 2/2-21, 4/2-56 and 2/1-51, characterized by very high values for seed yield, straw yield and harvest index, and high to moderate values for all other traits. Only 4/2-56 had very low values for N and protein content in the seed. One mutant line, 6/2-2, did not belong to any of the groups and differed from all other genotypes based on its very low seed and straw yield and very high values for nitrogen and protein content in the straw and neutral detergent fibre. All other mutants belonged to the second group, with low to moderate yield and moderate to high values for the other traits. Mutant lines with the highest seed and straw yield, as well as the best quality of seed and straw under different management systems, were identified and after additional evaluation will be submitted for official variety registration.

Keywords: wheat mutant lines • seed yield • straw mass • two-way cluster analysis

1 Introduction

The agricultural sector in the Republic of Macedonia is prioritized as one of the most important sectors of the Macedonian economy due to its importance for social security and poverty reduction. It provides sustenance to the majority of the population and accounted for 10.5% of GDP in 2015. Of the total working population 11% are engaged in agriculture and 40% of the population live in rural areas.

Cereals are the most important group of crops. In 2014, cereals were produced on 168,000 ha with total production of 625,921 t. Wheat is one of the most important cereal crops, mainly used for human consumption, followed by barley and maize. In 2014, the total production of wheat in Macedonia was 287,954 t (State Statistical Office, 2016). The production

[1] Email: mirjanajankulovska@yahoo.com

 Mutation Breeding, Genetic Diversity and Crop Adaptation to Climate Change (eds S. Sivasankar *et al.*)
DOI: 10.1079/9781789249095.0021

area, as well as the total production, has been decreasing in recent years, which results in a higher import rate for cereal grains. The decrease of winter wheat production will eventually result in reduced food security, since it is the essential crop for food supply in the country. Moreover, the demand for animal feed is huge, and this is mostly supplied from imports.

The value and need to increase cereal production, especially wheat, is widely recognized. The average wheat yield in Macedonia is approximately 3 t/ha, which is similar to the world average yield. Despite growing varieties with higher seed yield, farmers are interested in producing varieties that have good bread-making quality for human consumption, but also have high biomass yields with a quality that can be used as animal feed. For the farmers, maximizing yield is not the ultimate objective; profitability and risk management are more important criteria for growing a specific crop. Farmers are interested in dual-purpose cereal crops that can be exploited both as food and as feed. In that case, farmers can change the final purpose of the crop depending on circumstances. As a result, it is very important to develop adequate agronomic practices, including appropriate soil and water management systems that will maximize the yield of wheat (both natural and mutant varieties) used for food and feed. Delivering increased yields is a complex challenge that cannot be solved by a single approach. Three specific major challenges are: (i) to increase yield potential (maximize yield of given genotype under optimal conditions); (ii) to protect yield potential; and (iii) to increase resource use efficiency to ensure sustainability (Mukherjee, 2014). Sustainable production of food crops relies on germplasm improvement and genetic diversity (Pooja *et al.*, 2018). Also, optimal use of production parameters, such as seeding rate, nitrogen (N) application, coupled with appropriate genotypes may improve grain yield and quality, resulting in higher economic benefits (Bhatta *et al.*, 2017).

Nitrogen is an essential nutrient for proper growth and productivity of winter wheat (Fageria and Baligar, 2005). A high proportion of the applied N is not taken up by wheat, resulting in losses due to denitrification, surface runoff, volatilization and leaching (Mullen *et al.*, 2003; Fageria and Baligar, 2005). The loss of applied N contributes unnecessary costs for the farmers, in addition to creating negative environmental impacts such as surface and groundwater pollution (Bhatta *et al.*, 2017). Therefore, appropriate N management (i.e. rate, type and time) is essential for efficient utilization of fertilizer, as well as for improving grain yield. Many researchers detected an increase in wheat yield when N fertilizer was applied at different growth stages and dosages (Lloveras *et al.*, 2001; López-Bellido *et al.*, 2006; Velasco *et al.*, 2012). Moreover, varietal differences play a significant role in N uptake (Ortiz-Monasterio *et al.*, 1997), and it is therefore necessary to evaluate genotype-specific responses to N fertilization. Besides growth practice management, the decision to select which variety to grow is the most important agronomic decision a farmer makes to optimize yield and maximize net benefit (Bhatta *et al.*, 2017).

This study was therefore conducted to determine the effects of N fertilization on seed and straw yield and quality traits of ten winter wheat advanced mutant lines at two locations in Macedonia and to select the best performing lines for dual purpose growing.

2 Materials and Methods

2.1 Plant growth

Field experiments were conducted during the 2015–2016 growing season at two locations: near Skopje and near Veles (Gradsko), Macedonia. Ten advanced wheat mutant lines (2/2-21, 2/1-51, 4/1-194, 4/2-56, 5/1-8, 5/1-199, 6/2-2, 7/1-125, 7/1-143 and 8/2-37/1) were sown in a randomized block design with two replications for evaluation of their agronomic performance. Additionally, three different fertilizer levels were applied: (i) monoammonium phosphate (12:61) 27 g/m^2 + ammonium nitrate 34% 32 g/m^2; (ii) monoammonium phosphate (12:61) 33 g/m^2 + ammonium nitrate 34% 46 g/m^2; and (iii) monoammonium phosphate (12:61) 38 g/m^2 + ammonium nitrate 34% 53 g/m^2. Standard agronomic practices for wheat were applied during the vegetative growth season. At both locations, the experiment was conducted in two replications and the plot size was 1 m^2. Straw mass was determined on a half-plot basis and grain yield on a plot basis. The obtained values were converted to kg/ha. Harvest index (HI) was empirically calculated. Nitrogen (N) content in seed and straw was determined on ten randomly selected plants by micro-Kjeldahl method. Protein (Pr) content was estimated as N × 6.25. Neutral detergent fibre (NDF) and acid detergent fibre (ADF) were determined according to the procedures described by Goering and Van Soest, 1970.

2.2 Statistical analyses

To detect the effect of genotype, location and different fertilizer levels on seed yield, straw mass, harvest index, qualitative characteristics of the seed and straw, NDF and ADF, an analysis of variance was applied and the least significant difference (LSD) test was used for detecting the significant differences (at $p < 0.05$) between the treatments and their two-way interactions. The effect of the replication was not significant and the average value of both replications was used in the following model:

$$R_{ijr} = m + G_i + L_j + F_r + GL_{ij} + LF_{jr} + GF_{ir} + e_{ijr}$$

where R_{ijr} is the response variable, m is the grand mean, G is the effect of the genotype i, L is the effect of the location j, F is the effect of the fertilizer level r, GL_{ij} is the interaction between the genotype and the location, LF_{jr} is the interaction between the location and the fertilizer level, GF_{ir} is the interaction between the genotype and the fertilizer level and e_{ijr} is the random error.

The average values of the genotypes in all environments were standardized and normalized values (Z-score) based on the mean and standard deviation were obtained. The z-values were used to perform two-way cluster analysis in R 3.5.0 statistical package (R Core Team, 2013).

3 Results

The analysis of variance for the analysed traits in wheat mutants showed that the replication and three-way interaction (genotype × location × fertilizer level) did not have significant influence on any analysed trait. Therefore, the data for the replications were pooled together. The effects of the genotype, fertilizer level and genotype × location interaction had significant influence on all analysed traits. The effect of the location was not significant for N and protein content in the straw, while the genotype × fertilizer level had significant effect only on N and protein content in the seed. The fertilizer level × location interaction did not have significant influence on the variability of any trait (Table 21.1).

Considering the performance of mutant lines at both locations and under the three different fertilizer levels, it can be concluded that during the 2015–2016 growing season, both seed and straw yield were higher in Skopje (SK) than in Gradsko (GR) for all mutants, except for 2/2-21, which had higher seed and straw yield in Gradsko, compared with Skopje (Table 21.2). Under all three different fertilizer treatments, the mutant line 5/1-8 gave the highest seed and straw yield. The values for these traits increased with increasing fertilizer level in all mutants, except for 4/2-56, whose best performance was achieved under the second fertilizer level. The mutant line 2/2-21 had the highest harvest index at both locations, followed by 5/1-199 in Skopje and 4/2-58 under the second fertilizer level in Gradsko.

The analysed mutant lines had higher nitrogen and protein content in the seed in Gradsko, compared with Skopje. The increased application of N fertilizer led to increased nitrogen and protein content in the seed. The highest values for both traits at both locations were observed in the mutant line 7/1-125, under the third fertilizer level (Table 21.3). The same mutant line also had very high values for these traits under the second fertilizer level in Gradsko. It was followed by 8/2-137/1 and 5/1-199 in Gradsko, which also achieved a protein content > 15%.

Considering nitrogen and protein content in the straw, the mutant lines showed different values at the two locations, but again, generally, the values were highest under the third fertilization dosage. The same mutants had the highest N and protein contents at both locations. In Skopje, the best performing mutant lines were 6/2-2, 7/1-125, 4/2-56 and 5/1-8, successively, while in Gradsko, the highest value was in 4/2-56, followed by 5/1-8, 6/2-2, 7/1-125 and 2/1-51 (Table 21.3).

The NDF and ADF in the straw of wheat mutant lines were evaluated in order to receive information about the food intake, digestibility and, consequently, the energy intake for the straw of

Table 21.1. ANOVA for the analysed traits in wheat mutants.

Sources of variability	df	Mean squares								
		Seed yield	Straw yield	Nitrogen content in straw	Protein content in straw	Nitrogen content in seed	Protein content in seed	Harvest index	NDF	ADF
Genotype (G)	9	37,423,752.585**	14,354,780.869**	0.014**	0.564**	0.120**	4.699**	0.003**	44.994**	12.701**
Location (L)	1	136,715,415.00**	36,639,846.15**	0.003	0.126	0.050**	1.944**	0.015**	108.918**	15.130**
Fertilizer level (F)	2	10,871,078.067**	1,721,126.717**	0.608**	23.761**	0.154**	5.996**	0.002**	86.000**	18.880**
Genotype × Location (G × L)	9	14,473,121.889**	12,535,470.928**	0.006**	0.252**	0.003**	0.137**	0.002**	31.868**	6.486**
Genotype × Fertilizer level (G × F)	18	370,279.419	47,863.624	0.002	0.063	0.003**	0.129**	0.000	3.331	0.645
Fertilizer level × Location (F × L)	2	240,035.000	14,054.550	0.003	0.098	0.000	0.011	0.000	1.120	0.858
Error	18	167,418.833	38,906.494	0.001	0.032	0.001	0.020	0.000	1.685	1.150
Total	60									

*Significant at $p < 0.05$
**Significant at $p < 0.01$

Table 21.2. Mean values of seed yield, straw yield and harvest index of analysed mutant lines.

	Skopje			Gradsko		
Mutant line	Seed yield (kg/ha)	Straw yield (kg/ha)	Harvest index	Seed yield (kg/ha)	Straw yield (kg/ha)	Harvest index
2/1-51	10,698.33	7,531.67	0.59	9,858.33	9,228.67	0.52
2/2-21	11,791.67	7,205.00	0.62	13,921.67	9,845.67	0.59
4/1-194	10,203.00	8,758.67	0.54	3,938.33	3,790.33	0.51
4/2-56	12,465.00	9,469.00	0.57	11,705.00	8,668.33	0.57
5/1-199	12,198.33	7,584.00	0.62	7,353.33	6,667.33	0.52
5/1-8	15,163.33	11,048.00	0.58	11,503.00	9,522.67	0.55
6/2-2	6,631.67	5,190.00	0.56	4,873.33	4,925.67	0.50
7/1-125	10,563.33	8,520.67	0.55	5,941.67	4,730.00	0.55
7/1-143	12,393.33	10,086.67	0.55	4,016.67	3,303.33	0.55
8/2-137/1	10,706.67	8,764.00	0.55	9,513.33	7,846.67	0.55

G × L $LSD_{0.05}$ Seed yield 136.93
G × L $LSD_{0.05}$ Straw yield 66.01
G × L $LSD_{0.05}$ Harvest index 0.01

each genotype. The concept behind the detergent fibre analysis is that plant cells can be divided into less digestible cell walls (which contain cellulose, lignin and hemicellulose) and mostly digestible cell contents (starch and sugars). These two components can be separated by the use of two detergents: neutral detergent and acid detergent. Non-ruminants cannot digest cellulose, lignin and hemicellulose, while ruminants can partially digest cellulose and hemicellulose.

NDF is a good indicator of 'bulk' and thus feed intake (NDF = cellulose + lignin + hemicellulose), while ADF is an indicator of digestibility and thus energy intake (ADF = cellulose + lignin). The determination of both traits was performed following the procedure of Goering and Van Soest, 1970.

The mutant line 6/2-2 had the highest values for NDF in Gradsko, for all three fertilization dosages. In Skopje, NDF was the highest in 7/1-143 and 7/1-125, when the first fertilizer level was applied. ADF in Skopje was also the highest in 7/1-143, while in Gradsko the highest values were observed in mutant lines 2/1-51 and 4/2-56. Opposite to all other analysed traits, NDF and ADF generally decreased when the applied N quantity was increased (Table 21.4).

The fertilization treatment had a significant influence on all analysed traits (Table 21.5). Increasing the nitrogen dosage led to a significant increase (at level $p = 0.05$) of all traits, except NDF and ADF, where, on the contrary, a significant decrease of the values can be observed.

Seed and straw yield and harvest index were significantly higher in Skopje than in Gradsko (Table 21.6). All other traits had significantly higher values in Gradsko, except for the nitrogen and protein content in the straw, for which the difference was not significant.

The effect of the genotype was significant for all analysed traits (Table 21.7), indicating that the mutant lines differed in all analysed traits.

In order to classify the genotypes based on all analysed traits, a two-way cluster analysis was applied. According to their overall performance, at both locations and with the three different fertilization treatments, the mutant lines were classified in two main groups (Fig. 21.1). The first cluster consisted of mutants 5/1-8, 2/2-21, 4/2-56 and 2/1-5, characterized by very high values for seed yield, straw yield and HI, and high to moderate values for all other traits. Only 4/2-56 had very low values for N and protein content in the seed. The mutant line 6/2-2 did not belong to any of the groups, based on its very low seed and straw yield and very high values for nitrogen and protein content in the straw and NDF, and by these characteristics it differed from all other mutant lines. All other mutants belonged to the second group, with low to moderate yield and moderate to high values for the other traits.

Table 21.3. Nitrogen and protein content in the seed and the straw of analysed mutant lines.

		Skopje		Gradsko		Skopje		Gradsko	
Mutant line	Fertilization level	Nitrogen content in seed (%)	Protein content in seed (%)	Nitrogen content in seed (%)	Protein content in seed (%)	Nitrogen content in straw (%)	Protein content in straw (%)	Nitrogen content in straw (%)	Protein content in straw (%)
2/1-51	1	1.95	12.2	2.06	12.9	0.71	4.44	0.77	4.81
2/1-51	2	2.06	12.9	2.11	13.2	0.76	4.75	0.89	5.56
2/1-51	3	2.18	13.6	2.21	13.8	0.96	6.00	1.15	7.19
2/2-21	1	2.05	12.8	2.05	12.8	0.70	4.38	0.74	4.63
2/2-21	2	2.11	13.2	2.16	13.5	0.78	4.88	0.92	5.75
2/2-21	3	2.19	13.7	2.26	14.1	0.98	6.13	1.11	6.94
4/1-194	1	2.10	13.1	2.18	13.6	0.72	4.50	0.68	4.25
4/1-194	2	2.16	13.5	2.27	14.2	0.85	5.31	0.75	4.69
4/1-194	3	2.30	14.4	2.37	14.8	1.07	6.69	0.92	5.75
4/2-56	1	1.81	11.3	1.82	11.4	0.73	4.56	0.75	4.69
4/2-56	2	1.87	11.7	1.97	12.3	0.87	5.44	0.89	5.56
4/2-56	3	2.02	12.6	2.03	12.7	1.14	7.13	1.17	7.31
5/1-199	1	2.26	14.1	2.29	14.3	0.75	4.69	0.72	4.50
5/1-199	2	2.30	14.4	2.37	14.8	0.81	5.06	0.85	5.31
5/1-199	3	2.35	14.7	2.42	15.1	1.04	6.50	0.96	6.00
5/1-8	1	2.10	13.1	2.11	13.2	0.75	4.69	0.72	4.50
5/1-8	2	2.14	13.4	2.16	13.5	0.88	5.50	0.94	5.88
5/1-8	3	2.19	13.7	2.21	13.8	1.12	7.00	1.16	7.25
6/2-2	1	1.98	12.4	2.00	12.5	0.82	5.13	0.78	4.88
6/2-2	2	2.10	13.1	2.05	12.8	0.97	6.06	0.97	6.06
6/2-2	3	2.11	13.2	2.10	13.1	1.19	7.44	1.13	7.06
7/1-125	1	2.26	14.1	2.29	14.3	0.76	4.75	0.73	4.56
7/1-125	2	2.37	14.8	2.43	15.2	0.89	5.56	0.92	5.75
7/1-125	3	2.54	15.9	2.56	16	1.17	7.31	1.13	7.06
7/1-143	1	2.11	13.2	2.18	13.6	0.67	4.19	0.68	4.25
7/1-143	2	2.16	13.5	2.24	14	0.76	4.75	0.82	5.13
7/1-143	3	2.14	13.4	2.29	14.3	0.99	6.19	1.05	6.56
8/2-137/1	1	2.06	12.9	2.21	13.8	0.74	4.63	0.73	4.56
8/2-137/1	2	2.18	13.6	2.35	14.7	0.81	5.06	0.82	5.13
8/2-137/1	3	2.37	14.8	2.53	15.8	1.06	6.63	1.04	6.50

$G \times L\ LSD_{0.05}$ Nitrogen content in the seed 0.008
$G \times F\ LSD_{0.05}$ Nitrogen content in the seed 0.007
$G \times L\ LSD_{0.05}$ Protein content in the seed 0.047
$G \times F\ LSD_{0.05}$ Protein content in the seed 0.045
$G \times L\ LSD_{0.05}$ Nitrogen content in the straw 0.01
$G \times L\ LSD_{0.05}$ Protein content in the straw 0.06

Table 21.4. NDF and ADF of analysed mutant lines.

Mutant line	Fertilization level	Skopje		Gradsko	
		NDF (%)	ADF (%)	NDF (%)	ADF (%)
2/1-51	1	67.45	53.25	73.15	57.75
2/1-51	2	66.6	53.15	72.65	57.85
2/1-51	3	64.8	52.8	70.24	57.32
2/2-21	1	66.5	52.5	74.3	55.5
2/2-21	2	66.3	50.7	73.2	54.8
2/2-21	3	65.8	51.1	73.8	55.05
4/1-194	1	68.4	55.62	73.95	56.55
4/1-194	2	68.25	55.25	73.6	54.25
4/1-194	3	67.6	54.85	72.8	54.13
4/2-56	1	69.35	54.75	75.65	57.85
4/2-56	2	63.75	53.6	71.25	56.25
4/2-56	3	61.2	52.7	69.6	54.35
5/1-199	1	72.2	55.2	76.8	54
5/1-199	2	71.25	54.25	72.25	55.25
5/1-199	3	68.85	52.65	68.4	52.8
5/1-8	1	74.8	53.2	72.8	53.8
5/1-8	2	72.25	53.25	69.9	51.1
5/1-8	3	69.6	51.6	68.4	51.4
6/2-2	1	75.2	55.45	82.45	53.05
6/2-2	2	74.9	54.5	80.4	52.15
6/2-2	3	72.45	53.05	79.1	51.5
7/1-125	1	75.65	57.25	78.2	56.8
7/1-125	2	72.2	57	73.4	56.15
7/1-125	3	73.6	54.35	69.35	54.75
7/1-143	1	79.2	54.45	74	52.75
7/1-143	2	74.6	49.4	69.7	53.3
7/1-143	3	73.65	50.25	64.6	51
8/2-137/1	1	70.3	53.5	69.35	52.75
8/2-137/1	2	68.85	52.65	69.7	53.3
8/2-137/1	3	64.8	48.3	68.2	53.2

G × L $LSD_{0.05}$ NDF 0.434
G × L $LSD_{0.05}$ ADF 0.359

4 Discussion

The results from the present study revealed that the effects of the genotype, fertilizer level and genotype × location interaction had significant influence on all analysed traits. The effect of the location was not significant for N and protein content in the straw. N and protein content in the seed were significantly influenced by the interaction between the genotype and the fertilizer level, indicating different responses of the mutant lines to nitrogen fertilization treatment. The significant effect of genotype on all traits was observed, reflecting differences in yield potential, biomass production and quality characteristics of the mutant lines under study.

Considering that nitrogen is an essential element for crop development and biomass production, the absorption of N by plants plays an important role in their growth (Kaur *et al.*, 2015). It is one of the most required nutrients for wheat growing and is mostly not supplied in ideal quantity and period to guarantee high yield and good quality of the final product (Todeschini *et al.*, 2016). Identification of wheat genotypes that can efficiently use N is needed in order to sustain or increase yield and quality, while reducing the negative impacts of fertilizer on the environment (Hirel *et al.*, 2007; Foulkes *et al.*, 2009).

Table 21.5. The effect of fertilization treatment on analysed traits.

Fertilization level	Seed yield (kg/ha)	Straw yield (kg/ha)	Nitrogen content in straw (%)	Protein content in straw (%)	Nitrogen content in seed (%)	Protein content in seed (%)	Harvest index	NDF (%)	ADF (%)
1	9,065.40	7,318.85	0.73	4.58	2.09	13.08	0.55	73.49	54.80
2	9,714.00	7,685.20	0.86	5.36	2.18	13.62	0.55	71.25	53.91
3	10,536.50	7,898.90	1.08	6.73	2.27	14.18	0.57	69.34	52.86
$LSD_{0.05}$	218.13	105.16	0.015	0.094	0.012	0.075	0.005	0.69	0.57

Table 21.6. The effect of location on analysed traits.

Locations	Seed yield (kg/ha)	Straw yield (kg/ha)	Nitrogen content in straw (%)	Protein content in straw (%)	Nitrogen content in seed (%)	Protein content in seed (%)	Harvest index	NDF (%)	ADF (%)
Skopje	11,281.47	8,415.77	0.88	5.51	2.15	13.44	0.57	70.01	53.35
Gradsko	8,262.47	6,852.87	0.90	5.60	2.21	13.80	0.54	72.71	54.36
$LSD_{0.05}$	471.68	227.381	ns	ns	0.026	0.162	0.011	1.496	1.236

Table 21.7. Overall performance of mutant lines.

Mutant lines	Seed yield (kg/ha)	Straw yield (kg/ha)	N content in straw (%)	Pr content in straw (%)	N content in seed (%)	Pr content in seed (%)	HI	NDF (%)	ADF (%)
2/1-51	10,278.33	8,380.17	0.87	5.46	2.10	13.10	0.55	69.15	55.35
2/2-21	12,856.67	8,525.33	0.87	5.45	2.14	13.35	0.60	69.98	53.28
4/1-194	7,070.67	6,274.50	0.83	5.20	2.23	13.93	0.52	70.77	55.11
4/2-56	12,085.00	9,068.67	0.93	5.78	1.92	12.00	0.57	68.47	54.92
5/1-199	9,775.83	7,125.67	0.86	5.34	2.33	14.57	0.57	71.63	54.03
5/1-8	13,333.17	10,285.33	0.93	5.80	2.15	13.45	0.56	71.29	52.39
6/2-2	5,752.50	5,057.83	0.98	6.10	2.06	12.85	0.53	77.42	53.28
7/1-125	8,252.50	6,625.33	0.93	5.83	2.41	15.05	0.55	73.73	56.05
7/1-143	8,205.00	6,695.00	0.83	5.18	2.19	13.67	0.55	72.63	51.86
8/2-137/1	10,110.00	8,305.33	0.87	5.42	2.28	14.27	0.55	68.53	52.28
$LSD_{0.05}$	136.932	66.010	0.010	0.060	0.008	0.047	0.003	0.434	0.359

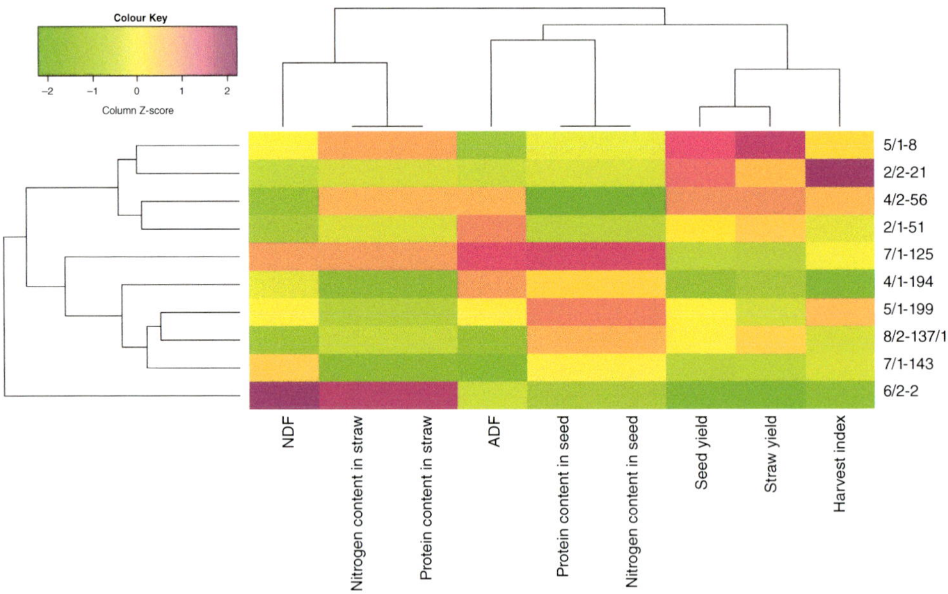

Fig. 21.1. Classification of mutant lines based on analysed traits.

The fertilization treatment in our study had significant influence on all analysed traits. Increasing the nitrogen dosage led to significant increase of all traits, except NDF and ADF, where, on the contrary a significant decrease of the values can be observed. The positive effect of N fertilization on wheat seed yield, biomass production, harvest index and other quantitative traits has been reported by other researchers (Viller and Guillaumes, 2010; Rahman *et al.*, 2011; Brezolin *et al.*, 2016; Ferrari *et al.*, 2016; Bhatta *et al.*, 2017; Costa *et al.*, 2018). Moreover, it was concluded that the complex relationships between the genotype, climate and management practices must be considered in defining efficient N supply management for achieving maximal yields (Costa *et al.*, 2018). For all mutant lines included in this study, the optimal N dosage was determined for both locations. The seed yield, straw yield, nitrogen and protein content in the seed and the straw and harvest index at both locations increased with increasing N dosage, except for the mutant line 4/2-56, for which the highest values for seed yield, straw yield and harvest index were achieved under the second fertilizer level. NDF and ADF decreased with the increase of N supply.

Cluster analysis was effective in classification of genotypes with similar traits (Khodadadi *et al.*, 2011; Ajmal *et al.*, 2013; Bhattarai *et al.*, 2017). In this study, it revealed two main groups, in which mutant lines were characterized by very high values for seed yield, straw yield and HI, and high to moderate values for all other traits, versus mutant lines with low to moderate yield and moderate to high values for the other traits. Only one mutant line did not belong to either of the groups, based on the very low seed and straw yield and very high values for nitrogen and protein content in the straw and NDF, and by these characteristics it differed from all other mutant lines.

5 Conclusion

Nitrogen fertilizer affected all analysed traits of the mutant lines included in this study, and genotypic response varied under the different testing environments (locations and different fertilizer treatments). Nitrogen treatment significantly increased seed yield, straw yield, harvest index, nitrogen and protein content in the seed and straw and significantly decreased NDF and ADF. The optimal management practices were established for each mutant line at the separate locations. The best performing mutant lines based on different traits were identified and will be recommended for official variety registration.

References

Ajmal, S.U. *et al.* (2013) Multivariate analysis of genetic divergence in wheat (*Triticum aestivum*) germplasm. *Pakistan Journal of Botany* 45(5), 1643–1648.

Bhatta, M. *et al.* (2017) Seeding rate, genotype, and topdressed nitrogen effects on yield and agronomic characteristics of winter wheat. *Crop Science* 57, 951–963.

Bhattarai, R.P. *et al.* (2017) Cluster analysis of Elite spring wheat (*Triticum aestivum* L.) genotypes based on yield and yield attributing traits under irrigated condition. *International Journal of Experimental Research and Review* 10, 9–14.

Brezolin, A.P. *et al.* (2016) The efficiency of wheat yields by nitrogen dose and fractionation. *African Journal of Agricultural Research* 11, 3440–3449.

Costa, J.S.P. *et al.* (2018) Single and split nitrogen dose in wheat yield indicators. *Brasileira de Engenharia Agrícola e Ambiental* 22(1), 16–21.

Fageria, N.K. and Baligar, V.C. (2005) Enhancing nitrogen use efficiency in crop plants. *Advances in Agronomy* 88, 97–185.

Ferrari, M. *et al.* (2016) Manejos e fontes de nitrogênio nos componentes de afilhamento de trigo. *Revista Brasileira de Ciências Agrárias* 11, 178–185.

Foulkes, M.J. *et al.* (2009) Identifying traits to improve the nitrogen economy of wheat: recent advances and future prospects. *Field Crops Research* 114, 329–342.

Goering, H.K. and Van Soest, P.J. (1970) *Forage Fiber Analysis (apparatus, reagents, procedures and some applications)*. USDA Agricultural Handbook No. 379. US Department of Agriculture, Washington, DC.

Hirel, B. *et al.* (2007) The challenge of improving nitrogen use efficiency in crop plants: towards a more central role for genetic variability and quantitative genetics within integrated approaches. *Journal of Experimental Botany* 58, 2369–2387.

Kaur, G. *et al.* (2015) Nitrogen levels effect on wheat nitrogen use efficiency and yield under field conditions. *African Journal of Agricultural Research* 10(23), 2372–2377.

Khodadadi, M. *et al.* (2011) Genetic diversity of wheat (*Triticum aestivum* L.) genotypes based on cluster and principal component analyses for breeding strategies. *Australian Journal of Crop Science* 5(1), 17.

Lloveras, J. *et al.* (2001) Bread-making wheat and soil nitrate as affected by nitrogen fertilization in irrigated Mediterranean conditions. *Agronomy Journal* 93, 1183–1190.

López-Bellido, L. *et al.* (2006) Fertilizer nitrogen efficiency in durum wheat under rainfed Mediterranean conditions: effect of split application. *Agronomy Journal* 98, 55–62.

Mukherjee, D. (2014) Water harvesting and resource conservation technique of Darjeeling Himalaya. *Himalayan Research Journal* 2(1), 72–79.

Mullen, R.W. *et al.* (2003) Identifying an in-season response index and the potential to increase wheat yield with nitrogen. *Agronomy Journal* 95, 347–351.

Ortiz-Monasterio, J.I. *et al.* (1997) Genetic progress in wheat yield and nitrogen use efficiency under four N rates. *Crop Science* 37, 898–904.

Pooja, D. *et al.* (2018) Genetic diversity based on cluster and principal component analyses for yield and its contributing characters in RILs of bread wheat (*Triticum aestivum* L.). *International Journal of Pure and Applied Bioscience* 6(2), 242–247.

R Core Team (2013): *R: A language and environment for statistical computing*. R Foundation for Statistical Computing, Vienna, Austria. Available at: http://www.R-project.org/ (accessed 2019)

Rahman, M.A. *et al.* (2011) Yield response and nitrogen use efficiency of wheat under different doses and split application of nitrogen fertilizer. *Bangladesh Journal of Agricultural Research* 36, 231–240.

State Statistical Office (2016) *Statistical Yearbook of the Republic of Macedonia*.

Todeschini, M.H. *et al.* (2016) Eficiência de uso do nitrogênio em cultivares modernas de trigo. *Bragantia* 75, 351–361.

Velasco, J.L. *et al.* (2012) Optimizing fertilizer nitrogen use efficiency by intensively managed spring wheat in humid regions: effect of split application. *Canadian Journal of Plant Science* 92, 847–856.

Viller, J.M. and Guillaumes, E. (2010) Use of nitrification inhibitor DMPP to improve nitrogen recovery in irrigated wheat on a calcareous soil. *Spanish Journal of Agricultural Research* 8, 1218–1230.

22 Induced Variation for Post-Emergence Herbicide Tolerance in Lentil

Sarvjeet Singh[1], S.R. Sharma, R.K. Gill and Shiv Kumar

Abstract

Lentil (*Lens culinaris* L. Medik.) is an important cool-season food legume but is a poor competitor to weeds because of a slow early growth rate. If weeds are left uncontrolled, they can reduce yield by up to 50%. Sensitivity of lentil to post-emergence herbicides warrants development of herbicide-tolerant cultivars. In the absence of natural variability, mutation breeding is a powerful tool to create variability for desired traits. Thus, 1000 seeds of a lentil genotype, LL1203, were exposed to gamma radiation (300 Gy, ^{60}Co) with the objective to induce herbicide tolerance. Seeds of all 530 surviving M_1 plants were harvested individually and divided in two parts to raise the M_2 generation in two different plots. Each plot was sprayed with imazethapyr (75 g/ha) and metribuzin (250 g/ha) herbicides 50 days after sowing, using water at 375 l/ha. Data on herbicide tolerance for individual M_2 plants were recorded after 14 days of herbicide spray on a 1–5 scale, where 1 = highly tolerant (plants free from chlorosis or wilting) and 5 = highly sensitive (leaves and tender branches completely burnt). For herbicide-tolerant M_2 plants, data were also recorded for pod and yield per plant. None of the M_2 plants showed a high level of tolerance to imazethapyr. However, 14 mutants having higher herbicide tolerance to metribuzin were selected. Two mutants ('LL1203-MM10', 'LL1203-MM7') recorded < 2.0 score, while six mutants recorded < 2.50 score as compared with the 3.13 score of the parent variety. The pods per plant and seed yield per plant of mutants 'LL1203-MM7' (383 and 12.4 g) and 'LL1203-MM10' (347 and 12.1 g) were higher than those of the parent genotype LL1203 (253 and 7.8 g). The study indicated that metribuzin-tolerant mutants have some other desirable traits that can be of use in lentil breeding.

Keywords: lentil • mutation • herbicides • tolerance • weeds

1 Introduction

Lentil (*Lens culinaris* Medik., $2n = 14$) is one of the oldest self-pollinated crop species, like barley, emmer wheat and pea (Harlan, 1992; Sandhu and Singh, 2007). It can flourish well on poor soils and adverse environmental conditions, as it has the ability of nitrogen fixation. Lentil is an underexploited crop and ground-breaking research has been made towards improving its agronomic and nutritional characteristics. Lentil cultivars have poor yields, due to weed infestation and their susceptibility to biotic and abiotic stresses. Lentil has a small leaf area with a slow growth rate in the early stages of crop growth, which makes it a poor competitor to weeds. If weeds are left uncontrolled, they can reduce lentil yield by up to 50% (Tanveer and Ali, 2003).

[1] Email: sarvjeetm@rediffmail.com

 Mutation Breeding, Genetic Diversity and Crop Adaptation to Climate Change (eds S. Sivasankar *et al.*)
DOI: 10.1079/9781789249095.0022

Therefore, weed management is important for obtaining maximum yield with good quality of produce. The high sensitivity of lentil to post-emergence herbicides such as amino acid synthesis-inhibitor families (e.g. imidazolinones and sulfonylureas), lipid-inhibitor families, cyclohexanediones and growth regulator-inhibitor families (e.g. phenoxyacetic acid and benzoic acid) have been observed (Mishra *et al.*, 2005; Singh *et al.*, 2014). Mutation breeding is useful to generate new variability for a particular trait if a sufficient level of variability is not present in available germplasm. Various mutagens have been used to generate new variability for herbicide tolerance and mutants with the desired level of tolerance have been selected in the M_2 generation and subsequent populations in different crops (Rizwan *et al.*, 2015, 2017). Induced mutagenesis is responsible for development of herbicide-tolerant cultivars in many crops; for example, the 'Coromup' variety of narrow-leaf lupin and 'Tracy-M' variety of soybean (Hartwig, 1987) for improved tolerance to the herbicide metribuzin. 'Tanjil-AZ-55' and 'Tanjil-AZ-33', two highly metribuzin-tolerant mutants of narrow-leaf lupin, have also been identified (Si *et al.*, 2009). Similarly, 'RH44', an imidazolinone herbicide-tolerant variety of lentil, was developed by mutagenesis and cross-breeding (Slinkard *et al.*, 2007). Herbicide-tolerant crop cultivars, including lentil, are required to manage weeds through the use of post-emergence herbicides. Such tolerant cultivars would have better chances of finding a place in a cereal-dominated cropping system as well as in further breeding programmes.

2 Materials and Methods

In the present study, 1000 pure seeds of a lentil genotype, LL1203, were exposed to gamma radiation (300 Gy or 30 kR, ^{60}Co) with the objective to induce herbicide tolerance. M_1 seeds were grown during rabi 2013–2014 in the Experimental Area of Pulses Section, Punjab Agricultural University, Ludhiana. Seeds of all 530 surviving M_1 plants were harvested individually and divided into two parts to raise the M_2 generation in two different plots. Each plot was sprayed with imazethapyr (75 g/ha) and metribuzin (250 g/ha) herbicides 50 days after sowing with a shoulder-mounted hand-operated knapsack sprayer using 375 l water/ha. Data on herbicide tolerance for individual M_2 plants were recorded after 14 days of herbicide spray on a 1–5 scale (Gaur *et al.*, 2013), where 1 = highly tolerant (plants free from chlorosis or wilting) and 5 = highly sensitive (leaves and tender branches completely burnt). Plants were scored again 45 days after spraying as they showed recovery. The mutant plants that were found to be green and healthy, without any symptoms of chlorosis and wilting after recovery, were categorized as tolerant while those with dried leaflets and completely dried plants were categorized as highly sensitive. The mutant plants that recovered after imazethapyr spray were named as 'LL1203-IM', while the mutants recovered after metribuzin spray were named as 'LL1203-MM' followed by a specific number. Data on herbicide-tolerant M_2 plants were also recorded for various yield components, namely plant height (cm), biomass per plant (g), pods per plant, 100-seed weight (g) and yield per plant (g). Data for five unsprayed random plants of the parental variety were also recorded and averaged for comparison with individual mutant plants.

3 Results

None of the M_2 plants showed a high level of tolerance when compared with the parent variety for imazethapyr herbicide. However, four mutant plants were identified with a higher number of pods per plant and yield per plant as compared with the parental variety (Table 22.1). The herbicide sensitivity score of these four mutant plants was higher, with a range of 2.25–3.00 as compared with a score of 2.00 for the parent variety (Fig. 22.1A). In the case of metribuzin herbicide, 14 mutants were identified that had a higher tolerance score than the parent variety. Among them, the mutant 'LL1203-MM10' had a score of 1.50, 'LL1203-MM7' scored 1.75; three other mutants recorded a score of 2.00, five mutants recorded a score of 2.25, while two mutants recorded 2.50 against the 3.13 score of the parent variety (Fig. 22.1B) (Table 22.2). Among the selected mutants, the biomass per plant ranged from 39.9 g in 'LL1203-MM7' to 25.2 g in 'LL1203-MM4'. The number of pods per plant ranged from 383 in 'LL1203-MM7' to 282 in 'LL1203-MM4' mutants. The maximum 100-seed weight (2.15 g) was observed in 'LL1203-MM6' and minimum (1.85 g) in 'LL1203-MM12'. The mean seed yield per plant of selected mutants was 10.29 g as compared with 7.86 g in the parent variety. The maximum seed yield per plant (12.38 g) was recorded in 'LL1203-MM7', while it was minimum (8.13 g) in 'LL1203-MM4' in the case of selected mutants (Table 22.2).

4 Discussion

Chemical and physical mutagenesis followed by selection of herbicide-tolerant mutants are the basis for selection of herbicide-tolerant plants, besides direct selection for natural variability. Absorption, metabolism and translocation are three crucial factors that affect herbicide movement to the active site and have a significant effect on expression of herbicide resistance/ susceptibility at the whole-plant level (Shaner

Table 22.1. Comparison of promising M_2 plants with their parent variety for imazethapyr herbicide tolerance and some other traits.

Mutant	Tolerance score	Plant height (cm)	Biomass per plant (g)	No. of pods per plant	100-seed weight (g)	Yield per plant (g)
LL1203-IM1	3.00	28.9	27.53	305	1.94	8.53
LL1203-IM2	2.50	31.2	28.37	291	2.04	9.05
LL1203-IM3	2.25	31.1	32.91	354	2.03	10.13
LL1203-IM4	2.50	30.8	29.11	274	2.01	8.86
Mean of mutants		30.5	29.48	306	2.01	9.14
STD DEV of mutants		1.08	2.37	34.41	0.04	0.69
Mean of 5 plants of parent variety		29.4	23.4	253	2.33	7.86
Range of mutants		28.9–31.2	27.53–32.91	274–354	1.94–2.04	8.53–10.13

Table 22.2. Comparison of promising M_2 plants with their parental variety for metribuzin herbicide tolerance and some other traits.

Mutants	Tolerance score	Plant height (cm)	Biomass per plant (g)	No. of pods per plant	100-seed weight (g)	Yield per plant (g)
LL1203-MM1	2.50	32.8	25.3	304	1.86	8.83
LL1203-MM2	2.25	31.7	29.5	315	2.04	9.50
LL1203-MM3	2.00	33.5	32.7	338	2.13	11.36
LL1203-MM4	2.50	29.7	25.2	282	2.02	8.13
LL1203-MM5	2.25	33.3	25.7	312	2.05	9.87
LL1203-MM6	2.00	31.4	30.8	316	2.15	10.84
LL1203-MM7	1.75	34.1	39.9	383	1.98	12.38
LL1203-MM8	2.25	29.7	28.8	307	2.13	9.30
LL1203-MM9	2.25	30.3	26.5	296	1.97	9.14
LL1203-MM10	1.50	32.6	36.8	347	1.97	12.05
LL1203-MM11	2.25	31.2	28.3	309	2.11	9.75
LL1203-MM12	2.00	31.6	31.9	313	1.85	10.21
LL1203-MM13	1.75	30.6	32.2	322	1.91	11.25
LL1203-MM14	1.75	31.7	33.3	366	1.86	11.43
Mean of mutants		31.7	30.5	322.1	2.00	10.29
STD DEV of mutants		1.38	4.37	27.45	0.10	1.28
Mean of 5 plants of parent variety		29.4	23.4	253	2.33	7.86
Range for mutants		29.7–34.1	25.2–39.9	282–383	1.85–2.15	8.13–12.38

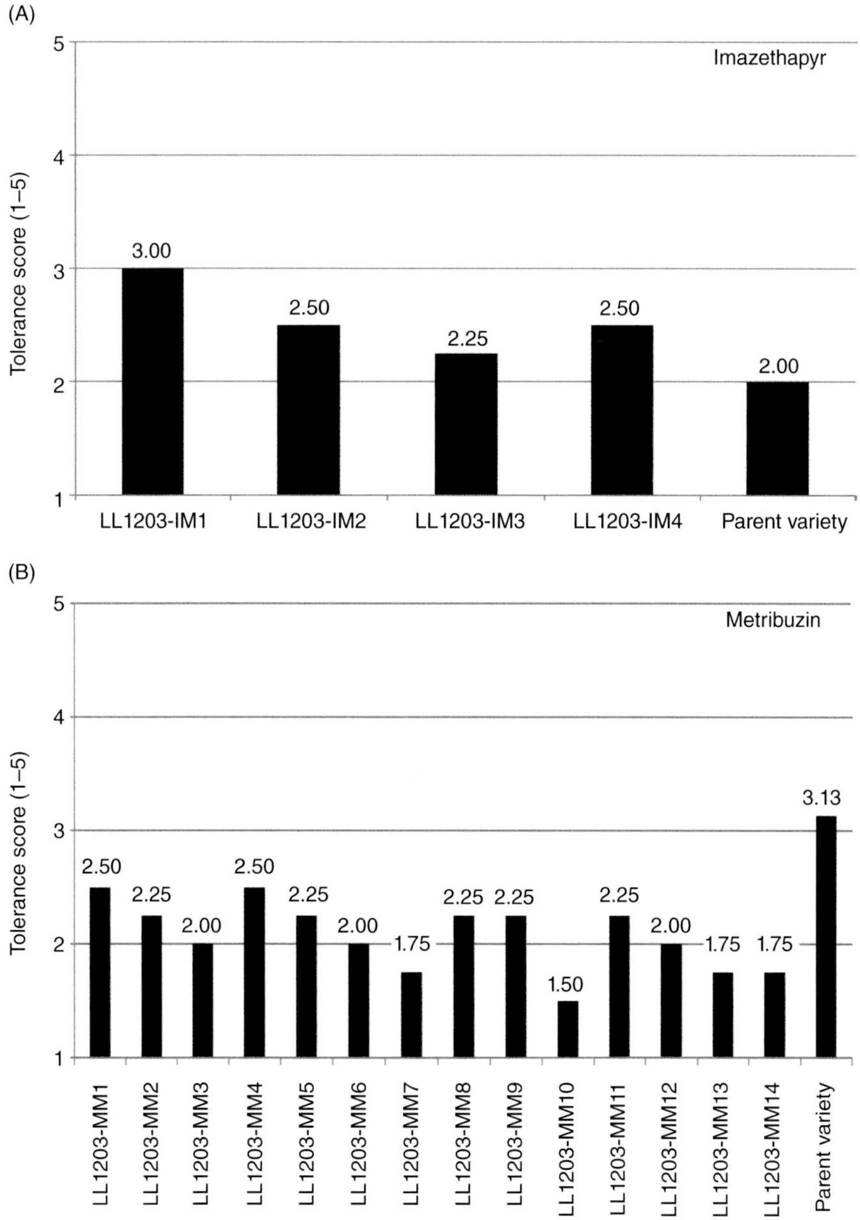

Fig. 22.1. Herbicide tolerance scores of selected lines. **(A)** Imazethapyr tolerance. **(B)** Metribuzin tolerance.

and Robson, 1985; Beyer *et al*., 1988). Results of the present study revealed that no mutant tolerant to imazethapyr was observed through mutation induction but positive mutations have been observed for other economically important traits such as biomass, pod number and seed yield per plant. Previous reports have suggested that imidazolinone and triazinone herbicides control weeds by inhibiting the enzyme acetolactate synthase (ALS) (Powles and Holtum, 1994) and photosystem II (Rutherford and Krieger-Liszkay, 2001). A similar study by Si *et al*. (2009) found a mutant, 'Tanjil-AZ-33', in narrow-leafed lupin which has more than a twofold yield increase compared with the parent cultivar 'Tanjil'. Imidazolinone-tolerant mutant in chickpea (Toker *et al.*, 2012) and a winter wheat cultivar 'Fidel' (Newhouse *et al.*, 1992) have also been reported. Herbicide-tolerant mutants have been used to develop commercial herbicide-tolerant cultivars in many crops, viz. imidazolinone-tolerant maize, rice, wheat, canola, sunflower and lentil; sulfonylurea-tolerant soybean and sunflower; cyclohexanedione-tolerant maize; and triazine-tolerant canola (Duke, 2005). Albino plants (chlorophyll-deficient mutations) were observed in the M_2 generation in the present study, as also reported earlier in chickpea (Toker *et al.*, 2012). Recovery of metribuzin-tolerant mutant plants indicated that herbicide tolerance can be induced through mutation in lentil. Different mutants were recovered with varied herbicide-tolerance scores, indicating that the trait was governed by different genes with major or minor effects. Recovery of mutants with enhanced expression of various traits indicated that variability for traits (plant height, biomass per plant, pods per plant and seed yield per plant) can also be induced through mutation.

5 Conclusion

Identification of an imidazolinone-tolerant lentil mutant must be given top priority to overcome weeds in winter-sown lentil, as weeds are known to be the most important factor that reduces seed yield (Tepe *et al.*, 2005). Identification and incorporation of herbicide-tolerant genes into commercial cultivars will be useful in devising an integrated weed management strategy. Thus, the tolerant mutants isolated in this present study can serve as potential material for solving the weed infestation problem in lentil for yield enhancement.

Acknowledgement

A grant received from a CGIAR Research Programme (CRP) on Grain Legumes through ICARDA, Rabat, Morocco, for carrying out these research activities is duly acknowledged. The travel grant awarded by the Joint FAO/IAEA Division of Nuclear Techniques in Food and Agriculture to the first author for presenting this work at the symposium is also duly acknowledged.

References

Beyer, E.M. *et al.* (1988) Sulfonylurea herbicides. In: Kearney, P.C. and Kaufmann, D.D. (eds) *Herbicides: Chemistry, Degradation, and Mode of Action*. Marcel Dekker, New York.

Duke, S.O. (2005) Taking stock of herbicide resistant crops ten years after introduction. *Pest Management Science* 61, 211–218.

Gaur, P.M. *et al.* (2013) Large genetic variability in chickpea for tolerance to herbicides imazethapyr and metribuzin. *Agronomy* 3, 524–536.

Harlan, J.R. (1992) *Crops and Man*, 2nd edn. American Society of Agronomy, Madison, Wisconsin.

Hartwig, E.E. (1987) Identification and utilization of variation in herbicide tolerance in soybean (*Glycine max*) breeding. *Weed Science* 35, 4–8.

Mishra, J.S. *et al.* (2005) Efficacy of herbicides against field dodder (*Cuscuta campestris*) in lentil, chickpea and linseed. *Indian Journal of Weed Science* 37, 220–224.

Newhouse, K.E. *et al.* (1992) Tolerance to imidazolinone herbicides in wheat. *Plant Physiology* 100, 882–886.

Powles, S.B. and Holtum, J.A.M. (eds) (1994) *Herbicide Resistance in Plants: Biology and Biochemistry*. Lewis Publ., Boca Raton, Ann Arbor, London and Tokyo.

Rizwan, M. *et al.* (2015) Development of herbicide resistant crops through induced mutations. *Advances in Life Sciences* 3, 1–8.

Rizwan, M. *et al.* (2017) Pre-breeding of lentil (*Lens culinaris* Medik.) for herbicide resistance through seed mutagenesis. *PLoS ONE* 12(2), e0171846.

Rutherford, A.W. and Krieger-Liszkay, A. (2001) Herbicide-induced oxidative stress in photosystem II. *Trends in Biochemical Science* 26, 648–653.

Sandhu, J.S. and Singh, S. (2007) History and origin. In: Yadav, S.S., McNeil, D. and Stevenson, P.C. (eds) *Lentil: An Ancient Crop for Modern Times*. Springer, Dordrecht, Netherlands, pp. 1–9.

Shaner, D.L. and Robson, P.A. (1985) Absorption, translocation, and metabolism of AC 252,214 in soybean (*Glycine max*), common cocklebur (*Xanthium strumarium*), and velvet leaf (*Abutilon theophrasti*). *Weed Science* 33, 469–471.

Si, P. *et al.* (2009) Improved metribuzin tolerance in narrow-leafed lupin (*Lupinus angustifolius* L.) by induced mutation and field selection. *Field Crops Research* 113, 282–286.

Singh, G. *et al.* (2014) Weed management in lentil with post-emergence herbicides. *Indian Journal of Weed Science* 46, 187–189.

Slinkard, A.E. *et al.* (2007) Lentil plants having increased resistance to imidazolinone herbicides. *US patent* 7232942.

Tanveer, A. and Ali, A. (2003) *Weeds and their Control*. Higher Education Commission, Islamabad, Pakistan.

Tepe, I. *et al.* (2005) Comparison of some winter lentil cultivars in weed–crop competition. *Crop Protection* 24(6), 585–589.

Toker, C. *et al.* (2012) Improvement in Imidazolinone resistance in *Cicer* species by induced mutation. *Plant Breeding* 131, 535–559.

23 Application of Mutation Techniques and Genotype × Environment Interaction for Grain Yield in Ion Beam Induced Mutant Rice Lines Tested in Multiple Locations in Malaysia

Mohd Rafii Yusop[1], Yusuff Oladosu, Abdul Rahim Harun, Asfaliza Ramli, Ghazali Hussin, Mohd Razi Ismail and Norhani Abdullah

Abstract

Genotype evaluation for stability and high yield in rice is an important factor for sustainable rice production and food security. These evaluations are essential, especially when the breeding objective is to release rice with high yields, adaptability and stability for commercial cultivation. To achieve this objective, this study was carried out to select high-yielding rice genotypes induced by ion beam irradiation. Seeds of the rice variety 'MR219' were subjected to different doses of 320 MeV carbon-ion beam irradiation to determine the optimum dose to produce high mutant frequency and spectrum. The optimum dose was 60 Gy. After several cycles of selection and fixation between 2009 and 2014 (M_0–M_6), six prospective lines with desirable characters were selected at the M_6 generation. The selected mutant lines along with other mutant varieties were then tested at five locations in two planting seasons to select high-yielding and stable genotypes. The experiment was conducted in a randomized complete block design with three replications across the locations and seasons. The pooled analysis of variance revealed highly significant differences ($p \leq 0.01, 0.05$) among genotypes, among locations and among genotypes by location by season (G×L×S interaction) for the yield traits except for seasons and genotype by season (G×S interaction). Based on univariate and multivariate stability parameters, rice genotypes were classified into three main categories. The first group comprised genotypes with high yield stability along with high yield per hectare. These genotypes include ML4 and ML6 and are widely adapted to diverse environmental conditions. One line exhibited high yield per hectare but low stability; this genotype (ML9) is suitable for specific environments. The last group had low yield per hectare and high stability and included 'MR220', 'Binadhan4' and 'Binadhan7'. This final group is more suitable for breeding specific traits or perhaps has yield component compensation. Hence, rice mutant lines ML4 and ML6 were recommended for commercial cultivation in Malaysia.

Keywords: rice • stability • GGE biplot • genotype × environment interaction • mutant line

1 Introduction

A genotype having high yield potential with a low degree of yield fluctuation across a wide range of environments is considered stable and desirable in any breeding programme, especially if the object is to release a commercial variety. However, the relative performance of a genotype varies due to genotype by environment interactions (G×E). The presence of G×E

[1] Email: mrafii@upm.edu.my

DOI: 10.1079/9781789249095.0023

reduces the correlation between phenotypes and genotypes and also complicates breeding work (Makumbi *et al.*, 2015). The stability and adaptability of a genotype during growth is affected by the G×E interaction. Thus, the value of both stability and adaptability seem to reduce when these interactions are highly significant (Liu *et al.*, 2017). Both G and G×E interactions should be measured for genotype evaluation (Yan *et al.*, 2000). Investigation of G×E interactions or phenotypic stability in rice has been achieved using several methods of analysis. Analysis of variance (ANOVA) is often used in statistical analysis for multi-environmental yield trials with the objective of identification of existence of G×E interactions. ANOVA measures the components of variation existing among random and fixed factors (e.g. genotype, replication, location, year) and the interactions between and within these factors. However, ANOVA has a limitation as it is unable to differentiate the significant differences of genotype in the non-additive term known as G×E (Zobel *et al.*, 1988). Aside from ANOVA, several other methods of analysing G×E data have been developed to reveal patterns of interaction, such as univariate and multivariate stability analysis. The recent and commonly used approach for identification of genotypes that are relatively stable in diverse environments is the GGE biplot; this method uses a graphical technique to display the interaction patterns and to picture the interrelationships among genotypes, environments and G×E. GGE biplots are constructed by singular value decomposition of environment-centred, or within-environment standardized, G×E data (Yan *et al.*, 2000). Both G (genotype) and G×E are displayed by the biplots, which are also applicable to cultivar assessment (Kang, 1993). GGE biplot analysis has been applied in numerous studies, including cultivar evaluation (Oladosu *et al.*, 2017a), mega-environment evaluation (Oladosu *et al.*, 2017b) and varietal stability assessment (Zhang *et al.*, 2016). The goal of this study was to evaluate advanced rice mutant lines for the G×E interaction using univariate and multivariate analysis. This model is useful in identifying highly stable and adapted genotypes for a specific environment or wide range of environments for crop production and release as commercial variety.

2 Materials and Methods

2.1 History of planting materials

The most cultivated rice variety in Malaysia is 'MR219', which covers about 90% of the cultivated area (Oladosu *et al.*, 2014). Due to its acceptability, efforts are constantly being made by small- and large-scale farmers on improving its yield potential. Therefore, its seeds were subjected to 320 MeV carbon-ion irradiation by a vertical beam line of AVF cyclotron at the Japan Atomic Energy Research Institute, Takasaki, in 2009. Fresh and healthy 'MR219' rice seeds were exposed to different doses of 0, 10, 20, 40, 60, 80, 100 and 120 Gy, respectively, to determine the optimum doses for the production of high mutant frequency and spectrum and the optimum dose was found to be 60 Gy. Ten thousand single M_1 plants per hill spaced in a 25 cm × 25 cm array were transplanted in the field to produce M_2 seeds. A total of 5250 plants were selected, from which two panicles per hill (i.e. from a single plant) were randomly harvested. About 5% of the M_2 families were selected for further screening in the M_3 generation. Superior plants were selected from superior progenies based on yield and yield-related traits (panicles per hill, number of grains per panicle). After several rounds of selection and fixation performed during the 2009–2014 seasons (M_0–M_6), six promising mutant lines with the required adaptive traits were recovered in the M_6 generation. For the G×E interaction, a set of 15 genotypes was used comprising six advanced mutant lines, six mutant varieties from abroad and three well-characterized released varieties. The advanced mutant lines (ML4, ML6, ML9, ML10, ML21 and ML24) were from preliminary studies of ion beam irradiation (Oladosu *et al.*, 2015). Three mutant varieties were obtained from the Bangladesh Institute of Nuclear Agriculture (BINA), viz. 'Binadhan4', 'Binadhan7' and 'Iratom', while the additional three mutant varieties 'VN121', 'VN124' and 'VN001' were supplied by the Vietnam Atomic Energy Institute (VINATOM). The released varieties ('MR219', 'MR220' and 'MR253') served as reference lines for validating the performance of advanced breeding lines in different environments. The field trials were conducted in five locations (Serdang, Tanjung Karang, Kota Serang Semut, Seberang Prai and Sekinchan) in two different cropping seasons. At each location, the experiment was laid out in a randomized complete block design with three replications. Plot size was 38 × 9 m^2, with a sub-plot size of 2 m^2 for each genotype in each replication. The optimum date for transplanting at each location was used according to the farmers' schedules. Fertilizer application was followed as prescribed by the Malaysian Agricultural Research and Development Institute (MARDI) starting from day 15 to day 75. The land was mechanically ploughed following the normal cultural practices of local farmers across all locations. Data analysis on yield (t/ha) was estimated from the weight of threshed grains from all panicles in 1.5 × 1.5 m^2, excluding border rows.

2.2 Analysis of variance (ANOVA) for genotype by environment interaction (G×E)

An analysis of variance (ANOVA) was calculated for grain yield, using the program SAS version 9.4, to determine the variation among the genotypes and the environment and to estimate the G×E interaction.

2.3 Stability analysis

G×E interaction was analysed using several methods of univariate and multivariate stability statistical analysis. The SAS program version 9.4 was used for the univariate stability parameter using SASG×E code developed by Dia *et al.* (2015), which is freely available. Univariate stability analysis parameters include: regression slope (b_i), deviation from regression (S^2_d), stability variance (σ_i^2), Wricke's ecovalance (W_i^2) and Kang's stability statistic (YS_i) (Dia *et al.*, 2015). The multi-environment trial (MET) data was analysed graphically using GGE biplot software for interpretation of the G×E interaction (Yan and Rajcan, 2002). This method consists of two concepts, the GGE concept (Yan *et al.*, 2000) and biplot concepts (Gabriel, 1971), and was used to analyse 15 genotypes visually. The GGE biplots were computed using the GGEBiplot GUI package of R statistical software in RStudio (CRAN, 2014; RStudio, 2014). The graphs describe: (i) mega-environment ('which-won-where' pattern of GGE); (ii) genotype evaluation; and (iii) tested environment.

3 Results and Discussion

3.1 Analysis of variance (G×E)

The pooled data for the analysis of variance over the ten locations are presented in Table 23.1. Locations (L), genotypes (G) and G×L×S were highly significant ($p \leq 0.01$) while G×L was significant at $p \leq 0.05$. However, seasons (S) and G×S were not significantly different. The partitioning of the total sum-of-squares percentage presented in Table 23.1 explains the variation percentage for yield per hectare. Locations accounted for the highest variation followed by genotype, at 33% and 20%, respectively. The significant effects of G×L×S interaction indicate the level at which genotypes responded differently due to environmental variation in a given

Table 23.1. Mean square and total sum-of-square percentage for yield traits.

		Yield/ha	
Source of variation	Degree of freedom	Means square	Total sum-of-square %
Reps (location)	10	12.49**	5.04
Locations (L)	4	78.68**	32.71
Seasons (S)	1	0.94^ns	0.04
Genotypes (G)	14	25.74**	19.56
G×L	56	5.39*	14.92
G×S	14	3.16^ns	1.79
G×L×S	60	5.75**	13.94
Error	290	3.58	11.99

**Highly significant ($p \leq 0.01$); *significant ($p \leq 0.05$); ns, not significant.

location. The results also show the changes in the comparative ranking of the genotypes from one environment to another, which indicates the necessity of testing rice varieties at multiple locations. This study illustrates the significant difficulties faced by plant breeders in selecting a new cultivar for release. However, analysis of variance does not explain details of the G×E interactions. Additional statistical analyses such as univariate and multivariate analysis are more useful for understanding and describing G×E interactions. Freeman (1973) reported that the main reason for genotype evaluation over a wide range of environments is to estimate their adaptability and stability.

3.2 Stability analysis

Univariate stability

When the regression coefficient value (b_i) of a genotype approximates unity ($b_i \approx 1$) ($p < 0.01$) and the trait mean is high, this is considered to represent a genotype that is stable across diverse environments. However, when the b_i value is associated with low trait mean, the genotype is poorly adapted to all environments. A regression coefficient value that is greater than unity describes a genotype with a high sensitivity to changes in environment, and a greater specificity of adaptability to a high-yielding environment. In this study, the b_i value ranged from –0.04 to 1.99 for yield/ha (Table 23.2). For yield/ha, the genotype ML9 is considered the best, due to its significant value for b_i ($p > 0.05$) being close to unity; this performance is followed by ML10. 'VN121' was considered the lowest in ranking. Eberhart and Russell (1966) proposed the regression coefficient (b_i) as a parameter of response, and deviations from regression (S^2_d) as a parameter of stability. Although high mean yield is a desirable attribute, it is not an indicator of yield stability (Eberhart and Russell, 1966). According to Eberhart and Russell (1966), a genotype is considered to be stable if the residual mean squares from the regression model on the environment index are small. A genotype is considered to be stable when deviation from regression (S^2_d) is not significantly different from zero. The genotypes with the highest yield/ha in this study were ML9, ML4, ML6 and ML10. Among these genotypes, only ML6 had a significant value of S^2_d.

Shukla's stability variance (σ_i^2) is a measure of stability rather than of mean performance, while Wricke's ecovalence (W_i^2) defines the contribution of each genotype to the G×E interaction sum of squares. According to these two

Table 23.2. Means of stability statistics.

Genotype	M	R	b_i	R	S^2_d	R	σ_i^2	R	W_i^2	R	YS_i	R
BINA4	5.83	8	1.37	10	0.94	1	0.92	2	9.89	2	8	5
BINA7	4.94	13	1.99	14	3.12*	12	3.49	5	29.95	5	1	10
IRATOM	3.6	15	1.52	11	2.75	5	2.58	4	22.8	4	–1	13
ML10	7.9	4	1.31	9	6.64	11	7.80***	14	63.55	14	9	3
ML21	7.37	6	1.78	13	3.33	6	4.01	7	33.93	7	6	6
ML24	7.49	5	0.22	6	3.99	7	7.12**	13	58.21	13	4	9
ML4	8.16	2	–0.11	4	1.02	2	1.59	3	15.04	3	16	1
ML6	8.04	3	4.52**	15	5.44**	15	5.02*	9	41.87	9	11	2
ML9	8.23	1	0.99*	8	6.35	10	8.69***	15	70.5	15	9	4
MR219	5.03	12	0.22	7	2.11	4	3.82	6	32.47	6	0	11
MR220	5.4	10	–0.12	3	1.84	3	0.89	1	9.63	1	5	7
MR253	4.86	14	–0.32	2	4.97**	14	6.06**	12	49.94	12	–8	15
VN001	5.54	9	1.64	12	4.37	9	5.71**	11	47.2	11	–2	14
VN121	6.88	7	–0.84	1	4.36	8	4.33	8	36.42	8	5	8
VN124	5.14	11	–0.04	5	4.72*	13	5.03*	10	41.88	10	0	12

Note: Means (corrected by least squares) (M); regression coefficient (b_i); deviation from regression (S^2_d); Shukla's stability variance (σ_i^2); Wricke's ecovalence (W_i^2); and Kang stability statistic (YS_i) for yield/ha: 15 rice genotypes tested in ten environments; *significant at 0.05 probability level; **significant at 0.01 probability level; ***significant at 0.001 probability level.

concepts, a genotype with low σ_i^2 and W_i^2 is considered to be stable. In this study, genotypes with the lowest stability variance are ranked top, and thus considered to be the most stable. Genotype 'MR220' > 'Binadhan4' > ML4 > 'Iratom' > 'Binadhan7' > 'MR219' > ML21 > 'VN121' > ML6 > 'VN124' > 'VN001' > 'MR253' > ML24 > ML10 > ML9 for yield/ha based on σ_i^2 and W_i^2 values (Table 23.2). According to Becker and Leon (1988), the range of variation indicates the level of interaction in the response of genotypes across environments. Genotypes with the lowest interaction variance are less responsive to the environment, while larger variances indicate environmental influence. However, it was very difficult to find the same pattern of response to validate this proposition and perhaps this makes selection difficult when considering different stability analyses on genotypes, due to shifts in genotype ranking.

Though the different stability statistics are indicative of the high, intermediate and low stability performance, the stability statistics per se were not informative and useful in selection unless they were integrated with yield capacity. Thus, efforts have been made to combine yield and stability parameters (ecovalence and stability variance) into a single selection criterion (Kang, 1993). Kang (1993) developed a yield-stability statistic (YS_i) to be used as a selection criterion when G×E interaction is significant and to demonstrate the benefit to growers of emphasizing stability of performance during selection for yield. Therefore, genotypes with YS_i value, i.e. greater than the mean, are considered stable. According to this concept (YS_i), the top six genotypes with highest yield/ha and YS_i across the entire ten environments are ML4, ML6, ML10, ML9, 'Binadhan4' and ML21, respectively. (Table 23.2).

Multivariate stability

A biplot is obtained by representing each genotype and environment as a point in a two-dimensional graph. These points are generally referred to as genotype markers and environment markers, respectively. A polygon view of the GGE biplot shown in Figs 23.1–23.3 explained 71% (PC1 = 51.47%, PC2 = 20.44%) of the genotype and genotype × environment variation for yield/ha. The results are presented in three sections below: (i) the first section represents the results on mega-environment analysis 'which-won-where', to identify the best genotypes for each environment; (ii) section two shows the genotype evaluation in respect to the overall site; and (iii) the final section deals with the environmental evaluation.

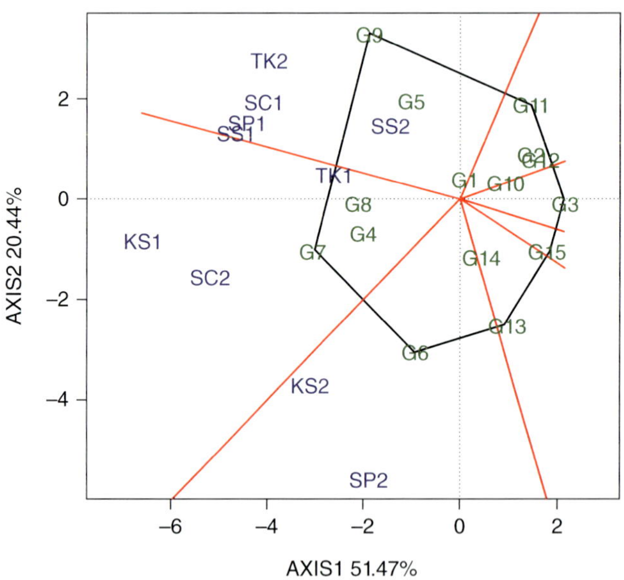

Fig. 23.1. 'Which-won-where' pattern for yield among the evaluated genotypes across the ten environments.

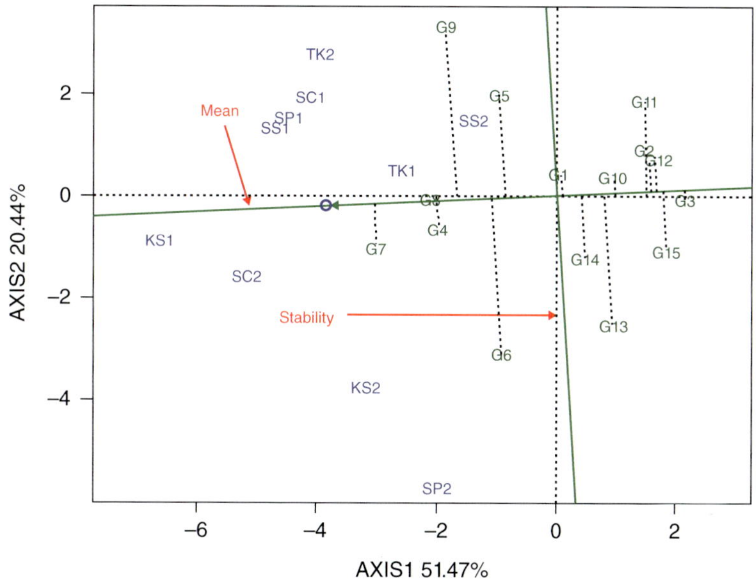

Fig. 23.2. Genotype stability determination for yield across the ten environments.

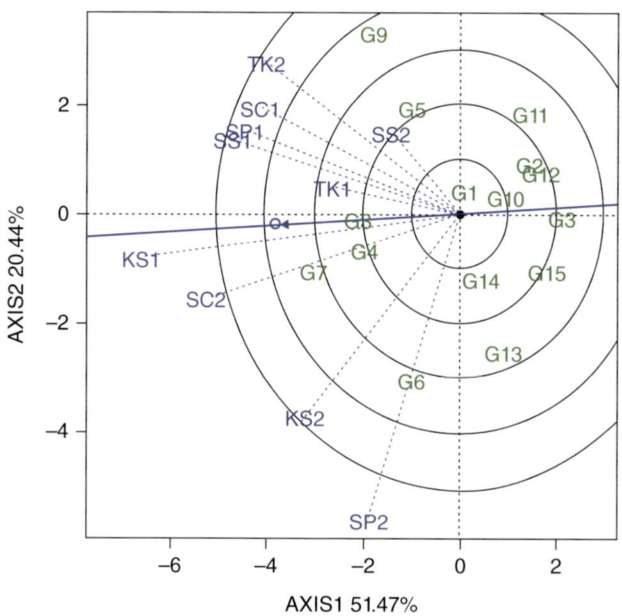

Fig. 23.3. Variation of the informativeness of test locations as environments for yield trait analysis.

MEGA-ENVIRONMENT ANALYSIS VIEW OF GGE BIPLOT (WHICH-WON-WHERE PATTERN). A mega-environment is defined as a set or group of locations that share the best cultivar across years (Yan and Rajcan, 2002). The mega-environment view is made up of an irregular polygon (Fig. 23.1) containing lines that run from the centre of the biplot perpendicular to the polygon. The polygon is constructed in such a way that it connects all genotypes that are far away from the centre of the biplot, thereby containing the entire genotype markers within the polygon. A line that originates from the centre of the biplot and perpendicularly intersects a polygon represents a hypothetical environment in which the two genotypes that are represented on either side of the polygon are said to perform equally. The perpendicular line that runs from the origin of the biplot to the polygon divides the biplot into sectors, each sector having its own winning cultivars or genotype. The winning cultivar is always located at the vertex, where two sides of the polygon meet and whose perpendicular lines form the boundary of that sector. If all environmental markers fall into one sector, this indicates that a single cultivar performs best in all environments. But if the environment marker falls on a different sector, this shows that a different cultivar won in a different location or environment, thus revealing the 'which-won-where' pattern of genotype × environment data. The seven rays shown in Fig. 23.1 divide the biplot into seven sections, and the ten environments fall into three sectors. The vertex genotype for each quadrant is the one that gave the highest yield for the environments that fall within that quadrant. The highest yield in the environments TK2, SC1, SP1, SS1 and SS2 is for the genotype ML9. The winning genotype in KS1, SC2 and TK1 was ML4, while ML24 performed best in KS2 and SP2. The other vertex genotypes ('VN001', 'VN124', 'Iratom' and 'MR253') are poorest in all ten environments.

GENOTYPE EVALUATION (MEAN VS STABILITY VIEW OF THE GGE BIPLOT). The mean vs. stability view of GGE biplot (Fig. 23.2) is denoted for the average environment coordinate (AEC) aspect, which is based on singular value partitioning (SVP = 1) (Yan *et al.*, 2007). The genotype comparison is facilitated by the view based on mean performance and stability across environments within a mega-environment. The arrow marked 'mean' on the AEC abscissa gives the direction of genotypes ranked in increasing order according to the higher trait performance, while the perpendicular line marked 'stability' on the AEC ordinate measures the stability for each genotype; this is facilitated by the genotype vector connecting the genotype to the AEC abscissa. The shorter the genotype vector, the higher is the stability, and vice versa. In Fig. 23.2, the 15 genotypes are ranked according to their means of yield and yield stability. The closer a genotype is to the concentric circles, the higher its mean yield. Genotypes that are further away in either direction from the biplot origin have a greater G×E interaction and reduced stability (Yan and Rajcan, 2002). The genotypes on the righthand side of the perpendicular line performed better than overall mean performance. Therefore, stable genotypes should have both high mean and short vector length from AEC. For broad selection, the ideal genotypes should have both high mean yield and high stability. In the biplot, genotypes ML4 (G7), ML6 (G8) and ML10 (G4) were the best as they are close to the concentric circle and have the shortest vector from the AEC. On the other hand, for specific selection, the ideal genotypes are those that have high mean yield but low stability and respond best to particular environments. Therefore, ML6 (G8) is regarded as the most stable genotype across all environments.

ENVIRONMENT EVALUATION. Two factors that contribute to the identification of the ideal location include the ability to differentiate the genotype and being a representative of the target region (Zhang *et al.*, 2016). Discrimination signifies the ability of location to maximize the variance among genotypes. The representing ability refers to a location being representative of other locations included in the study (Yan and Tinker, 2006). High yields and stable performances of genotypes could be effectively screened at an ideal location. The graphical display of test environment evaluation shows the environmental markers that correspond to the centre of origin of the biplot for different locations. The environmental evaluation can be classified into three main categories as follows.

1. The first is the environmental marker that has a very short vector, i.e. is closer to the biplot origin (TK1). This indicates that all genotypes perform equally in this environment and its use is of little value because it provides little or no information enabling genotype discrimination or identifying genotype differences.
2. The second is the environmental marker with a long vector and large angle from the AEC abscissa (TK2 and SP2). Although this cannot be

used to select superior genotypes, it can be used in culling unstable genotypes.

3. The third type of environmental marker is that with a long vector and small angle, closer to the AEC abscissa (KS1 and SC2). This is regarded as an ideal environment (Yan and Rajcan, 2002). This is very useful because it represents many environments and can, therefore, be a representative of a mega-environment.

The GGE biplot vector view used in determining the discriminative ability of test locations is presented in Fig. 23.3. The longer the location vector, the higher is the discriminative ability of the location. Therefore, among the ten locations in this study, KS1 and SC2 were the most informative (discriminating) and representative of test locations.

4 Conclusion

Based on univariate and multivariate stability parameters (b_i, S^2_d, σ_i^2, W_i^2, YS_i and GGE biplot), rice genotypes were classified into three main categories. The first group were genotypes with high yield stability along with high yield/ha. These genotypes include ML4 and ML6 and they are widely adapted to diverse environmental conditions. The second group exhibited high yield/ha but low stability and is represented by the single genotype (ML9), which is suitable for specific environments (SS2, TK2, SC1, SP1 and SS1). The last group comprises genotypes with low yield per hectare but high yield stability, which includes genotypes 'MR220', 'Binadhan4' and 'Binadhan7'. This final group is more suitable for breeding specific traits or perhaps has yield component compensation such as the capacity to recover rapidly from stresses and capacity for stress tolerance. The performance of these varieties in comparison with the nationally recommended variety 'MR219' indicates the importance of specific adaptation. In general, breeding for the stability of performance under variable conditions is a complex and difficult task, because yield and yield components are greatly influenced by different factors such as environment, which has a direct or indirect effect. Therefore, evaluation of varieties under different environments and adoption of simultaneous selection for yield and stability is a reliable selection criterion for any plant breeding programme. The identification of improved genotypes with ability to provide stable high yield across variable environments and their release for cultivation by farmers will enable farmers to reap high yield and stable income.

Acknowledgement

The authors would like to acknowledge the financial support of IAEA's Coordinated Research Project Integrated Utilization of Cereal Mutant Varieties in Crop/Livestock Production Systems for Climate Smart Agriculture (D23030). We also appreciate the support of the Malaysian Nuclear Agency and Malaysian Agricultural Research and Development Institute for research materials and research facilities.

References

Becker, H.C. and Leon, J. (1988) Stability analysis in plant breeding. *Plant Breeding* 101, 1–23.

CRAN (2014) Comprehensive R Archive Network for the R programming language. Available at: http:// cran.r-project.org/web/packages/available_packages_by name.html#available-packages-A (accessed 25 September 2018)

Dia, M., *et al.* (2015) Analysis of genotype × environment interaction (G×E) using SAS programming. Available at: http://cuke.hort.ncsu.edu/cucurbit/ wehner/software.html (accessed July 2018)

Eberhart, S.T. and Russell, W.A. (1966) Stability parameters for comparing varieties 1. *Crop Science* 6, 36–40.

Freeman, G.H. (1973) Statistical methods for the analysis of genotype-environment interactions 2. *Heredity* 31, 339.

Gabriel, K.R. (1971) The biplot graphic display of matrices with application to principal component analysis. *Biometrika* 58, 453–467.

Kang, M.S. (1993) Simultaneous selection for yield and stability in crop performance trials: consequences for growers. *Agronomy Journal* 85, 754–757.

Liu, Z. *et al.* (2017) Stability analysis of seven agronomic traits for soybean [(*Glycine max* (L.) Merr.] Tokachi nagaha and its derived cultivars using the AMMI model. *Plant Production Science* 20, 499–506.

Makumbi, D. *et al.* (2015) Agronomic performance and genotype × environment interaction of herbicide-resistant maize varieties in Eastern Africa. *Crop Science* 55, 540–555.

Oladosu, Y. *et al.* (2014) Genetic variability and selection criteria in rice mutant lines as revealed by quantitative traits. *Scientific World Journal* 2014, article 190531. doi: 10.115/2014/190531

Oladosu, Y. *et al.* (2015) Genetic variability and diversity of mutant rice revealed by quantitative traits and molecular markers. *Agrociencia* 49, 249–266.

Oladosu, Y. *et al.* (2017a) Genotype × environment interaction and stability analyses of yield and yield components of established and mutant rice genotypes tested in multiple locations in Malaysia. *Acta Agriculturae Scandinavica, Section B – Soil & Plant Science* 67, 590–606.

Oladosu, Y. *et al.* (2017b) Assessing the representative and discriminative ability of test environments for rice breeding in Malaysia using GGE Biplot. *International Journal of Science Technology and Research* 6, 8–16.

RStudio (2014) RStudio: Integrated development environment for R (Computer software v0.98.1074). Available at: http://www.rstudio.org/ (accessed 25 September 2018)

Yan, W. and Rajcan, I. (2002) Biplot analysis of test sites and trait relations of soybean in Ontario. *Crop Science* 42, 11–20.

Yan, W. and Tinker, N.A. (2006) Biplot analysis of multi-environment trial data: principles and applications. *Canadian Journal of Plant Science* 86, 623–645.

Yan, W. *et al.* (2000) Cultivar evaluation and mega-environment investigation based on the GGE biplot. *Crop Science* 40, 597–605.

Yan, W. *et al.* (2007) GGE biplot vs. AMMI analysis of genotype-by-environment data. *Crop Science* 47, 643–653.

Zhang, P.P. *et al.* (2016) GGE biplot analysis of yield stability and test location representativeness in proso millet (*Panicum miliaceum* L.) genotypes. *Journal of Integrated Agriculture* 15, 1218–1227.

Zobel, R.W. *et al.* (1988) Statistical analysis of a yield trial. *Agronomy Journal* 80, 388–393.

24 Impact of Mulch-Based Cropping Systems Using Green Mulch and Residues on the Performance of Advanced Mutant Lines of Maize (*Zea mays* (L.)) Under Infested Field with the Parasitic Weed *Striga asiatica* (L.) Kuntze in Madagascar

Harimialimalala Jhonny Rabefiraisana[1], Abdelbagi Mukhtar Ali Ghanim, Alice Andrianjaka, Berthe Rasoamampionona, Ljupcho Jankuloski, Mbolatiana Alinà Razafindrasoa, Ratsimiala Ramonta Isabelle, Ivan Ingelbrecht, Nirina Hanitriniaina Ravelonjanahary and Noronirina Victorine Rakotoarisoa

Abstract

In Madagascar, cereal yields remain insufficient due to various biotic and abiotic constraints, including *Striga asiatica*, a parasitic weed that has contributed to decreased maize yield up to 100%. This work aims to assess the impact of the practice of two cropping systems on the maize crop infested by *S. asiatica*. PLATA maize seed of the putative tolerant mutant line from the M_5 generation after gamma irradiation at 100, 200 and 300 Gy and of the sensitive parent variety were grown in fields naturally infested or artificially inoculated with one pinch of around 3000 ready-to-germinate seeds of *S. asiatica*. The cropping system (SCV) is a community of plants that is managed by a farm unit to achieve various human goals. The residue of *Stylosanthes* sp. legumes was used as mulch SCVm and the legume cowpea was planted with the host plant for the intercropping system SCVv. Results have shown that the use of mulch, either residue SCVm or green mulch SCVv, minimizes *S. asiatica* infestation on maize plants. The SCV reduces significantly the number of emerging *Striga* plants with an emergence of 1.33 ± 0.36 for SCVm, 4.33 ± 0.27 for SCVv and 15.00 ± 1.08 for the control. Moreover, M_5 lines have shown significant differences in plant survival rate of 50.57 ± 1.25% to 80.00 ± 0.91%, versus 13.50 ± 0.47% for the parent variety. Yields of the parent and M_5 lines on SCVm are, respectively, 3.46 ± 0.02 t/ha and 4.64 ± 0.01 t/ha, and 2.30 ± 0.04 t/ha and 3.61 ± 0.05 t/ha for SCVv, while that of the control plot remains low, varying from 0.50 ± 0.04 t/ha to 2.29 ± 0.01 t/ha. Cover increases soil humidity and delays the development of *S. asiatica* and infection of the host plant, thus improving host plant yield. These results demonstrate the benefit of the integrated approach of mutation breeding and cultural practice to ensure more durable crop production under heavy *Striga* infestation.

Keywords: mutation breeding • gamma ray • maize crop • *Striga asiatica* • cropping system

1 Introduction

Maize is one of the most important agronomic crops in the world (Fakorede, 2001) and the major substitute for rice after cassava in Madagascar (USAID, 2013). It is also used as animal feed and in industry (Fussell, 1999).

The witchweeds, *Striga asiatica*, are parasitic plants that pose a major biotic constraint to cereal production in Madagascar. They can completely

[1] Email: nyjahmalayah@gmail.com

DOI: 10.1079/9781789249095.0024

destroy the yield of cereal crops such as rice, maize and sorghum in the mid-west, north-east and south-west regions of Madagascar (Sallé and Raynal-Roques, 1989; Andrianaivo *et al.*, 1998). Severe crop damage (60–100%) has been reported in cases of heavy infestation with this parasitic weed in Madagascar (Elliot *et al.*, 1993). When *Striga* infestation is too severe, farmers abandon heavily infested fields and migrate to new ones (Sallé and Raynal-Roques, 1989). *Striga* is typically found in dry, infertile soils in semi-arid tropical grasslands and savannahs (Spallek *et al.*, 2013) and reduces crop yields by extracting water, nutrients and photosynthetates from the root system of the host plant, resulting in stunting and yield reduction (Nail *et al.*, 2014). Several control options against *S. asiatica* can be used by farmers, including cultural measures such as normal hand pulling, delayed sowing, trap crops, herbicide application, use of nitrogen fertilizer and cereal–legume rotation (Elliot *et al.*, 1993). The actual use of these is limited by the reluctance of farmers to accept them for both biological and socio-economic reasons (Oswald, 2005). Control using chemical fertilizers and pesticides may cause significant surface and groundwater contamination and potential health risks (Guziejewski *et al.*, 2012). We therefore tried to address these problems by using advanced lines from the M_5 generation derived from mutants, combined with cropping systems using green mulch and residues. First, inducing mutation via gamma-ray irradiation has proved to be a promising and viable approach to improve many agronomically important traits by shortening the growing period, improving suitability for rotation and increasing tolerance or resistance to abiotic and biotic stresses in major crop plants (Bibi *et al.*, 2009). During the past 75 years, around 2672 mutant varieties of rice, wheat, cotton and sunflower have been released officially for commercial cultivation and have made a major economic impact in many countries (Maluszynski *et al.*, 2000). On the other hand, there are many benefits to adopting cropping systems such as intercropping with nitrogen-fixers to enhance soil fertility and increase abiotic-stress tolerance (George, 2012). Studies on grasslands have shown that multispecies plots produced 15% higher yields than monocrops (Prieto *et al.*, 2015). Mulching is an important agronomic practice, beneficial in conserving soil moisture, suppressing weeds, improving soil fertility and modifying the soil's physical environment (Yoo-Jeong *et al.*, 2003). This work aims at assessing the impact of the practice of two cropping systems on advanced mutant lines of maize infested by *S. asiatica* in the mid-west region of Madagascar.

2 Materials and Methods

2.1 Experimental design

Experiments were conducted in the 2015 summer cropping season in a field of the National Centre for Applied Research and Rural Development station (FOFIFA) in Kianjasoa (49° 22′ 62.1″ S, 19° 03′ 20.3″ E). It is situated in the mid-west region of Madagascar, where *S. asiatica* is a serious limitation to cultivation of cereals such as rice and maize. Three advanced mutant maize lines at M_5 generation that were selected as putatively tolerant to *S. asiatica* were provided from the Plant Breeding and Biotechnology Unit (UBAP) at the Faculty of Sciences in Antananarivo-Madagascar and were used in this study. These advanced lines were derived from induced mutation by gamma-ray irradiation at 100, 200 and 300 Gy followed by mass selection and then individual and genealogy selections from the M_2 plant in the field. *S. asiatica* seeds for this experiment were collected from mature healthy *Striga* plants in Kianjasoa during 2014. Two types of cropping systems were used during the experiment: (i) the residue of *Stylosanthes* sp. was used as mulch cover (SCVm); or (ii) *Vigna unguiculata* (cowpea) was planted with the host plant in an intercropping system (SCVv). The experimental design included three plots (A, B, C) which represented the two different cropping systems and one control plot of monocrop maize (Table 24.1). Plots were spaced at 1 m and subdivided into three sub-plots of 10 m × 10 m, which corresponded to three replications per plot. The experiment was laid out as a split-plot design.

For each 10 m × 10 m sub-plot, the number of maize holes along the length was 20 and 14,

Table 24.1. Plot design experiment.

Cropping system	Tested genotype
Plot A: control	Parent (PLATA 0 Gy) Mutant lines (Mutant-PL1, Mutant-PL2, Mutant-PL3)
Plot B: intercropping (SCVv)	Parent (PLATA 0 Gy) Mutant lines (Mutant-PL1, Mutant-PL2, Mutant-PL3)
Plot C: mulch cover (SCVm)	Parent (PLATA 0 Gy) Mutant lines (Mutant-PL1, Mutant-PL2, Mutant-PL3)

according to the width. In fact, 280 holes per subplot were made, making a total of 840 holes for each plot. Maize was planted at a row-to-row distance of 75 cm and a plant-to-plant distance of 25 cm within the rows, with a depth of 20 cm. Prior to planting, the *Striga* seeds were mixed with fine sand at a ratio of 2:98 (seed:sand) so that one pinch would deliver around 3000 viable *Striga* seeds. One pinch of *S. asiatica* preconditioned seeds was introduced and scattered in each hole; then before sowing, the hole was covered with a pile of soil 10 cm thick. Seeding was done on 16 November 2015. For the mulch cover (SCVm), spreading of the cover or crop residues (*Stylosanthes* sp.) was carried out after sowing. The thickness of this cover reached 10 cm to cover the ground. Concerning the intercropping (SCVv), the cowpea seeds were sown 2 weeks after maize planting, in order to ensure proper development of the host plant before the cover plant gained space or covered the host plant. Cowpea seeds were sown between the maize plants.

2.2 Data collection and analysis

The number of emerged *Striga* plant per maize plant per 0.5 m^2 for each plot was counted from 80 days after sowing (80 DAS) until maturity. Plant height was measured, the survival rate of maize plants was recorded and maize grain yield (t/ha) was calculated. The percentage damage was calculated by the following formula:

$$\textit{Percentage damage} = \frac{\textit{total number of burned leaves per plant}}{\textit{total number of green leaves per plant} + \textit{total number of burned leaves per plant}} \times 100$$

The results of the experiments were analysed using R Software version 3.4.1. The effects of experimental factors were evaluated by the analysis of variance (ANOVA), and comparisons between means were carried out using Tukey HSD test at the significance level of $p \leq 0.05$. Relationships between the parameters were highlighted by the Pearson correlation test.

3 Results

Results of the analysis of variance have shown that *Striga* count, maize plant height, survival rate of maize plant, percentage damage and yield were influenced by the genotypes and cropping system (Table 24.2; Fig. 24.1). The data clearly indicated significant differences between the treatments and genotypes, whatever the variable considered (Table 24.2).

At 80 DAS, the first *Striga* emergence was observed only in the control plot. The difference in number of *Striga* shoots counted per 0.5 m^2 in the control plot and the cropping system (SCVv and SCVm) is statistically significant ($p \leq 0.05$). The *Striga* number in the mulch cover plot (SCVm) was significantly lower than in the intercropping plot (SCVv), which was significantly lower than that counted from the control plot. The SCVm and SCVv treatments give shade to the ground, thereby reducing its temperature and retaining moisture. As a result, they limit the germination and even the growth of *Striga* plants. According to Akobundu (1987) and Singh *et al.* (2008), mulching can inhibit weed seed germination by shading and in some cases through the release of allelopathic substances. Our results corroborate the findings of Fasil and Verkleij (2007), who reported that intercropping maize with cowpea can significantly reduce the emergence of *Striga*. The genotypes tested have a significant effect on the emerged *Striga* because, in the SCVv, SCVm and control plots, the number of *S. asiatica* plants that emerged in the mutant plots (Mutant-PL1, Mutant-PL2 and Mutant-PL3) were significantly lower than that in the

Table 24.2. Means with standard errors of number of *Striga* count at 80 DAS per 0.5m^2, *Striga* count at maturity per 0.5m^2, maize plant height, percentage damage, survival rate of maize plant and maize grain yield (45 maize plants per plot chosen randomly were used for the measurements for each variable).

Plot	Genotype	*Striga* count at 80 DAS (/0.5 m^2)	*Striga* count at maturity (/0.5 m^2)	Maize plant height (cm)	Percentage damage (%)	Survival rate of maize plant (%)	Maize grain yield (t/ha)
A: Control							
	Parent (PLATA)	7.11 ± 0.41a	15.00 ± 1.08a	177.63 ± 0.97f	69.25 ± 1.25a	13.50 ± 0.47h	0.50 ± 0.04f
	Mutant-PL1	2.14 ± 0.48b	7.30 ± 0.45c	211.23 ± 4.44e	43.75 ± 0.63b	50.57 ± 1.25e	2.29 ± 0.01e
	Mutant-PL2	2.30 ± 0.29b	7.24 ± 0.25cd	208.13 ± 3.24e	43.25 ± 0.75b	51.33 ± 0.49e	2.18 ± 0.01e
	Mutant-PL3	2.45 ± 0.19b	7.41 ± 0.26c	202.31 ± 3.20e	43.50 ± 1.19b	52.76 ± 0.98e	2.26 ± 0.03e
B: Intercropping (SCVv)	Parent (PLATA)	0.00 ± 0.00c	9.78 ± 0.21b	230.91 ± 8.15d	42.75 ± 0.95b	24.20 ± 0.98g	2.30 ± 0.04e
	Mutant-PL1	0.00 ± 0.00c	4.33 ± 0.27d	248.16 ± 5.16bcd	36.00 ± 1.08c	69.66 ± 0.47c	3.58 ± 0.03cd
	Mutant-PL2	0.00 ± 0.00c	4.38 ± 0.36d	264.00 ± 2.16b	36.25 ± 0.48c	63.65 ± 0.46d	3.53 ± 0.01cd
	Mutant-PL3	0.00 ± 0.00c	4.47 ± 0.35d	244.34 ± 3.43cd	36.50 ± 0.65c	65.62 ± 0.22d	3.61 ± 0.05c
C: Mulch cover (SCVm)	Parent (PLATA)	0.00 ± 0.00c	6.00 ± 0.71cd	257.37 ± 1.15bc	41.75 ± 0.25b	28.26 ± 0.27f	3.46 ± 0.02d
	Mutant-PL1	0.00 ± 0.00c	1.37 ± 0.37e	314.44 ± 0.74a	27.75 ± 1.11d	73.66 ± 0.23b	4.51 ± 0.00ab
	Mutant-PL2	0.00 ± 0.00c	1.40 ± 0.36e	332.30 ± 1.42a	27.50 ± 0.87d	76.08 ± 0.77b	4.48 ± 0.01b
	Mutant-PL3	0.00 ± 0.00c	1.33 ± 0.36e	321.23 ± 2.50a	28.00 ± 0.41d	80.00 ± 0.91a	4.64 ± 0.01a

Different letters show significant statistical differences between the treatments according to Tukey's test ($p \leq 0.05$); values within columns with the same letter are not significantly different.

Fig. 24.1. (A) Maize plant with mulch cover system (SCVm). **(B)** Maize plant with intercropping system (SCVv). **(C)** *Striga asiatica* plant on the control plot. **(D)** *Striga asiatica* plant on maize plant. **(E)** Ear of maize mutant cover by mulch cover system (SCVm) and control. **(F)** Ear of maize mutant cover by cowpea (SCVv) and control.

parent variety. In agreement with this, Ezeaku and Gupta (2004) reported that the genetic differences between sorghum cultivars affect the time of parasite attachment, with tolerant varieties showing later attachment and later parasite emergence than sensitive cultivars. In 2010, Reda pointed out that the observed reduction and delay in *Striga* emergence may be attributed to reduced germination and reduced haustorium initiation and attachment. In our results, we can explain that the mutant lines (Mutant-PL1, Mutant-PL2, Mutant-PL3) had several advantages over their parent for several traits. Our mutant lines (Mutant-PL1, Mutant-PL2, Mutant-PL3) appeared to use a mechanism of resistance that delays and inhibits *Striga* proliferation, perhaps as a consequence of lower production of germination stimulant. Ejeta (2007) identified various components of resistance, such as low germination stimulation, low haustorial initiation, mechanical barriers and antibiosis. This finding agrees with Graves (1995), who had reported that mutant alleles at the *lgs* (low germination stimulant) locus drastically reduce *Striga* germination stimulant activity. Vogler (1996) reported that *lgs*-determined inheritance of low *Striga* germination stimulant activity in *Sorghum* corresponds to inheritance of a recessive allele. Arnaud *et al.* (1999) demonstrated that the use of fully tolerant or resistant genotypes reduces infection and prevents further build-up of the *Striga* seed bank.

In this experiment, the percentage damage varied from 27.50 ± 0.87% to 69.25 ± 1.25%. This percentage damage for both cropping plots (SCVv and SCVm) was significantly lower than that assessed from the control plot. Mutant-PL1, Mutant-PL2 and Mutant-PL3 had low percentage damage as compared with the parent, whatever the cropping system. The emerged *Striga* was significantly and positively correlated with the percentage damage ($r = 0.842$) (Table 24.3). The possible reason for the lowest and highest value of percentage damage in all treatment is that all the Mutant-PL1, Mutant-PL2, Mutant-PL3 and parent (PLATA) plants did not have equal access to the nutrient resources. Another important reason for these types of results is that the Mutant-PL1, Mutant-PL2, Mutant-PL3 are mainly controlled and regulated by genetics rather than the cropping system. Graves (1995) reported that the infection inevitably results in reduced growth, yield and quality of crops and host damage becomes noticeable even before the parasite emerges from the soil.

The maize plant height in the mulch cover plot was significantly lower than in the intercropping plot, which was significantly lower than that measured from the control plot (Table 24.2). The maximum heights of all the Mutant-PL1, Mutant-PL2, Mutant-PL3 and parent (PLATA) plants in the SCVm plot were higher (257.37 ± 1.15 cm to 332.30 ± 1.42 cm) than those of the SCVv plot (230.91 ± 8.15 cm to 264.00 ± 2.16 cm) and for the control plot (177.63 ± 0.97 cm to 211.23 ± 4.44 cm). The differences are statistically significant ($p \leq 0.05$) between the treatments. On the other hand, in the cropping systems, the mutant lines were significantly ($p \leq 0.05$) taller than the parent variety, which is consistent with the observation that the parent becomes stunted when infected by *S. asiatica* but the tolerant genotypes maintain their normal size, and the *Striga* emergence was significantly and negatively correlated with plant height ($r = -0.638$) (Table 24.3). The cropping system (SCVm and SCVv) combined with Mutant-PL1, Mutant-PL2 and Mutant-PL3 improved plant growth. Midega *et al.* (2013) observed that intercropping can suppress weeds that reduce the competition between cultivated plants and weeds for water and nutrition and thus improves the growth of cultivated plants. The availability of nutrients plays a very important role in plant development. This agrees with the report of Pappa *et al.* (2012) that intercropping systems with legumes can supply nitrogen, not only as companion plants but also post crop. Smil (2001) and Tsubo (2003) found that the fertilization benefits for the cereal crop when associated with a nitrogen-fixing leguminous crop can be ascribed to nitrogen excretion and nodule decomposition of the latter crop during the growing period. Ahonsi *et al.* (2004) demonstrated striking benefits from mulch applications, including nutrient recycling, conservation of moisture, maintenance of uniform soil temperature, reduction of soil erosion and compaction from heavy rain and increase of water penetration.

The mulch cover system (SCVm) had a high rate of plant survival for all of the genotypes tested, followed by the intercropping system (SCVv), with the lowest plant survival rate being in the control plot (Table 24.2). Regarding the

Table 24.3. Pearson's correlation coefficients (r) describing association of emerged *Striga*, maize plant height, percentage damage, survival rate of maize plant and maize grain yield tests.

		Pearson's r	p	VS-MPR[a]
Emerged *Striga*	Percentage damage	0.842	< 0.001	5.234e+26
Emerged *Striga*	Maize plant height	–0.638	< 0.001	1.166e+11
Emerged *Striga*	Survival rate of maize plant	–0.864	< 0.001	2.216e+30
Emerged *Striga*	Maize grain yield	–0.913	< 0.001	6.959e+39
Percentage damage	Maize plant height	–0.433	< 0.001	9816
Percentage damage	Survival rate of maize plant	–0.838	< 0.001	3.785e+24
Percentage damage	Maize grain yield	–0.764	< 0.001	4.465e+14
Maize plant height	Survival rate of maize plant	0.426	< 0.001	6780
Maize plant height	Maize grain yield	0.709	< 0.001	1.069e+15
Survival rate of maize plant	Maize grain yield	0.847	< 0.001	7.545e+27

[a]Vovk-Sellke Maximum p-Ratio. Based on the p-value, the maximum possible odds in favour of H_1 over H_0 equals $1/(-e\ p \log(p))$ for $p \leq .37$ (Sellke *et al.*, 2001).

genotype effect, the Mutant-PL1, Mutant-PL2, Mutant-PL3 lines had a higher plant survival rate than their parent (PLATA). The percentage damage was significantly and negatively correlated with survival rates of maize plants (r = –0.838) (Table 24.3). Ratnadass (2012) revealed that biological nitrogen fixation, green manure and organic matter reincorporated into the soil led to increased soil fertility, humidity conservation and microbiological stimulation. All these ensure long-term productivity.

The grain yields of mutant lines in the mulch cover plot were significantly higher than those in the intercropping plot, which were significantly higher than those calculated from the control plot (Table 24.2). The maize grain yields of mutant lines in all cropping systems are high, ranging from 2.30 ± 0.04 t/ha to 4.64 ± 0.01 t/ha, compared with those of the parent variety, ranging from 0.50 ± 0.04 t/ha to 3.46 ± 0.02 t/ha. The percentage damage was significantly and negatively correlated with maize grain yield (r = –0.764) (Table 24.3). The maize grain yield augmentation in the cropping system plots compared with the control can be attributed to nutrient resource availability combined with the ability of the mutant lines to counter *S. asiatica* development. Aliyu (2006) pointed out that higher grain yield may also be attributed to the effectiveness of the cropping system, which not only reduces the *S.* seed bank but also increases the nitrogen supply to the host crop. Moreover, cropping systems (SCVm and SCVv) have improved maize grain yield. Cardinale *et al.* (2007) stated that intercropping had been shown to produce 1.7 times more biomass than single-species monocropping and to be 79% more productive than monocropping systems. Saito (1982) reported that maize–legume intercropping proved more productive than sole maize cropping.

4 Conclusion

The maize mutant lines showed superiority in facing *S. asiatica* attack, whatever the kind of cropping system. In brief, mulch cover (SCVm) performance was significantly better than the intercropping system (SCVv). Both SCVm and SCVv cropping systems have shown a positive impact on agronomic performance. Cover increases soil humidity and delays *S. asiatica* development and attack of the host plant. Our mutant lines use various components of resistance to *Striga* development. The combination of the use of both cropping system (SCVv and SCVm) and the mutant maize lines improves and increases maize grain yield. These results demonstrate the benefit of an integrated approach of mutation breeding and cultural practice to ensure more durable crop production under heavy *Striga* infestation in Madagascar.

Acknowledgements

The authors thank the International Atomic Energy Agency, Joint FAO/IAEA Division of Nuclear Techniques in Food and Agriculture, Austria, for support of this project.

References

Ahonsi, M.O. *et al.* (2004) Effects of ALS-inhibitor herbicides, crop sequence, and fertilization on natural soil suppressiveness to *Striga hermonthica*. *Agriculture, Ecosystems & Environment* 104(3), 453–463.

Akobundu, I.O. (1987) *Weed Science in the Tropics – Principles and Practices*. Wiley, Chichester, UK.

Aliyu, B.S. (2006) Effect of intra- and inter-row mixing of sorghum with two varieties of cowpea on host crop yield in a *Striga hermonthica* infested field. *African Journal of Agricultural Research* 1(2), 024–026.

Andrianaivo, A.P. *et al.* (1998) *Biologie et gestion du* Striga *à Madagascar*. Direction de la protection des végétaux (DPV), Ministère de l'agriculture, Antananarivo, Madagascar.

Arnaud, M.C. *et al.* (1999) Physiology and histology of resistance to *Striga hermonthica* in *Sorghum bicolor* var. *Framida*. *Australian Journal of Plant Physiology* 26, 63–70.

Bibi, S. *et al.* (2009) Genetic differentiation of rice mutants based on morphological traits and molecular marker (RAPD). *Pakistan Journal of Botany* 41(2), 737–743.

Cardinale, B.J. *et al.* (2007) Impacts of plant diversity on biomass production increase through time because of species complementarity. *Proceedings of the National Academy of Sciences of the United States of America* 104(46), 18123–18128.

Ejeta, G. (2007) *Striga* resistance in *Sorghum*: exploitation of the intricate host-parasite biology. *Crop Science* 47, S216–S227.

Elliot, P.C. *et al.* (1993) Weeds in rice in Madagascar. *International Rice Research Notes*, pp. 53–54.

Ezeaku, I.E. and Gupta, S.C. (2004) Development of sorghum populations for resistance to *Striga hermonthica* in the Nigerian Sudan Savanna. *African Journal of Biotechnology* 3(6), 324–329.

Fakorede, M.A.B. (2001) *Revolutionizing Nigerian Agriculture with Golden seed*. Inaugural lecture series, Obafemi Awolowo University Press Ltd, Ile-Ife, Nigeria.

Fasil, R. and Verkleij, J.A. (2007) Cultural and cropping systems approach for *Striga* management – a low cost alternative option in subsistence farming. In: Ejeta, G. and Gressel, J. (eds) *Integrating New Technologies for Striga Control: Towards Ending the Witch-hunt*. World Scientific Publishing Co., Singapore, pp. 229–240.

Fussell, B. (1999) *The Story of Maize*, 2nd edn. North Point Press, New York.

George, S.J. (2012) A sustainable agricultural landscape for Australia: a review of interlacing carbon sequestration, biodiversity and salinity management in agroforestry systems. *Agriculture Ecosystems & Environment* 163(12), 28–36.

Graves, J.D. (1995) Host-plant responses to parasitism. In: Press, M.C. and Graves, J.D. (eds) *Parasitic Plants*. Chapman & Hall, London, pp. 206–225.

Guziejewski, D. *et al.* (2012) Square wave adsorptive stripping voltametric determination of diazinon in its insecticidal formulations. *Environmental Monitoring & Assessment* 184(11), 6575–6582.

Maluszynski, M. *et al.* (2000) Officially Released Mutant Varieties – The FAO/IAEA Database. In: *Mutation Breeding Review* 12. International Atomic Energy Agency (IAEA), Vienna.

Midega, C.A.O. *et al.* (2013) Effects of mulching, N-fertilization and intercropping with *Desmodium uncinatum* on *Striga hermonthica* infestation in maize. *Crop Protection* 44, 44–48.

Nail, K. *et al.* (2014) *Striga asiatica. Harvest Choice Pest Geography* 2, 1–6.

Oswald, A. (2005) *Striga* control – technologies and their dissemination. *Crop Protection* 24, 333–342.

Pappa, V.A. *et al.* (2012) Legumes intercropped with spring barley contribute to increased biomass production and carry-over effects. *Journal of Agricultural Science* 150, 584–594.

Prieto, I. *et al.* (2015) Complementary effects of species and genetic diversity on productivity and stability of sown grasslands. *Nature Plants* 1(4), 15033.

Ratnadass, A. (2012) Plant species diversity for sustainable management of crop pests and diseases in agro-ecosystems: a review. *Agronomy for Sustainable Development* 32, 273–303.

Reda, F. *et al.* (2010) Virulence study of *Striga hermonthica* populations from the Tigray region (northern Ethiopia). *World Journal of Agricultural Sciences* 6(6), 676–682.

Saito, S.M.T. (1982) The nitrogen relationship of maize/bean associations. In: Graham, P.H. and Harris, S.C. (eds) *Biological Nitrogen Fixation*. Papers presented at workshop at CIAT, 9–13 March 1981. Centro Internacional de Agricultura Tropical, Cali, Colorado, pp. 631–639.

Sallé, G. and Raynal-Roques, A. (1989) Le *Striga*. *La recherche* 206(20), 45–52.

Sellke, T. *et al.* (2001) Calibration of *p* values for testing precise null hypotheses. *The American Statistician* 55, 62–71.

Singh, D. *et al.* (2008) Evaluation of spatial arrangements of soybean (*Glycine max*) and pigeon pea (*Cajanus cajan*) in intercropping system. *Haryana Agricultural University Journal of Research* 38(1/2), 33–36.

Smil, V. (2001) *Feeding the World: A Challenge for the Twenty-first Century*. MIT Press, Cambridge, Massachusetts.

Spallek, T. *et al.* (2013) The genus *Striga*: a witch profile. *Molecular Plant Pathology* 3(9), 1–9.

Tsubo, M.E. (2003) Productivity of maize-bean intercropping in a semi-arid region of South Africa. *Water SA* 29(4), 381–388.

USAID (2013) *BEST Analysis – Madagascar*. US Agency for International Development (USAID) Office for Food for Peace, Washington, DC (prepared by Fintrac Inc.).

Vogler, R.K. (1996) Inheritance of low production of *Striga* germination stimulant in sorghum. *Crop Science* 36, 1185–1191.

Yoo-Jeong, Y. *et al.* (2003) Effect of organic mulches on soil bacterial communities one year after application. *Biology and Fertility of Soils* 38, 273–281.

25 Groundnut Mutants with End-of-season Drought Tolerance for the Marginal Dry Lands of North Kordofan State, Sudan

Elgailani Abdalla[1], Tarig Ahmed, Omar Bakhit, Yasir Gamar, Salih Elshaikh, Yasir Mohammed, Abdellatif Sulaiman and Hatim Mardi

Abstract

Groundnut (*Arachis hypogaea* L.), produced in the traditional small-scale rainfed sector of Western Sudan, accounts for 80% of the total annual groundnut acreage, producing 70% of the total production. Low productivity of groundnut is a characteristic feature in North Kordofan State, which is characterized as the most vulnerable state to the impact of climate change. Terminal drought stress resulting from reduction in rainfall amount and distribution at the end of the season is the most deleterious drought period, as it coincides with groundnut pod filling and maturation periods. High and stable yields under subsistence farming conditions in North Kordofan State could be realized only by using adapted high-yielding, drought-tolerant genotypes. Mutation induction by gamma-rays of 200 and 300 Gy was utilized to irradiate 500 dry seeds of the Spanish-type groundnut genotypes, Barberton, Sodari, ICGV 89104, ICGV 86743, ICGV 86744 and ICG 221, aiming at increasing the chances of obtaining genotypes with the desired drought-tolerant traits. Mutants were selected from the M_3 plants using visual morphological traits. Groundnut mutants at the M_4 and M_5 generations, advanced by single seed descent, were evaluated for end-of-season drought tolerance. A terminal drought period of 25 days was imposed after 60 days from planting, using a rainout shelter. Mutants that survived 25 days of terminal drought stress were further evaluated for agronomic performance under rainfed field conditions. The groundnut mutant, Barberton-b-30-3-B, produced 1024 kg/ha, a significantly higher mean pod yield over 12 seasons compared with 926 kg/ha for 'Gubeish', the widely grown released check cultivar, showing overall yield advantage of 11%. Under 5 years of participatory research, Barberton-b-30-3-B was ranked the best with yield increment of 21% over 'Gubeish' under the mother trials. The GGE biplot analysis for 12 and five seasons, respectively, showed that Barberton-b-30-3-B was stable and produced a good yield in both high and low rainfall situations. Hence, Barberton-b-30-3-B was found to be a suitable mutant for sustainable profitable yields in the marginal dry lands of North Kordofan State and was officially released as 'Tafra-1' by the National Variety Release Committee during its second meeting of April 2018.

Keywords: groundnut mutants • end-of-season drought tolerance • North Kordofan • Sudan

1 Introduction

Groundnut (*Arachis hypogaea* L.) is an important oilseed crop in the rainfed sector of Western Sudan which contributes about 80% of Sudan's total groundnut acreage, producing 70–80% of the total production, depending on the expansion or reduction of groundnut areas under the

[1] Email: elgailani_ers@hotmail.com

 Mutation Breeding, Genetic Diversity and Crop Adaptation to Climate Change (eds S. Sivasankar *et al.*)
DOI: 10.1079/9781789249095.0025

irrigated sector in the central clay plains. North Kordofan State is generally classified as one of the areas of Sudan that are most vulnerable to climatic change (NAPA, 2007). In the past three decades, drought was recorded twice in every three years. Long-term rainfall data analysis has shown a drastic reduction in season length due to early cessation of rains (Adam, 2002). Growing seasons have been reduced by about 30 days at the northern boundary of the crop-producing zone and by 24 days at the southern boundary. The expected humid period favourable for crop growth around El-Obeid in North Kordofan State is generally reduced to less than 55 days (FAO, 2006). Drought, among several abiotic factors, is the main limiting factor that affects groundnut production in the rainfed sector in Sudan. Moderate water deficits (drought episodes) during the pre-flowering phase in groundnut increased subsequent rates of crop growth and pod growth (Nageswara Rao *et al.*, 1988). Drought at early growth stages could change root distribution patterns of the crop, and the change in root distribution patterns was more pronounced with long durations of drought (Thangthong *et al.*, 2017). Terminal drought stress resulting from a reduction in rainfall amount and distribution at the end of the growing season is the most deleterious drought period for groundnut as it coincides with pod filling and maturation periods. Hence, crop varieties with long and medium maturity durations (90–120 days) are rendered unsuitable for achieving sustainable crop production under the prevailing shifts in season length, frequent terminal drought incidences and unreliable rainfall patterns. New generations of early-maturing, drought-tolerant groundnut varieties are required as additional marginal areas need to be cultivated in North Kordofan State to compensate for a reduction in the cultivated groundnut area as great as 85%, arising from the latest division of States which created West Kordofan State and deprived North Kordofan State of large annual cultivated groundnut areas.

Several statistical methods have been used for the analysis and interpretation of multi-environment trial data (Yates and Cochran, 1938; Crossa and Cornelius, 1997; Gauch and Zobel, 1997; Yan *et al.*, 2000, 2007). The genotype main effect plus G×E interaction (GGE) biplot methodology is considered effective in identifying the best-performing and stable genotype from a multi-environment trial data (Yan and Tinker, 2006). The GGE biplot analysis utilizes both genotype (G) and genotype × environment interaction (GEI) effects and graphically displays GEI in a two-way table (Yan *et al.*, 2000). It helps visual evaluation of the relationships among the test environments, genotypes and the GEI and has been recommended as the best method for the analysis of GEI (Yan and Kang, 2003). Plant breeders widely use the GGE biplot in the variety evaluation of many crops (Yan *et al.*, 2000; Yan and Hunt, 2001; Yan and Rajcan, 2002). The main objective of this study was to conduct field experiments to evaluate and select groundnut mutants with stable high yield, aiming at stabilizing groundnut production in the marginal rainfed areas of North Kordofan State under the prevailing abiotic pressures augmented by climate change.

2 Materials and Methods

Studies on mutation induction and evaluation of mutants were conducted on rainfed sandy soil at the research farm of El-Obeid Research Station, North Kordofan State, Sudan. Dry uniform seeds of six groundnut cultivars, viz. Sodari, Barberton, ICGV 89104, ICGV 86744, ICGV 86743 and ICG 221, all belonging to the Spanish-type groundnut (*Arachis hypogaea* L. subsp. *fastigiata* var. *vulgaris*), were irradiated in 1999 in Pretoria, South Africa, using gamma-rays at dose levels of 200 and 300 Gy. Five hundred M_1 seeds of each genotype were grown under controlled irrigation conditions. M_2 seeds of individual plants were harvested separately and divided into basal (b) and apical (a) seeds. Basal and apical seeds from the dose of 200 Gy were used in the drought screening nursery, while seeds from the dose of 300 Gy were advanced by the bulk pedigree method to M_5 for further handling and evaluation. The pedigree breeding method, using visual morphological plant characters, was used in selecting the mutants and generation advancement. Visible aerial plant morphological characters were growth habit, number of branches, stem colour, branching pattern, leaf size and shape, leaf colour, flower colour and maturity duration, while visible underground plant characters included pod length, pod size, pod reticulation, pod beak, pod constriction, basal or profuse pod setting and peg length. Plants grown from M_3 and M_4 seeds were subjected to a terminal drought-stress period of 25 days using a sprinkler irrigation facility and a mechanical rainout shelter. The sprinkler irrigation facility was used to relieve any drought stress during the first 2 months of crop age, while the mechanical rainout shelter was used to impose terminal drought stress of 25 days' duration after 60 days of optimum crop growth. High harvest index and high dry-matter production were found to make a moderate direct contribution to high pod yield under terminal drought stress, whereas high leaf relative water content (LRWC), low specific leaf area (SLA), low canopy temperature (CT) and low leaf senescence (LSENS) were found to make an indirect contribution to high pod yield under terminal drought stress (Abdalla *et al.*, 2007, 2008). Promising mutants were incorporated into the rainfed groundnut breeding programme since the 2006–2007 cropping season. Barberton-b-30-3-B, a promising mutant with terminal drought tolerance, was tested together with 11 genotypes in the national rainfed groundnut variety trial. Three more mutants were introduced into the national rainfed variety trial in the 2010–2011 cropping season. The national rainfed groundnut variety trial usually accommodated 12 entries, including at least the latest released variety as a control. The 12 genotypes including mutants were arranged in a randomized complete block design with five replications. Plots contained five rows of 5 m length with 60 cm spacing between rows and 20 cm between holes where two seeds were sown. Plots were harvested after 85 days from planting. Total dry matter per plot and total dry pod weight per plot were recorded and the harvest index was calculated for each plot as the percentage pod yield divided by total biological yield. Daily rainfall data were recorded at El-Obeid research farm, where field experiments were continuously conducted from the 2006–2007 cropping season.

A participatory research approach, where farmers are engaged in the management of experiments side by side with researchers, was initiated as a new research protocol supported by the International Fund for Agricultural Development (IFAD) through the Seed Development project in seven selected villages in Sheikan and Elrahad localities from the 2013–2014 cropping season. Participatory research or client-driven research is increasingly recognized and practised as an integral part of modern innovation systems and is considered a proven complement to conventional non-participatory research approaches when used appropriately and expertly (Ashby and Lilja, 2004; Ceccarelli and Grando, 2007). Implemented activities in selected villages were jointly supervised by the research team and the farmers. The degree of farmer involvement differed according to the type of participatory research. Full on-station trials were completely researcher-managed experiments. Mother trials, each composed of a single full replication, were grown in farmers' fields but jointly supervised and managed by researchers and farmers, whereas baby trials, at the other extreme, were represented by four sets, with each set comprising three genotypes randomly selected from a complete replication, and were wholly managed by farmers. From each village, two community members (male and female) were selected to constitute a gender-balanced participatory research committee at the village level (PRCV). The sum of the PRCV constituted the participatory research committee at each locality level (PRCL). The participatory research committee at the State level (PRCS) was constituted from the participatory research committees of two localities (PRCL), the liaison head of the extension team of each locality and the participating research team of scientists from the El-Obeid Research Station. Pod yield was recorded from the mother trials and the baby trials. Genotypes under test were exclusively ranked by the farmers for the baby trials, whereas ranking of the mother trials was performed by the State-level participatory research committee through group discussions. Data analysis was carried out using MSTATC and GENSTAT. The GENSTAT statistical package provides a descriptive tool for exploring the genotype-by-environment interaction to identify superior genotypes that are widely or specifically adapted (Yan, 2001; Yan and Tinker, 2006).

3 Results

3.1 Pod yield

Pod yield (kg/ha) of 12 genotypes evaluated at El-Obeid research farm during the 2006–2010 cropping seasons is shown in Table 25.1. Differences between genotypes were not significant during this period. Pod yield of 12 groundnut genotypes including four mutants continuously tested during 2013–2017 cropping seasons is shown in Table 25.2. Significant differences between genotypes were observed in the 2017 season and the combined analysis over five seasons. The mutant Barberton-b-30-3-B significantly outyielded the widely cultivated released rainfed check variety, 'Gubeish'. Differences between the same genotypes tested in farmers' fields through the participatory research protocol (mother and baby trials) were significant in all the individual and combined analysis of variances, as shown in Table 25.2. Barberton-b-30-3-B significantly outyielded 'Gubeish' in the individual and combined mean over three locations at the level of the mother trials, whereas it significantly exceeded 'Gubeish' in two seasons at the level of the baby trials. Mean pod yield of three selected genotypes and the mutant Barberton-b-30-3-B continuously tested at the El-Obeid research farm since the 2006–2007 cropping season showed that the mutant Baberton-b-30-3-B significantly recorded the highest mean pod yield among these genotypes (Table 25.3).

Daily rainfall recorded at the El-Obeid research farm during the 2000–2017 cropping seasons is shown in Table 25.4. The total effective rainfall was calculated from the daily rainfall received after planting. Pod yield of four genotypes, including the mutant Barberton-b-30-3-B, and the rainfall data recorded during the cropping seasons from 2006 to 2017 at the El-Obeid research farm were used to study the performance of these genotypes in different rainfall environments using the GGE biplot model of GENSTAT as shown in Fig. 25.1 for the 'which-won-where' pattern (see section 4.1, below) for the total and effective rainfall environments, respectively. The GGE biplot for identifying the most stable genotype is presented in Fig. 25.2 for the total and the effective rainfall environments, respectively. Pod yield and rainfall data during the 2013–2017 cropping seasons were used to study the performance of 12 genotypes including four mutants using the GGE biplot 'which-won-where' pattern as presented in Fig. 25.3 for the total and effective rainfall environments, respectively. The comparison GGE biplot for identifying high-yielding, stable genotype is presented in Fig. 25.4 for the total and effective rainfall environments, respectively.

Table 25.1. Pod yield (kg/ha) of 12 genotypes tested at the El-Obeid research farm (on-station) during seasons 2006–2010.

	Season					
Genotype	2006	2007	2008	2009	2010	Mean
ICG 221	1178	1518	1396	867	978	1187 a
ICGV 92121	982	1130	1441	989	1180	1144 a
RF 1012	996	1403	1409	833	1194	1167 a
ICGV 86744	1084	1621	1464	833	1036	1208 a
ICGV 93255	1153	1306	1144	789	984	1075 ab
ICGV 89171	1082	1562	1227	856	896	1125 a
ICGV 89104	1271	1637	1381	811	969	1214 a
Barberton-b-30-3-B	1214	1528	1205	798	1241	1198 a
S-White	1281	1486	1376	778	1061	1196 a
ICGV 93269	857	1165	1119	733	848	944 b
SODIRI	1134	1363	1393	756	1007	1130 a
Gubeish	1124	1457	1442	722	898	1128 a
SE ±	124	131	126	84	111	50*
CV %	21	20	21	23	24	22

Mean values with different letters are significantly different at that level; *Significantly different at 0.05 probability level; **highly significantly different at 0.01 probability level

Table 25.2. Pod yield (kg/ha) of 12 genotypes tested at on-station and farmer fields (mother and baby trials), seasons 2013–2017.

	Pod yield (kg/ha)													
	On-station						Mother				Baby			
Genotype	2013	2014	2015	2016	2017	Mean	2015	2016	2017	Mean	2015	2016	2017	Mean
Barberton-b-16-2	444	778	533	1222	678	731	917	796	854	836	515	978	624	706
ICGV 93420	600	911	544	1044	849	790	778	930	891	866	552	822	546	640
ICGV 86744-b-1	400	866	700	1000	1000	789	833	914	856	868	414	956	470	613
ICGV 86744	544	844	744	1067	667	813	840	884	842	855	443	1033	479	652
RF 0955	578	900	778	1089	611	791	764	825	875	821	549	836	477	621
ICGV 89171	498	978	644	1056	736	782	806	801	847	819	559	878	437	625
ICGV 89104	522	933	700	900	556	722	813	790	864	822	624	922	503	683
Barberton-b-30-3-B	511	1011	776	1200	1049	909	965	767	972	901	782	867	655	768
ICGV 93261	376	900	722	1244	800	808	993	712	833	846	777	911	482	723
ICGV 89104-a-23-1	411	722	667	1089	878	753	799	684	903	795	383	1000	622	668
ICGV 92121	489	789	678	1078	533	713	445	606	482	511	324	278	270	291
Gubeish	456	889	600	956	667	714	696	690	850	745	354	1378	535	756
SE ±	57	76	77	77	112*	82**	78**	59**	56**	38**	74**	67**	55**	33**
CV %	26	19	25	16	33	24	24	34	27	30	31	22	43	37

*Significantly different at 0.05 probability level; **highly significantly different at 0.01 probability level

Table 25.3. Mean pod yield (kg/ha) of four genotypes continuously tested at the El-Obeid research farm during 2006–2017 cropping seasons.

	Season												
Genotype	2006	2007	2008	2009	2010	2011	2012	2013	2014	2015	2016	2017	Mean
ICGV 89171	1082	1562	1227	856	896	496	1231	498	978	644	1056	736	939^{bc}
ICGV 89104	1271	1637	1381	811	969	484	1262	522	933	700	900	556	952^{ab}
Barberton-b-30-3-B	1214	1528	1205	798	1241	564	1177	511	1011	785	1200	1049	1024^{a}
Gubeish	1124	1457	1442	722	898	680	1223	456	889	600	956	667	926^{c}
SE ±	124	131	126	84	111	66	105	57	76	77	77	124*	47*
CV %	21	20	21	23	24	27	19	26	19	26	16	33	5

Mean values with different letters are significantly different at that level; *Significantly different at 0.05 probability level; **highly significantly different at 0.01 probability level

Table 25.4. Monthly effective rainfall (Total[1]) and total seasonal rainfall (Total[2]) (mm) at El-Obeid research farm, 2000–2017 cropping seasons.

	Year																		Mean	
Month	2000	2001	2002	2003	2004	2005	2006	2007	2008	2009	2010	2011	2012	2013	2014	2015	2016	2017	OBD	FAO
July	35	70	20	120	80	57	191	201	85	92	184	63	186	48	79	25	69	126	102	98
Aug	110	145	75	125	145	92	137	190	135	164	84	140	250	275	80	62	146	141	147	111
Sept	23	51	2	0	0	52	49	40	43	44	51	58	10	65	71	110	58	049	46	62
Oct	18	4	0	6	2	0	52	9	0	0	78	31	2	0	52	23	0	0	16	15
Total[1]	186	270	97	251	227	201	429	440	263	300	379	292	448	388	282	220	273	316	311	286
Total[2]	198	270	115	357	227	332	556	647	317	307	423	326	448	420	317	220	313	406	344	318

Total[1] = Effective rainfall; Total[2] = Total seasonal rainfall; OBD, El-Obeid; FAO, Food and Agriculture Organization of the United Nations

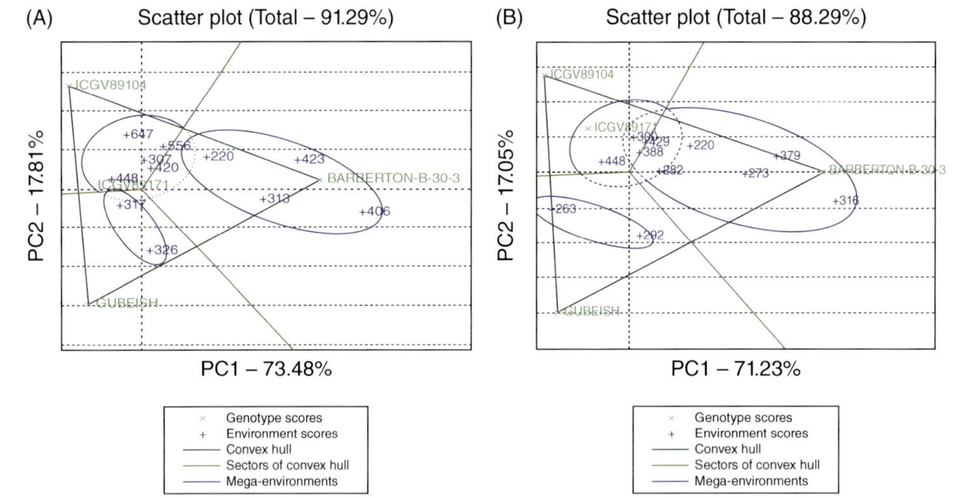

Fig. 25.1. Which-won-where pattern for four groundnut genotypes. **(A)** Genotypes evaluated for 12 years (2006–2017) under total-rainfall amount. **(B)** Genotypes evaluated for 12 years (2006–2017) under effective-rainfall amount.

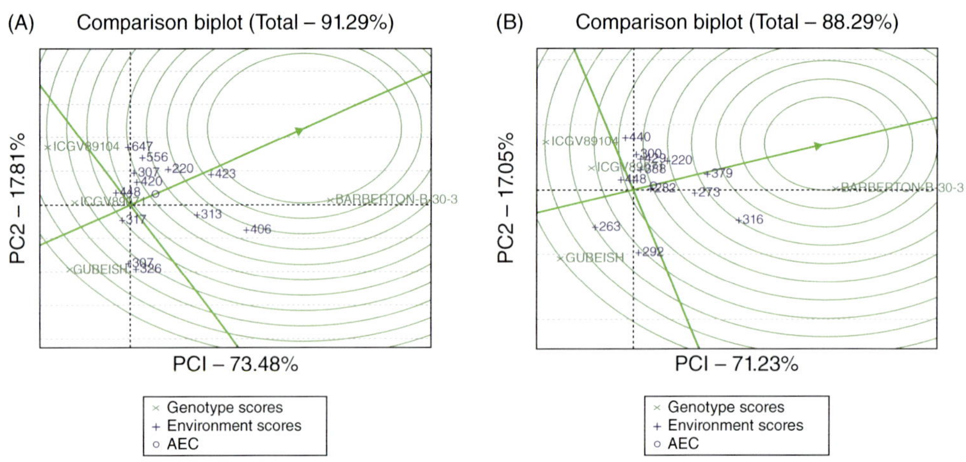

Fig. 25.2. A comparison GGE biplot showing the best groundnut genotype based on mean performance and stability across 12 environments. **(A)** Total-rainfall environments (2006–2017). **(B)** Effective-rainfall environments (2006–2017).

3.2 Harvest index (%)

Harvest index calculated as the percentage of pod yield divided by the total biological yield (pod yield and hay yield) for all genotypes tested during seasons 2006–2017 is presented in Tables 25.5 and 25.6. The mutant Barberton-b-30-3-B significantly exceeded all genotypes in the 2006 season and the combined mean of the test periods of 2006–2010 and 2013–2017.

3.3 Participatory research

Besides on-station evaluation, mutants in the national rainfed groundnut variety trial were also evaluated in farmers' fields using participatory research tools. Participatory research committees at village levels (PRCV) and the locality levels (PRCL) in Sheikan and Rahad localities evaluated the groundnut genotypes through field observation and group discussions.

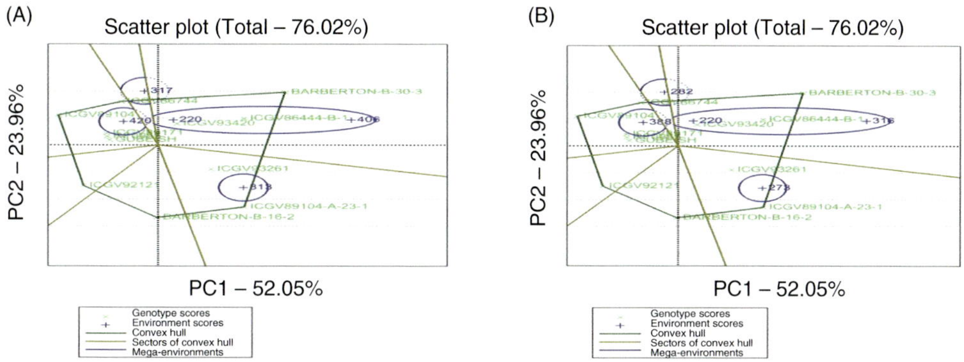

Fig. 25.3. The which-won-where pattern for 12 groundnut genotypes evaluated for 5 years. **(A)** (2013–2017) under total-rainfall amount. **(B)** (2013–2017) under effective-rainfall amount.

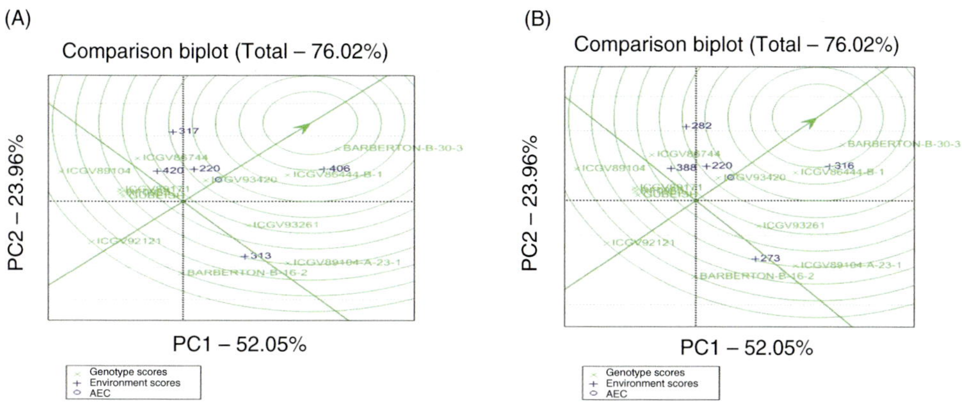

Fig. 25.4. A comparison GGE biplot showing the best groundnut genotype. **(A)** Based on mean performance and stability across five total-rainfall. **(B)** Based on mean performance and stability across five effective-rainfall.

Ranking of genotypes from full on-station researcher-managed trials, mother trials and baby trials is presented in Table 25.7. The mutant Barberton-b-30-3-B was ranked better than the released check variety 'Gubeish' on ten out of 11 occasions.

3.4 Economic analysis

A partial cost benefit analysis was utilized to calculate the economic returns of genotypes and mutants from the on-station, mother and baby trials, as shown in Tables 25.8, 25.9 and 25.10. Barberton-b-30-3-B recorded the highest economic return at all levels of participatory research.

4 Discussion

4.1 Pod yield (kg/ha)

There was a strong positive correlation ($r = 0.90$) between total and effective rainfall amounts recorded during the 2000–2017 cropping seasons as shown in Table 25.4. Hence, total effective rainfall could be reliably used to explain adaptation of genotypes to rainfall, instead of the total rainfall that includes considerable rainfall amounts recorded during April, May and June, outside the groundnut growth period that extends from July to October. Grain yield showed a non-significant positive correlation with total rainfall (0.47) and effective rainfall (0.35). Daily

Table 25.5. Harvest index (%), on-station at El-Obeid research farm, 2006–2010.

Genotype	Harvest index					
	2006	2007	2008	2009	2010	Mean
ICG 221	40	57	43	44	44	46
ICGV 92121	39	54	47	43	43	45
RF 1012	42	64	42	48	49	49
ICGV 86744	37	57	35	46	46	44
ICGV 93255	41	57	40	45	45	46
ICGV 89171	34	55	37	47	47	44
ICGV 89104	39	57	37	49	49	46
Barberton-b-30-3-B	55	68	47	50	50	54
S-White	41	56	41	46	46	46
ICGV 93269	39	58	39	47	47	46
SODIRI	39	56	39	47	47	46
Gubeish	41	63	45	54	54	51
SE ±	3*	3	4	2	2	1**
CV %	19	12	20	11	10	14

* = Significant; ** = highly significant

Table 25.6. Harvest index (%), on-station at El-Obeid research farm, 2011–2017.

	Harvest index							
Genotype	2011	2012	2013	2014	2015	2016	2017	Mean
Barberton-b-16-2	25	51	34	29	32	50	38	37
ICGV 93420	18	50	39	38	28	40	48	38
ICGV 86744-b-1	29	47	34	29	32	38	44	35
ICGV 86744	37	39	38	30	35	42	41	37
RF 0955	27	45	38	34	39	45	45	40
ICGV 89171	35	36	35	31	25	33	47	34
ICGV 89104	35	36	36	30	37	40	42	37
Barberton-b-30-3-B	35	36	37	34	39	49	50	42
ICGV 93261	35	37	35	29	32	42	46	37
ICGV 89104-a-23-1	24	47	35	30	35	44	49	30
ICGV 92121	34	43	36	29	34	42	48	38
Gubeish	24	48	35	31	32	39	44	36
SE ±	2	15	2	2	4	3	4	1*
CV %	20	11	12	13	26	18	19	18

rainfall recorded at El-Obeid research farm during 2000–2017 showed that total rainfall during October with amounts below 20 mm occurred 12 times (70%) in the last 18 years and 7 times (58%) during the study period of the last 12 years, thus rendering end-of-season drought a frequent drought phenomenon (Table 25.4). It is worth mentioning that there was little deviation between the actual mean rainfall during the months of July, August, September and October during 2000–2017 and the expected total calculated from a model using rainfall data of the past 60 years (FAO, 2006). The model identified the humid period suitable for crop growth at the test site (research farm of the El-Obeid Research Station) as 55 days. A research site with these characteristics is considered marginal for groundnut production. Hence, groundnut genotypes with good performance under this environment increase the chances of expanding groundnut cultivation to similar marginal environments and could stabilize groundnut production in the newly demarcated State of North Kordofan.

Table 25.7. Participatory research, variety ranking, on-station, mother and baby trials, 2013–2017.

Genotype	Ranking													
	On-station						Mother				Baby			
	2013	2014	2015	2016	2017	Mean	2015	2016	2017	Mean	2015	2016	2017	Mean
Barberton-b-16-2	9	10	11	2	7	8	3	6	7	3	7	4	2	2
ICGV 93420	1	4	10	8	4	3	9	1	3	2	5	11	4	7
ICGV 86744-b-1	11	7	5	9	2	6	5	2	6	2	9	5	10	8
ICGV 86744	3	8	3	6	7	3	4	3	10	6	8	2	8	5
RF 0955	2	5	1	4	9	2	10	4	4	5	6	9	9	8
ICGV 89171	6	2	8	7	6	4	7	5	9	7	4	8	11	8
ICGV 89104	4	3	5	11	10	5	6	7	5	5	3	6	6	3
Barberton-b-30-3-B	5	1	2	3	1	1	2	8	1	1	1	10	1	1
ICGV 93261	12	5	4	1	5	3	1	9	11	7	2	7	7	4
ICGV 89104-a-23-1	10	11	7	4	3	7	8	11	2	7	10	3	3	4
ICGV 92121	7	9	6	5	11	8	12	12	12	9	12	12	12	10
Gubeish	8	6	9	10	8	10	11	10	8	8	11	1	5	5

Table 25.8. Mean pod yield, gross benefit and net benefit (SDG) per hectare of groundnut genotypes, El-Obeid full trials, 2013–2017 seasons.

Genotype	Grain yield (kg/ha)	Gross benefit (SDG/ha)	Net benefit (SDG/ha)	Ranking
Barberton-b-16-2	731	5190	3215	11
ICGV 93420	790	5664	3689	5
ICGV 86744-b-1	789	5602	3627	6
ICGV 86744	813	5772	3797	2
RF 0955	791	5695	3720	4
ICGV 89171	782	5552	3577	7
ICGV 89104	722	5198	3223	10
Barberton-b-30-3-B	909	6475	4500	1
ICGV 93261	808	5736	3762	3
ICGV 89104-a-23-1	753	5210	3236	8
ICGV 92121	713	5205	3230	9
Gubeish	714	5097	3123	12

Table 25.9. Mean pod yield, gross benefit and net benefit (SDG) per hectare of mother trials of groundnut genotype combined over seasons 2015–2017.

Genotype	Grain yield (kg/ha)	Gross benefit (SDG/ha)	Net benefit (SDG/ha)	Ranking
Barberton-b-16-2	836	5936	3960	5
ICGV 93420	866	6333	4358	2
ICGV 86744-b-1	868	6248	4273	3
ICGV 86744	855	6113	4138	4
RF 0955	821	4530	2555	11
ICGV 89171	819	5884	3839	7
ICGV 89104	822	5914	3943	6
Barberton-b-30-3-B	901	6219	4534	1
ICGV 93261	846	5680	3705	8
ICGV 89104-a-23-1	795	5580	3606	9
ICGV 92121	511	3898	1923	12
Gubeish	745	5383	3408	10

Table 25.10. Baby trials, mean yield, gross benefit and net benefit (SDG) per hectare of baby trials of groundnut genotypes combined over seasons 2015–2017.

Genotype	Grain yield (kg/ha)	Gross benefit (SDG/ha)	Net benefit (SDG/ha)	Ranking
Barberton-b-16-2	706	4977	3002	4
ICGV 93420	640	4402	2427	8
ICGV 86744-b-1	613	4189	2214	10
ICGV 86744	652	4409	2434	7
RF 0955	621	4197	2223	9
ICGV 89171	625	4061	2086	11
ICGV 89104	683	4550	2575	6
Barberton-b-30-3-B	768	5226	3251	1
ICGV 93261	723	4600	2625	5
ICGV 89104-a-23-1	668	4984	3009	3
ICGV 92121	291	2044	69	12
Gubeish	756	5190	3215	2

The GGE biplot model provides breeders with a complete and visual evaluation of all aspects of the data by creating a biplot that simultaneously represents both mean performance and stability, identifies mega-environments and identifies the best cultivar in each environment (Yan and Tinker, 2006). A mega-environment is a group of environments in which a single genotype or a group of similarly performing genotypes are specifically adapted and excel in performance. The purpose of mega-environment analysis is to divide a target crop region into meaningful sub-regions so that repeatable G×E interaction can be explored and exploited. Suitability of genotypes for particular environments was carried out using the 'which-won-where' function of the GGE biplot (Yan and Tinker, 2006). The 'which-won-where' function of a GGE biplot is an extended use of the 'pair-wise comparison' function (Yan *et al.*, 2007). A polygon is first drawn on genotypes that are furthest from the biplot origin so that all other genotypes are contained within the polygon. Then perpendicular lines to each side of the polygon are drawn, starting from the biplot origin. Genotypes located on the vertices of the polygon performed either the best or the poorest in one or more environments (Yan and Hunt, 2002). Four groundnut genotypes were evaluated at the El-Obeid research farm for 12 consecutive seasons (2006–2017). The total rainfall ranged from 220 to 647 mm and the total effective rainfall varied from 220 to 448 mm. Based on the GGE analysis using the total rainfall amount, the first two principal components explained 91.3% of the total interaction variation for grain yield, implying that they sufficiently explained GGE. The total rainfall environments segregated into distinct sectors of the 'which-won-where' biplot, with each sector having different genotypes performing best (Fig. 25.1A). The winner genotypes were Barberton-b-30-3-B, ICGV 89104 and 'Gubeish'. A winner genotype is the vertex genotype for each sector that has the largest value among all genotypes in a mega-environment falling within that sector (Yan *et al.*, 2000). The mutant Barberton-b-30-3-B was the best in the mega-environment with total rainfall of 423, 406, 316 and 220 mm, whereas ICGV 89104 was the best genotype in the mega-environment with total rainfall of 647, 556, 448, 420 and 307 mm. 'Gubeish' was the best in the mega-environment with only two total rainfalls of 326 and 316 mm. The mutant Barberton-b-30-3-B could be grown in below-average as well as above-average rainfall environments, whereas the genotype ICGV 89104 could be suitable for average and above-average rainfall environments. The widely cultivated check variety could perform better only in environments with average rainfall. The GGE biplot analysis for the effective rainfall amount is presented in Fig. 25.1B. Based on the GGE analysis, the first two principal components explained 88.3% of the total interaction variation. The winner genotypes were Barberton-b-30-3-B, ICGV 89104 and 'Gubeish'. The mutant Barberton-b-30-3-B was the best in the mega-environment with rainfall of 429, 316, 379, 388, 300, 282, 273 and 220 mm, whereas ICGV 89104 was the best genotype in the mega-environment with 448, 440, 429, 388 and 300 mm. 'Gubeish' was the best in the mega-environment with total rainfalls of only 292 and 263 mm.

The genotype comparison GGE biplot for identifying the ideal genotype across environments is presented in Fig. 25.2 for the total and the effective rainfall environments. Genotype ranking based on genotype-focused scaling assumes that stability and mean yield are equally important. Thus, an ideal genotype should have the highest mean performance and be absolutely stable (Yan, 2002; Yan and Tinker, 2006). According to Yan and Tinker (2006), such an ideal genotype is defined as having the greatest vector length of the high-yielding genotype with zero GEI, as represented by an arrow pointing to it. Using the ideal genotype as the centre, concentric circles were drawn to help visualize the distance between each genotype and the ideal genotype, where a genotype is identified as more desirable if it is located closer to the ideal genotype. In this study, Barberton-b-30-3-B represented the ideal genotype which combined both high average yield with yield stability across both total and effective rainfall environments.

The GGE biplots showing 'which-won-where' functions or which genotypes are best for which environment for pod yield for 12 groundnut genotypes continuously evaluated for five consecutive seasons (2013–2017) are presented in Fig. 25.3 for total and effective seasonal rainfall amounts. Based on the GGE analysis, the first two principal components explained 76% of the total interaction variation. The 12 total and effective rainfall environments segregated into distinct sectors of 'which-won-where' biplot, each sector having different genotypes performing best. The vertices (winner) genotypes were Barberton-b-30-3-B, ICGV 89104-a-23-1, Barberton-b-16-2, ICGV 92121, ICGV 89104 and ICGV 86744, as explained by Yan *et al.*, 2000.

The genotype comparison GGE biplot for identifying the ideal genotype across environments is presented in Fig. 25.4, for total and effective rainfall environments. The ideal genotype which combined both high average yield with yield stability across both total and effective rainfall environments was represented by Barberton-b-30-3-B.

4.2 Harvest index (%)

The mutant Barberton-b-30-3-B recorded the highest mean harvest index (HI) in this study. Previous studies have shown positive genetic correlation of harvest index with high pod yield under terminal drought stress (Nagda *et al.*, 2001; Abdalla *et al.*, 2008) and these authors also reported significant positive correlation between HI and pod yield in groundnut, while Arjunan *et al.* (1999) reported positive but non-significant correlation. Nautiyal *et al.* (2002) stated that selection for high HI could result in groundnut cultivars with improved performance under rainfed agriculture. The possibility of improving groundnut pod yield through selecting for high HI has also been reported in several path coefficient analysis studies (Mathews *et al.*, 1988; Bera and Das, 2000; Nagda *et al.*, 2001). Based on HI, Barberton-b-30-3-B was predicted to be the best drought-tolerant mutant.

4.3 Participatory research

Differences between genotypes were significant at all participatory research levels where the mutant Barberton-b-30-3-B exceeded the control cultivar ('Gubeish') by 27% under full on-station trials, and by 21% under the mother trials. Ranking of genotypes by the participatory research committee and farmers is in line with the agronomic superiority and the high economic returns of the mutant Barberton-b-30-3-B as shown in Tables 25.8, 25.9 and 25.10 for the on-station, mother and baby trials, respectively. Participatory research or demand-driven research is complemented by conventional research and facilitated by the inclusion of farmers' ideas and their quality preferences, thus improving the quality of research that could eventually enhance adoption (Ashby and Lilja, 2004; Ceccarelli and Grando, 2007). The mutant Baberton-b-30-3-B has biological and economic leverage over all genotypes, a result that supports releasing it for the low-rainfall drylands of North Kordofan State.

5 Conclusion

Results have shown that mutant Baberton-b-30-3-B, which exceeded 'Gubeish' (the latest released variety) by 11% over all environments, is a suitable candidate for high yield and wide stability in both effective and total rainfall scenarios. The mutant Barberton-b-30-3-B evaluated under 5 years of participatory research also exceeded the control cultivar ('Gubeish') by 27% under full on-station trials, and by 21% under the mother trials. Farmers and the participatory research committee ranked Barberton-b-30-3-B as the best among all genotypes. This ranking goes in line with the agronomic superiority and high economic return of the mutant Barberton-b-30-3-B and supports releasing it for the low-rainfall drylands of North Kordofan State. Barberton-b-30-3-B was officially released as 'Tafra-1' by the National Variety Release Committee during its second meeting of April 2018 as the first groundnut mutant in Sudan.

References

Abdalla, E.A. *et al.* (2007) Easily measurable morpho-physiological traits as selection criteria for terminal drought tolerance in groundnut (*Arachis hypogaea* L.). *Gezira Journal of Agricultural Science* 5(2), 21–34.

Abdalla, E.A. *et al.* (2008) Performance of nine groundnut mutants and their parents under terminal drought stress conditions in western Sudan. *Gezira Journal of Agricultural Science* 6, 6(1), 18–22.

Adam, H.S. (2002) *Agro-climatology*. University of Gezira Printing and Publication House, Wad Medani, Sudan.

Arjunan, A.V. *et al.* (1999) Path analysis of characters contributing to drought resistance in groundnut. *Madras Agricultural Journal* 86, 36–39.

Ashby, J.A. and Lilja, N. (2004) Participatory research: does it work? Evidence from participatory plant breeding. In: *New Directions for a Diverse Planet*. Proceedings of the 4th International Crop Science Congress, Brisbane, Australia, 26 September–1 October 2004. Available at: www.cropscience.org.au (accessed 2019).

Bera, S.K. and Das, P.K. (2000) Path coefficient analysis of groundnut at different locations and years. *Agricultural Science Digest* 20, 9–12.

Ceccarelli, S. and Grando, S. (2007) Decentralized – participatory plant breeding: an example of demand driven research. *Euphytica* 155, 349–360.

Crossa, J. and Cornelius, P.L. (1997) Sites regression and shifted multiplicative model clustering of cultivar trial sites under heterogeneity of error variances. *Crop Science* 37, 406–415.

FAO (2006) *Local Climate Estimator*. Environment and Natural Resources Working Papers No. 20. FAO, Rome.

Gauch, H.G. Jr and Zobel, R.W. (1997) Identifying mega-environments and targeting genotypes. *Crop Science* 37, 311–326.

Mathews, R.B. *et al.* (1988) The physiological basis for yield differences between four genotypes of groundnut (*Arachis hypogaea* L.) in response to drought. I. Dry matter production and water use. *Experimental Agriculture* 24, 191–202.

Nagda, S.N. *et al.* (2001) Gene effects for specific leaf area and harvest index in three crosses of groundnut (*Arachis hypogaea* L.). *Annals of Applied Biology* 139, 301–306.

Nageswara Rao, R.C. *et al.* (1988) Effect of water deficit at different growth phases of peanut. II. Response to drought during pre-flowering phase. *Agronomy Journal* 80, 431–438.

NAPA (2007) *The national adaptation plan of action*. Higher Council for Environment and Natural Resources, Ministry of Environment and Physical Development, Republic of the Sudan.

Nautiyal, P.C. *et al.* (2002) Moisture-deficit-induced changes in leaf-water content, leaf carbon exchange rate and biomass production in groundnut cultivars differing in specific leaf area. *Field Crops Research* 74, 67–79.

Thangthong, N. *et al.* (2017) Root distribution patterns of peanut genotypes with different drought resistance levels under early-season drought stress. *Journal of Agronomy and Crop Science* 204, 111–122.

Yan, W. (2001) GGE biplot: a Windows application for graphical analysis of multi-environment trial data and other types of two-way data. *Agronomy Journal* 93, 1111–1118.

Yan, W. (2002) Singular-value partition for biplot analysis of multi-environment trial data. *Agronomy Journal* 94, 990–996.

Yan, W. and Hunt, L.A. (2001) Interpretation of genotype-by-environment interaction for winter wheat yield in Ontario. *Crop Science* 41, 19–25.

Yan, W. and Hunt, L.A. (2002) Biplot analysis of diallel data. *Crop Science* 42, 21–30.

Yan, W. and Kang, M.S. (2003) *GGE Biplot Analysis: A Graphical Tool for Breeders, Geneticists, and Agronomists*. CRC Press, Boca Raton, Florida.

Yan, W. and Rajcan, I. (2002) Biplot evaluation of test sites and trait relations of soybean in Ontario. *Crop Science* 42, 11–20.

Yan, W. and Tinker, N.A. (2006) Biplot analysis of multi-environment trial data: principles and application. *Canadian Journal of Plant Science* 86, 623–645.

Yan, W. *et al.* (2000) Cultivar evaluation and mega-environment investigation based on the GGE biplot. *Crop Science* 40, 597–605.

Yan, W. *et al.* (2007) GGE biplot vs. AMMI analysis of genotype-by-environment data. *Crop Science* 47, 643–655.

Yates, F. and Cochran, W.G. (1938) The analysis of group experiments. *Journal of Agricultural Science* 28, 556–580.

26 Induction of Variability for Yield Components in Indian Mustard (*Brassica juncea* L. Czern & Coss) under Acidic Soil Regime of Jharkhand

Arun Kumar[1], Binay Kumar Agarwal, Rajesh Kumar, Sanjay J. Jambhulkar, Varsha Rani and Zille Ali. Haider

Abstract

Indian mustard (*Brassica juncea* L.) is the most important oilseed crop of the state of Jharkhand in India, where 78% of the cultivable soil is acidic, causing a sizeable yield reduction. Potential seed yield from such soils cannot be realized within existing varieties and therefore a mutation breeding approach has been followed to isolate mutants tolerant to acidic soil. Three doses of gamma-rays (900 Gy, 1000 Gy and 1100 Gy) and a combined treatment of gamma irradiation and 0.3% EMS were used for induction of mutation in the varieties 'Shivani' and 'Pusa Bold'. A total of 139,720 M_2 plants (75,760 of 'Shivani' and 63,960 of 'Pusa Bold') were screened under acidic soil conditions (pH 4.8). A wide spectrum of variability for tolerance to soil acidity, earliness, seed colour, seed yield and yield components, and morphological traits was observed in the M_2 generation. True-breeding mutants for different traits were confirmed in the M_3 generation. Mutations were recorded in 'Shivani' and 'Pusa Bold', respectively, for secondary branch number (38 and 24), siliquae per plant (1223 and 562) and single plant seed yield (45.49 g and 34.84 g). In addition, a large spectrum of variability for morphological characters was identified.

Keywords: induced mutagenesis • *Brassica juncea* • gamma radiation • combination treatment • soil acidity tolerance

1 Introduction

Against the total domestic demand of 25.88 million tonnes (Mt) of vegetable oil, India is able to meet hardly 10.52 Mt (40%) through its domestic production. The remaining 15.35 Mt (59.31%) is obtained through import with an investment of 75 billion rupees (Rs 74996 crore) during 2017–2018. Rapeseed-mustard (*Brassica rapa*, *B. juncea* and *B. napus*) shared 24.98% of the nation's total oilseed basket with an average production of 1304 kg/ha as against 1974 kg/ha for the rest of the world. In India, approximately 85% of Brassicaceae-derived vegetable oil is produced from Indian mustard (*B. juncea*) and it ranks second in the country and first in the state of Jharkhand with an average yield of 766 kg/ha. Acidic soil, which accounts for > 70% of the total geographical area of Jharkhand state (Agarwal *et al.*, 2012), is the major limiting abiotic factor for productivity of rapeseed-mustard. Potential seed yield cannot be realized from such

[1] Email: arun.bau4@yahoo.co.in

 Mutation Breeding, Genetic Diversity and Crop Adaptation to Climate Change (eds S. Sivasankar *et al.*)
DOI: 10.1079/9781789249095.0026

soil. Therefore, development of genotypes tolerant to acid soil is essential.

Mutation breeding is an important method to enhance the spectrum of variability for characters of agronomic and economic significance, which is a prerequisite for any crop improvement programme. Mutation breeding is accomplished by chemical or physical treatments followed by selection for heritable changes in target traits. Mutagenesis has been widely used as a potent method for enhancing variability for crop improvement (Subudhi *et al.*, 1991). A number of genotypes tolerant to soil acidity have been developed in soybean, barley and sorghum using this tool (Foy *et al.*, 1992, 1993; Foy, 1996). This method has been used successfully in the genetic improvement of qualitative and quantitative characters in oleiferous *Brassica* species (Robbelen, 1990; Jambhulkar, 2007, 2015). Mutation breeding has been successfully used to develop a large number of high-yielding varieties of oilseeds and other crops (Maluszynski *et al.*, 1995; Bhatia *et al.*, 1999; Chopra, 2005).

The present work on mutation breeding on Indian mustard (*B. juncea* L.) started in the Department of Genetics and Plant Breeding, Birsa Agricultural University, Ranchi (Jharkhand), India, in 2014–2015 as a collaboration with the Board of Research in Nuclear Sciences, Department of Atomic Energy (DAE), Government of India, through project 35/14/21/2014-BRNS/0401 aimed to induce genetic variability for yield components by employing gamma-rays alone and in combination with ethyl methanesulfonate (EMS) to develop novel mutants that can perform better under acidic soil condition.

2 Materials and Methods

2.1 Area of investigation

The area of study is situated in the eastern part of India that includes the city of Ranchi, which is the capital of Jharkhand state. Geographically, the area is located between 23° 17′ N latitude and 85° 10′ E longitude at an altitude of 625 m above mean sea level. The total geographical area of the state is 797,142 km^2 of which only 8.31% has neutral soil; the rest of the soil is affected by soil acidity ranging from slightly to extremely acid.

2.2 Materials and mutagenesis treatment

Well-filled seeds (250 g per treatment) of two varieties, namely 'Shivani' and 'Pusa Bold', were used for mutation induction. Seeds were irradiated with 900 Gy, 1000 Gy and 1100 Gy doses of gamma-rays using ^{60}Co source at Bhabha Atomic Research Centre (BARC), Trombay, Mumbai. For combination treatment, all doses of gamma-rays together with 0.3% EMS were used. An aqueous solution of EMS (0.3% v/v) was made in phosphate buffer (0.5 M sodium phosphate, pH 7.0) and treatment time was 8 h post irradiation. Untreated seeds and seeds soaked in buffer solution and distilled water separately served as control. Thus, in total nine treatments of each variety, including three controls, viz. 900 Gy, 1000 Gy, 1100 Gy, 900 Gy + 0.3% EMS, 1000 Gy + 0.3% EMS, 1100 Gy + 0.3% EMS, 0 Gy(C1), buffer-soaked (C2) and water-soaked (C3), were constituted for the present study.

2.3 M_1 generation

Seeds of each treatment were sown in the field in a non-replicated trial during rabi 2014–2015 in rows 5 m long, with spacing of 30 cm between rows and 10 cm between plants. Each treatment was sown in 800 rows, along with ten rows of each control side by side. Recommended agronomic practices were followed to raise the crop. At the end of the M_1 generation, 3493 M_1 plants consisting of 1894 plants of 'Shivani' and 1599 plants of 'Pusa Bold' were harvested separately on a single-plant basis which served as the source of the M_2 generation raised during rabi 2015–2016. Observations on germination percentage, growth and development were recorded from seedling to maturity.

2.4 M_2 generation

During rabi 2015–2016, the M_2 generation was raised on a single plant progeny basis from 3493 individual M_1 plants and was grown under natural acidic soil condition (pH 4.8) along with a single row of controls. The total plant population was 139,720, out of which 75,760 were derived from 'Shivani' and 63,960 from 'Pusa Bold'. Selection was made on the basis of plant height, earliness, long silique, basal branching, seed colour, large seed size, number of branches, compact plant type and high yield attributes. In total, 304 single plants from 'Shivani' and 229 from 'Pusa Bold' were selected as putative mutants for different characters. Mutation frequency of each visible mutant in the M_2 generation was calculated as suggested by Gaul (1958).

2.5 M_3 generation

All the selected putative mutants were grown in plant-to-progeny rows along with the parent to study the performance of agronomic characters and selection of mutants true to type in the M_3 generation under acidic soil during 2016–2017. Each M_2 plant progeny was grown in three-row plots of 2 m length. Standard spacing of 30 cm between rows and 10 cm between plants was maintained. The control was sown in two-row plots at intervals of six M_3 progenies. Observations of traits on putative mutants were recorded from seedling to maturity. True-breeding mutants were subjected to further study.

3 Results

Results obtained from this investigation are presented below under two main subheads:

1. Mutation frequency observed in M_2 generation
2. Performance of selected mutants in M_3 generation for yield and yield components

3.1 Mutation frequency

The effects of different treatments of mutagens on cv. 'Shivani' and cv. 'Pusa Bold' are presented in Table 26.1. Frequency of mutants induced by different doses of gamma-rays and combination treatments for 'Shivani' are presented in Table 26.2a and for 'Pusa Bold' in Table 26.2b and Figs 26.1 and 26.2. Out of six treatments, the combination of 100 Gy + 0.3% EMS had the highest mutation frequency of 1.5%, followed by 1100 Gy + 0.3% EMS with 1.3% in 'Shivani'; and 1100 Gy + 0.3% EMS had 1.8% followed by 900 Gy + 0.3% EMS with 1.26% in 'Pusa Bold'. It is not clear whether the differences between varieties or between gamma-ray doses are significantly different. The highest frequencies were obtained in the combination treatments and the differences can be attributed entirely to the chemical mutagen. The gamma-rays may damage DNA by direct action or indirectly through free-radical formation, whereas the chemical mutagen, EMS, normally acts as a base-substituting agent. One of the possibilities in the combined treatment could be restriction of repair process of damaged DNA in the embryo, leading to accumulation of mutations and a high frequency of base substitution, which was partly suggested by Mondal *et al.* (2007). An increase in the frequency of mutants with an increase in mutagen dose in mustard was also observed by Sangsiri *et al.* (2005).

Among the different types of morphological variants, the most common mutant forms had an altered frequency of primary and secondary branches (in 'Shivani' 0.17% and in 'Pusa Bold' 0.12%) whereas other morphological mutants, viz. early flowering (0.09%), high silique density (0.022%), long shoots in comparison with parent (0.017%), tall (0.017%), large seed size (0.015%), dwarf (0.008%) and cream flower colour mutants (0.005%), had comparatively lower frequency in 'Shivani' (Table 26.2a) (Fig. 26.2). In 'Pusa Bold', various morphological mutants, viz. tall (0.038%), high silique density (0.03%), early (0.03%), long main shoot length (MSL) (0.027%), large seed (0.02%) and yellow seed coat mutant (0.008%), were recorded in relatively lower frequency (Table 26.2b, Fig. 26.2). Similar results were also observed by Girija *et al.* (2013) in *B. juncea* when treated with gamma-rays and EMS. Morphological characters play an important role as phenotypic markers to identify and maintain the purity of genotype as well as in linkage mapping (Jambhulkar, 2007). Dwarfing genes in *Brassica* spp. are useful in increasing seed yield by reducing lodging and increasing harvest index. Dwarf mutants compared with their parents have been isolated in rapeseed-mustard using physical and chemical mutagens (Chauhan and Kumar, 1986; Das and Rahman, 1988; Zanewich *et al.*, 1991; Rai and Singh, 1993), whereas Devlin *et al.* (1992)

Table 26.1. Effect of different treatments of mutagen on Indian mustard cv. 'Shivani' and cv. 'Pusa Bold'.

		M_1		M_2		M_3	
Treatment by variety	Code	Seed treated	M_1 plants harvested	Number of plants	Number of variants	Mutation frequency (%)	True breeding mutants
'Shivani'							
Control 1 (0 Gy)	SC1	100g	100	–	–	–	–
Control 2 (Water soaked)	SC3	100g	100	–	–	–	–
900 Gy	S1	250g	301	12,040	27	0.224	7
1000 Gy	S2	250g	683	27,320	36	0.132	13
1100 Gy	S3	250g	521	20,840	44	0.211	5
900 Gy + 0.3% EMS	S4	250g	140	5,600	60	1.071	10
1000 Gy + 0.3% EMS	S5	250g	90	3,600	54	1.5	2
1100 Gy + 0.3% EMS	S6	250g	159	6,360	83	1.317	12
Total			**1,894**	**75,760**	**304**	**0.401**	**49**
'Pusa Bold'							
Control 1 (0 Gy)	PBC1	100g	100	–	–	–	–
Control 2 (Water soaked)	PBC3	100g	100	–	–	–	–
900 Gy	PB1	250g	428	17,120	31	0.181	13
1000 Gy	PB2	250g	432	17,280	36	0.208	16
1100 Gy	PB3	250g	509	20,360	34	0.167	7
900 Gy + 0.3% EMS	PB4	250g	83	3,320	42	1.265	10
1000 Gy + 0.3% EMS	PB5	250g	61	2,440	24	0.984	5
1100 Gy + 0.3% EMS	PB6	250g	86	3,440	62	1.802	10
Total			**1,599**	**63,960**	**229**	**0.358**	**61**

Table 26.2a. Frequency of mutants induced by different doses of gamma-rays and combination treatments in M_2 generation of cv. 'Shivani'.

Character	Different dose of gamma-rays +/– 0.3% EMS						
	900 Gy	1000 Gy	1100 Gy	900 Gy + 0.3%EMS	1000 Gy + 0.3% EMS	1100 Gy + 0.3% EMS	Total
Appressed	0	2	0	5	0	4	11
Bold Seed	1	0	1	4	4	1	11
Cream Flower Colour	0	0	0	0	1	3	4
Semi-Dwarf	0	0	0	0	3	1	4
Dwarf	0	0	1	2	1	2	6
Early	4	6	10	14	13	21	68
Late	2	3	0	0	0	0	5
Profuse Branch	9	16	18	25	26	32	126
Long Silique	0	0	2	0	0	2	4
MSL[a]	4	2	3	0	0	4	13
Base Branching	1	1	0	0	0	0	2
Semi Appressed	0	0	0	1	1	0	2
Small Silique	1	0	0	2	0	2	5
Plant Height	1	2	3	3	1	3	13
White Flower	1	0	0	0	2	1	4
Yellowish Green Leaf	0	0	1	0	0	0	1
Large Leaf	0	1	0	0	0	0	1
Chlorophyll Mutation	0	0	1	1	0	3	5
Erect Plant	0	0	0	0	1	0	1
Red Seed Coat	0	0	0	0	0	1	1
High Silique Density	3	3	4	3	1	3	17
Total	27	36	44	60	54	83	304

[a]Length of the main fruiting axis or inflorescence

Table 26.2b. Frequency of mutants induced by different doses of gamma-rays and combination treatments in M_2 generation of cv. 'Pusa Bold'.

	Different dose of gamma-rays +/– 0.3% EMS						
Character	900 Gy	1000 Gy	1100 Gy	900 Gy + 0.3% EMS	1000 Gy + 0.3% EMS	1100 Gy + 0.3% EMS	Total
Appressed	0	1	1	0	1	2	5
Bold Seed	3	2	3	1	0	4	13
Brown Red Seed Coat	0	0	1	0	0	0	1
Cream Flower Colour	0	2	0	1	0	0	3
Dwarf	0	0	4	2	1	0	7
Early	2	3	1	5	3	5	19
Late	2	0	0	0	0	0	2
Profuse Branch	9	15	9	11	8	26	78
Long Silique	3	1	1	1	1	1	8
MSL[a]	2	2	2	5	0	6	17
Base Branching	3	3	0	1	0	0	7
Purple Leaf	0	0	0	1	1	1	3
Semi Appressed	0	2	0	2	1	1	6
Small Silique	0	0	1			0	1
Plant Height	5	3	4	6	1	5	24
Twisted Plant	1	0	2	0	0	1	4
White Flower	0	0	2	0	0	0	2
Yellow Seed Coat	0	0	0	2	3	0	5
Yellowish Green Leaf	0	0	0	1	1	0	2
High Silique Density	1	2	3	3	3	10	22
Total	31	36	34	42	24	62	229

[a] Length of the main fruiting axis or inflorescence

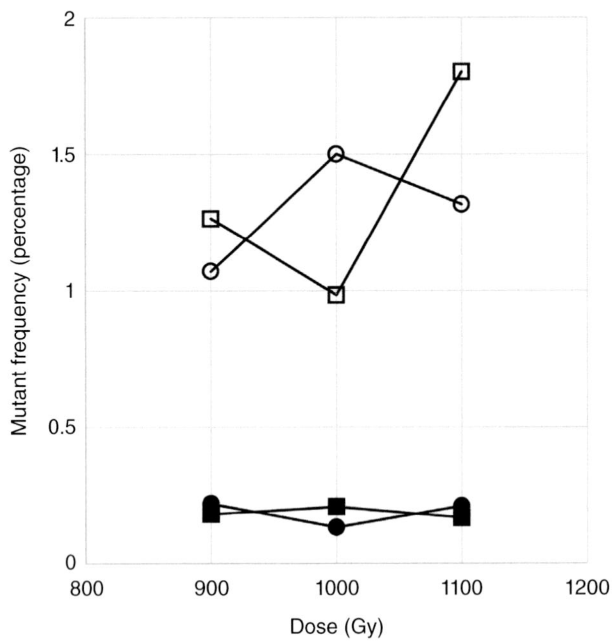

Fig. 26.1. Dose dependence of M_2 mutation frequency. Round symbols, cv. 'Shivani'; square symbols, cv. 'Pusa Bold'; Solid symbols, gamma-rays alone; open symbols, gamma-rays plus 0.3% v/v EMS.

isolated an elongated internode mutation in *B. rapa*. A novel mutant with a yellow-white flower colour was observed in the male sterile progeny derived from commercial *B. napus* hybrid CO22 isolated by Yu *et al.* (2004). Yellow-seeded rapeseed-mustards are more desirable than brown- or black-seeded types because of their thinner seed coat, higher oil content, higher protein and lower fibre content than the brown-seeded varieties with the same genetic background (Stringam *et al.*, 1974; Woods, 1980; Xiao, 1982).

3.2 Performance of selected mutants in M_3 generation for yield and yield components

In the M_3 generation, 49 mutants from 'Shivani' and 61 mutants from 'Pusa Bold' were found to be true breeding (Table 26.1). Observations for major yield-contributing traits were recorded from all the true-breeding mutants under acidic soil conditions (pH 4.8). Results of yield-contributing traits of some selected true-to-type mutants of 'Shivani' and 'Pusa Bold' are presented in Table 26.3.

Early and late mutants

Out of 17 true-breeding mutants of 'Shivani', three mutants matured in 92–97 days as compared with a duration of 105–110 days in their parent. Similarly, in 'Pusa Bold', one mutant matured in 99 days as compared with a duration of 115–120 days in its parent (Table 26.3). Early-maturing genotypes with comparable yield are the most rewarding to fit into the multiple-cropping system, especially in those areas where irrigation water is limited. Early-flowering mutants of *B. juncea* were reported by Nayar and George (1969).

Tall and dwarf mutants

Significantly dwarf and tall mutants were isolated from 'Shivani' with heights of 57 cm and 205 cm, respectively, and from 'Pusa Bold' with heights of 55 cm and 196 cm, respectively (Table 26.3). The dwarfing gene in *Brassica* sp. is useful in increasing seed yield by reducing lodging and increasing harvest index. Similar results were reported in rapeseed-mustard using physical and chemical mutagens (Chauhan and Kumar,

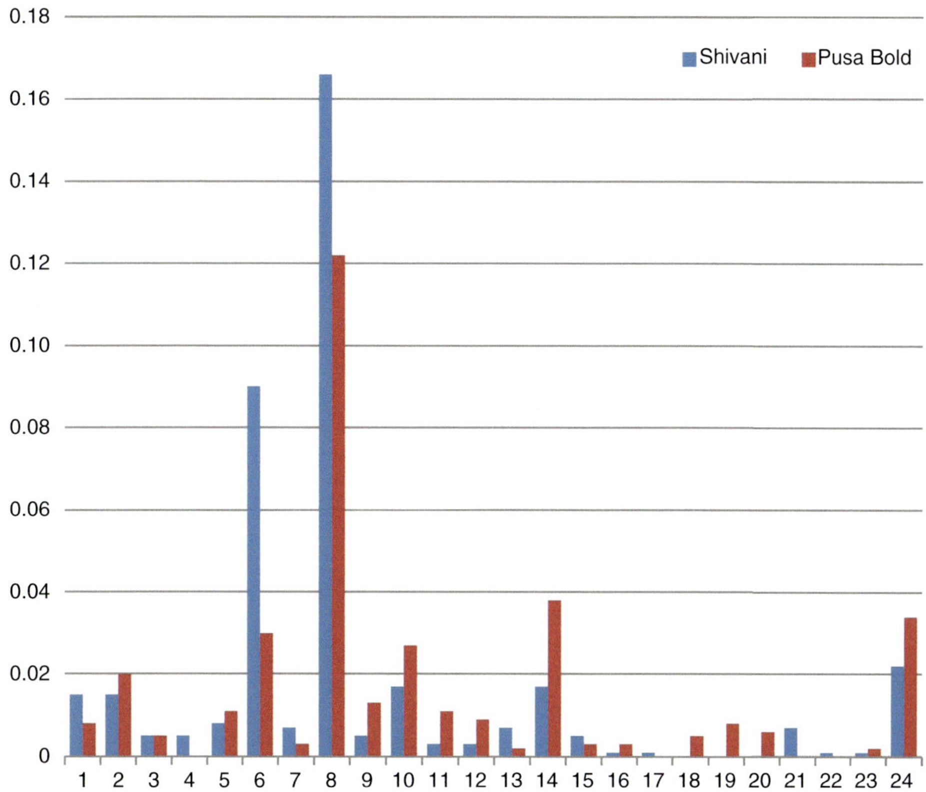

Fig. 26.2. Percentage of various mutant classes in the M_2. The *y* axis is the percentage occurrence and the *x* axis represents the traits presented in Table 26.2. They are: 1, Appressed; 2, Bold Seed; 3, Cream Colour Flower; 4, Semi-Dwarf; 5, Dwarf; 6, Early; 7, Late; 8, Profuse Branching; 9, Long Silique; 10, MSL; 11, Base Branching; 12, Semi Appressed; 13, Small Silique; 14, Plant Height; 15, White Flower; 16, Yellowish Green Leaf; 17, Large Leaf; 18, Purple Leaf; 19, Yellow Seed Coat; 20, Twisted Plant; 21, Chlorophyll Mutation; 22, Erect Plant; 23, Brown/Red Seed Coat; 24, High Density Silique.

Table 26.3. Range of variation for major yield contributing traits of selected mutants of Indian mustard in the M_3 generation under field conditions at low soil pH (4.8).

	'Shivani'		'Pusa Bold'	
Characters	Control	Mutant lines	Control	Mutant lines
Days to Maturity	105–110	92–115**	115–120	99–121*
Plant Height (cm)	130–135	57–205**	135–140	55–196**
Primary Branch	3–4	0–14**	4–5	0–7*
Secondary Branch	4–5	0–38**	5–6	0–24**
Main Shoot Length (cm)	50–55	13–123**	55–60	15–97**
Silique/Plant	130–140	6–1223**	140–150	7–562**
Silique Length (cm)	3.5–3.8	2.9–5.0*	3.5–4.0	2.1–6.2**
No. of Seed/Silique (g)	10–12	7–18**	12–13	2–16**
1000 Seed Weight (g)	3.5–4.0	2.9–6.2**	5.0–6.0	4.1–8.5**
Seed Yield/Plant (g)	4.5–5.0	0.45–45.59**	5.0–7.0	0.62–34.84**

*, **indicate significance at 5% and 1% based on *t*-test

1986; Das and Rahman, 1988; Zanewich *et al.*, 1991; Rai and Singh, 1993).

Profuse secondary branches

The maximum numbers of 38 and 24 secondary branches occurred, respectively, in a mutant of 'Shivani' and in a mutant of 'Pusa Bold' (Table 26.3), compared with 5–6 in the parents. Profuse secondary branching is a desirable yield attribute. Secondary branching in Indian mustard is one of the major yield-contributing traits and is directly correlated with number of siliques per plant and seed yield per plant. Mutants with more branches were reported by Chauhan and Kumar (1986) and Javed *et al.* (2003).

Main shoot length

A wide range of variation for main shoot length (MSL) ranging from 13 cm to 123 cm was observed in 'Shivani' mutants and in 'Pusa Bold' mutants, ranging from 15 cm to 97 cm. Mutants with a long main fruiting axis were also reported by Naz and Islam (1979) and Shah *et al.* (1990).

Siliques per plant

The number of siliques per plant ranged between 6 and 1223 in 'Shivani' mutants and 7 to 562 in 'Pusa Bold' mutants. The mutants bearing higher siliques were also coupled with most of the desirable yield-contributing characters leading towards high seed yield per plant (Table 26.3).

Silique length and seeds per silique

Silique length varied from 2.9 cm with seven seeds (S6-15-63) to 5.0 cm with 17 seeds (S3-15-17) as compared with 3.5–3.8 cm with 10–12 seeds in the parent 'Shivani'. Though the maximum of 18 seeds were found in a 4 cm silique of 'Shivani' (S1-15-3), seed size in this mutant was small. However, in 'Pusa Bold', the longest silique (6.2 cm) was observed in the mutant PB2-14-8 (Table 26.3).

Test weight (1000-seed weight)

A wide spectrum of variation for test weight (1000-seed weight) was observed in mutants of both the varieties. Test weight in 'Shivani' mutants varied from 2.9 g to 6.2 g as compared with 3.5–4.0 g in the parent. In 'Pusa Bold' mutants, the test weight ranged from 4.1 g to 8.5 g as against 5.0–6.0 g in the parent (Table 26.3).

Seed yield per plant

Yield is a complex character which is determined by several component traits and governed by polygenes with small additive effect. Direct selection on the basis of seed yield is not advisable. In the present investigation, selected mutants had better yield-contributing characters compared with their parents and therefore resulted in higher seed yield per plant. Selection of mutants with desirable yield-contributing characters, without compromising economic end product quality, is always advocated for the development of novel genotypes. The seed yield per plant under acidic soil conditions varied greatly, ranging from 0.46 to 45.49 g per plant in the mutants of 'Shivani' and 0.62 to 34.84 g per plant in the mutants of 'Pusa Bold' as compared with 4.5–5.0 g and 5.0–7.0 g in the parents 'Shivani' and 'Pusa Bold', respectively. The highest seed yield of 45.49 g per plant was observed in mutants of 'Shivani'. In 'Pusa Bold', the maximum seed yield for the mutants derived was 34.84 g per plant (Table 26.3).

Induced mutations have been successfully employed to isolate novel mutants with desirable economic traits such as plant height, number of siliques per plant, number of seeds per silique, seed weight, seed yield and oil content (Chauhan and Kumar, 1986; Rehman *et al.*, 1987; Mahla *et al.*, 1990, 1991; Robbelen, 1990; Shah *et al.*, 1990, 1998, 1999; Rehman, 1996; Javed *et al.*, 2003; Jambhulkar, 2007, 2015).

4 Conclusion

Induced mutagenesis, particularly through a combination of gamma-rays with EMS, was successful in creating variation in *Brassica juncea* for a wide spectrum of heritable variability for plant characteristics of economic and academic significance. The frequency of induced mutations was greatly increased in the combination of mutagens compared with gamma-rays alone. Out of six different doses of mutagens, a combination of 1000 Gy + 0.3% EMS for 'Shivani' and 1100 Gy + 0.3% EMS for 'Pusa Bold' was found to be most effective. Forty-nine mutants of 'Shivani' and 61 of 'Pusa Bold' were found to be true breeding in the M_3 generation for tolerance to soil acidity with improved yield-contributing traits including seed yield per plant. These mutants may be used straight away or in future breeding programmes for the development of improved genotypes of mustard suitable for acidic soil.

Acknowledgements

The authors acknowledge with gratitude The Board of Research on Nuclear Sciences (BRNS), BARC, DAE, Govt. of India, Mumbai for financial support and encouragement through a project sanctioned vide No 35/14/21/2014-BRNS/0401. This work is part of the project, funded by the BRNS, Department of Atomic Energy (DAE), Govt of India. We also express our sincere thanks to Birsa Agricultural University, Ranchi (Jharkhand), India, for providing all kinds of facilities, guidance and support to carry out the present research project. The authors are highly grateful to and express deep sense of gratitude to International Atomic Energy Agency (IAEA), Vienna, Austria, for acceptance of the paper and for financial support for participation in the FAO/IAEA International Symposium on Plant Mutation Breeding and Biotechnology scheduled during 27–31 August 2018 at IAEA, Vienna, Austria.

References

Agarwal, B.K. *et al.* (2012) *Soil Acidity and Nutrient Status in Central and North Eastern Plateau Zone (Zone IV) of Jharkhand and its Management for Crop Production*. Status Report for the Govt of Jharkhand.

Bhatia, C.R. *et al.* (1999) Oilseed cultivars development from induced mutations and mutation altering fatty acid composition. *Mutation Breeding Review* 11, 1–36.

Chauhan, Y.S. and Kumar, K. (1986) Gamma ray induced chocolate B seeded mutant in *Brassica campestris*. *Current Science* 55, 410.

Chopra, V.L. (2005) Mutagenesis: investigating the process and processing the outcome for crop improvement. *Current Science* 89, 353–359.

Das, M.L. and Rahman, A. (1988) Induced mutagenesis for the development of high yielding varieties in mustard. *Journal of Nuclear Agriculture and Biology* 17, 1–4.

Devlin, P.F. *et al.* (1992) Photophysiology of the elongated internode (ein) mutant of *Brassica rapa*: einmutant lacks a detectable phytochrome B-like polypeptide. *Plant Physiology* 100, 1442–1447.

Foy, C.D. (1996) Tolerance of barley cultivars to an acid aluminium toxic subsoil related to mineral element concentrations in their shoots. *Journal of Plant Nutrition* 19, 1361–1380.

Foy, C.D. *et al.* (1992) Tolerance of soybean germplasm to an acid tatum subsoil. *Journal of Plant Nutrition* 15, 527–547.

Foy, C.D. *et al.* (1993) Tolerance of Sorghum genotypes to an acid aluminium toxic tatum subsoil. *Journal of Plant Nutrition* 161, 97–127.

Gaul, H. (1958) Present aspect of induced mutations in plant breeding. *Euphytica* 7(3), 275–279.

Girija, M. *et al.* (2013) Gamma rays and EMS induced flower color and seed mutants in cowpea (*Vigna unguiculata*). *Advanced Applied Science Research* 4, 134–139.

Jambhulkar, S.J. (2007) Mutagenesis: generation and evaluation of induced mutations. In: Gupta, P.C. (ed.) *Advances in Botanical Research. Vol. 45: Rapeseed Breeding*. Academic Press, Elsevier, USA, pp. 418–428.

Jambhulkar, S.J. (2015) Induced mutagenesis and allele mining. In: Kumar, A., Banga, S.S., Meena, P.D. and Kumar, P.R. (eds) *Brassica Oilseeds Breeding and Management*. CAB International, Wallingford, UK, pp. 53–67.

Javed, M.A. *et al.* (2003) Development of high yielding mutants of *Brassica campestris* L. cv. Toria selection through gamma rays irradiation. *Asian Journal of Plant Sciences* 2, 192–195.

Mahla, S.V.S. *et al.* (1990) Effect of mutagen on yield and its component characters in mustard. *Haryana Agriculture University Journal Research* 20, 259–264.

Mahla, S.V.S. *et al.* (1991) Mutagen induced polygenic variability in some mustard (*Brassica juncea* L.) varieties and their hybrids. *Journal of Oilseed Research* 8, 173–177.

Maluszynski, M. *et al.* (1995) Application of in vivo and in vitro mutation techniques for crop improvement. *Euphytica* 85, 303–315.

Mondal, S. *et al.* (2007) Induction of genetic variability in a disease-resistant groundnut breeding line. *BARC New Letter* 285, 237–246.

Nayar, G.G. and George, K.P. (1969) X-ray induced early flowering, appressed pod mutant in *Brassica juncea* Coss. In: *Proceedings, Symposium on Radiation and Radiomimetic Substances in Mutation Breeding*, Sept. 26–29, Bombay, India, pp. 409–413.

Naz, R.M.A. and Islam, R.Z. (1979) Effect of irradiation of *Brassica trilocularis* (yellow sarson). *Pakistan Journal of Agricultural Research* 17, 87–93.

Rai, B. and Singh, D. (1993) A note on potential sources of dwarfing genes in Indian rapeseed (*Brassica campestris* L. Prain). *Indian Journal of Genetics* 53, 153–156.

Rehman, A. (1996) New mutant cultivar. *Mutation Breeding Newsletter* 42, 27.

Rehman, A. *et al.* (1987) Promising mutants in *Brassica campestris*. *Mutation Breeding Newsletter* 29, 14–15.

Robbelen, G. (1990) Mutation breeding for quality improvement – a case study for oilseed crops. *Mutation Breeding Review* 6, 1–43.

Sangsiri, C.W. *et al.* (2005) Gamma radiation induced mutations in mungbean. *Science Asia* 31, 251–255.

Shah, S.A. *et al.* (1990) Induction and selection of superior genetic variables of oilseed rape, *Brassica napus* L. *The Nucleus* 7, 37–40.

Shah, S.A. *et al.* (1998) Use of induced genetic variability for the improvement of oilseed rape (*Brassica napus* L.). In: *Proceedings of 2nd International Symposium New Genetical Approaches to Crop Improvement II.* Nuclear Institute of Agriculture, Tandojam, Pakistan, pp. 229–237.

Shah, S.A. *et al.* (1999) Evolution of high yielding and early flowering variety of rapeseed *(Brassica napus L.)* through in-vivo mutagenesis. In: *Proceedings of 3rd International Symposium New Genetical Approaches to Crop Improvement III.* Nuclear Institute of Agriculture, Tandojam, Pakistan, pp. 47–53.

Stringam, G.R. *et al.* (1974) Chemical and morphological characteristics associated with seed coat colour in rapeseed. In: *Proceedings of 4th International Rapeseed Congress*, Giessen, West Germany, pp. 99–108.

Subudhi, P.K., Mohapatra, B.K. and Sinha, S.K. (1991) use of pollen traits for early detection of induced micromutations in wheat. *Indian Journal of Genetics* 51(1), 107–111.

Woods, D.L. (1980) The association of yellow seed coat with other characters in mustard *Brassica juncea*. *Cruciferae Newsletter* 5, 23–24.

Xiao, D. (1982) Analysis of the correlation between seed coat colour and oil contents of *Brassica napus* L. *Acta Agronomica Sinica* 8, 24–27.

Yu, C.Y. *et al.* (2004) The discovery of novel flower color mutation in male sterile rapeseed (*Brassica napus* L.). *Yi Chuan* 26, 330–332.

Zanewich, K.P. *et al.* (1991) Dwarf mutant of Brassica: responses to applied gibberellins and gibberellin content. *Plant Growth Regulation* 10, 121–127.

Section 3.

Mutation Induction Techniques for Enhanced Genetic Variation

27 The Barley *chloroplast mutator* (*cpm*) Mutant, an Extraordinary Source of Plastome Variability

Alejandra Landau[1], Franco Lencina, María Elizabeth Petterson, María Gabriela Pacheco, Susana Costoya, Vanina Brizuela and Alberto Prina

Abstract

The plastome is usually considered a highly conserved genome. Compared with the nuclear genome, it is small and has different genetic rules. Through different molecular methods (TILLING, candidate gene sequencing, amplicon massive sequencing and plastome re-sequencing) applied to barley *chloroplast mutator* (*cpm*) seedlings, we detected more than 60 polymorphisms affecting a wide variety of plastid genes and several intergenic regions. The genes affected belonged mostly to the plastid genetic machinery and the photosynthetic apparatus, but there were also genes like *matK*, whose functions are so far not clearly established. Among the isolated mutants, we found the first *infA* gene mutant in higher plants, two mutants in *ycf3* locus and the first *psbA* gene mutant in barley. The latter is used in breeding barley cultivars where PSII is tolerant to toxic herbicides. Most of the molecular changes were substitutions, and small indels located in microsatellites. However, particular combinations of polymorphisms observed in the *rpl23* gene and pseudogene suggest that, besides an increased rate of mutations, an augmented rate of illegitimate recombination also occurred. Although a few substitutions were observed in the mitochondria of *cpm* plants, we have not yet determined the implications of the *cpm* for mitochondrial stability. The spectrum of plastome polymorphisms highly suggests that the *cpm* gene is involved in plastid DNA repair, more precisely taking part in the mismatch repair system. All results show that the *cpm* mutant is an extraordinary source of plastome variability for plant research and/or plant breeding. This mutant also provides an interesting experimental system in which to investigate the mechanisms responsible for maintaining plastid stability.

Keywords: barley chloroplast mutator • chloroplast TILLING • plastome polymorphisms • plastid DNA instability

1 Introduction

The chloroplast genome or plastome is usually considered as a highly conserved genome in most higher plants (Palmer, 1985, 1990; Wolfe *et al*., 1987; Clegg, 1993; Wicke *et al*., 2011). Its current variability is scarce and artificial induction of plastome mutants is far from being a routine practice, as compared with nuclear genes (Prina *et al*., 2012). Plastome genes have been little used in breeding and their functionalities have been difficult to investigate. There exist exceptions to the rule of the genetically conserved plastome, which are hypothesized to occur due to failures of DNA replication, recombination and repair (DNA-RRR) systems (Zhang *et al*., 2016). One striking example is the barley *chloroplast mutator* (*cpm*) mutant, first identified by the

[1] Email: landau.alejandra@inta.gob.ar

 Mutation Breeding, Genetic Diversity and Crop Adaptation to Climate Change (eds S. Sivasankar *et al*.)
DOI: 10.1079/9781789249095.0027

presence of a few chlorophyll-deficient longitudinal stripes on the leaves of M_4 and M_5 plants coming from a combined mutagenic treatment of X-rays and sodium azide (Prina, 1992). Prina (1992) analysed chlorophyll deficiencies (CDs) in progenies of reciprocal crosses and backcrosses and determined that the CDs were cytoplasmically transmitted and that their induction was controlled by a single nuclear mutator gene, when homozygous recessive. In addition, it was observed that once CDs were induced, they were expressed independently of the nuclear constitution. All these results suggested that the CDs were a consequence of a nuclear mutator gene mutant that produced plastome instability. This experimental material identified a gene that was called the barley *chloroplast mutator*. It was the first mutator genotype reported in monocots to induce a wide spectrum of cytoplasmically inherited CDs.

Chloroplasts, as well as mitochondria, are multi-copy DNA organelles. Depending on the species, the number of chloroplasts per cell ranges from a few tens to over 100 (Lopez-Juez and Pyke, 2005). Moreover, chloroplast genome copy number is very high in leaf tissue, with upwards of 10,000 or more copies of the chloroplast DNA per leaf cell (Morley and Nielsen, 2016).

When a plastid genome mutant arises, it is a single mutation to a single genome in a single plastid in a single cell. Many stochastic events lead to this mutation being fixed, lost, or continuing segregation (sorting out), which occurs more slowly and under more relaxed rules than those that regulate the segregation of nuclear gene mutants (Birky, 2001). This means that, after a mutational event, plastome mutant sectors are in the beginning restricted to a much smaller portion of the plastid genome than nuclear gene mutations. Plastid, cells and plants carrying plastome mutant sectors may carry a whole range of predominance of either wild-type or mutant genomes. Thus, a plant cell can carry more than one type of allele in its plastids, and these cells and the plants carrying them are called heteroplastomic or heteroplastidic (Prina *et al.*, 2012).

In this chapter, we summarize the most important results of the molecular analysis of the *cpm*, which we consider to be an extraordinary source of plastome variability.

2 Materials and Methods

2.1 Isolation and characterization of mutants

The isolation of *cpm* and its breeding behaviour were described in Prina (1992). The isolation, characterization and analyses of the *infA* gene mutants were described in Prina *et al.* (2003) and Landau *et al.* (2007, 2011). The characterization and analysis of the *ycf3* mutant were described in Landau *et al.* (2009). The isolation and molecular analyses of the *psbA* mutant were described in Rios *et al.* (2003).

2.2 Chloroplast TILLING

The chloroplast TILLING (Targeting Induced Local Lesions in Genomes) approach was described in Landau *et al.* (2016).

2.3 Massive amplicon sequencing of plastome of six cytoplasmic lines (CL)

Plant material

Genomic DNA was isolated from seedlings of six CLs according to Saghai-Maroof *et al.* (1984). Two of these mutants, CL1 and CL4 (Fig. 27.1), were described in Prina (1996). CL1 has a positional variegated pattern mainly evident on the first leaf when fully grown. It presents a green top and the green continues toward the bottom only in the surroundings of the midrib, showing diffuse borders between the albino and the green zones. CL4 is a virescent type (light-green young tissues that tend to be normal green when fully grown). At the second leaf stage it shows positional variegation of the *virido/xantha* pattern (Fig. 27.1B) (Prina, 1996); all of the leaves tend to be normal green in later stages. The CL4 mutant phenotype is manifest only by a transient light-green colour observed mainly on the bottom of growing leaves. CL5, CL7C and CL9 have phenotypes with some similarities to the former lines. CL13 has transversal bands of chlorotic tissue (Fig. 27.1D).

Fig. 27.1. Cytoplasmic lines (CLs), derived from the *cpm* mutant, whose plastomes were sequenced by massive amplicon sequencing (NGS). **(A)** Normal green-albino *CL1*. **(B)** Virescent *CL4, CL5, CL9*. **(C)** Virido-xantha *CL7C*. **(D)** Transversal bands *CL13*.

Long PCR

Primers were designed by Primer 3 software to obtain 14 amplicons of ca. 6–14 kb with an overlap of ca. 1 kb in order to amplify the complete chloroplast genome of barley (136 kb, according to the reference sequence GenBank NC_008590). LongAmp Taq DNA Polymerase (New England Biolabs) was used and the reactions were set up according to the manufacturer's protocol.

Sequencing

The mutants were sequenced by Illumina technology for next-generation sequencing. The polymorphisms were found performing de novo assembly with SPAdes and Mauve software.

Mapping against the reference sequence was also performed with CLC and Galaxy (Bowtie for Illumina y BWA for Illumina). The identification of variants was done with GATK software.

2.4 Mitochondrial TILLING

The plant material, DNA isolation and the TILLING protocol were the same as for cpTILLING, described in Landau *et al.* (2016). For the PCR, most primers were designed based on the mitochondrial genome of wheat (*Triticum aestivum*) (GenBank: AP_008982.1) because the mitochondrial genome sequence of barley was not yet available. The size of the amplicons was between 0.9 and 1.5 kb and comprised the genes *18S rRNA*, *26S rRNA*, *rps7*, *rps2*, *rps3*, *ccmFN*, *cob*, *cox1*, *nad2*, *nad5*, *nad7* and *matR*. The amplicons also included some intergenic regions located in the surroundings of those genes. The PCR reaction was performed as described in Landau *et al.* (2016).

3 Results

3.1 Forward genetics: sequencing of candidate genes based on phenotypes

Visual selection

Four mutants isolated among the selfed progeny of chloroplast mutator (*cpm*/*cpm*) homozygotes were described by Prina (1996). Due to their mode of inheritance, they were designated as cytoplasmic lines (CLs). The molecular analysis of two of them is presented here.

INFA MUTANTS. The seedlings of a CL called CL2 were described by Prina (1996) as carrying a curious positional pattern of variegation (Fig. 27.2) mainly expressed in the first two leaves. After performing several biochemical analyses and electronic microscopy observations, we concluded that the CL2 mutant had a delayed plastid protein synthesis during embryogenesis (Prina *et al.*, 2003). The candidate gene proposed as defective for this phenotype was the *infA* gene, which encodes IF1 protein (translation initiation factor 1). This gene was sequenced in CL2 and in wild-type barley and a missense mutation was found in the mutant (Table 27.1) (Landau *et al.*, 2007). Later, two other mutants with the same phenotype as CL2 (CL2-like 1 and CL2-like 2) were isolated from new *cpm*/*cpm* pools and we found two different missense mutations in *infA* gene, one in each seedling, both also transitions (see Table 27.1). In another experiment, the *cpm* mutator mutant was introduced by crosses to the nuclear genome of plants carrying the original CL2 plastome mutant. In those progenies we looked for reversions to the wild-type green phenotype, and one seedling carrying a normal green stripe on a CL2 phenotype in the first leaf was found (Fig. 27.2). The DNA was isolated from both kinds of tissues and

Fig. 27.2. *infA* gene mutants. **(A)** Wild-type barley (left) and *albo-viridis* CL2 (right). **(B)** CL2 seedlings carrying the *cpm*. The arrow points to a seedling with a normal green stripe that shows a phenotypic reversion.

the *infA* gene was sequenced. Interestingly, in the greener tissue it was found that the original substitution observed in CL2 seedlings was accompanied by another transition missense mutation, which was interpreted as having a compensatory effect of the original CL2 mutation, turning the *albo-viridis* tissue into homogeneous normal green. This was not a real genetic reversion, nor was it explicable as a re-emergence by sorting out of the wild-type plastome which had been at low copy number. It was a phenotypic reversion (Landau *et al.*, 2011) attributable to a compensatory effect of a suppressor mutation.

YCF3 MUTANT. Another *cpm*-induced mutant, called CL3, was described by Prina (1996) as a homogeneous *viridis* (light-green) type (Fig. 27.3). After several physiological and biochemical analyses, we concluded that CL3 seedlings were more sensitive to high temperatures than wild-type barley. We observed that the PSI was altered in CL3 seedlings but none of the PSI proteins encoded by the plastome seemed to be responsible for this syndrome. Thus, we proposed two chaperones involved in PSI assembly as candidate genes, *ycf3* and *ycf4*. Both genes were sequenced and we found two point mutations in intron 1 of *ycf3* gene (insertions T 150 + T 528 C). When analysed we found that intron 1 was unspliced at high temperature in CL3 (Fig. 27.3) and the chaperone was almost missing at this temperature. We concluded that the lack of YCF3 chaperone in the assembly of PSI at high temperature is the cause of CL3 phenotype (Landau *et al.*, 2009).

Table 27.1. *infA* gene mutants and the molecular changes found at the DNA level and the predicted changes at the protein level.

Mutant	DNA change	Amino acid change
CL2	T 157 C	Ser 52 Pro
CL2-like 1	T 97 C	Phe 32 Leu
CL2-like 2	A 185 G	Asp 61 Gly
Double mutant with compensatory effect	T 157 C + A 178 G	Ser 52 Pro + Met 59 Val

Selection with herbicides

PSBA MUTANT. The *psbA* mutant was isolated after a treatment with the PSII toxic herbicide atrazine. It was determined that this mutant has a missense mutation in *psbA* gene (A 790 G; Ser 264 Gly), which encodes D1 protein (Rios *et al.*, 2003).

3.2 Reverse genetics

Chloroplast TILLING

Since the forward genetics approaches were very laborious and time consuming in the identification of the mutated genes based on phenotypes, we decided to adapt the TILLING technique (which has been widely used for nuclear genes)

Fig. 27.3. WT and CL3 mutant barley seedlings 6 days after germination. **(A)** At 18°C. **(B)** At 32°C.

to the chloroplast genome. Instead of using mutagens for the generation of a mutagenized population, we used the *cpm* as a biological mutagen. We performed cpTILLING on two populations that were homozygous for the *cpm* nuclear gene mutation during different numbers of generations of self-pollination (Landau *et al.*, 2016).

At least 61 different mutational events were discovered; most of them were transitions and some indels (large or mononucleotide indels) (Table 27.2). Most of the indels were insertions and consisted of one or two nucleotides localized in microsatellites (mononucleotide repeats of 9–10 bases). In addition, four large indels (one insertion of 15 bp and three deletions of 45, 79 and 620 bp) were detected, all of them having direct repeat sequences flanking the inserted/deleted fragments which ranged from 7 to 25 bp. The largest deletion, of 620 nucleotides, was found in the coding sequence of the *psbA* gene (Landau *et al.*, 2016).

A peculiar situation was observed for the *rpl23* gene. This gene is located in both inverted repeats (IRs) of the chloroplast genome, i.e. there are two copies of *rpl23* per plastid genome and there is also a pseudogene in the large single copy (LSC). A high rate of molecular changes was found in the gene, all of them corresponding to five polymorphic differences between the gene and pseudogene that already exist in nature, suggesting that these molecular changes more probably arose from recombination than from mutational changes (for details see Landau *et al.*, 2016; Lencina *et al.* 2019).

Mitochondrial TILLING

Some of the chloroplast mutators described so far also affected mitochondria (Martinez-Zapater *et al.*, 1992; Sakamoto *et al.*, 1996; Kuzmin *et al.*, 2005). Therefore, we investigated the mitochondrial genome of *cpm* plants with a strategy similar to the cpTILLING approach. The mitochondrial genome of barley is almost four times bigger (around 500 kb) than the plastome but it has many fewer genes. The genes analysed were: *18S rRNA*, *26S rRNA*, *rps7*, *rps2*, *rps3*, *ccmFN*, *cob*, *cox1*, *nad2*, *nad5*, *nad7*, *atp1*, *nad4*, *nad7*, *nad9* and *matR*. Using this technique, we found 18 different mutational events, all heteroplasmic and most of them located in introns or intergenic regions (Table 27.3).

3.3 Massive amplicon sequencing of other CLs

In order to obtain more information about the plastome polymorphisms induced by *cpm*, we

Table 27.2. Type and number of polymorphisms found through cpTILLING in two groups of seedlings that carried *cpm* for different numbers of generations. Group A consists of two families that carried *cpm* during generations 12–17. Group B consists of four families that carried *cpm* for five generations.

	Group A (*n* = 182)	Group B (*n* = 122)	Total (*n* = 304)
Transitions	23	20	43
Transversions	1	2	3
Indels	11	4	15
No. of mutational events	35	26	61

Table 27.3. Type and number of polymorphisms found through mtTILLING in two groups of seedlings that carried the *cpm* during different numbers of generations. Group A consisted of two families that carried the *cpm* during 12–17 generations. Group B consisted of four families that carried the *cpm* for five generations.

	Group A (*n* = 182)	Group B (*n* = 122)	Total (*n* = 304)
Transitions	8	–	8
Transversions	1	–	1
Indels	5	4	9
No. of mutational events	14	4	18

conducted a massive amplicon sequencing in the other six cytoplasmic lines (CLs). Some relevant polymorphisms were found in more than one CL, like a T insertion in a microsatellite region of the *matK* gene, which was observed in CL4, CL5, CL7C and CL9, all of them coming from a common pool of *cpm/cpm* plants and therefore the insertion in *matK* could arise from the same mutational event. Besides, two point mutations which were observed in two different genes belonging to the ATPase complex were observed in CL13.

4 Discussion

The barley chloroplast mutator mutant *cpm* (Prina, 1992) induces a wide spectrum of cytoplasmically inherited chlorophyll-deficient (CD) mutants. In addition to the CD mutants, it was subsequently demonstrated that the *cpm/cpm* gene malfunction produces a wide range of molecular changes in the plastome, mostly point mutations, substitutions and small indels located in microsatellites (Landau *et al.*, 2007, 2009, 2011, 2016; Prina *et al.*, 2009, 2012). A peculiar case was observed for the *rpl23* gene in which five polymorphisms were detected in different combinations. Interestingly, they corresponded to the five differences that already exist between the gene and pseudogene in wild-type barley. In this way, it was hypothesized that these molecular changes were probably not due to mutation but were a result of recombination events involving the two copies of the *rpl23* gene and pseudogene. All the results from the different molecular analyses showed a spectrum of molecular changes that resembled the footprint of the malfunction of a gene involved in mismatch repair (Landau *et al.*, 2016). Some preliminary results obtained by TILLING analysis of the mitochondrion, which detected polymorphisms mostly located in introns and intergenic regions in *cpm* seedlings, suggested that the genetic stability of the mitochondrion could also be affected in *cpm* seedlings. If this is true, it could be concluded that the nuclear encoded CPM protein would be dually targeted to both the chloroplast and the mitochondrion. Nevertheless, from this analysis we cannot rule out the presence of large rearrangements in the mitochondrion of *cpm* seedlings, because a TILLING strategy is not the most suitable technique to detect them.

Among the most interesting plastome mutants isolated from *cpm* seedlings we would mention the following.

1. CL2 was the first *infA* gene mutant (encoding the translation initiation factor 1) described in higher plants (Prina *et al.*, 2003; Landau *et al.*, 2007). Until then, *infA* gene functions had only been deduced from experimental results in bacteria (Croitoru *et al.*, 2004, 2006). We could also isolate a second mutation in the *infA* gene, which had a compensatory effect on the first one, a fact that definitively demonstrated the relationship between the *infA* gene and the *albo-viridis* phenotype observed (Landau *et al.*, 2011).

2. CL3 was described by Prina (1996) as a homogeneous *viridis* (light-green) type. CL3 seedlings were more sensitive to high temperatures than wild-type and it was demonstrated that this sensitivity was caused by two point mutations found in intron 1 of *ycf3* gene, which inhibited the splicing of the intron 1 at high temperature and, consequently, the lack of YCF3 chaperone in the assembly of PSI (Landau *et al.*, 2009).

3. The *psbA* gene mutant was isolated after azide treatment of *cpm* seedlings. When the *psbA* gene was sequenced it was found that it carries a point mutation well known in other plant species, but this was the first barley mutant tolerant to PSII toxic herbicides (Rios *et al.*, 2003). This experimental material, after crosses and selection for agronomic characteristics, is ready to enter a barley breeding programme.

Through the use of next-generation sequencing technology, we obtained the sequences of the whole chloroplast genomes of six other CLs and detected several polymorphisms located in the plastomes. We found a polymorphism in *matK* gene, whose function has not yet been clearly established, in several of the CLs analysed that curiously presented similar phenotypes. This gene encodes a maturase that is involved in its own splicing and in other group II introns splicing. In CL4, CL5, CL7C and CL9 the maturase would be truncated because of the frame shift. Other interesting polymorphisms were found in two genes belonging to the ATPase complex, which were observed in CL13, and it remains to be investigated whether they could be the cause of the CL13 phenotype.

5 Conclusion

All these results show that the barley chloroplast mutator mutant (*cpm*) not only induces a wide spectrum of cytoplasmically inherited chlorophyll-deficient mutations, but also induces an enormous quantity of plastome polymorphisms mostly consisting of substitutions and small indels located in microsatellites. The particular combinations of five polymorphisms observed in the *rpl23* gene suggest that an increased rate of illegitimate recombination (or gene conversion) also occurred in *cpm* plants. This peculiar landscape of molecular changes strongly suggests that the *cpm* gene is involved in plastid DNA repair, maybe taking part in the plastome mismatch repair system.

Although a few polymorphisms were found in mitochondria, we cannot yet conclude firmly that *cpm* also affects mitochondria.

Among the isolated plastome mutants there are some novel ones: the first *infA* gene mutant reported in higher plants, others found in *matK* gene and in the ATPase complex mutant. The *psbA* gene mutant can also be useful in breeding commercial varieties by conferring a high level of tolerance to atrazine.

The spectrum of plastome polymorphisms observed in *cpm* seedlings, mostly corresponding to point mutations that widely affect different portions of the plastome, makes the *cpm* an extraordinary source of variability that otherwise is extremely scarce in nature and its artificial induction has so far been poorly handled. This outburst of plastome variability can be very useful for plant research and/or plant breeding.

References

Birky, C.W. Jr (2001) The inheritance of genes in mitochondria and chloroplasts: laws, mechanisms, and models. *Annual Review of Genetics* 35, 125–148.

Clegg, M.T. (1993) Chloroplast gene sequences and the study of plant evolution. *Proceedings National Academy of Sciences USA* 90, 363–367.

Croitoru, V. *et al.* (2004) Generation and characterization of functional mutants in the translation initiation factor IF1 of *Escherichia coli*. *European Journal of Biochemistry* 271, 534–544.

Croitoru, V. *et al.* (2006) RNA chaperone activity of translation initiation factor IF1. *Biochimie* 88, 1875–1882.

Kuzmin, E.V. *et al.* (2005) A mitochondrial mutator system in maize. *Plant Physiology* 137, 779–789.

Landau, A. *et al.* (2007) Two infA gene mutations independently originated from a mutator genotype in barley. *Journal of Heredity* 98, 272–276.

Landau, A. *et al.* (2009) A cytoplasmically inherited barley mutant is defective in photosystem I assembly due to a temperature-sensitive defect in ycf3 splicing. *Plant Physiology* 151, 1802–1811.

Landau, A. *et al.* (2011) A second infA plastid gene point mutation shows a compensatory effect on the expression of the cytoplasmic line 2 (CL2) syndrome in barley. *Journal of Heredity* 102, 633–639.

Landau, A. *et al.* (2016) Plastome mutations and recombination events in barley chloroplast mutator seedlings. *Journal of Heredity* 107(3), 266–273.

Lencina, F. *et al.* (2019) The rpl23 gene and pseudogene are hotspots of illegitimate recombination in barley chloroplast mutator seedlings. *Scientific Reports* 9(9960).

Lopez-Juez, E. and Pyke, K.A. (2005) Plastids unleashed: their development and their integration in plant development. *International Journal of Developmental Biology* 49, 557–577.

Martinez-Zapater, J.M. *et al.* (1992) Mutations at the Arabidopsis CHM locus promote rearrangements of the mitochondrial genome. *Plant Cell* 4, 889–899.

Morley, S. and Nielsen, B. (2016) Chloroplast DNA copy number changes during plant development in organelle DNA polymerase mutants. *Frontiers in Plant Science* 7, 57.

Palmer, J.D. (1985) Comparative organization of chloroplast genomes. *Annual Review of Genetics* 19, 325–354.

Palmer, J.D. (1990) Contrasting modes and tempos of genome evolution in land plant organelles. *Trends in Genetics* 6, 115–120.

Prina, A.R. (1992) A mutator nuclear gene inducing a wide spectrum of cytoplasmically inherited chlorophyll deficiences in barley. *Theoretical and Applied Genetics* 85, 245–251.

Prina, A.R. (1996) Mutator-induced cytoplasmic mutants in barley: genetic evidence of activation of a putative chloroplast transposon. *Journal of Heredity* 87, 385–389.

Prina, A.R. *et al.* (2003) A cytoplasmically inherited mutant controlling early chloroplast development in barley seedlings. *Theoretical and Applied Genetics* 107, 1410–1418.

Prina, A.R. *et al.* (2009) Genetically unstable mutants as novel sources of genetic variability: the chloroplast mutator genotype in barley as a tool for exploring the plastid genome. In: Shu, Q.Y. (ed.) *Induced Plant Mutations in the Genomics Era. FAO*, Rome, pp. 255–256.

Prina, A.R. *et al.* (2012) Mutation induction in cytoplasmic genomes. In: Shu, Q.S., Foster, B.P. and Nakagawa, H. (eds) *Plant Mutation Breeding and Biotechnology*. FAO/IAEA, Rome, pp. 201–206.

Rios, R.D. *et al.* (2003) Isolation and molecular characterization of atrazine tolerant barley mutants. *Theoretical and Applied Genetics* 106, 696–702.

Saghai-Maroof, M.A., Soliman, K.M., Jorgensen, R.A. and Allard, R.W. (1984) Ribosomal DNA spacer-length polymorphism in barley: Mendelian inheritance, chromosomal location, and population dynamics. *Proceedings National Academy of Sciences USA* 81, 8014–8019.

Sakamoto, W. *et al.* (1996) Altered mitochondrial gene expression in a maternal distorted leaf mutant of Arabidopsis induced by chloroplast mutator. *Plant Cell* 8, 1377–1390.

Wicke, S. *et al.* (2011) The evolution of the plastid chromosome in land plants: gene content, gene order, gene function. *Plant Molecular Biology* 76, 273–297.

Wolfe, K.H. *et al.* (1987) Rates of nucleotide substitution vary greatly among plant mitochondrial, chloroplast, and nuclear DNAs. *Proceedings National Academy of Sciences USA* 84, 9054–9058.

Zhang, J. *et al.* (2016) Coevolution between nuclear encoded DNA replication, recombination and repair genes and plastid genome complexity. *Genome Biology and Evolution* 8(3), 622–634.

28 Progress of Mutant Resource Development and TILLING on Starch Biosynthesis in Wheat

Huijun Guo[1], Yongdun Xie, Linshu Zhao, Hongchun Xiong, Jiayu Gu, Shirong Zhao and Luxiang Liu

Abstract

Induced mutations have been widely utilized for the development of plant mutant germplasm and varieties since 1927 and have contributed to genetic diversity enhancement and food security in the world. Mutant resources are essential for gene identification and functional characterization by forward and reverse genetic strategies. The publishing of annotated wheat reference genomes is greatly promoting the progress of wheat functional genomic research. Mutant resources of a broad spectrum and diversified wild- types will be the prerequisites in this process, in part due to the polyploid nature of wheat. This review describes the progress of mutant resource development derived from the winter wheat cultivar 'Jing411'. The segregating M_2 population has been used for mining functional mutant alleles of key genes involved in starch biosynthesis and could be further used for allele mining of any other target genes. The morphological mutant resources developed from various mutagens have been, and are going to be, used to develop genetic populations for gene mapping and the genetic analysis of biological functions.

Keywords: wheat • mutant resource • morphological mutant • TILLING • mutation allele

1 Introduction

The induced mutation technique induces mutation at higher frequency than spontaneous mutation and has been utilized to develop plant mutant germplasm and varieties for over 60 years. This method has played an important role in increasing genetic diversity and maintaining food security in the world. Several different mutagens have been used to induce mutations, including gamma-rays, X-rays, space mutagens, heavy-ion beams, chemical mutagens, etc. Targeting Induced Local Lesion in Genomes (TILLING) (McCallum *et al.*, 2000) is a reverse genetic approach that combines mutation induction with a high-throughput detection technique to discover mutations of target genes. After nearly 20 years, this approach has been used widely, for example to demonstrate the mutation characteristics of gamma-rays, which generate fewer point mutations and more knockout mutations than ethyl methanesulfonate (EMS) treatment (Sato *et al.*, 2006), to annotate gene functions (Zhai *et al.*, 2016) and to identify mutants for target traits (Slade *et al.*, 2012). The Cel1 enzyme digestion method to detect mutations, on which the procedure depends, has been refined for use with a variety of detection platforms, such as denaturing HPLC (Colasuonno *et al.*, 2016; McCallum *et al.*, 2000), high-resolution melting curve analysis (Botticella *et al.*, 2011; Ishikawa *et al.*, 2010), sequencing (Tsai *et al.*, 2011;

[1] Email: guohuijun@caas.cn

DOI: 10.1079/9781789249095.0028

Krasileva *et al.*, 2017), capillary electrophoresis (Suzuki *et al.*, 2008), agarose gels (Raghavan *et al.*, 2007; Dong *et al.*, 2009) and non-denaturing polyacrylamide gel electrophoresis (Uauy *et al.*, 2009).

Many mutated populations for the TILLING approach have been developed by the single seed descent (SSD) method in rice (Till *et al.*, 2007), maize (Till *et al.*, 2004), tomato (Minoia *et al.*, 2010) and other species. The genetic background of wheat is quite complicated, and wheat species may be diploid, tetraploid or hexaploid. Hexaploid wheat (bread wheat) is further classified into spring and winter wheat according to the growth habit, and TILLING populations based on some wild-types have been established (Dhaliwal *et al.*, 2015). At least four populations of tetraploid wheat have been developed, which have been used to mine novel alleles of *waxy* and *SBEII* genes (Slade *et al.*, 2005; Botticella *et al.*, 2011), involved in starch metabolism, and *ε-LCY* and *β-LCY* genes (Richaud *et al.*, 2018) in carotenoid metabolism. At least seven populations of spring bread wheat have been established and one of them has been used to uncover hidden mutations in the genome by the exon capture method (Krasileva *et al.*, 2017). However, progress in winter bread wheat was not as fast as in spring bread wheat and fewer mutated populations have been established.

Morphological mutants were essential for the development of gene discovery and gene functional research in model plants such as *Arabidopsis* and rice. *MONOCULM 1* (*MOC1*), controlling rice tillering, was cloned from a single tiller rice mutant (Li *et al.*, 2003). Meanwhile, many scientists have developed their own mutant resources which provide specific resources for their research. Bread wheat has a huge genome, which is about 40 and 140 times those of rice and *Arabidopsis*, respectively, and its gene function research is far behind both model plant species. The development of a large-capacity mutant resource that includes various morphological variations will provide the material not only for gene function research, but also for breeding and improvement. Several mutant resources have already been reported (Rakszegi *et al.*, 2010; Bovina *et al.*, 2014; Dhaliwal *et al.*, 2015), but the few mutants suggest that these resources will not be sufficient for a systematic approach to wheat functional genomics.

Bread wheat is a staple food in China. The planting area is more than 20 million hectares each year, of which the greatest area is for winter wheat. It is hence of importance to build a large-capacity morphological mutant library to enhance the basis for wheat genetic research in China. 'Jing411' is a winter wheat (*Triticum aestivum* L.) variety and has been planted in large areas for many years. It has a high yield and good adaptability potential and has been used as a parent in crosses from which several new varieties have been bred. Using 'Jing411' as the wild-type, an M_2 population for reverse genetic research was developed to mine functional mutant alleles of starch biosynthesis genes by TILLING; a large morphological mutant resource was developed for forward genetic research by irradiation with gamma-rays (200 and 250 Gy), heavy-ion beams (Guo *et al.*, 2007) and other mutagens (Guo *et al.*, 2008) through field and laboratory screening.

2 The Wheat TILLING Population and its Utilization in Mining Mutant Alleles of Starch Biosynthetic Genes

The seeds of 'Jing411' were treated with 0.5%, 1.0% and 1.5% EMS solution (sodium phosphate buffer, pH 7.0) for 4 or 6 h, and an M_2 population including more than 3000 individual plants was developed by SSD. In order to assess the mutation frequency and practicability for novel allele mining, the known gene *TaAGP.L-B1* was selected as the target to mine point mutations by TILLING. *TaAGP.L-B1* encodes the large subunit of adenosine diphosphate glucose phosphorylase (AGPase), the key enzyme in starch biosynthesis. Eighteen point mutations were identified from 1218 individual M_2 plants including both transitions (72%) and transversions (22%), and seven out of the 18 mutations were located in exon regions. The mutation density was up to 1/35 kb. The transition mutation in line E3-1-3 (C2531T) was predicted to be a functional allele. The *TaAGP.L-B1* gene expression levels at 12, 18 and 24 days after flowering were all significantly reduced in line E3-1-3 compared with the wild-type 'Jing411' and for two other mutant lines (E051-6 and E422) which were predicted to have no impact on gene function. Grain starch content was significantly decreased. The results of gene expression and grain starch content confirmed that the mutation in line E3-1-3 affected gene function of *TaAGP.L-B1*, and the M_2 population carried additional mutations which could be further used to TILL functional alleles of this and other target genes. The details were reported in Guo *et al.* (2017b).

SSIV (starch synthase IV) is one of the isoforms of soluble starch synthase, but knowledge of its function in wheat growth and development is limited. Mutant alleles of *TaSSIVb-D* were identified in this population. Fifty-four point mutations were identified from the M_2 resource by TILLING, including a nonsense mutation C2077T (Q269-Stop) in line E054-3 and five missense mutations which were predicated to be functional alleles. The *TaSSIVb-D* gene was expressed in leaves, and its expression levels at seeding, elongation and heading stages were most significantly reduced in the nonsense mutant and missense mutant (line E1137). Starch granule number in the chloroplast was significantly decreased in both mutant lines, E054-3 and E1137, which also differed for photosynthesis parameters and PSII efficiency (Guo *et al.*, 2017a). The results indicated that function of the gene *TaSSIVb-D* in wheat was identical with that of *OsSSIVb* in rice, which is involved in leaf starch granule biosynthesis. Potential functional point mutations of the genes *TaSSIVb-A* and *TaSSIVb-B* have been identified and double and triple mutants of *TaSSIVb* have been developed. Synergistic and/or interaction effects of the three homeologous genes will be analysed, and their genetic function in wheat leaf and grain will be further elucidated. The results furthermore confirmed that the M_2 resource carries ample mutations in different genes.

3 A Morphological Mutant Resource for Wheat

The ^{7}Li (30–50 Gy with 43 MeV initial energy) and ^{12}C (50–70 Gy with 80 MeV initial energy) heavy-ion beams, mixed partial field (which is used to simulate the cosmic rays in space environment; it is generated from E2 beam lines of LINAC of Beijing Electron Positron Collider and consists of $\pi^{\pm}$, π^0, $\mu^{\pm}$, $e^{\pm}$, γ, p and other particles) (Guo *et al.*, 2008), space mutagenesis through the satellites Shijian 8 (15 days in orbit with an orbital inclination of 63° at perigee altitude of 180 km, apogee altitude of 469 km) and Shijian 10 (operated 12 days in a circular orbit of 252 km from the ground with an orbital inclination of 42.9° and an average orbital period of 93 min), gamma-rays and EMS treatment were utilized to treat the seeds. Morphological variants from seedling stage until maturity were screened in M_2 segregating populations, including the above-mentioned M_2 resource for TILLING. A phenotypic mutant resource with broad mutation spectrum on various traits has been established, derived from wild-type 'Jing411', which contains a broad range of morphological mutants. The mutant phenotypes included alterations to plant architecture, leaf, spike, seed, root, quality, stress resistance and other traits.

Among 1387 plant architecture mutants, 737 were dwarf or semi-dwarf (61 carrying *RhtD1b* and four carrying *RhtB1b* allele) (Xiong *et al.*, 2016), 292 were taller than the parental plants (Fig. 28.1A), 143 had fewer tillers, 50 had more tillers than the parent (Fig. 28.1B), 75 had altered tiller angle (larger or smaller), 73 were affected in seedling growth habit and 17 had altered epicuticular wax. Leaf morphology was affected by 382 mutants, including 18 yellow-green leaf, 89 necrotic leaf, 49 flag leaf angle (larger or smaller), 19 flag leaf size and 207 leaf shape. Spike morphology was affected by 395 mutants, including 34 big spike, 256 compact spike, 97 spelta-like spike, four degenerate spike and four with altered awns. Seed morphology was affected by 740 mutants, including nine red seed, 64 with higher thousand-grain weight (TGW), 390 with lower TGW, 46 with higher grain hardness and 231 of altered grain shape. Quality traits were affected by 361 mutants, including 346 with altered protein content (higher or lower) and 15 with altered wet gluten content. Stress resistance was affected by 111 mutants, including 45 with improved powdery mildew resistance (immune, highly resistant or medium resistant) and 66 being salt tolerant. In addition, 161 mutants had altered root morphology, including 89 mutants with altered heading date (earlier or later), 15 with altered senescence and 802 with various other phenotypes.

Fig. 28.1. Mutants affecting plant height and tiller number. **(A)** Plant height. **(B)** Tiller number.

4 Utilization and Future Prospects of the Wheat Mutant Resource

TILLING has been used to mine alleles for many years and the mutation density ranged according to species and population as reported in the literature. Many mutations have been identified, but most of them were silent mutations or located in introns or non-conservative domains, which would not lead to gene functional variation. However, molecular biologists and breeders would prefer functional and favourable alleles. The M_2 resource described here carried functional mutations for the four target genes, implying that the resource could be used to mine functional mutations of other target genes and the mutants could be further used to elucidate gene function, which will promote functional genomics research in the post-genomics era for reverse genetics.

It has been proved that mutation irradiation can induce DNA mutations such as substitutions and deletions, which result in phenotypic variation. Meanwhile, the mutation characteristics of each mutagen are different. For example, gamma-rays are believed to induce more large deletions than substitutions; while EMS treatment induces more substitutions; and heavy-ion beams induce substitution, small insertion-deletions and large deletions (Fitzgerald *et al.*, 2015; Du *et al.*, 2017). In the morphological mutant resource, for a specific trait, mutants have been induced by different mutagens and they might carry different types of mutation, which would greatly benefit wheat functional genetics and breeding. Taking plant height as an example, the wild-type 'Jing411' carried a wild-type allele at *RhtD1a* and *RhtB1a*, as determined by KASP assays and genetic segregation in populations derived from different crosses. We found not only dwarf mutants as mutant alleles of *RhtD1b* and *RhtB1b* allele (Xiong *et al.*, 2016), but also unknown mutated dwarfing genes that were induced by different mutagens. Mapping of the unknown dwarfing gene is ongoing.

With the rapid development of sequencing technology, the cost has been decreased dramatically. Several different strategies to fine-map mutant genes have been developed based on sequencing, such as MutMap, MutRenSeq, MutChromSeq, Exome sequencing, etc., which make gene mapping much faster. Given that the annotated wheat reference genome is available from the International Wheat Genome Sequencing Consortium (IWGSC, 2018), the morphological mutant resource described here has provided abundant plant material which makes it possible to use MutMap and other mapping strategies for fine-mapping and mutant gene isolation in wheat, which would promote wheat functional gene research for both forward and reverse genetics.

Acknowledgement

This work is supported by the National Key Research and Development Program (2016YFD0 102100) and NSFC project (31771791) of China.

References

Botticella, E. *et al.* (2011) High resolution melting analysis for the detection of EMS induced mutations in wheat SBEIIa genes. *BMC Plant Biology* 11, 156.

Bovina, R. *et al.* (2014) Development of a TILLING resource in durum wheat for reverse- and forward-genetic analyses. *Crop and Pasture Science* 65, 112–124.

Colasuonno, P. *et al.* (2016) DHPLC technology for high-throughput detection of mutations in a durum wheat TILLING population. *BMC Genetics* 17, 43.

Dhaliwal, A.K. *et al.* (2015) An ethylmethane sulfonate mutant resource in pre-green revolution hexaploid wheat. *PLoS ONE* 10, e0145227.

Dong, C. *et al.* (2009) A modified TILLING method for wheat breeding. *The Plant Genome* 2, 39.

Du, Y. *et al.* (2017) Identification of substitutions and small insertion-deletions induced by carbon-ion beam irradiation in *Arabidopsis thaliana*. *Frontiers in Plant Science* 8, 1851.

Fitzgerald, T.L. *et al.* (2015) An assessment of heavy ion irradiation mutagenesis for reverse genetics in wheat (*Triticum aestivum* L.). *PLoS ONE* 10, e0117369.

Guo, H. *et al.* (2007) Biological effects of high energy ^{7}Li ion beams implantation on wheat. *Plant Mutation Reports* 1, 31–35.

Guo, H. *et al.* (2008) Mutagenic effects of mixed particle field irradiation on wheat. *Scientia Agricultura Sinica* 41, 654–660.

Guo, H. *et al.* (2017a) Novel mutant alleles of the starch synthesis gene *TaSSIVb-D* result in the reduction of starch granule number per chloroplast in wheat. *BMC Genomics* 18, 358.

Guo, H. *et al.* (2017b) Development of a high-efficient mutation resource with phenotypic variation in hexaploid winter wheat and identification of novel alleles in the *TaAGP.L-B1* gene. *Frontiers in Plant Science* 8, 1404.

Ishikawa, T. *et al.* (2010) High-resolution melting curve analysis for rapid detection of mutations in a Medaka TILLING library. *BMC Molecular Biology* 11, 70.

IWGSC (2018) Shifting the limits in wheat research and breeding using a fully annotated reference genome. *Science* 361, eaar7191.

Krasileva, K.V. *et al.* (2017) Uncovering hidden variation in polyploid wheat. *Proceedings National Academy of Sciences USA* 114, E913–E921.

Li, X.Y. *et al.* (2003) Control of tillering in rice. *Nature* 422, 618–621.

McCallum, C.M. *et al.* (2000) Targeting Induced Local Lesions IN Genomes (TILLING) for plant functional genomics. *Plant Physiology* 123, 439–442.

Minoia, S. *et al.* (2010) A new mutant genetic resource for tomato crop improvement by TILLING technology. *BMC Research Notes* 3, 69.

Raghavan, C. *et al.* (2007) Rapid method for detecting SNPs on agarose gels and its application in candidate gene mapping. *Molecular Breeding* 19, 87–101.

Rakszegi, M. *et al.* (2010) Diversity of agronomic and morphological traits in a mutant population of bread wheat studied in the Healthgrain program. *Euphytica* 174, 409–421.

Richaud, D. *et al.* (2018) Identification of Lycopene epsilon cyclase (LCYE) gene mutants to potentially increase beta-carotene content in durum wheat (*Triticum turgidum* L. ssp. *durum*) through TILLING. *PLoS ONE* 13, e0208948.

Sato, Y. *et al.* (2006) Mutant selection from progeny of gamma ray irradiated rice by DNA heteroduplex cleavage using brassica petiole extract. *Breeding Science* 56, 179–183.

Slade, A.J. *et al.* (2005) A reverse genetic, nontransgenic approach to wheat crop improvement by TILLING. *Nature Biotechnology* 23, 75–81.

Slade, A.J. *et al.* (2012) Development of high amylose wheat through TILLING. *BMC Plant Biology* 12, 69.

Suzuki, T. *et al.* (2008) MNU-induced mutant pools and high performance TILLING enable finding of any gene mutation in rice. *Molecular Genetics and Genomics* 279, 213–223.

Till, B.J. *et al.* (2004) Discovery of induced point mutations in maize genes by TILLING. *BMC Plant Biology* 4, 12.

Till, B.J. *et al.* (2007) Discovery of chemically induced mutations in rice by TILLING. *BMC Plant Biology* 7, 19.

Tsai, H. *et al.* (2011) Discovery of rare mutations in populations: TILLING by sequencing. *Plant Physiology* 156, 1257–1268.

Uauy, C. *et al.* (2009) A modified TILLING approach to detect induced mutations in tetraploid and hexaploid wheat. *BMC Plant Biology* 9, 115.

Xiong, H. *et al.* (2016) Enhancement of dwarf wheat germplasm with high-yield potential derived from induced mutagenesis. *Plant Genetic Resources* 16, 74–81.

Zhai, S. *et al.* (2016) Genetic analysis of phytoene synthase 1 (Psy1) gene function and regulation in common wheat. *BMC Plant Biology* 16, 228.

29 Gamma-rays in the Development of Rice Lines Tolerant to Aryloxyphenoxypropionate Herbicides

Alexander de Andrade[1], Augusto Tulmann-Neto, Fernando Adami Tcacenco, Rubens Marschalek, Adriana Pereira, Antonio Mendes de Oliveira Neto, Klaus Konrad Scheuermann, Ester Wickert and José Alberto Noldin

Abstract

The aryloxyphenoxypropionate (APP) herbicides are graminicides with excellent control of many grass weed species, including weedy rice (*Oryza sativa* L.). These herbicides block fatty acid biosynthesis by inhibition of the enzyme acetyl-CoA carboxylase (ACCase) and cause death of the plant. Through induced mutation of rice seeds with gamma-rays, rice lines resistant to APP have been developed. Plant dose–response assays confirmed resistance to the APP herbicides quizalofop-p-ethyl and haloxyfop-p-methyl. The carboxyl-transferase (CT) domain fragments of ACCase from the resistant line and the susceptible control were sequenced and compared. A point mutation was detected in the amino acid position 2027. Results indicated that resistance to APP herbicides is a consequence of an altered ACCase enzyme that confers resistance. APP-resistant rice provides an option to improve the efficiency of weed management in rice crops.

Keywords: ACCase inhibitors • acetyl-CoA carboxylase • quizalofop-p-ethyl • haloxyfop-p-methyl

1 Introduction

Induced mutation is used with great success by different breeding programmes for developing new cultivars. The method commonly used for promoting mutation in rice is seed treatment with ionizing radiation, such as gamma-rays (Tulmann-Neto *et al.*, 2011).

The Epagri rice breeding programme has been working since 1985 in collaboration with the Center of Nuclear Energy in Agriculture (CENA/USP (University of São Paulo)) for the development of new cultivars through mutation with gamma-rays. The partnership allowed the development of two mutant rice cultivars, SCS114 'Andosan' (Ishiy *et al.*, 2006) and SCS118 'Marques' (Schiocchet *et al.*, 2014). These mutant cultivars showed good resistance to lodging, high yield potential and long grains with good quality.

Different research groups have sought to improve the induced mutation technique as a tool to develop cultivars that show tolerance to herbicides. In paddy rice production systems, weedy rice (red rice) is the main limiting factor that reduces rice yield and can severely reduce rice production worldwide. The use of herbicide-resistant rice cultivars may improve weed control, reducing costs and labour associated with manual removal of red rice (roguing). Herbicide-resistant rice is particularly useful where rice is direct seeded.

[1] Email: alexanderandrade@epagri.sc.gov.br

 Mutation Breeding, Genetic Diversity and Crop Adaptation to Climate Change (eds S. Sivasankar *et al.*)
DOI: 10.1079/9781789249095.0029

Herbicides inhibiting acetyl-coenzyme A carboxylase (ACCase) are very effective to control grass weeds, including weedy rice in paddy rice production systems. ACCase inhibitors affect the enzyme by blocking fatty acid biosynthesis, causing plant death. This herbicide resistance in rice is conferred by a single point mutation resulting in an amino acid substitution of the carboxyl transferase domain of ACCase, as has been reported in many grass weed species (Wenger *et al.*, 2012; Kaundun, 2014; Li *et al.*, 2014; Andrade *et al.*, 2018). In this work, the development of mutant rice lines with resistance to ACCase-inhibiting herbicides developed through gamma-ray irradiation is reported.

2 Materials and Methods

A mutation breeding programme for rice was initiated in 2010 at Epagri to develop herbicide-tolerant lines. Epagri's commercial rice cultivar, 'Sabbore' (susceptible to ACCase-inhibiting herbicides), was selected to create the mutant population. The irradiation procedure was performed at CENA/USP, Piracicaba, São Paulo, with 250 Gy of gamma-rays from a cobalt-60 source, using 500 g of 'Sabbore' seeds. Thereafter, the seeds were sown in trays with sandy soil at Epagri, in Itajaí, Santa Catarina. Seedlings were individually transplanted to the field, at the two-to-three leaf stage, forming the M_1 population. Plants were harvested and eight to ten seeds per plant were selected to generate an M_2 population with approximately 200,000 M_2 plants. These plants were screened for herbicide resistance (Fig. 29.1).

The M_2 plants were grown under field conditions. Fifteen days after plant emergence at the three-to-four leaf stage (V_3–V_4), quizalofop-p-ethyl was applied at 75 g a.i./ha. About 15 days after herbicide application, surviving rice plants were transplanted to individual pots in a greenhouse. The progenies of these putative mutants (M_3 generation) were again tested under field conditions with application of quizalofop-p-ethyl at 75 g a.i./ha to confirm if the mutant lines were resistant to the herbicide.

Total genomic DNA was extracted from leaf tissues of the resistant (R) and susceptible (S) rice lines using the protocol described by Doyle and Doyle (1990). DNA quantity and quality were measured in a biophotometer (Eppendorf, Germany) and a work solution was prepared at a concentration of 20 ng DNA/ml. Thirty individual plants from each population were sequenced. Primers were designed to amplify regions in the CT domain known to be involved in sensitivity to ACCase herbicides. Eight sets of primers were generated through the NCBI/Primer-BLAST tool (Andrade *et al.*, 2018). The

Fig. 29.1. **(A)** Plants were screened for APP herbicide resistance under field conditions. **(B)** Effect of herbicide 15 days after application with quizalofop-p-ethyl, 75 g a.i./ha.

primers were designed based on the choroplastic ACCase sequences generated for *Oryza sativa* subsp. *japonica* cv. 'Nipponbare' (locus identifier LOC_ Os05g229401) deposited at the Rice Genome Annotation Project. The amplified DNA was further subjected to sequencing PCR with the BigDye (Applied Biosystems). All PCR reactions contained 40 ng of DNA, PCR buffer 1X, 1.2 mM of $MgCl_2$, 0.2 mM of each dNTP, 0.4 μM of each primer, 1.5 U of *Taq* DNA polymerase, in a final volume of 25 μl. The reactions were carried out in a Veriti Thermal Cycler (Applied Biosystems, USA) with the following programme: (i) an initial denaturation step for 5 min at 95°C; (ii) 35 cycles of 45 s at 95°C, 45 s at 62°C and 1 min at 72°C; and (iii) a final extension step of 7 min at 72°C. The amplified DNA was further subjected to PCR sequencing. The PCR reactions contained 2 μl of Big Dye, 3 μl of Save Money buffer, 0.2 μl of each primer (forward or reverse), 50 ng of DNA and sterile water to a final volume of 20 μl. PCR was performed on a Veriti Thermal Cycler with the following programme: initial denaturation for 5 min at 96°C, 35 cycles of 15 s at 96°C, 15 s at 62°C and 4 min at 60°C. The samples were purified with the BigDye XTerminator Kit (Applied Biosystems) following the manufacturer's protocol. The PCR products that resulted from the sequencing PCR with the forward and reverse primers (two replicates each, total of four sequences per accession) were sequenced in an ABI 3130 Genetic Analyzer (Applied Biosystems). The sequences obtained were analysed with the software MEGA 6.0.

3 Results

Fifteen days after herbicide application, six individual M_2 plants were selected (Fig. 29.2). Among the selected tolerant plants, four presented very high sterility and two had normal growth. Seeds from the two selected tolerant plants with normal growth were cultivated under field conditions for herbicide screening and seed propagation. Both progenies (M_3 generation) resulting from the M_2 selected plants were identified as tolerant to APP herbicides and named as resistant (R) mutants (Fig. 29.3). No injury symptoms were observed in any of them. No segregants were observed and a highly homogeneous population in agronomic and resistance traits was produced. No differences were identified in the evaluations between the

Fig. 29.2. The progenies of these putative mutants (M_3 generation) were again tested under field conditions with application of quizalofop-p-ethyl (75 g a.i./ha) to confirm if the mutant lines were resistant to the herbicide.

susceptible (S) and the two R mutants under field conditions. Both selected mutants presented similar growth patterns.

The full DNA sequence obtained from the two plants was aligned with the sequence deposited for the *Oryza sativa* subsp. *japonica* cv. 'Nipponbare' (LOC_Os05g22940.1) at the Rice Genome Annotation Project site and also with the sequence for the rice cultivar 'Sabbore' (S line) that was submitted to gamma irradiation. Sequencing of the carboxyl-transferase region ACCase gene revealed one single base pair change (transversion G > T) that was found in the coding region of the ACCase gene at position 2027. This change causes an amino acid change in the ACCase protein. In the *Oryza sativa* subsp. *japonica* cv. 'Nipponbare', as well as in the S line, used as the parental population for the irradiation treatments described earlier, the amino acid at this position is a tryptophan, whereas in the R line obtained in the present study the amino acid at the same position is a cysteine.

4 Discussion

The mutation technique has been successfully used in the Epagri rice breeding programme. Rice seeds have been treated with gamma-rays, sodium azide or ethyl methanesulfonate (EMS). Over time it has been observed that the use of different doses of several mutating agents allows researchers to obtain new populations of mutant plants continuously. This procedure is the basis for the success of the mutation breeding programme. Gamma-ray treatment has been used at 250–350 Gy. It was observed that higher levels applied to plants in the M_1 generation induced more physiological disorders such as reduction in plant height and sterility. The pre-germinated rice M_2 plants that were screened at a density greater than 60 plants/m^2 had a larger number of false positive plants. In the M_2 generation, several treatments showed a greater increase in genetic variability for traits like earliness, plant height, plant type and spikelet fertility.

The origin and mechanism of the herbicide resistance was revealed by sequencing the acetyl-CoA carboxylase gene. The results confirmed the hypothesis that a target-site mutation is the mechanism of resistance to APP herbicides in this Epagri R rice line (Fig. 29.3). The replacement of amino acids in specific positions of the ACCase protein may cause different degrees of sensitivity to herbicides because of difficulties in their access to and/or the fit within the enzyme active site (Osuna *et al.*, 2012).

Resistance can be conferred by several mechanisms, including reduced target-site sensitivity, target-site amplification, sequestration of the herbicide away from the target site, increased rate of herbicide detoxification, or decreased rate of herbicide activation. The last two are considered

Fig. 29.3. (A) Rice plants with FOP-tolerant mutation of ACCase gene were not affected by foliar spray of quizalofop-p-ethyl (75 g a.i./ha) herbicide. **(B)** Rice plants without the tolerant mutation were killed by the same herbicide spray.

to be metabolism-based resistance (Devine *et al.*, 1997). The target-site mutation occurs due to a single key-point mutation in the carboxyl transferase (CT) domain of ACCase. Studies with different species have shown that resistance to ACCase inhibitors is conferred by target-site mutations at ACCase codon positions 1781, 1999, 2027, 2041, 2078, 2088 and 2096 (Powles and Yu, 2010; Délye *et al.*, 2011; Papapanagiotou *et al.*, 2015). The level of resistance depends on the herbicides, recommended field rates, weed species, plant growth stages, specific amino acid changes and the number of gene copies and mutant ACCase alleles (Kaundun, 2014).

Herbicides that inhibit the enzyme ACCase will be an additional tool for management of weedy rice in commercial rice production. Potential advantages include effective control of weedy rice, which is one of the most important weeds in rice production, in rotation with the Clearfield® Production Systems (BASF), available and extensively used in 70–80% of the rice area in Brazil. Herbicide-resistant rice provides an option to improve the efficiency of weed management. In areas where weedy rice is a problem, the availability of ACCase herbicide resistance may result in a gain in rice production. This mutant has a different mechanism of herbicide resistance and will provide an option to control weedy rice that is resistant to imidazolinone herbicides. Previously, there were restrictions on the use of APP herbicides in rice due to the limited selectivity and the injury to rice plants. Any of the rice lines described here is suitable to be developed as a rice cultivar or hybrid for use in commercial rice production as part of a weed-control system. The planting of multiple herbicide-resistant rice in rotation has been suggested as one of the methods to prevent the selection and spread of herbicide-resistant gene(s) among weedy-rice populations (Beckie and Tardif, 2012).

5 Conclusion

Induced mutation using gamma-rays is suitable for the development of mutant rice lines with tolerance to APP herbicides.

References

Andrade, A. *et al.* (2018) Development of rice (*Oryza sativa*) lines resistant to aryloxyphenoxypropionate herbicides through induced mutation with gamma rays. *Plant Breeding* 137, 364–369.

Beckie, H. and Tardif, F.J. (2012) Herbicide cross resistance in weeds. *Crop Protection* 35, 15–28.

Délye, C. *et al.* (2011) 'Universal' PCR assays detecting mutations in acetylcoenzyme A carboxylase or acetolactate synthase that endow herbicide resistance in grass weeds. *Weed Research* 51, 353–362.

Devine, M.D. *et al.* (1997) Physiological, biochemical and molecular aspects of herbicide resistance based on altered target sites. In: Roe, R.M. *et al.* (eds) *Herbicide Activity: Toxicology, Biochemistry and Molecular Biology*. IOS, Amsterdam, pp. 159–185.

Doyle, J.J. and Doyle, J.L. (1990) Isolation of plant DNA from fresh tissue. *Focus* 12, 13–15.

Ishiy, T. *et al.* (2006) Rice mutant cultivar SCS114 Andosan. *Plant Mutation Reports* 1(2), 25.

Kaundun, S.S. (2014) Resistance to acetyl-CoA carboxylase-inhibiting herbicides. *Pest Management Science* 70(9), 1405–1417.

Li, L. *et al.* (2014) Target-site mechanism of ACCase-inhibitors resistance in American sloughgrass (*Beckmannia syzigachne* Steud.) from China. *Pesticide Biochemistry and Physiology* 110, 57–62.

Osuna, M.D. *et al.* (2012) Resistance to ACCase inhibitors in *Eleusine indica* from Brazil involves a target site mutation. *Planta Daninha* 30, 675–681.

Papapanagiotou, A.P. *et al.* (2015) ACCase mutations in *Avena sterilis* populations and their impact on plant fitness. *Pesticide Biochemistry and Physiology* 123, 40–48.

Powles, S.B. and Yu, Q. (2010) Evolution in action: plants resistant to herbicides. *Annual Review of Plant Biology* 61, 317–347.

Schiocchet, M.A. *et al.* (2014) SCS118 Marques – New rice cultivar obtained through induced mutation. *Crop Breeding and Applied Biotechnology* 14, 68–70.

Tulmann-Neto, A. *et al.* (2011) Genetic improvement of crops by mutation techniques in Brazil. *Plant Mutation Reports* 2, 24–37.

Wenger, J. *et al.* (2012) Acetyl-CoA carboxylase inhibitors. In: Krämer, W. *et al.* (eds) *Modern Crop Protection Compounds*. Wiley-VCH, Weinheim, pp. 447–477.

30 National Repository of EMS Induced Mutants of an Upland Rice Cultivar Nagina 22: Progress Update on Characterization and Utilization

Amitha Mithra V. Sevanthi[1], Prashant Kale, Chandra Prakash, M.K. Ramkumar, Neera Yadav, V. Sureshkumar, Yugandhar Poli, Nitasha Grover, Motilal Behera, Ansuman Khandual, P. Jeyaparakash, Meera K. Kar, Muthurajan Raveendran, S. Manonmani, S. Gopala Krishnan, Satendra K. Mangrauthia, N. Sarla, M.S. Sheshshayee, Trilochan Mohapatra, Ashok K. Singh, Nagendra K. Singh and Rameshwar P. Sharma

Abstract

The Indian initiative for creating mutant resources in rice has generated 87,000 mutants in the background of a popular drought- and heat-tolerant upland cultivar, Nagina 22 (N22), through EMS mutagenesis. So far, 541 macro-mutants from this resource have been identified, maintained in the mutant garden and characterized in detail based on 44 descriptors pertaining to distinctness, uniformity and stability (DUS) of rice and other agronomic parameters. The similarity index of the mutants was more than 0.6 for nearly 90% of the mutants with respect to DUS descriptors, further establishing the validity of the mutants. The available high-quality sequence resource of N22 has been improved by reducing the gaps by 0.02% in the coding sequence (CDS) region. This was made possible using the newly synthesized whole-genome data of N22 which helped to remove 9006 'Ns' and replace 12,746 existing nucleotides with the accurate ones. These sequence and morphological details have been updated in the mutant database 'EMSgardeN22'. Further, 1058 mutants have been identified for low-P tolerance, tolerance to sheath blight, blast, drought, heat, higher photosynthetic efficiency and agronomic and root traits from this resource. A novel herbicide-tolerant (imazethapyr) mutant earlier identified and characterized from this resource is now being used in introgressing the herbicide-tolerant trait in eight major rice varieties in India. Further, robust and simpler screening systems have been tested for studying low-P tolerance of the mutants. A grain-size mutant, heat-tolerant mutant, drought-tolerant mutant, stay-green mutant and low-P tolerant and water-use efficient high-root-volume mutants have been characterized at morphological and molecular levels. A brief account of all these mutants, the entire mutant resource and the elaborate trait-based screenings is presented in this chapter.

Keywords: EMS induced mutant resource • Nagina 22 • mutant database • rice • herbicide tolerance

[1] Email: amithamithra.nrcpb@gmail.com

DOI: 10.1079/9781789249095.0030

1 Introduction

Ethyl methanesulfonate (EMS) mutagenized populations are amenable to TILLING, a popular reverse genetic approach, owing to their near-complete nature of genome saturation (McCallum *et al.*, 2000; Till *et al.*, 2007). The long and cumbersome nature of the TILLING workflow has been made easier by approaches based on next-generation sequencing (Tsai *et al.*, 2011; Guo *et al.*, 2014). The amenability of EMS mutagenized populations to forward genetics has been well established in the current decade with the multiple mapping strategies based on bulked segregant analysis (BSA) and high-throughput short-read sequencing (Abe *et al.*, 2012; Fekih *et al.*, 2013; Takagi *et al.*, 2013; Zou *et al.*, 2017; Jiao *et al.*, 2018). Further, these populations have the advantage of being free from restrictions that apply to transgenic materials in countries like India and the members of the European Union. Hence, EMS mutagenized populations are still in vogue in major crop species such as rice, wheat, sorghum, soybean and canola and are being extensively used in basic and applied aspects of plant sciences (Abe *et al.*, 2012; Mohapatra *et al.*, 2014; Dhaliwal *et al.*, 2015; Wang *et al.*, 2016; Shoba *et al.*, 2017; Song *et al.*, 2017; Jiao *et al.*, 2018).

India is the second largest producer of rice (*Oryza sativa* L.) next to China and the third largest exporter next to Thailand and Vietnam. Nearly 95% of the rice produced in India is consumed within the country, making it the second most consumed food grain (Singha, 2013). Hence rice improvement has remained one of the flagship research fields in the country. India participated in the international rice genome sequencing project, contributing to sequencing of the long arm of chromosome 11 (IRGSP, 2005). Subsequently, functional genomics in rice was initiated with the development of many mapping populations followed by identification of many major quantitative trait loci (QTLs) and genes for important traits such as drought tolerance, salinity tolerance, heat tolerance, resistance to various biotic stresses such as sheath blight and bakane disease, yield traits such as grain number and grain size, and quality traits such as aroma (Swamy and Sarla, 2008; Deshmukh *et al.*, 2010; Pandit *et al.*, 2010; Shanmugavadivel *et al.*, 2013, 2017; Pachauri *et al.*, 2014; Yadav *et al.*, 2015; Fiyaz *et al.*, 2016; Prakash *et al.*, 2016; Richa *et al.*, 2016; Verma *et al.*, 2017). Mutant resources are the important components of functional genomics, and in rice alone nearly 25 such resources are available (Sevanthi *et al.*, 2018). A network of research institutions generated the mutant resources 'National Repository of EMS induced Mutants in rice in the background of an upland rice cultivar Nagina 22' in India (Mohapatra *et al.*, 2014; Mithra *et al.*, 2016). An account of this mutant resource has been recently elaborated by Sevanthi *et al.* (2018). Further improvements made in this resource, including the Nagina 22 (N22) sequencing information, stability analysis and the extensive screening done for various traits and identification of mutants, characterization of the mutants at morphological, genetic and molecular level as well as their utilization in breeding, are presented in this chapter.

2 Materials and Methods

2.1 Plant materials and phenotyping

Mutant generation and screening for various traits are as detailed in Mohapatra *et al.* (2014) and Sevanthi *et al.* (2018). In brief, for maintenance and rejuvenation of the entire set of 87,000 mutants, they were grown in research farms by all the partner institutes in India. For morphological descriptors, 541 mutants maintained in the mutant garden were raised at two locations in India, (New Delhi and Cuttack, Odisha) during kharif 2017 (June–September). However, at the New Delhi location, the crop lodged at the reproductive stage and hence data pertaining to that period could not be recorded (Sevanthi *et al.*, 2018). Subsequently, in the kharif season of 2018 a total of 25 post-anthesis DUS descriptors, except for decorticated grain characteristics, in the 541 mutants grown in the mutant garden were recorded according to Shobha Rani *et al.* (2004) and are included in this chapter. Stability analysis for each of those traits as well as the complete set of mutants was done as described in Sevanthi *et al.* (2018).

2.2 Improving genic region sequencing information of N22

N22 and four of its mutants were subjected to whole-genome sequencing (WGS) using Illumina sequencing chemistry (both HiSeq 1000 and MiSeq) and assembled to obtain the sequencing information of the 66,638 genes using the reference of 'Nipponbare', pseudomolecule V 7.0 (Sevanthi *et al.*, 2018). Additional WGS data of N22 was generated employing Illumina HiSeq 1000 with paired-end sequencing technology and used to improve the coding sequence (CDS) coverage. Library construction was done using TrueSeq DNA PCR-Free LT kit (Illumina, Singapore). The average insert size was 100 bp. The good-quality reads were mapped to CDS sequences (final consensus sequences as reported in Sevanthi *et al.*, 2018). Alignment file format processing was done using SAMtools (Li *et al.*, 2009). VarScan v. 2.4.3 (Koboldt *et al.*, 2012) was used for variant calling. Improved CDS consensus sequences were generated using bcftools available under the SAMtools suite of programs (Li, 2011). The final nucleotide compositions in the older and the current version were counted using a python script.

2.3 Screening the mutant resources

The mutant resources were screened for multiple biotic and abiotic stresses, low-P and -N tolerance, herbicide tolerance, and for various physiological and agronomic traits according to standard screening protocols. The references for the same are provided in Table 30.1.

2.4 Characterization and utilization of the identified mutant resources

A brief account of the mutants so far characterized and published is provided below. The utilization of the herbicide-tolerant mutant and other promising mutants in varietal development and evaluation programmes is also provided in brief.

3 Results

3.1 Characterization of mutant garden

Comparison of the mutants based on the similarity index calculated over the 25 post-anthesis traits revealed that only 68 of the 541 mutants (12.6%) had an index lower than 0.6, while 323 mutants (60%) had a score > 0.7 and 243 mutants (45%) showed more than 0.8 (Fig. 30.1A). This indicated high similarity of the mutants to the N22, establishing their identity as mutants of N22. Among the 25 post-anthesis DUS characteristics recorded, seven traits, most of them related to awns (stem length excluding panicle, presence of awns in panicle, colour of awns, length of the longest awn, distribution of the awns, attitude of panicle branches and sterile lemma colour), showed very low similarity index (< 0.5) while five traits (anthocyanin colouration of lemma apex, attitude of leaf blade, curvature of main axis of the panicle, time to maturity and grain width) showed moderate similarity index (> 0.5 to < 0.70) compared with the N22 (Fig. 30.1B). The trait expression values for the rest of the 13 traits were very similar to N22 in most of the mutants. When the DUS data obtained from this location in the previous season (Sevanthi *et al.*, 2018) as well as the current season were compared with those from National Rice Research Institute, 36 DUS traits were common. Comparisons based on these 36 traits revealed that 18 of the traits behaved similarly across both locations, while others behaved very differently (Fig. 30.1C). Among the pre-anthesis traits, only six out of 16 traits (37.5%) were highly variable between the two environments, while in the post-anthesis traits 12 out of 20 traits (60%) were variable between the locations.

3.2 Improving CDS details of N22

The details of CDS sequence generation and processing are elaborated in Sevanthi *et al.* (2018). Further N22 sequences were generated to the tune of 16.3× coverage with the total number of reads being 69,765,084. After quality check, a total of 64,025,944 reads were retained for further mapping and variant calling. As this data is from N22 itself, but not the mutants, it was directly mapped on the previously assembled N22 reference genome (Sevanthi *et al.*, 2018). This exercise could not only fill the gaps of 9006 nucleotides and improve the coverage by 0.01% but also changed the nucleotide composition (Fig. 30.2). This makes sense, as

Table 30.1. Details of the EMS induced mutant resources subjected to screening for various traits and the number of promising mutants obtained.

Trait category	Trait	Number of mutants screened	Number of promising mutants identified	Screening procedure
Abiotic stresses	Drought tolerance including alternate wet and dry conditions (based on grain yield or survival in PEG 6000)	2,680	36	Lima *et al.* (2015) and IRRI standard evaluation procedure
	Salinity tolerance (based on survival)	3,000	2	IRRI standard evaluation procedure
	Heat tolerance (based on grain yield)	15	4	Panigrahy *et al.* (2011) and Poli *et al.* (2013)
Biotic stresses	Sheath blight tolerance	2,570	5	Bashyal *et al.* (2017)
	Nematode resistance	3	1	–
	Blast (mutants with no spots or disease lesion score 1)	100,000	60	Growing at two different disease hot-spot locations, Hazaribagh, Odisha, and Cunnoor, Tamil Nadu
Plant nutrition	Low-P tolerance (Tiller number and grain yield)	4,000	144	Mutants were screened in low-P (Olsen P 1.8 kg/ha) and normal condition (60 kg/ha; Olsen P, 24 kg/ha) Panigrahy *et al.* (2014) and Yugandhar *et al.* (2017a)
	Low-N tolerance (Tiller number and grain yield)	20	6	–
Herbicide tolerance	Imazethapyr tolerance	100,000	1	Shoba *et al.* (2017)
	Glufosinate tolerance	100,000	0	2% Basta
Physiological traits	Photosynthesis	1,070	12	Photosynthetic rate and prior selection based on high biomass and yield
	Water use efficiency (using $\Delta^{13}C$ as surrogate)	680	60	Reddy *et al.* (2018)
	Epicuticular wax	1,500	3	Sevanthi *et al.* (2018)
	Stay-green trait	20,000	20	Visual examination and dark-induced senescence assay (Talla *et al.*, 2016)
Morphological traits	Biomass	23,250	60	Visual examination
	Short root	1,000	2	Hydroponics
	Plant height	23,250	18 tall and 509 semi-dwarf and dwarf mutants	Visual examination
	High yield	3,250	44	Single plant yield
	High tillering	8,250	10	Visual examination
	Grain type mutants	5,500	20	Visual examination
	Panicle architecture	3,500	30	Visual examination
	Strong culm mutants	518	11	Visual examination

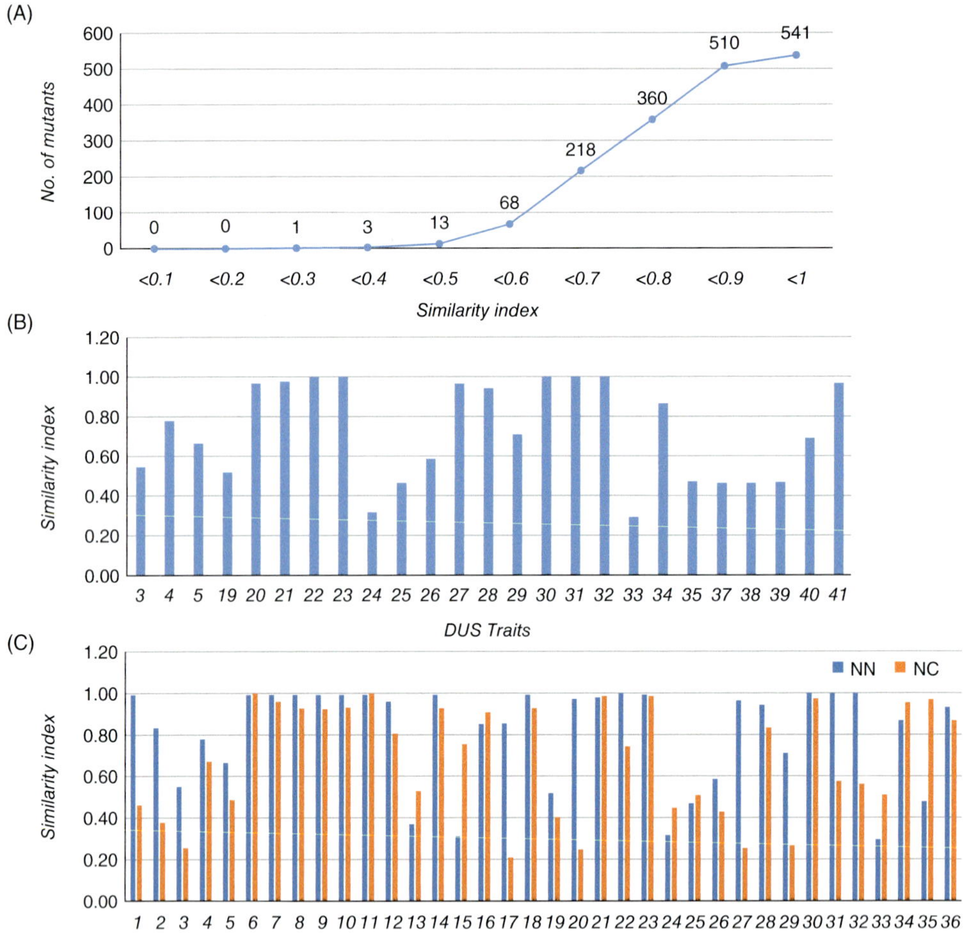

Fig. 30.1. Mutant and trait-wise similarity indices of the 541 mutants of the mutant garden. **(A)** Distribution of similarity indices of the 541 mutants based on 25 post-anthesis traits at New Delhi location; 88% of the mutants had similarity index more than 0.6. **(B)** Trait-wise similarity index for the 25 post-anthesis traits at New Delhi location. **(C)** Comparison of similarity indices between New Delhi (NN) and Cuttack (NC) locations for 36 traits (Data from Sevanthi *et al.* (2018) has also been considered for this comparison). DUS trait codes: 1-Basal leaf: sheath colour, 2-Culm: attitude, 3-Flag leaf: attitude of blade (late observation), 4-Grain: length, 5-Grain: width, 6-Leaf: anthocyanin colouration, 7-Leaf sheath: anthocyanin colouration, 8-Leaf: anthocyanin colouration of auricles, 9-Leaf: auricles, 10-Leaf: collar, 11-Leaf: colour of ligule, 12-Leaf: intensity of green colour, 13-Leaf: length of blade, 14-Leaf: ligule, 15-Leaf: Pubescence of blade surface, 16-Leaf: shape of ligule, 17-Leaf: width of blade, 18-Leaf: anthocyanin colouration of collar, 19-Lemma: anthocyanin colouration of apex, 20-Lemma: anthocyanin colouration of area below apex, 21-Lemma: anthocyanin colouration of keel , 22-Male sterility, 23-Panicle: presence of secondary branching, 24-Panicle: attitude of branches, 25-Panicle: awns, 26-Panicle: curvature of main axis, 27-Panicle: exertion, 28-Panicle: secondary branching, 29-Spikelet: colour of tip of lemma, 30-Stem: anthocyanin colouration of internodes, 31-Stem: anthocyanin colouration of nodes, 32-Stem: intensity of anthocyanin colouration of nodes, 33-Stem: length (excluding panicle; excluding floating rice), 34-Stem: thickness, 35-Sterile lemma: colour, 36-Time of heading (50% of plants with panicles)(days), 37-Panicle: colour of awns (late observation), 38-Panicle: distribution of awns, 39-Panicle: length of longest awn, 40-Time: maturity (days), 41-Lemma and palea: colour.

	'A's	'C's	'G's	'T's	Bases replaced	Gap filled (N)	Total length of the CDS
Old assembly	26609536	26154187	28257312	25538072	106559107	6770704	113329811
New assembly	26607666	26154645	28262544	25543258	106568113	6761698	113329811

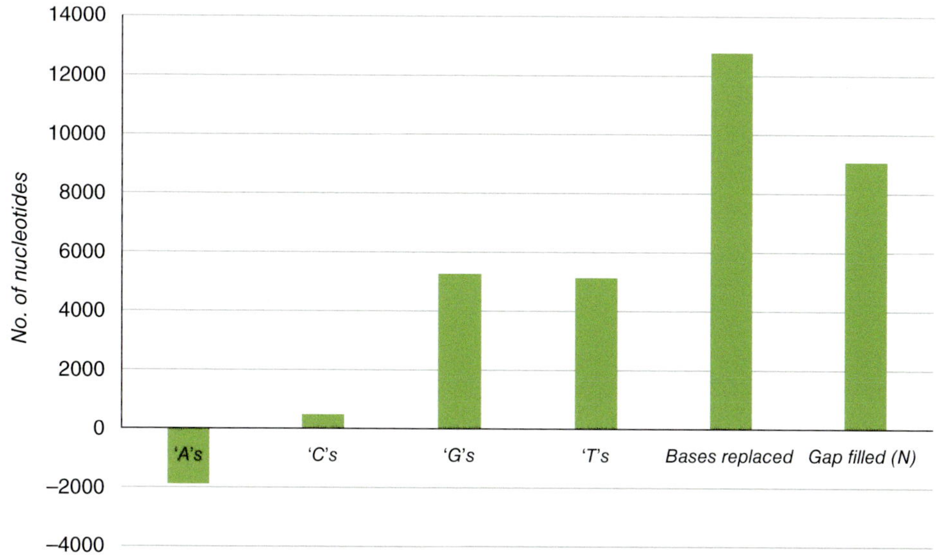

Fig. 30.2. Improvement of the CDS assembly of the Nagina 22 sequences over that of Sevanthi *et al.* (2018). The improvement was made possible by generating whole genome sequencing data of the WT at 16.3x depth using Illumina platform.

in the earlier version, the contributions of the mutant reads were much higher than N22. A total of 12,746 nucleotides were replaced with different bases (Fig. 30.2). Thus the total improvement of the N22 CDS sequences accounted for 21,752 nucleotides representing 0.02% of the CDS regions. This exercise increased the number of genes with complete coverage (no 'N's) by 60.

3.3 Screening of the mutant resources for specific traits

The entire or partial resource has been screened for herbicide tolerance, low-P tolerance, biotic stresses (rice blast and sheath blight), abiotic stresses (such as drought, heat and salinity), physiological traits (such as water use efficiency (with $\Delta^{13}C$ as surrogate), epicuticular wax content and photosynthetic rate) and morphological traits (such as plant height, root architecture, biomass, stay-green trait and single plant yield). The details of screening, reference for screening procedure and the number of promising mutants obtained are given in Table 30.1.

3.4 Characterization of the mutants for specific traits

Though the mutant resources were screened for a broad range of traits (Table 30.1), only a few of them were taken up for detailed characterization. A brief account of the mutants characterized is given here. The details of the same are published elsewhere.

Abiotic stresses

A set of 1100 mutants were subjected to drought stress at 30 DAS by withholding water at the National Institute for Plant Biotechnology, New Delhi, and the promising mutants were tested in both 25% (w/v) of 6000 PEG and in pot culture. The mutants performing better in both the conditions were short-listed as drought-tolerant mutants (Lima *et al.*, 2015). One of these drought-tolerant mutants, *ewst1*, was found to be similar to N22 except for chalky endosperm, decorticated grain colour and grain weight. This mutant was characterized and the details that follow were described by Lima *et al.* (2015). Mutant *ewst1* exhibited maximum root length and better performance for relative water content, cell membrane stability and chlorophyll concentration under drought stress than the wild-type (WT). It also had a higher number of xylem and phloem cells, variable and smaller size of central meta-xylem cells and a higher number of closed stomata under drought stress. Comparative genome-wide microarrays for the mutant and the WT revealed changes in gene expression for genes involved in exocytosis, tryptophan biosynthesis, protein phosphorylation and signalling. Enrichment of the upstream *cis*-acting regulatory elements of the differentially expressed genes (DEGs) revealed that most of the genes were found to be enriched for those elements that were regulators of the α-amylase genes in rice (Lima *et al.*, 2015).

The WT (N22) has pale green leaves and low chlorophyll content even when grown under ambient growth conditions (Sevanthi *et al.*, 2018; Sinha *et al.*, 2018). Five mutants which exhibited better drought tolerance as well as darker green leaves than the WT, both at vegetative and maturity stages, were selected and tested for their stay-green phenotype using dark-induced senescence assay (Panigrahy *et al.*, 2011). They were found to be functional stay-green types and they performed well under heat stress. Under heat stress (40°C), all four mutants were found to accumulate comparatively lower amounts of antioxidants (Panigrahy *et al.*, 2011). The best-performing one among these five mutants, NH219, was found to have a better yield compared with the WT even under 44°C (Poli *et al.*, 2013). Genetic analysis of this mutant with a heat-sensitive genotype, IR64, using 70 F_2 segregants, revealed a marker, RM 229 on chromosome 11, to be linked with higher yield under high temperature, while delayed senescence was associated with a marker on chromosome 2 (Poli *et al.*, 2013).

Plant nutrition

A set of 4000 N22 mutants were screened under field conditions both under normal P (60 kg/ha; Olsen P indicator 24 kg/ha) and low P in a field facility (no P added for 38 years, Olsen P 1.8 kg/ha) available at the Indian Institute of Rice Research, Hyderabad (Fig. 30.3) (Panigrahy *et al.*, 2014; Yugandhar *et al.*, 2017a, 2018a). Out of these 4000 mutants, 144 were found to be either low-P tolerant (lpt) or low-P sensitive (lps) as compared with the WT (N22), based on their grain yield and tiller number under low-P conditions. Panigrahy *et al.* (2014) identified three mutants (NH776, NH710 and NH719) as lpt mutants which gave higher grain yield in low-P conditions than other mutants and N22. A positive correlation was observed between activity of acid phosphatases (ACPs) measured during flowering stage and grain yield in a low-P field. The lpt mutants had higher ACP activity and P content than other mutants and N22. Further, a set of 67 mutants were screened in low-P, AWD (alternate wetting and drying) and normal conditions. Based on phenotypic and genotypic data, five mutants (NH686, NH787, NH363, NH669 and NH355) were selected as lpt mutants for low-P conditions and four markers (RM19696, RM263, RM3688 and RM1942) showed association with grain yield in all three conditions (Yugandhar *et al.*, 2017b). A lps mutant, NH101, was characterized for various morpho-agronomic, biochemical and molecular traits in both –P and +P conditions in pot soil experiments. Seminal and adventitious roots, panicle length and unfilled spikelet/panicle, antioxidant enzymes (SOD, POD and APX), and the relative expression levels of the genes involved in the maintenance of inorganic phosphate (Pi) homeostasis (MPH), i.e. OsPHR2, SPX1/2 and OsPT4, 6 and 8, showed a significant increase in the NH101 mutant in Pi-deprived condition compared with N22. Other traits such as number of tillers and filled spikelets per panicle, yield, Pi content and externally secreted ACPs, activity of CAT, and the relative expression levels of MPH genes (i.e. *OsmiR399a*, *OsPHO1;2*, *OsIPS1*, *OsPAP10a*,

Fig. 30.3. Screening Nagina 22 mutants in low-P tolerance field (Olsen P 1.8) facility at Indian Institute of Rice Research, Hyderabad, India.

OsPT2, 9 and *10*) showed a significant reduction in NH101 compared with N22 (Yugandhar *et al.*, 2018a).

As field screening is too exhaustive in terms of land, labour and time, four different screening methods, namely, hydroponics without sand (H), hydroponics with sand (HS), large pots with soil (PS) and glasses with soil (GS), were compared to determine the most reliable and simplest method for screening rice mutants and germplasm for low-P tolerance (Yugandhar *et al.*, 2017a). HS was identified as the most reliable method when low-P soil is not available, while GS is the next best method in soil experiments. Yet another simple screening method was standardized for identification of lpt and lps mutants in hydroponics under laboratory conditions (Yugandhar *et al.*, 2018b). In this study it was demonstrated that growing 14-day-old seedlings in hydroponics for 22 days with +2P (double P; 0.64 mM NaH_2PO_4) can help to identify low-P tolerant genotypes based on their final (36th day) shoot dry weight. This method can be used for rapid screening of genotypes under laboratory conditions.

Herbicide tolerance

Field screening of one lakh (100,000) M_2 EMS induced mutants of N22, especially created for this purpose, was undertaken using imazethapyr (commercial name: Pursuit™). The herbicide (2.5 ml) was diluted to 1 l with water and used as a spray on 30-day-old seedlings and one seedling was identified as a promising imazethapyr-tolerant mutant (Shoba *et al.*, 2017). This mutant has been designated as ‘herbicide tolerant mutant (HTM) Robin’. Genetic and molecular analysis revealed that the herbicide tolerance trait was under the control of monogenic dominance with eight non-synonymous mutations in the Acetohydroxy Acid Synthase (*AHAS*) gene of the mutant compared with the WT (Shoba *et al.*, 2017). Our mutant is different from the Clearfield™ mutant as the former does not have the A122T change responsible for the latter (Livore *et al.*, 2007; Shoba *et al.*, 2017).

Physiological traits

As a novel strategy, above-ground biomass, grain yield and $\Delta^{13}C$ as a surrogate for water use efficiency (WUE) were measured for 680 M_2 mutants grown under semi-irrigated aerobic conditions to identify contrasting pairs of mutants for WUE and root traits (Reddy *et al.*, 2018). Based on these evaluations in M_2 and subsequently in M_3, 150 promising mutants were identified and evaluated for root traits in the M_4 generation according to Sheshshayee *et al.* (2011). This exercise showed a positive correlation between WUE and high root biomass. Finally, two high-volume root mutants and one low-volume root mutant were selected and, along with the WT, were subjected to targeted sequencing of 112 genes implicated in WUE, drought tolerance or drought adaptive traits, root phenotype and photosynthesis. One of the high root mutants had non-synonymous variations in three genes, namely *Hox10*, *Zeaxanthin oxidase* and *citrate synthase*. Further genetic analysis of the mutant with the WT and selective phenotyping revealed that the high-volume root segregants invariably had the *Hox10* allele of the mutant type (Reddy *et al.*, 2018).

One of the five stay-green mutants (NH162), mentioned earlier (section 3.4), was further characterized as it had higher photosynthetic rates during senescence. The mutant leaves were darker green than those of the WT under dark-induced senescence. Further, the WT leaves treated with a solution of BA (6-benzyl adenine, an analogue of cytokinin) at 2 mg/ml had chlorophyll content similar to the mutant even after 72 h, indicating that cytokinin plays a key role in the process of senescence by regulating genes involved in chlorophyll inter-conversion, thereby maintaining the chlorophyll a/b ratio (Talla *et al.*, 2016).

Morphological traits

A short-grain mutant (Ngangkham *et al.*, 2018) and a high-tillering dwarf mutant (Kulkarni *et al.*, 2014) were characterized at genetic and molecular levels by crossing them with a contrasting genotype, IR64, which has longer grains and semi-dwarf moderate tillering architecture. In the case of the short-grain mutant, through BSA and graphical genotyping of the mutant type individuals, the causal gene was initially mapped to the telomeric region of the short arm of chromosome 5. Using advanced generations of this population, a backcross mapping population with IR64 and a large F_2 population derived from a cross with Pusa 1121, another long-grain genotype, the gene was fine-mapped to a 250 Kb region. As this region contained a known grain size regulator, *SRS3*, sequencing of this gene was undertaken in the WT, mutant and 88 more rice accessions. This exercise revealed a novel allele of this gene designated as TEMS5032, owing to C-to-T transition in the mutant, which created a stop codon in this Kinesin 13 family gene. Though the transcription of this gene remained unaffected in the mutant compared with the WT, 13 of 25 cell-cycle genes analysed showed differential expression (Ngangkham *et al.*, 2018).

Genetic analysis of a high-tillering and dwarf mutant was made by crossing it with IR64 and generating a fairly large mapping population. Mapping by the BSA approach revealed the presence of a major QTL on chromosome 4 for both plant height and number of tillers. Fine-mapping of this QTL region using progeny of a single heterozygous F_2 plant and additional markers narrowed down this QTL to a 984 Kb interval. Bioinformatics analysis of this region revealed the presence of a gene annotated as *CAROTENOID CLEAVAGE DIOXYGENASE7* (*CCD7*), which is the closest homologue of Arabidopsis *MAX3*, a strigolactone biosynthetic gene involved in axillary branching (Kulkarni *et al.*, 2014). This is allelic to *HTD1* in rice but with a different mutation (Zou *et al.*, 2005). In the sixth exon of this gene, a double mutation (CC to AA) has been induced in our mutant, leading to creation of a stop codon in the place of serine (Kulkarni *et al.*, 2014).

3.5 Utilization of the identified mutants

Some of the promising mutants identified have been directly utilized in breeding programmes. They include two lpt mutants, a high Fe- and Zn-content mutant, and the heat-tolerant mutant. The lpt mutants are also utilized as check varieties for screening of rice germplasm for low-P tolerance. The heat-tolerant mutant is also being used as tolerant check in the heat tolerance trials. Further, the HTM-Robin mutant is being extensively used in varietal development through marker-assisted backcross breeding (Shoba *et al.*, 2017). Six out of the eight rice varieties chosen for introgression of the HT trait are in advanced breeding generations.

4 Discussion

Induced mutations have added immense value to both basic and applied sciences (Parry *et al.*, 2009; Suprasanna *et al.*, 2015). So far more than 3200 crop varieties have been released all over the world, with 822 varieties in rice alone (Suprasanna *et al.*, 2015; https://mvd.iaea.org, accessed April 2021). Advances in molecular techniques have revived this 90-year-old science with vigour (Dhaliwal *et al.*, 2015; Abe *et al.*, 2012; Li, S. *et al.*, 2016, 2018; Li, G. *et al.*, 2017). We have established the 'National Repository of EMS Induced Mutants of Nagina 22' with 87,000 mutant resources, a fairly large one compared with the other mutant resources. This is the largest resource in rice among the physical/chemical induced resources and the second largest one when compared with the T-DNA-based resources (Miyao *et al.*, 2003; Wu *et al.*, 2005; Zhang *et al.*, 2006; Sevanthi *et al.*, 2018). This resource was established with the three objectives of maintenance, characterization and utilization of induced mutants in rice. Besides the 541 mutants that have been phenotyped for multiple traits, 1058 mutants for various traits have been identified (Table 30.1). Barring the 704 mutants identified based on visual examination, the rest of the 354 mutants were identified after appropriate screening procedures, some of them (such as herbicide tolerance and blast tolerance) involving large-scale field evaluations (Table 30.1) (Fig. 30.3). The mutants identified would serve as a valuable repository for breeding and gene characterization.

Comparison of the mutant similarity indices showed that nearly 87.5% of the mutants had stability score > 0.6 (Fig. 30.1A). This result was similar to that observed in the previous season in both New Delhi and Cuttack locations, where 92.3% and 89.18% of the mutants had similarity indices > 0.6 (Sevanthi *et al.*, 2018). Over the locations, 50% of the traits showed similar behaviour while the remaining 50% showed highly variable behaviour.

Of all the traits so far studied, only imazethapyr herbicide tolerance has been taken to the stage of utilization in practical breeding by introgressing this trait in a number of major varieties pertaining to the country's different rice ecosystems (Shoba *et al.*, 2017). As development of dual herbicide-tolerant varieties is now recommended in almost all crop species to delay development of herbicide-resistant super weeds, we are in the process of testing the resources with other herbicides (Bonny, 2016; Fartyal *et al.*, 2018).

Another major trait that has been investigated in detail includes low-P tolerance. One of the major outcomes is the standard protocols developed in rice for evaluation of low-P tolerance of the rice germplasm including mutants (Panigrahy *et al.*, 2014; Yugandhar *et al.*, 2017a, 2018a). The major QTL and subsequently gene known for P use efficiency (higher grain yield under deficient P) in rice is the PUP1 QTL on chromosome 12 comprising the *Pstol* gene encoding a protein kinase (Wissuwa *et al.*, 1998; Gamuyao *et al.*, 2012). Most of the traditional drought-tolerant rice cultivars are akin to 'Kasalath' in retaining the PUP1 locus, which is absent in the irrigated rice cultivars (Chin *et al.*, 2011). As N22 is a traditional rice cultivar, we conducted a local BLAST of *Pstol* gene sequence with our WGS data (sequenced at 162× depth) generated from N22 to determine whether or not this gene is present in the N22 genome. These results showed the presence of this gene. Further, we looked at the de novo assembly of the N22 sequence data, which clearly indicated that the entire 452 Kb PUP1 QTL region was present in N22 except for a 3 Kb region. This could be either due to lack of sequence coverage or due to absence of this region altogether in the N22 genome. Since the *Pstol* gene was present in our WT genome, we compared the nucleotide and amino acid sequences of N22 with that of 'Kasalath', which carries a functional allele, and found that there were 31 nucleotide and 16 amino acid substitutions. We have not yet studied whether these changes have rendered N22 ineffective for low P tolerance. However, it is known from our earlier low-P tolerance screening studies that N22 is highly sensitive to low P supply (Panigrahy *et al.*, 2014; Yugandhar *et al.*, 2017a). Thus the mutants obtained from our resource with improved performance under low-P conditions have the potential to yield both *Pstol*-based and novel candidates for improving efficiency of P use in rice.

The experience so far gained with respect to mapping some of the mutants has suggested that all the causal genes have been novel alleles of the known genes with different degrees of phenotype effect (Kulkarni *et al.*, 2014; Ngangkham *et al.*, 2018; Reddy *et al.*, 2018; Sevanthi *et al.*, 2018). The availability of this resource along with the high-quality sequencing resource of N22 is expected to aid in the identification of useful mutants and genes for crop improvement within and outside the network.

5 Conclusion

Development of appropriate facilities and rapid high-throughput protocols are necessary for better exploitation of the mutant resource generated. As currently the available sequencing resource includes only the CDS data, it would be worthwhile to generate the high-quality WGS data to facilitate NGS-based mapping strategies and identification of causal genes in the useful mutants.

Acknowledgement

The contribution of Dr Amolkumar Solanke, National Institute for Plant Biotechnology, New Delhi, and Dr Someshwar Bhagat in screening the 100,000 mutants in the Hazaribagh location for rice blast is sincerely acknowledged. The financial support provided by the Department of Biotechnology, New Delhi, through Project No. BT/PR9264/AGR/02/406/2007 and BT/PR10787/AGIII/103/883/2014 is also sincerely acknowledged.

References

Abe, A. *et al.* (2012) Genome sequencing reveals agronomically important loci in rice using MutMap. *Nature Biotechnology* 30, 174–178.

Bashyal, B.M. *et al.* (2017) Screening and identification of new sources of resistance to sheath blight in wild rice accessions. *Indian Journal of Genetics and Plant Breeding* 77(3), 341–347.

Bonny, S. (2016) Genetically modified herbicide-tolerant crops, weeds, and herbicides: overview and impact. *Environmental Management* 57(1), 31–48.

Chin, J.H. *et al.* (2011) Developing rice with high yield under phosphorus deficiency: Pup1 sequence to application. *Plant Physiology* 156(3), 1202–1216.

Deshmukh, R. *et al.* (2010) Identification of candidate genes for grain number in rice (*Oryza sativa* L.). *Functional & Integrative Genomics* 10(3), 339–347.

Dhaliwal, A.K. *et al.* (2015) An ethylmethane sulfonate mutant resource in pre-green revolution hexaploid wheat. *PLoS ONE* 10(12), e0145227.

Fartyal, D. *et al.* (2018) Developing dual herbicide tolerant transgenic rice plants for sustainable weed management. *Scientific Reports* 8(1), 11598.

Fekih, R. *et al.* (2013) MutMap+: Genetic mapping and mutant identification without crossing in rice. *PLoS ONE* 8, 1–10.

Fiyaz, R.A. *et al.* (2016) Mapping quantitative trait loci responsible for resistance to Bakanae disease in rice. *Rice* 9(1), 45.

Gamuyao, R. *et al.* (2012) The protein kinase Pstol1 from traditional rice confers tolerance of phosphorus deficiency. *Nature* 488(7412), 535.

Guo, L. *et al.* (2014) Application of re-sequencing to rice genomics, functional genomics and evolutionary analysis. *Rice* 7, 1–10.

International Rice Genome Sequencing Project (2005) The map-based sequence of the rice genome. *Nature* 436, 793–800.

Jiao, Y. *et al.* (2018) Efficient identification of causal mutations through sequencing of bulked F2 from two allelic bloomless mutants of *Sorghum bicolor*. *Frontiers in Plant Science* 8, 1–11.

Koboldt, D.C. *et al.* (2012) VarScan 2: somatic mutation and copy number alteration discovery in cancer by exome sequencing. *Genome Research* 22(3), 568–576.

Kulkarni, K.P. *et al.* (2014) A substitution mutation in OsCCD7 cosegregates with dwarf and increased tillering phenotype in rice. *Journal of Genetics* 93(2), 389–401.

Li, G. *et al.* (2017) The sequences of 1,504 mutants in the model rice variety kitaake facilitate rapid functional genomic studies. *Plant Cell* 29, 1218–1231.

Li, H. (2011) A statistical framework for SNP calling, mutation discovery, association mapping and population genetical parameter estimation from sequencing data. *Bioinformatics* 27(21), 2987–2993.
Li, H. *et al.* (2009) The Sequence Alignment/Map format and SAMtools. *Bioinformatics* 25(16), 2078–2079.
Li, S. *et al.* (2016) Frequency and type of inheritable mutations induced by γ rays in rice as revealed by whole genome sequencing. *Journal of Zhejiang University-SCIENCE B* 17(12), 905–915.
Li, S. *et al.* (2018) High-resolution melting-based TILLING of γ ray-induced mutations in rice. *Journal of Zhejiang University-SCIENCE B* 19(8), 620–629.
Lima, J.M. *et al.* (2015) Physiological, anatomical and transcriptional alterations in a rice mutant leading to enhanced water stress tolerance. *AoB Plants* 7, lv023.
Livore, A. *et al.* (2007), inventors; Instituto Nacional de Tecnologia Agropecuaria (INTA), assignee. Rice plants having increased tolerance to imidazolinone herbicides. United States patent application US 10/569,576.
McCallum, C.M. *et al.* (2000) Targeted screening for induced mutations. *Nat biotechnol* 18(4), 455–457.
Mithra, S.A. *et al.* (2016) DBT propelled national effort in creating mutant resource for functional genomics in rice. *Current Science* 110(4), 543–548.
Miyao, A. *et al.* (2003) Target site specificity of the Tos17 retrotransposon shows a preference for insertion within genes and against insertion in retrotransposon-rich regions of the genome. *The Plant Cell* 15(8), 1771–1780.
Mohapatra, T. *et al.* (2014) EMS induced mutants of upland rice variety Nagina22: generation and characterization. *Proceedings Indian National Science Academy* 80, 163–172.
Ngangkham, U. *et al.* (2018) An EMS-induced new sequence variant, TEMS5032, in the coding region of SRS3 gene leads to shorter grain length in rice (*Oryza sativa* L.). *Journal of Applied Genetics* 59(4), 377–389.
Pachauri, V. *et al.* (2014) Identification of candidate genes for rice grain aroma by combining QTL mapping and transcriptome profiling approaches. *Cereal Research Communications* 42(3), 376–388.
Pandit, A. *et al.* (2010) Combining QTL mapping and transcriptome profiling of bulked RILs for identification of functional polymorphism for salt tolerance genes in rice (*Oryza sativa* L.). *Molecular Genetics and Genomics* 284, 121–136.
Panigrahy, M. *et al.* (2011) Heat tolerance in rice mutants is associated with reduced accumulation of reactive oxygen species. *Biologia Plantarum* 55(4), 721.
Panigrahy, M. *et al.* (2014) Hydroponic experiment for identification of tolerance traits developed by rice Nagina 22 mutants to low-phosphorus in field condition. *Archives of Agronomy and Soil Science* 60(4), 565–576.
Parry, M.A. *et al.* (2009) Mutation discovery for crop improvement. *Journal of Experimental Botany* 60, 2817–2825.
Poli, Y. *et al.* (2013) Characterization of a Nagina22 rice mutant for heat tolerance and mapping of yield traits. *Rice* 6, 36.
Prakash, C. *et al.* (2016) Unraveling the molecular basis of oxidative stress management in a drought tolerant rice genotype Nagina 22. *BMC Genomics* 17(1), 774.
Reddy, S.H. *et al.* (2018) Allele-specific analysis of single parent backcross population identifies HOX10 transcription factor as a candidate gene regulating rice root growth. *Physiologia Plantarum* 166(2), 596–611.
Richa, K. *et al.* (2016) Functional characterization of novel chitinase genes present in the sheath blight resistance QTL: qSBR11-1 in rice line tetep. *Frontiers in Plant Science* 7, 244.
Sevanthi, A.M. *et al.* (2018) Whole genome characterization of a few EMS-induced mutants of upland rice variety Nagina 22 reveals a staggeringly high frequency of SNPs which show high phenotypic plasticity towards the wild-type. *Frontiers in Plant Science* 9, 1179.
Shanmugavadivel, P.S. *et al.* (2013) Mapping quantitative trait loci (QTL) for grain size in rice using a RIL population from Basmati × indica cross showing high segregation distortion. *Euphytica* 194, 401–416.
Shanmugavadivel, P.S. *et al.* (2017) High resolution mapping of QTLs for heat tolerance in rice using a 5K SNP array. *Rice* 10, 28.
Sheshshayee, M.S. *et al.* (2011) Phenotyping for root traits and their improvement through biotechnological approaches for sustaining crop productivity. In: Costa de Oliveira, A. and Varshney, R.K. (eds) *Root Genomics*. Springer, Berlin, pp. 205–223.
Shoba, D. *et al.* (2017) Development and genetic characterization of a novel herbicide (Imazethapyr) tolerant mutant in rice (*Oryza sativa* L.). *Rice* 10, 10.

Shobha Rani, N. *et al.* (2004) *National Guidelines for the Conduct of Tests for Distinctiveness, Uniformity and Stability*. Directorate of Rice Research, Rajendranagar, 6–13.

Singha, K. (2013) Paddy processing mills in India: an analysis. *Rice Research* 1, 114. doi: 10.4172/jrr.1000115

Sinha, S.K. *et al.* (2018) Transcriptome analysis of two rice varieties contrasting for nitrogen use efficiency under chronic N starvation reveals differences in chloroplast and starch metabolism-related genes. *Genes* 9(4), 206.

Song, J. *et al.* (2017) Next-generation sequencing from bulked-segregant analysis accelerates the simultaneous identification of two qualitative genes in soybean. *Frontiers in Plant Science* 8, 1–11.

Suprasanna, P. *et al.* (2015) Induced mutations and crop improvement. In: Bahadur, B., Venkat Rajam, M., Sahijram, L. and Krishnamurthy, K. (eds) *Plant Biology and Biotechnology*. Springer, New Delhi.

Swamy, B.M. and Sarla, N. (2008) Yield-enhancing quantitative trait loci (QTLs) from wild species. *Biotechnology Advances* 26(1), 106–120.

Takagi, H. *et al.* (2013) QTL-seq: rapid mapping of quantitative trait loci in rice by whole-genome resequencing of DNA from two bulked populations. *Plant Journal* 74, 174–183.

Talla, S.K. *et al.* (2016) Cytokinin delays dark-induced senescence in rice by maintaining the chlorophyll cycle and photosynthetic complexes. *Journal of Experimental Botany* 67, 1839–1851.

Till, B.J. *et al.* (2007) Discovery of chemically induced mutations in rice by TILLING. *BMC Plant Biology* 7, 19.

Tsai, H. *et al.* (2011) Discovery of rare mutations in populations: TILLING by sequencing. *Plant Physiology* 156, 1257–1268.

Verma, R.K. *et al.* (2017) Mapping of QTLs for grain yield and its component traits under drought stress in elite rice variety of Assam, India. *International Journal of Current Microbiology and Applied Sciences* 6(6), 1443–1455.

Wang, Y. *et al.* (2016) Fine mapping and whole-genome re-sequencing identify the seed coat color gene in Brassica rapa. *PLoS ONE* 11, 1–14.

Wissuwa, M. *et al.* (1998) Mapping of QTLs for phosphorus deficiency tolerance in rice (Oryza sativa L.). *Theoretical and Applied Genetics* 97, 777–783.

Wu, J.L. *et al.* (2005) Chemical-and irradiation-induced mutants of indica rice IR64 for forward and reverse genetics. *Plant Molecular Biology* 59, 85–97.

Yadav, S. *et al.* (2015) Identification of QTLs and possible candidate genes conferring sheath blight resistance in rice (*Oryza sativa* L.). *Springerplus* 4(1), 175.

Yugandhar, P. *et al.* (2017a) Comparing hydroponics, sand, and soil medium to evaluate contrasting rice Nagina 22 mutants for tolerance to phosphorus deficiency. *Crop Science* 57(4), 2089–2097.

Yugandhar, P. *et al.* (2017b) Identifying markers associated with yield traits in Nagina22 rice mutants grown in low phosphorus field or in alternate wet/dry conditions. *Australian Journal of Crop Science* 11(5), 548.

Yugandhar, P. *et al.* (2018a) Nagina 22 mutants tolerant or sensitive to low P in field show contrasting response to double P in hydroponics and pots. *Archives of Agronomy and Soil Science* 64(14), 1–13.

Yugandhar, P. *et al.* (2018b) Increased catalase activity and maintenance of Photosystem II distinguishes high-yield mutants from low-yield mutants of rice var. Nagina22 under low-Phosphorus stress. *Frontiers in Plant Science* 9, 1543.

Zhang, J. *et al.* (2006) RMD: a rice mutant database for functional analysis of the rice genome. *Nucleic Acids Research* 34(Suppl. 1), D745–D748.

Zou, J. *et al.* (2005) Characterizations and fine mapping of a mutant gene for high tillering and dwarf in rice (*Oryza sativa* L.). *Planta* 222, 604–612.

Zou, T. *et al.* (2017) OsLAP6/OsPKS1, an orthologue of Arabidopsis PKSA/LAP6, is critical for proper pollen exine formation. *Rice* 10, 53.

31 Radiation-Induced Mutations in Genetic Enhancement and Development of New Crop Varieties in Black Gram (*Vigna mungo* (L.) Hepper)

Souframanien Jegadeesan[1] and Kandali Sreenivasulu Reddy

Abstract

Black gram (*Vigna mungo* (L.) Hepper), popularly known as urdbean or mash or black gram, is a grain legume rich in protein (25–28%), widely cultivated in the Indian subcontinent and to a lesser extent in Thailand, Australia and other Asian and South Pacific countries. Genetic improvement in this crop is hindered due to the narrow genetic base. As genetic variability is a prerequisite for any crop improvement programme, induced mutations provide an important source for generating variability. Radiation (gamma, X-rays and neutron) induced mutants were identified for various morphological and biochemical traits, creating a pool of genetic variability. These mutants were used in a cross-breeding programme to develop high-yielding, disease-resistant varieties in black gram. The effective blend of mutation and recombination breeding at the Bhabha Atomic Research Centre has resulted in the release of five black gram varieties (TAU-1, TAU-2, TPU-4, TU94-2 and TU-40) by incorporating desirable traits like large seed, wider adaptability, resistance to disease and improved quality. These varieties have been developed from mutants directly or by using them in cross-breeding programmes. For example, a black gram variety, NO.55, was irradiated with gamma-rays and electron beams to obtain a large number of mutants. The large-seed mutants, UM-196 and UM-201, were used in cross-breeding with the elite cultivar T-9 for developing the high-yielding varieties TAU-1, TAU-2, TPU-4, TU94-2 and TU-40. TAU-1 has become the most popular variety in Maharashtra state, occupying the maximum area under black gram cultivation. Induced mutations will continue to play an increasing role in generating genetic variability for various traits as a major component of environmentally sustainable agriculture.

Keywords: black gram • electron beam • gamma-rays • mutation • recombination breeding

1 Introduction

Globally, grain legumes are the second most important group of crops. Among legumes, pulses are important in India, as most of the dietary protein is derived from them. Global pulse production is around 73.21 million tonnes, of which about 18.3 million tonnes were produced annually in India during 2013–2014. Among pulses, black gram (*Vigna mungo*) is one of the most important crops with its origin in the Indian subcontinent. Black gram is a small annual plant and is commonly known in India as urdbean; genetic variability is low in this inbreeding crop and there is a narrow genetic base among the released cultivars (Gupta *et al.*, 2005). As genetic variability is essential for any crop improvement programme, induced mutations provide an important source for variability.

[1] Email: souf@barc.gov.in

 Mutation Breeding, Genetic Diversity and Crop Adaptation to Climate Change (eds S. Sivasankar *et al.*)
DOI: 10.1079/9781789249095.0031

Radiations, including gamma-rays, X-rays, fast neutrons and thermal neutrons, have been widely applied to induce mutations and have contributed to great progress in plant breeding. Crop improvement through mutation breeding has resulted in the development of improved varieties that are directly used for commercial cultivation or in recombination breeding by hybridizing mutants among themselves, with their derivatives and cultivars. In the past 80 years, physical mutagens, mostly ionizing radiations (gamma-rays, X-rays, neutrons, etc.), have been used widely for inducing mutations and more than 70% of mutant varieties were developed using physical mutagenesis (Mba, 2013). Over 232 different crops and plant species have been subjected to mutation breeding, including various major crops such as wheat, rice, grapefruit, rapeseed, sunflower, cotton and banana (IAEA, 2017). The Mutant Variety Database (MVD) of the Joint FAO/IAEA Division indicates that more than 3281 mutant varieties with improved characters have been officially released, including direct mutant, mutant × mutant, mutant × variety, use of mutant trait (e.g. semi-dwarf) and radiation-induced translocation in breeding programmes (IAEA, 2017). Most of those varieties were developed using ionizing radiation, mainly gamma-rays (64%), followed by X-rays (22%). Mutation is regarded as random and success in obtaining a desired mutant trait depends on three factors: the efficiency of mutagenesis, the starting plant material and mutant screening (Hase *et al.*, 2012). The frequency with which the desired mutants appear depends, in part, on the efficiency of the mutagenesis. Among different ionizing radiations, gamma-rays have been commonly used and numerous mutants have been produced in black gram (Souframanien and Pandey, 2006; Souframanien *et al.*, 2016). Mutation breeding in black gram was initiated in 1960 and mutations affecting various morphological and physiological characters were obtained. Subsequently these mutants were used in cross-breeding programmes to develop new black gram varieties. Apart from the conventional electromagnetic radiations, like X-ray and gamma-ray, the electron beam is now an alternative source of energy to induce mutations. In the present study, we report the use of electron beams for induction of chlorophyll and morphological mutants in black gram and also their effectiveness and efficiency in comparison with gamma-rays in black gram.

2 Materials and Methods

2.1 Developing mutant population

Genetically pure, uniform and dry seeds (12% moisture content) of black gram variety TU94-2 were irradiated with four different doses (200, 300, 400 and 500 Gy) of ^{60}Co gamma-rays at the Bhabha Atomic Research Centre (BARC), Trombay, India. For electron beam irradiation, seeds were treated with 7.5 MeV electron beam with four different doses (200, 300, 400 and 500 Gy) from the Electron Accelerator Facility at Raja Ramanna Centre for Advanced Technology (RRCAT), Indore, India. Aluminium scatterers were used to achieve low doses (~ 8 Gy/s) in a field of 15 cm × 15 cm. Treated seeds along with untreated control of the variety (TU94-2) were sown in the experimental field facility at Trombay to raise the M_1 generation. Observations on injury and lethality were recorded. M_1 plants were harvested separately to raise plant progeny rows in the M_2 generation with recommended package of practices. Observations on chlorophyll and viable morphological mutations were recorded from the day of emergence until plants attained physiological maturity. Data on biological abnormalities such as injury and lethality in the M_1 generation and mutation frequencies in the M_2 generation were used to determine mutagenic effectiveness and efficiency according to the formulae suggested by Konzak *et al.* (1965).

Selection of desirable characteristics was made in the M_2 generation and the selected plants were tagged. Each tagged plant was individually harvested and advanced to the M_3 generation. Selection in the M_3 generation was followed by selecting individual rows of homozygous lines with desirable characters. Morphological mutants identified were advanced to subsequent generations to study the true breeding nature of the mutants. Plants producing high yield and having synchronized maturity in a row were identified and bulked. Selected mutants were used in a cross-breeding programme to develop F_1 generations. Selections were made from segregating generations of the progenies

(F_2 to F_5) derived from these crosses. Selected lines were multiplied and uniform lines were bulked. High-yielding, disease-resistant lines were evaluated in the station trials and All India Coordinated Research Project trials of ICAR along with a check variety. The best performing lines were released for commercial cultivation.

2.2 Identifying a yellow mosaic virus-resistant mutant

The M_2 population was screened in the yellow mosaic virus (YMV) hotspot area at Trombay. Each plot consisted of a 3 m row with ten plants per metre and 30 cm between the rows. Under field conditions, resistant plants were identified as those that did not show any yellowing of leaves or pods during the growth period, while susceptible plants showed various grades of yellowing depending on the stage of infection. For scoring, any appearance of yellow colour on the leaves was scored as a susceptible plant, while those that remained green until maturity were scored as resistant. Normal cultural practices were followed, except that there was no insecticide spraying. LBG17, a cultivar highly susceptible to YMV, was used as a refuge every five rows.

2.3 Identifying low-oligosaccharide mutant

Mature seeds (M_3) of black gram were soaked in distilled water overnight at room temperature. About ten seeds were ground in 10 ml of 80% ethanol in a mortar with a pestle. The slurry was transferred to a centrifuge tube and kept on a shaker for about 5 h. The slurry was centrifuged at 12,000 rpm (20,913 ×*g*) at 24°C for 20 min. The clear supernatant was collected into a fresh tube. About 1.5 ml of the extract was dried in a vacuum concentrator at 45°C. The dried residues were suspended in 75 μl of acetonitrile:water (65:35 v/v). The contents were centrifuged at 12,000 rpm (20,913 ×*g*) for about 10 min and 4 μl of the clear supernatant was injected into the HPLC column. Each sample was extracted in five replications and each replication was measured once for quantification of soluble sugar concentrations. HPLC analysis was carried out with JASCO HPLC System using a carbohydrate column (30 cm × 3.9 mm) and acetonitrile:water (65:35 v/v) as the mobile phase. The flow rate was adjusted to 1.0 ml/min. Known concentrations of the samples (4 μl) were injected and the sugars in the various samples were identified by comparison with the corresponding standard samples. The sugar concentrations in the various genotypes were calculated from the standard calibration curve of the individual sugars and were expressed as milligrams per gram of seed meal.

3 Results

3.1 Chlorophyll mutations

In this study, a wide spectrum of chlorophyll mutations showed that the M_2 generation of the mutagen treated population of TU94-2 included *albino*, *xantha*, *chlorina* and *viridis* classes of chlorophyll mutants. In these experiments, the frequency of chlorophyll mutations in gamma-ray and electron beam treated populations ranged from 3% to 5% and 4% to 11%, respectively. Among different types of chlorophyll mutations, *xantha* was the most frequent (2% and 4%) followed by *chlorina* and *albino* in both gamma-ray and electron beam treated populations. In these experiments, the overall frequency of chlorophyll mutations in the electron beam treated population was higher (33%) compared with the gamma-ray treated population (17%). Chlorophyll mutation frequency rate was higher (8%) in the electron beam treated population than that of the gamma-ray (4%) treated population (Fig. 31.1). In the present study, there was a dose-dependent increase in the spectrum and frequency of chlorophyll mutations in both gamma-ray (except 400 Gy) and electron beam treated M_2 populations.

3.2 Morphological mutations

A wide spectrum of morphological mutations was observed in the M_2 generation of gamma-ray and electron beam treated populations. The spectrum of morphological mutations (dwarf, tall, leaf size and shape, increased branch

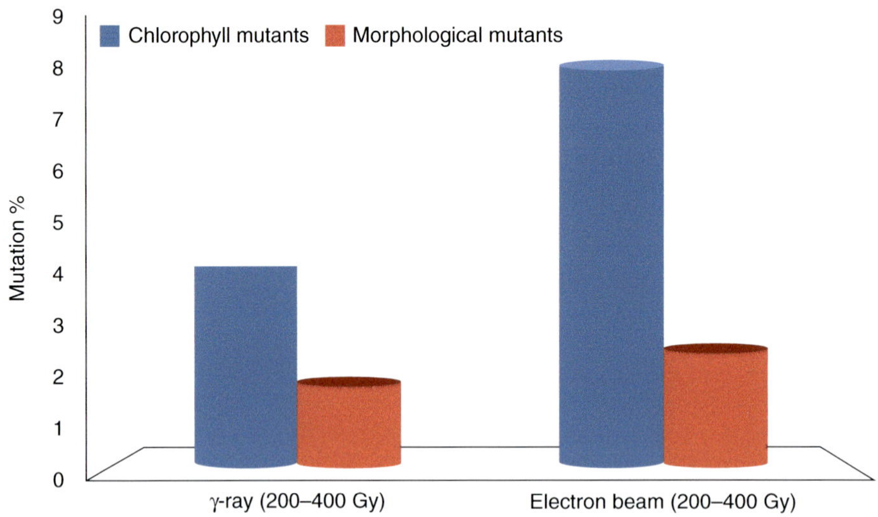

Fig. 31.1. Mutant frequency in the M_2 generation of TU94-2 black gram treated with gamma-rays (6840 M_2 plants) and electron beam at 200, 300 and 400 Gy (6605 M_2 plants). No error bar shown as the data was not replicated.

number, branch position, stem girth, early and late flowering, increased pod number, pod and seed colour, synchronous maturity, increased primary and lateral root length mutants) observed was common to both gamma-ray and electron beam treated populations (Fig. 31.2). Overall morphological mutation frequency was higher (10%) in the electron beam treated population compared with that of the gamma-ray (7%) treated population (Fig. 31.1).

As the mutagen dose increased, for both gamma-rays (up to 400 Gy) and the electron beams, the spectrum and frequency of morphological mutations appeared to increase.

3.3 Synchronous maturity and yellow mosaic virus-resistant mutant

Most black gram varieties have an indeterminate growth habit and asynchronous floral maturity. In order to develop a black gram genotype suitable for mechanical harvesting, synchronous maturity of pods is required. In this study, a mutant of determinate type, synchronized maturity and top bearing mutants were identified as suitable for mechanical harvesting (Fig. 31.3). Black gram mutants resistant to yellow mosaic virus (YMV) disease were identified under field screening using the infector row method. Resistant plants were tagged and harvested individually and advanced to the M_3 generation. Selection in the M_3 generation was conducted by selecting individual rows of homozygous lines with YMV resistance and advanced to subsequent generations to study the true breeding nature of the mutants. Lines showing resistance to YMV disease were identified and bulked and evaluated under field conditions with susceptible check (Fig. 31.4).

3.4 A black gram mutant with a reduced content of raffinose family oligosaccharides

HPLC methodology was standardized for the separation and quantification of sucrose and raffinose family oligosaccharides (RFOs) in black gram. Using this protocol, sugars in various samples were identified by comparison of their retention times with that of standard samples. The sugar contents in the various genotypes of black gram were calculated from the corresponding standard calibration curves and expressed in milligrams per gram of seed meal. The HPLC profile of the black gram samples revealed the presence of sucrose, raffinose, stachyose,

Fig. 31.2. Morphological mutations induced by gamma-rays and electron beam in black gram. **(a)** Crinkle leaf mutant. **(b)** Narrow leaf mutant. **(c)** Top pod bearing mutant. **(d)** Dwarf mutant. **(e)** Seed colour mutants. P, parent; M, mutant.

verbascose and ajugose. In TU94-2 (parent) the total RFO content was 58.3 ± 17.5 mg/g with the individual RFO content for raffinose, stachyose, verbascose and ajugose found to be 1.0 ± 0.22 mg/g, 37.27 ± 1.07 mg/g, 19.70 ± 0.89 mg/g and 0.33 ± 0.12 mg/g, respectively. Black gram mutant TU43-1, which recorded low total RFOs (26.64 ± 7.28 mg/g), also recorded lower contents of verbascose (13.95 ± 0.25 mg/g), sucrose (3.28 ± 0.87 mg/g) and ajugose (0.35 ± 0.11 mg/g). Low stachyose content was observed in TU1-820-1-5 (11.13 mg/g). Low verbascose content was observed in TU43-1 (13.95 ± 0.25 mg/g). These low stachyose and verbascose mutants are particularly valuable in developing new black gram genotypes for food and feed with improved digestibility and functionality. TU51 recorded a low stachyose content (15.11 ± 0.47 mg/g) compared with its parent TU94-2 (37.27 ± 1.07 mg/g). Low stachyose and verbascose content of 15.85 ± 1.40 mg/g and 17.02 ± 2.30 mg/g, respectively, was observed in the TU55-1 mutant. The black gram mutant TU1-820-1-5 recorded high verbascose (31.02 ± 0.74 mg/g) and lowest raffinose (0.18 ± 0.05 mg/g) content among the black gram genotypes studied. Black gram mutants with reduced RFOs having low stachyose and verbascose identified in this study may be used as valuable and unique germplasm for future breeding and genetic research.

3.5 Mutants in cross-breeding and development of varieties

Black gram mutants having variation for morphological traits such as plant height, number of branches per plant, days to first flowering, clusters per plant, pods per plant, seeds per pod, seed yield per plant, 100-seed weight were identified and advanced to further generations to study their true breeding nature. Mutants with desirable characters and high yield were either released for cultivation directly or used in hybridization with other mutants or varieties. The large-seed mutants, UM-196 (dark green leaf mutant) and UM-201 were used in cross-breeding with the elite cultivar T-9 for developing high-yielding varieties TAU-1, TAU-2 and TPU-4. So far, five varieties have been released for cultivation which were developed through induced mutation and mutant derivatives through cross-breeding. Mutation breeding has made a significant contribution in increasing the production of black gram in India. Five of the seven mutant varieties of black gram

Fig. 31.3. Electron beam induced synchronous maturity in black gram.

Fig. 31.4. Field evaluation of black gram mutant resistant to yellow mosaic disease.

released in India have been developed at the Nuclear Agriculture & Biotechnology Division of the Bhabha Atomic Research Centre (BARC), Mumbai. The variety TAU-1, developed at BARC in collaboration with the Dr Panjabrao Deshmukh Krishi Vidyapeeth Agricultural University, Akola, has become the most popular variety in Maharashtra and has covered most of the area under black gram cultivation in that state.

4 Discussion

Chlorophyll mutations offer one of the most reliable indices for the assessment of genetic effects of mutagenic treatments. Genotypic differences in response to induction of chlorophyll mutations can be observed as the frequency of induced chlorophyll mutations in the M_2 generation. The overall frequency of chlorophyll mutations in the electron beam treated population was higher compared with the gamma-ray treated population, suggesting higher efficiency of electron beam treatment. A similar trend was reported in chickpea (Joshi-Saha *et al.*, 2015). In this study, more viable chlorophyll mutations (*chlorina* and *viridis*) and lethal mutants (*albino* and *xantha*) were observed at higher doses of electron beam irradiation in comparison with gamma-rays. A similar trend of progressive increase in mutation frequency of chlorophyll mutations was observed with increasing doses of gamma-rays in black gram (Makeen *et al.*, 2013) and other *Vigna* species (Ignacimuthu and Babu, 1988). Khan and Tyagi (2010) also reported a reduction in chlorophyll mutation frequency in one variety of soybean but an increase in another for the same mutagen at a range of doses; however, statistical support for this is not provided. About 250–300 loci might be involved in the breakdown of the chlorophyll apparatus in barley (Swami Nathan, 1957). Among different types of chlorophyll mutations, *xantha* was the most frequent (2% and 4%) followed by *chlorina* and *albino* in both gamma-ray and electron beam treated populations. Previous studies using gamma-rays showed *xantha* as the most frequently observed chlorophyll mutations in black gram (Lal *et al.*, 2009; Makeen *et al.*, 2013). Induced mutations have been used to generate genetic variability and have been successfully utilized to improve yield and yield components in black gram (Kundu and Singh, 1981; Singh and Singh, 2001; Souframanien and Pandey, 2006; Vanniarajan *et al.*, 2017). These reports suggest that mutagenesis is a potential tool to be used for crop improvement. In the present study, there was a dose-dependent increase in the spectrum and frequency of morphological mutations in both gamma-ray (until 400 Gy) and electron beam treated M_2 populations. Differential induction of morphological mutation in different doses of gamma-rays has been reported in mung bean (Mishra and Singh, 2014) and chickpea (Kharkwal, 2000). The frequency of morphological mutations increased with higher doses of electron beam. A similar trend was observed in electron beam treated M_2 populations of adzuki bean (Luo *et al.*, 2012). Viable morphological mutation frequency was higher (2.4%) in the electron beam experiment in comparison with gamma-rays (1.7%) (Fig. 31.2). A similar mutation rate has been observed in cockscomb (*Celosia* spp.) seeds treated with electron beam (Wang *et al.*, 2006).

Induced mutants are utilized directly for varietal development or in recombination breeding. In breeding programmes, hybridization provides many possibilities for generating new combinations of characters which can be selected in the segregating populations. In contrast, with induced mutations it is possible to improve single traits without causing extensive disruption in the genome. Induced mutations have been successfully used in cross-breeding programmes. In black gram hybridization, induced mutations and a combination of the two led to the isolation of cultures with large seed size and higher yield (Pawar *et al.*, 1988). Similarly, the rational use of induced plant type mutations affecting canopy structure in cross-breeding programmes led to a selection that has given significantly higher yield over the check cultivars in black gram (Pawar *et al.*, 2000). Sustained induced mutagenesis in black gram using gamma-rays and X-rays resulted in a wide spectrum of mutants affecting various traits. Traits like large seed size, increased harvest index, semi-dwarf habit, earliness, new plant types, improved seed quality and enhanced disease resistance were induced and incorporated into different crops (Souframanien, 2017). The effective blend of mutation and recombination breeding resulted in the release of seven black gram varieties for commercial cultivation in India. Among these, five varieties have been released by BARC. These varieties were developed by incorporating desirable agronomic features such as large seed in TPU-4 and high harvest index and wider adaptability in TAU-1. Yellow mosaic virus resistance in TU94-2 and TU-40 was also introgressed into these varieties from an induced mutant line.

5 Conclusion

The current study showed the existence of a significant amount of genetic variability in irradiation-derived mutants as compared with their original parent. Among the various doses, 400 Gy gamma-rays and 300 Gy electron beam were more desirable and resulted in low plant damage and high genetic effects. Mutants identified in this study could serve as a source of variability for crop improvement. Sustained improvement of plant productivity will be necessary to meet the growing demand and challenges of climate changes. Induced mutations will continue to play an increasing role in developing crop varieties with traits such as improved protein and starch quality, enhanced uptake of specific metals, deeper rooting system, and resistance to drought, diseases and salinity as a major component of environmentally sustainable agriculture. The impact of crop varieties derived from mutation breeding around the world demonstrates the potential of induced mutagenesis and its breeding strategies for crop improvement.

References

Gupta, S. *et al.* (2005) Pedigree analysis of urd bean varieties released in India. *Indian Journal of Pulses Research* 18, 153–155.

Hase, Y. *et al.* (2012) Development of an efficient mutagenic technique using ion beam: towards more controlled mutation breeding. *Plant Biotechnology* 29, 193–200.

IAEA (2017) IAEA Mutant Variety Database. International Atomic Energy Agency, Vienna. Available at: http://mvd.iaea.org/ (accessed April 2021)

Ignacimuthu, S. and Babu, C.R. (1988) Radio sensitivity of the wild and cultivated urd and mung beans. *Indian Journal of Genetics and Plant Breeding* 48, 331–342.

Joshi-Saha, A. *et al.* (2015) Identification of novel mutants through electron beam and gamma irradiation in chickpea (*Cicer arietinum* L.). *Journal of Food Legumes* 28, 1–6.

Khan, M.H. and Tyagi, S.D. (2010) Studies on effectiveness and efficiency of gamma rays, EMS and their combination in soybean (*Glycine max* (L.) Merrill). *Journal of Plant Breeding and Crop Science* 2, 55–58.

Kharkwal, M.C. (2000) Induced mutations in chickpea (*Cicer arietinum* L.) IV. Types of macro-mutations induced. *Indian Journal of Genetics and Plant Breeding* 60, 305–320.

Konzak, C.F. *et al.* (1965) Efficient chemical mutagenesis. *Radiation Botany* 5(Suppl.), 49–70.

Kundu, S.K. and Singh, D.P. (1981) EMS-induced variability in black gram. *Crop Improvement* 8, 71–72.

Lal, G.M. *et al.* (2009) Mutagenic sensitivity in early generation in black gram. *Asian Journal of Agricultural Science* 1, 9–11.

Luo, W.X. *et al.* (2012) Effects of electron beam radiation on trait mutation in azuki bean (*Vigna angularis*). *African Journal of Biotechnology* 11, 12939–12950.

Makeen, K. *et al.* (2013) Study of chlorophyll and macromutations induced by gamma rays and sodium azide in urd bean (*Vigna mungo* L. Hepper). *African Journal of Agricultural Research* 8, 5958–5961.

Mba, C. (2013) Induced mutations unleash the potentials of plant genetic resources for food and agriculture. *Agronomy* 3, 200–231.

Mishra, D. and Singh, B. (2014) Studies on effectiveness and efficiency of gamma rays in green gram (*Vigna radiata* (L.) Wilczek). *SABRAO Journal of Breeding and Genetics* 46, 34–43.

Pawar, S.E. *et al.* (1988) Genetic improvement of pulse crops: induced mutations and their use in cross breeding. In: *Proceedings, FAO/IAEA Workshop on Improvement of Grain Legume Production Using Induced Mutations*, Vienna, pp. 235–249.

Pawar, S.E. *et al.* (2000) *Genetic improvement of black gram using induced mutations*. Paper presented in DAE-BRNS Symposium on the use of Nuclear Techniques in crop improvement, Dec 6–8, 2000, Mumbai, pp. 170–174.

Singh, M. and Singh, V.P. (2001) Genetic analysis of certain mutant lines of urdbean for yield and quality traits in M_4 generation. *Indian Journal of Pulses Research* 14, 60–62.

Souframanien, J. (2017) The role of radiation induced mutations in crop improvement. In: *Proceedings of National Symposium on Applications of Radioisotopes and Tracer Techniques in Agriculture and Environment*, 16–17 February 2017, TNAU, Coimbatore, pp. 10–24.

Souframanien, J. and Pandey, R.N. (2006) Mutational and biotechnological approaches for urdbean improvement: a review. *Indian Journal of Pulses Research* 19, 155–164.

Souframanien, J. *et al.* (2016) Comparative effectiveness and efficiency of electron beam and ^{60}Co γ-rays in induction of mutations in black gram [*Vigna mungo* (L.) Hepper]. *Journal of Food Legumes* 29, 1–6.

Swami Nathan, M.S. (1957) Swedish mutation work in relation to plant breeding – a review. Special Symposium on Genetics and Plant Breeding in South Asia. *Indian Journal of Genetics and Plant Breeding* 17, 276–295.

Vanniarajan, C. *et al*. (2017) Gamma rays induced urd bean [*Vigna mungo* (L.) Hepper] mutants with YMV resistance, good batter quality and bold seeded type. *Legume Research* 42(1), 25–31.

Wang, D. *et al*. (2006) Plant growth of E-beam irradiated or ion implanted seeds of ornamental Celosia. *Journal of Radiation Research and Radiation Processing* 24, 188–192.

32 Development of New Bread Wheat Resistant Mutants for Ug99 Rust Disease (*Puccinia graminis* f. sp. *Tritici*).

Abdulwahid Saif[1], Aref Al-Shamiri and Abdulnour Shaher

Abstract

M_3 derived mutants from two bread wheat varieties, namely, 'Giza 186' and 'Saha 93', were screened for resistance to the rust Ug99 at two locations in Njoro (Kenya) and in Tihama (Yemen). At Tihama, two mutants of 'Giza 186' (G-M2- 2010-1-28 and G-M2- 2010-41- 52) and four mutants of 'Saha 93' (S-M2- 2010-16-12, S-M2- 2010-21-13, S-M2- 2010-22-14 and S-M2- 2010-27-15) were seen to be resistant at both seedling and adult stages while their parents were resistant at seedling stage and susceptible at adult stage. In Kenya, the resistance score of the mutants was slightly different from those obtained at Tihama. The mutants G-M2- 2010-1-28 and G-M2- 2010-41-52 were stable in their level of resistance recorded at Tihama, but only two mutants of 'Saha 93' (S-M2- 2010-16-12 and S-M2- 2010-27-15) were resistant at both growth stages. S-M2- 2010-22-14 and S-M2- 2010-21-13 were resistant at the seedling stage while susceptible at adult stage. Further selection on these mutants for yield potential, agronomic performance and yellow rust disease resistance, as well as on selected mutants of both 'Giza 186' and 'Saha 93', at M_5–M_6 stages identified superior mutant lines compared with the two parents 'Saha 93' and 'Giza 186'. These included the line Erra-010-GM2w-41-52-40, which ranked first in yield (3768 kg/ha), followed by the lines Erra-010-SwM2-16-12-19, Erra-010-GM2w-1-28-18 and Erra-010-SwM2-22-14-6. Moreover, it can be concluded that Erra-010-GM2w-41-52-40 and Erra-010-SwM2-16-12-19 are highly recommended for their resistance to stem and yellow rust diseases as well as for yield potential and preference by farmers. Therefore, efforts are in progress to increase their seeds for dissemination over a wide range of farmers and wheat areas where rust diseases are an epidemic, and for registration of the lines as improved mutant varieties.

Keywords: resistant mutant • bread wheat • Ug99 rust

1 Introduction

Yellow rust, leaf rust and stem rust caused by *Puccinia striformis*, *P. recondita* and *P. graminis tritici* are the major diseases that can attack seedlings and adult plants of wheat in Yemen. Yellow rust is the major one affecting wheat-production areas and the inoculum can survive throughout the year, due to different sowing dates. March and August are the maximum yearly peaks in appearance of yellow rust. Infection can continue throughout crop growth and its impact can be seen through to harvest (Watt, 1975; GTZ, 1985; Kamal *et al.*, 1985). Wheat is considered one of the most important cereal crops in Yemen. It occupies about 14% of the total area of cereals and 21% of total cereal production, ranking second among cereal crops

[1] Email: Amozaid@yahoo.com

DOI: 10.1079/9781789249095.0032

after sorghum. The total wheat production of 149,000 t constitutes only 9% of the total wheat consumption in the country. The area under wheat cultivation increased from 72,000 ha in 1988 to about 150,000 ha in 2010, with grain production of 250,000 t, which gradually increased until 2014 (MAI, 2015). The reason for this increase could be the introduction of improved high-yielding wheat varieties and quality support by the government to expand the area under the wheat crop. Average yield was not in excess of 1.5 t/ha. It is grown mainly in the Central highland, Northern highland, Southern highland and Eastern high plateau regions in two cycles: the winter season and summer season. In the mid-altitude areas over 1000 m above sea level, wheat is grown in one cycle, October–November. Generally, the productivity of wheat per hectare is low due to several biotic and abiotic stresses. Severe epidemics of stripe rust were recorded in Yemen in 1991, 1993, 1994 and 1996 when yield losses ranged from 10% to 50% (AREA, 1997). In the case of stem or black rust, the causal pathogen is *Puccinia graminis* f. sp. *tritici*. Infection occurs through spores transmitted by wind. The disease is among the most threatening diseases of wheat, especially for long-duration cultivars under favourable conditions, 20–30°C and high humidity.

Scientists have developed several new wheat varieties with resistant properties they hope will help eradicate the virulent wheat rust Ug99 that has spread from Africa to Asia and is devastating the Iranian wheat crop (Singh *et al.*, 2011). Seven races belonging to the Ug99 lineage are now known and have spread to various wheat-growing countries in the eastern African highlands, as well as Zimbabwe, South Africa, Sudan, Yemen and Iran. Because of the susceptibility of 90% of wheat varieties grown worldwide, the Ug99 group of races was recognized as a major threat to wheat production and food security (Singh *et al.*, 2011). The wind-borne fungus moving across countries was first detected in Uganda in 1999, hence the name Ug99. It spread east and north-east through Africa and into south-west Asia, as far as the north of Iran (Singh *et al.*, 2008). There are now 11 known races of Ug99. They are all closely related and are believed to have evolved from a common ancestor, but differ in their virulence/avirulence profiles and the countries in which they have been detected (Hodson *et al.*, 2012). Global wind models indicate that the fungus may next spread to Iran's neighbouring countries, including Pakistan, Afghanistan and India, while in Yemen it was detected in the summer of 2006 in farmers' fields and Ug99 nurseries. However, this stem rust race was not detected in the winter season of 2006 as determined by collecting stem rust samples and analysing them in foreign laboratories in the USA and Canada. Different Ug99 nurseries have been planted with different lines to confirm presence of the disease. The confirmation of appearance of Ug99 in Yemen was by growing wheat at the Tihama Research Station in 2007. After that, ICARDA sent nurseries each year for evaluation against Ug99 disease and for growing them in different wheat-production areas, such as Taiz, Sana'a, Mareb, Sayuon, Dhamar and Al Kod, in order to monitor Ug99 appearance. In the 2008 summer season, one of many different wheat nurseries at Dhamar Research Station Farm contained 11 lines of wheat, including one mutant that differed in its level of resistance. In this context, it is worth mentioning that under the INT/5/150 Project 'Responding to the Transboundary Threat of wheat Black Stem Rust' supported by the IAEA, several activities have been carried out. In total, 12 mutant variants of M_2 wheat derived from 'Sonalika' variety along with the parent were sent to Njoro (Kenya) in 2009 for screening for resistance to stem rust disease. Out of these, two mutants were identified (namely, Erra-2009-WM2-1 and Erra-2009-WATs) as having medium resistance to Ug99. This result was reported by Davendor, the focal point for implementing CIMMYT nurseries in Kenya, at the 2nd technical meeting of 2010, held in Kenya under the INT/5/150 Project (Davendor, 2010). Another activity implemented under the INT/5/150 project consisted of two sets, each composed of 13 mutant variants of 'Giza 186' and ten mutants of 'Saha 93' along with the parents, planted in a disease hotspot at Tihama in Yemen and Njoro in Kenya for screening for Ug99 disease resistance. This chapter illustrates two types of results: screening of mutants in Tihama and Njoro for stem rust disease resistance; and advanced yield trials of selected mutants for yield potential and resistance to yellow rust disease in farmers' fields.

2 Materials and Methods

The study consists of two cycles.

2.1 Cycle I: Initial screening

Two sets of M_3 wheat mutant variants along with their parents, each consisting of 13 and ten mutants derived from 'Giza 186' and 'Saha 93' varieties, respectively, were planted in Tihama in Yemen and Njoro in Kenya at hotspots for Ug99 disease. The planting was done on 10 December 2012 and 5 August 2012 under irrigation. Each mutant was sown in two rows that were 3 m long and 30 cm apart; and a spreader variety (highly susceptible) was sown after every eight rows. Natural conditions in both locations were favourable for the development of Ug99 disease in terms of temperature (20–25°C) and high humidity, and allowed the breeder to screen mutants against the disease. The mutant variants were scored for stem rust reaction twice, on 22 January 2012 and 15 February 2012, in Tihama, using a modified Cobb method at seedling stage (10–25-day-old plants), and later at the start of spike emergence. The same steps were followed for scoring resistance of mutants in Njoro, Kenya. Scoring was done on 15 August 2012 and 22 September 2012 based on type of infection of the plant, and the following system was used for scoring of disease response: Immune (0), Resistant (R), Moderately Resistant (MR) and Susceptible (S). The wheat (*T. aestivum* L.) cultivars 'Giza 186' and 'Saha 93' are high-yielding introduced varieties which are being improved for rust disease resistance using mutation techniques.

2.2 Cycle II: Advanced yield trial

Two mutants of 'Giza 186' and four mutants of 'Saha 93' along with their parents were planted on 12 July 2014 and 22 July 2015 in an advanced yield trial at Shibam (Yemen) with three replications aimed at evaluating mutant lines for yellow rust disease resistance and yield potential. Each mutant was planted in six rows that were 6 m long and 30 cm apart. The following data were recorded: days to flowering and maturity, plant height, seeds per spike, 1000-kernel weight, grain yield and score for yellow rust disease resistance using a modified Cobb scale: Resistance, Medium Resistance and Susceptible. Data were collected for all of the above attributes and pooled analysis was done using the LSD0.05 method for separating the means of mutants and to select the best lines.

3 Results

3.1 Cycle I: Initial screening

Resistance scores of M_3 mutant variants grown in Tihama for stem rust disease screening showed that two mutants of 'Giza 186' (G-M2-2010-1-28 and G-M2- 2010-41-52) and four of 'Saha 93' (S-M2- 2010-16-12, S-M2- 2010-21-13, S-M2- 2010-22-14 and S-M2- 2010-27-15) were immune to Ug99 disease, while the score for the untreated parent ranged from 0 and 50S at both seedling and spike emergence stages (Table 32.1). Meanwhile, the reaction of mutants exposed to stem rust disease at Njoro differed slightly from that obtained in Tihama. The same two mutant variants (G-M2- 2010-1-28 and G-M2- 2010-41-52), derived from 'Giza 186', were resistant and varied from a score of MR to immune. Only two mutants of 'Saha 93' showed resistance to Ug99 disease, namely S-M2- 2010-16-12 and S-M2- 2010-27-15. The score for the untreated parent ranged from 0 to 30S at both seedling and spike emergence stages (Table 32.2). The two mutants S-M2- 2010-21-13 and S-M2-2010-22-14 that appeared to be resistant in Tihama were susceptible in Njoro. The rest of the mutants in both locations varied in susceptibility to stem rust reaction (Table 32.2). In this context, it is worth mentioning that the similarity of the results for stem rust reaction of some mutants at both locations might be due to the occurrence of the same, or related, races of Ug99. Stem rust testing nurseries of different lines planted by ICARDA in different countries, including Yemen, Kenya and Ethiopia, showed that the related races for Ug99 (Sr24-, Sr31+, Sr36+) are common in Kenya, Ethiopia and Yemen but with differences of virulence (Figueroa *et al.*, 2016).

Table 32.1. Name/ Pedigree of M_3 mutant variants of 'Giza 186' and 'Saha 93' varieties (Season 2012; Tihama location, Yemen).

			Stem rust reaction	
Order number	Name of parent line	Mutated variant	25/01/2012	22/02/2012
1	Giza186	G-M2- 2010-1-28	0	0
2	Giza186	G-M2- 2010-2-41	MS	20S
3	Giza186	G-M2- 2010-7-42	5S	10S
4	Giza186	G-M2- 2010-13-44	5S	20MS
5	Giza186	G-M2- 2010-21-45	5MS	10MS
6	Giza186	G-M2- 2010-23-46	5S	20S
7	Giza186	G-M2- 2010-31-47	0	30S
8	Giza186	G-M2- 2010-33-48	5S	20S
9	Giza186	G-M2- 2010-35-49	0	20S
10	Giza186	G-M2- 2010-41-52	0	0
11	Giza186	G-M2- 2010-43-55	5S	20MS
12	Giza186	G-M2- 2010-49-56	10S	40S
13	Giza186	G-M2- 2010-49-43	0	10S
14	Giza186	Untreated	0	40S
15	Saha 93	S-M2- 2010-7-1	0	0
16	Saha 93	S-M2- 2010-7-2	TR	30S
17	Saha 93	S-M2- 2010-9-3	0	20S
18	Saha 93	S-M2- 2010-9-5	TR	20S
19	Saha 93	S-M2- 2010-13-6	TR	20S
20	Saha 93	S-M2- 2010-16-12	0	0
21	Saha 93	S-M2- 2010-21-13	0	0
22	Saha 93	S-M2- 2010-22-14	0	0
23	Saha 93	S-M2- 2010-27-15	0	0
24	Saha 93	S-M2- 2010-29-16	5S	20S
25	Saha 93	Untreated	0	30S

3.2 Cycle II: Advanced yield trial

Results of this trial showed significant yield differences between mutants and their parents over the years. The differences due to their interactions were also significant (Table 32.3). Most of the mutant lines tended to give a comparatively higher yield than their parents, with the exception of Erra-010-SwM2-27-15-36 which, among the mutants, had the lowest yield (3023 kg/ha) but was not significantly different from the parent 'Giza 186', which had a yield of 3056 kg/ha. 'Saha 93' gave a yield of 2944 kg/ha (Table 32.4). The highest yield was recorded by Erra-008-GM2w-4-52-40, which attained a yield of 3768 kg/ha. Erra-010-GM2w-1-28-18 and Erra-010-SwM2-16-12-19 had similar grain yields but were significantly different with respect to their parents (Table 32.4). They gave 3579, 3566 and 3508 kg/ha, respectively, while the yields obtained from the two parents, 'Saha 93' and 'Giza 186', were 2944 and 3056 kg/ha, respectively. Generally, all tested mutant lines, on average, had a yield advantage and their yields ranged between 3426 and 3768 kg/ha with the exception of Erra-010-SwM2-27-15-36, which had the lowest yield among the mutants. These findings are supported by Amer (2000), Moshref (2001) and Abdulwahid and Aref (2007), who reported an increase in yield for induced wheat mutants tested in different locations, compared with their respective parents. Regarding plant height, Erra-008-GM2w-41-52-40 attained the highest value (74 cm) while the lowest was recorded for Erra-010-SwM2- 27-15-36 (59 cm). The rest of the mutants gave averages which ranged between 64 and 66 cm, while the parents 'Saha 93' and 'Giza 186' attained 58 and 61 cm, respectively (Table 32.4). Similar observations were reported by Abo-Hegazi (1978), Budak (1995), Moshref (2001) and Abdulwahid and

Table 32.2. Name/Pedigree of M_3 mutant variants of 'Giza 186' and 'Saha 93' varieties (Season 2012; Njoro, Kenya).

			Stem rust reaction	
Order number	Name of parent line	Mutated variant	15/08/2012	22/9/2012
1	Giza186	G-M2- 2010-1-28	MR	MR
2	Giza186	G-M2- 2010-2-41	30MS	40MS
3	Giza186	G-M2- 2010-7-42	30MS	40MSS
4	Giza186	G-M2- 2010-13-44	40MS	50MS
5	Giza186	G-M2- 2010-21-45	20MS	40MS
6	Giza186	G-M2- 2010-23-46	5MS	40MS
7	Giza186	G-M2- 2010-31-47	30MS	30MS
8	Giza186	G-M2- 2010-33-48	20MS	40MS
9	Giza186	G-M2- 2010-35-49	40MS	50MS
10	Giza186	G-M2- 2010-41-52	0	0
11	Giza186	G-M2- 2010-43-55	50MS	20MS
12	Giza186	G-M2- 2010-49-56	10S	40S
13	Giza186	G-M2- 2010-49-43	0	10S
14	Giza186	Untreated	0	50S
15	Saha 93	S-M2- 2010-7-1	20S	30S
16	Saha 93	S-M2- 2010-7-2	20S	30S
17	Saha 93	S-M2- 2010-9-3	0	20S
18	Saha 93	S-M2- 2010-9-5	5S	20S
19	Saha 93	S-M2- 2010-13-6	5S	30S
20	Saha 93	S-M2- 2010-16-12	0	0
21	Saha 93	S-M2- 2010-21-13	0	20S
22	Saha 93	S-M2- 2010-22-14	0	20S
23	Saha 93	S-M2- 2010-27-15	0	0
24	Saha 93	S-M2- 2010-29-16	5S	20S
25	Saha 93	Untreated	0	30S

Table 32.3. Average of pooled error for studied parameters over seasons and its significance.

Source of variation	df	Days to flowering	Plant height (cm)	Seeds per spike	1000-kernel weight (g)	Yield (kg/ha)
Year	1	959,207*	0.840**	2.3922*	30.43*	280.56*
Year(rep)	4	8.493	32.92	15.92	1.467	88,928
Mutants	7	521.18*	1,114.58*	1,243.12*	369.013*	886,490*
Mutant × year	7	39.69*	70.66**	40.20*	27.490*	551,794*
Error	28	1.917	19.09	0.2886	4.603	40,123

Analysis of variance results: **significant at 0.01, *significant at 0.05

Aref (2008, 2012), who found an increase in plant height in induced mutants when different doses of gamma-rays were used. The resistance score of the mutants with respect to yellow rust disease ranged from susceptible to resistant, including the parents. Erra-010-GM2w-41-52-40 and Erra-010-SwM2-16-12-19 were resistant to yellow rust disease, while the mutants Erra-010-GM2w-1-28-18 and Erra-010-SwM2-22-14-6 were medium in terms of their level of resistance. The parents 'Saha 93' and 'Giza 186' were susceptible, with scores ranging between 40S and 50S (Table 32.4). Based on these findings, it is obvious that the mutation induction technique is a valuable tool in creating variation within targeted varieties/cultivars to allow breeders to select for many traits, including disease resistance. For the number of days to heading and maturity, significant

Table 32.4. Grain yield, plant height and resistance score of wheat mutant lines for yellow rust disease tested at Shibam during the 2014 and 2015 seasons.

	Yield (kg/ha)			Plant height (cm)			
Name /Pedigree	2014	2015	Average	2014	2015	Average	YR reaction
Erra-010-GM2w-41-52-40	3710	3826	3768	72	76	74	R
Erra-010-GM2w-1-28-18	3372	3786	3579	63	69	66	MR
Erra-010-SwM2-16-12-19	3268	3864	3566	61	67	64	R
Erra-010-SwM2-22-14-6	3471	3544	3508	65	69	67	MR
Erra-010-SwM2-21-13-13	3318	3534	3426	61	68	65	S
Erra-010-SwM2-27-15-36	2946	3100	3023	57	61	59	40S
Parent Saha 93 (Untreated)	2884	3003	2944	56	60	58	40S
Parent Giza 186 (Untreated)	2993	3119	3056	59	63	61	50S
Average	3245	3472		62	67		
LSD at 0.05 for Mutant			456			5	
LSD at 0.05 for Years			212			3	
LSD0.05 for year x mutant			312			5	

differences were recorded among all tested mutants and their parents. The duration until heading and maturity ranged between 50–56 and 96–104 days, respectively. The parents 'Saha 93' and 'Giza 186' were late compared with the mutants, and they ranged from 60–62 and 104–112 days, respectively (Table 32.5). This result is more or less in agreement with the findings of Reddy and Gupta (1989), Moshref (2001) and Abdulwahid and Aref (2008, 2012), who found some early-maturity mutants from Triticale and wheat crops using gamma radiation.

The average number of seeds per spike varied within the tested mutants. The highest value was recorded from Erra-008-GM2w-41-52-40 (53 seeds) and the lowest was obtained from Erra-010-SwM2-27-15-36 (38 seeds) with the difference being significant at 0.05 *p* value. The rest of the mutants gave averages which ranged between 44 and 46 seeds per spike, while the two parents ranged between 39 and 43 seeds per spike. Generally, most of the mutants tended to give higher values of seeds per spike compared with their parents (Table 32.6). These results agree with the findings of Amer (2000) and Abdulwahid and Aref (2007, 2012).

With respect to 1000-kernel weight, Erra-008-GM2w-41-52-40 had the highest value (48 g) and significantly differed from the others, including the parents. This was followed by Erra-010-SwM2-22-14-6, Erra-010-SwM2-16-12-19 and Erra-010-GM2w-1-28-18 for which averages ranged between 42 g and 44 g. The mutant Erra-010-SwM2-27-15-36 gave the lowest value (38 g) compared with the other mutants, whereas no difference was found for values attained by the parents, 'Saha 93' and 'Giza 186', as shown in Table 32.6. These results are in agreement with those of Amer (2000) and Abdulwahid and Aref (2012).

4 Conclusion

According to the scores for the resistance of mutant lines to stem and yellow rust diseases, as well as yield potential and the preference of farmers, it can be concluded that Erra-010-GM2w-41-52-40 and Erra-010-SwM2-16-12-19 are the best mutants, and it is recommended to increase their seeds for disseminating over a wide range of farmers and wheat areas where rust diseases are a problem, and to register them as improved mutant varieties. The other mutants, S-M2- 2010-22-14 and S-M2-2010-21-13, that were resistant to stem rust disease in Tihama but susceptible in Njoro, might be used as stock for breeding of wheat for stem rust disease resistance.

Table 32.5. Days to flowering and maturity of wheat mutant lines tested at Shibam for the 2014 and 2015 seasons.

	Days to flowering			Days to maturity		
Name/Pedigree	2014	2015	Average	2014	2015	Average
Erra-010-GM2w-41-52-40	54	58	56	104	104	104
Erra-010-GM2w-1-28-18	49	51	50	96	100	98
Erra-010-SwM2-16-12-19	50	58	54	94	102	98
Erra-010-SwM2-22-14-6	50	54	52	94	98	96
Erra-010-SwM2-21-13-13	50	56	53	97	100	99
Erra-010-SwM2-27-15-36	54	58	56	100	104	102
Parent Saha 93 (Untreated)	58	62	60	103	105	104
Parent Giza 186 (Untreated)	60	64	62	110	114	112
Average	53	58		100	103	
LSD at 0.05 for Mutant			4			6
LSD at 0.05 for Years			4			5
LSD0.05 for year x mutant			6			5

Table 32.6. Number of seeds/spike and 1000-kernel weight of wheat mutant lines tested at Shibam during the 2014 and 2015 seasons.

	Seeds/spike			1000-kernel weight (g)		
Name/Pedigree	2014	2015	Average	2014	2015	Average
Erra-010-GM2w-41-52-40	50	56	53	46	50	48
Erra-010-GM2w-1-28-18	44	48	46	40	44	42
Erra-010-SwM2-16-12-19	44	46	45	40	46	43
Erra-010-SwM2-22-14-6	42	48	45	44	44	44
Erra-010-SwM2-21-13-13	42	46	44	40	42	41
Erra-010-SwM2-27-15-36	36	40	38	40	38	39
Parent Saha 93 (Untreated)	38	40	39	38	38	38
Parent Giza 186 (Untreated)	41	45	43	36	40	38
Average	42	46		41	43	
LSD at 0.05 for Mutant			5			4
LSD at 0.05 for Years			4			4
LSD0.05 for year x mutant			3			5

References

Abdulwahid, A. and Aref Al-Shamiri (2007) *Improvement of some important crops in Yemen through induced mutations*. Seventh Arab Conference on the Peaceful Use of Atomic Energy, Sana'a, Yemen. 4–8 Dec, pp. 256–278.

Abdulwahid, A. and Aref Al-Shamiri (2008) Evaluation of yield and agronomic characters of local wheat mutants under rainfed condition. *Yemeni Journal of Agricultural Research and Studies* 18, 13–27.

Abdulwahid, A. and Aref Al-Shamiri (2012) different mutant lines of local barley variety (*T. Hordeum vulgare* cv. Bakkur) with improved quantitative and qualitative traits under rainfed conditions. Eleventh Arab Conference on the Peaceful Uses of Atomic Energy, Sudan, December, pp. 245–276.

Abo-Hegazi, A. (1978) Studies on the effect of Gamma rays on some characteristics of Egyptian, and Mexican varieties of wheat in a breeding program for their improvement. *Isotope and Radiation Research* (Suppl.) *10*, 2 139–157.

Amer, I. (2000) *Mutation breeding in wheat crop*. 40th Science week, Syrian Arab Republic, pp. 243–254.
AREA (1997) *Survived study on wheat rusts disease in Yemen*. Report made by specialized personnel. Agricultural Research Authority (AREA), Yemen.
Budak, N. (1995) Plant height response of semi-dwarf and non-semi-dwarf wheat to the environment. *Crop Sci.* 447–499.
Davendor (2010) Responding to the trans-boundary threat of wheat black stem rust (Ug99) 2nd Coordination Meeting of INT/5/150, Ngoro-Kenya.
Figueroa, M. *et al.* (2016) Changing the game: using integrative genomics to probe virulence mechanisms of the stem rust pathogen *Puccinia graminis* f. sp. *Tritici*. *Frontiers in Plant Science* 7, 205.
GTZ (1985) *Plant Protection Project* (1985) *Annual Report*. GTZ, Sana'a.
Hodson, D.P. *et al.* (2012) *Tracking the Wheat Rust Pathogens*. Borlaug Global Rust Initiative Technical Workshop Proceedings, 1–4 September 2012, Beijing.
Kamal, M. *et al.* (1985) *Research in Plant Pathology in Yemen*. Final Report of the FAO Plant Pathologist. April 1976 and Feb.1985, Agri-Res. Authority, Taiz, YAR.
MAI (2015) *Statistical Year Book* (1988–2014). General Administration of Statistics, Ministry of Agriculture and Irrigation, Yemen.
Moshref, T. (2001) Yield performance and stability of new bread wheat genotypes under various environments. *Journal of Agricultural Science (Mansoura University)* 26, 575–583.
Reddy, V. and Gupta, P. (1989) Induced mutations for crop improvement. *Mutant Breeding* 7, 1–41.
Singh, R.P. *et al.* (2008) Will stem rust destroy the world's wheat crop? *Advances in Agronomy* 98, 270–309.
Singh, R.P. *et al.* (2011) The emergence of Ug99 races of the stem rust fungus is a threat to wheat world production. *Annual Review of Phytopathology* 49, 465–481.
Watt, M. (1975) *Crop Protection in the Tihama Region, YAR*. FAO, *Rome*.

33 Determination of Radiosensitivity of *Coffea arabica* var. 'Venecia' Seeds to Gamma-ray Irradiation

Reina Céspedes, Noel Arrieta, Miguel Barquero[1], Ana Abdelnour, Nielen Stephan and Ingelbretch Ivan

Abstract

Coffee is one of the most commercially available raw materials, being the tropical product with the highest market value in the world. In Costa Rica it is the third most important product for agricultural exports and provides the main income for many families in the country. However, coffee is under threat due to coffee leaf rust disease (CLR). Mutation breeding in coffee is a promising approach to develop new varieties resistant to CLR. As a new technology for coffee, basic tests related to mutation induction need to be done. The plant material used was *Coffea arabica* var. 'Venecia' seeds, with a moisture content of 27.3%. The applied irradiation doses were 0, 80, 100, 120, 140, 160 and 180 Gy. For each treatment, three replicates of 200 g were used, with a seed number range of 765–808 units per replicate. The irradiated seeds were planted on the same day. Eighty days after treatment the number of seedlings was quantified, the hypocotyl height and radicle length were measured and the opening of cotyledons was determined for each dose. The effects of the radiation doses on seed germination frequency were recorded. At the dose of 80 Gy, germination was reduced over the control by 9.65%, at 100 Gy by 34.06%, at 120 Gy by 52.76%, at 140 Gy by 60.24%, at 160 Gy by 65.56% and at 180 Gy by 75.40%. Seedling growth was affected and a delay in opening of the cotyledons was observed at higher doses. This radiosensitivity test, based on seed germination as compared with unirradiated control, revealed that the LD_{50} for the variety tested is in the range 100–120 Gy experimentally, and according to the regression is 125 ± 30 Gy. This dose will be used for further bulk experiments and is of great importance, because the LD_{50} is considered as the range where the appearance of useful mutations in breeding programmes is favoured. The establishment of these parameters is a necessary advance to continue with measurements of genetic and phenotypical parameters to implement mutation breeding in coffee looking for new sources of resistance against CLR.

Keywords: LD_{50} • coffee • Gy • coffee leaf rust • ionizing radiation

1 Introduction

Coffea arabica is the most widely cultivated species of *Coffea* and the only tetraploid species (2× = 44) in the genus (Lim, 2013). The two main species cultivated throughout the tropical world are *Coffea arabica* and *Coffea canephora*, which represent 70% and 30%, respectively, in world production (Perrois *et al*., 2015). Over the past 50 years, both production and consumption of coffee have risen considerably. Now over 70 countries produce coffee. Some of the coffee-producing countries have seen considerable benefits through higher yields and growing volumes of sales. But many, especially smallholders, who produce most of the world's coffee, are also

[1] Email: mbarquero@icafe.cr

 Mutation Breeding, Genetic Diversity and Crop Adaptation to Climate Change (eds S. Sivasankar *et al.*)

DOI: 10.1079/9781789249095.0033

facing growing challenges from climate change and more difficult natural growing conditions (FAO, 2015.) One of the main problems is attacks by pests and diseases, as coffee growth is affected by microorganisms such as fungi, bacteria, viruses and nematodes (Canet *et al.*, 2016). Among these diseases, coffee leaf rust (CLR), caused by *Hemileia vastatrix* (*Hv*), represents the biggest threat to coffee production worldwide and ranks amongst the most serious fungal diseases in history (Silva *et al.*, 2018). It is important to mention that the fungus causing this disease is of high mutability, quickly breaking the coffee resistance genes. The existing physiological rust races have already broken almost all the *SH* resistance genes (Del Grossi *et al.*, 2013). Many coffee cultivars considered resistant in the past are now presenting susceptibility. For a perennial crop like coffee, where a cultivar needs a lifespan of at least 15 years, durable resistance to leaf rust is very important to be a successful cultivar (Hiroshi *et al.*, 2007).

For all the reasons above, the induction of genetic variation by different methods and techniques becomes a basic tool in the hands of plant breeders for the development of new cultivars with tolerance to abiotic stress, resistance to pests and diseases and improvement of quality and agricultural yield (Bermúdez *et al.*, 2016).

While mutations may occur spontaneously, they can also be induced artificially. Artificially induced mutations can be created by physical mutagens, such as X-rays, gamma-rays and neutrons, and by chemical mutagens in plant mutation breeding. Gamma-rays are ionizing radiation and are used in inducing mutations in seeds and other planting materials (Beyaz and Yildiz, 2017). Gamma-rays have proved to be more economical and effective compared with other ionizing radiations because of their easy availability and power of penetration. Previous studies have revealed that seed exposure to high doses of gamma-rays disturbs protein synthesis, hormone balance, leaf gas exchange and enzyme activity. The morphological, structural and functional changes depend on the strength and duration of gamma doses of exposure (Marcu *et al.*, 2013).

Mutant individuals present negative changes at an increasing rate as the irradiation dose increases, for which reason it is very important to know the lethal dose (LD_{50}). The LD_{50} corresponds to the amount of irradiation absorbed where 50% of the population that has been exposed survives, a proportion considered as the range where the appearance of useful mutations in breeding programmes is favoured (Espino *et al.*, 2013).

At present, there is a lack of information on the effects of gamma irradiation on *Coffea arabica*. The following research intends to contribute with the compilation of information in this regard, by performing a determination of the lethal dose (LD_{50}) of gamma irradiation in seeds of *Coffea arabica* var. 'Venecia' and evaluating the effects on growth.

2 Materials and Methods

2.1 Plant material

The plant material used corresponds to *Coffea arabica* var. 'Venecia' seeds, from the experimental farm of ICAFE 'La Palmira' in the coffee region of Pérez Zeledón. Dry seed (4.2 kg) was obtained from the Costa Rican Coffee Institute with a moisture content of 27.3%.

2.2 Gamma irradiation

The equipment used for the irradiation process was an Ob-Servo Ignis irradiator, with a cobalt 60 radioactive source, and an activity of 4.4×10^{14} Bq (Becquerel). Seven irradiation treatments were performed, at 0, 80, 100, 120, 140, 160 and 180 Gy. The dose rate corresponded to 0.8 Gy/s.

2.3 Radiosensitivity test

For each treatment, three replicates of 200 g each were used, with the number of seeds in each sample varying from 765 to 808. The irradiated seeds were planted in 71 cm × 39 cm boxes divided into three sections, one for each replica, and then covered with a 6 cm substrate layer. The treatments were irrigated daily for 13 weeks and germination of the seeds was evaluated weekly.

2.4 Measurement of hypocotyl, cotyledons and radicle

For measurement of the variables a sample of each treatment was taken, the sample size being defined by the treatment with the lower number of plants. A measurement of the length of the hypocotyl and the radicle was made. For plants with cotyledons, width and length were measured.

2.5 Statistical analysis

For each of the applied doses a weekly seed germination curve was carried out. The LD_{50} was obtained by linear regression. An analysis of variance (ANOVA) was carried out to determine the effects of the irradiation on the parameters, cotyledon width, cotyledon length, hypocotyl length and radicle length.

3 Results

With the seed germination weekly data, the germination curves of the treatments were determined. Seed germination started on the 7th week after irradiation. In the last evaluation, the maximum germination rate occurred in the non-irradiated treatment (0 Gy). As shown in Fig. 33.1, there is a seed germination pattern. As the dose of irradiation increased, the seed germination decreased, a behaviour that was maintained throughout the evaluation.

At the end of the evaluation, after the 13th week, the total numbers of seeds germinated by the treatments were obtained according to each dose. A linear regression was obtained with the predicted values (Fig. 33.2). The same decreasing pattern by increasing the radiation doses was observed. It is estimated that the LD_{50} is 125 ± 30, according to the linear regression.

A one-way analysis of variance was done with each of the variables: cotyledon width, cotyledon length, radicle length and hypocotyl length (Fig. 33.3). For each treatment we worked with a sample of 126 randomly selected seedlings and this number was defined by the treatment with lowest germination (180 Gy). For each of the variables there were significant differences.

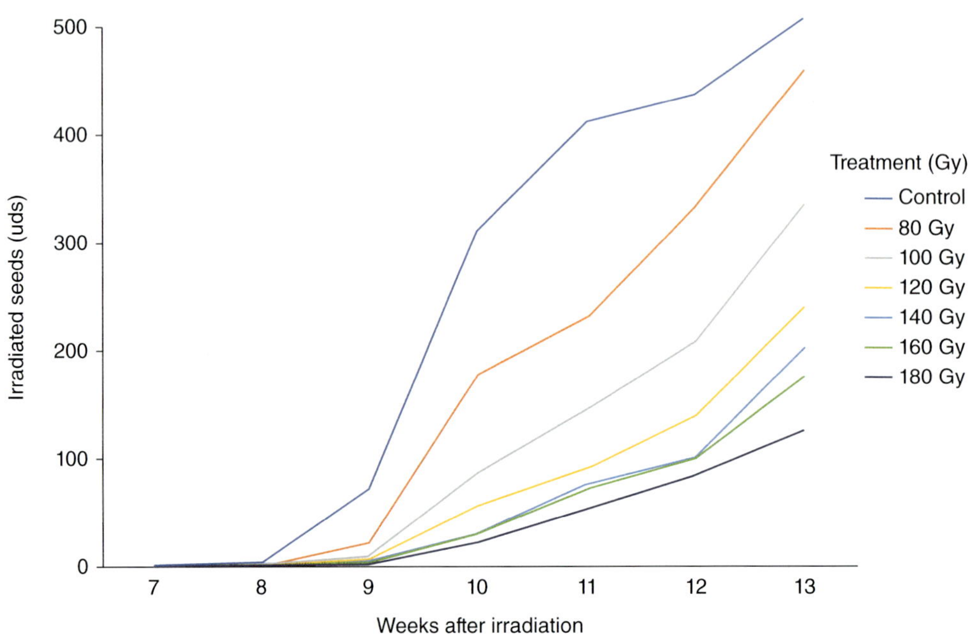

Fig. 33.1. *Coffea arabica* var. 'Venecia' seed germination curves from 7th to 13th week for each dose of irradiation.

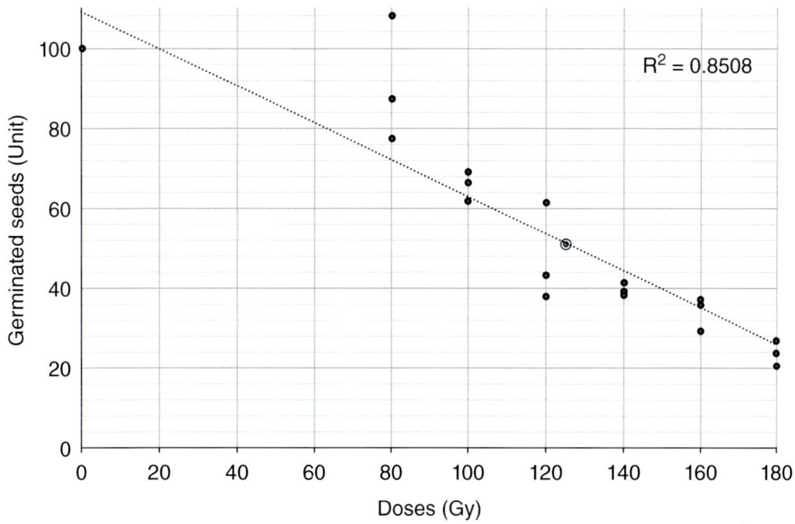

Fig. 33.2. Determination of the LD_{50} of *Coffea arabica* var. 'Venecia' seeds (125 ± 30 Gy).

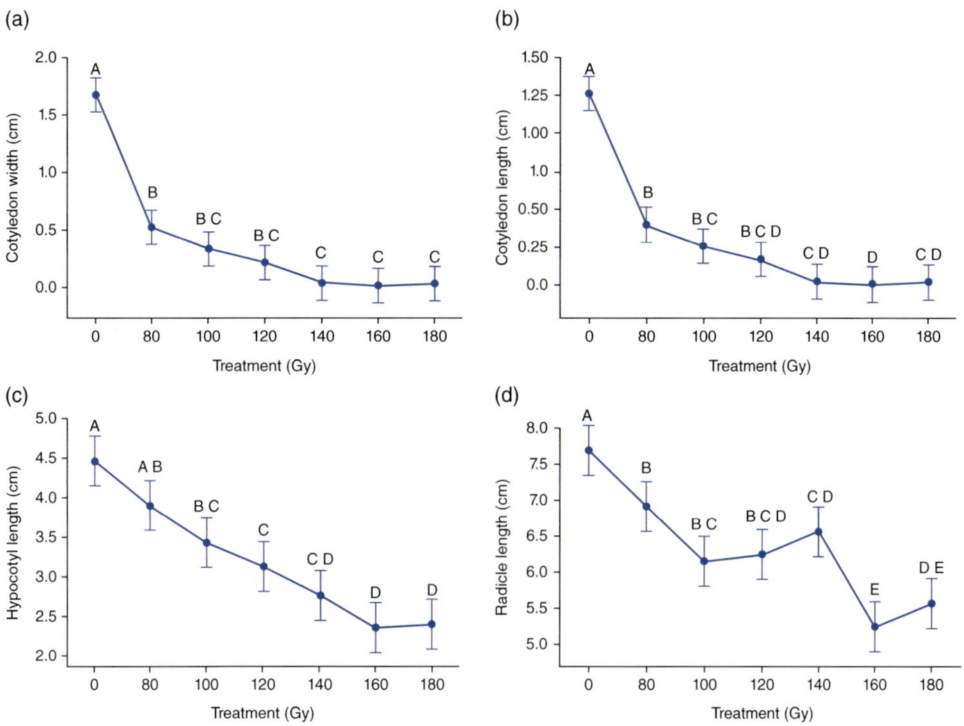

Fig. 33.3. Effect of radiation doses on *C. arabica* var. 'Venecia' growth. **(a)** Cotyledon width. **(b)** Cotyledon length. **(c)** Hypocotyl length. **(d)** Radicle length. Bars mean 95% of confidence interval (n = 126).

4 Discussion

Coffee seed germination decreased as the radiation dose increased and all irradiated treatments showed growth inhibition. These results are in accordance with the findings of previous researchers who reported that the seed germination potential of different crops decreased by increasing the irradiation dose. Growth inhibition induced through high-dose irradiation has been attributed to cell cycle arrest in the G2/M phase during somatic cell division and/or to a variety of damages in the entire genome. In fact, the inhibition of mitosis may be transient but also the radiological lesion that occurs at the gene and chromosomal level can be lethal for dividing cells, which are highly sensitive to radiation (Iglesias *et al.*, 2010). Processes such as auxin destruction, changes of the ascorbic acid content and physiological and biochemical disturbances could induce the inhibition of plant germination and development (Marcu *et al.*, 2013) as shown by the obtained data (Fig. 33.1). It is important to mention that coffee seeds usually require between 45 and 60 days to germinate. One of the most critical factors in germination, both in the laboratory and in the soil, is the endocarp, also known as parchment. Parchment removal allows the seed to germinate within 14–21 days. Its presence retards germination by limiting the availability of oxygen to the embryo (Guevara *et al.*, 1997). In our case, the removal of the endocarp was not done, which explains the need to perform an evaluation for a longer period.

The lethal dose may vary depending on the species, type and dose of irradiation (Aparna *et al.*, 2013). It was found that the LD_{50} in the 'Venecia' variety is around 120 Gy, 125 ± 30 Gy according to the linear regression (Fig. 33.2). It is important to consider that the determination of the lethal dose can vary according to several factors. One of these is the water content in the plant material. Ionizing radiation can cause the breaking of covalent bonds and the breakdown of water molecules. This produces the formation of free radicals that can damage the different organelles of the cell. Additionally, these radicals can cause damage to the DNA molecule, which can sometimes induce inversions, translocations and changes in the structure of the chromosomes (Álvarez *et al.*, 2017). It was determined that the moisture content of the treated seed was 27.3%, being a reference for further investigations. The same decreasing pattern by increasing the radiation doses was observed. It is estimated that the LD_{50} is 125 ± 30 Gy, according to the linear regression (Fig 33.2).

As a result of the treatments, it was possible to observe significant differences in growth in all the evaluated variables: cotyledon width, cotyledon length, hypocotyl length and radicle length. The effect of growth reduction was observed in all the variables of the irradiated treatments compared with non-irradiated (Fig. 33.3). Aparna *et al.* (2013) reported that radiation increases plant sensitivity to gamma-rays and this may be caused by the reduced amount of endogenous growth regulators, especially cytokinins, as result of breakdown, or lack of synthesis, due to radiation. They noticed that treating seeds with high rates of gamma radiation reduced germination, with a corresponding decline in the growth of plants. This also explains the reduction in the appearance and growth of the cotyledons. Also, the reduction of root length at higher doses of gamma-rays is presumably produced by the decreasing mitotic activity in meristematic tissues (Aparna *et al.*, 2013).

5 Conclusion

With this research it was possible to determinate the lethal dose (LD_{50}) of gamma irradiation in *Coffea arabica* var. 'Venecia' seeds. This is around 120 Gy, specifically 125 ± 30 Gy, according to the linear regression.

The effect of irradiation dose on the development of *Coffea arabica* var. 'Venecia' was to reduce the width of the cotyledon and the length of the cotyledon, hypocotyl and radicle. It was possible to determine the existence of significant differences in the evaluated variables. In the same way, the number of germinated seeds per treatment reduced as the radiation dose increased.

References

Álvarez, A. *et al.* (2017) Dosis óptima de irradiación gamma con Co60 para inducción de mutagénesis en pastos. *Nova Scientia* 9(2), 65–82.

Aparna, M. *et al.* (2013) Impact of gamma rays on the seed germination and seedling parameters of groundnut (*Arachis hypogaea* L.). *Asian Journal of Experimental Biological Sciences* 4(1), 61–68.

Bermúdez, I. *et al.* (2016) Mutagénesis in vitro en suspensiones celulares embriogénicas de banano cv. Grande naine (Musa AAA). *Biotecnología Vegetal* 16(2), 103–111.

Beyaz, R. and Yildiz, M. (2017) The use of gamma irradiation on plant mutation breeding. In: Jurić, S. (ed.) *Plant Engineering*. IntechOpen, London. doi: 10.5772/intechopen.69974.

Canet, G. *et al.* (2016) *La situacion y tendencias de la produccion de cafe en America Latina y el Caribe*. Instituto Interamericano de Cooperacion para la Agricultura (IICA) y Centro de Investigacion y Asistencia en Tecnologia y Diseno del Estado de Jalisco A.C. (CIATEJ).

Del Grossi, L. *et al.* (2013) Rust resistance in Arabic coffee cultivars in Northern Paraná. *Brazilian Archives of Biology and Technology* 56(1), 27–33.

Espino, A. *et al.* (2013) Determinación de la dosis letal (DL50) con Co60 en vitroplántulas de *Agave tequilana* var. Azul. *Revista Fitotecnia Mexicana* 36(4).

FAO (2015) *FAO Statistical Pocketbook*. Food and Agriculture Organization of the United Nations, Rome.

Guevara, E. *et al.* (1997) Efecto del sustrato y su condición hídrica sobre la germinación de semilla de café caturra. *Agronomía Costarricense* 21(2), 207–216.

Hiroshi, G. *et al.* (2007) Resistance to leaf rust in coffee carrying SH3 gene and other SH genes. *Brazilian Archives of Biology and Technology* 50(5), 753–757.

Iglesias, L. *et al.* (2010) Efecto de radiaciones gamma en *Abies religiosa* (Kunth) Schltd. Et Cham. *Revista Chapingo: Serie Ciencias forestales y del ambiente* 16(1), 5–12.

Lim, T.K. (2013) *Edible Medicinal and Non-Medicinal Plants*, Vol. 5. Springer Science, Dordrecht.

Marcu, D. *et al.* (2013) Gamma radiation effects on seed germination, growth and pigment content, and ESR study of induced free radicals in maize (*Zea mays*). *Journal of Biological Physics* 39, 625–634.

Perrois, C. *et al.* (2015) Differential regulation of caffeine metabolism in *Coffea arabica* (Arabica) and *Coffea canephora* (Robusta). *Planta* 241, 179–191.

Silva, D. *et al.* (2018) Population genomic footprints of host adaptation, introgression and recombination in coffee leaf rust. *Molecular Plant Pathology* 19(7), 1742–1753.

34 Determination of the Optimal Conditions for Mutagenesis Induction in a Commercial Arabica Coffee Variety

Andrés Gatica-Arias[1], Jorge Rodríguez-Matamoros, Ana Abdelnour-Esquivel and Marta Valdez-Melara

Abstract

Low genetic diversity and autogamous reproduction limit genetic improvement of *Coffea arabica* L. As a consequence, susceptibility to biotic and abiotic stresses increases. Induced mutagenesis is an alternative strategy for increasing genetic variability and for the development of varieties tolerant or resistant to biotic and abiotic factors. In the present study, the effect of three mutagenic agents (NaN_3, EMS and ^{60}Co gamma-rays) on survival of Arabica coffee zygotic embryos was evaluated. The zygotic embryos were immersed for 10 min in a solution of NaN_3 (0, 2.5, 5.0, 7.5, 10.0, 12.5, 15.0 and 20.0 mM) or for 2 h in a solution of EMS (0, 0.5, 1, 1.5, 2, 4 and 6% v/v) or irradiated with 0, 20, 40, 60, 80 or 100 Gy. As the concentration or dose of the applied mutagen increased, survival decreased. The LD_{50} values for sodium azide, EMS and ^{60}Co were 12.5 mM (51.6%), 1% v/v (48.3%) and 40 Gy (50.0%), respectively. Our results indicated that coffee zygotic embryos are suitable for chemical and physical mutagenesis and this offers an alternative for the genetic improvement of agriculturally important traits in coffee.

Keywords: lethal dose • zygotic embryos • EMS • sodium azide • gamma rays • mutations

1 Introduction

Coffee is an important beverage around the world and is one of the main export products for many countries in Africa, Asia and Latin America (Mishra and Slater, 2012; Davis *et al.*, 2019). Almost 60% of global coffee production relies on *Coffea arabica* L. because of its quality, aromatic characteristics and low caffeine content (Mishra and Slater, 2012; Ahmed *et al.*, 2013; Davis *et al.*, 2019). The main cultivated varieties of *C. arabica* L. are derived from 'Typica' or 'Bourbon' coffee, which has resulted in low genetic diversity among varieties of this species (Mishra and Slater, 2012). The autogamous mode of reproduction and the mode of dispersion from the centre of origin have also contributed to this genetic uniformity (Mendonça, 2014). As a consequence, growth and production of *C. arabica* L. are threatened by biotic and abiotic stresses.

Induced mutagenesis offers a promising alternative for the development of varieties resistant to these stresses, since it could accelerate the spontaneous mutation process that occurs naturally in plants and increase the pool of genes available for genetic improvement (Gressel *et al.*, 2006). Exposure of vegetative or reproductive plant tissues and organs to ultraviolet (UV),

[1] Email: andres.gatica@ucr.ac.cr

 Mutation Breeding, Genetic Diversity and Crop Adaptation to Climate Change (eds S. Sivasankar *et al.*)

DOI: 10.1079/9781789249095.0034

X-ray, gamma radiation, sodium azide (NaN_3) and ethyl methanesulfonate (EMS) causes heritable changes in the DNA (Mba *et al.*, 2010). Physical and chemical agents with mutagenic properties have been used in plant breeding programmes to generate novel mutants with improved yield and quality traits, as well as resistance against biotic and abiotic stresses (Ali *et al.*, 2014; Wannajindaporn *et al.*, 2014). Moreover, mutants are not subject to the regulations imposed on genetically modified organisms. In this sense, coffee mutant varieties could be approved and accepted by consumers. In the present study, a procedure was developed for the treatment of zygotic embryos of *C. arabica* L. var. 'Caturra' with NaN_3, EMS and gamma-rays (^{60}Co).

2 Materials and Methods

2.1 Plant material

Mature coffee fruits (*C. arabica* L. var. 'Caturra') were acquired from the Coffee Research Center (CICAFE) of Costa Rica. Zygotic embryos were used as plant material in experiments to optimize conditions for mutagenesis.

2.2 Disinfection of zygotic embryos

Mature fruit from genetically homogeneous mother plants maintained in the greenhouse or in the field were collected and the pulp, mucilage and parchment were removed by hand. A uniform seed stock was prepared by selecting disease-free seeds and removing any small, shrivelled or damaged seeds. The seeds were immersed in distilled water with two drops of Tween 20 for 30 min with orbital rotation, then disinfected in 3.5% (v/v) sodium hypochlorite for 1 h and finally rinsed three times with sterile distilled water. The seeds were soaked in sterile distilled water for 48 h before the zygotic embryos were excised in a laminar flow cabinet with the aid of tweezers and a scalpel. The zygotic embryos were maintained in a solution of citric acid and ascorbic acid (100 mg/ml each; pH 5.6) and disinfected one last time with 0.01% w/v NaOCl for 5 min. Finally, the zygotic embryos were washed three times with sterile distilled water.

2.3 Mutagen dose determination

Disinfected zygotic embryos were incubated in conical tubes (15 ml Falcon) with 10 ml of Murashige and Skoog (MS) medium (Murashige and Skoog, 1962) complemented with either 0.0, 2.5, 5.0, 7.5, 10.0, 12.5, 15.0 or 20.0 mM NaN_3 adjusted to pH 3.0 with 0.1 M K_2HPO_4–KH_2PO_4 buffer for 10 min, or EMS (0, 0.5, 1, 1.5, 2, 4 and 6% v/v) for 2 h. The cultures were placed in the dark at 26 ± 2°C on a horizontal platform at 150 rpm throughout the experiment.

For the irradiation experiments, the excised zygotic embryos were cultured in Petri dishes containing 20 ml of germination medium (GER) consisting of full-strength MS salts supplemented with BAP (benzyladenine) at 1 mgL, GA_3 (gibberellic acid) at 1 mgL, sucrose at 30 g/l and Gelrite™ at 2.5 g/l, pH 5.6. Then the zygotic embryos were irradiated with ^{60}Co gamma rays (0, 20, 40, 60, 80 and 100 Gy; exposure time 0, 17 s, 35 s, 53 s, 70 s and 88 s, respectively) using an ObServo Ignis (Institute of Isotopes Co., Budapest, Hungary). The treated zygotic embryos were washed three times for 1 min with 10 ml of MS medium (pH 5.6). Finally, 10 ml of MS medium were added and the cultures were maintained in the dark for 24 h with constant shaking (100 rpm) at 26 ± 2°C. All reagents were purchased from Sigma-Aldrich® unless otherwise specified. In all cases, the pH of the medium was adjusted to 5.6 with HCl (1 N) or NaOH (1 N) before autoclaving for 21 min at 121°C. Each treatment was carried out in duplicate, with 30 embryos per treatment. After mutagenic treatment, the zygotic embryos were incubated in the dark at 37°C for 24 h in a solution of 1% w/v TTC (triphenyl tetrazolium chloride; Phytotechnology Laboratories®) and cell viability was observed using a stereoscope (SM2 8000; Nikon, Tokyo, Japan). The survival rate was calculated as: (number of surviving explants ÷ number of treated explants) × 100. The data was subjected to one-way analyses of variance (ANOVA) and the significance of the differences between means was analysed using Fisher's least significant difference (LSD) test at the 0.05 probability level. Statistical tests were processed in the Minitab® 16.1.0 statistical package.

2.4 Effect of antioxidants on germination of zygotic embryos

After mutagenic treatment of zygotic embryos with 1% v/v EMS as described above, the embryos were treated as follows: (a) mutagenized embryos were immersed in a 0.1% w/v citric acid/ascorbic acid antioxidant solution (pH 5.6) for a 16 h imbibition period (treatment AS); (b) mutagenized embryos were cultured in Petri dishes containing 20 ml of GER medium supplemented with activated charcoal at 300 mg/l (treatment AC); and (c) combination of treatments a + b (treatment AS + AC). Controls with no mutagenic agent or antioxidant agent were included. Each treatment consisted of two replicates composed of 25 zygotic embryos. Finally, the zygotic embryos were cultured in Petri dishes containing 20 ml of GER medium. The cultures were maintained at 26 ± 2°C under a 16 h photoperiod using fluorescent bulbs of 30 μmol/m²/s light intensity (Philips F32T8/DX PDT, 32 Watt). After 15 days of culture, the percentage of oxidation and germination of the zygotic embryos was determined.

2.5 Bulk mutagenesis

Zygotic embryos were disinfected and excised according to the protocol described above. For each treatment, 400 zygotic embryos were treated with NaN_3 (12.5 mM), EMS (1% v/v) or ^{60}Co gamma-rays (40 Gy) as described above. A control with no mutagenic treatment was also included. After exposure to the mutagenic agents, the zygotic embryos were cultured in Petri dishes containing 20 ml of GER medium supplemented with activated charcoal at 300 mg/l (treatment AC). The cultures were maintained for 7 days in the dark at 26 ± 2°C and afterwards under a 16 h photoperiod using fluorescent bulbs of 30 μmol/m²/s light intensity. Each treatment was performed in triplicate. After 4 weeks of culture, germination was calculated as: (number of germinated embryos × 100) ÷ total number of embryos.

3 Results

3.1 Mutagen dose determination

The steps and time required for mutation induction in zygotic embryos of *C. arabica* L. cv. 'Caturra' are described in Fig. 34.1. The first step in the procedure was the selection, preparation and disinfection of the seeds. Zygotic embryos were then excised and treated with either NaN_3, EMS or ^{60}Co gamma rays. After the mutagenic treatment, the zygotic embryos were washed and incubated in the dark for 24 h. Finally, the treated zygotic embryos were cultured on GER medium for 4 weeks.

Survival of zygotic embryos in MS medium without NaN_3 (at pH 3.0 or pH 5.6) was greater than 0% (Fig. 34.2A). Sodium azide reduced survival at concentrations at or above 2.5 mM (95%), 5.0 mM (91.6%), 7.5 mM (86.6%), 10.0 mM (70.0%), 12.5 mM (51.6%), 15.0 mM (45.0%) and 20.0 mM (40.0%) (Fig. 34.2A). A negative correlation between NaN_3 concentration and survival rate was observed (R^2 = 0.9366). NaN_3-treated zygotic embryos showed decreased cell viability in comparison with the non-treated control, as indicated by the colouration generated by the TTC assay (Fig. 34.2B). The estimated LD_{50} for sodium azide was 12.5 mM (SE ± 7.1).

Regarding the EMS treatment, compared with the non-treated control (0 mM EMS; 100.0%), survival of zygotic embryos was significantly reduced ($p < 0.05$) when EMS was applied in increasing concentrations from 0.5% (63.3%) to 6.0% (10.0%) (Fig. 34.3A). A negative correlation was observed between EMS concentration and survival rate (R^2 = 0.7898). The estimated LD_{50} value was 1% EMS (SE ± 28.1). TTC staining of the zygotic embryos exposed to EMS showed clear differences in the red colour as the concentration increased (Fig. 34.3B).

Survival of zygotic embryos irradiated with gamma rays (^{60}Co) decreased significantly ($p <$ 0.05) from 100.0% (0 Gy) to 11.61% with 100 Gy (Fig. 34.4A). A negative correlation between the gamma-ray dose and survival rate was observed (R^2 = 0.9193). Gamma-ray doses higher than 40 Gy severely affected cell viability, as shown by TTC staining of the zygotic embryos (Fig. 34.4B). The LD_{50} value for ^{60}Co was 40 Gy (SE ± 10.2).

3.2 Effect of antioxidants on germination of zygotic embryos

In order to minimize oxidation of coffee zygotic embryos exposed to the different muta-

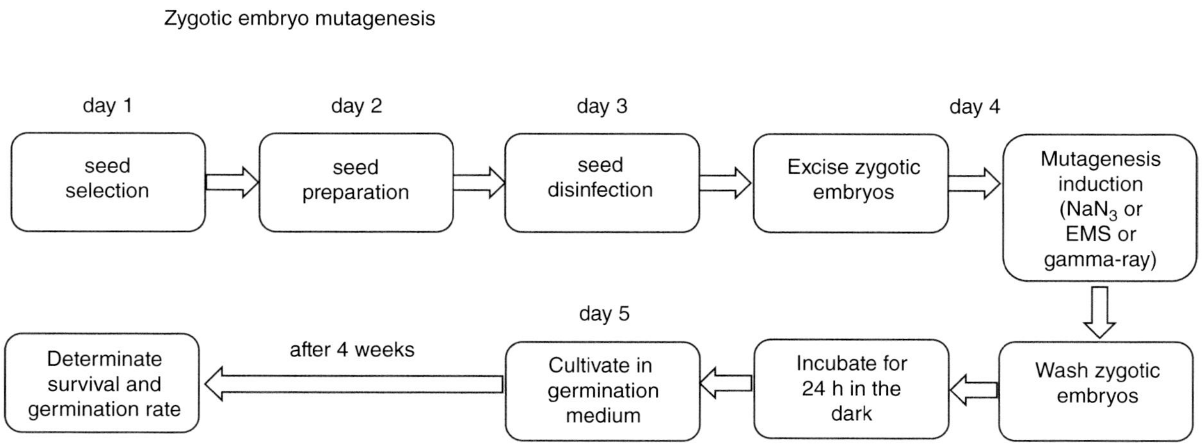

Fig. 34.1. Steps and time required to produce a mutant M_1 population from coffee (*C. arabica* L. var. 'Caturra') zygotic embryos using chemical or physical mutagenesis.

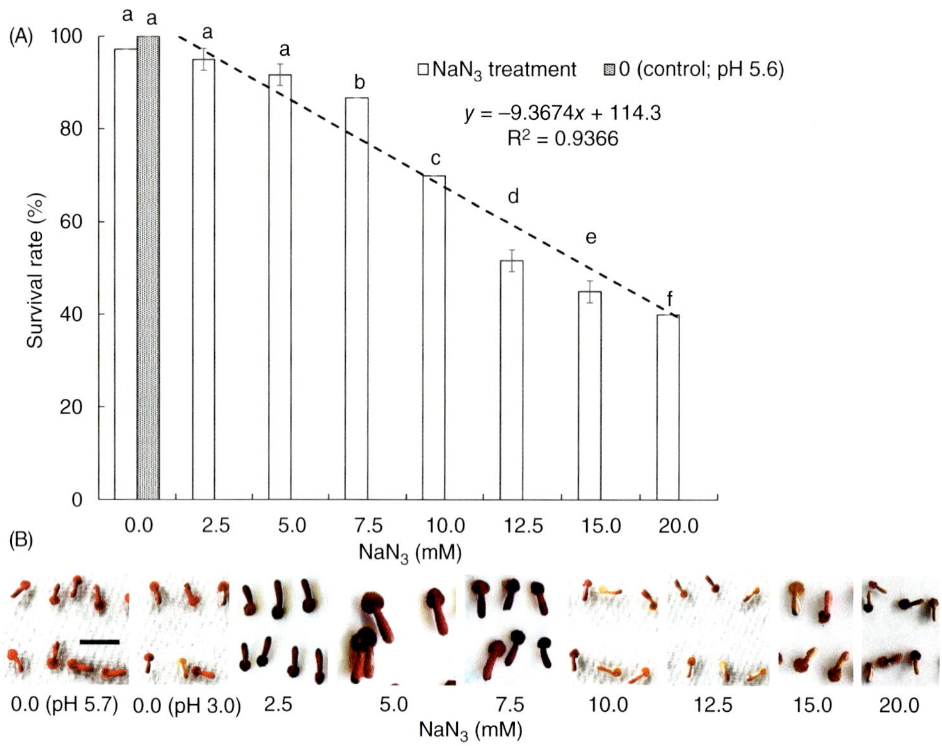

Fig. 34.2. Effect of NaN_3 concentration on survival and viability of coffee (*C. arabica* L. var. 'Caturra') zygotic embryos. **(A)** Survival rate (%) versus NaN_3 concentrations. Each value represents the mean ± SD of two repetitions. Linear regression trend is shown. Values followed by the same letter do not differ significantly ($p = 0.05$). **(B)** Zygotic embryos treated with different NaN_3 concentrations and incubated in TTC at 37°C for 24 h (red colour indicates viability of plant tissue). Bar = 1 cm.

genic agents, the effects of two antioxidant agents were evaluated. Table 34.1 shows the percentage of oxidation of EMS-treated and non-treated zygotic embryos with or without treatment with antioxidant agents. One-way ANOVA showed significant differences among the treatments ($p = 0.05$). After 15 days, oxidation of the non-mutagenized zygotic embryos cultured in GER medium reached 6.0%, whereas germination of embryos treated with AS, AC and AS + AC was 18.0%, 22.0% and 1.0%, respectively. Treatment of EMS mutagenized zygotic embryos with AC decreased oxidation by 22% compared with the GER medium, AS and AS + AC treatments (Table 34.1). A decrease in germination after exposure of zygotic embryos to mutagenic agents was observed (Fig. 34.5). However, there was also a negative effect on germination caused by exposure to the antioxidant solution of citric acid and ascorbic acid at 0.1% w/v (AS treatment) (Fig. 34.5). This shows how the antioxidant solution was not able to alleviate the stress to the embryos caused by the different mutagenic treatments and, in turn, affected the germination process.

3.3 Bulk mutagenesis

After 4 weeks of culture on GER medium supplemented with activated charcoal at 300 mg/l (treatment AC), germination of zygotic embryos treated with NaN_3 (12.5 mM), EMS (1% v/v) or ^{60}Co gamma-rays (40 Gy) was 18.4%, 42.3% and 44.6% , respectively, compared with 93.8% germination of the non-treated zygotic embryos.

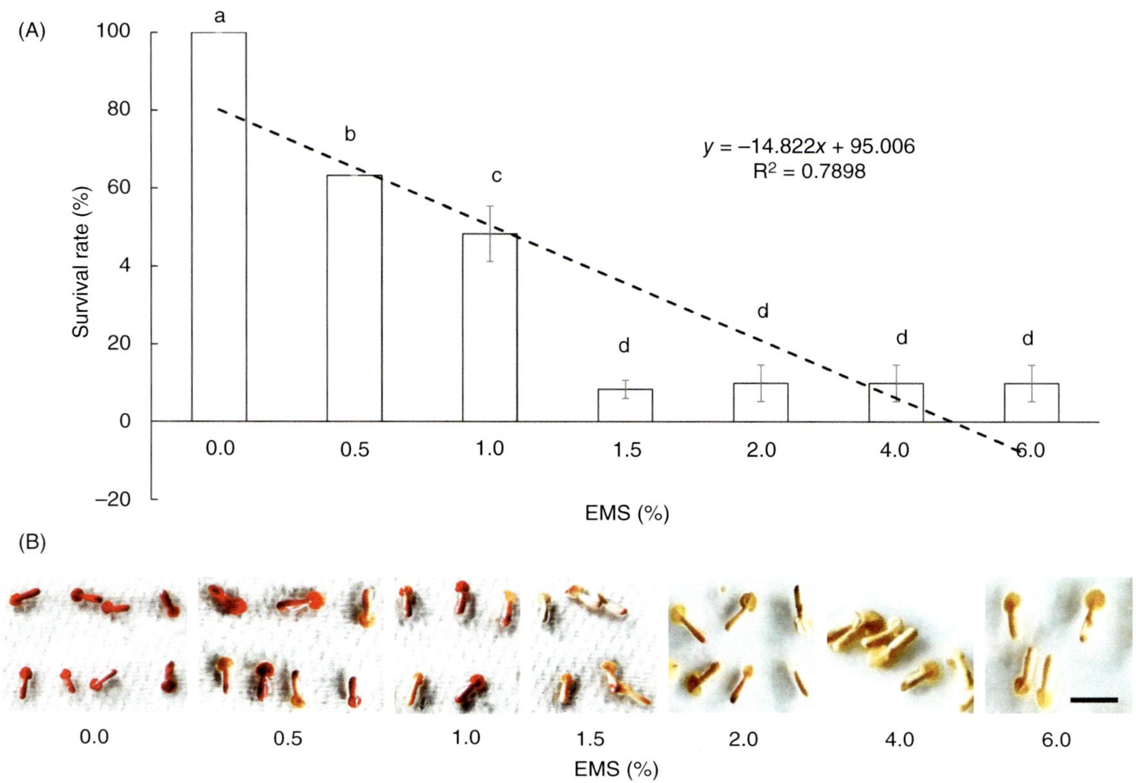

Fig. 34.3. Effect of EMS concentration on survival and viability of coffee (*C. arabica* L. var. 'Caturra') zygotic embryos. **(A)** Survival percentage versus EMS concentrations. Each value represents the mean ± SD of two repetitions. Linear regression trend is shown. Values followed by the same letter do not differ significantly ($p = 0.05$). **(B)** Zygotic embryos treated with different EMS concentrations after incubation with TTC at 37°C for 24 h (red colour indicates viability of the tissue). Bar = 1 cm.

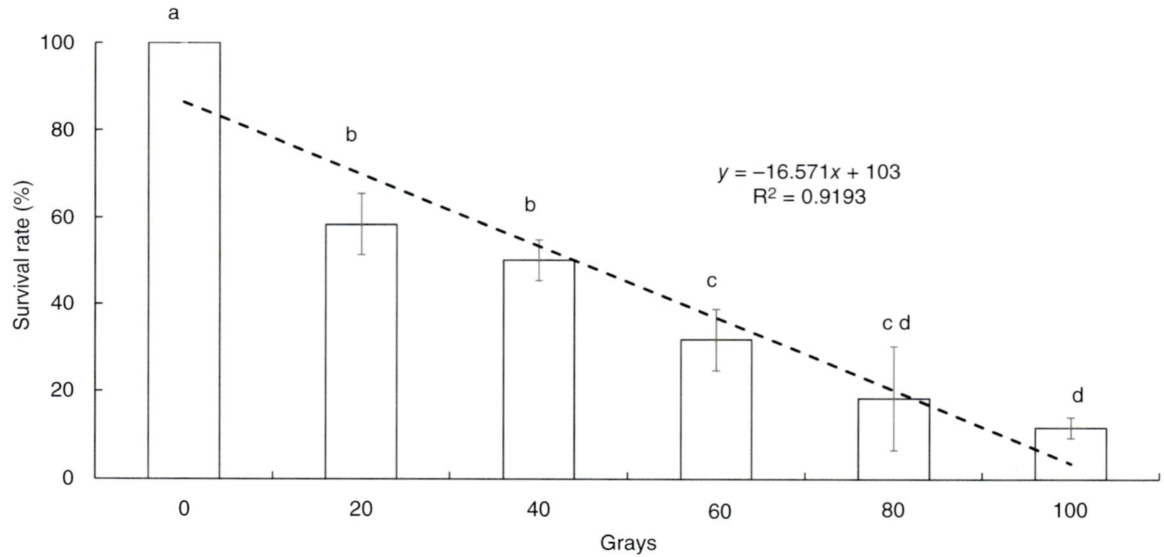

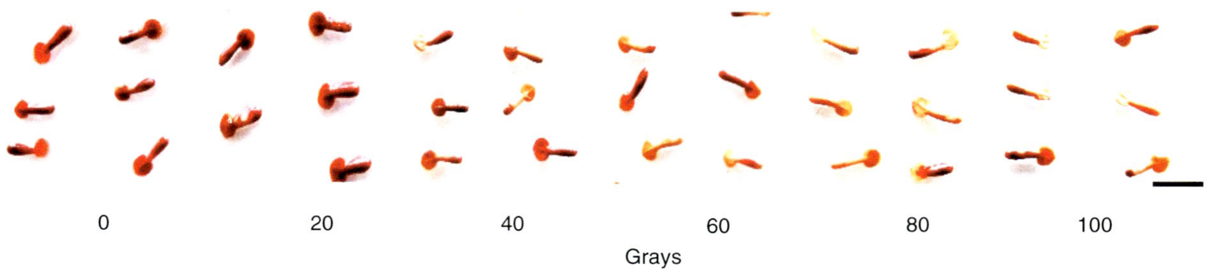

Fig. 34.4. Effect of ^{60}Co dose on survival and viability of coffee (*C. arabica* L. var. 'Caturra') zygotic embryos. **(A)** Survival percentage versus ^{60}Co dose. Each value represents the mean ± SD of two repetitions. Linear regression trend is shown. Values followed by the same letter do not differ significantly ($p = 0.05$). **(B)** Zygotic embryos treated with different ^{60}Co doses after incubation with TTC at 37°C for 24 h (red colour indicates viability of the tissue). Bar = 1 cm.

Table 34.1. Effect of antioxidant agents on oxidation and germination of zygotic coffee embryos (*C. arabica* L. var. 'Caturra') treated with 1% v/v EMS.

Treatment	EMS treatment	Oxidation (%)	Germination (%)
GER medium	No	6 a	94 a
AS		18 a	0 b
AC		22 a	78 a
AS + AC		10 a	2 b
GER medium	Yes (1% v/v)	100 b	0 b
AS		100 b	0 b
AC		78 b	22 b
AS + AC		100 b	0 b

Values followed by the same letter do not differ significantly; (P = 0.05); AS, Antioxidant solution; AC, Activated charcoal; AS + AC, Antioxidant solution plus activated charcoal.

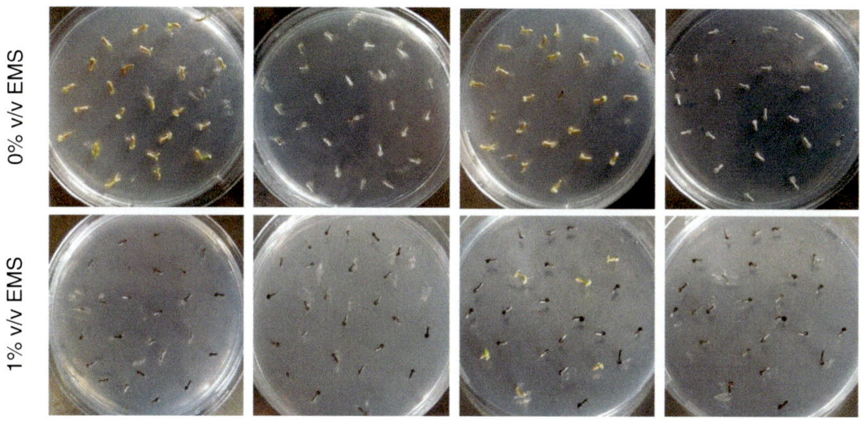

Fig. 34.5. Germination of zygotic coffee embryos (*C. arabica* L. var. 'Caturra') after exposure to mutagenic agents and antioxidant treatment.

4 Discussion

4.1 Mutagen dose determination

In coffee, mutation induction using embryogenic callus (Bolívar-González *et al.*, 2018), seedlings (Dada *et al.*, 2018) and seeds (Vargas-Segura *et al.*, 2019) has been reported. Nevertheless, to the best of our knowledge, the conditions for physical and chemical mutagenesis in zygotic embryos of *C. arabica* L. var. 'Caturra' have been not described. The first step in a mutagenesis programme involves the selection of a mutagenic agent and dose (concentration and duration of treatment) appropriate for the crop and cultivar (Roychowdhury and Tah, 2011). The response of the material to exposure to these agents can depend on several factors, such as the concentration of the mutagen, the pH, the temperature, the duration of exposure and the type of material treated (Srivastava *et al.*, 2011). To avoid an excessive loss of experimental material, preliminary studies should be carried out to determine the mean lethal dose of the mutagenic agents, which is the concentration of the mutagen at which the highest frequency of mutations is obtained (Kangarasu *et al.*, 2014). In our study, the influence of the concentration of NaN_3 or EMS and the dose of gamma-rays on zygotic embryo survival and germination rate were evaluated. In this sense, as the concentration of the applied mutagen increased, survival decreased.

The reduction in viability and germination of plant material in mutagenic treatments may be due to a delay or inhibition of biological and physiological processes necessary for germination, which include enzymatic activity, hormonal imbalance and inhibition of the mitotic process (Khan *et al.*, 2009). Many studies have shown an inverse relationship between sodium azide concentration and percentage survival of treated material. Azide ions play a significant role in mutation by interacting with enzymes and DNA in cells. Azide anions are strong inhibitors of cytochrome oxidase, which inhibits the oxidative phosphorylation process. In addition, they are strong inhibitors of the proton pump and alter the potential of the mitochondrial membrane. These effects can hinder the biosynthesis of ATP, and the resulting decrease in ATP availability can slow the germination rate (Khan *et al.*, 2009). It has been observed that low concentrations of sodium azide are usually optimal for inducing mutations. However, this depends on the species, since some species are very sensitive to sodium azide and others, such as *Arabidopsis thaliana*, show no mutagenic effects (Shu *et al.*, 2011). Similar doses to those used in this work (5 mM, 10 mM and 15 mM) were tested on embryogenic cell suspensions of a banana cultivar and it was determined that cell viability decreased with the increase in mutagen concentration (Xu *et al.*, 2011). At the lower concentration, cell survival was approximately 50%. Likewise, Muthusamy *et al.* (2007) treated embryogenic peanut calluses (*Arachis hypogaea* L.) with NaN_3 (0, 1, 2, 3, 4 and 5 mM) for 30 min and observed that the percentage survival remained unchanged (100%) at concentrations below 4 mM. At 4 mM and 5 mM, survival decreased to 60.4% and 43.2%, respectively. In cotton, somatic embryo maturation and germination decreased as the concentration of NaN_3 increased (Ganesan *et al.*, 2005).

On the other hand, the reduction in embryo viability at increased concentrations of the EMS may be due to cytotoxic effects of the alkylating agents, which may affect the ability of the cells to divide. This may also be affected by the phase of the cell cycle, the activity of the repair mechanisms, or by variations in the absorption of EMS influenced by the cell position (Jankowicz-Cieslak *et al.*, 2012). In our study, the LD_{50} for EMS was set at a concentration of 1% v/v for 2 h. A similar LD_{50} has been reported in investigations with other species. Oat seeds were treated with 0.8%, 0.9%, 1%, 1.2%, 1.4%, 1.6%, 1.8% and 2% EMS, and it was determined that the optimal dose was 0.9%, at which seed survival was 37% (Chawade *et al.*, 2010). Concentrations higher than 1.6% EMS significantly reduced survival to approximately 0%. These results were similar to those obtained in our work. In rice, approximately 50% survival of seeds was achieved in treatment with 0.5% EMS. These authors also observed a decrease in viability to 0% in seeds exposed to 1.5% and 2% EMS (Benjavad and Shahrokhifar, 2012). Banana *in vitro* shoots were treated with 0.25%, 0.5%, 1.0% and 1.5% EMS for 4 h; and it was demonstrated that treatment with 1.5% EMS was lethal (Jankowicz-Cieslak *et al.*, 2012).

As in the present study, previous reports have demonstrated the negative effect of increasing concentrations of sodium azide and EMS on survival of treated material (Talebi *et al.*, 2012; Gandhi *et al.*, 2014; Lee *et al.*, 2017; Aslam *et al.*, 2018; Vargas-Segura *et al.*, 2019).

Regarding treatment with gamma-rays, in research conducted in 1962, similar results were obtained with coffee seeds (var. 'Padang') exposed to gamma radiation. After 10 weeks, 61% of the seeds irradiated with 8 kr (70.16 Gy) and 89% of the seeds irradiated with 4 kr (35.08 Gy) had germinated (Moh, 1962). Studies of the effect of ionizing radiation on other species have shown similar results. Embryogenic calli of peanuts (*Arachis hypogaea* L.) were treated with 0, 10, 20, 30, 40 and 50 Gy and it was demonstrated that at doses below 40 Gy, survival was 100%; however, at 40 Gy and 50 Gy, the percentage survival decreased to 72.4% and 50.2%, respectively (Muthusamy *et al.*, 2007). On the other hand, embryogenic calluses of sugarcane were treated with the same doses and it was observed that below 40 Gy the percentage of regeneration decreased by 59.32%. As the dose was increased, survival and yields decreased and most calluses became necrotic (Ali *et al.*, 2014). It was determined that for somatic embryogenic cell suspensions of date palm, a dose of 20–30 Gy, depending on the genotype used, favoured the induction of mutations (Moha, 2012). The reduction in survival with higher doses of radiation may be due to increased chromosomal damage caused by the radiation itself. High doses of radiation are harmful to the plant genome and cause severe DNA damage, leading to a large number of mutations, most of which are lethal (Benjavad and Shahrokhifar, 2012; Moha, 2012).

The low survival rate of plants exposed to certain doses of gamma radiation has also been attributed to a physiological imbalance that affects the integrity of several important macromolecules, such as endogenous auxins (Ramesh *et al.*, 2012).

4.2 Effect of antioxidants on germination of zygotic embryos

Oxidation of mutagenized tissues is due to clastogenic effects (chromosomal damage) and damage to the cellular structure triggered by physical and chemical mutagenic agents and by reactive oxygen species generated in conditions of environmental stress (Khan *et al.*, 2009). Previously, it was noted that activated carbon in different concentrations did not affect the development of pistachio (*Pistacia vera*) embryo axes (Benmahioul *et al.*, 2012). These authors explained that the addition of activated carbon to the culture medium not only decreases oxidation of the explants by the adsorption of different harmful compounds, but also often has a promoter effect on growth and organogenesis. Moreover, it was reported that in oil palm embryos cultured in medium with growth regulators (NAA 0.1 mg/l, BAP 0.1 mg/l and AG3 0.1 mg/l) and activated carbon at 2 g/l, the survival was higher in treatments with added activated carbon (Suranthran *et al.*, 2011). They attributed this effect to the adsorption of unwanted toxic exudates that accumulate in culture media, and to the adsorption of growth-inhibiting substances produced by the decomposition of culture media during autoclaving, such as 5-(hydroxymethyl)-2-furaldehyde (HMF), an inhibitory compound formed by hydrolysis of sucrose.

It was mentioned that ascorbic acid is an antioxidant used to control the oxidation of phenols and its use in tissue culture reduces oxidation levels. It is also involved in cell division and elongation (Kariyama and Nisyawati, 2013). Moreover, banana shoots exposed to different concentrations of ascorbic acid (50, 100 and 200 mg/l) showed decreased development, an effect similar to that observed in our work (Chikezie, 2012). Likewise, it was pointed out that ascorbic acid at concentrations of 5 mg/l to 150 mg/l considerably reduced the budding capacity of banana explants (apices). This author explained that the rapid oxidation of ascorbic acid to dehydroascorbic acid may be responsible for the significant reduction in the sprouting response of the explants, since dehydroascorbic acid at low concentration inhibits the activity of several enzymes *in vitro*, including fructose-1, 6-biphosphatase and hexokinase (Chikezie, 2012). It was also shown that citric acid at 150 mg/l and ascorbic acid at 100 mg/l suppressed the development of guava apices compared with the control (Tagelsir *et al.*, 2006).

5 Conclusion

We investigated the effect of NaN_3, EMS and ^{60}Co gamma rays on survival of Arabica coffee zygotic embryos. In conclusion, as the concentration of the applied mutagen increased, survival decreased. These observations indicate that NaN_3, EMS and ^{60}Co gamma-rays negatively affect the growth and development of coffee zygotic embryos. The optimal concentration and duration of mutagenic treatment will be used to generate a NaN_3, EMS and ^{60}Co gamma-rays mutant population in coffee plants and offers a rapid and economical alternative to improve agronomic important traits.

Acknowledgements

This study was financed by the University of Costa Rica, the Ministerio de Ciencia, Tecnología y Telecomunicaciones (MICITT) and the Consejo Nacional para Investigaciones Científicas y Tecnologicas (CONICIT) (project No. 111-B5-140; FI-030B-14).

Author contributions

A.G.-A conceived the project, designed and coordinated the experiments, analysed data and wrote the paper; J.R.-M. designed and performed the experiments and analysed data; A.A-E. and M.V-M. discussed the results and edited the paper.

Conflict of interest

On behalf of all authors, the corresponding author states that there is no conflict of interest. All authors read and approved the final manuscript.

References

Ahmed, W. *et al.* (2013) Somatic embryogenesis of a coffee (*Coffea arabica* L.) hybrid using leaf explants. *Journal of Horticultural Science and Biotechnology* 88(4), 469–475.

Ali, A. *et al.* (2014) Effect of sodium azide on seed germination and related agro-metrical traits in M1 lentil (*Lens culinaris* Medik.) generation. *World Journal of Agricultural Sciences* 10(3), 95–102.

Aslam, M. *et al.* (2018) Result of chemical mutagenesis on quantitative as well as qualitative traits of maize (*Zea mays* (L.)) *International Journal of Pure and Applied Biosciences* 6(1), 12–15.

Benjavad, A. and Shahrokhifar, B. (2012) Ethyl methane sulphonate (EMS) induced mutagenesis in Malaysian rice (cv. MR219) for lethal dose determination. *American Journal of Plant Science* 3, 1661–1665.

Benmahioul, B. *et al.* (2012) Influence of activated charcoal on *in vitro* embryo germination and growth of plantlets of pistachio (*Pistacia vera* L.). *International Journal of Science Nature* 3(3), 613–616.

Bolívar-González, A. *et al.* (2018) Responses of Arabica coffee (*Coffea arabica* L. var. Catuaí) cell suspensions to chemically induced mutagenesis and salinity stress under *in vitro* culture conditions. *In Vitro Cell Developmental Biology Plant* 54, 576.

Chawade, A. *et al.* (2010) Development and characterization of an oat TILLING-population and identification of mutations in lignin and β-glucan biosynthesis genes. *BMC Plant Biology* 10, 86–99.

Chikezie, U.N.Y. (2012) Effect of ascorbic acid on blackening and sprouting of *Musa* spp shoot tips. *Journal of Biotechnology and Bioinformatics* 2, 11–17.

Dada, K.E. *et al.* (2018) Biological effect of gamma irradiation on vegetative propagation of *Coffea arabica* L. *African Journal of Plant Science* 12(6), 122–128.

Davis, A.P. *et al.* (2019) High extinction risk for wild coffee species and implications for coffee sector sustainability. *Science Advances* 5, 1–9.

Gandhi, E.S. *et al.* (2014) The effect of ethyl methane sulphonate and diethyl sulphate on chilli (*Capsicum annuum* L.) in M1 generation. *International Letters of Natural Sciences* 5, 18–23.

Ganesan, M. *et al.* (2005) Mutagenic effect of sodium azide on somatic embryo regeneration and root growth of cotton (*Gossypium hirsutum* L. cv. SVPR2). *Journal of Agricultural Science and Technology* 1, 365–380.

Gressel, J. *et al.* (2006) Agriculture: the selector of improbable mutations. *PNAS* 103, 12215–12216.

Jankowicz-Cieslak, J. *et al.* (2012) Induction, rapid fixation and retention of mutations in vegetatively propagated banana. *Plant Biotechnology Journal* 10, 1056–1066.

Kangarasu, S. *et al.* (2014) Determination of lethal dose for gamma rays and ethyl methane sulphonate induced mutagenesis in cassava (*Manihot esculenta* Crantz.). *International Journal of Science Research* 3(1), 3–6.

Kariyama, K. and Nisyawati (2013) Effect of ascorbic acid, activated carbon and light duration on explant browning of banana cultivar Barangan (*Musa acuminata* L.) *in vitro* culture. *International Journal of Recent Research of Applied Studies* 16(1), 118–123.

Khan, S. *et al.* (2009) Sodium azide: a chemical mutagen for enhancement of agronomic traits of crop plants. *Environment & We, an International Journal of Science and Technology* 4, 1–21.

Lee, D.K. *et al.* (2017) Determination of the optimal condition for ethylmethane sulfonate-mediated mutagenesis in a Korean commercial rice, *Japonica* cv. Dongjin. *Applied Biological Chemistry* 60(3), 241–247.

Mba, C. *et al.* (2010) Induced mutagenesis in plants using physical and chemical agents In: Davey, M.R. and Anthony, P. (eds) *Plant Cell Culture: Essential Methods*, 1st edn. Wiley-Blackwell, Chichester, UK, pp. 111–130.

Mendonça, A.P. (2014) *Desempenho de seleções clonais de Coffea arabica, em Campinas*, SP. Master's Thesis. Campinas, BR, Instituto Agronômico.

Mishra, M.K. and Slater, A. (2012) Recent advances in the genetic transformation of coffee. *Biotechnology Research International* 2012, Article 580857. doi: 10.1155/2012/580857

Moh, C.C. (1962) The use of radiation-induced mutations in crop breeding in Latin America and some biological effects of radiation in coffee. *International Journal of Applied Radiation and Isotopes* 13, 467–475.

Moha, S. (2012) In vitro mutagenesis for improving date palm (*Phoenix dactylifera* L.). *Emirates Journal of Food Agriculture* 24(5), 400–407.

Murashige, T. and Skoog, F. (1962) A revised medium for rapid growth and bioassays with tobacco tissue cultures. *Physiologia Plantarum* 15, 473–497.

Muthusamy, A. *et al.* (2007) Effects of mutagens on somatic embryogenesis and plant regeneration in groundnut. *Biologia Plantarum* 51(3), 430–435.
Ramesh, H.L. *et al.* (2012) Effect of gamma radiation on morphological and growth parameters of mulberry variety M5. *International Journal of Science and Nature* 3(2), 447–452.
Roychowdhury, R. and Tah, J. (2011) Germination behaviors in M_2 generation Dianthus after chemical mutagenesis. *International Journal of Advanced Science and Technology Research* 1(2), 448–454.
Shu, Q.Y. *et al.* (eds) (2011) *Plant Mutation Breeding and Biotechnology*. CABI, Wallingford, UK.
Srivastava, P. *et al.* (2011) Mutagenic effects of sodium azide on the growth and yield characteristics in wheat (*Triticum aestivum* L. em. Thell.). *Asian Journal of Plant Science* 10, 190–201.
Suranthran, P. *et al.* (2011) Effect of plant growth regulators and activated charcoal on *in vitro* growth and development of oil palm (*Elaeis guineensis* Jacq. var. Dura) zygotic embryo. *African Journal of Biotechnology* 10(52), 10600–10606.
Tagelsir, M.I. *et al.* (2006) Enhancement of growth and control of browning of tissue cultures of Guava (*Psidium guajava* L.). *Journal of Science Technology* 7(1), 74–83.
Talebi, A.B. *et al.* (2012) Ethyl methane sulphonate (EMS) induced mutagenesis in Malaysian rice (cv. MR219) for lethal dose determination. *American Journal of Plant Science* 3, 1661–1665.
Vargas-Segura, C. *et al.* (2019) Sensitivity of seeds to chemical mutagens, detection of DNA polymorphisms and agro-metrical traits in M1 generation of coffee (*Coffea arabica* L.). *Plant Tropical Biology* 22, 451–464.
Wannajindaporn, A. *et al.* (2014) Evaluation of genetic variability in *in vitro* sodium azide-induced *Dendrobium* 'Earsakul' mutants. *Genetics Molecular Research* 13(3), 5333–5342.
Xu, C.X. *et al.* (2011) The effect of ethyl methane sulphonate (EMS) and sodium azide (NaN_3) on plant regeneration capacity of an embryogenic cell suspension of 'Yueyoukang 1' (*Musa*, AAA), a banana cultivar resistant to Fusarium wilt. *Acta Horticulturae* 897, 301–302.

35 Virulence Genes of New Population of Coffee Rust (*Hemileia vastatrix*) Affecting Coffee Variety 'Lempira', in Honduras; Resistant and Susceptible Varieties

Yonis Morales[1] and Rolando Grajeda[2]

Abstract

The coffee variety 'Lempira', released in Honduras in 1998, was classified 100% resistant to races I and II of coffee rust identified by Portugal's Centre for Research into Coffee Rusts (Centro de Investigação das Ferrugens do Cafeeiro) (CIFC) in 1997. However, since 2007, the disease has been reported in seed foundation plots and producer farms, the most recent epidemic report being in April 2016 in Vegas de Jalan, Juticalpa Olancho, affecting 210 ha. Since this variety constitutes 45% of the cultivated area under coffee in the country, there is a need to identify the virulence genes of the new strain and to determine the resistance and susceptibility of other cultivated varieties. For these purposes, mass samples of rust were inoculated on leaf discs of the differential clones 1343/269, 110/5, 147/1, 152/3, 33/1, 419/20, 832/1 and 832/2, together with 87/1, 1006/10, 420/10 and 420/2 from the Federal University of Vicosa, as well as on the two main cultivated resistant varieties ('Parainema' and 'IHCAFE-90'), and seven promising genotypes, under controlled temperature conditions and relative humidity. After 20–60 days of inoculation, seven virulence genes were identified (*v1*, *v2*, *v4*, *v5*, *v6*, *v7*, *v9*), of which *v1*, *v4*, *v6*, *v7* and *v9* had not been reported in Honduras previously. It is inferred that this rust population arose by recombination of race *v5* with *v6*, *v7* or *v9*. Races with 3, 4, 5, 6 or 7 virulence determinants were identified as the most complex and aggressive strains described but they lacked the *v3* and *v8* determinants. In addition, it was found that 'Parainema', 'H27', 'T5296-170', 'Central American', 'Pacamara yellow' and 'Anacafe-14' are resistant because they possess the *SH8* gene, absent from 'Lempira'. 'IHCAFE-90' and 'Obatá' showed 20% susceptibility, and 'Ruiru 11' was susceptible. The results reveal the diversity of rust virulence genes in Honduras and emphasize the importance of the *SH3* and *SH8* genes as sources of resistance.

Keywords: differential clones • virulence • genotype • race • *SH3* and *SH8* genes

1 Introduction

Coffee rust is the main fungal disease that affects coffee cultivation in all areas where plantations are established worldwide. This implies that every producer must have the economic resources and the basic knowledge for efficient management of the pathogen, before the levels of incidence and severity surpass the control capacity of the fungicides available on the market and the plant loses its capacity to regenerate foliar tissues and enters a process of regressive death of fruits, bandolas and eventually the whole plant. Due to the great variability of coffee prices, which represent the main source of income for producers, several inter-institutional

[1] Email: ymorales71@gmail.com; [2] Email: rolandograjeda01@gmail.com

 Mutation Breeding, Genetic Diversity and Crop Adaptation to Climate Change (eds S. Sivasankar *et al.*)

DOI: 10.1079/9781789249095.0035

cooperation efforts have been established in Honduras for the development and release of new varieties of coffee that offer genetic resistance to coffee rust and are more productive, better adapted to climate and soil conditions, and have better or equal cup quality as traditional susceptible varieties. One of the main varieties released is 'Lempira', which since 1998 has been adopted and cultivated in all the coffee regions of the country located at altitudes from 700 m to above 1500 m above mean sea level. This has allowed the renewal of 65% of the coffee plantations in Honduras, where 'Lempira' is grown on 45% of the coffee cultivated area. It has allowed producers with limited economic resources to grow coffee, especially those who manage small areas between 0.7 and 3.5 ha, which represent 95% of the producers in the country. However, by the end of 2015, reports had been growing that the rust was affecting this variety to the degree of causing severe defoliation and the loss of harvest and plantations, putting at risk a large part of the national production and the sustainability of thousands of coffee-producing families. As this was the first experience of loss of resistance to rust documented in Honduras, there was a need to know which virulence genes were present in this new rust population, and to identify resistance or susceptibility in hitherto released varieties in order to make the best decision when opting for renewal of affected plantations. For this, the present research work was carried out by inoculating a mass sample of rust collected from a seed-production plot of the 'Lempira' variety on plants that differentiate the virulence genes of the pathogen. A susceptible variety was also included as a control. By inoculation of rust in the resistant varieties released and available to date, it was possible to identify all the virulence genes present in the new rust population that are capable of affecting the 'Lempira' variety, as well as the potential genes responsible for resistance in these varieties. Resistant materials have been identified in the process described below.

2 Materials and Methods

2.1 Selection and conservation of the coffee rust population

To characterize the diversity of specific virulence genes that make up the new rust population affecting 'Lempira', a mass sample was collected from 200 highly infected leaves free of hyperparasitic fungi (Fig. 35.1a). The leaves were selected from 50 randomly distributed plants from the seed-production plot established at the Carlos Alberto Bonilla Research Centre of Campamento Olancho. Uredospores were collected in gelatine capsules and then stored in a 250 ml beaker covered completely with aluminium foil. This was placed inside a desiccator with 30% sulfuric acid and maintained at a temperature of 5°C in a refrigerator for later use in the inoculation studies (Zambolim and Chaves, 1974).

2.2 Selection of plants differentiating virulence genes of *Hemileia vastatrix*

To characterize the diversity of genes present and absent in the new rust, the existing population in the field was evaluated by inoculating a representative mass sample. Well-differentiated young leaves were chosen, and discs of approximately 1 inch (2.5 cm) in diameter were inoculated using uredospores collected from 12 differential clones (1343/269, 110/5, 147/1, 152/3, 33/1, 419/20, 832/1, 832/2, 87/1, 1006/10, 420/10 and 420/2), all from the Federal University of Vicosa and managed in the greenhouse in the Honduran Coffee Institute (IHCAFE). These differentials (Fig. 35.2) were strategically selected to identify the nine virulence genes (*v1* to *v9*). Recombination among these gives rise to many possible genotypes of the fungal pathogen (*Hemileia vastatrix*), known as races.

2.3 Virulence and resistance tests

For the virulence test, two inoculations (repetitions) were made with the 12 selected clones, the first to obtain a preliminary notion of the genes present in the rust population and the second to validate the previous results, especially in the differential clones where only flecking was presented at the beginning. This is because there is the possibility that the fungus did not thrive until sporulation, due to some non-genetic limiting conditions such as low relative humidity or the development stage of the leaf tissue used. To determine resistance in the ten varieties released as resistant to rust, one trial

Fig. 35.1. Inoculation technique of rust masses and validation strategy of resistance or susceptibility in resistant varieties and plants that differentiate resistance genes to identify virulence genes present in rust populations emerging in Honduras. **(a)** Collection of coffee rust spores in gelatine capsule from an infected leaf in 'Lempira' mother lines. **(b)** Direct inoculation of mass population sample of rust that affects variety 'Lempira'. **(c)** Virulence test in differential plants, accompanied by six 'Caturra' discs as a control located in the centre of the growth chamber. **(d)** Test of resistance in commercial varieties and with apparent resistance in the field.

Fig. 35.2. Selection of differential plants for the determination of virulence genes present in the population of coffee rust affecting the variety 'Lempira' in the Olancho Honduras region.

was conducted with three repetitions, inoculating 13 leaf discs of the genotype of interest (differential plant or variety) in each repetition and using six leaf discs of the variety 'Caturra' as a susceptible check and indicator of favourable conditions for development of the fungus (Fig. 35.1). These 'Caturra' plants were managed under greenhouse conditions to avoid contamination with spores from the environment and to guarantee their quality and reliability for inoculation. In the test for resistance of cultivated varieties and those in the process of being released, ten representative plants were used per variety. The inoculation technique used was the one proposed by D'Oliveira (1958). This consists of depositing the uredospores on the underside of the leaf with the aid of a brush, followed by a sprinkling of distilled water using a manual atomizer, 48 h of incubation in the dark at a temperature of 20–22°C, and subsequent transfer of the inoculated leaves to the incubation room at the same temperature, with humidity between 60% and 80% (modified from Zambolim and Chaves, 1974). This provided favourable conditions for spore germination and the development of the disease if the material evaluated was susceptible. The resistance or susceptibility behaviour of differentiating clones and varieties was evaluated during an observation period of 60 days, which is considered sufficient time for the uredospores to germinate and colonize leaf discs by forming small pustules that indicate susceptibility of the genotype (D'Oliveira and Rodrigues, 1960).

3 Results

3.1 Virulence test

After a period of 20–60 days of evaluation it was determined that the new rust population that affects the 'Lempira' variety possesses the seven virulence genes, *v1*, *v2*, *v4*, *v5*, *v6*, *v7* and *v9*, confirmed by the development of rust pustules in the differentials 110/5, 152/3, 419/20, 1006/10 and 420/2. This shows that the fungus possesses at least the virulence genes complementary to the resistance genes present in the affected differentiating plant (Table 35.1).

3.2 Test of resistance to population of Roya Olancho region

Of the ten varieties released as resistant and validated in this study with the new emerging rust population in Honduras, only seven showed a complete or vertical resistance to rust and two showed that only 80% of their progeny are really resistant ('IHCAFE-90' and 'Obatá'). In addition it

Table 35.1. Identification of virulence genes present in the rust population that affects the 'Lempira' variety in Honduras, 2017.

Differential clones	Physiological group and resistance genes to *Hemileia vastatrix*	Susceptible (S) or Resistant (R) to the sample of coffee rust	Virulence genes (v) identified (*Hemileia vastatrix*)								
			1	2	3	4	5	6	7	8	9
1343/269	R(SH6)	R**									
110/5	J(SH4,5)	S				4	5				
147/1	T(SH1,3,4,5)	R									
152/3	Y(SH2,4,5)	S		2		4	5				
33/1	G(SH3,5)	R									
419/20	3(SH5,6,9)	S					5	6			9
832/1	A(SH5,6,7,8,9)	R									
832/2	A(SH5,6,7,8,9)	R									
87/1	C(SH1,5)	R*									
1006/10	L(SH1,2,5)	S	1	2			5				
420/10	1(SH5,6,7,9)	S					5	6	7		9
420/2	2(SH5,8)	R									
Virulence genes found in the new population of coffee rust			1	2		4	5	6	7		9

* Indicates plants with resistance genes not yet identified

was discovered that 'Ruiru 11' is highly susceptible, despite being considered as being resistant to the disease in Kenya, its country of origin. All the results of resistance and susceptibility were validated with the cultivar 'Caturra' since it was always susceptible in all the tests where it was inoculated, either with differentiating plants or commercial varieties (Fig. 35.3 and Table 35.2).

4 Discussion

Of the seven genes identified, *v1*, *v4*, *v6*, *v7* and *v9* had not been reported in Honduras. When considering the potential genetic recombination in each biological cycle of the fungus, it is inferred that the evaluated rust population is composed of both simple races of two gene combinations (*v5* combined with *v6*, *v7* or *v9*) and races with 3, 4, 5, 6 and 7 genes. The latter constitute the most complex and aggressive populations described to date. However, they lack the *v3* and *v8* genes, thus leaving the resistance genes *SH3* and *SH8* as one of the most viable options for the development of future varieties with rust resistance in the country, preferably by fixing the *SH3* gene, which to date has not been overcome by any rust population existing in the Central American region.

The complete resistance reaction observed in the differential clones 832/1 and 832/2 confirms that these Timor hybrids can still be considered as donors of resistance genes in future breeding programmes for coffee. It should be kept in mind, however, that the reserve of resistance genes that these offer is becoming smaller, due to the emergence of more complex rust populations capable of overcoming resistance genes.

Fig. 35.3. Witness reaction on leaf discs of 'Caturra' (susceptible) and resistant varieties.

Table 35.2. Validation of resistant varieties against new rust isolates that affect the variety 'Lempira'.

No.	Variety of coffee	Susceptible (S) or Resistant (R) to the sample of coffee rust
1	Parainema	R
2	T5296-170	R
3	Anacafe-14	R
4	F1 Centroamericano	R
5	Pacamara Amarillo	R
6	H27	R
7	IHCAFE-90	80% R
8	Obatá	80% R
9	Batian	R
10	Ruiru 11	S
11	Caturra	S

5 Conclusions and Recommendations

This study reveals the increasing reduction of sources of available genetic resistance for the genetic improvement of coffee for rust resistance. This is especially so in Honduras, where currently seven of the nine available resistance genes have been overcome by the pathogen through the emergence of new and aggressive breeds, leaving the *SH3* gene as the only useful gene for present coffee improvement projects. Recurrent use creates a high genetic uniformity for resistance for future improved varieties, a factor that is limiting when new pathogen variants arise to defeat this gene, with the ability to affect all the derivative varieties of the resistant parent. Therefore, it is essential to identify and create new resistance genes through mutations induced by various means and in many susceptible varieties in order to guarantee genetic diversity for resistance to coffee rust and thus reduce the risk of epidemics that can affect a nation or continent producing coffee. This is important to ensure the productive and economic stability of the sector worldwide.

References

D'Oliveira, B. (1958) As Ferrugens do cafeiro. *Revista do Café Portugués* 3, 39–61.

D'Oliveira, B. and Rodrigues, C.J. Jr (1960) O Problema das ferrugens do cafeeiro. Determinacao da resistencia a *Hemileia vastatrix* em Coffea arabica. *Revista do Café Portugués* 8, 5–87.

Zambolim, L. and Chaves, G.M. (1974) Efeito de baixas temperaturas e do binômio temperatura-umidade relativa sobre a viabilidade dos uredosporos de *Hemileia vastatrix* Berk. et Br. e *Uromyces phaseoli* Arth. *Experientiae* 17, 151–184.

36 Mutagenesis of *in vitro* Explants of *Coffea* spp. to Induce Fungal Resistance

Souleymane Bado[1], Fatemeh Maghuly, Vitor Varzea and Margit Laimer

Abstract

Coffee is one of the most valuable commodity tree crops worldwide. However, it suffers from several devastating diseases and pests, for example coffee leaf rust and coffee berry borer, whose impact is being amplified by changing climatic conditions. Development of new adapted varieties remains a laborious effort by conventional breeding due to the long juvenile period in tree crops. Plant cell/tissue culture represents the ultimate method to produce large amounts of true-to-type healthy plants and of explants for mutation breeding. In fact, mutation induction combined with *in vitro* cell/tissue culture techniques has proved to be effective for developing improved cultivars of perennial crops. Prior to mutation breeding, cell and tissue radiosensitivity tests to various mutagens need to be performed, so that optimal treatments can be applied for large population development. Thus, different *in vitro* explants (plantlet, leaf, callus, embryogenic callus, globular and torpedo stage embryos) of *Coffea arabica* and *Coffea canephora* were exposed to different gamma-ray doses (0, 10, 15, 20, 40, 60 and 80 Gy). After 9–21 weeks incubation, a radiosensitivity test was conducted on the different explants and LD_{50} doses corresponding to 50% of viability or survival of callus, embryogenic callus, globular and torpedo stage embryos and 50% growth reduction (GR_{50}) of shoot were also determined. Callus explants showed a relatively high radio-resistance (LD_{30}–LD_{50} 50–100 Gy) in comparison with entire plantlets or embryos (LD_{30}–GR_{50} 8–46 Gy). Globular embryo development into plantlets and also leaf area of irradiated plantlets were more severely affected by irradiation than other explants. It was possible to confirm the relative radio-resistance of unicellular explants compared with multicellular explants. Estimation of optimal mutation induction dosage range for various *in vitro* explants is important for tree crops, especially for coffee improvement.

Keywords: coffee • mutation induction • *in vitro* mutagenesis • radiosensitivity test • cell/tissue culture

1 Introduction

Coffee is one of the most important tree crop plants worldwide regarding its social and economic contributions. Coffee is one of the most important beverages, with over 2.25 billion cups consumed every day (Labouisse *et al.*, 2008), and a great contributor to the economies of tropical and subtropical developing countries, which constitute the main production areas, as it is one of their key export and cash crops. Coffee provides cash income for over 75 million smallholders in these regions. Furthermore, coffee represents the second most traded commodity after oil (Pietro, 2015). As a crop plant, coffee is cultivated across most continents in over 11 million

[1] Email: souleymane@hotmail.com

 Mutation Breeding, Genetic Diversity and Crop Adaptation to Climate Change (eds S. Sivasankar *et al.*)
DOI: 10.1079/9781789249095.0036

hectares which produced about 120.6 million 60 kg bags during 2009–2010 (Déchamp *et al.*, 2015).

Despite its importance, coffee breeding has been a slow, long, costly and time-consuming process (about 30 years to release a variety) because of the long juvenile period and limited potential for germplasm improvement due to a very narrow genetic base of the two most used species, *Coffea arabica* (tetraploid, $2n = 44$) and *C. canephora* (diploid, $2n = 22$) (Davis *et al.*, 2006; Kumar *et al.*, 2016). Additionally, absence of pest resistance in the preferred *C. arabica* or existing resistances have shown limitations with the emergence of new diseases and pests under changing climate conditions (Kumar *et al.*, 2016). Most hybrid cultivars are derived from spontaneous mutations, which are rare events, consequently limiting breeding opportunities. Thus, the potential application of mutation breeding to broaden genetic variability for improvement of crops like *Coffea* spp. with little application of improvement through biotechnology is evident.

Since the pioneering attempt of Carvalho in the 1950s using X-ray irradiated seed (Carvalho *et al.*, 1967) it has taken more than a half century to see a publication presenting preliminary results of research activities of mutation breeding programmes in coffee, as a consequence of Coordinated Research Project (CRP) D22005 initiated by the Joint FAO/IAEA Division, and its organized capacity building on mutation induction for coffee improvement in Latin America (Dada *et al.*, 2014, 2018; Bolívar-González *et al.*, 2018). Coffee offers a wide range for mutation induction as multicellular explants (seed, *in vitro* and *in vivo* cuttings, whole plant and grafting) and unicellular explants (somatic and gametic cells and pollen). Cell/tissue cultures have been used for multiplication of coffee planting material to produce true-to-type, though heterogeneous, cultivars. This was linked to the generation of some somaclonal variation (Etienne and Bertrand, 2003; Landey, 2013). However, no improved cultivar has been officially identified and released from those somaclones. *In vitro* cell/tissue culture in combination with induced mutation has proved to be a valuable method to broaden genetic variability (Van Harten and Broertjes, 1989). It is the optimal method for producing a large population size in a heterogeneous crop species, such as tree crops, in a short time period. It also offers the chance to accelerate mutation breeding when using single (somatic or gametic) explants through early screening and selection of improved lines. But as a prerequisite, homogeneous and uniform explants for efficient mutation induction must be obtained.

Prior to mutation induction, radiosensitivity tests need to be performed to determine the optimal dose treatment for mutation induction. In this study we determined the radiosensitivity of different *in vitro* explants to gamma irradiation. The data provide useful information in optimizing irradiation treatments for mutation induction in coffee.

2 Materials and Methods

Plant material of different genotypes of *Coffea* spp. was available in the form of various explants. Coffee can be exposed to gamma irradiation as shoot, leaf disc, somatic calli, embryogenic calli, globular embryo and cotyledonary embryo. A radioactive cobalt-60 (gamma) source with a dose rate of 94.08 Gy/min from the Plant Breeding and Genetics Laboratory of the Joint FAO/IAEA Division at Seibersdorf, Austria, was used to induce mutations.

2.1 Irradiation of shoots/cuttings

Cuttings of young fresh shoots of *C. canephora* cv. 'Quillou' were prepared in Magenta boxes with five cuttings per box, with three boxes for a total of 15 cuttings per dose. The cuttings were irradiated at dose range of 0, 10, 15, 20, 40 and 60 Gy. After incubation for about 9 weeks on culture medium according to Zamarripa (1993), the irradiated shoots were evaluated for the number of emerged nodes, shoot length, number of roots, longest root length, leaf area and secondary formation.

2.2 Irradiation of leaf discs

Young shoots of *C. canephora* cv. 'Quillou' with green leaves were selected for leaf disc irradiation.

Five shoots were prepared per dose. After gamma irradiation, the leaves were isolated and cut into leaf discs and placed on media for callogenesis and then on embryogenesis media (Etienne *et al.*, 2005). After incubation for 24 weeks, a radiosensitivity test was evaluated for callogenesis and embryogenesis in response to gamma irradiation (doses used were 0, 10, 15, 20, 40 and 60 Gy).

2.3 Irradiation of somatic callus

Prior to mutagenesis, calli were induced on leaf discs of *C. arabica* cv. 'Pacamara'. Leaf discs with 100% callusing response were selected for gamma irradiation on solid culture media (Etienne *et al.*, 2005). After exposure of calli to 0, 10, 15, 20 and 40 Gy, they were assessed for survival and ability to induce embryogenic calli.

2.4 Irradiation of embryogenic callus

As for callus irradiation, somatic embryogenic calli were induced from *C. arabica* cv. 'Java' and *C. canephora* cvs 'Niaoulli' and 'Quillou', and then exposed to gamma irradiation. Following preliminary studies, a wider dose range was applied: 0, 10, 15, 20, 40, 60 and 80 Gy. The irradiated calli were evaluated for survival and their ability to form embryos.

2.5 Irradiation of globular and cotyledonary embryos

Embryogenic calli of *C. arabica* cv. 'Java' were subjected to somatic embryo formation. The formed embryos were sorted into globular and cotyledonary/torpedo stages which were exposed to gamma irradiation (0, 10, 15, 20, 40 and 60 Gy). Ten explants of *C. arabica* cv. 'Java' per replication were prepared, with three replications per dose, to assess the radiosensitivity of the globular and cotyledonary embryos. Survival rate was used to determine the mutation induction dose of these two stages of embryo.

The optimal dose for mutation induction in *Coffea* spp. was determined as growth reduction at 50% (GR_{50}) and 30% (GR_{30}) for shoot/cutting irradiation, or lethality dose at 50% (LD_{50}) and 30% (LD_{30}) for leaf disc, somatic calli, embryogenic calli, globular and cotyledonary embryos gamma irradiation (Mba *et al.*, 2010; Bado *et al.*, 2015). The collected data were subjected to analysis of variance (ANOVA) and least significant differences (LSD) of means (5% level) using GenStat 9.2.

3 Results

Analysis of variance of the number of emerged nodes, shoot length, number of roots, longest root length and leaf area on shoots grown *in vitro* from cuttings of *C. canephora* cv. 'Quillou' exposed to gamma irradiation was significant ($p < 0.05$) among gamma irradiation doses (Table 36.1). The results showed that gamma irradiation inhibited the growth of shoots or cuttings at doses above 20 Gy. These treatments were lethal for root formation and more pronounced for secondary induction, with less than 30% root or secondary induction at 40 Gy. None of the explants had roots, and consequently no secondary induction, at 60 Gy (data not shown). No growth was observed with 60 Gy. The data collected from each parameter were normalized according to Mba *et al.* (2010) and the optimal dose for mutation induction using coffee cuttings was estimated to be between 19 and 46 Gy, respectively, for GR_{30} and GR_{50} (Table 36.2 and Fig. 36.1). Thus, leaf area was severely affected by gamma irradiation followed by number of formed roots, and the longest root length saw similar responses, whereas number of emerged nodes and length of shoot growth were relatively radioresistant with GR_{50} above 40 Gy (Fig. 36.1).

The effects of gamma irradiation on callogenesis induction rate on leaf discs of *C. canephora* cv. 'Quillou' increased over time. After incubation for 9 weeks, all treatments below 20 Gy exhibited almost 100% callus induction rate. The treatments above 20 Gy inhibited callus formation on leaf discs, with 36% and 7% reached at 24 weeks incubation, respectively, for 40 Gy and 60 Gy (Fig. 36.2). Successively, induced calli were evaluated for their ability to develop into somatic embryogenic calli. The results showed a progressive inhibition of embryogenesis with the increase of gamma irradiation treatments.

Table 36.1. Analysis of variance for effects of gamma irradiation on *Coffea canephora* cv. 'Quillou' shoot and root growth.

		Number of emerged nodes		Leaf area (cm²)		Number of roots		Longest root length		Shoot growth length	
Source of variation	DF	Means square	F-value	Means square	F-value	Means square	F-value	Means square	F-value	Means square	F-value
Dose	5	19.333	70.81*	33.182	23.90*	57.511	34.25*	20.3172	22.02*	0.83458	25.13*
Residual	84	0.2730		1.388		1.679		0.9225		0.03321	
Total	89										

DF = Degree of Freedom; *denotes significant differences at 5% probability level

Table 36.2. Means and standard deviation of various growth parameters in the responses of *Coffea canephora* cv. 'Quillou' shoot and root to gamma irradiation.

Dose (Gy)	Number of emerged nodes	Leaf area (cm²)	Number of roots	Longest root length (cm)	Shoot growth length (cm)
0	2.73 ± 0.46a	3.19 ± 1.21a	5.27 ± 1.39a	2.73 ± 0.83a	0.55 ± 0.19ab
10	2.73 ± 0.46a	3.74 ± 1.93a	4.47 ± 1.55ab	3.02 ± 1.02a	0.63 ± 0.25a
15	2.87 ± 0.64a	3.19 ± 1.13a	4.53 ± 1.64ab	2.53 ± 0.95a	0.60 ± 0.25ab
20	2.60 ± 0.51a	2.12 ± 1.16b	3.73 ± 1.39bc	2.19 ± 1.23ab	0.51 ± 0.19ab
40	1.47 ± 0.74b	0.82 ± 0.72c	2.13 ± 1.06c	1.04 ± 1.17bc	0.37 ± 0.06c
60	0.00 ± 0.00c	0.00 ± 0.00d	0.00 ± 0.00d	0.00 ± 0.00c	0.00 ± 0.00d
CV%	25.3	54.2	38.6	50.1	41.0
LSD	0.38	0.84	0.94	0.70	0.13
GR_{30}–GR_{50}	22–41.5	19–27	20–34	25–34	39–46

*Means followed by the same letter do not differ significantly at $p = 0.05$ according to the Least Significance Difference.

However, 20 Gy exhibited a stimulation effect contrary to 60 Gy which was completely lethal in embryogenesis. The optimal dose range for mutation induction in coffee using leaf disc irradiation was estimated between 30 Gy and 36 Gy, respectively, for LD_{30} and LD_{50} (Fig. 36.2).

Somatic calli of *C. arabica* cv. 'Pacamara' and embryogenic calli of *C. arabica* cv. 'Java' and *C. canephora* cvs 'Niaoulli' and 'Quillou' were exposed to gamma irradiation and the effects were assessed as survival rate. Analysis of variance of survival rate of the two types of callus was not significant ($p < 0.05$) among irradiation treatments. All treated calli showed a change in colour as a response to irradiation, compared with yellow control. The callus developed somatic embryos until exposed to a 40 Gy dose. Further, embryogenic calli survived until 80 Gy, but exhibited formation of globular as well as cotyledonary embryos only at doses below 40 Gy. However, the optimal dose for mutation induction in coffee through calli irradiation was estimated as between 55 Gy and 100 Gy, respectively, for LD_{30} and LD_{50} (Fig. 36.3), and for embryogenic calli above 80 Gy for LD_{30}.

Two successive stages of embryo development in culture, namely globular and cotyledonary (or torpedo) embryos, were subjected to gamma irradiation. Analysis of variance of survival rate showed significance ($p < 0.05$) among the treatments (Table 36.3). Increasing the applied dose of gamma irradiation diminished the survival rate and the ability of embryo differentiation to plantlets (Table 36.4). All doses above 20 Gy were lethal for globular embryos, whereas 50% of cotyledonary embryos survived at 40 Gy. Thus, the optimal dose for mutation induction was established between 8–20 Gy and 17–30 Gy, respectively, for LD_{30} and LD_{50} for globular and cotyledonary embryos. These results showed

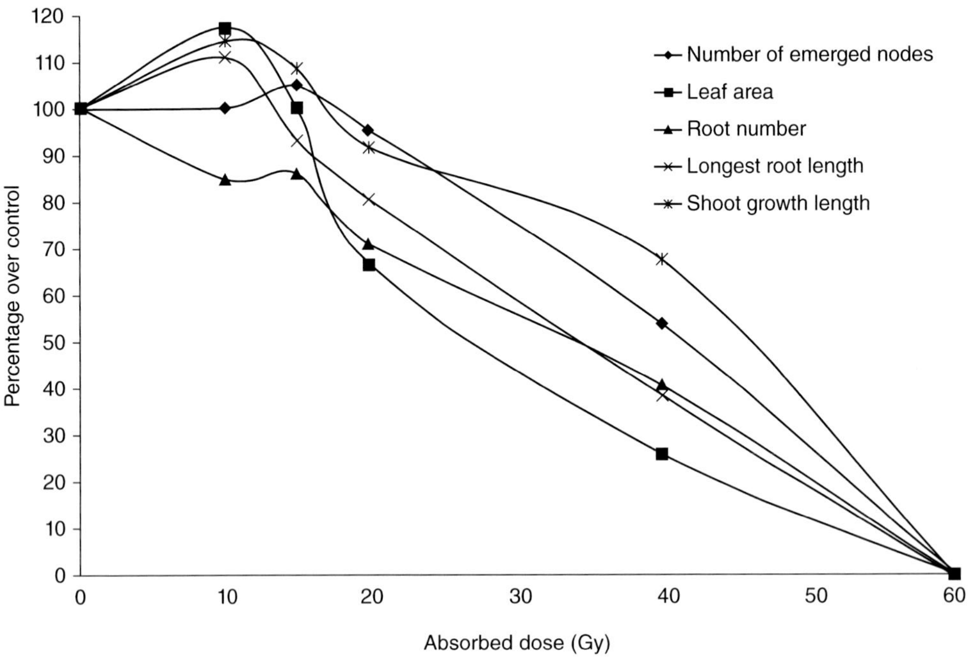

Fig. 36.1. Effects of gamma irradiation on *Coffea canephora* cv. 'Quillou' shoot and root growth after nine weeks.

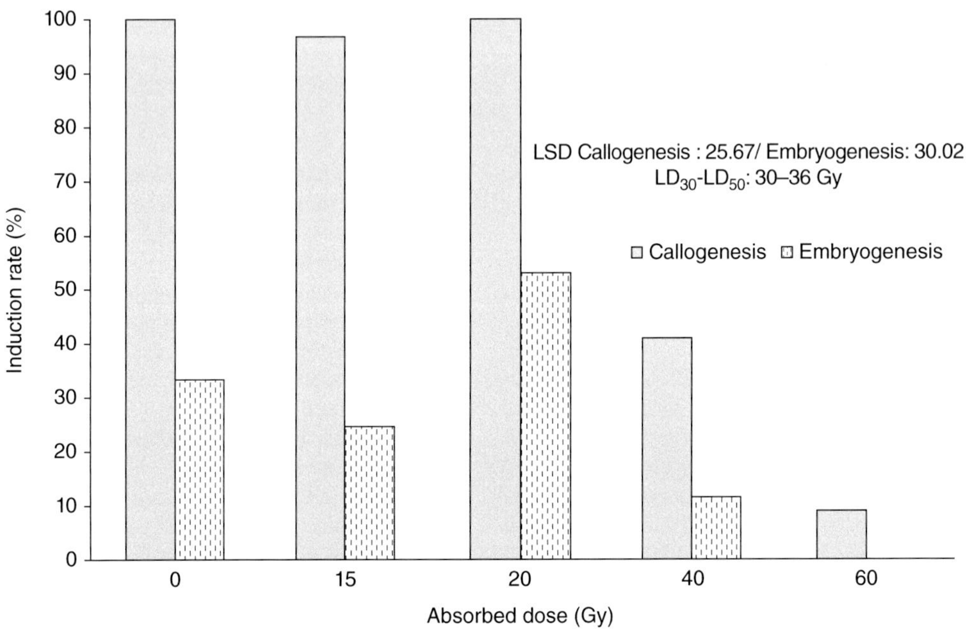

Fig. 36.2. Effects of gamma irradiation on *Coffea canephora* cv. 'Quillou' leaf discs in response to callogenesis and somatic embryogenesis after 24 weeks incubation.

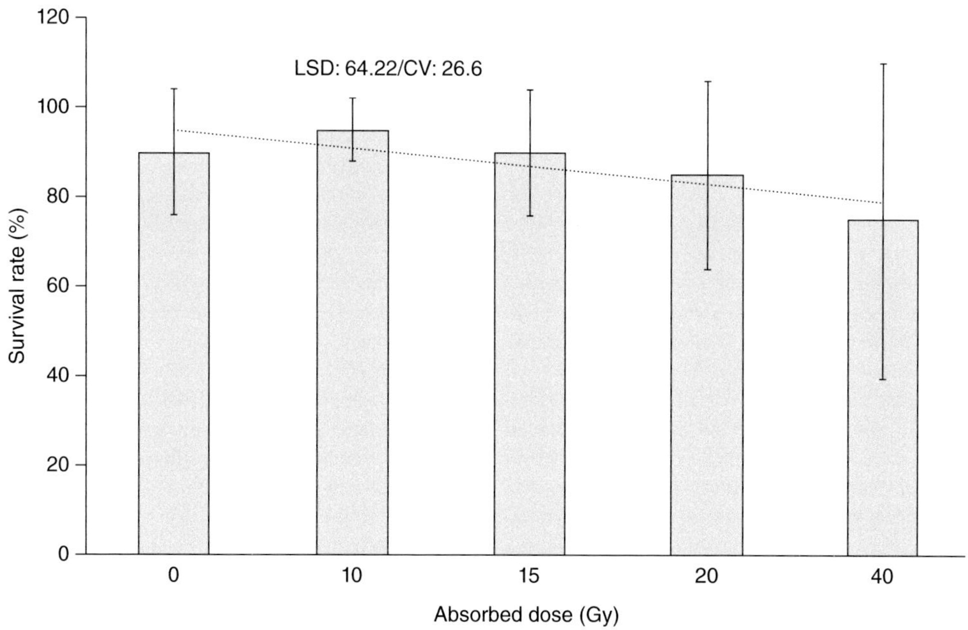

Fig. 36.3. Effects of gamma irradiation on survival rate of *Coffea arabica* cv. 'Pacamara' callus and curve of the optimal dose estimation ($y = -4x + 99$; $R^2 = 0.6957$; standard error of regression coefficient, STEYX = 4.19).

Table 36.3. Analysis of variance for effects of gamma irradiation doses on *in vitro* survival rate of globular and cotyledonary embryos of *Coffea arabica* cv. 'Java'.

Source of variation	DF	Globular embryo		Cotyledonary embryo	
		Means square	F-value	Means square	F-value
Replication	2	80.00	3.69	126.7	0.75
Dose	4	4816.67	222.31*	4193.3	24.91*
Residual	8	21.67		168.3	
Total	14				

DF = Degree of Freedom; *denotes significant differences at 5% probability level

Table 36.4. Effects of gamma irradiation on survival rate of globular and cotyledonary embryos of *Coffea arabica* cv. 'Java'.

Dose (Gy)	Globular embryo	Cotyledonary embryo
0	83.33 ± 5.77a	93.33 ± 5.77a
15	75.0 ± 7.07a	86.67 ± 5.77a
20	40.0 ± 10.00b	46.67 ± 25.17b
40	0.0 ± 0.00c	50.00 ± 10.00b
60	0.0 ± 0.00c	0.00 ± 0.00c
CV%	11.6	23.4
LSD	8.76	24.48
LD_{30}–LD_{50}	**8.04–20.68**	**17.12–30.59**

the relative radio-resistance of cotyledonary embryo to gamma irradiation in comparison with globular embryo.

4 Discussion

In mutation induction, radiosensitivity tests are performed with the purpose of selecting the optimal dose for a specific genotype in a species. This corresponds to the gamma treatment which will provide the mutant trait with a low-mutational-load genetic background at a frequency that can be detected in a mutant population. This is especially important in vegetatively propagated crops, such as coffee, with a long juvenile period and, consequently, difficulty in restoring an elite genetic background by backcrossing (Bado *et al.*, 2017). Therefore, the determination of the optimal dose for mutation induction in such crop species is highly recommended. Irradiation of *in vitro* cuttings of *C. canephora* cv. 'Quillou' showed significant growth reduction with increasing gamma irradiation dose (Fig. 36.1). Leaf area was most affected by higher gamma irradiation. Roots were more severely affected than shoot development, even if the number and length of roots and the number and length of emerged shoots were similar. However, the optimal dose was established between 19 Gy and 46 Gy for mutation induction in *in vitro* coffee shoots (Table 36.2). This is relatively high in comparison with *in vivo* vegetative cuttings of three cultivars of *C. arabica*, which were estimated at between 12 Gy and 15 Gy when considering the sprouting success percentage in response to gamma irradiation (Dada *et al.*, 2018). In general, the effects of ionizing irradiation are a function of the energy absorbed in the exposed tissue; consequently, high doses may result in a higher mutation frequency. These findings are in the same range as those for other vegetatively propagated crops such as cassava at between 6 Gy and 25 Gy (Owoseni *et al.*, 2006; Ndofunsu *et al.*, 2015), potato at 30 Gy (Al-Safadi and Arabi, 2003, 2007), sweet potato at 45 Gy (Asare and Akama, 2014) and yam between 20 Gy and 50 Gy (Amano and Tsugawa, 1985), for gamma irradiation of *in vitro* stem cuttings. Our results corroborate the reports of bud/cutting-mutagenized tree crops, for example apple and citrus with the optimal dose established between 20–80 Gy and 16–40 Gy for scion/bud-cutting irradiation (Spiegel-Roy and Kochba, 1973; Hearn, 1984; Gulsen *et al.*, 2007; Kodym *et al.*, 2012).

Leaf disc irradiation results in mutation in a single cell which then may multiply during callogenesis and become manifest in the resulting embryos. The fact that somatic or gametic embryos originate from a single cell prevents chimera formation among regenerated plantlets. This makes them ideal explants for mutation induction and allows early screening during callogenesis or embryogenesis in comparison with shoot mutagenesis, for which chimera dissolution by repeated sub-culturing is required. Embryogenesis in the calli produced was more radiosensitive to gamma irradiation than callogenesis, which may be explained by the ability to induce non-differentiated consecutive differentiated cell formation (Fig. 36.2). However, the responses of leaf discs showed a higher radiosusceptibility to gamma irradiation compared with shoots, when comparing explants of the same *Coffea* spp. cv. 'Quillou'. This confirms the differential response of explants to the same mutagens reported by Bado *et al.* (2016) using various explants of potato.

The survival rate of calli induced prior to irradiation showed that they had more resistance to gamma radiation than leaf disc explants. Thus, the mutation induction estimated is much higher than the optimal doses for other propagules (Fig. 36.3). However, our findings consolidate the report Kodym *et al.* (2012) on optimal doses on vegetative and perennial crops; for example, in apple, callus irradiation was between 50 Gy and 100 Gy, whereas in citrus callus irradiation showed higher radio-resistance with 120–200 Gy and 240 Gy, respectively, reported by Spiegel-Roy and Kochba (1973) and Gonzaga *et al.* (2011), which may reflect the variability found within and between species, irradiation source and dose rate, and irradiation conditions. These reports of a higher mutation induction dose for single explants are supported by our results of embryogenic calli of three cultivars, indicating that coffee embryogenic callus was more radio-resistant than other explants in the present study. The survival and formation of embryos by 80 Gy irradiated embryogenic calli showed that the LD_{50} is much higher.

Nevertheless, the next stages of somatic embryogenesis formation, i.e. globular and cotyledonary embryos, were more radiosensitive to gamma irradiation with mutation induction between 8.00–20.00 Gy and 17.00–30.00 Gy for LD_{30} and LD_{50}, respectively, for globular and cotyledonary embryos compared with callus and embryogenic callus (Table 36.4). This can be explained by the higher resistance of single-cell explants compared with multi-cellular explants (Bado *et al.*, 2017). This goes along with similar results of mutation induction dose range observed in the present study with shoot irradiation for multi-cellular explants (Table 36.2 and Fig. 36.1). Gamma irradiation of globular-stage somatic embryos of cassava has revealed a much higher optimal dose of 50 Gy for mutation induction (Joseph *et al.*, 2004), indicating a probable species-related response of *in vitro* explants and also gamma irradiation dose rate used. However, the higher radio-resistance observed with cotyledonary embryos, compared with globular embryos, may be explained by their advanced stage in plantlet formation (Table 36.4), where cotyledonary embryos were quite advanced in their development compared with the early differentiation stages, which would be expected to make them less vulnerable to gamma rays.

5 Conclusion

It has been proved that *in vitro* cell/tissue culture combined with mutation induction is effective in improvement of crop plants, especially for those with long juvenile periods and heterogeneous crop species, e.g. tree crops. This chapter presents pioneer work where various *in vitro* explants have been considered and subjected to mutagenesis. Therefore, this research provides optimal doses for improvement of *Coffea* spp. by mutation breeding using multi-cellular (shoot/cutting, globular and cotyledonary/torpedo embryos) or single-cell explants (leaf disc, callus and embryogenic callus). The evaluation of the different putative mutants generated in various *in vitro* explants will allow the determination of the effectiveness of explants for mutation induction by genotypic and phenotypic analysis of traits such as coffee leaf rust. Thus, the most efficient *in vitro* explant for mutation induction in *Coffea* spp. will be identified.

Acknowledgements

This work was supported by the Food and Agriculture Organization of the United Nations and the International Atomic Energy Agency through their Joint FAO/IAEA Program of Nuclear Techniques in Food and Agriculture as Coordination Research Project D22005. We would like to thank Keji Emmanuel Dada and Enock Sapey for valuable support.

References

Al-Safadi, B. and Arabi, M.I.E. (2003) *In vitro* induction, isolation and selection of potato mutants resistant to late blight. *Journal of Genetics and Breeding* 57, 359–364.

Al-Safadi, B. and Arabi, M.I.E. (2007) *In vitro* induction, isolation and selection of potato mutants tolerant to salinity. *Advances in Horticultural Science* 20, 127–132.

Amano, E. and Tsugawa, H. (1985) Effect of gamma-irradiation on garlic and nagaimo (seedling). *Japan Journal of Breeding* (Suppl.) 35(2), 216–217.

Asare, P.A. and Akama, C.K. (2014) Reaction of sweet potato (*Impomea batatas* L.) to gamma irradiation. *Journal of Applied Sciences* 14(17), 2002–2006.

Bado, S. *et al.* (2015) Plant mutation breeding: current progress and future assessment. *Plant Breeding Reviews* 39, 23–88.

Bado, S. *et al.* (2016) *In vitro* methods for mutation induction in potato (*Solanum tuberosum* L.). *African Journal of Biotechnology* 15(39), 2132–2145.

Bado, S. *et al.* (2017) Plant mutation breeding for the improvement of vegetatively propagated crops: successes and challenges. *CAB Reviews* 12(028), 1–21.

Bolívar-González, A. *et al.* (2018) Responses of Arabica coffee (*Coffea arabica* L. var. Catuaí) cell suspensions to chemically induced mutagenesis and salinity stress under in vitro culture conditions. *In Vitro Cellular & Developmental Biology – Plant* 54, 1–14.

Carvalho, A. *et al.* (1967) Genetic relationships of selected *Coffea* species. *Ciência e Cultura (Brasil)* 19(1), 161–165.

Dada, K.E. *et al.* (2014) Radio sensitivity testing in coffee (*Coffea arabica*) as a prelude to coffee improvement through mutation breeding. In: *ASIC Conference (25th Conference on Coffee Science)*, 8–13 September, Colombia, pp. 177–178.

Dada, K.E. *et al.* (2018) Biological effect of gamma irradiation on vegetative propagation of *Coffea arabica* L. *African Journal of Plant Science* 12(6), 122–128.

Davis, A.P. *et al.* (2006) An annotated taxonomic conspectus of the genus Coffea (Rubiaceae). *Botanical Journal of the Linnean Society* 152(4), 465–512.

Déchamp, E. *et al.* (2015) Coffee (*Coffea arabica* L.). *Methods in Molecular Biology* (Clifton, N.J.) 1224, 275–291.

Etienne, H. and Bertrand, B. (2003) Somaclonal variation in *Coffea arabica*: effects of genotype and embryogenic cell suspension age on frequency and phenotype of variants. *Tree Physiology* 23, 419–426.

Etienne, H. *et al.* (2005) Somatic embryogenesis protocol: coffee (*Coffea arabica* L. and *C. canephora* P.). In: Jain, S.M. and Gupta, P.K. (eds) *Protocols for Somatic Embryogenesis in Woody Plants*. Springer, Dordrecht, The Netherlands, pp.167–179.

Gonzaga, D.L. *et al.* (2011) Radiossensibilidade de dois tipos de propágulos de citros. *Bragantia, Campinas* 70(1), 13–18.

Gulsen, O. *et al.* (2007) Development of seedless and Mal Secco tolerant lemons through budwood irradiation. *Scientia Horticulturae* 112, 184–190.

Hearn, C.J. (1984) Development of seedless orange and grapefruit cultivars through seed irradiation. *Journal of the American Society for Horticultural Science* 109, 270–273.

Joseph, R. *et al.* (2004) Induced mutations in cassava using somatic embryos and the identification of mutant plants with altered starch yield and composition. *Plant Cell Reports* 23, 91–98.

Kodym, A. *et al.* (2012) Methodology for physical and chemical mutagenic treatments. In: Shu, Q.Y., Forster, B.F. and Nakagawa, H. (eds) *Plant Mutation Breeding and Biotechnology*. CABI, Wallingford, UK, and FAO, Rome, pp. 169–180.

Kumar, P.K.V. *et al.* (2016) Coffee. In: Mani, M. and Shivaraju, C. (eds) *Mealybugs and their Management in Agricultural and Horticultural Crops*. Springer, India, pp. 643–655.

Labouisse, J.P. *et al.* (2008) Current status of coffee (*Coffea arabica* L.) genetic resources in Ethiopia: implications for conservation. *Genetic Resources and Crop Evolution* 55, 1079–1093.

Landey, R.B. (2013) Influence of micropropagation through somatic embryogenesis on somaclonal variation in coffee (*Coffea arabica)*: assessment of variations at the phenotypical, cytological, genetic and epigenetic level. *Végétal Biologie. Université Montpellier II - Sciences et Techniques du Languedoc*, 2013 [In English].

Mba, C. *et al.* (2010) Induced mutagenesis in plants using physical and chemical agents. In: Davey, M.R. and Anthony, P. (eds) *Plant Cell Culture: Essential Methods*. Wiley, Chichester, UK, pp. 111–130.

Ndofunsu, D. *et al.* (2015) *In vitro* gamma radio-sensitivity test in Congolese cassava, *Manihot esculenta* Crantz accession. *Academia Journal of Biotechnology* 3(1), 001–005.

Owoseni, O. *et al.* (2006) Radio-sensitivity and *in vitro* mutagenesis in African accessions of cassava, *Manihot esculenta* Crantz. *Plant Mutation Reports* 1(2), 32–36.

Pietro, G. (2015) *FAO Coffee Pocketbook*. Food and Agriculture Organization of the United Nations, Rome. Available at: http://www.fao.org/3/a-i4985e.pdf (accessed April 2019)

Spiegel-Roy, P. and Kochba, J. (1973) Mutation breeding in citrus. In: *FAO/IAEA Induced Mutations in Vegetative-Propagated Plants*. IAEA, Vienna, pp. 91–105.

Van Harten, A.M. and Broertjes, C. (1989) Induced mutations in vegetatively propagated crops. *Plant Breed Reviews* 6, 55–91.

Zamarripa, A. (1993) Étude et development de l'embryogenèse en millieu liquid du caféier (*Coffea canephora* P., *Coffea arabica* L. et l'hybrid Arabusta). Thèse de doctorat. École National Sepérieure Agronomique, Rennes, France.

Section 4.

Mutation Breeding in Vegetatively Propagated and Ornamental Crops

37 Somaclonal Variation in Clonal Crops: Containing the Bad, Exploring the Good

Nicolas Roux[1], Rachel Chase, Ines Van den houwe, Chih-Ping Chao, Xavier Perrier, Jean-Pierre Jacquemoud-Collet, Julie Sardos and Mathieu Rouard

Abstract

Somaclonal variation describes random cellular changes in plants regenerated through tissue culture. It occurs in certain crops that undergo micropropagation and has been recorded in different explant sources, from leaves and shoots to meristems and embryos. In banana (*Musa* spp.), a clonal crop conserved *in vitro*, somaclonal variation has been observed after prolonged periods in tissue culture, resulting from an increase in subcultures performed on a given clone. According to scientific literature, variants, or off-types, often show characteristics such as abnormal growth and flower or fruit defects in frequencies ranging from 1% to 32%. This variation poses a problem for gene bank managers, whose mandate is to maintain the genetic integrity of their collections for research and breeding. In the case of the Bioversity International *Musa* Germplasm Transit Centre (ITC), stress during the *in vitro* process is minimized by various techniques and plants are regenerated after 10 years, making it a long and costly process. Identifying somaclonal variation at an early stage would be an ideal solution; however, this requires suitable molecular markers. Recent studies revealed that techniques such as direct DNA sequencing and single nucleotide polymorphisms (SNPs) are able to detect the underlying factors of somaclonal variation and are becoming more accessible. On the other hand, somaclonal variation can be beneficial as it allows the natural development of new varieties and supplies genetic stocks used for future genetic studies. Harnessing the diversity of somaclones is easier, faster and cheaper compared with other methods of crop improvement, although it is also less predictable. So far, variants of crops such as apple, strawberry, potato and banana have been successfully adopted into global markets. In this chapter, we will discuss how to minimize the adverse effects of somaclonal variation while maximizing its benefits for greater crop diversity, with a particular focus on banana.

Keywords: clonal crops • somaclonal variation • crop improvement • mutation breeding • genetic stocks

1 Introduction

Vegetative or clonal propagation is an asexual reproduction in which successive mitosis of specialized vegetative propagules (as bulbs, corms, tubers, cuttings, buds and apomictic seeds) develops new plants and results in a clonal population. Clonal crops such as potatoes, yams, sweet potatoes, banana and cassava complement maize, rice, wheat, legumes, vegetables and livestock and provide income, nutrition and food security for around 300 million poor people in developing countries (Thiele *et al.*, 2017).

A clone is usually considered to be genetically uniform material derived from a single individual that is vegetatively propagated either

[1] Email: n.roux@cgiar.org

 Mutation Breeding, Genetic Diversity and Crop Adaptation to Climate Change (eds S. Sivasankar *et al.*)
DOI: 10.1079/9781789249095.0037

in vivo or *in vitro*. *In vitro* rapid mass propagation (or micropropagation) has the main advantage of increasing the propagation rate. But tissue culture *in vitro* may also create undesired variation, also called somaclonal variation. The term 'somaclonal variation' was first introduced by Larkin and Scowcroft (1981) to describe the variation observed among plants regenerated after passing through tissue culture or cell culture. Somaclonal variations were first recorded in potato, sugarcane, rice and maize in the 1970s and 1980s (Karp, 1995). There appear to be two types of somaclonal variation: heritable and epigenetic (Skirvin *et al.*, 1994; Kaeppler *et al.*, 2000).

- Heritable variation is stable through the sexual cycle or repeated asexual propagation. Somaclonal variation can involve either single or multiple genes and can be due to alterations in DNA sequence, genes, chromosomes or entire sets of chromosomes.
- Epigenetic variation may be unstable even when asexually propagated. In this case, somaclonal variation involves mechanisms of gene silencing or gene activation that were not due to chromosome aberrations or sequence change.

According to reviews by Skirvin *et al.* (1994) and Krishna *et al.* (2016), the factors affecting somaclonal variation can be of different origins, such as:

- the tissue culture environment (e.g. temperature or light);
- the culture medium, including growth regulators such as auxins or kinetin;
- the explant source (tissue with preformed shoots are more stable than other types of explants such as adventitious buds with undifferentiated tissues);
- the plant genotype (within a species, the frequency of somaclonal variations usually occur at higher ploidy levels);
- the number of subcultures; and
- the time spent in tissue culture *in vitro* (not regenerated).

Crops that are propagated by tissue culture are more likely to display somaclonal variation. Among all the reviews written on somaclonal variation, Bairu *et al.* (2011) gave a useful list of 180 examples of plants with the sources of variation and reference. The list showed that most of the somaclonal variation occurs in species with breeding limitations (such as clonal crops) and are due to the preferred method of propagation via tissue culture. A selection of species from Bairu *et al.* (2011) is given in Table 37.1.

2 Challenges of Somaclonal Variation

2.1 In commercial production

Commercial micropropagation was initiated in the 1970s and 1980s, when the number of commercial laboratories undertaking these activities grew significantly. There were high expectations during that period and then in the 1990s many failures occurred due to the production of off-types (Reuveni *et al.*, 1993; Skirvin *et al.*, 1994). Even though many laboratories reduced their commercial tissue culture operations,

Table 37.1. A selection of five crops showing somaclonal variation (from Bairu *et al.*, 2011).

Species	Common name	Source of variation	Detection method	Reference
Allium sativum L.	Garlic	Genotype, colchicine treatment	Morphology, isozyme patterns	Chomatova *et al.* (1990)
Musa spp.	Banana	Number of subcultures	Morphology, RAPD, microsatellite markers	Ray *et al.* (2006)
Solanum tuberosum L.	Potato	Embryogenic culture	Morphology	Rietveld *et al.* (1991)
Fragaria L.	Strawberry	6-benzylaminopurine	Morphology, RAPD	Biswas *et al.* (2009)
Saccharum L.	Sugarcane	Callus culture	Morphology, susceptibility to red-rot disease	Singh *et al.* (2008)

such companies still exist (especially for ornamental plants) which follow best practices to avoid the high rate of somaclonal variation. Commercial companies multiply few diverse accessions at high rates.

2.2 In gene banks

For *in vitro* gene banks the situation is different from commercial laboratories since the goal is to conserve the highest possible diversity and maintain a limited number of plantlets per accession. To limit the number of subcultures and reduce as much as possible the manpower needed to maintain a high quantity of accessions, the cultures are maintained under slow growth conditions. For example, at Bioversity's International *Musa* Germplasm Transit Centre (ITC), accessions can be maintained for 1 year on average without subculturing if the conditions include reduced temperature and light, growth regulator in the medium and with minimal replication to maintain healthy germplasm. Gene banks multiply their accessions regularly and at rates depending on demand. Quality management systems are generally put in place to ensure that the distributed material is true to type. Somaclonal variants are not true-to-type accessions and therefore cannot be distributed by gene banks – this amounts to a loss of conserved/available genetic diversity. Gene banks must therefore detect somaclonal variants or 'off-types', eliminate them from the active collection and replace them from the original source if possible.

2.3 Detection and solutions

Commercial companies as well as gene banks have been investigating how to limit, as much as possible, the production of somaclonal variants. Based on the causes listed above, the following recommendations are made to limit the production of off-types/variants (Smith *et al.*, 1992).

- Select the ideal genotype or accession that shows relative stability as starting material or mother plant.
- Minimize stress through explant sources, regeneration techniques and culture environment.
- Limit subculture cycles and regenerate plants regularly. Restrict multiplication to approximately 1000 plants, which corresponds to around ten subcultures from initiation. If we consider a multiplication rate of two at each subculture, we should have, from each explant, $2^{10} = 1024$ plantlets after ten subcultures.

At the ITC, in order to perform the third recommendation above we estimated that we reach 1000 plants per meristem after ten subcultures, which corresponds to 10 years (as accessions are subcultured once per year on average).

During the early 2000s, the ITC initiated the so-called Field Verification exercise, which was put in place to field-verify all accessions that had been *in vitro* at the ITC for more than 10 years. The first step of the ongoing exercise is that three to five plantlets of accessions available for distribution and detected as virus-free are sent to the field, i.e. the field collection of the USDA Research Station in Puerto Rico (USDA-ARS-TARS). At least three plants per accession are grown in the field and 34 morphological descriptors are recorded (based on the minimum descriptors TAG, 2010) together with a set of ten standard photos agreed upon by a panel of taxonomists called the Taxonomic Advisory Group (TAG). All morphological data and photos, plus any comments from the USDA curator, are compiled in the *Musa* Germplasm Information System (MGIS) database (Ruas *et al.*, 2017). The data is then shared with the TAG panel, where each expert gives their opinion on the true-to-type nature of each ITC accession to determine its genetic integrity.

In parallel, the ITC collection (i.e. 1566 accessions) is being genotyped, using flow cytometry to determine the ploidy and using 19 SSR markers to record the genomic constitution (based on methods in Hippolyte *et al.*, 2012, and Christelová *et al.*, 2017). The entire process is depicted in Fig. 37.1.

The Field Verification process has enabled the detection of mislabelled or misclassified accessions, but not somaclonal variants. It was only after the publication of the whole *Musa* genome sequence (D'hont *et al.*, 2012) that we could map SNP markers to detect some cases of somaclonal variants. By aligning the genotyping by sequencing (GBS) data to the referenced sequenced genome, it was possible to visualize the distribution by chromosome of the ploidy

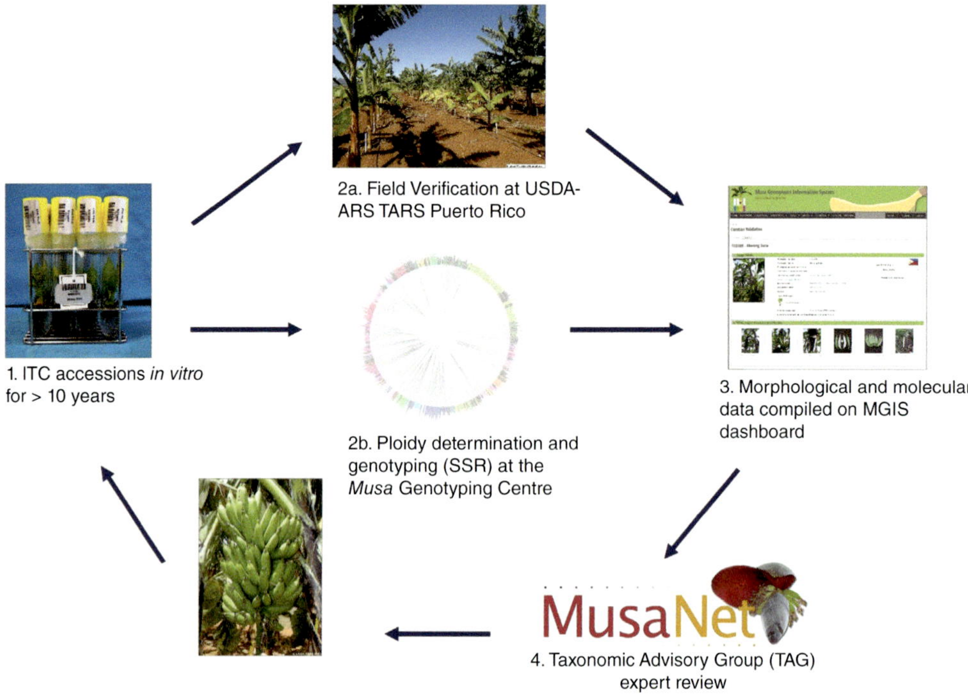

Fig. 37.1. The Field Verification process of ITC banana accessions.

ratio on biallelic SNP data, the ploidy ratio being the ratio for a given SNP, between the number of reads observed on the major allele divided by the total number of reads obtained for that given SNP (e.g. an SNP with A allele for 15 reads and C for 30 reads will have a ploidy ratio of 30/(30 + 15) = 0.666). For a typical diploid profile, all chromosomes are expected to exhibit a peak centred around 0.50 (Fig. 37.2a), and around 0.66 for a triploid (Fig. 37.2b); ratios of 0.75 are expected for AA×AB allotetraploids (Fig. 37.2c).

For most of the accessions, we have found that all the chromosomes matched with the overall expected ploidy. Nevertheless, we detected in some accessions that one or a few chromosomes had unexpected ploidy ratios. For example, when comparing the profile of the mother plant of plantain cultivar 'Ihitisim' (AAB) NGA-124 (originating from IITA field collection in Onne, Nigeria) maintained in the field but never propagated *in vitro*, with its daughter plant 'Ihitisim' (AAB) ITC0121 introduced *in vitro* at ITC in 1986, we detected that the chromosome profiles are not identical. For the accession NGA-124, the peaks of all 11 chromosomes are at a major allele frequency of approximately 0.66, as expected of a triploid (Fig. 37.3a). For the accession ITC0121, the peak of chromosome 3 is centred around 0.75, which corresponds to the expectation of a tetraploid profile for that specific chromosome (Fig. 37.3b).

These preliminary observations need more investigation on a larger number of accessions, but this method may only detect somaclonal variations that are due to aneuploidy or large chromosomal aberrations. Consequently, a routine somaclonal variation detection pipeline could be put in place in order to improve and accelerate the process of determining genetic fidelity of accessions maintained *in vitro* not only in gene banks, but also in commercial tissue culture laboratories.

3 Benefits of Somaclonal Variation

3.1 In crop improvement

Even though most somaclonal variants are of no value as they have deleterious traits, in some

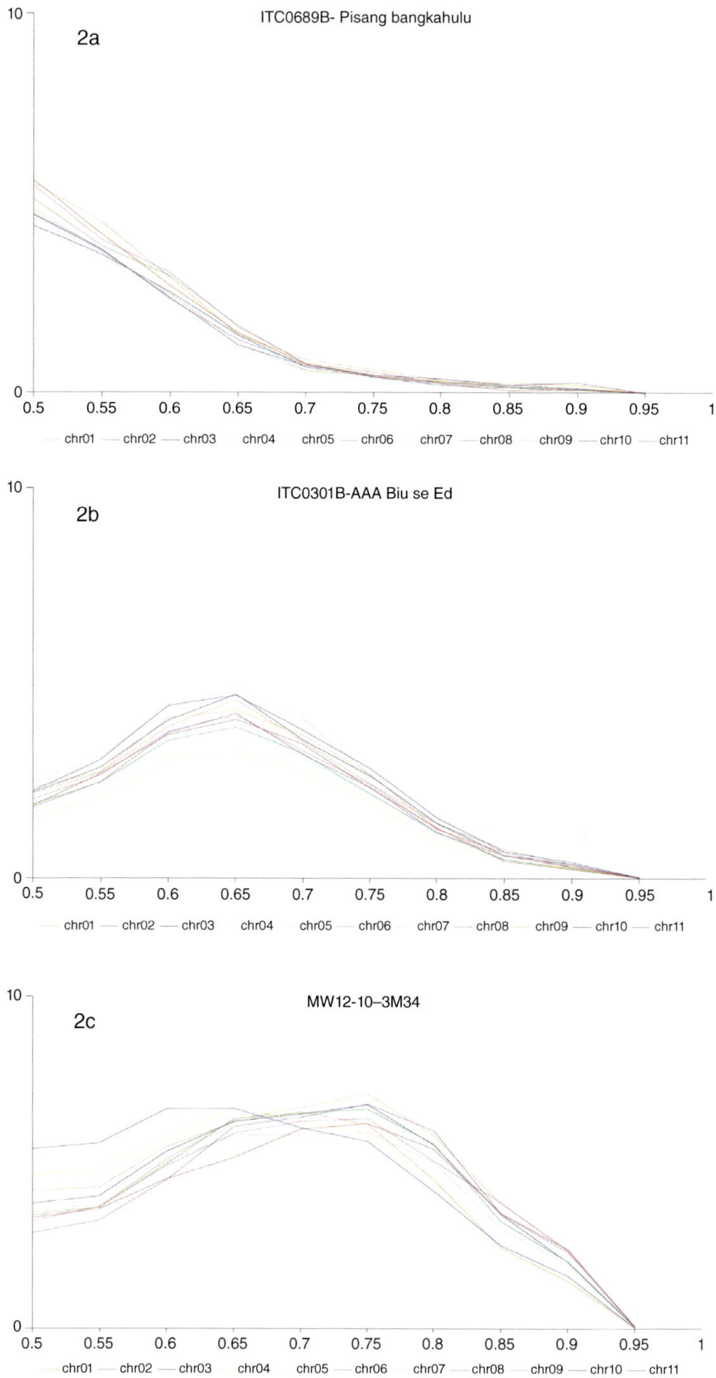

Fig. 37.2. Distribution for the 11 chromosomes of the ploidy ratio on biallelic SNP data. *X*-axis: consecutive classes of ploidy ratio calculated for a given SNP as the ratio of the number of reads with the major allele on the total number of reads obtained for that given SNP, ratio which varies between 0.5 and 1 according to this definition. *Y*-axis: frequency (in %) of SNPs observed along the chromosome for a given class of ploidy ratio.

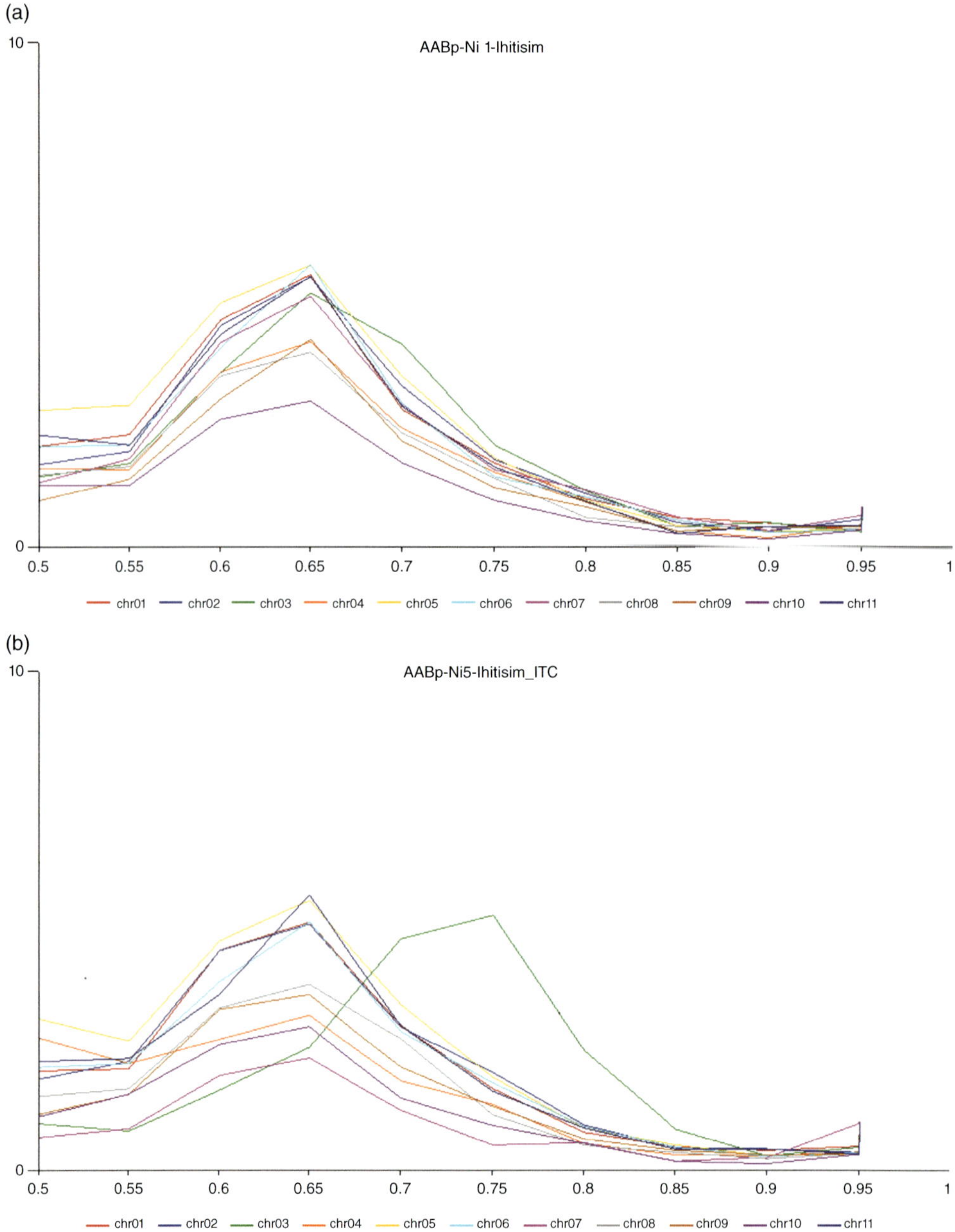

Fig. 37.3. **(a)** Ploidy distribution per chromosome of NGA-124 'Ihitisim' (AAB). All chromosomes are forming a peak at ratio 0.66 (2/3) which is expected from a triploid genotype. **(b)** Ploidy distribution per chromosome of ITC0121 'Ihitisim' (AAB). All chromosomes are forming a peak at ratio 0.66 except for chromosome 3 (in green) with a peak at ratio 0.75 corresponding to tetraploid profile.

cases they do produce improved traits and are even considered as an additional crop improvement method, very often compared with a mutation induction technique even though there are no physical or chemical mutagens involved. The advantage is that, when the variants are stable, they represent another source of genetic diversity that is important for crop improvement as they provide new variability for breeders. Some examples are listed in Table 37.2.

3.2 The banana case study: 'Formosana' (GCTCV-218)

Researchers at the Taiwan Banana Research Institute (TBRI) have been investigating since the early 1980s how to obtain resistant varieties to Fusarium wilt disease caused by the fungus *Fusarium oxysporum* f. sp. *cubense* tropical race 4 (FOC TR4). No fungicide can kill this pathogen, so selection of resistant varieties is the fastest way to prevent further damage. Conventional breeding for triploid cultivars such as the Cavendish, the subgroup representing 90% of the banana cultivated area in Taiwan, is difficult as they are sterile. Researchers from TBRI found that among the millions of plants distributed from tissue culture to farmers, about 3% displayed variation in size or colour of the pseudostem and leaves and in the shape of leaves and fruits (Hwang, 1986). In 1984, TBRI decided to initiate a project to screen Cavendish plantlets for FOC TR4 resistance that included four steps:

Step 1: The establishment of a FOC TR4-infested test plot
Step 2: Healthy plants are planted at high density with plants of a susceptible variety (Fig. 37.4a)
Step 3: Screening for somaclonal variants with resistance to FOC TR4 (Fig. 37.4b)
Step 4: Recurrent selection of an improved type (Fig. 37.4c).

'Formosana' was not the first somaclonal variant detected to be tolerant to FOC TR4; nevertheless, the general wilt incidence was 4.1%, which is significantly lower than the resistant 'Tai Chiao' with 9.5% and the 'Giant Cavendish' with 29.6% (Hwang and Ko, 2004). In addition, most of the agronomical characteristics of 'Formosana' (i.e. stronger pseudostem, thicker leaves, better hand formation and more uniform hand size and higher yield) make it a superior cultivar even if it takes one additional month to produce fruit as compared with its progenitor 'Pei Chiao'. Starting in 2000, 'Formosana' was distributed to farmers and was very well received for its resistance to FOC TR4 and its fruit quality. Commercial planting was possible after only 6 years of research. Although there are now millions of plants produced commercially, it is still not known what kind of variation occurred in the genome. With the sequenced information now available, new molecular markers and tools have been developed that could help us understand this variation.

3.3 Use of mutants as genetic stocks

Genetic stocks, broadly defined as plants or populations generated and/or selected for genetic studies, represent a unique and growing class of extremely valuable germplasm which, depending on crop, type of genetic stock and user community, may represent genetic resources of either transient or long-lasting value. Genetic stocks can be divided into three general groups: cytological stocks (e.g. chromosome addition/substitution, aneuploids, amphiploids), mutants (e.g. induced/insertion mutants, tilling populations) and germplasm sets (e.g. mapping populations,

Table 37.2. Selection of desirable traits and development of some commercially exploited varieties through somaclonal variation in different horticultural crops (from Krishna *et al.*, 2016).

Species	Common name	Improved characteristic of somaclone	Reference
Malus × domestica Borkh.	Apple	Resistance to *Erwinia amylovora*	Chevreau *et al.* (1998)
Musa acuminata L.	Banana	GCTCV clones; resistance to *Fusarium* wilt	Hwang *et al.* (1992); Hwang and Ko (2004)
Rubus fruticosus L.	Blackberry	Thornless var. 'Lincoln Logan'	Hall *et al.* (1986)
Mangifera indica L.	Mango	Resistant to *Colletotrichum gleosporiensis*	Litz *et al.* (1991)
Ipomoea batatas L. lam.	Sweet potato	Tolerant to salinity	Anwar *et al.* (2010)

Fig. 37.4 **(a)** (top left): FOC TR4 infested test plot. **(b)** (top right): Shorter 'Formosana' (left) and original 'Formosana' (right). **(c)** (bottom): Comparison of GCTCV-218 'Formosana' (left) and parental Giant Cavendish 'Pei Chao' (right) planted in an infested field.

parental lines, reference germplasm). Any genetic stock collection can represent from a few lines to tens of thousands of lines and therefore can potentially offer a challenge, as well as a burden, to gene bank managers from the standpoint of storage and maintenance. Another challenge with genetic stock collections is the rapidly changing technology used to develop new genetic stocks, which may make older collections obsolete. Therefore, the gene bank manager is faced with having to predict the long-term value, and hence the need for long-term maintenance, of any given collection. Despite the contrasting options of long-term value for some collections versus short-term value for others, there is no question that genetic stock collections should be preserved and that the global system, including CGIAR gene banks, needs to play a role in their preservation.

In April 2010, a group of experts met in Bologna, Italy, for a Genetic Stocks Management Workshop organized in the framework of the System-wide Genetic Resources Programme (SGRP) to develop a curator decision tree (Fig. 37.5) and

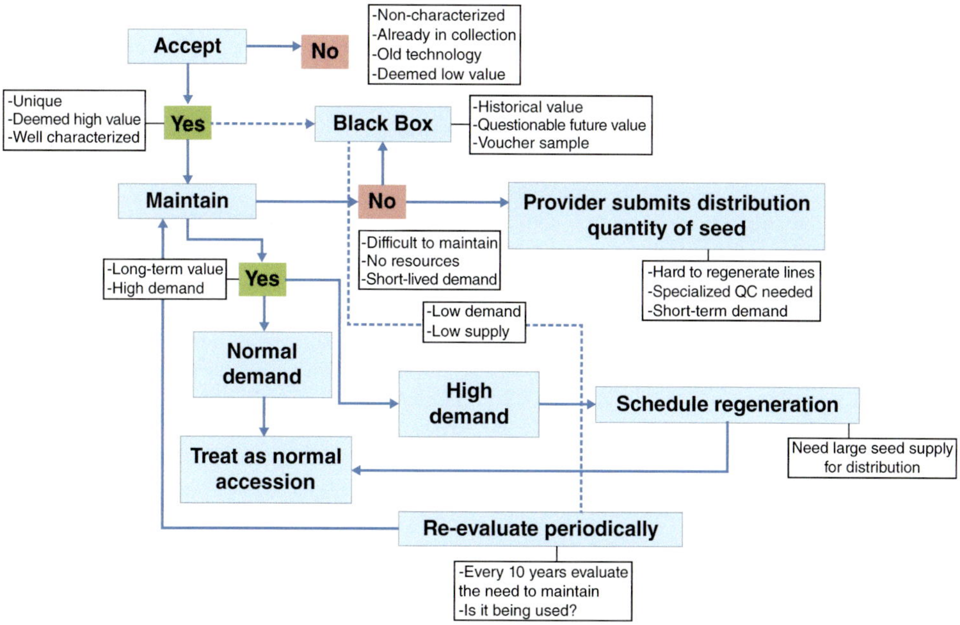

Fig. 37.5. Decision tree offering specific examples for handling genetic stocks in gene banks (SGRP, 2011).

recommendations on the management of genetic stock collections.

Recommendations on genetic stocks (modified from SGRP, 2011) include the following.

- Genetic stock collections are a valuable genetic resource in need of attention from the international community to ensure conservation and access to a wider community.
- An inventory needs to be made of where genetic stock collections are located and who is responsible for the maintenance and distribution of these stock collections.
- There should be involvement of the different crop communities to highlight the urgency of safeguarding genetic stock collections.
- User communities should be a key part of the effort to inventory, collect and safeguard genetic stock collections for target crops.
- A database system is needed which can accommodate data from genetic stock collections.
- There should be regular workshops involving curators and gene bank managers, breeders and researchers to ensure proper identifying, prioritizing and care of genetic stock collections.
- Clear policy rules need to be used when exchanging genetic stocks. Under the Plant Treaty such material can be considered PGRFA (plant genetic resources for food and agriculture) Under Development and is subject to accessibility restriction.
- The international community (CGIAR, FAO, IAEA) should actively support the conservation of genetic stocks of value and importance, as they are tools which can further the mission of sustainably increasing and improving livelihoods.

4 Conclusion

For mass propagation and gene banks wanting to multiply on a large scale and distributing worldwide true-to-type material, somaclonal variants are not desirable and represent a challenge that needs to be overcome using adapted methodologies. Thanks to the sequencing of a great number of crops since the beginning of this century, early detection pipelines could help gene banks avoid propagating or distributing off-types and could improve our understanding of the causes of somaclonal variation occurrence.

Nevertheless, somaclonal variation has proved to be an interesting source of variation and has been used as a genetic improvement methodology, often considered similar to mutation breeding, although no physical or chemical mutagens are believed to be involved. However, our knowledge on the cause of somaclonal variation remains incomplete. As mutants, somaclonal variants even if not directly used as improved germplasm can be used as genetic stocks to finally understand the pathway of many traits, such as the resistance to FOC TR4, as obtained by TBRI on their 'Giant Cavendish' tissue-culture variant (GCTCV) clones through somaclonal variation.

The Joint FAO/IAEA Division with its laboratories and facilities at Seibersdorf, Austria, and its Mutant Variety Database with information on more than 3000 plant mutant varieties (cultivars) is in an ideal position to coordinate the collection of genetic stocks to be distributed worldwide for the benefit of mankind, similar to what the CGIAR centres do within the Genebank Platform.

Acknowledgements

Many thanks to our colleagues from the Bioversity International *Musa* Genetic Resources Team, USDA-TARS-ARS in Puerto Rico, the *Musa* Genotyping Centre at the Institute for Experimental Botany, Czech Republic, CIRAD, TBRI, MusaNet and the Joint FAO/IAEA Division. We are grateful also to the CGIAR Research Programme on Roots, Tubers and Bananas and the CGIAR Genebank Platform, both funded by CGIAR donors.

References

Anwar, A. *et al.* (2010) Assessment of somaclonal variation for salinity tolerance in sweet potato regenerated plants. *African Journal of Biotechnology* 9, 7256–7265.

Bairu, M.W. *et al.* (2011) Somaclonal variation in plants: causes and detection methods. *Plant Growth Regulation* 63, 147–173.

Biswas, M.K. *et al.* (2009) Development and evaluation of *in vitro* somaclonal variation in strawberry for improved horticultural traits. *Scientia Horticulturae* 122, 409–416.

Chevreau, E. *et al.* (1998) Fire blight resistance and genetic trueness-to-type of four somaclonal variants from the apple cultivar Greensleeves. *Euphytica* 104, 199–205.

Chomatova, S. *et al.* (1990) Protein complex and esterase isoenzyme patterns of *Allium sativum* L. cultivars and clones-regenerants. *Biologia Plantarum* 32, 321–331.

Christelová, P. *et al.* (2017) Molecular and cytological characterization of the global Musa germplasm collection provides insights into the treasure of banana diversity. *Biodiversity and Conservation* 26(4), 801–824.

D'hont, A. *et al.* (2012) The banana (*Musa acuminata*) genome and the evolution of monocotyledonous plants. *Nature* 488, 213–217.

Hall, H.K. *et al.* (1986) Germplasm release of 'Lincoln Logan', a tissue culture-derived genetic thornless 'Loganberry' . *Fruit Varieties Journal* 40, 134–135.

Hippolyte, I. *et al.* (2012) Foundation characteristics of edible Musa triploids revealed from allelic distribution of SSR markers. *Annals of Botany* 109(5), 937–951.

Hwang, S.-C. (1986) Variation in banana plants propagated through tissue culture. *Journal of Chinese Society for Horticultural Science* 32, 117–125.

Hwang, S.-C. and Ko, W.-H. (2004) Cavendish banana cultivars resistant to Fusarium wilt acquired through somaclonal variation in Taiwan. *Plant Disease* 88(6), 580–588.

Hwang, S.-C. *et al.* (1992) Breeding for resistance to Fusarium wilt of Cavendish banana by using tissue culture method. *Special Publication Taichung District Agricultural Improvement Station* 29, 229–237.

Kaeppler, S.M. *et al.* (2000) Epigenetic aspects of somaclonal variation in plants. *Plant Molecular Biology* 43, 179–188.

Karp, A. (1995) Somaclonal variation as a tool for crop improvement. *Euphytica* 85, 295–302.

Krishna, H. *et al.* (2016) Somaclonal variations and their applications in horticultural crops improvement. *3 Biotech* 6, 54.

Larkin, P.J. and Scowcroft, W.R. (1981) Somaclonal variation – a novel source of variability from cell cultures for plant improvement. *Theoretical and Applied Genetics* 60(4), 197–214.

Litz, R.E. *et al.* (1991) Mango somatic cell genetics. *Acta Horticulturae* 291, 133–140.

Ray, T. *et al.* (2006) Genetic stability of three economically important micropropagated banana (*Musa* spp.) cultivars of lower Indo-Gangetic plains, as assessed by RAPD and ISSR markers. *Plant Cell, Tissue and Organ Culture* 85, 11–21.

Rietveld, R.C. *et al.* (1991) Somaclonal variation in tuber disc-derived populations of potato. *Theoretical and Applied Genetics* 82, 430–440.

Reuveni, O. *et al.* (1993) Factors influencing the occurrence of somaclonal variation in micropropagated bananas. *Acta Horticulturae* 336, 357–384.

Ruas, M. *et al.* (2017) MGIS: managing banana (*Musa* spp.) genetic resources information and high-throughput genotyping data. *Database (Oxford)*, 2017, bax046. doi: 10.1093/database/bax046

SGRP (2011) *Report prepared by Nicolas Roux, Mathieu Rouard and Dave Ellis for the Global Public Goods Programme Phase 2 of the System-wide Genetic Resources Programme. Genetic Stocks Management Workshop, Bologna, Italy 28–29 April 2010*. System-wide Genetic Resources Programme, Rome.

Singh, G. *et al.* (2008) In vitro induction and characterization of somaclonal variation for red rot and other agronomic traits in sugarcane. *Euphytica* s60, 35–47.

Skirvin, R.M. *et al.* (1994) Sources and frequency of somaclonal variation. *HortScience* 29(11), 1232–1237.

Smith, M. *et al.* (1992) *Banana tissue culture: Getting it right.* Banana Industry Protection Board of Queensland Annual Report 1992, pp. 30–32.

TAG (2010) Minimum Descriptors for Musa. MusaNet, available at: www.musanet.org (accessed 2019).

Thiele, G. *et al.* (2017) Roots, tubers and bananas: planning and research for climate resilience. *Open Agriculture* 2, 350–361.

38 Induced Mutations for Generating Bananas Resistant to Fusarium Wilt Tropical Race 4

Joanna Jankowicz-Cieslak[1], Florian Goessnitzer, Sneha Datta, Altus Viljoen, Ivan Ingelbrecht and Bradley J. Till[2]

Abstract

Bananas are a staple for more than 400 million people. Additionally, more than 16.5 million tonnes are exported, making it both an important food security and a cash crop. Productivity of Cavendish-type bananas is threatened by both abiotic and biotic stresses. The fact that triploid bananas are sterile, parthenocarpic and obligate vegetatively propagated makes them particularly susceptible to diseases, including Fusarium wilt caused by *Fusarium oxysporum* f. sp. *cubense* (Foc) tropical race 4 (Foc TR4). This is because continual clonal propagation has led to loss of genetic diversity. Additionally, lack of meiosis limits methods for breeding. Foc TR4 has been devastating Cavendish bananas in South-east Asia but has recently also been reported from Queensland in Australia, the Middle East and Mozambique, thus threatening global banana production. To address this, we are performing mutagenesis of *in vitro* propagated bananas to broaden the genetic diversity in order to find new alleles conferring disease resistance. We have developed methods for efficient induction of mutations in isolated apical meristems from shoot tips using chemical mutagens and ionizing radiation. Mutation discovery methods have been adapted to recover mutations including single point mutations and large deletions spanning millions of base pairs. We have created approximately 5000 mutated lines for forward-genetic screens to identify TR4 resistance in greenhouse-evaluated material. A subset of ca. 500 *in vitro* plantlets was subjected to glasshouse-based screening using a virulent *F. oxysporum* isolate. To date, 23 lines showing altered resistance responses to Foc TR4 have been identified.

Keywords: Musa • TR4 • LC-WGS • EMS • gamma irradiation • Panama disease

1 Introduction

Numerous factors have negative impacts on global food security, including population growth, reduction in soil quality, climate change and a growing demand for meat-based diets (Godfray *et al.*, 2010; Foley *et al.*, 2011). A multifaceted approach is thus needed to increase food production over the next decades. This includes implementing sustainable agronomic practices and reducing food losses, improving infrastructures and promoting healthier food. A fundamental component of food security is genetic improvement of crops to increase yields (Duvick, 2005; Ceccarelli *et al.*, 2010) and increased resilience of plants to pests and diseases, which are spreading more rapidly due to changes in climate and increasing global commerce and travel (Gautam *et al.*, 2013; Velásquez *et al.*, 2018).

[1] Email: .jankowicz@iaea.org; [2] Email: till.brad@gmail.com

 Mutation Breeding, Genetic Diversity and Crop Adaptation to Climate Change (eds S. Sivasankar *et al.*)

DOI: 10.1079/9781789249095.0038

Banana (*Musa* spp.) is the fourth most important food crop in developing countries and among the top ten food commodities for South-east Asia, Africa and Latin America (FAOSTAT, 2010; World Agriculture, 2018). Most edible varieties are clones from the inter- and intraspecific hybridization of diploid seeded *Musa acuminata* and *Musa balbisiana* (Simmonds and Shepherd, 1955; Perrier *et al.*, 2011). The vast majority of edible bananas are sterile, parthenocarpic and triploid. The Cavendish banana comprises approximately 45% of all bananas worldwide. Large-scale and commercial propagation of Cavendish cultivars are accomplished through tissue culture-based mass clonal production, propagation and maintenance systems developed since the 1970s (Gowen, 1995). This has resulted in a genetic bottleneck with Cavendish bananas planted throughout the world being nearly identical genetically.

A lack of genetic diversity in agricultural crops may result in severe losses due to pathogens and pests. For banana, this first occurred in the early and mid-20th century when Fusarium wilt (Panama disease), caused by the soil-borne fungus *Fusarium oxysporum* f. sp. *cubense* (Foc), devastated the banana export industry that consisted of Gros Michel (Stover, 1962). Cavendish bananas replaced Gros Michel (Ploetz, 2005, 2006) as they were found to be resistant to Foc race 1, the strain that affects Gros Michel. However, a new strain called Foc tropical race 4 (TR4) began to severely affect Cavendish bananas in five Asian countries and in the Northern Territory of Australia, and is now spreading globally (Ploetz, 2015; Viljoen *et al.*, 2020). Reports of its spread to Jordan, Pakistan, Lebanon, Oman and Mozambique since 2013 have raised serious alarms as no commercial replacement for the Cavendish banana resistant to TR4 had yet been found (Molina *et al.*, 2009; García-Bastidas *et al.*, 2013; Ordonez *et al.*, 2015). The use of conventional methods such as chemical and cultural control to eradicate TR4 has been largely unsuccessful. Its spread thus threatens both the global export trade and local markets, and therefore it can negatively impact on commercial economies and the livelihoods of small growers.

Traditional methods of breeding to create new allelic combinations are challenging in vegetatively propagated crops, especially those unable to produce sufficient seed. This is due to their reduced ability to undergo meiosis, recombination and independent assortment. Alternative methods exist to introduce new allele diversity and genetic material. Transgenic approaches have been used since the 1990s to introduce novel genes and regulatory elements into plants (Tester and Langridge, 2010). More recently, genome editing approaches have been described that allow precise modification of DNA sequences within the plant (Zhang *et al.*, 2018). These methods require pre-knowledge of the gene sequence to be altered, or a novel sequence to be introduced. A powerful alternative is forward genetics. Here, random mutagenesis is performed that results in novel allele combinations. Phenotypic selection is applied to recover plants with the desired improved trait without the need for *a priori* knowledge of gene function (Jankowicz-Cieslak and Till, 2015). When used as a breeding method, the process is referred to as plant mutation breeding. An effective approach to generate allele diversity randomly is to mutagenize plant cells with chemicals or ionizing radiation (Mba *et al.*, 2010). This approach dates back to the 1920s, with Lewis John Stadler's pioneering work with X-ray irradiation of cereal crops (Stadler, 1928a,b). Since then, more than 3360 mutant crop varieties have been registered with improved yields, nutritional quality and resistance to biotic and abiotic stresses (Jankowicz-Cieslak and Till, 2015; MVD, 2021).

The type and dosage of mutagen used affects the density and spectrum of induced mutations in treated material. The chemical mutagen ethyl methanesulfonate (EMS), for example, induces single base-pair (bp) point mutations at a high density (Kurowska *et al.*, 2011; Till *et al.*, 2018). In many species EMS causes nearly 100% G:C to A:T transition changes at densities ranging from one mutation per 23,000 bp to one million base pairs. Achievable mutation densities depend on factors such as mutagen used, dosage, genotype and ploidy of the plant. Point mutations can cause missense changes, affect RNA splice sites and lead to the gain or loss of stop codons. EMS is thus an excellent choice to induce single-gene effects creating a range of allele types from hypomorphic to loss-of-function (Henikoff *et al.*, 2004). It has been extensively used for the reverse genetics method known as TILLING. Technologies for the discovery of newly induced point mutations include enzymatic mismatch cleavage, whole genome sequencing, amplicon sequencing and exome-capture sequencing (Till *et al.*, 2003; Tsai *et al.*, 2011; Henry *et al.*, 2014; Burkart-Waco *et al.*, 2017; Gupta *et al.*, 2017; Krasileva *et al.*, 2017).

Mutagenesis using ionizing radiation is an alternative to chemical mutagenesis. The most commonly used method is gamma irradiation. The IAEA's Mutant Variety Database (MVD) lists 1703 officially released mutant crops produced by gamma irradiation (MVD, 2021). Treatment of plant cells with gamma irradiation can result in a broader spectrum of mutations that includes point mutations, large insertions and deletions (indels), and loss of whole chromosomes (Henry *et al.*, 2015; Li *et al.*, 2015). A common feature of many reports on gamma irradiation is the accumulation of large indels that span tens of thousands to millions of base pairs resulting in copy number variation or complete loss of, tens to hundreds of genes (Jankowicz-Cieslak and Till, 2015). Irradiation dosage of treated tissues should be optimized so that a sufficient number of new mutations accumulate, but also that the resulting material maintains suitable fitness for propagation. The most accurate measure for optimization is direct interrogation of the DNA sequence. Low-coverage whole genome sequencing (LC-WGS) has been described that allows a rapid evaluation of plant genomes for larger indel events by measuring copy number variation. This has successfully been applied to Arabidopsis, poplar and rice to uncover gamma-induced mutations ranging from small deletions to loss of whole chromosomes (Henry *et al.*, 2015; Tan *et al.*, 2015). Once optimized, mutagenesis can be applied on a larger scale so that the desired trait improvements can be recovered.

In 2015, the Joint FAO/IAEA Programme initiated a mutation breeding programme for the genetic improvement of Cavendish bananas for resistance to Fusarium wilt TR4 in China, the Philippines, Iran and Mozambique to help address the global threat of Fusarium wilt to banana production. The focus of the work presented in this report was to develop a banana mutation breeding pipeline to recover mutants with resistance to Foc TR4. A pilot-scale project was undertaken to evaluate different mutation induction and mutant screening methodologies with the goal of producing a streamlined pipeline for higher-throughput screens. Owing to the polyploid and parthenocarpic nature of Cavendish-type bananas, we aimed to broaden the genetic diversity by inducing a range of different allele types in order to increase the chances of producing useful phenotypes in the absence of crossing. Therefore, we optimized conditions for EMS and gamma-irradiation mutagenesis. Optimizations were performed by direct evaluation of DNA mutation events using enzymatic mismatch cleavage for EMS-induced point mutations and LC-WGS for gamma-induced indels. Resulting mutant populations were subjected to net-house-based quarantine screening using Foc TR4, whereby variations in resistance responses were recovered. This suggests that Foc TR4 resistance can be increased using induced mutations and that physical and chemical mutation induction techniques are highly effective to introduce novel genetic variation in triploid banana.

2 Materials and Methods

2.1 Plant material

In vitro banana (*Musa acuminata*) plantlets of the Cavendish (AAA) cultivars 'Grande Naine' and 'Williams' were obtained from Du Roi Laboratory (https://www.duroilab.co.za/, accessed 22 July 2021). Plantlets were maintained *in vitro* through mitotic propagation of shoot tips in liquid S27 media as previously described (Jankowicz-Cieslak *et al.*, 2012; Jankowicz-Cieslak and Till, 2016).

2.2 Mutagenic treatment

Shoot tips of *in vitro* 'Grande Naine' and 'Williams' bananas were isolated and placed in a Petri dish containing a few drops of water. For physical mutagenesis, 10–15 shoot tips were bulked in a Petri dish and subjected to 20 Gy, 30 Gy, or 40 Gy gamma irradiation using a cobalt-60 source (2.144 Gy/s; Seibersdorf, Austria). For chemical mutagenesis, bulk treatment was performed as previously described (Jankowicz-Cieslak *et al.*, 2012; Jankowicz-Cieslak and Till, 2016). EMS treatment (1%) was used to prepare 2000 lines for phenotypic assessment. Each shoot tip was used as the starting material for subsequent cultures, and thereafter referred to as a mutant line (Jankowicz-Cieslak *et al.*, 2012; Jankowicz-Cieslak and Till, 2016; Datta *et al.*, 2018). A minimum of three rounds of meristematic isolation and bisection were carried out to ensure removal of mutation-induced genotypic heterogeneity (mosaicism/chimerism).

2.3 DNA isolation and mutation discovery

DNA was extracted from leaf material using the DNeasy Plant Mini Kit (Qiagen, Cat. No. 69106) according to the manufacturer's guidelines. Primer design and enzymatic mismatch cleavage for EMS-induced mutations followed standard methods optimized for PAGE-gel TILLING assays (Jankowicz-Cieslak *et al.*, 2012). For next-generation sequencing, genomic DNA was assessed for quality and quantity using NanoDrop, agarose gel electrophoresis and Qubit fluorimeter (Duitama *et al.*, 2017). Library preparation, sequencing and bioinformatic analysis of data were performed as described by Datta *et al.* (2018). To evaluate the efficiency of physical irradiation, a whole genome sequencing approach was adapted using low-coverage reads for the detection of large genomic insertions, deletions and chromosomal aberrations (Henry *et al.*, 2014; Tan *et al.*, 2016). Sequencing data was grouped into 100 kilobase (kb) bins for analysis, and indels were scored when three or more bins showed variation in the number of reads. Thus, the minimum indel size detectable with this approach is 300,000 bp. This approach was selected owing to the fact that gamma irradiation can produce a broader spectrum of induced mutations than EMS, and LC-WGS allows cataloguing large induced mutations at a much lower cost than deeper sequencing. To develop this for triploid bananas the mutant variety 'Novaria', created using gamma irradiation, was used as a positive control. The methods used for mutation induction, chimera dissolution, DNA evaluation and resulting data on accumulation of induced mutations have been previously published (Jankowicz-Cieslak *et al.*, 2012; Jankowicz-Cieslak and Till, 2016; Datta *et al.*, 2018). Sequencing was carried out on 'Williams' material treated with 20 and 40 Gy gamma irradiation.

2.4 Foc TR4 resistance screening

Mutagenized plantlets were produced and maintained in the FAO/IAEA laboratories in Seibersdorf, Austria. To evaluate the efficacy of mutation breeding for disease resistance, a subset of over 500 mutated lines was selected. This material was sent to the Taiwan Banana Research Institute (TBRI), which was contracted to provide a service for banana disease screening. TBRI has established efficient protocols for screening for response to Foc TR4 (Promusa, 2021). Briefly, the *in vitro* plantlets were first acclimatized in greenhouse conditions and then hardened off for one month. A virulent Foc TR4 isolate was used to prepare inoculum for pathogenicity testing, and sterilized sandy soil was inoculated to obtain a concentration of approximately 4×10^3 colony-forming units/g soil. One-month-old banana plantlets were transplanted into pots containing the inoculum. Disease symptoms, such as leaf discoloration or the longitudinal splitting of pseudostems, were recorded monthly. The severity of the inner corm infection was rated as follows: 0 = without any symptoms; 1 = less than 10% of the surface discoloured; 2 = 10–20% surface discoloured; 3 = 30% surface discoloured; and 4 = severe wilting of plants.

3 Results

The EMS treatment of 'Grande Naine' resulted in the recovery of a high density of one mutation per 57 kb induced G:C to A:T changes that were stably inherited for at least six generations. All mutations recovered were heterozygous and pedigree analysis suggested rapid dissolution of chimeric sectors within three or fewer subcultures when employing meristematic isolation followed by longitudinal bisection (Jankowicz-Cieslak *et al.*, 2012).

Gamma-ray dosages of 20, 30 and 40 Gy were selected based on growth reduction tests. Low-coverage whole genome sequencing for recovery of induced copy number mutations was adapted for triploid banana using the mutant variety 'Novaria' as a positive control. Sequence analysis revealed 'Novaria' to contain multiple deletions, including one of approximately 3.8 million base pairs spanning over 100 genes (Fig. 38.1).

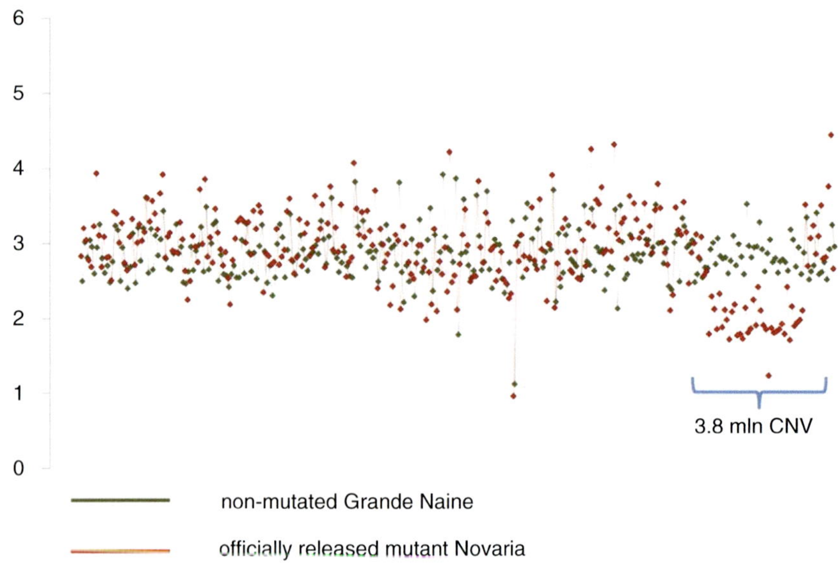

Fig. 38.1. Relative sequence read coverage (RSRC) plots of a 3.8 million base pair deletion in chromosome 5 of mutant 'Novaria' (red) compared with non-mutated control (green). Modified from Datta *et al.* (2018).

This method was then applied to newly gamma-irradiated 'Williams' material. Overall, single sample sequencing revealed 20 large indels across the banana (AAA) genome with a size range of approximately 300,000 to 6.8 million base pairs. Analysis of single mutant lines revealed the accumulation of mutations on one to three chromosomes per line. In some lines no mutations were observed. No statistical difference could be observed in the accumulation of detected indels when comparing 20 Gy and 40 Gy treated material when assayed by LC-WGS (Fig. 38.2) (Datta *et al.*, 2018).

Mutagenized material was shipped to TBRI where disease screening was carried out (Figs 38.3 and 38.4). Plantlets irradiated at a dosage of either 30 or 40 Gy were less infected with Foc TR4 than those treated by EMS. Based on the disease screening, 40 candidate mutants were selected with rating 0 (13 plants) and 1 (27 plants).

4 Discussion

The use of induced mutations remains an important tool for crop breeding. Today there are over 3360 officially released mutant crop varieties in the Mutant Variety Database (MVD, 2021). Of these, 63% have been directly released without further introgressions, suggesting the efficacy of mutagenizing elite cultivars (Jankowicz-Cieslak and Till, 2015; Jankowicz-Cieslak *et al.*, 2017). However, only 12% of the released varieties are vegetatively propagated. This may be in part due to the fact that the major consumed crops are seed crops but may also be due to the extra challenges associated with clonally propagated crops. According to the MVD, only three mutant banana varieties have been officially released. All three were produced by treatment with gamma irradiation. The variety 'Al-Beely' showed 30% higher yield and was released in 2007 in Sudan, while 'Klue Hom Thong KU1' with larger bundle was approved in 1985 for release in Thailand. 'Novaria', demonstrating early maturity and improved fruit quality, was released in Malaysia in 1995 and is reported as the most commercially successful (MVD, 2021). Additional mutation breeding activities in banana have been reported for which there are no varieties reported in the MVD. Gamma irradiation has been used to improve agronomic characteristics in a banana cultivar with resistance to Foc TR4 (Smith *et al.*, 2006). EMS was used to

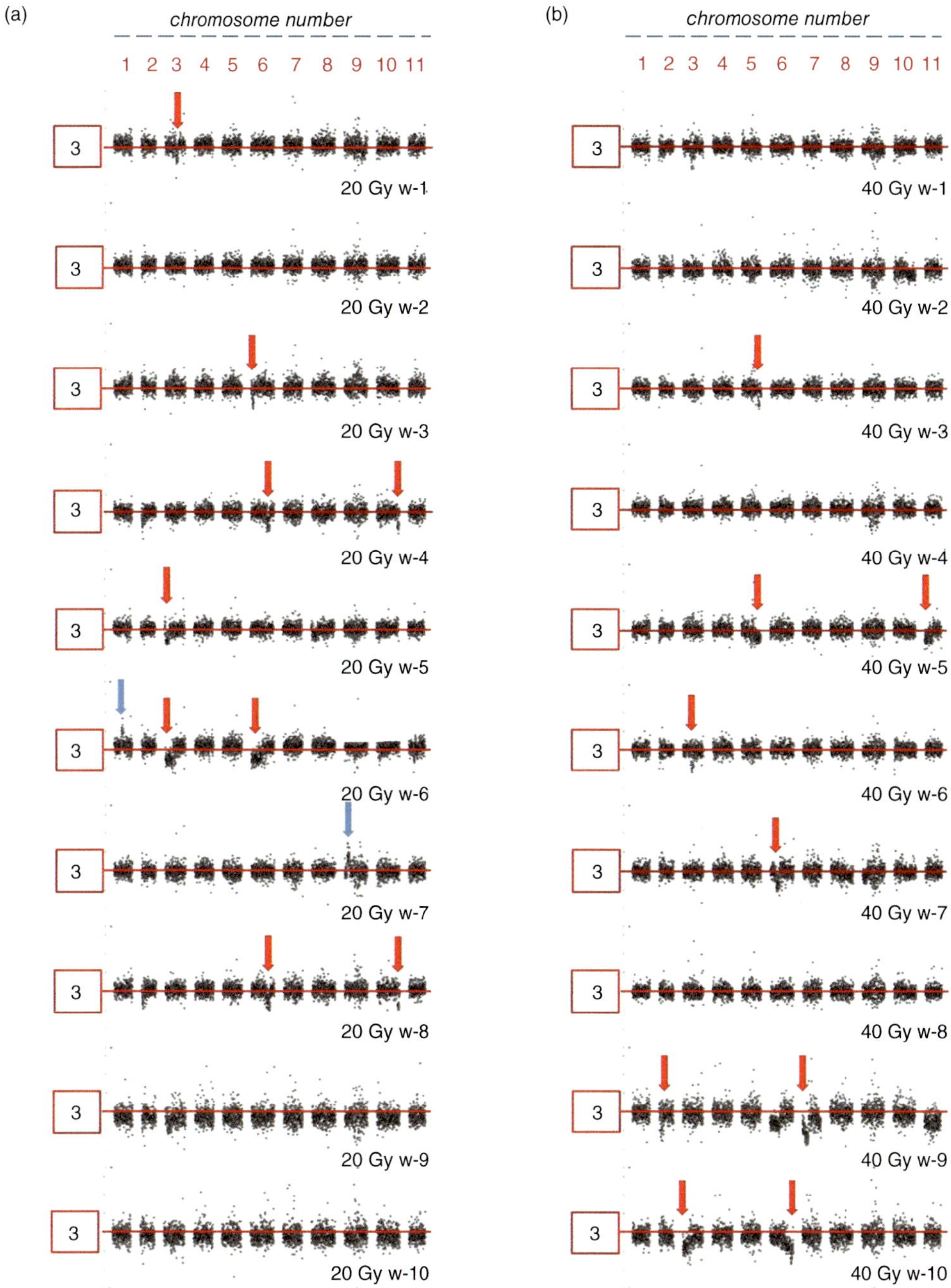

Fig. 38.2. RSRC plots of newly mutagenized Cavendish cultivar 'Williams'. Data from ten lines mutagenized at 20 Gy and 40 Gy (left and right panels, respectively) are shown. Arrows mark selected regions of selected visually detectable putative CNV. Red and blue arrows represent deletions and insertions, respectively. Modified from Datta *et al.* (2018).

Fig. 38.3. A pipeline for screenhouse-based mutant pre-screening and selection of 'Grande Naine' bananas with enhanced resistance to Foc TR4. An *in vitro* mutant population was generated **(a)** and plantlets were hardened *in vivo* **(b)**. One-month-old banana seedlings ready for inoculation **(c)** with Foc TR4 inoculum were prepared **(d, e)**. Inoculated plants **(f)** were scored for symptoms **(g)** and mutants with identified resistance were transplanted to bigger pots for shoot proliferation **(h)** and subsequent *in vitro* re-establishment **(i)**.

improve tolerance to *Fusarium oxysporum* f. sp. *Cubense* in AAA banana (Bhagwat and Duncan, 1998).

One contributing factor to the underrepresentation of vegetatively propagated crops in the MVD may be that their reduction in meiosis complicates breeding when compared with seed-propagated crops. For example, obligate vegetatively propagated crops such as some triploid bananas have completely lost their ability to undergo meiosis. On the one hand, this makes such crops ideal candidates for using induced mutations, as there are few alternatives for generating novel allelic combinations in their genomes. On the other hand, the act of inducing mutations is made much more complex and time consuming. A typical route is performing mutagenesis on tissue culture propagules or cell cultures. The former is advantageous in that establishment of cultures of multicellular tissues can be easier than cell cultures, but it is disadvantageous in that mutagenesis of multicellular tissues results in a genetic chimera or mosaic, where each cell accumulates a different set of mutations. Steps must be taken to dissolve chimeric sectors so that plants derived from these materials are genotypically homogeneous with all tissues harbouring the induced mutations (and are thus capable of expressing the altered trait). The work we have carried out with chemical and physical mutagenesis of banana shows that at least for AAA-type bananas, chimeric sectors can be quickly resolved via meristematic isolation followed by longitudinal bisection of meristems (Jankowicz-Cieslak and Till, 2016). This follows on the observation of Roux *et al.*

Fig. 38.4. Examples of disease symptoms observed 3 months post inoculation. Corm with **(a, b)** no symptoms and with **(c)** mild and **(d)** severe symptoms.

(2001) when following the passage of synthetic aneuploidy in banana tissue cultures that were treated with colchicine. In addition to dissolution of chimeras, the number and type of mutations (density and spectrum) serve as an indicator for a successful mutation breeding programme. It is only within recent years that data are accumulating on the effect that gamma irradiation has on the DNA of crops (Henry *et al.*, 2015; Li *et al.*, 2015, 2016). In AAA bananas, we found that approximately 70% of lines treated with 20 Gy and 60% treated with 40 Gy harboured large deletions (Datta *et al.*, 2018). Single sample analysis suggests that mutated bananas (where mutations are detected) accumulate mutations on one to three chromosomes per line. This may be an underrepresentation, as analysis of biological replicates resulted in the discovery of 18 putative indel events on four chromosomes of a 40 Gy treated 'Williams' sample (Datta *et al.*, 2018). Further work is necessary to learn if this can be improved or if it represents near optimal mutation accumulation. Much more is known, however, about chemical mutagenesis. Large data sets on the effect of EMS on plant genomes started to accumulate after the advent of TILLING (Kurowska *et al.*, 2011; Till *et al.*, 2018). More recently, exome capture sequencing technologies have allowed the recovery of more than 10 million EMS-induced mutations (Henry *et al.*, 2014; Krasileva *et al.*, 2017; Hussain *et al.*, 2018). What has emerged from these studies is a clear expectation of the type of mutation induced (mostly G:C to A:T transitions). The density of induced point mutations varies depending on ploidy, with the frequency of mutation events increasing as ploidy increases. This is presumably because polyploids harbour more functional copies of genes and thus can accumulate more mutations before lethality is reached. Our studies of EMS mutagenized triploid bananas resulted in an estimation of mutation density of 1/57 kb, which is higher than reported densities for diploid plants but lower than that reported for tetraploids. This exactly meets expectation and suggests that EMS mutagenesis in bananas is highly maximized.

Another factor affecting the success of mutation breeding of vegetatively propagated plants surrounds how the mutations are created. Induced mutations, such as those caused by EMS, result in accumulation of heterozygous lesions whereby only one copy of a gene is altered. It is thought that most deleterious point mutations are recessive and thus the majority of mutations would have to be made homozygous before a phenotype is expressed. This is accomplished in seed-propagated plants through self-fertilization or cross-fertilization with sibling plants. How, then, could point mutations work for vegetatively propagated crops? Dominant phenotypes will be observable and there are examples of dominant phenotypes arising from point mutations (Sharma *et al.*, 2014). There may also be the possibility to observe the more frequent recessive mutations. This is owing to the very fact that obligate vegetatively propagated species have no means to expunge deleterious alleles that occur naturally. Over time, such plants would accumulate deleterious alleles in gene copies to the point where only one functional copy is remaining. Thus, plants can become functionally haploid at many loci. The process of accumulating such deleterious mutations over time was postulated by Muller, and has been termed Muller's ratchet (Muller, 1932, 1964; Felsenstein, 1974). Observation of high heterozygosity and accumulation of heterozygous predicted deleterious alleles has been observed in vegetatively propagated crops such as polyploid bananas and potato (Potato Genome Sequencing *et al.*, 2011; Jankowicz-Cieslak *et al.*, 2012). Therefore, induction of point mutations in a single copy of a gene may result in the expression of otherwise recessive traits. The same phenomenon holds true for gamma-induced mutations. However, as observed in our work, such treatments can also induce deletion of millions of base pairs spanning more than 100 genes. Indeed, in the mutant variety 'Novaria', we observed deletion of 189 genes, including those involved in regulating gene expression, protein phosphorylation and cellular biogenesis. Large genomic deletions, insertions and translocations have been reported for gamma irradiation. These may have more profound and genome-wide impacts on gene expression and function, resulting in a wider spectrum of phenotypic consequences (Mileyko *et al.*, 2008; DeBolt, 2010; Żmieńko *et al.*, 2014). From this perspective, it is interesting to consider that gamma irradiation will produce multigenic effects. This may explain how induced mutations can produce stable novel phenotypes that are characterized in traditional breeding activities as being complex traits. For example, more than 1300 registered mutant varieties are associated with a variation in yield. How the novel induced mutations in these varieties correspond to yield QTL in non-mutagenized material that is harbouring presumably minor functional variations remains unexplored. Further study of the recently induced mutations may reveal gene functions and networks previously lost or hidden due to thousands of years of human selection.

Plant diseases are predicted to remain major components of yield loss in future decades. Global climate change and variation are predicted to result in a wider spread of diseases. Increasing global travel by human beings can also cause diseases to spread to new geographical locations. Indeed, recent investigations have revealed Foc TR4 in new regions such as the greater Mekong subregion in Laos, Myanmar and Vietnam (Zheng *et al.*, 2018). Cavendish varieties occupy approximately 40% of the total global area of production (Ploetz, 2015). Thus, a global programme is needed to prevent and manage this devastating disease and also to look into ways to improve disease tolerance. Mutation breeding is one approach that has proved to be a successful tool in crop disease resistance.

For the over 3360 mutants officially registered in the MVD, 558 are improved for resistance to biotic stresses, including improved resilience to *Fusarium* pathogens (MVD, 2021). Among most recently registered mutants with improved disease resistance are soybean and mandarin cultivars. Both were registered in 2017 and a sweet orange variety was released in 2016. The DT2010 soybean cultivar was developed in Vietnam by a cross of two mutant varieties, DT2008 × DT99, and combines some improved characteristics of the two mutant parent varieties, which are high yields, resistance to rust and drought tolerance (MVD, 2021). The *Citrus reticulata* 'Blanco' mutant is a product of direct use of an induced mutation and was registered under the name 'NIAB Kinnow' (MVD, 2021). It exhibits better climatic adaptability in different localities of Pakistan compared with the local cultivar. The tree of this Mandarin variety is highly vigorous, large, and spreading with dense foliage. 'NIAB Kinnow' has shown moderate to high resistance to citrus canker, scab and withertip diseases as well as low incidence of major insect pests. Based on its dynamic yield, high quality and disease resistance potential, 'NIAB Kinnow' is expected to enhance and sustain citrus production in the country. The sweet orange mutant, *Citrus sinensis* L. Osbeck registered under the name 'IAC2014', was generated via gamma-ray treatment of buds. Around 1000 M_1V_1 plants were obtained, which were vegetatively propagated up to M_1V_4, and approximately 7600 plants were planted in the field. One mutant clone showed seedless fruits and greater tolerance to citrus canker. These recent examples not only show that mutation breeding can work effectively for disease tolerance but also point out the fact that the programme can be implemented successfully in the case of vegetatively propagated crops. Therefore, we can expect similar successes in vegetatively propagated banana.

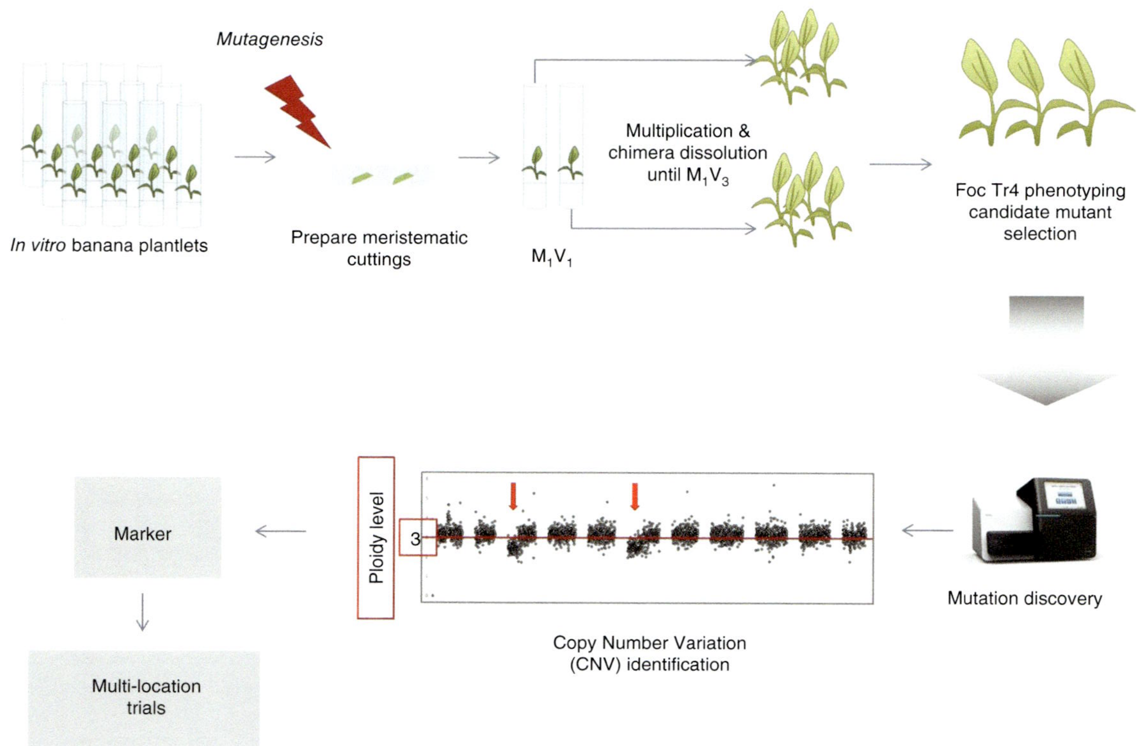

Fig. 38.5. Schematic diagram of banana mutagenesis, mutant phenotyping and recovery of mutations via molecular tools. Shoot tips isolated from banana plantlets propagated and maintained by shoot apical meristem culture are subjected to EMS or gamma irradiation, chimera dissolution and Foc TR4 phenotyping. Lines identified with desired phenotypes are then sequenced and SNP discovery or chromosomal dosage analysis is carried out to detect SNPs (EMS) or indels (gamma). Knowledge of induced mutations is useful to monitor the purity of material and stability of mutations through successive rounds of vegetative propagation. The mutations thus serve as a marker to track and monitor lines in the laboratory and in multi-location field trials. Sequenced mutations also allow improved plants to be uniquely identified. This may be important to protect farmers' rights. Knowledge of DNA sequence variation will also provide a list of putative candidate genes causative for the improved phenotype. Modified from Datta *et al.* (2018).

Mutation breeding projects typically involve the production and screening of thousands of mutant lines (Mba *et al.*, 2010). This can be especially challenging in vegetatively propagated crops. Dissolution of chimeras and maintaining and propagating disease-free material can require extra resources when compared with mutation breeding projects in seed-propagated crops. In addition, technical aspects need to be defined and optimized. This includes determining the stages at which plants can be phenotyped. In the case of the mutant banana population described here, we decided to conduct the first round of screening at the 1-month stage so that interesting material could be identified quickly. Given the large number of resources required for such a project, we decided to apply genotypic screening of a subset of mutant material prior to application of disease phenoytping. This is advantageous in that the effect of the mutagenic treatment is confirmed before additional resources are allocated. Based on this pilot work, we propose a pipeline for mutation breeding for disease resistance in banana (Fig. 38.5). Our pilot scale screen produced between 18.2 and 26.1% of material showing improved resistance. It should be noted that, due to our limited sample size, our pilot screen lacked a mock treated control for disease symptoms. The observed percentages are much higher than expected and all material needs to be validated and confirmed in field trials before conclusions can be drawn.

5 Conclusion

In summary, efficient pipelines for mutagenesis and recovery of induced mutations remain an important tool, especially in vegetatively propagated crops where optimized mutagenesis and phenotypic screening are key components for generating traits such as Foc TR4 resistance. In this report, we described methods for the induction of single nucleotide and large indel mutations in *in vitro* banana cultures. Glasshouse TR4 screening assays were applied and promising mutants expressing improved tolerance were recovered. Evaluation of this promising material is ongoing and will include whole genome sequence evaluation and also multi-location trials to assess performance under different environments.

Acknowledgements

The authors thank Chih-Ping Chao and Shih-Hung Huang of the Taiwan Banana Research Institute for providing the fee-based banana disease screening service described in this work. The Food and Agriculture Organization of the United Nations and the International Atomic Energy Agency through their Joint FAO/IAEA Programme of Nuclear Techniques in Food and Agriculture supported this work. This work is part of IAEA Coordinated Research Projects D24012 and D22005.

References

Bhagwat, B. and Duncan, E.J. (1998) Mutation breeding of banana cv. Highgate (*Musa* spp., AAA Group) for tolerance to *Fusarium oxysporum* f. sp. *cubense* using chemical mutagens. *Scientia Horticulturae* 73, 11–22.

Burkart-Waco, D. *et al.* (2017) Next-generation sequencing for targeted discovery of rare mutations in rice. In: *Biotechnologies for Plant Mutation Breeding*. Springer, Cham, pp. 323–340.

Ceccarelli, S. *et al.* (2010) Plant breeding and climate changes. *The Journal of Agricultural Science* 148, 627–637.

Datta, S. *et al.* (2018) Induction and recovery of copy number variation in banana through gamma irradiation and low coverage whole genome sequencing. *Plant Biotechnology Journal* 16, 1644–1653.

DeBolt, S. (2010) Copy number variation shapes genome diversity in Arabidopsis over immediate family generational scales. *Genome Biology and Evolution* 2, 441–453.

Duitama, J. *et al.* (2017) Deep assessment of genomic diversity in cassava for herbicide tolerance and starch biosynthesis. *Computational and Structural Biotechnology Journal* 15, 185–194.

Duvick, D.N. (2005) The contribution of breeding to yield advances in maize (*Zea mays* L.). *Advances in Agronomy* 86, 83–145.

FAOSTAT (2010) Food and Agriculture Data. Available at: http://www.fao.org/faostat/en/#home (accessed 7 September 2021).

Felsenstein, J. (1974) The evolutionary advantage of recombination. *Genetics* 78, 737–756.

Foley, J.A. *et al.* (2011) Solutions for a cultivated planet. *Nature* 478, 337–342.

García-Bastidas, F. *et al.* (2013) First report of *Fusarium oxysporum* f. sp. *cubense* tropical race 4 associated with Panama disease of banana outside Southeast Asia. *Plant Disease* 98, 694.

Gautam, H.R. *et al.* (2013) Climate change and its impact on plant diseases. *Current Science* 105, 1685–1691.

Godfray, H.C.J. *et al.* (2010) Food security: the challenge of feeding 9 billion people. *Science* 327, 812–818.

Gowen, S. (1995) *Bananas and Plantains*. Chapman & Hall, London.

Gupta, P. *et al.* (2017) Next-generation sequencing (NGS)-based identification of induced mutations in a doubly mutagenized tomato (*Solanum lycopersicum*) population. *Plant Journal* 92, 495–508.

Henikoff, S. *et al.* (2004) TILLING. Traditional mutagenesis meets functional genomics. *Plant Physiology* 135, 630–636.

Henry, I.M. *et al.* (2014) Efficient genome-wide detection and cataloging of EMS-induced mutations using exome capture and next-generation sequencing. *Plant Cell* 26(4), 1382–1397.

Henry, I.M. *et al.* (2015) A system for dosage-based functional genomics in poplar. *Plant Cell* 27(9), 2370–233.

Hussain, M. *et al.* (2018) Identification of induced mutations in hexaploid wheat genome using exome capture assay. *PLoS ONE* 13, e0201918.

Jankowicz-Cieslak, J. and Till, B.J. (2015) Forward and reverse genetics in crop breeding. In: Al-Khayri, J.M, Jain, S.M. and Johnson, D.V. (eds) *Advances in Plant Breeding Strategies: Breeding, Biotechnology and Molecular Tools.* Springer, New York, pp. 215–240.

Jankowicz-Cieslak, J. and Till, B.J. (2016) Chemical mutagenesis of seed and vegetatively propagated plants using EMS. *Current Protocols in Plant Biology* 1(4), 617–635.

Jankowicz-Cieslak, J. *et al.* (2012) Induction, rapid fixation and retention of mutations in vegetatively propagated banana. *Plant Biotechnology Journal* 10, 1056–1066.

Jankowicz-Cieslak, J. *et al.* (2017) Mutagenesis for crop breeding and functional genomics. In: Jankowicz-Cieslak, J., Tai, T.H., Kumlehn, J. and Till, B.J. (eds) *Biotechnologies for Plant Mutation Breeding: Protocols*. Springer, Cham, pp. 3–18.

Krasileva, K.V. *et al.* (2017) Uncovering hidden variation in polyploid wheat. *Proceedings of the National Academy of Sciences* 114, E913–E921.

Kurowska, M. *et al.* (2011) TILLING – a shortcut in functional genomics. *Journal of Applied Genetics* 52, 371–390.

Li C. *et al.* (2015) Genome-wide characterization of cis-acting DNA targets reveals the transcriptional regulatory framework of opaque2 in maize. *Plant Cell* 27, 532–545.

Li S. *et al.* (2016) Frequency and type of inheritable mutations induced by γ rays in rice as revealed by whole genome sequencing. *Journal of Zhejiang University Sci B* 17, 905–915.

Mba C. *et al.* (2010) Induced mutagenesis in plants using physical and chemical agents. In: Davey, M.R. and Anthony, P. (eds) *Plant Cell Culture: Essential Methods*. Wiley, Chichester, UK, pp. 111–130.

Mileyko, Y. *et al.* (2008) Small-scale copy number variation and large-scale changes in gene expression. *Proceedings National Academy of Sciences* 105, 16659–16664.

Molina, A.B. *et al.* (2009) Recent occurrence of *Fusarium oxysporum* f. sp. *cubense* tropical race 4 in Asia. *Acta Horticulturae* 828, 109–116.

Muller, H.J. (1932) Some genetic aspects of sex. *The American Naturalist* 66, 118–138.

Muller, H.J. (1964) The relation of recombination to mutational advance. *Mutation Research* 1, 2–9.

MVD (2021) Mutant Variety Database. Joint FAO/IAEA Mutant Variety Database. Available at: https://mvd.iaea.org/ (accessed 22 July 2021)

Ordonez, N. *et al.* (2015) Worse comes to worst: bananas and panama disease—when plant and pathogen clones meet. *PLOS Pathogens* 11, e1005197.

Perrier, X. *et al.* (2011) Multidisciplinary perspectives on banana *(Musa* spp.) domestication. *Proceedings National Academy of Sciences* 108, 11311–11318.

Ploetz, R.C. (2005) Panama disease: an old nemesis rears its ugly head – Part 1. The beginnings of the banana export trades. *Plant Health Progress.*

Ploetz, R.C. (2006) Panama disease: an old nemesis rears its ugly head – Part 2. The Cavendish era and beyond [Online]. *Plant Health Progress* doi:10.1094/PHP-2006-0308-01-RV.

Ploetz, R.C. (2015) Management of Fusarium wilt of banana: a review with special reference to tropical race 4. *Crop Protection* 73, 7–15.

Potato Genome Sequencing *et al.* (2011) Genome sequence and analysis of the tuber crop potato. *Nature* 475, 189–195.

Promusa (2021) Taiwan Banana Research Institute. Available at: https://www.promusa.org/Taiwan+Banana+Research+Institute+-+TBRI (accessed 22 July 2021)

Roux, N. *et al.* (2001) Effectiveness of three micropropagation techniques to dissociate cytochimeras in *Musa* spp. *Plant Cell, Tissue and Organ Culture* 66, 189–197.

Sharma, S. *et al.* (2014) A dominant mutation in the light-oxygen and voltage2 domain vicinity impairs phototropin1 signaling in tomato. *Plant Physiology* 164, 2030–2044.

Simmonds, N.W. and Shepherd, K. (1955) Taxonomy and origins of cultivated bananas. *Journal of the Linnean Society of Botany (London)* 55, 302–312.

Smith M.K. *et al.* (2006) Towards the development of a Cavendish banana resistant to race 4 of fusarium wilt: gamma irradiation of micropropagated Dwarf Parfitt (*Musa* spp., AAA group, Cavendish subgroup). *Australian Journal of Experimental Agriculture* 46(1), 107–113.

Stadler, L.J. (1928a) Genetic effects of X-rays in maize. *Proceedings National Academy of Sciences USA* 14, 69–75.

Stadler, L.J. (1928b) Mutations in barley induced by X-rays and radium. *Science* 68, 186–187.

Stover, R.H. (1962) *Fusarium Wilt (Panama Disease) of Bananas and Other Musa Species*. Commonwealth Mycological Institute, Kew, UK.

Tan, E.H. *et al.* (2015) Catastrophic chromosomal restructuring during genome elimination in plants. *eLife Sciences* 4, e06516.

Tan, E. *et al.* (2016) Chromosome dosage analysis in plants using whole genome sequencing. *Bio-protocol* 6(13), e1854.

Tester, M. and Langridge, P. (2010) Breeding technologies to increase crop production in a changing world. *Science* 327, 818–822.

Till, B.J. *et al.* (2003) Large-scale discovery of induced point mutations with high-throughput TILLING. *Genome Research* 13, 524–530.

Till, B.J. *et al.* (2018) TILLING: the next generation. *Advances in Biochemical Engineering/Biotechnology* 164, 139–160.

Tsai, H. *et al.* (2011) Discovery of rare mutations in populations: TILLING by sequencing. *Plant Physiology* 156, 1257–1268.

Velásquez, A.C. *et al.* (2018) Plant-pathogen warfare under changing climate conditions. *Current Biology* 28, R619–R634.

Viljoen, A. *et al.* (2020) Occurrence and spread of the banana fungus *Fusarium oxysporum* f. sp. *cubense* TR4 in Mozambique. *South African Journal of Science* 116, 11–12.

World Agriculture (2018) *Climate Change, Crop Plant Diseases and Future Food Production*. Available at: http://www.world-agriculture.net/article/climate-change-crop-plant-diseases-and-future-food-production (accessed 6 April 2021)

Zhang, Y. *et al.* (2018) Applications and potential of genome editing in crop improvement. *Genome Biology* 19, 210.

Zheng, S.-J. *et al.* (2018) New geographical insights of the latest expansion of *Fusarium oxysporum* f. sp. *cubense* Tropical Race 4 Into the Greater Mekong Subregion. *Frontiers in Plant Science* 9, 457.

Żmieńko, A. *et al.* (2014) Copy number polymorphism in plant genomes. *Theoretical and Applied Genetics* 127, 1–18.

39 Induction and Selection of Mandarin Mutants with Fruits Containing Low Number of Seeds

Marinês Bastianel[1], Vera L.N.P. Barros, Augusto Tulmann Neto, Paulo S. Souza, Rose M. Pio and Rodrigo R. Latado

Abstract

The Brazilian citrus industry has a worldwide presence for production and export of sweet orange juice, but it has little contribution to the production of fruits for the fresh fruit market. One requirement of this market is the production of seedless fruits. The Fremont IAC 543 mandarin produces fruits with good commercial qualities, large numbers of seeds (10–12), and plants with resistance to Alternaria brown spot (ABS), an important disease present in several countries. The objective of this work was to induce and select mutants of Fremont IAC 543 mandarin with seedless fruits or fruits with a low number of seeds, using gamma-ray induced mutagenesis. *In vivo* buds were irradiated with doses of 20 and 30 Gy of gamma-rays. After irradiation and grafting of 2000 *in vivo* buds with each mutagenic dose, 4000 plants were produced and planted in an experimental field. During development of these plants, they were pruned several times allowing only the development of M_1V_4 branches or more advanced ones (without new grafting). A total of 32 branches were selected during the harvesting period because they produced seedless fruits and nine mutant clones were selected after another vegetative multiplication. Fruit and juice qualities, including seed number of the fruits, were evaluated in a further experiment including six mutants and a control. The results obtained showed that all mutants produced fruits with a lower number of seeds (between 3.7 and 9.1 seeds per fruit) in relation to the control (22.0 seeds per fruit), but without the existence of other alterations (fruit metric and chemical characteristics of the juice). All selected mutants (nine) are participating in advanced agronomic evaluation experiments, with a greater number of replicates and several local checks, in order to evaluate commercial yield, presence of chimeras, disease resistance and organoleptic quality of the fruits.

Keywords: citrus • mutant • seed • breeding

1 Introduction

Mandarin cultivation is an important activity in Brazilian citriculture, despite the fact that the country is recognized worldwide as a major producer of oranges and exporter of not-from-concentrate (NFC) orange juices and frozen concentrate orange juice (FCOJ). The production of mandarins in Brazil gave the country eighth place in the ranking of producing countries in 2016 (FAO, 2019). Mandarin is the main citrus species commercialized in the fresh fruit market in several countries.

Currently, ‘Ponkan’ is the most produced mandarin (*Citrus reticulata*) in Brazil, followed by ‘Murcott’ tangor (*C. reticulata* × *C. sinensis*)

[1] Email: mbastianel@ccsm.br

DOI: 10.1079/9781789249095.0039

and 'do Rio' mandarin (willowleaf mandarin). Brazilian consumers differ from the majority of consumers in other countries because they still prefer to purchase and consume large fruits (equal to or greater than 7–8 cm in diameter) and without the requirement of seedlessness. However, this situation may soon be changed, because there is an increased supply of imported citrus fruits from other countries, with excellent characteristics of shape, colour, taste and the absence of seeds. Thus, there is a great opportunity in Brazil for cultivation of new citrus varieties, with preference for those that have characteristics that meet consumer demands. Other opportunities can also arise, with the possibility of increasing the country's exports of fresh fruits.

One of the main problems of mandarin production in some countries, including Brazil, is the Alternaria brown spot disease (ABS). It has been costly to produce fruits in Brazil's main producing regions because the two main mandarin varieties ('Ponkan' and 'Murcott') are highly susceptible to the disease's causal agent, the fungus *Alternaria alternata* (Michelin *et al.*, 2016). Nowadays, new mandarin varieties with ABS resistance are in great demand by Brazilian growers.

The 'Fremont' mandarin was obtained by crosses between clementine (*Citrus* × *clementina*) and 'Ponkan' mandarins, conducted by P.C. Reece (Florida, USA) and selected by J.R. Furr (California, USA), with commercial release in 1964 (Pio *et al.*, 2006). It has fruits with a mid-season period of maturation, smooth peel, easy-to-peel fruit (which facilitates easy handling during harvest and post-harvest) and fruits with large numbers (10–12) of seeds, an undesirable characteristic. The resistance of Fremont plants to ABS disease makes it an interesting variety for orchard diversification (Pacheco *et al.*, 2017).

The selection of spontaneous or induced mutants has contributed significantly to obtaining new citrus varieties in Brazil and other countries (Spiegel-Roy, 1990; Spina *et al.*, 1991; Latado *et al.*, 2001). Mutant clones of 'Pêra' sweet orange with different characteristics were induced using 40 Gy of gamma rays in buds (Tulmann Neto *et al.*, 1996). The spectrum of selected mutants presented plants with compact size, fruits with low seed numbers, fruits with alteration in harvesting period (early or late harvest) and/or greater tolerance to citrus canker disease (Tulmann Neto *et al.*, 1996). As a practical result, the 'IAC 2014' variety, a sweet orange mutant with higher tolerance to citrus canker on leaves and fruits, and with seedless fruits (Latado *et al.*, 2005), was commercially released in 2016 in Brazil.

The objective of this work was to induce and select mutants of Fremont IAC 543 mandarin with seedless fruits or fruits with low number of seeds using gamma-ray induced mutagenesis.

2 Materials and Methods

2.1 Experiment: radiosensitivity determination

A preliminary experiment was carried out with the objective of evaluating the radiosensitivity of Fremont IAC 543 mandarin to gamma-rays. For this, several budwoods were irradiated with doses of 10, 20, 30, 40 and 50 Gy using the cobalt-60 (^{60}Co) source of Centro de Energia Nuclear na Agricultura (CENA/USP) in Piracicaba, São Paulo. Non-irradiated buds were used as experimental control. The irradiated and control buds were grafted onto Rangpur lime plants. Forty buds and plants were used per treatment, using a completely randomized design.

Approximately 60 days after grafting, evaluations were performed for two parameters: (i) percentage of developed buds; and (ii) length of developed branches (M_1V_1). The obtained data were used in analysis of variance using the F and Tukey tests ($p < 1\%$), in addition to linear regression analysis.

2.2 Experiment: induction and selection of mutants

An experiment was carried out with the irradiation of buds with 20 and 30 Gy of gamma-rays, with selection based on the earlier experiment. After irradiation and grafting of 2000 buds with each mutagenic dose, 4000 plants were produced and planted in the experimental field. During development of these plants, they were pruned several times, allowing only the development of M_1V_4 branches or more advanced ones (without new grafting).

Planting was done at intervals of 1 m in rows 6 m apart, which is not appropriate to evaluate plant yield but is enough for selection of mutants during 3–4 years.

Selection of mutant clones for lower number of seeds in fruits was carried out during the fruiting period, using the following methodology. All fruits of each plant were cut transversally in a middle region of the fruit to verify the presence of seeds (Fig. 39.1). Branches containing seedless fruits and some buds were selected and excised for multiplication and formation of new plants. Three grafted plants were produced from buds of each selected branch.

After vegetative propagation of putative mutants by grafting, a second selection was carried out in plants kept in the greenhouse in order to confirm the mutations for lower number of seeds in fruits.

2.3 Evaluation of selected mutant clones

A field experiment was established in 2014 at Cordeirópolis, Sao Paulo, Brazil (22° 28' 49" S, 47° 27' 21" W; 649 m altitude), a region with an average temperature of 20.2°C and annual rainfall of 1305 mm (Cwa climate, according to Köppen classification). Trees were planted at 4 m × 7 m spacing (within and between rows, respectively) and the grove was not irrigated. The experimental design was in randomized blocks, with six mutated clones and a control, with five replications (blocks) and three plants per plot. Fertilization, disease and pest management were performed according to conventional local recommendation.

The variables analysed were related to fruit quality. For this, 20 fruits were randomly collected from each tree for individual fruit metric evaluations (mass, length and width), juice content (%) and for chemical analysis of juice. Total soluble solids (TSS), acidity (%) and ratio (TSS/acidity) were estimated according to Schinor *et al.* (2013). The number of seeds in each fruit was also evaluated, followed by calculation of the average number of seeds per fruit of each mutant.

3 Results and Discussion

3.1 Experiment: radiosensitivity determination

Physiological effects due to exposure of buds to gamma-rays were observed a few days after the irradiation. These effects were mostly observed when higher doses of the mutagen (40 and 50 Gy) were used. The most common effects were the presence of deformed leaves (the first ones developed) and leaves with chlorotic spots.

Except for the dose of 50 Gy, the other doses did not cause significant reductions in shoot length after 60 days of development and the lengths ranged from 13.2 to 15.6 cm (Table 39.1). On the other hand, irradiation of buds with a dose of 50 Gy caused a significant reduction in shoot length (4.9 cm) and in the

Fig. 39.1. Method of selection of mandarin mutants for lower number of seeds in fruits.

Table 39.1. Shoot length (cm) and percentage of developed shoots after 60 days of bud irradiation with several doses of gamma-rays.

Dose (Gy)	Shoot length (cm ± SE)	Developed shoots (%)
0	14.9 ± 0.8a	98
10	13.3 ± 0.6a	88
20	15.6 ± 0.5a	100
30	14.6 ± 0.5a	95
40	13.2 ± 0.8a	98
50	4.9 ± 0.3b	62
F test	11.12**	
CV (%)	47.99	

SE = Standard errors. Means followed by the same letter in the column did not differ statistically by Tukey's test ($p < 0.05$).

number of developed shoots (survival rate 62%) (Table 39.1).

Linear regression analysis confirmed a tendency for a gradual reduction of shoot length with the increase of mutagenic doses.

Radiosensitivity experiments are conducted to evaluate the physiological effects of distinct doses on the growth rate of propagules, expressed as growth reduction (GR) or plant lethality (lethal dose, LD). Sensitivity to a certain dose of mutagen can vary between species, cultivars and propagules such as buds, shoot apexes and seeds (Broertjes and Van Harten, 1988). The doses selected and used in the present study for the mutagenic treatment of citrus buds were lower than those used by other authors with the same type of propagule, specifically, 40 and 50 Gy of gamma-rays (Spiegel-Roy and Kochba, 1973; Tulmann Neto *et al.*, 1996; Gulsen *et al.*, 2007).

3.2 Experiment: induction and selection of mutants

Irradiation of a large number of buds (4000) with two distinct doses of mutagen (2000 for each dose) resulted in an average loss of almost 26% of irradiated buds after grafting, while the mean loss of control-grafted buds (non-irradiated buds) was approximately 5%.

During development of the plants derived from irradiated buds, they were pruned several times to allow only the development of M_1V_4 branches or more advanced ones (without new grafting). The first selection of mutants containing seedless fruits was performed during the harvesting period (Fig. 39.1) and resulted in the selection of 32 branches, in a total of 1340 evaluated plants (2.4%).

After vegetative multiplication of the 32 putative mutant branches by means of new grafting, a second selection was carried out in greenhouse conditions and it was observed that only nine mutant clones maintained the previously selected characteristic, namely, fruits containing a low number of seeds (0.7% of the total).

The absence of seeds is one of the most important traits in the fresh fruit citrus market. For this reason, most efforts of citrus breeding programmes are directed to the development of seedless cultivars. There are several reports on the use of induced mutagenesis for selection of citrus mutants in several countries and, among the objectives pursued, obtaining mutants with low number of seeds is the main one. Hensz (1977, 1985) reported works involving mutagenesis in grapefruit in Texas (USA). The same was done in Israel (Spiegel-Roy *et al.*, 1985, 1990), Italy (Russo *et al.*, 1981; Starrantino *et al.*, 1988) and Brazil (Tulmann Neto *et al.*, 1996; Latado *et al.*, 2005).

3.3 Evaluation of six selected mutant clones

After vegetative propagation of buds of six mutated clones, fruit and juice qualities and seed number of the fruits were evaluated in another experiment. The results obtained showed that all mutants produced fruits with a reduced number of seeds (between 3.7 and 9.3) in relation to the control (22.0), but demonstrated that they were not entirely seedless (Table 39.2). Clone #5.3 showed fruits with the least number of seeds (3.7) (Fig. 39.2). New evaluations will be carried out in different locations and during more years to verify the stability and repeatability of these results.

Evaluations of fruit metric and chemical characteristics of the juice (Tables 39.2 and 39.3) demonstrated that mutant clones apparently did not undergo other changes (mutations) in fruits and juice, which may be interesting if

Table 39.2. Physical characteristics of the fruits of several mutants and control plants of Fremont IAC 543 mandarin.

Genotype	Mass (g ± SE)	Length (cm ± SE)	Width (cm ± SE)	H/W ratio (± SE)	Juice content (% ± SE)	Number seeds/ fruit (± SE)
Control	103.0 ± 6.6	5.4 ± 0.1	6.3 ± 0.2	0.86 ± 0.0	44.0 ± 1.2	22.0 ± 5.1
#5.1	104.4 ± 6.1	5.4 ± 0.1	6.2 ± 0.2	0.86 ± 0.01	44.3 ± 4.9	5.4 ± 2.3
#5.2	96.1 ± 5.7	5.3 ± 0.1	6.2 ± 0.2	0.86 ± 0.01	45.0 ± 3.2	4.2 ± 2.1
#5.3	94.6 ± 4.1	5.2 ± 0.1	6.0 ± 0.1	0.86 ± 0.01	44.6 ± 1.0	3.7 ± 1.9
#9.1	102.9 ± 3.1	5.3 ± 0.1	6.3 ± 0.1	0.84 ± 0.01	44.4 ± 4.2	6.4 ± 2.9
#9.2	93.8 ± 1.3	5.1 ± 0.04	6.0 ± 0.1	0.85 ± 0.02	45.1 ± 2.6	9.3 ± 2.4
#9.3	87.1 ± 9.6	5.0 ± 0.2	5.8 ± 0.3	0.86 ± 0.01	45.6 ± 3.8	8.1 ± 2.1

SE = Standard errors.

Fig. 39.2. Fruits harvested from mutant #5.3 (left) and control (right), of Fremont IAC 543 mandarin.

Table 39.3. Chemical characteristics of the fruit juices of several mutants and control plants of Fremont IAC 543 mandarin.

Genotype	TSS (°brix ± SE)	Acidity (% ± SE)	TSS/Acid ratio (± SE)
Control	10.36 ± 0.61	0.91 ± 0.05	11.43 ± 0.35
#5.1	9.71 ± 0.89	0.97 ± 0.07	10.02 ± 0.82
#5.2	10.47 ± 0.55	0.96 ± 0.04	10.89 ± 0.24
#5.3	9.84 ± 0.56	0.86 ± 0.05	11.39 ± 0.02
#9.1	10.18 ± 0.72	0.91 ± 0.11	11.29 ± 0.61
#9.2	9.98 ± 0.55	0.96 ± 0.14	10.48 ± 1.05
#9.3	10.40 ± 0.79	1.00 ± 0.01	10.50 ± 0.68

SE = Standard errors.

some of them are commercially grown in the future. The variables related to size, shape and juice content showed similar results among different genotypes (mutants and control). A slightly larger change was observed in fruit mass (Table 39.2), mainly in clone #9.3 (87.1 g per fruit), which produced lighter fruits than control plants (103.0 g per fruit).

There were also no drastic changes in chemical characteristics of the juices in fruits from mutants, with total soluble solids (TSS) ranging from 9.7 to 10.5, acidity close to 1.0 and ratio (TSS/acidity) ranging from 10.0 to 11.4 (Table 39.3).

Small changes in fruit and juice characteristics can be caused by genetic components (mutations) as well as by growing conditions. They should not cause great concern to breeders, because they can be corrected or overcome by actions such as changes in rootstocks, nutritional management of the orchard, pruning and thinning of young fruits.

The selected nine mutants are currently in advanced agronomic evaluation experiments which involve a greater number of replicates and several locales (regions), in order to evaluate commercial yield, presence of chimeras, disease resistances and organoleptic quality of the fruits (Fig. 39.3).

4 Conclusion

The induction and selection of Fremont mandarin mutants with fruits containing low number of seeds is possible, using gamma-ray induced mutagenesis.

Fig. 39.3. Plant of Fremont mandarin mutant clone #9.2 (Cordeirópolis, SP).

References

Broertjes, C. and Van Harten, A.M. (1988) *Applied Mutation Breeding in Vegetatively Propagated Crops*. Elsevier Science, Amsterdam.

FAO (2019) Available at: http://www.fao.org/economic/est/est-commodities/citrus-fruit/en/ (accessed 14 January 2019)

Gulsen, O. *et al.* (2007) Development of seedless and *Mal Secco* tolerant lemons through budwood irradiation. *Scientia Horticulturae* 112, 184–190.

Hensz, R.A. (1977) Mutation breeding and the development of the 'Star Ruby' grapefruit. *Proceedings of the International Society of Citriculture* 2, 582–585.

Hensz, R.A. (1985) 'Rio Red', a new grapefruit with a deep-red color. *Journal of the Rio Grande Valley Horticultural Society* 38, 75.

Latado, R.R. *et al.* (2001) Mutantes de laranja 'Pêra' com número reduzido de sementes obtidos através de mutações induzidas. *Revista Brasileira de Fruticultura* 23, 339–344.

Latado, R.R. *et al.* (2005) Caracterização agronômica de mutantes de laranjeira Pêra sem sementes ou com alteração no período de maturação de frutos. *Laranja* 26, 109–120.

Michelin, T.H.V. *et al.* (2016) Reação de híbridos de citros à inoculação com *Alternaria alternata*. *Summa Phytopathologica* 42, 313–320.

Pacheco, C.A. *et al.* (2017) Fremont - IAC 543: tangerine with potential for the Brazilian market. *Revista Brasileira de Fruticultura* 39, 1–6.

Pio, R.M. *et al.* (2006) Características da variedade Fremont quando comparadas com as das tangerinas Ponkan e Clementina nules. *Revista Brasileira de Fruticultura* 28, 222–226.

Russo, F. *et al.* (1981) Mutation applied for citrus improvement. *Proceedings of the International Society of Citriculture* 1, 68–70.

Schinor, E.H. *et al.* (2013) Sunki mandarin vs Poncirus trifoliata hybrids as rootstocks for pera sweet orange. *Journal of Agricultural Science* 5, 190–200.

Spiegel-Roy, P. (1990) Economic and agricultural impact of mutation breeding in fruit trees. *Mutation Breeding Review* 5, 1–26.

Spiegel-Roy, P. and Kochba, J. (1973) Mutation breeding in citrus. In: FAO/IAEA *Induced Mutations in Vegetative-Propagated Plants*. IAEA, pp. 91–105.

Spiegel-Roy, P. *et al.* (1985) Seedless induced mutant in lemon (*Citrus limon*). *Mutation Breeding Newsletter* 26, 1.

Spiegel-Roy, P. *et al.* (1990) Seedless induced mutant in highly seeded lemon (*Citrus limon*). *Mutation Breeding Newsletter* 36, 11.

Spina, P. *et al.* (1991) Use of mutagenesis at the Instituto Sperimentale per Lagrumicoltora Acireale: results and prospects for the future. In: *FAO/IAEA International Symposium on the Contribution of Plant Mutation Breeding to Crop Improvement*, 1990, Vienna, Áustria, IAEA, pp. 257–261.

Starrantino, A. *et al.* (1988) The genetic stability of three seedless clonal selections obtained by gamma rays of the seedy 'Monreal' Clementine. *Proceedings of the International Society of Citriculture* 2, 175–181.

Tulmann Neto, A. *et al.* (1996) Indução e seleção de mutantes em laranja 'Pêra' mediante emprego da radiação gama. *Pesquisa Agropecuária Brasileira* 31, 743–752.

40 ^{60}Co Gamma Irradiation-Induced Mutation in Vegetatively Propagated *Philodendron erubescens* 'Gold'

Dayani Karunananda[1], Ramya Ranathunga and Wathsala Abeysinghe

Abstract

Philodendron erubescens 'Gold', an ornamental plant and a popular climber with brilliant greenish yellow leaves, is used in indoor gardening and landscaping. It is commonly propagated through vegetative cuttings, thus incorporation of new traits through conventional breeding is impracticable. As commercial floriculture always demands novel varieties, this study was carried out to induce mutation in *P. erubescens* 'Gold' leaves using gamma-ray irradiation. Rooted cuttings ($n = 200$) of *P. erubescens* 'Gold' were subjected to 70 Gy, 100 Gy and 150 Gy gamma-rays and recovered on a propagator. Surviving shoots were transferred to pots. Regenerated shoots were multiplied vegetatively and ten M_1 lines were maintained as M1-1 to M1-10 for 12 generations (M_1V_{12}) to evaluate growth and morphological variations along with their genetic stability. Of all 70 Gy and 100 Gy treated cuttings, 24 and two, respectively, survived after 6 months. Most of the irradiated plants had lost regeneration ability except for two M_1 plants, which also showed comparatively reduced growth (one leaf in 45 days). Only one regenerated M_1 plant showed morphological variation in its leaves and it was multiplied and maintained as lines. Several variations, including characteristics of leaves (shape, size, colour), stems (internodal length and branching) and plant stature, were observed among M_1 lines and in subsequent vegetative generations. Leaves had three different colour patches, but neither the colour nor its distribution pattern was uniform or stable. The M1-4 line showed the highest stability of colour distribution in leaves; the colour composition of its leaves ranged as 0–10% dark bluish green, 60–90% strong yellow green and 10–30% brilliant greenish yellow throughout the 12 generations. This study demonstrates that gamma irradiated *P. erubescens* 'Gold' line M1-4 can be a promising mutant to develop as a new *Philodendron* cultivar.

Keywords: mutants • survival rate • regeneration • morphological variation • growth retardation

1 Introduction

With the improvement of the living standard of people, use of floricultural products has increased tremendously throughout the world. Unlike the food crop sector, floriculture is changing rapidly; within a short time period, demand for these products has changed greatly. The floriculture market always expects novel products and continuous adding of new items is needed to remain competitive within the industry. Therefore, all flower and foliage breeders develop new varieties with different characters. Not only cut flowers but also pot plants and cut foliage are in significant demand from consumers. Countries in the tropical part of the world are the leading

[1] Email: dayani.karunananda@gmail.com

 Mutation Breeding, Genetic Diversity and Crop Adaptation to Climate Change (eds S. Sivasankar *et al.*)
DOI: 10.1079/9781789249095.0040

producers of foliage varieties as pot plants and cut leaves; warm tropical climates enable year-round production of greens. The Sri Lankan floriculture industry is growing fast and is based on foliage and potted plant production, which has a high potential to expand further (Padmini and Kodagoda, 2017).

Philodendrons are climbing herbs belonging to the family Araceae and they show epiphytic growth. Several philodendron types are popular as cut foliage as well as pot plants in both local and foreign markets, which attracts the attention of producers as an easy income generator. Each philodendron cultivar on the market has its unique colour and shape. *Philodendron erubescens* 'Gold' is the most common philodendron type used in indoor decoration and landscape gardening in Sri Lanka. It is a climber with brilliant greenish yellow (151 D in the RHS colour chart) (RHS, 2003) young leaves and strong yellow green (144 A in the RHS colour chart) mature leaves. Leaves are more or less lanceolate with pinkish-green medium-length petioles. It is a fast-growing climber with a good shade tolerance and is well adapted to indoors with training and pruning. No variegated types are recorded in philodendrons, but leaf variegation is an important factor influencing the popularity of ornamental plants (Koh and Davies, 2001). As a member of the family Araceae, *P. erubescens* 'Gold' produces an inflorescence spadix surrounded by a spathe (Chouteau *et al.*, 2006), but this is not usually observed under domestic cultivation. Therefore, it is commonly multiplied by vegetative propagation and it is very unlikely to develop new varieties through conventional breeding. Genetic variation in such ornamentals takes place using spontaneous mutations, but the frequency of desirable mutations is extremely low (Dai and Magnusson, 2012). Induced mutation provides the opportunity to create genetic variation in vegetatively propagated plants and develop new varieties (Schum and Preil, 1998).

The application of induced mutation has been practised since the 1930s (Broertjes and Van Harten, 1988) and is routinely practised by floriculture breeders. However, mutation induction in vegetatively propagated crops is comparatively difficult compared with seed crops, as mutation takes place in cells that may have regeneration ability but are also surrounded by other tissues. Chimeric shoots can be isolated and propagated to obtain variegated plants (Bado *et al.*, 2017). Although irradiation of *in vitro* cultures has become the most popular method of generating mutants in vegetatively propagated crops, irradiation of rooted cuttings has the longer history, as it was practised even before the introduction of tissue culture (Ahloowalia *et al.*, 2004). *In vitro* mutagenesis is convenient; it is easy to handle a large number of plantlets after mutation and so facilitate rapid multiplication of mutants. Irradiation of rooted cuttings avoids the high cost involved in tissue culture maintenance and the burden of development of protocols for culture regeneration, including acclimatization. The death rate of vegetatively propagated plants subjected to irradiation is comparatively higher than for seeds and these plants show higher shoot-growth retardation after irradiation (Park and Andersen, 1990; Dwimahyani and Widiarsih, 2010). Therefore, the number of plants that remain to evaluate is low. A number of flower varieties have been developed in different countries through mutation breeding. The most commonly changing visible character in new mutants is flower colour (Datta, 1997). Variations in leaves are also visible (Abdullah *et al.*, 2009) but improvement of foliage ornamentals through mutational breeding is not common. Variations of foliage that have been created by mutagenesis are chlorophyll mutations and changing of leaf architecture (leaf shape, internodal length, etc.) but mild mutations cannot be recognized easily by eye. However, changes in foliage shape and colour have been recorded by several authors. Abdullah *et al.* (2009) recorded chlorophyll and leaf shape mutation in *Curcuma alismatifolia*. Variation of leaf shape of *Buddleia davidii* was recorded by Dai and Magnusson (2012). Koh and Davies (2001) obtained variegated leaves in *Tillandsia fasciculate* Swartz var. *fasciculate* for ornamental purposes.

We have not found any records of mutation breeding of *Philodendron* species but there is frequent production of mutated varieties of *Anthurium*, which belongs to the same family (Puchooa and Sookun, 2003), suggesting the potential for mutant development of *P. erubescens* 'Gold'. Therefore, this study was initiated to induce desirable genetic variation such as variegated leaf colour, different shape and compact structure of *P. erubescens* 'Gold' and to develop novel foliage cultivars using gamma irradiation as a mutation tool.

2 Materials and Methods

The experiment was conducted at the Horticulture Research and Development Institute (HORDI), Gannoruwa, Sri Lanka, during 2008–2016. Rooted cuttings of *P. erubescens* 'Gold' were used for the experiment and the ^{60}Co gamma irradiation source located at HORDI was used to irradiate plant samples.

Pseudostem cuttings of *P. erubescens* 'Gold' 10–15 cm long with three to four nodes were rooted in a sand bed for 4 weeks. In the radiosensitivity test, 3-week-old rooted cuttings were subjected to irradiation with gamma-ray doses of 0, 25, 50, 75 and 100 Gy. Thirty cuttings were used for each dose and together with the control they were immediately planted in the same sand bed that had been used to root cuttings previously. Survival number was recorded for 3 months at 7-day intervals for each treatment. Relative humidity within the propagator was maintained at 65–67% and the temperature was around 25°C. A fertilizer solution consisting of 20% nitrogen, 20% phosphorus, 20% potassium and a trace of MgO was sprayed at 1-week intervals at the rate of 10 g/l. Survival percentages were plotted against irradiation doses at the end of the third month after irradiation to characterize radiosensitivity of the rooted cuttings of *P. erubescens* 'Gold'. The LD_{50} value was calculated from the plot at each time of record keeping.

After the radiosensitivity test, three doses were selected for mass irradiation, including two doses around the LD_{50} (70 Gy and 100 Gy) and one higher dose (150 Gy). Six hundred visually uniform philodendron cuttings were rooted in the sand beds of a propagator and maintained for 2 weeks. Rooted cuttings were carefully uprooted and subjected to gamma irradiation in ten batches for each dose, which consisted of 20 cuttings per batch. Irradiated cuttings were replanted in sand beds as lines according to completely randomized design (CRD). Conditions were maintained as for the radiosensitivity test and the same nutrient and cultural treatments were provided. Dead cuttings were removed, and survival numbers were counted at 2-week intervals for 6 months. Plants that showed bud breaking were potted separately in a compost mixture and watering and fertilization were done regularly. Shoots emerging from growing cuttings were excised carefully and propagated in the sand medium. Ten growing shoots were selected from them, maintained as M_1 plant lines (M1-1 to M1-10) and evaluated for internode length, leaf size and growth (time taken to produce a leaf) and the number of colours in a leaf. Colours present in leaves were recorded by comparing with the Royal Horticulture Society (RHS) colour chart (RHS, 2003). A SPAD-502 chlorophyll meter was used to take readings in ten randomly selected colour patches of each M1 line. The SPAD meter reading for each colour patch was recorded separately and variances were analysed.

In the first vegetative generation, chlorophyll content of each colour patch of all M_1 lines was measured and compared with chlorophyll content of mature and young leaves of the mother plant. To determine the chlorophyll content of different colour patches, patches in each colour were separated using scissors and total chlorophyll was extracted into 96% methanol. The absorbances of each solution at 665, 649 and 470 nm wavelengths were measured spectrophotometrically and chlorophyll a, chlorophyll b, carotene and xanthophylls in each colour patch were quantified according to the method described by Lichtenthaler and Wellburn (1983).

Shoots descended from each M_1 line were maintained through 12 vegetative generations as V_1–V_{12} populations. Plants were evaluated for leaf colour, uniformity of colour and colour distribution in leaves throughout the 12 generations. Readings were taken for the first two leaves that emerged after transplanting. Colours present in the leaves were determined by comparing with the RHS colour chart. SPAD values for each area were recorded separately. The area occupied by each colour was determined by analysing digital photographs using Adobe Photoshop software.

Statistical evaluations were performed using SAS statistical package (SAS portable 9.1 software). One-way variance analysis (ANOVA) was implemented by general linear model procedures to test the differences among different lines; then means were separated using LSD test at an alpha 0.05 level.

3 Results

Development of novel plant types using mutagenesis is widely practised in floriculture as it is the easiest and fastest method of generating new traits (Datta, 1997). The advantage of the use of mutation in floriculture compared with food crops is that whatever morphological changes take place and are visible, they could be desirable as novel traits in ornamentals. Therefore, mutation breeding plays a very important role in floriculture breeding (Maluszynski *et al.*, 1995) and the results of this study are of value for the tropical foliage industry.

3.1 Radiosensitivity test

In the radiosensitivity test, the death rate increased with the increasing dose of gamma-ray and nothing regenerated until 3 months after the irradiation. The irradiated shoots showed progressive death and those that survived remained dormant until the end of the experiment. Therefore, the LD_{50} was calculated only using survival at the end of the third month (Fig. 40.1) and 50% death was estimated to be at 81 ± 20 Gy.

Determination of doses for large-scale irradiation was done using the results of the radiosensitivity test. Significant variation was observed in survival, bud breaking and shoot production after gamma irradiation of *P. erubescens*

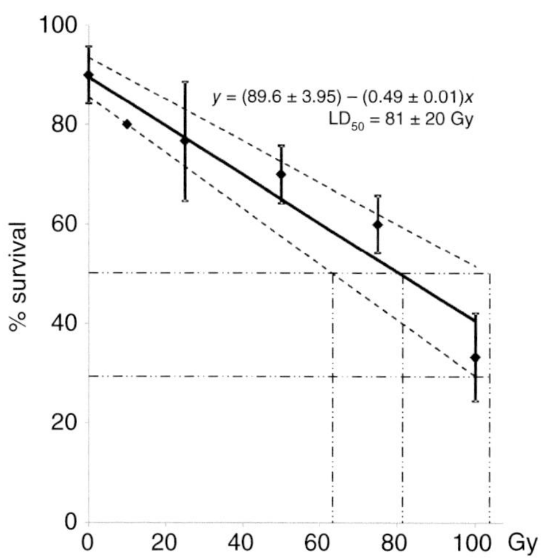

Fig. 40.1. Determination of LD_{50}.

'Gold' rooted cuttings. In this experiment, early and higher death rate was observed at 150 Gy and all those plants died within a month. Most of the rooted plants, particularly those from 70 and 100 Gy treatments, appeared to be struggling to survive and eventually died in the propagator.

From the very beginning of mutation breeding, doses that lead to 50% lethality (LD_{50}) have often been chosen as the ideal dose for irradiation (Oldach *et al.*, 2008). However, in this experiment the lethal dose was determined by the death of the plant at the third month, but it was realized that, although those irradiated cuttings had survived, they had lost their regeneration ability. Hence the use of survival percentage to determine effective dose was not useful. Therefore, consideration of other factors such as growth, size of leaf and shoot, bud breaking and shoot elongation would be important to determine the effective dose for irradiation. Several authors have used a growth parameter GR_{50} in place of LD_{50} as an effective dose (Dai and Magnusson, 2012). The main reason for the use of other growth parameters would be the loss of regenerating ability due to mutation.

The mass irradiation experiment revealed that irradiation of rooted cuttings results in only very few mutants. The death rate of vegetative planting material is very high after mutagenesis as compared with seeds. Bado *et al.* (2017) pointed out that the success of the mutation treatment of vegetatively propagated crops relies on the size of the population. As induced mutation is a random event, the use of a large population size is desirable (Bado *et al.*, 2017). Therefore, it is required that a large number of cuttings are irradiated and the use of lower doses is recommended. Going for a lower percentage of death of about 20% as effective dose was suggested by Oladosu *et al.* (2016) to obtain a large number of mutants. Similarly, higher irradiation doses reduce the regeneration capacity of shoots. Selection at lower doses such as LD_{25} and LD_{35} will help to obtain low DNA damage with desirable mutations and higher regeneration rates from treated plants (Jain, 2006). However, Brunner (1995) showed that the mutation percentage increased with the increasing irradiation dose, hence the effective dose is difficult to predict and needs to be determined experimentally.

3.2 Regeneration of irradiated cuttings

Twenty-four shoots from 70 Gy and two from 100 Gy treated shoots survived to the sixth month after treatment and two shoots (rooted cutting numbers 70-2-10 and 70-6-3) from 70 Gy treated shoots started bud breaking and regeneration. Figure 40.2(a) shows the survival of irradiated shoots against time and Fig. 40.2(b) shows the bud breaking of shoots at the seventh month after irradiation. One bud from 70-6-3 and two buds from 70-2-10 produced shoots and these showed a significant delay in leaf production and growth (Table 40.1). Swaroop *et al.* (2015) observed the same results in irradiated stem cuttings of bougainvillea with 85 days delay in sprouting of 20 Gy treated cuttings.

According to Park and Andersen (1990), bud dormancy of mutated *Passiflora* cuttings became a frequent drawback in the development of a novel mutant through induced mutation. The growth rate of mutated plants was significantly reduced, due to both genetic and epigenetic aberrations due to irradiation (Sparrow and Gunckel, 1955). Ghani *et al.* (2013) showed that the number of plants which recover in the hardening of irradiated *Gerbera* is reduced with the increasing dose, presumably due to forced stress by gamma-rays. During this phase, plant growth and development are temporarily suspended and recovery takes a long time.

Irradiation of vegetative cuttings or plants results in chimeras. Mutated plants can be obtained through intrasomatic selection and multiplication of such chimeric cells. In this scenario, *in vitro* multiplication techniques are frequently used over conventional methods (Datta, 2018). Mutations in cells that have regeneration ability get the chance to multiply that variation (Van Harten, 2002). In seed irradiation,

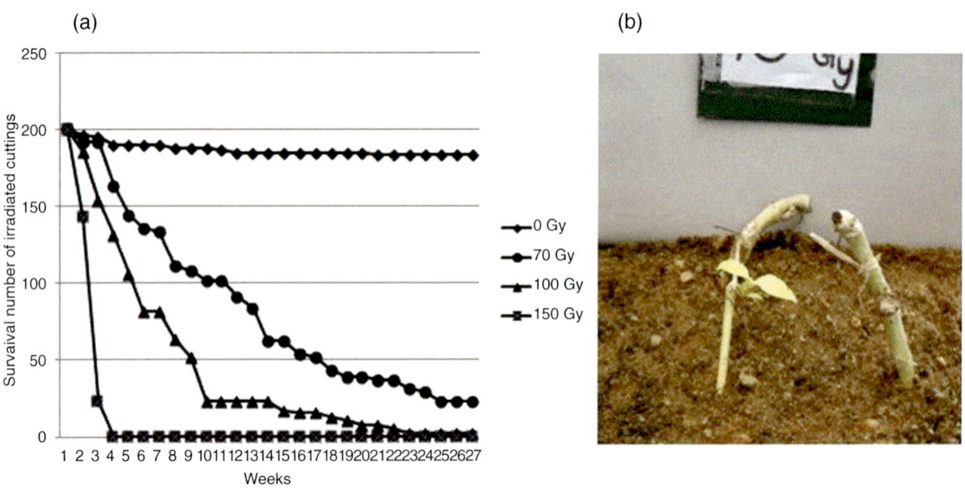

Fig. 40.2. (a) Survival of irradiated shoots against time. **(b)** Bud breaking of shoots at 7th month after irradiation.

Table 40.1. Growth performance of two regenerated shoots against the control during first year.

Parameter	70-2-10	70-6-3	Control
Time taken to first leaf initiation (weeks)	22	28	2
Number of new shoots from planted cutting	2	1	2.4
Time taken to produce a leaf (days, average)	24.5	53.4	9.1
Colour of new leaf	Light yellow	Mixed (variegated)	Greenish yellow
Internode length (cm)	1.1	0.58	

zygotes are produced that carry variation and the resultant plants carry the same variation in every cell. This is the main difference between irradiation of rooted cuttings versus cell cultures and seeds, hence only chimeras can be expected from rooted cuttings. The use of cuttings with inactive axillary buds provided an additional advantage in this experiment, as the buds in each node literally have the ability to regenerate and produce new shoots. It is possible that variations in such buds may be induced by the mutagenesis and therefore those buds may develop as putative mutants.

3.3 Chimera production from irradiated cuttings

Regenerated shoots of 70-2-10 were almost identical to control plants morphologically but the shoot of 70-6-3 was a chimera and manifested some clear variations in leaves, stem and plant stature. Importantly, most of the leaves possessed colour patches belonging to three different colours. The whole shoot had a different phenology from that of the mother plant; this is therefore a sectorial chimera distinct from the mother shoot. Although it possessed variegations, almost all the leaves in the M_1 of 70-2-10 were crooked and aberrant from the beginning. The prominent abnormalities of these leaves included differences in shape, size and leaf margin (Fig. 40.3a).

Hewawasam *et al.* (2004), Dwimahyani and Widiarsih (2010) and Dai and Magnusson (2012) reported growth retardation and leaf aberration in mutated plants. Plant survival and the degree of reduction of height depend on the nature and extent of chromosomal damage created by induced mutations (Sax, 1942; Lea, 1947). Increasing chromosomal damage may increase such negative effects in mutational breeding (Hewawasam *et al.*, 2004).

Both M_1 lines obtained in this experiment had the ability to regenerate new shoots when the top part was excised. However, multiplication of 70-2-10 was stopped after obtaining the fourth vegetative shoot, due to its similarity to control plants. The chimeric shoot of 70-6-3 took more than 45 days to produce a new leaf, but it had the ability to regenerate new side shoots when apical dominance was removed by dissociating a growing apical bud.

3.4 Production of new phenotypes from sectorial chimera

The M_1 line, which was maintained in the protected house, started production of shoots with completed leaves after about 3 years. Twenty-seven dissociated shoots from M_1 plants successfully regenerated in a nursery bed (Fig. 40.3b). This is in agreement with the observation of Ghani *et al.* (2013). Shoots have taken a long time to recover from stress created by irradiation.

New plants obtained by M_1 chimeric shoot propagation were morphologically dissimilar to each other and plants with aberrant leaves were not carried over for further selections. Table 40.2 gives the different characters for ten selected shoots dissociated from the M_1 plant (Fig. 40.3(b) and Table 40.2).

As shown in Table 40.2, growth (leaf emergence) was clearly reduced in the M_1 plant, which was accelerated a little in the M1-1 to M1-10 vegetative lines. Initiation of new shoots was comparatively slow in some lines including the M_1. Physical removal of the top part of the plant along with the growing apical bud in the M_1 generation induced the production of side shoots. Leaf size (length and width) was significantly reduced in mutated shoots with respect to the control line.

However, shoots with complete leaves were selected as vegetative lines; hence, incomplete and aberrant leaves were not observed in selected vegetative lines from the M_1 plant. Reduction of plant height and growth after exposure to irradiation was clearly observed and it was noted for several subsequent vegetative generations. Chromosomal damage associated with non-chromosomal damage due to irradiation effects may be the main reason for growth reduction (Sparrow and Gunckel, 1955) but DNA repair may correct such errors. Gill *et al.* (2015) showed that DNA repair by the plant corrected the irradiation damage of DNA caused by ultraviolet and ionizing radiation. Although DNA repair may be the reason for correction of plant height, leaf aberrations and growth speed of the shoots in this experiment, the damage and healing

Fig. 40.3. **(a)** Mother plant. **(b)** M_1 plant of 70-2-10. **(c)** Isolation of M1 vegetative line.

Table 40.2. Characters of selected vegetative lines developed from dissociated shoots of an M_1 chimera against the mother plant.

	Mother plant	M1	M1-1	M1-2	M1-3	M1-4	M1-5	M1-6	M1-7	M1-8	M1-9	M1-10
No. of shoots/ year (*n* = 2) (Study period = 5 years)	8.6 ± 0.76	6.8 ± 0.82	2.3 ± 0.44	5.1 ± 0.70	3.4 ± 0.54	6.2 ± 1.05	3.2 ± 0.64	9.4 ± 1.40	4.1 ± 0.58	3.8 ± 0.78	11.4 ± 1.62	6.4 ± 0.71
Av. internodal length (cm) (*n* = 12)	3.4 ± 0.37	0.95 ± 0.12	1.51 ± 0.18	2.3 ± 0.21	2.39 ± 0.12	1.88 ± 0.23	2.38 ± 0.12	1.7 ± 0.18	1.48 ± 0.14	3.35 ± 0.28	2.5 ± 0.13	1.12 ± 0.15
Av. leaf length (cm) (*n* = 12)	18.02 ± 0.98	7.4 ± 0.32	12.4 ± 0.46	13.3 ± 0.58	11.4 ± 0.34	14.3 ± 0.66	13.3 ± 2.67	9.3 ± 1.04	10.2 ± 0.87	14.4 ± 2.09	13.8 ± 1.25	6.6 ± 0.59
Av. leaf width (cm) (*n* = 12)	11.61 ± 0.77	3.54 ± 0.31	4.68 ± 0.37	5.63 ± 0.42	3.43 ± 0.23	8.09 ± 0.81	10.2 ± 0.53	3.73 ± 0.35	4.38 ± 0.43	5.51 ± 0.62	6.58 ± 0.73	3.91 ± 0.27
Leaf emergence interval (day) (*n* = 2) (Study period = 5 years)	8 ± 0.76	46.2 ± 4.9	32.4 ± 4.15	29.8 ± 2.56	37.6 ± 2.46	23.2 ± 1.63	26.5 ± 3.06	32 ± 2.53	28.6 ± 1.01	34.8 ± 2.65	19 ± 2.06	17 ± 1.10
Colours in a leaf (no.) (*n* = 10)	1 ± 0	2.7 ± 0.15	2.4 ± 0.16	2.6 ± 0.16	2.5 ± 0.16	2.9 ± 0.1	1.8 ± 0.2	2.1 ± 0.17	2.7 ± 0.15	2.6 ± 0.16	2.3 ± 0.26	2.4 ± 0.26

were not uniform among shoots, as reflected in different phenological traits in different shoots.

Leaf variegation has never been observed in *P. erubescens* 'Gold', in contrast to the range of colours from light green when immature to greenish yellow when mature (but only one colour appears in a leaf at a time). Young leaves are always brilliant greenish yellow (151 D in RHS colour chart) and become light green when mature (144 A in the RHS colour chart). Although the mother plant had monochrome leaves, three prominent colours were observed in a single leaf of all M_1 lines: dark bluish green (131 A), strong yellow green (144 A) and brilliant greenish yellow (151 B–D) according to the RHS colour chart (Fig. 40.4a). This leaf variegation is the most prominent character of the M_1 generation and its subsequently derived lines. Generally, the chlorophyll content is responsible for the green colour of leaves and variations in chlorophyll content are visible as colour variations. Therefore, these results indicate that the chlorophyll content of mutant leaves varies across the leaf lamina. Analysis of SPAD meter value and chlorophyll a and b contents shows significant variation among the three type of colour patches (Fig. 40.4b and Table 40.3).

The colour patch distribution pattern varied among the plants and leaves and some possessed only one or two colours. Even though the spreading of colour patches in leaves varied in mutants, chlorophyll content and SPAD meter readings of the patches belonging to the same colour were consistent (Table 40.3). Among these three common colours present in mutants, strong yellow green and brilliant greenish yellow were present in the mother plant as well as in the control, but dark bluish-green was not found in control. Further, there were shoots with complete bluish-green leaves and the growth of such shoots was faster than in variegated shoots. Several authors have reported colour and photosynthesis rate reduction after irradiation, due to chlorophyll mutations (Gustafsson, 1942; Hasbullah *et al.*, 2012). In contrast, the chlorophyll content of some of these chimeric shoots and colour patches increased in this experiment. Similar results had been obtained by Huang *et al.* (2013) in mutated rice.

The *OsDET1* mutation that Huang *et al.* (2013) described was a G-to-C single base substitution which resulted in an increased chlorophyll content of this rice mutant, and which might be fundamental for enhanced photo responsiveness.

Colours play an important role in amenity horticulture, where plants having multiple colours are vitally important to be attractive in the market. Therefore, the development of a plant line from this new chimera is useful to introduce as a new philodendron variety for the floriculture industry. But drastic changes in colour with time may not be tolerated in the industry; therefore, the establishment of the colour distribution pattern is needed. Results of the study on quantification of each colour in the leaves using Adobe Photoshop software revealed that the colour distribution of leaves of all these lines is not stable and varies from leaf to leaf and from generation to generation. Not only the colour distribution pattern but also the number of colours in the leaves varied with the vegetative generation in some of the lines. Changing of this colour distribution pattern is common in some plants; for example, *Acalypha wilkesiana* shows variation of colour distribution from leaf to leaf. Frank and Chitwood (2016) described chimeras with changing colour pattern and non-patterned colour in a sectorial chimera, which originated with highly active transposons. Such variants are popularly used in landscaping along with other chimeras.

With the passing of generations, M1-9 and M1-5 lines eventually produced monochrome leaves. These plant lines were well established and grew vigorously in later generations. Lines with multicoloured leaves showed retarded growth. Furthermore, a prominent character of these multicoloured lines was unstable distribution of colour patches in leaves. The colour distribution pattern and the area covered by each colour varied among the leaves and the generations. This variability was significantly low in the M1-4 line. The graphs in Fig. 40.5 show the colour distribution of second and third leaves of every vegetative generation of ten selected lines. In M1-4, the amount of each colour in the leaves did not show much variation throughout the 12 generations. Among the variegated lines only M1-4 showed stability, hence it can be considered as a promising ornamental line.

(a) (b)

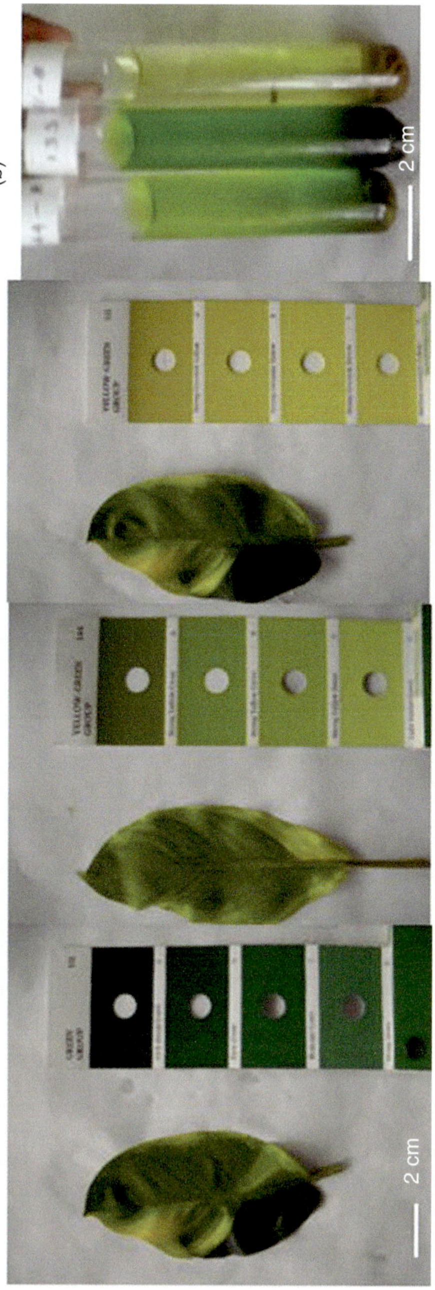

Fig. 40.4. **(a)** Comparison of colour patches with RHS colour chart. **(b)** Chlorophyll extraction of three different colour patches.

Table 40.3. SPAD meter values and chlorophyll a and b contents of three different leaf-colour patches (n = 12).

Colour of the leaf/leaf patch	Number in RHS colour chart	SPAD meter reading	Chlorophyll-a content (mg/g)	Chlorophyll-b content (mg/g)
Dark bluish green	131 A	76.8 ± 7.6	11.67 ± 1.1	21.83 ± 1.2
Strong yellow green	144 A-C	48.2 ± 7.3	8.68 ± 0.08	13.47 ± 0.55
Brilliant greenish yellow	131 B-D	14.4 ± 3.6	5.28 ± 0.34	1.77 ± 0.69
Control young leaves	131 D	19.6 ± 8.3	4.47 ± 1.2	2.91 ± 2.4
Control mature leaves	144 A-C	42.2 ± 11.3	5.6 ± 2.1	6.43 ± 5.2

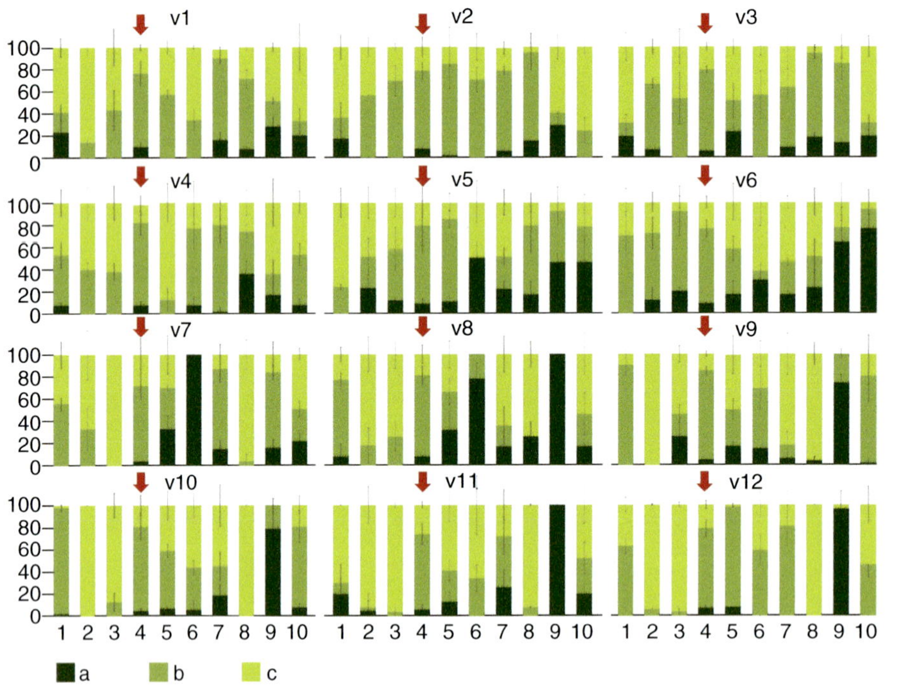

Fig. 40.5. Area occupied by different colour (%) in leaves of different M1 lines (M1-1 to M1-10) through 12 vegetative generations (V_1 to V_{12}). Arrow, M1-4 line; **(a)** dark bluish green; **(b)** strong yellow green; **(c)** brilliant greenish yellow.

4 Conclusion

In the present study, variation was induced in *P. erubescens* 'Gold' through ^{60}Co gamma irradiation treatment. The LD_{50} of *P. erubescens* 'Gold' rooted cuttings is around 85 Gy, but a high loss of regeneration ability limited the development of novel mutants. Significant delays or losses of bud breaking, shoot emergence and stem elongation were prominent in the mutants; however, ^{60}Co gamma-rays induced unstable variegation (sectorial mutation with changing colour pattern) in the leaves of M_1 lines. From the studied M_1 lines, the most stable colour distribution in leaves was in M1-4 and this was observed for 12 vegetative generations. This study demonstrates that the gamma irradiated *P. erubescens* 'Gold' line M1-4 can be used as a potential mutant to develop a new philodendron cultivar.

References

Abdullah, T.L. *et al.* (2009) Changes in flower development, chlorophyll mutation and alteration in plant morphology of *Curcuma alismatifolia* by gamma irradiation. *American Journal of Applied Sciences* 6(7), 1436–1439.

Ahloowalia, B.S. *et al.* (2004) Global impact of mutation-derived varieties. *Euphytica* 135, 187–204.

Bado, S. *et al.* (2017) Plant mutation breeding for the improvement of vegetatively propagated crops: successes and challenges. *CAB Reviews 2017* 12(28), 1–20.

Broertjes, C. and Van Harten, A.M. (1988) *Applied Mutation Breeding for Vegetatively Propagated Crops*. Elsevier, Amsterdam, Oxford, New York, Tokyo.

Brunner, H. (1995) Radiation induced mutations for plant selection. *Applied Radiation and Isotopes* 46(6–7), 589–594.

Chouteau, M. *et al.* (2006) A comparative study of inflorescence characters and pollen-ovule ratios among the genera *Philodendron* and *Anthurium* (Araceae). *International Journal of Plant Science* 167(4), 817–829.

Dai, W. and Magnusson, V. (2012) Morphological variations in *Buddleia* induced by gamma irradiation. *HortScience* 47(1), 81–83.

Datta, S.K. (1997) *Ornamental Plants. Role of Mutation.* Daya Publishing House, Delhi.

Datta, S.K. (2018) Breeding of new ornamental varieties: Rose. *Current Science* 114(6),1194–1206.

Dwimahyani, I. and Widiarsih, S. (2010) The effects of gamma irradiation on the growth and propagation of *in-vitro* Chrysanthemum shoot explants (cv. *Yellow Puma*). *Atom Indonesia* 36(2), 45–49.

Frank, M.H. and Chitwood, D.H. (2016) Plant chimeras: the good, the bad, and the 'Bizzaria'. *Developmental Biology* 419, 41–53.

Ghani, M. *et al.* (2013) Induction of novel variants through physical and chemical mutagenesis in Barbeton daisy (*Gerbera jamesonii* Hook.). *Journal of Horticultural Science & Biotechnology* 88(5), 585–590.

Gill, S.S. *et al.* (2015) DNA damage and repair in plants under ultraviolet and ionizing radiations. Review Article. *The Scientific World Journal*, Article ID 250158, Hindawi Publishing Corporation. doi: 10.1155/2015/250158

Gustafsson, A. (1942) The plastid development in various types of chlorophyll mutations. *Hereditas* 28(3–4), 483–492.

Hasbullah, N.A. *et al.* (2012) Physiological responses of callus from *Gerbera jamesonii* gamma irradiation. *International Journal of Brazilian Archives of Biology and Technology* 55(3), 411–416.

Hewawasam, W. *et al.* (2004) New phenotypes of *Crossandra infundibuliformis* var. Danica through In-vitro culture and induced mutation. *Tropical Agriculture Research* 16, 253–270.

Huang, J. *et al.* (2013) Mutation of OsDET1 increases chlorophyll content in rice. *Plant Science* 210, 241–249.

Jain, S.M. (2006) Mutation-assisted breeding for improving ornamental plants. *Acta Horticulturae* 714, 85–98.

Koh, Y.C. and Davies, J.F.T. (2001) Mutagenesis and in vitro culture of *Tillandsia fasciculata* Swartz var. fasciculata (Bromeliaceae). *Scientia Horticulturae* 87, 225–240.

Lea, D.E. (1947) *Action of Radiation on Living Cells*. Cambridge University Press, Cambridge, UK.

Lichtenthaler, H.K. and Wellburn, R. (1983) Determinations of total carotenoids and chlorophylls a and b of leaf extracts in different solvents. *Biochemical Society Transactions* 11, 591–592.

Maluszynski, M. *et al.* (1995) Application of in vivo and in vitro mutation techniques for crop improvement. *Euphytica* 85, 303–315.

Oladosu, Y. *et al.* (2016) Principle and application of plant mutagenesis in crop improvement: a review. *Biotechnology & Biotechnological Equipment* 30(1), 1–16.

Oldach, K.H. *et al.* (2008) Identification of a chemically induced point mutation mediating herbicide tolerance in annual medics (*Medicago* spp.). *Annals of Botany* 101, 997–1005.

Padmini, S.M.P.C. and Kodagoda, T.D. (2017) Present status and future scope of floriculture industry in Sri Lanka and its potential in women empowerment. *Sri Lanka Journal of Social Sciences* 40(1), 31–40.

Park, N.B. and Andersen, A.S. (1990) Gamma ray treatment of *Passiflora caerulea* cuttings: mutation and rooting data. *Die Gartenbauwissenschaft* 55(2), 63–65.

Puchooa, D. and Sookun, D. (2003) Induced mutation and *In Vitro* culture of *Anthurium andreanum.* In: LaTourette, J.A. and Bachraz, D.Y. (eds) *Proceedings of the Sixth Meeting of Agricultural Scientists, Réduit, Mauritius*, 8–9 May 2003. Food and Agricultural Research Council, Réduit, Mauritius, pp. 17–28.

RHS (2003) *Colour Chart*. Royal Horticultural Society, London.

Sax, K. (1942) The mechanism of X-ray effects on cells. *Journal of General Physiology* 25, 533–537.

Schum, A. and Preil, W. (1998) Induced mutations in ornamental plants. In: Jain, S.M. *et al*. (eds) *Somaclonal Variation and Induced Mutations in Crop Improvement*, Kluwer Academic Publishers, Dordrecht, Boston, London.

Sparrow, A.H. and Gunckel, J.E. (1955) The effect on plants of chronic exposure to gamma radiation from radio cobalt. *Proceedings International Conference on Peaceful Uses of Atomic Energy, Geneva* 12, 52–59.

Swaroop, K. *et al*. (2015) Effect of different doses of gamma rays for induction of mutation in bougainvillea cv. Mahatma Gandhi. *Indian Journal of Agricultural Sciences* 85(9), 1245–1247.

Van Harten, A.M. (2002) Mutation breeding of vegetatively propagated ornamentals. In: Vainstein, A (ed.) *Breeding for Ornamentals: Classical and Molecular Approaches*. Kluwer Academic Publishers, Dordrecht, pp. 105–127.

Section 5.

Induced Genetic Variation for Crop Improvement in the Genomic Era

41 The Power of Next-generation Sequencing and Machine Learning for Causal Gene Finding and Prediction of Phenotypes

Anna S. Sowa, Lisa Dussling, Jörg Hagmann and Sebastian J. Schultheiss[1]

Abstract

The wide application of next-generation sequencing (NGS) has facilitated and accelerated causal gene finding and breeding in the field of plant sciences. A wide variety of techniques and computational strategies is available that needs to be appropriately tailored to the species, genetic architecture of the trait of interest, breeding system and available resources. Utilizing these NGS methods, the typical computational steps of marker discovery, genetic mapping and identification of causal mutations can be achieved in a single step in a cost- and time-efficient manner.

Rather than focusing on a few high-impact genetic variants that explain phenotypes, increased computational power allows modelling of phenotypes based on genome-wide molecular markers, known as genomic selection (GS). Solely based on this genotype information, modern GS approaches can accurately predict breeding values for a given trait (the average effects of alleles over all loci that are anticipated to be transferred from the parent to the progeny) based on a large training population of genotyped and phenotyped individuals (Crossa *et al.*, 2017). Once trained, the model offers great reductions in breeding speed and costs. We advocate for improving conventional GS methods by applying advanced techniques based on machine learning (ML) and outline how this approach can also be used for causal gene finding.

Subsequent to genetic causes of agronomically important traits, epigenetic mechanisms such as DNA methylation play a crucial role in shaping phenotypes and can become interesting targets in breeding pipelines. We highlight an ML approach shown to detect functional methylation changes sensitively from NGS data.

We give an overview about commonly applied strategies and provide practical considerations in choosing and performing NGS-based gene finding and NGS-assisted breeding.

Keywords: genomic selection • machine learning • next-generation sequencing • bioinformatics • DNA methylation

1 Introduction

Humans have applied plant breeding and plant selection since agriculture arose, a practice which is estimated to date back about 11,000 years to the domestication of the fig (Kislev *et al.*, 2006). In the simplest terms, plant selection is a coevolutionary process between humans and plants. Breeding involves combining parental plants to obtain progeny with the best characteristics. Thomas Andrew Knight is considered to be the first person to use wide cross for the introduction of disease resistance when he produced the 'Siberian Bittersweet' as a scab-resistant variety of apple

[1] Email: sebastian.schultheiss@computomics.com

 Mutation Breeding, Genetic Diversity and Crop Adaptation to Climate Change (eds S. Sivasankar *et al.*)
DOI: 10.1079/9781789249095.0041

in the 1820s (Nabours, 1930). This work was well regarded by more well-known breeders such as Mendel, whose own interest in inheritance, fertilization and natural hybridization most likely stemmed from an initial interest in crop improvement (van Dijk *et al.*, 2018). The best characteristics are in themselves a variable trait, but most often include: higher yield; ease of cultivation and harvest; tolerance to environmental stress and resistance against pests; and better storage and processing characteristics. Plants, of course, are not only improved for dietary consumption but also for features relevant to fibre, fuel, shelter, landscaping and a variety of other human activities.

Over the years, plant breeding has expanded far beyond the fields where crops grow. Initially, plant selection was beneficial to individual farmers and communities. Depending on environment and desired trait, different locations used phenotype selection to parse out desired plants for further cultivation. Over time, as farming became more commercial, seed companies and research institutions began to have a stake in plant selection. Over the years, although a smaller percentage of the population is involved in direct agriculture, the number of stakeholders has expanded. The group of involved parties now includes chemical manufacturers, non-governmental global organizations such as FAO, IAEA and IRRI, biotech companies and charitable organizations worldwide.

Since the goal of plant domestication is to meet evolving human needs, the commercialization of those needs is not a surprise. Over the years, humans subjected plants to pressures that account for increase in fruit size, synchronization of growth and flowering, loss of seed dispersal, changes in plant architecture, and other qualities that have resulted in more effective cultivation, higher yields and more valuable products (Asano *et al.*, 2011). In the 21st century, however, food security became the most important goal in plant breeding. Producing enough food to feed global demand requires the knowledge of both biotech and rural farmers to come together in an innovative and progressive way. Currently, farming is facing a lot of pressure, including climate change, invasive pests, new diseases, decreasing investment in research and finally the lack of appropriate public knowledge about agriculture and food sciences. This is in part due to the success of agriculture. Today, the Bureau of Labor Statistics reports that only 2% of people in the USA are involved in food production, down from almost two-thirds only half a century ago.

Crop sciences and the involved stakeholders are all driven by different factors. Over the years, heightened food production has always been a uniting goal. Although it is hard to pinpoint, seed sales began as early as 1743 with Vilmorin Seeds selling vegetable seed; and the company from which Mendel himself obtained seeds, Benary, was founded by Ernst Benary in Erfurt, Thuringia, in 1842 (Benary, n.d.). One of the earliest commercial plant breeding companies for grain seeds in England, Gartons Agricultural Plant Breeders, was founded in the late 19th century by John Garton. William Farrer released an early fungus-resistant wheat variety in 1903. In 1908, the concept of heterosis (hybrid vigour) was introduced; and in the 1920s, statistical methods tried to analyse environmental effects on gene selection. Finally, in 1933 came the breakthrough of cytoplasmic male sterility developed in maize. These early breeding techniques were the beginning of the methods of crop selection we see today (Schlegel, 2018).

The 20th century ushered in a rush of genetic-guided improvements in crop science using phenotypic, pedigree and performance data (Crossa *et al.*, 2017). These measures equated to big advantages for the industry, but signs of grain yield stagnation in some crops, especially in stress-prone regions, is already evident (Ray *et al.*, 2012). Genetic technologies hold great promise for increasing the effectiveness and speed of plant selection.

In the following paper, we highlight current genetics-based techniques in plant breeding and point out some possibilities for moving the field forward using novel next-generation sequencing (NGS) and machine learning (ML) methods.

2 Overview of Current Methods

The most important part of plant breeding is selection among genetically diverse parents. Even today, selection is still often based on phenotype. Breeders choose the 'best' crops under their breeding goal, using their knowledge, expectations and observed phenotype. They propagate those crops to increase the proportion of desirable traits. The earliest form of this process was selecting naturally occurring variants from the wild, exemplified by the derivation of maize from teosinte (Doebley *et al.*, 1995). Further scientific discovery gave breeders the ability to follow markers linked to traits and map them to areas of the plant genome. Some initial DNA markers were polymorphisms in restriction enzyme sites and simple sequence repeats (Kan and Dozy, 1978). They were superseded by single nucleotide polymorphisms (SNPs), which could be identified through high-throughput genotyping. With the use of NGS to identify large numbers of SNP–phenotype correlations, many gene-trait associations have been identified.

2.1 Quantitative trait loci (QTL)

When many markers are genotyped, a genetic map can be built based on the frequency of the recombination events between these markers. When a quantitative trait is evaluated, a model can be built which explains some of the observed trait variation according to marker segregation and implies that those markers are linked to genes underlying the trait variation. This results in quantitative trait loci (QTL) maps, by which a breeder can get an idea of how many loci are controlling a trait, their relative importance and approximate position in the genome. This analysis is limited to the alleles segregating in the specific analysed cross and the amount of the variance explained by the alleles at any one locus is dependent on the allelic variation at any other loci that may contribute to trait variation. In cases where the QTLs have a large, dominant effect, it may be possible to directly use them for the selection of useful markers. More often, however, more work is necessary and the ultimate desired outcome of this method is the identification of a locus or loci (and polymorphism(s)) responsible for the phenotypic diversity.

While thousands of marker-trait associations have been reported in different plant species, far fewer examples of successfully exploiting mapped QTL have been reported in the literature. Key lessons learned from applying markers in plant breeding include the following (Bernardo, 2008).

1. The purpose of detecting QTL should be clearly defined before embarking on QTL mapping.
2. Procedures for marker-based selection depend on the number of QTL.
3. Estimates of QTL effects for complex traits are often inconsistent.
4. Gain per unit cost and time rather than gain per cycle should be considered.
5. Exact phenotyping is crucial.

2.2 Expression QTL (eQTL)

With time, the increase in the number of genome wide-association studies (GWAS) showed that the majority of variants are found in non-coding regions of the genome and are likely involved in gene regulation (Nica and Dermitzakis, 2013). The analysis of these regulation-specific variants is called expression QTL (eQTL). Variation in gene expression is known to influence many traits in plants. Studies in the tomato, for example, have used eQTL analysis to identify hotspots regulating patterns related to plant defence and photosynthesis (Ranjan *et al.*, 2016). eQTLs are especially relevant in plants, since closely related plant species sometimes have only minor coding divergence but develop vastly different characteristics, demonstrating the importance of gene expression patterns (Koenig *et al.*, 2013). In Arabidopsis, for example, eQTL studies have demonstrated that elements whose physical location coincides with the location of the regulated gene (*cis*-eQTLs) have an effect on local expression levels, while elements whose location is in a different position from the gene being investigated (*trans*-eQTLs) have more of a global influence on expression (Breseghello and Coelho, 2013). eQTL analysis in plants has the potential to reveal a lot of information on gene regulatory networks and help to identify key transcriptional regulators.

The availability of whole genome sequencing (WGS) changed the perspective on plant genotyping and identification of DNA markers from fragment-based polymorphisms to sequence-based SNPs. These technologies are also time consuming and provide information on targeted individuals, which have limited use in plant breeding and gene discovery. Another limiting factor for plant breeding is time and cost.

In general, causal gene finding (QTL, eQTL, WGS or others) has a lot of potential for research purposes but offers little direct impact for breeders, although it has yielded rewards such as the identification of translation factors involved in virus resistance (Sanfaçon, 2015). Identifying and manipulating one trait at a time has become straightforward, but unlikely to result in useful varieties. The challenges are thus found in trying to improve multiple traits of interest simultaneously, which are intertwined due to pleiotropic effects, linkage or population genetic structure (Breseghello and Coelho, 2013).

NGS has stepped in to handle some of these challenges. In contrast to human genetics, where many of these methods were first developed, plant biotechnology has to deal with relatively large genome sizes, the increase in gene copy numbers or polyploidy and the large number of offspring in plants, which all increase the computational power needed for discovery, sequencing and genotyping of thousands of markers in a single step. The hardware requirements are usually one or two orders of magnitude above those for human genetics.

2.3 Sequencing and genotyping of a subset of the genome

NGS has the potential to improve on the previously described techniques to make causal gene finding easier and more pertinent for breeders. The first step in NGS-based marker development involves library construction. Since sequencing whole genomes is still costly for large crop genomes, several techniques generate partial, reduced representation of the genome by using restriction enzymes (reduced-representation libraries, complexity reduction of polymorphic sequences, restriction-site associated DNA sequencing (RADseq), sequence-based polymorphic marker technology, multiplexed shotgun genotyping and genotyping-by-sequencing (GBS)), or by using sequence capture libraries without enzymatic digestion (molecular inversion probes, solution hybrid selection and microarray-based genomic selection) (Ray and Satya, 2014).

RADseq and GBS have already proved useful for plant breeding purposes, for example studying anthracnose disease resistance in *Lupinus angustifolius* L. (Yang *et al.*, 2012; Glaubitz *et al.*, 2014). Not needing any *a priori* genome sequence information is a major benefit of RADseq and, if reference genomes are available, missing data can be reliably imputed (Browning *et al.*, 2018). These methods were used to identify QTLs for anthocyanin pigmentation of fruit (Ray and Satya, 2014) and GBS was useful for development of high-density SNP maps in wheat and barley (Poland *et al.*, 2012).

2.4 Bulked segregant analysis (BSA)

Rather than cutting down the number of markers, a technique called bulked segregant analysis (BSA) reduces the number of sequencing libraries required to identify (causal) genetic markers that are linked to mutant phenotypes (Zou *et al.*, 2016). It was originally designed to find genes that have large effects but has recently been modified for analysing minor causal alleles, i.e. QTL with relatively small effects as well as linked and interacting QTL (Ehrenreich *et al.*, 2010). Since many traits with agronomic value are genetically complex, affected by many genes and the environment, naïve analysis requires a large number of individuals. BSA reduces this complexity by considering only the phenotypes at the extreme ends of the spectrum and bulks the individuals of each end into one pool. Bulk analysis has been adapted to work well in plant populations and has the potential to reduce cost noticeably. As an example, considering a population with 500 individuals with two extremes, bulk analysis will only amount to 2/500th or 0.4% of the cost of whole population analysis, although it will require a greater read depth (Zou *et al.*, 2016). BSA has been successfully applied to study leaf rust in wheat, salt resistance in cotton, bean common mosaic virus in the common bean, *Phytophthora* root rot in pepper and photosynthetic traits in poplar, as well as salt- and cold-tolerance in rice, among many others (Zou *et al.*, 2016).

In more detail, gene finding by forward genetics works by first applying a mutagenic agent like a chemical or radiation to an organism of interest and inducing random genetic variants. Mutants with a trait of interest are isolated and are crossed in a defined scheme to a known genetically diverged strain to retrieve genetic markers and to generate a recombinant mapping population. Often, multiple independent alleles of the mutant phenotype are present in the population so it is important to determine if the phenotype of interest is monogenetically controlled and to show that the independent mutant alleles are allelic. In this case, close linkage can be a problem if you have only one mutant allele. At the causal locus of the phenotype, and also at linked genetic markers surrounding it, the mutant plants have a different allele frequency than the non-mutagenized individuals and therefore they can be analysed in a pool. Thus, in a typical scenario, it is only necessary to sequence two pools of individuals and genotype the genetic markers along the genome to identify the region harbouring the causal locus (the mapping interval). For an easily detectable difference in allele frequencies, correct phenotyping is crucial. Any wrongly phenotyped individual will reduce the differentiating signal between mutant and wild-type pools. The width of the mapping interval depends on many factors, most importantly the number (and thus the density) of genetic markers and the recombination frequency of the species. The width of the interval itself is determined by the number of recombination events and, for any given number of recombination events, the number of markers affects the interval size only because of the distance between the recombination events and the markers nearest to the end of the interval. In practice, most recombination events are further apart than markers in even low-density maps. This means that the number of recombination events is much more important than the number of markers. Early attempts with genome-wide but targeted genotyping, and nowadays with reduced representation sequencing (see above), are easily capable of identifying a sufficient number of genetic markers. Since the genetic markers are, however, likely to be different from the causal mutation, an additional targeted sequencing step is still required. This two-step process has become unnecessary due to affordable whole genome sequencing, since it simultaneously enables the detection of typically abundant genetic markers, the calculation of their allele frequencies in the pools and the identification of the causal genetic change in the mapping interval (Schneeberger *et al.*, 2009; Schneeberger, 2014). It even allows backcrossing the mutant lines to the non-mutagenized progenitors (isogenic mapping populations), in which case only the induced mutations serve as genetic markers, given that a mutagen with a moderate to high effect is applied (Abe *et al.*, 2012). Even the need for a suitable reference genome can be compensated by specific methodologies (Nordström *et al.*, 2013) and NGS can also generate a reference genome and annotation *de novo* with constantly decreasing effort.

Once genetic markers have been identified with BSA, it is useful to narrow down the candidate causal mutations (Schneeberger, 2014) by screening them for mutation patterns that are typical for the mutagenesis method used (e.g. the widely used chemical ethyl methanesulfonate mainly induces C-to-T substitutions), rank them by their impact on gene function in coding regions (SIFT) (Vaser *et al.*, 2016), or screen the prevalence of a mutation in other species (PROVEAN) (Choi *et al.*, 2012). In addition, mutations closer to the 'peak' of the allele frequency difference in the mapping interval can be considered to be more likely causal.

2.5 Genomic selection and ML-based genomic selection

Once relevant molecular markers in linkage disequilibrium with QTL markers are identified, they can be used for marker-assisted selection. During such a selection, only plants or animals with the desirable allele associated with the trait of interest are considered. The method has been successfully applied for traits controlled by a single or a few QTLs with major impact on trait expression, such as the SCMV-resistance genes *Scm1* and *Scm2* in maize (Quint *et al.*, 2002). However, it has been proved inadequate for complex quantitative traits such as grain yield in cereals (Doebley *et al.*, 2006), which are determined by a large number of minor QTLs that also interact with each other. Genomic selection (GS) has been developed to overcome the limitations of both conventional and marker-assisted breeding. GS aims to estimate the genetic value of an individual based on a large set of markers across the whole genome. This requires a prediction model, which is generated from the genotypic and phenotypic data of a large training population. This model can then be used to determine the genomic estimated breeding value of any individual according to their genomic profile. In turn, the model is used to predict the breeding value for genotypes from the breeding population with the highest performance potential and select them as a parent for hybridization or next-generation advancement. This is applied to individuals that have been genotyped but not yet phenotyped (Crossa *et al.*, 2017). GS uses the predicted values as a substitute of phenotyping in a few selection cycles, thus reducing the cost and time for breeders in variety development. For instance, in maize, a breeder can test-cross half of the available lines and use GS for the other half, thus saving 50% of the test plots and also time. This efficiency is also observed in the second cycle of selection, which uses the trained population from the previous cycle to predict the doubled haploid lines. Based on GS, these lines can go directly to the second stage of evaluations (Crossa *et al.*, 2017). GS has the potential to be useful for tracking complex traits with low heritability or for simple traits with high heritability, but the usefulness of GS is limited by genotyping costs and unclear user guidelines for how breeding programmes should be modified to make optimal use of the predictive ability of GS and its space- and cost-saving potential.

Initially, the application of GS was also severely limited by the availability of data. The phenotype of a large number of individuals of the training population has to be determined for accurate and precise predictions. Similarly, the genotype of each target individual has to be identified, ideally in a comprehensive, fast and unbiased way. Genome-wide coverage of the markers can be achieved at low cost per base through NGS. Today, high-throughput genome-wide genotyping strategies such as GBS are available to genotype large numbers of individuals at low costs and within a short time frame. Without the need for array design or probe design from a reference genome, GS becomes feasible even for orphan crops. Furthermore, missing information can be imputed based on information from a reference population (Bhat *et al.*, 2016).

GS can be applied to maintain the genetic diversity of the breeding programme, accelerate the breeding cycle and increase genetic gain beyond the results of conventional or marker-assisted approaches. A variety of models, both parametric and non-parametric, has been developed in recent years.

The standard parametric approach is based on best linear unbiased prediction (BLUP). Genomic BLUP assumes equal variance between the loci and an additive impact of all genomic markers on the phenotype. The genetic merit of an unknown individual is predicted based on its genetic distance to individuals in a reference population by calculating a kinship matrix. Genomic BLUP is an effective, accurate and widely used method to predict genetic and phenotypic values, provided the underlying assumptions are met. However, the predictive range of BLUP-based methods is limited by the performance of the reference population. Novel marker combinations in individuals that outperform previous generations are not identified (Meuwissen *et al.*, 2001; Endelman, 2011).

We propose non-parametric approaches such as support vector regression (SVR) as viable alternatives to BLUP, since they do not require the use of a kinship matrix. SVR is a supervised learning method that generates a prediction model from the training data. Larger datasets are required, which theoretically increases computational complexity significantly, potentially beyond the point of feasibility. SVR and other kernel methods solve this problem by using the kernel trick, a mathematical equivalence of the kernel function to the naïve scalar product calculation, which makes the method feasible and highly efficient for high-dimensional data (Bishop, 2006). However, the selected kernel function has to fit the reference data and appropriate parameters have to be determined to prevent overfitting and ensure predictive performance (Crossa *et al.*, 2017; Wang *et al.*, 2018). Cross-validation with the training population is used to mitigate overfitting and accurately assess the performance.

2.6 Making ML-based GS useful for causal gene finding

A trained machine-learning predictor offers more than just the prediction of phenotypes of unknown individuals. Studying the internal workings of a trained SVR weight vector has the potential to reveal causal loci as well as environmental conditions with great impact on the phenotype of interest. During the training process, underlying patterns of the dataset are learned and stored within the model. The predictor has to determine the impact of all possible marker combinations on the phenotype. For this, each marker state and effect is weighted according to the training population. If further information such as environmental conditions or specific treatments are available, they can be integrated easily to increase the predictive performance. Once the model is trained, predictions are made by combining the modelled effects of the markers present in any unknown individual.

At the same time, the model information can be extracted to gain further insight into the reference population. Combinations of markers that affect the predicted phenotype the most are of considerable interest. In the case of an additive model, the sum of all weighted marker effects determines the final phenotype prediction, which we can determine. Features with strong effects on the phenotype have higher weights and are thus candidates for causal genes. An analysis of feature contributions revealed both beneficial features for high-performing maize crosses as well as disadvantageous contributions to low yield (Suazo-López *et al.*, 2014).

Comprehensive models are required to profile and predict complex and non-linear interactions between genotype, environment and phenotype. However, the extraction of information from non-linear models is more challenging and requires further processing to generate sensible and interpretable results. Several feature-importance measures have been developed and successfully applied to both classifiers and regressors (Vidovic *et al.*, 2016), for instance to the precise and efficient classification, ranking and visualization of DNA sequence properties (Sonnenburg *et al.*, 2007). Similarly, relevant markers can be identified from non-linear models on a genome-wide scale, assessed and visualized according to their importance.

Compared with BLUP-based GS, Computomics' ML-based approaches have several advantages, including: (i) higher prediction accuracy compared with BLUP methods; (ii) integration of environmental data; (iii) ability to transfer learned marker-phenotype associations from one population to another; (iv) the potential for causal gene discovery; (v) evaluation of the potential of new genetic material *in silico*; and (vi) the ability to reliably predict millions of hybrid crosses without any existing phenotype data of the parents, evaluating new genetic material.

2.7 Epigenetic causes of phenotypes

Explaining phenotypes solely by genetic changes did reveal many causes for disease and phenotypes, but also stayed behind expectations, since many association studies yielded weak or inconsistent results (Boyle *et al.*, 2017). The findings of epigenetic changes that affect phenotypes (so-called epialleles) in many different plant species steered attention towards the most widely analysed epigenetic mark, DNA methylation of cytosine bases. One of its functions in plants is to silence the expression of transposable elements, which are densely methylated, and to regulate stress responses and developmental processes, for example of important traits such as flowering time (Yaish *et al.*, 2011). Different extents of methylation in or near gene promoters can sporadically affect gene expression. One landmark example is the African oil palm *Elaeis guineensis*, where the efficiency of propagation is often reduced by the occurrence of a homeotic floral phenotype known as 'mantling'. This dramatically reduces oil yield. Ong-Abdullah *et al.* (2015) found that a drastic decrease in DNA methylation in mantled plants near a single splice acceptor site affects alternative splicing and is causal for this severe phenotype. Epialleles have been identified in other species as well (Weigel and Colot, 2012) and typically consist of a single genomic region with an altered DNA methylation profile. Thus, detecting and pinpointing such changes is possible with sequencing the genome-wide methylation landscape with the commonly used bisulfite-sequencing protocol. These causal methylation differences can be short genomic regions and only involve methylation in specific sequence contexts (CG, CHG, CHH, H: A, C or T), so a sensitive and accurate computational identification of differentially methylated regions is crucial. Typical specialized software only focuses on differences at single sites, which are enriched in gene bodies and variation of them is biologically less meaningful (Becker and Weigel, 2012), or these tools do not perform rigorous statistical tests on regions. Our developed tool, MethylScore, improves these shortcomings and was specifically designed for plant methylation pattern. MethylScore utilizes machine learning to reduce the bias towards less important genic methylation variance; it is efficient for large populations and suitable for whole genome as well as targeted sequencing data. Its underlying methodology successfully identified functional methylation changes caused by salt stress (Wibowo *et al.*, 2016) and was widely used for screening DNA methylation in plant populations (Hagmann *et al.*, 2015), during development (Daccord *et al.*, 2017; Tedeschi *et al.*, 2019) and across species (Seymour *et al.*, 2014).

Although it is a matter of debate whether and for how long epigenetic changes are stable across generations (Heard and Martienssen, 2014), current research is focusing on ways of making such changes heritable, so that they can soon become interesting targets for breeding.

3 Conclusion

We have presented a variety of NGS-based methods for gene finding and breeding that are ready to be used today. In some cases, breeding systems have to be adapted to take full advantage of these benefits. Depending on the breeding goals and trait of interest, a broader method like GS might be the right choice for the breeder, while in other cases the identification of a causal mutation as the most reliable marker could be desirable. Gene editing, for instance, requires a precise understanding of the genomic basis for the desired phenotypes to successfully deliver a useful new trait.

Our approach of interpretable ML techniques combines these two goals and makes this a versatile tool both for breeders looking for variety performance and for researchers looking for a gene editing target. With epigenetic analysis, we have only scratched the surface of potential traits that could be regulated even in existing lines through differential methylation. MethylScore provides a robust statistical analysis method for this type of analysis.

To reach the goal of more rapid breeding cycles and faster adaptation of plants to new climate conditions, the use of these techniques becomes inevitable and invaluable.

References

Abe, A. *et al.* (2012) Genome sequencing reveals agronomically important loci in rice using MutMap. *Nature Biotechnology* 30, 174–178.

Asano, K. *et al.* (2011) Artificial selection for a green revolution gene during japonica rice domestication. *Proceedings of the National Academy of Sciences of the United States of America* 108(27), 11034–11039.

Becker, C. and Weigel, D. (2012) Epigenetic variation: origin and transgenerational inheritance. *Current Opinion in Plant Biology* 15(5), 562–567.

Benary Seeds (n.d.) History. Available at: benary.com/company/history (accessed 6 April 2021)

Bernardo, R. (2008) Molecular markers and selection for complex traits in plants: learning from the last 20 years. *Crop Science* 48(5), 1649.

Bhat, J.A. *et al.* (2016) Genomic selection in the era of next generation sequencing for complex traits in plant breeding. *Frontiers in Genetics* 7(DEC), 1–11.

Bishop, C.M. (2006) *Pattern Recognition and Machine Learning*. Springer, New York.

Boyle, E.A. *et al.* (2017) An expanded view of complex traits: from polygenic to omnigenic. *Cell* 169(7), 1177–1186.

Breseghello, F. and Coelho, A.S.G. (2013) Traditional and modern plant breeding methods with examples in rice (*Oryza sativa* L.). *Journal of Agricultural and Food Chemistry* 61(35), 8277–8286.

Browning, B.L. *et al.* (2018) A one-penny imputed genome from next-generation reference panels. *American Journal of Human Genetics* 103(3), 338–348.

Choi, Y. *et al.* (2012) Predicting the Functional Effect of Amino Acid Substitutions and Indels. *PLoS ONE* 7(10), e46688.

Crossa, J. *et al.* (2017) Genomic selection in plant breeding: methods, models, and perspectives. *Trends in Plant Science* 22(11), 961–975.

Daccord, N. *et al.* (2017) High-quality de novo assembly of the apple genome and methylome dynamics of early fruit development. *Nature Genetics* 49(7), 1099–1106.

Doebley, J.F. *et al.* (2006) The molecular genetics of crop domestication. *Cell* 127(7), 1309–1321.

Doebley, J. *et al.* (1995) Teosinte branched1 and the origin of maize: evidence for epistasis and the evolution of dominance. *Genetics* 141(1), 333–346.
Ehrenreich, I.M. *et al.* (2010) Dissection of genetically complex traits with extremely large pools of yeast segregants. *Nature* 464(7291), 1039–1042.
Endelman, J.B. (2011) Ridge regression and other kernels for genomic selection with R package rrBLUP. *The Plant Genome Journal* 4(3), 250–255.
Glaubitz, J.C. *et al.* (2014) TASSEL-GBS: a high-capacity genotyping by sequencing analysis pipeline. *PLoS ONE* 9(2), e90346.
Hagmann, J. *et al.* (2015) Century-scale methylome stability in a recently diverged Arabidopsis thaliana lineage. *PLoS Genetics* 11(1), e1004920.
Heard, E. and Martienssen, R.A. (2014) Transgenerational epigenetic inheritance: myths and mechanisms. *Cell* 157(1), 95–109.
Kan, Y.W. and Dozy, A.M. (1978) Polymorphism of DNA sequence adjacent to human beta-globin structural gene: relationship to sickle mutation. *Proceedings of the National Academy of Sciences* 75(11), 5631–5635.
Kislev, M.E. *et al.* (2006) Early domesticated fig in the Jordan Valley. *Science* 312(5778), 1372–1374.
Koenig, D. *et al.* (2013) Comparative transcriptomics reveals patterns of selection in domesticated and wild tomato. *Proceedings of the National Academy of Sciences of the United States of America* 110(28), E2655–E2662.
Meuwissen, T.H.E. *et al.* (2001) Prediction of total genetic value using genome-wide dense marker maps. *Genetics* 157(4), 1819–1829.
Nabours, R.K. (1930) Plant hybridization before Mendel. *Science* 72(1863), 274–275.
Nica, A.C. and Dermitzakis, E.T. (2013) Expression quantitative trait loci: present and future. *Philosophical Transactions of the Royal Society of London. Series B, Biological Sciences* 368(1620), 20120362.
Nordström, K.J.V. *et al.* (2013) Mutation identification by direct comparison of whole-genome sequencing data from mutant and wild-type individuals using k-mers. *Nature Biotechnology* 31, 325–330.
Ong-Abdullah, M. *et al.* (2015) Loss of Karma transposon methylation underlies the mantled somaclonal variant of oil palm. *Nature* 525, 533–537.
Poland, J.A. *et al.* (2012) Development of high-density genetic maps for barley and wheat using a novel two-enzyme genotyping-by-sequencing approach. *PLoS ONE* 7(2), e32253.
Quint, M. *et al.* (2002) Development of RGA-CAPS markers and genetic mapping of candidate genes for sugarcane mosaic virus resistance in maize. *Theoretical and Applied Genetics* 105, 355–363.
Ranjan, A. *et al.* (2016) eQTL regulating transcript levels associated with diverse biological processes in tomato. *Plant Physiology* 172(1), 328–340.
Ray, D.K. *et al.* (2012) Recent patterns of crop yield growth and stagnation. *Nature Communications* 3, 1293.
Ray, S. and Satya, P. (2014) Next generation sequencing technologies for next generation plant breeding. *Frontiers in Plant Science* 5, 367.
Sanfaçon, H. (2015) Plant translation factors and virus resistance. *Viruses* 7(7), 3392–3419.
Schlegel, R.H.J. (2018) *History of Plant Breeding*, 1st edn. CRC Press, Boca Raton, Florida.
Schneeberger, K. (2014) Using next-generation sequencing to isolate mutant genes from forward genetic screens. *Nature Reviews Genetics* 15(10), 662–676.
Schneeberger, K. *et al.* (2009) Simultaneous alignment of short reads against multiple genomes. *Genome Biology* 10(9), R98.
Seymour, D.K. *et al.* (2014) Evolution of DNA methylation patterns in the Brassicaceae is driven by differences in genome organization. *PLoS Genetics* 10(11), e1004785.
Sonnenburg, S. *et al.* (2007) Accurate splice site prediction using support vector machines. *BMC Bioinformatics* 8(Suppl. 10), S7.
Suazo-López, F. *et al.* (2014) Support vector machines with features contribution analysis for agricultural yield prediction. *Annual Research & Review in Biology* 4(24), 4246–4258.
Tedeschi, F. *et al.* (2019) EFFECTOR OF TRANSCRIPTION factors are novel plant-specific regulators associated with genomic DNA methylation in Arabidopsis. *New Phytologist* 221(1), 261–278.
van Dijk, P.J. *et al.* (2018) How Mendel's interest in inheritance grew out of plant improvement. *Genetics* 210(2), 347–355.
Vaser, R. *et al.* (2016) SIFT missense predictions for genomes. *Nature Protocols* 11, 1–9.
Vidovic, M.M.-C. *et al.* (2016) Feature Importance Measure for Non-linear Learning Algorithms. Available at: http://arxiv.org/abs/1611.07567 (accessed 2019)

Wang, X. *et al.* (2018) Genomic selection methods for crop improvement: current status and prospects. *Crop Journal* 6(4), 330–340.

Weigel, D. and Colot, V. (2012) Epialleles in plant evolution. *Genome Biology* 13(10), 249.

Wibowo, A. *et al.* (2016) Hyperosmotic stress memory in Arabidopsis is mediated by distinct epigenetically labile sites in the genome and is restricted in the male germline by DNA glycosylase activity. *ELife* 5.

Yaish, M.W. *et al.* (2011) The role of epigenetic processes in controlling flowering time in plants exposed to stress. *Journal of Experimental Botany* 62(11), 3727–3735.

Yang, H. *et al.* (2012) Application of next-generation sequencing for rapid marker development in molecular plant breeding: a case study on anthracnose disease resistance in *Lupinus angustifolius* L. *BMC Genomics* 13(1), 318.

Zou, C. *et al.* (2016) Bulked sample analysis in genetics, genomics and crop improvement. *Plant Biotechnology Journal* 14(10), 1941–1955.

42 Ion Beam Mutagenesis – an Innovative and Effective Method for Plant Breeding and Gene Discovery

Tomoko Abe[1], Hiroyuki Ichida, Yoriko Hayashi, Ryouhei Morita, Yuki Shirakawa, Kotaro Ishii, Tadashi Sato, Hiroki Saito and Yutaka Okumoto

Abstract
We have developed a unique technology for mutation induction of plants using energetic ion beams at the RI Beam Factory (RIBF) of Rikagaku Kenkyūjo (RIKEN) (Institute of Physical and Chemical Research). Ion beams effectively induce mutations at relatively low doses without severely inhibiting growth. The irradiation treatment can be given to various plant materials and mutation can be induced in a short time, between seconds and a few minutes. The linear energy transfer (LET) of ions depends on the nuclide and velocity. Since LET value affects the mutation frequency, it is an important parameter to determine the most effective irradiation condition in mutagenesis. We determined the most effective dose in each LET for mutation induction in imbibed rice seeds. Subsequently, we analysed the mutated DNA responsible for the phenotype in morphological mutants. Most of the mutations were small deletions of less than 100 bp. Irradiations of C-ions and Ne-ions are effective for plant breeding because of the very high mutation rate and sufficient energy to disrupt a single gene. On the other hand, all mutations induced by Ar-ion (290 keV/μm) irradiation were large deletions ranging from 176 bp to approximately 620 kb. The average number of mutations in the target exon regions was 7.3, 8.5 and 4.3 per M_3 mutant plant in C-ions, Ne-ions and Ar-ions, respectively. The number of mutations induced by heavy-ion irradiation was relatively small. We could identify six responsible genes for eight mutants induced by C-ion and Ne-ion irradiations and two responsible genes for four mutants induced by Ar-ion irradiation. Three of these were genes not previously described.

Keywords: heavy-ion beam • carbon ion • argon ion • LET

1 Introduction

The world is facing a serious food and energy crisis. Plant mutation breeding has played a great role to overcome such crisis and maintain world stability. The breeding of new mutant varieties provides higher yield potential, more productive biomass energy use, better adaptation to climate change and variability. New techniques are needed to achieve faster and more effective breeding. We have developed a unique technology for mutation induction using energetic heavy-ion beams at the RI Beam Factory (RIBF). This development was achieved through an efficient synergic link between agricultural science and accelerator physics. At relatively low doses, ion beams induce mutations at a high rate without severely inhibiting growth. The irradiation treatment can be given to various plant materials and quick irradiations lasting between

[1] Email: tomoabe@riken.jp

 Mutation Breeding, Genetic Diversity and Crop Adaptation to Climate Change (eds S. Sivasankar *et al.*)
DOI: 10.1079/9781789249095.0042

seconds and a few minutes are enough to induce mutations. We have collaborated with flower companies and public agricultural experiment stations since 1996 on the potential of practical breeding using heavy-ion beams. As a result, 30 new cultivars were generated by collaborative research, including sterile *Verbena* 'Temari Coral Pink' (Kanaya *et al.*, 2008) which became the first plant registered by Suntory Flowers Ltd under the Seed and Seedling Law, a new flower colour of *Petunia* 'Safinia Rose Veined', tearless onion 'Smile ball' (Kato *et al.*, 2016) and short-culm Japanese barnyard millet 'Nebari-ko No. 2' (Nakajo *et al.*, 2009). Over 70 new varieties of plants and microbes have been created by ion-beam irradiation in Japan. Mutation breeding by ion-beam irradiation was mainly performed using flowers and ornamental plants. Recently, beneficial variants have been grown for various crops (Table 42.1) (Nakajo *et al.*, 2009; Cabanos *et al.*, 2012; Endo *et al.*, 2016; Kato *et al.*, 2016; Sawada *et al.*, 2016; Ishikawa *et al.*, 2018). The ion beam is thus a highly efficient tool for improving crops through mutation breeding (Abe *et al.*, 2012, 2015).

The effect of linear energy transfer (LET) value on mutation rate is an important factor to determine the most effective irradiation condition for mutagenesis. We investigated the effect of LET on mutation induction using the model plant *Arabidopsis thaliana*. The most effective LETs for mutation induction and lethality were 30 keV/μm and 290 keV/μm, respectively (Kazama *et al.*, 2008). This highly mutation-efficient LET for dry seeds was designated as $LETmax^{dry}$. Subsequently, we have reported detailed analyses on the molecular nature of DNA alterations induced by heavy-ion irradiation with $LETmax^{dry}$ using morphological mutants. The majority of mutations were small deletions ranging from several to several tens of base pairs which were sufficient to disrupt a single gene (Kazama *et al.*, 2011). On the other hand, irradiation with 290 keV/μm ions showed a different mutation spectrum from that at $LETmax^{dry}$. The proportion of small deletions was low, while that of large deletions and rearrangements was high (Hirano *et al.*, 2012, 2015).

Rice is an important plant not only as a staple crop but also as a model for genomic research. It is well known that the radiosensitivity (Myttenaere *et al.*, 1965) and storage effects (Constantin *et al.*, 1970) of dry seeds are affected by their water content. In order to eliminate those effects, we have applied ion-beam irradiation to imbibed seeds. In this report, we evaluate the effect of LET on mutation induction on imbibed rice seeds. Subsequently, we have analysed the mutated genes corresponding to morphological mutants.

2 Materials and Methods

2.1 Plant materials and heavy-ion beam irradiation

Rice seeds (*Oryza sativa* L. cv. 'Nipponbare') were soaked in water for 3 days at 28°C in the dark before irradiation. Germinated seeds placed in Petri dishes in a single layer were irradiated with carbon ions (C-ions) (135 MeV per nucleon), neon ions (Ne-ions) (135 MeV per nucleon) and argon ions (Ar-ions) (95 MeV per nucleon) in the RIBF, with a range of doses from 2.5 Gy to 80 Gy. The LET ranges of C-ions, Ne-ions and Ar-ions were 23–70 keV/μm, 63–100 keV/μm and 290 keV/μm, respectively. The LET values of these ions were calculated at the surface of the seeds. LET was adjusted by the range shifter of an automatic irradiation system for biological samples (Ryuto *et al.*, 2008). Seedlings were grown in a greenhouse. Survival rates and plant height were observed at 2 weeks after irradiation. Surviving seedlings were transplanted to a paddy field to obtain M_2 seeds. Mutation frequency was calculated based on the numbers of M_1 lines that showed chlorophyll-deficient mutants (CDM) in the M_2 generation. Seed fertility was evaluated using the percentage of fertile seeds per panicle.

2.2 Isolating the rice mutants

Mutant screening was conducted in the M_2 generation. We isolated the gibberellin-related dwarf zebra and shortened-plastochron mutants from the mutant population grown in a greenhouse at approximately 3 weeks after planting. Other mutants were isolated from the population grown in a paddy field.

Table 42.1. New varieties or parent plants developed in various crops using ion-beam irradiation.

Plant	Breeding goal	Original variety	Treated plant material	Treatment:Ion/LET (keV/µm)/Dose (Gy)	No. mutants/ selection population	New variety
Lettuce	Low polyphenol oxidase	Round	Protoplast	C/23/2	1/242 calli	
Lettuce	High yield	Papa Lettuce	Cultured leaf	H/0.45/3	4/98 M_2 plants	Fortuna
Onion	Tearless and non-pungent	Sapporo-ki	Dry seed	Ne/63/20	18/197 M_2 bulbs	Smile Ball
Midi Tomato	Parthenocarpy	Hanano-suzu	Dry seeds	H/0.45/10	16/400 M_1 plants	Lucina
Edible Chrysanthemum	Late flowering and big flower	Etenraku	Cultured floral petal	C/23/10	43/859 regenerated plants	Kiku-Meigetsu
Japanese barnyard millet	Short culm	Mojappe	Dry seed	C/23/20	17/906 M_2 plants	Nebarikko 2 go
Rice	Low cadmium uptake	Koshihikari	Dry seed	C/80/40	3/2592 M_2 plants	Koshihihari-kan 1 go
Peanut	Reducing major allergens	Nakateyutaka	Dry seeds	C, N/23, 30/100	17/11335 M_2 seeds	

2.3 DNA extraction and PCR analysis

DNA was extracted from leaves of the morphological mutants (M_3 generation) using the DNeasy® Plant Kit (Qiagen, Hilden, Germany). Mutations in the responsible genes were characterized using polymerase chain reaction (PCR). PCR was performed on the relevant genes using primer pairs amplifying approximately 1 kb. When the entire coding region of the responsible gene was amplified, we assumed that small deletions or base substitutions had occurred. In this situation, PCR products were purified and sequenced. If the whole or a part of the coding region could not be amplified, we assumed structural variants such as large deletion, inversion or translocation had occurred. In this case, we designed other primers and conducted a detailed analysis including TAIL-PCR for amplifying flanking sequence of the structural variants.

2.4 Whole exome sequencing and genetic linkage analysis

The details of whole exome capturing in rice have been described elsewhere (Ichida *et al.*, 2019). Briefly, target capturing probes that cover 91.0% of all coding sequences (34,040,553 bases of target exon regions within 373,245,519 bases of the rice genome) defined in IRGSP 1.0 annotations (Kawahara *et al.*, 2013) were designed and synthesized as a SeqCap EZ Developer Library (Roche Sequencing Solutions, Pleasanton, California, USA). The designed probes spanned 300,746 exon regions in the rice genome and were expected to cover a total of 41.75 Mb. Massively parallel (next-generation) sequencing was done using Illumina HiSeq 2500 and 4000 instruments (Illumina, San Diego, California, USA). The resulting rice exome sequencing reads were processed with an in-house sequencing data analysis pipeline built on the Hokusai parallel computing platform operated by Advanced Centre for Computing and Communication, RIKEN (Ichida *et al.*, 2019). Raw sequencing reads were mapped to the reference 'Nipponbare' sequences using BWA (Li and Durbin, 2009) and then subjected for variant calling with three distinct algorithms, GATK (McKenna *et al.*, 2010), Pindel (Ye *et al.*, 2009) and Bedtools (Quinlan and Hall, 2010).

We conducted an improved genetic linkage analysis using both F_2 progeny derived from a cross between 'Gimbozu' and each mutant plant, and *mPing* sequence-characterized amplified region (SCAR) markers (Monden *et al.*, 2009) for three mutants, namely 3-14 (ripening failure), 4-13 (weak growth) and 6-62 (dwarf). For this analysis, we used 20 F_2 progenies for 3-14, 24 F_2 progenies for 4-13 and 20 F_2 progenies for 6-62. We conducted a conventional linkage analysis for the two virescent mutants (Ne-1779 and Ar-5-160). The F_2 population derived from a cross between each mutant and the non-mutated cultivar, such as 'Gimbozu' or 'Nipponbare', was grown and the genetic linkage between each phenotype and the mutations identified by exome sequencing was determined. We used 13 F_2 progenies for Ne-1779 and nine for Ar-5-160.

3 Results

3.1 Effects of LET on survival and mutation induction

The survival rate of M_1 plants decreased as the dose increased. The irradiation with higher LET required a lower dose to decrease the survival rate (Fig. 42.1). Table 42.2 shows the effect of heavy-ion irradiation at different LET values on mutation rate (Hayashi *et al.*, 2008, 2009, 2016). There was no practical difference in the mutation frequency among C-ion, Ne-ion and Ar-ion irradiations. The LET of 50–70 keV/μm at 15 Gy was the most effective for CDM induction in the irradiation with C-ions and Ne-ions. In addition, the mutation frequency obtained from Ar-ions (290 keV/μm) was comparable to C-ions and Ne-ions at a lower dose of irradiation.

3.2 Candidate gene-based mutation analysis by PCR and sequencing

We first isolated ten previously reported rice mutants (dwarf, zebra, early-flowering, shortened plastochron and tall) from mutagenized populations by C-ion and Ne-ion irradiations. We performed PCR and sequencing analysis of the candidate genes to reveal each mutation type and the size of mutation. Sequencing analysis revealed

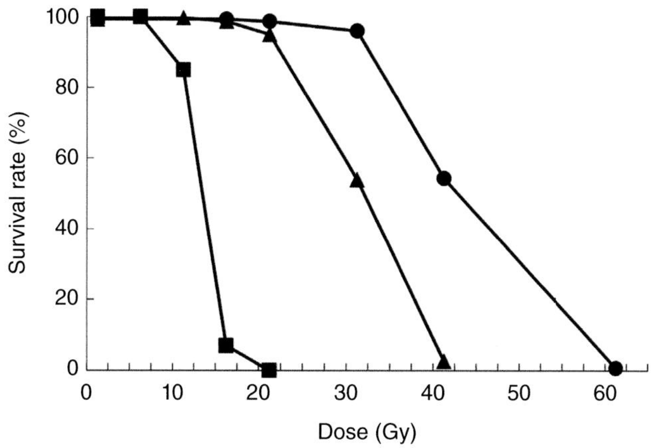

Fig. 42.1. Effect of LET on survival rate at 2 weeks after irradiation of imbibed seeds. Circles, triangles and squares indicate C-ions with 30 keV/μm, C-ions with 50 keV/μm and Ar-ions with 290 keV/μm, respectively.

Table 42.2. Highest mutation frequency under each irradiation treatment.

Ion	LET (keV/μm)	Dose (Gy)	Number of M_2 lines	Mutation frequency (%)
C	23	30	121	4.96
	30	20	446	3.59
	40	20	378	5.29
	50	15	1958	6.43 ± 0.20
Ne	63	15	2257	6.61 ± 0.20
	70	15	146	6.85
	80	15	145	3.45
Ar	290	2.5	686	6.10
		5	1374	6.92 ± 0.85

The seed fertility of M_1 plants was > 50%.

that all mutations tested were deletions (Table 42.3). Nine mutants possessed small deletions (< 100 bp) and the remaining harboured a 72.3 kb deletion. Next, we tested seven mutants induced by Ar-ion irradiation. PCR and sequencing analysis showed that all mutations induced by Ar-ion irradiation were also deletions (Table 42.3). All mutations were large deletions (≥ 100 bp) ranging from 176 bp to approximately 620 kb. Ar-ion irradiation could induce larger deletions than C-ion or Ne-ion beams.

3.3 Whole exome sequencing analysis of ion beam-induced rice mutants

We applied whole exome capturing in a total of 12 rice mutants induced by heavy-ion beam irradiations. Four mutants each from C-ion, Ne-ion and Ar-ion irradiations were subjected to whole exome sequencing and variant detection. We then compared the types and numbers of mutations detected in these conditions. In the eight C-ion and Ne-ion mutants, 28 single nucleotide variants (SNVs), 22 small deletions (1 to 18, 81 and 96 bp), three large deletions (739 bp, 46.4 kb and 102.2 kb), seven insertions (1–15 bp) and three inversions (299 kb, 897 kb and 3.3 Mb) were detected by the pipeline (Table 42.4). There were six SNVs, seven small deletions, 14 large deletions and two inversions in 29 homozygous mutated genes in C-ion and Ne-ion mutants. In the four Ar-ion (290 keV/μm) mutants, nine SNVs, three small deletions (1, 5 and 72 bp), four large deletions (103 bp, 552 bp, ca. 900 bp, and 133.6 kb) and two insertions (1 and 9 bp)

Table 42.3. Mutations identified in the mutants induced by C-, Ne- and Ar-ion irradiation.

Ion	Dose (Gy)	LET (keV/μm)	Phenotype	Mutated gene	Mutation type	Size (bp)
C	15	50	High tillering dwarf	*D10*	Deletion	3
			Zebra	*ZN1*	Deletion	22
			Dwarf	*KAO*	Deletion	2
			Dwarf	*GID2*	Deletion	72,348
Ne	15	63	Early flowering (1 month)	*OsHY2*	Deletion	2
			Dwarf	*GID2*	Deletion	4
			Dwarf	*GA3ox2*	Deletion	12
			Dwarf & Short grain	*D1*	Deletion	3
			Dwarf & Short grain	*SRS3*	Deletion	1
		70	Dwarf & Short grain	*D11*	Deletion	25
Ar	2.5	290	Early flowering (20 days)	*Ghd7*	Deletion	Approx. 620,000
	5	290	Shortened plastochron	*PLA1*	Deletion	176
			Dwarf	*GID2*	Deletion	2,627
			Early flowering (20 days)	*Ghd7*	Deletion	65,534
			Tall	*EUI1*	Deletion	47,930
			Dwarf & Short grain	*D1*	Deletion	22,148
			Dwarf & Short grain	*D2*	Deletion	117,089

were identified. In both conditions, nearly or more than half of the mutations detected were SNVs, followed by small (< 100 bp) and large (≥ 100 bp) deletions. There were one SNV, one small deletion and eight large deletions in ten homozygous mutated genes in Ar-ion mutants. In the homozygous mutant gene, the proportion of large deletions induced by Ar-ion irradiation was increased from that induced by C-ion and Ne-ion irradiations. The average number (±SE) of mutations in the target exon regions was 7.3 ± 1.3, 8.5 ± 1.7 and 4.3 ± 0.2 per M_3 mutant in C-ion, Ne-ion and Ar-ion, respectively.

3.4 A combinatorial approach to efficient identification of responsible genes by whole exome sequencing and genetic linkage analysis

We applied a combinational approach by whole exome sequencing and conventional genetic linkage analysis to the previously isolated 12 mutant lines. Table 42.5 shows examples in which the causative genes were successfully identified by this approach. The responsible genes for the 3-14 mutant (*Os04g0475600*) and the 4-13 mutant (*Os04g0662900*) were annotated as '2-OG-Fe(II) oxygenase domain containing protein' and 'RNA polymerase III transcriptional repressor, MAF1 domain containing protein', respectively, in the rice annotation database RAP-DB (available at https://rapdb.dna.affrc.go.jp/, accessed in 2019). The dwarf phenotype observed in 6-62 mutant was linked to a 102 kb deletion on chromosome 9 and this deletion caused the loss of seven genes (*Os09g0240200*, *Os09g0240350*, *Os09g0240500*, *Os09g0240850*, *Os09g0240975*, *Os09g0241100* and *Os09g0241200*). In the case of Ne-1779 and Ar5-160 mutants, 46.4 kb and 133.6 kb deletions were linked to the virescent phenotype. The 46.4 kb deletion contained *Os10g0421800*, encoding pentatricopeptide repeat containing protein, which is presumably a rice orthologue of *Arabidopsis PROTON GRADIENT 3* (Yamazaki *et al.*, 2004). The 134 kb deletion contained *Os09g0135400*, encoding a tetratricopeptide repeat containing protein, which is presumably a rice orthologue of *Arabidopsis CLUMPED CHLOROPLASTS 1* (Yang *et al.*, 2011). We identified five responsible genes from seven mutants induced by C-ion and Ne-ion irradiation, and two responsible genes from four mutants induced by Ar-ion irradiation by genetic linkage analysis. In addition, exome sequencing analysis indicated that the eight mutants, namely 3-14 (ripening failure), 3-51 and 6-62 (dwarf), 4-13 (weak growth), 5-12 (stripe) (Ge *et al.*, 2017),

Table 42.4. Numbers and types of heavy-ion induced mutations identified from whole exome sequencing.

		No. Mutated regions[a]							
Ions	Mutant ID	SNV[b]	Small insertion	Small deletion	Large deletion	Inversion duplication	Total	Per plant[c]	Mutated genes homozygous/total[a]
C ion	3-14	5	0	3	0	0	8		1/4
	3-51	4	2	5	0	0	11		1/1
	4-13	3	0	1	0	2	6		1/1
	5-12	2	1	1	0	0	4		2/3
	Total	14	3	10	0	2	29	7.25	5/9
Ne ion	6-62	5	2	2	1	0	10		11/13
	7-30	1	0	3	0	0	4		1/1
	7-3B	5	2	5	1	0	13		5/6
	Ne-1779	3	0	2	1	1	7		7/8
	Total	14	4	12	3	1	34	8.5	24/28
Ar ion	Ar-G1263	1	1	2	0	0	4		0/1
	Ar-G976	3	0	1	0	0	4		1/3
	Ar5-160	2	1	0	1	0	4		7/8
	Ar5-287	3	0	0	2	0	5		2/5
	Total	9	2	3	3	0	17	4.25	10/17

[a]Homozygous and heterozygous mutation;
[b]SNV, single nucleotide variant;
[c]average number of mutations in the target exon regions per M_3 plant

Table 42.5. Responsible gene identified by whole exome sequencing and genetic linkage analysis.

Mutant ID	Phenotype	Ion•Dose•LET	No. of mutated genes[a]	Causative gene			
				Os ID	Description	Mutation (size)	Gene symbol, orthologue
3–14	Poor filling	C•15•50	1	*Os04g0475600*	2OG-Fe(II) oxygenase domain containing protein	Deletion (1 bp)	NEW
4–13	Weak	C•15•60	1	*Os04g0662900*	RNA polymerase III transcriptional repressor, MAF1 domain containing protein	Inversion (897 kb)	NEW
5–12	Stripe	C•15•60	2	*Os01g0109300*	Similar to predicted protein	Deletion (18 bp)	NEW
6–62	Dwarf	Ne•15•63	11	*Os09g0240200*	Zinc finger, B-box domain containing protein	Deletion (102 kb)	NEW
				Os09g0240350	Hypothetical gene		
				Os09g0240500	Similar to sulfate transporter 4.1		
				Os09g0240850	Hypothetical conserved gene		
				Os09g0240975	Hypothetical gene		
				Os09g0241100	Similar to nucleotide binding		
				Os09g0241200	Hypothetical protein		
Ne-1779	Virescent	Ne•15•63	7	*Os10g0421800*	Pentatricopeptide repeat domain containing protein	Deletion (46 kb)	*Arabidopsis* protein gradient-regulation3
7-30	Pale green	Ne•15•70	1	*Os08g0163400*	RNA polymerase sigma-70 domain containing protein	Deletion (12 bp)	OsSIG1
Ar5-160	Virescent	Ar•5•290	7	*Os09g0135400*	Tetratricopeptide repeat containing protein	Deletion (134 kb)	*Arabidopsis* clumped chloroplasts1
Ar-G976	Early flowering	Ar•5•290	1	*Os03g0309200*	Similar to phytochrome B	Deletion (5 bp)	PHY B

[a]Number of homozygous mutated genes

7-30 (pale green), 7-3B (lesion mimic) and Ne-1799 (virescent), possessed a mutated gene, which had been reported elsewhere (Ichida *et al.*, 2019). We found a 12 bp deletion within the *Os08g0163400* gene in 7-30 mutant, which encodes a plastid RNA polymerase sigma factor, and it has been reported that the disruption of *Os08g0163400* causes pale green phenotype (Tozawa *et al.*, 2007). We also found a 5 bp deletion in *Os03g0309200*, encoding a rice phytochrome B, which was reported as a responsible gene for the early flowing phenotype, and therefore we concluded that Ar-G976 (early flowering) is a *phyB* mutant (Takano *et al.*, 2005).

4 Discussion

4.1 Mutation frequency in C-, Ne- and Ar-ion irradiation

Dose–response curve of survival rate and plant height were important factors to determine the optimal dose for ion-beam mutagenesis. The optimal dose for mutation induction that did not reduce the survival rate but decreased plant height was lower than the shoulder dose in the survival curve (Hayashi *et al.*, 2008; Kazama *et al.*, 2008; Yamaguchi *et al.*, 2009). Shoulder doses in the survival curve were 20 Gy in 50 keV/μm C-ion and 10 Gy in Ar-ion (Fig. 42.1). The optimal dose was lower than shoulder dose in rice (Table 42.2). The highest mutation frequency by C-ion and Ne-ion irradiations was obtained with LET of 50–70 keV/μm in rice. By contrast, the $LETmax^{dry}$ of C-ions were 30 keV/μm (Kazama *et al.*, 2008) and 50 keV/μm (Murai *et al.*, 2013) in *Arabidopsis* and diploid einkorn wheat (*T. monococcum*), respectively. The LETmax^dry values in other plant species have to be determined to enable the efficient use of ion beam mutagenesis. Mutation frequencies in C-ion and Ne-ion irradiations were almost the same as those of Ar-ion irradiation with LET of 290 keV/μm. In *Arabidopsis* mutagenesis, the effective dose was 400 Gy for $LETmax^{dry}$ and 50 Gy for Ar-ion irradiation, while LET of 290 keV/μm showing lower mutation frequency. In rice, the effective dose for C-ion and Ne-ion irradiation was 15 Gy and the optimum dose for Ar-ion irradiation was 5 Gy. This dose difference was small compared with that in *Arabidopsis*. The mutation frequency of C-ion and Ne-ion irradiation was in the same range for Ar-ion irradiation with LET of 290 keV/μm. Dose is proportional to the LET and the number of particles. The particle number of C-ions and Ar-ions in the above irradiation conditions passing through a 10 $μm^2$ layer of cells was estimated to be 190 ions and 11 ions, respectively. One Ar-ion might therefore give sufficient energy to produce the same mutation effect as 17 C-ions. The effect of heavy-ion dose on lethality increased according to the increase of LET. The highest relative biological effectiveness for lethality of dry seeds occurred at 251–305 keV/μm for Ar-ions (Tanaka *et al.*, 1997; Morishita *et al.*, 2003). Ar-ion irradiation was also more effective at decreasing the survival rate than C-ions in rice (Fig. 42.1).

4.2 Mutated genes induced by C-, Ne- and Ar-ion irradiation

The small deletions (< 100 bp) occupied 90.0% of total mutations induced by C-ion and Ne-ion irradiations in rice by PCR and sequencing analyses (Table 42.3). Kazama *et al.* (2011) reported that 80% of the mutations caused by C-ion with LETmax^dry were small DNA alterations such as base substitutions and small deletions in *Arabidopsis*. A similar tendency was observed in the present study, suggesting that the LETmax^dry irradiation could induce small DNA alterations in a wide range of plant species. The deletions induced by Ar-ion were much larger in size than those induced by C-ion and Ne-ion irradiation. Hirano *et al.* (2012) reported that more than half of the deletions induced by Ar-ion irradiation are large deletions (> 100 bp) in *Arabidopsis*. The proportion of the large deletions (> 100 bp) in our study was even higher than that reported for *Arabidopsis*.

Whole exome sequencing analysis showed that a theoretical number of mutations per total number of nucleotides in the target exon regions was 2.1, 2.5 and 1.3 × 10^{-7} bp in M_3 plants for C-ion, Ne-ion and Ar-ion irradiations, respectively. The exome sequencing analysis of ethyl methanesulfonate (EMS) mutagenized rice plants in a total of 39 Mb target regions detected 70 to 508 mutations in each line, which corresponds to a mutation rate of 1.8×10^{-6} to 1.3×10^{-5} bp in M_2 plants (Henry *et al.*, 2014). The number of mutations induced by ion-beam irradiation was approximately 10–100 times lower than that of EMS-treated plants. It should be noted that whole exome sequencing is prone to misidentifying large deletions and insertions due to its principle that utilizes hybridization-based capture and regions of read depth discontinuity in the genome. Therefore, these results may contain some systematic bias in both conditions derived from this limitation. It may be possible that some of the large insertions and deletions as well as some structural changes such as chromosomal rearrangements would not have been identified, especially in the Ar-ion irradiated samples. This could explain why there is no major difference in the types of induced mutations between C-ion, Ne-ion and Ar-ion mutants, while the target-gene approach by PCR and Sanger sequencing showed high frequency of large deletions in Ar-ion mutants. Development of improved algorithms to detect large indels and structural variants from exome sequencing datasets would be key to solving this problem.

4.3 Responsible gene identified by whole exome sequencing and genetic linkage analysis

We applied whole exome sequencing for the rapid and efficient identification of responsible genes in mutants (Table 42.5). This approach utilizes exome sequencing to systematically identify the induced mutations in exon regions, which are more likely to cause altered phenotypes than promoter and intergenic regions, and identifies the responsible mutations for the altered phenotype by linkage analysis with a segregating population (Ichida *et al.*, 2019). This approach often makes it possible to identify responsible genes even without linkage analysis after exome sequencing analysis, since the number of mutations induced by heavy-ion beams is relatively small, and often only one mutation is found near the mapped location. We identified six responsible genes from eight mutants induced by C-ion and Ne-ion irradiation and two responsible genes from four mutants induced by Ar-ion irradiation. Such high success rates demonstrated that whole exome sequencing, combined with genetic linkage analysis, is an effective method to identify responsible genes or mutations. We could not determine the responsible genes for the two mutants induced by C-ion and Ne-ion irradiation, and one mutant induced by Ar-ion irradiation, since there were no mutations linked to each mutant's phenotype. It is possible that the causative mutation is located in non-coding regions, such as promoter regions or intergenic regions, which were not captured in exome sequencing analysis. Whole genome sequencing (WGS) would be better than exome sequencing from this point of view, because WGS has the potential to detect mutations generated in both coding- and non-coding DNA regions, although the cost was much higher at the time of the start of our study. We isolated the two early-flowering mutants for which causative mutations would be located in intergenic regions by WGS analysis.

It was unexpected but interesting that the responsible mutations identified with these approaches were mostly deletions of sizes between 1 bp and 134 kb, and only one mutant (4-13) was caused by an inversion of 897 kb. The genes identified by PCR and sequencing also showed that all 17 tested were deletions (Table 42.3). This could reflect the fact that deletions have a higher chance to inactivate genes by causing deleterious in-frame deletions, frame shifts, filler-DNA insertion at double-strand break sites and non-synonymous substitutions and nonsense mutations in amino acid sequences, whereas SNVs could inactivate genes only when non-synonymous substitutions at an active site or other important residues or nonsense mutations in amino acid sequences occurred.

5 Conclusion

We determined the most effective heavy-ion irradiation conditions for rice mutagenesis. There was no practical difference in the mutation frequency among C-ion, Ne-ion and Ar-ion irradiations to rice imbibed seeds. In C-ion and Ne-ion irradiations, most mutations were small deletions of less than 100 bp. Thus, C-ion and Ne-ion irradiation can efficiently generate knockout mutations of a target gene and can be applied to mutation breeding and reverse genetics. On the other hand, all mutations induced by Ar-ion irradiation were large deletions. The average number (±SE) of mutations in the target exon regions was 7.3 ± 1.3, 8.5 ± 1.7 and 4.3 ± 0.2 per M_3 mutant in C-ion, Ne-ion and Ar-ion, respectively. Since the number of mutations induced by heavy-ion irradiations was relatively small, we could identify eight responsible genes from 12 mutants induced by C-ion, Ne-ion and Ar-ion irradiations.

Recently, we determined the LETmax^dry of C-ions in dry rice seeds to be LETs of 30 keV/μm, similar to that in *Arabidopsis*. When we adjusted water content of the seeds to 13% before irradiation, mutation frequency was 14.3% at 175 Gy (Hayashi *et al.*, 2017; Ichida *et al.*, 2019). We recommend that irradiation of dry seeds has a higher mutation frequency than imbibed seeds. Beneficial variants have been isolated in various plant species, such as salt-resistant rice, high-yield rice (Morita *et al.*, submitted), low polyphenol-oxidase lettuce (Sawada *et al.*, 2016), low-allergen peanut (Cabanos *et al.* 2012), high-yield nori (an edible seaweed) (Niwa *et al.*, 2011) and high oil-content algae (Ota *et al.*, 2013) and *Euglena* (Yamada *et al.*, 2016). We have built a new beam line to increase available ion species with higher LET and wider energy range (Fukunishi *et al.*, 2015). We aim at advances in examination of the effects of physical factors (e.g. ion species, LET and dose) on DNA-mutated regions with detection using whole genome sequencing and elucidating the mechanism of mutagenesis with ion beams. In addition, the combination of ion-beam induced mutants and genome sequencing technology may enable discovery of genes and thus lead to a new field in biology, 'mutagenomics' (Katano *et al.*, 2016; Koide *et al.*, 2018; Yamatani *et al.*, 2018).

Acknowledgements

This research is supported by grants, JSPS through the 'Funding Program for Next Generation World-Leading Researchers' (NEXT Program, GR096), MEXT KAKENHI 'Genome Science' (GR096), CSTI, Cross-ministerial Strategic Innovation Promotion Program (SIP), 'Technologies for creating next-generation agriculture, forestry and fisheries' (funding agency: Bio-oriented Technology Research Advancement Institution, NARO). We are grateful to the staff at the Research Resource Centre, RIKEN Brain Science Institute, for their technical assistance in DNA sequencing and target enrichment experiments. The irradiation experiments were conducted at RIBF operated by the RIKEN Nishina Centre and the Centre for Nuclear Study (CNS) at the University of Tokyo. Bioinformatics analysis was performed using the Hokusai parallel computing system under the project numbers Q16208 and Q17208.

References

Abe, T. *et al.* (2012) Ion beam mutagenesis. In: Shu, Q.Y., Forster, B.P. and Nakagawa, H. (eds) *Plant Mutation Breeding and Biotechnology*. CABI, Wallingford, UK, pp. 99–106.

Abe, T. *et al.* (2015) Ion beam breeding and gene discovery for function analyses using mutants. *Nuclear Physics News* 25, 30–34.

Cabanos, C.S. *et al.* (2012) Heavy-ion beam irradiation is an effective technique for reducing major allergens in peanut seeds. *Molecular Breeding* 30, 1037–1044.

Constantin, M.J. *et al.* (1970) Effects of modifying factors on the response of rice seeds to gamma-rays and fission neutrons. *Radiation Botany* 10, 539–549.

Endo, S. *et al.* (2016) Breeding of new edible Chrysanthemum cultivar 'Yamaen K4'. *Bulletin of Agricultural Research Yamagata* 8, 93–102.

Fukunishi, N. *et al.* (2015) New high-energy beam transport line dedicated to biological applications in RIKEN RI beam factory. *Proceedings 13th International Conference on HIAT* 2015, 42–44.

Ge, C. *et al.* (2017) Single-point mutation of a histidine-aspartic domain-containing gene involving in chloroplast ribosome biogenesis leads to white fine stripe leaf in rice. *Science Reports* 7, 3298.

Hayashi, Y. *et al.* (2008) Effects of ion beam irradiation on mutation induction in rice. In: *Proceedings of 18th International Conference on Cyclotrons and Their Applications (Cyclotron 2007)*, pp. 222–224.

Hayashi, Y. *et al.* (2009) LET-dependent effect of C-ion beam irradiation on mutation induction in rice. *RIKEN Accelerator Progress Report* 24, 285.
Hayashi, Y. *et al.* (2016) Effect of Ar-ion beam irradiation on imbibed seed of rice. *RIKEN Accelerator Progress Report* 49, 256.
Hayashi, Y. *et al.* (2017) Optimisation of mutation induction by heavy-ion beam irradiation in rice with seed water content change. *RIKEN Accelerator Progress Report* 50, 27.
Henry, I.M. *et al.* (2014) Efficient genome-wide detection and cataloging of EMS-induced mutations using exome capture and next-generation sequencing. *Plant Cell* 26, 1382–1397.
Hirano, T. *et al.* (2012) Molecular nature of mutations induced by high-LET irradiation with argon and carbon ions in *Arabidopsis thaliana*. *Mutation Research* 735, 19–31.
Hirano, T. *et al.* (2015) Comprehensive identification of mutations induced by heavy-ion beam irradiation in *Arabidopsis thaliana*. *Plant Journal* 82, 93–104.
Ichida, H. *et al.* (2019) Targeted exome sequencing of unselected heavy-ion beam-irradiated populations reveals less-biased mutation characteristics in the rice genome. *Plant Journal* 98, 301–314.
Ishikawa, S. *et al.* (2018) Ion-beam irradiation, gene identification, and marker-assisted breeding in the development of low-cadmium rice. *Proceedings National Academy of Sciences* 109, 19166–19171.
Kanaya, T. *et al.* (2008) Heavy-ion beam-induced sterile mutants on verbena (*Verbena* x *hybrida*) with an improved flowering habit. *Plant Biotechnology* 25, 91–96.
Katano, M. *et al.* (2016) Suppressor screen and phenotype analyses revealed an emerging role of the monofunctional peroxisomal enoyl-CoA hydratase 2 in compensated cell enlargement. *Frontiers in Plant Science* 7, 132.
Kato, M. *et al.* (2016) Production and characterization of tearless and non-pungent onion. *Science Reports* 6, 23779.
Kawahara, Y. *et al.* (2013) Improvement of the *Oryza sativa* Nipponbare reference genome using next generation sequence and optical map data. *Rice* 6, 1–10.
Kazama, Y. *et al.* (2008) LET-dependent effects of heavy-ion beam irradiation in *Arabidopsis thaliana*. *Plant Biotechnology* 25, 113–117.
Kazama, Y. *et al.* (2011) Characterization of highly efficient heavy-ion mutagenesis in *Arabidopsis thaliana*. *BMC Plant Biology* 11, 161.
Koide, Y. *et al.* (2018) Lineage-specific gene acquisition or loss is involved in interspecific hybrid sterility in rice. *Proceedings National Academy of Sciences* 115, E1955–E1962.
Li, H. and Durbin, R. (2009) Fast and accurate short read alignment with Burrows-Wheeler transform. *Bioinformatics* 25, 1754–1760.
McKenna, A. *et al.* (2010) The Genome Analysis Toolkit: a MapReduce framework for analyzing next-generation DNA sequencing data. *Genome Research* 20, 1297–1303.
Monden, Y. *et al.* (2009) High potential of a transposon *mPing* as a marker system in *japonica* x *japonica* cross in rice. *DNA Research* 16, 131–140.
Morishita, T. *et al.* (2003) Dose response and mutation induction by ion beam irradiation in buckwheat. *Nuclear Instruments and Methods in Physics Research B* 206, 565–569.
Murai, K. *et al.* (2013) A large-scale mutant panel in wheat developed using heavy-ion beam mutagenesis and its application to genetic research. *Nuclear Instruments and Methods in Physics Research B* 314, 59–62.
Myttenaere, C. *et al.* (1965) Radiosensitivity of rice seed in relation to water content and free radicals. *Radiation Botany* 5, 443–451.
Nakajo, S. *et al.* (2009) Breeding of new Japanese barnyard millet cultivars with low amylose content and short culm, 'Hie-Iwate 1', 'Hie-Iwate 2', 'Hie-Iwate 3'. *Millet Newsletter* 24, 12–19.
Niwa, K. *et al.* (2011) Mutation breeding in the marine crop *Porphyra yezoensis* (Bangiales, Rhodophyta): cultivation experiment of the artificial red mutant isolated by heavy-ion beam mutagenesis. *Aquaculture* 314, 182–187.
Ota, S. *et al.* (2013) Phenotypic spectrum of *Parachlorella kessleri* (Chlorophyta) mutants produced by heavy-ion irradiation. *Bioresource Technology* 149, 432–438.
Ryuto, H. *et al.* (2008) Heavy-ion beam irradiation facility for biological samples in RIKEN. *Plant Biotechnology* 25, 119–122.
Quinlan, A.R. and Hall, I.M. (2010) BEDTools: a flexible suite of utilities for comparing genomic features. *Bioinformatics* 26, 841–842.
Sawada, R. *et al.* (2016) Low polyphenol oxidase mutant induced by $^{12}C^{6+}$ ion beam irradiation to protoplasts of lettuce (*Lactuca sativa* L.). *Horticulture Research (Japan)* 15, 347–353.

Takano, M. *et al.* (2005) Distinct and cooperative functions of phytochromes A, B, and C in the control of deetiolation and flowering in rice. *Plant Cell* 17, 3311–3325.

Tanaka, A. *et al.* (1997) Effects of heavy ions on the germination and survival of *Arabidopsis thaliana*. *International Journal of Radiation Biology* 72, 121–127.

Tozawa, Y. *et al.* (2007) The plastid sigma factor SIG1 maintains photosystem I activity via regulated expression of the *psaA* operon in rice chloroplasts. *Plant Journal* 52, 124–132.

Yamada, K. *et al.* (2016) Efficient selective breeding of live oil-rich *Euglena gracilis* with fluorescence-activated cell sorting. *Science Reports* 6, 26327.

Yamaguchi, H. *et al.* (2009) Mutagenic effects of ion beam irradiation on rice. *Breeding Science* 59, 169–177.

Yamatani, H. *et al.* (2018) Impairment of Lhca4, a subunit of LHCI, causes high accumulation of chlorophyll and the stay-green phenotype in rice. *Journal of Experimental Botany* 69, 1027–1035.

Yamazaki, H. *et al.* (2004) PPR motifs of the nucleus-encoded factor, PGR3, function in the selective and distinct steps of chloroplast gene expression in *Arabidopsis*. *Plant Journal* 38, 152–163.

Yang, Y. *et al.* (2011) *CLUMPED CHLOROPLASTS 1* is required for plastid separation in Arabidopsis. *Proceedings National Academy of Sciences* 108, 18530–18535.

Ye, K. *et al.* (2009) Pindel: a pattern growth approach to detect break points of large deletions and medium sized insertions from paired-end short reads. *Bioinformatics* 25, 2865–2871.

43 Using Ionizing Radiation for Improving the Development and Yield of Agricultural Crops

Stanislav Geras'kin[1], Roman Churyukin, Polina Volkova and Sofiya Bitarishvili

Abstract

The response of barley seedlings was studied after gamma irradiation of seeds with doses in the range of 2–50 Gy. It was shown that stimulation of plant growth occurred in the dose range of 16–20 Gy. The influences of the dose rate, the quality of seeds and their moisture on the manifestation of radiation effects were investigated. We studied, under controlled conditions, the activities of metabolic and antioxidant enzymes and observed an increase in their activity in the range of doses that cause stimulation of seedling growth. We showed that changes in the balance among different classes of phytohormones were probably involved in the acceleration of plant growth after irradiation of seeds using stimulating doses. Gamma irradiation of barley seeds significantly influenced the development of plants during the growing season. After irradiation with stimulating doses, we observed a reduction in the duration of the initial stages of ontogenesis; the phase of full ripeness occurred 5–7 days earlier than in the controls. The manifestation of the effect of irradiation depended on the conditions in which the plants developed. During the growing season of 2014, which was a dry year, plants originating from the irradiated seeds showed an increase in the number of productive stems, which led to an increase in yield by 34–38%; during the optimal 2015 season, an increase in the number of grains per spike caused an increase in yield by 8–29%. Therefore, our field study has shown that at least some hormetic effects can occur in the field. Irradiation of seeds can increase field germination, stimulate the growth and development of plants and increase their resistance to unfavourable environmental conditions. A more complete understanding of the underlying mechanisms of hormesis is needed to exploit its potential benefits in crop production.

Keywords: barley • germination • enzymes • phytohormones • hormesis

1 Introduction

Currently agriculture faces many challenges associated with global climate change and the gap between food demand and the availability of agricultural land. As a result, there is now growing understanding that 'business as usual' in agriculture will be insufficient to meet these challenges. This emphasizes the need for development of new technologies like radiation-induced growth stimulation, aimed to increase plant resistance to unfavourable conditions and yields.

It is well known that low doses of gamma-rays can stimulate cell division, growth and development in plants (Cedergreen *et al.*, 2007; Calabrese and Blain, 2009; Jan *et al.*, 2012). Although a conclusive explanation for the

[1] Email: stgeraskin@gmail.com

 Mutation Breeding, Genetic Diversity and Crop Adaptation to Climate Change (eds S. Sivasankar *et al.*)

DOI: 10.1079/9781789249095.0043

stimulation effects of gamma-rays has not become available yet, some research (Kim *et al.*, 2004; Jan *et al.*, 2012) supports a hypothesis that changes in enzyme activities, phytohormonal balance and an increase in the antioxidant capacity of cells are involved in this process. Several studies have shown improvement not only in root and shoot elongation, but also in biomass growth and harvested yield in plants exposed to different kinds of stress (Velini *et al.*, 2008; Cedergreen *et al.*, 2009). From an agricultural point of view, stimulatory effects on harvestable plant traits possess the potential for improving crop production and quality. However, a high reproducibility of effects is an essential prerequisite for implementing new agricultural practices. With this in mind, in this study an attempt was made to elucidate the radiation-induced changes at the early stages of plant development and to identify how the advantage obtained in the early stages of ontogenesis can be realized during the vegetation season under field conditions.

2 Materials and Methods

2.1 Irradiation of seeds

The spring barley (*Hordeum vulgare* L.) cultivars 'Nur' (first reproduction, elite, super elite) and 'Grace' (first reproduction) were used in our experiments. 'First reproduction' is the first generation of plants grown from the elite seeds; 'elite seeds' are the seeds obtained from the plants grown from the original seeds; and 'super elite seeds' are the seeds of the first reproduction step that are produced by the manufacturer of a cultivar. The seeds were irradiated with gamma-rays of ^{60}Co (GUR-120, RIRAE, Obninsk) at room temperature. The radiation doses were assessed using DKS-101 dosimeter (Politehphorm-M, Russia) and were confirmed using a thermoluminescent dosimeter. The following doses were applied: 0 (the control), 2, 4, 6, 8, 10, 13, 16, 20, 25 and 50 Gy at dose rates of 20, 60, and 350 Gy/h. Dry seeds with a moisture content of 13–15% were used for these experiments. A portion of the seeds was irradiated after increasing the moisture content to 40% by soaking in distilled water for 18 h.

2.2 Seed germination

Both irradiated and control seeds were placed in rolls of filter paper. Each roll was put in a container with distilled water (200 ml) and seedlings were grown in a thermostat (MIR-254, Sanyo, Japan) maintained at 20°C. A seed was considered as germinated when the radicle elongated to 2–3 mm. A germination rate (percentage of normally developed seedlings after the 7th day) and lengths of shoot and root were evaluated after the 7th day of germination. All experiments were performed in four replicates of 400 seeds for each dose.

2.3 Enzymatic activity analyses

Activities of enzymes (superoxide dismutase (EC 1.15.1.1, SOD), catalase (EC 1.11.1.6, CAT), guaiacol peroxidase (EC 1.11.1.7, GPOX), pyruvate kinase (EC 2.7.1.40, PK), glucose-6-phosphate dehydrogenase (EC 1.1.1.49, G6PD), malate dehydrogenase (EC 1.1.1.37, MDH) and shikimate dehydrogenase (EC 1.1.1.25, SKDH) in seedlings after 3, 5 and 7 days of germination were evaluated. The choice of this time range is based on the fact that the radiation-induced boost in cell division and seedling growth seems to be limited to a short time frame (about 6–7 days) after irradiation (Gudkov, 1991). The assessment of enzyme activities was carried out according to protocols detailed in Geras'kin *et al.* (2017).

After weighing, a sample of fresh plant material (two seedlings per sample) was homogenized in 1 ml of potassium phosphate buffer solution using a mortar and a pestle. The pH of buffer solution depended on the enzyme that was analysed in the moment. Only green parts of the seedlings (shoots) were used for enzyme extraction. The homogenates were centrifuged using MiniSpin Plus (Eppendorf, Germany) at 14,500 rpm for 10 min at 4°C. Supernatants were used for enzyme assays. Enzyme activities were evaluated using a 'NanoDrop-2000' (Thermo Fisher Scientific, USA) spectrophotometer in accordance with the recommendations of Bisswanger (2004). Thirty extracts were used for each enzyme assay for each experimental condition after 3, 5 and 7 days of germination (90 extracts for each enzyme in total). The measured activities were recalculated in a specific activity per gram of green shoots.

2.4 Quantitative analysis of phytohormones

The quantitative analysis of the selected phytohormones (indol-3-acetic acid (IAA), indolyl-3-butyric acid (IBA), zeatin, and abscisic acid (ABA)) was carried out from the 4th to 7th day of germination according to protocols detailed in Bitarishvili *et al.* (2018). The analysis of the phytohormonal extracts was carried out on a Nexera LC-30 high-performance liquid chromatograph (Shimadzu, Japan) coupled with an SPD-M20A diode array detector (Shimadzu). The analysis was performed in triplicate (three samples of plant material per one experimental point) and each sample was analysed in two technical replicates to eliminate instrumental errors.

2.5 Field study

Plants from irradiated (8, 16, 20 and 50 Gy at dose rate 60 Gy/h) and control seeds were sown in the same field. Four randomly distributed plots were used for each dose. The phase of ontogenesis was considered complete if it was manifested in no less than 75% of the plants. The crop was harvested at a stage of complete ripeness after 95–98, 103–106 and 99–100 days of growing, respectively, in 2014, 2015 and 2016. The following parameters were evaluated: the height of the plants, the number of stems (tillering capacity), the average number of productive stems per plant, the number of grains per ear, the average straw weight of 100 plants, the average weight of 100 plant ears and of 1000 seeds. The evaluation of parameters was carried out according to protocols detailed in Churyukin and Geras'kin (2017). At least 400 plants were analysed for each dose.

3 Results

3.1 Morphological and biochemical effects of seed irradiation

As the first step, an attempt was made to elucidate the radiation-induced changes at the early stages of plant development. To this end, in three independent experiments with two barley cultivars we investigated the response of seeds to irradiation within the dose range from 2 to 50 Gy. This range covers both stimulatory and inhibitory doses, which is essential to prove the possible existence of a hormetic effect. The root and shoot lengths exceeded the corresponding control level in the dose range from 10 to 20 Gy (Fig. 43.1). It should be stressed that in three independent experiments the maximum shoot and root lengths were observed at the same dose of 20 Gy. Therefore, this shape of curve should not be considered as accidental, but rather it reflects qualitative patterns of a barley plant's response to irradiation. Interestingly, the maximum manifestation of the hormetic effect on germination, growth and yield of durum wheat was observed also under irradiation of seeds with dose of 20 Gy (Saleh and El-Shoney, 1974; Melki and Marouani, 2010).

Conditions of exposure such as dose rate, seed quality, moisture content, etc. may have a significant influence on the growth and development of plants. Indeed, irradiation of seeds with either low or high dose rates showed no positive effect on seedling growth (Fig. 43.2). Excessive moisture content usually increases the sensitivity of seeds to irradiation (Gudkov, 1991). Indeed, irradiation of seeds with an enhanced moisture content, even using the optimal dose rate, resulted in a statistically significant suppression of both root and shoot growth.

To test the possible influence of seed quality on manifestation of the radiobiological effect, we compared the results of gamma irradiation of the first reproduction, elite and super-elite seeds. The control seeds of the first reproduction had significantly reduced root length when compared with the elite and super-elite. However, irradiation improved root growth of the first reproduction seeds to the level that is typical for the elite and super-elite seeds (Fig. 43.3). These findings imply that hormesis is more probable when seed quality is low. Thus, irradiation of seeds may stimulate plants to use their genetic potential up to the limit, but not to bypass it.

The observed morphological effects of seed irradiation raise a question about the biological sense and the mechanisms underlying their origin and maintenance. Stimulation of growth is impossible without changes in the operation of key enzymes. We studied the effect of gamma irradiation on the activity of enzymes that

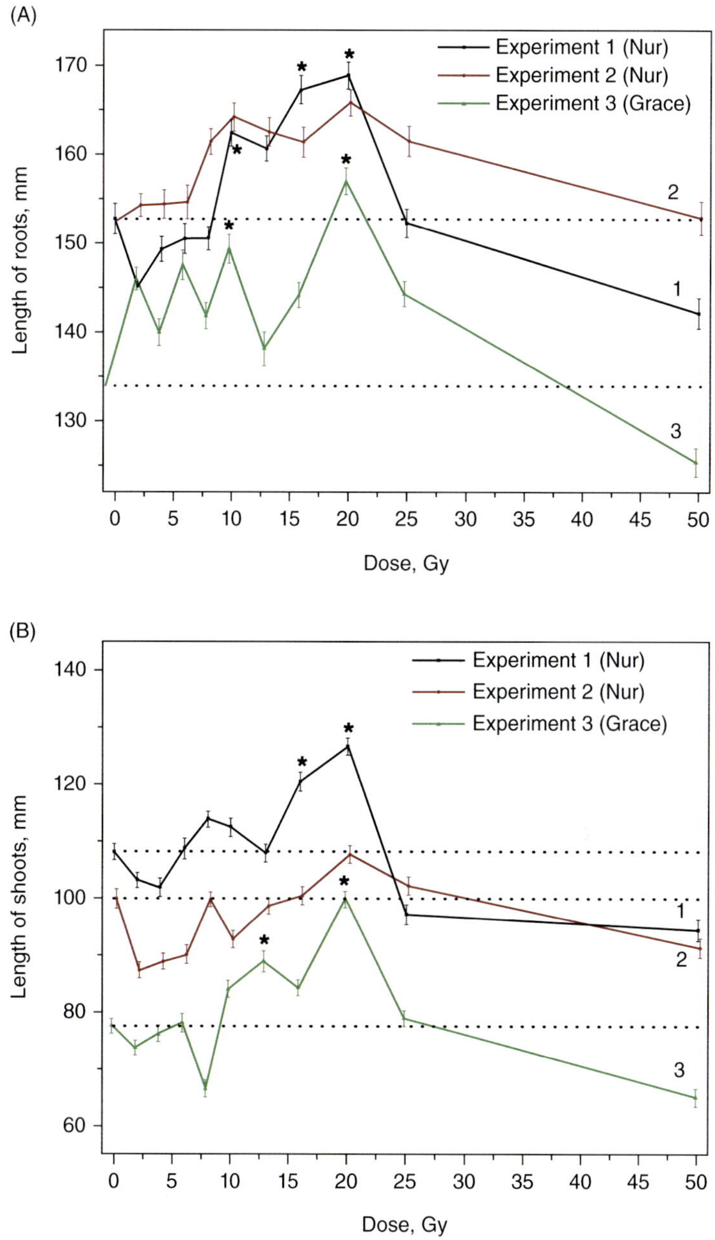

Fig. 43.1. (A) Length of roots and **(B)** length of shoots, depending on gamma-radiation dose (dose rate of 60 Gy/h). Means of four replicates of 400 seeds for each dose ± SE. Dotted line = mean values in control; *significant difference ($p < 0.05$) from the control level, *t*-test.

play key roles in the antioxidant system (SOD, CAT, GPOX), metabolism of glucose (PK), the Krebs cycle (MDH), the shikimate biosynthetic pathway (SKDH) and the oxidative pentose phosphate pathway (G6PD). These biochemical processes certainly contribute to the stimulation of seedling growth that we observed in the range of 16–20 Gy. The irradiated plants

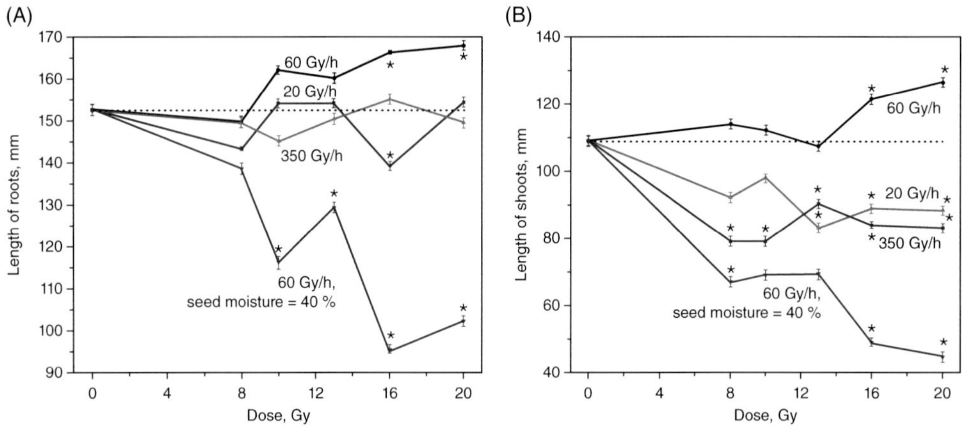

Fig. 43.2. (A) Length of roots and **(B)** length of shoots, depending on absorbed dose, dose rate and moisture content ('Nur' cultivar). Means of four replicates of 400 seeds for each dose ± SE. Dotted line = mean values in control; *significant difference ($p < 0.05$) from the control level, *t*-test.

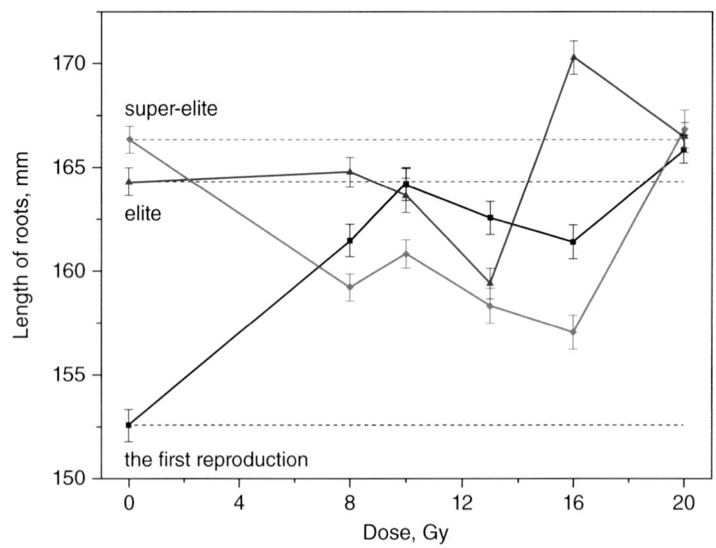

Fig. 43.3. Root lengths depending on seed quality when irradiated with a dose rate of 60 Gy/h ('Nur' cultivar). Means of four replicates of 400 seeds for each dose ± SE. Dotted line = mean values in control.

exhibited significant changes in enzymatic activity, depending on the dose and time after irradiation. In general, changes in the enzyme activities were in good agreement with the results of the morphological traits assessment. The majority of the enzymes studied increased their activity in the range of doses that caused stimulation of seedling development (Geras'kin *et al.*, 2017).

Growth and development of plants are under strict regulation by phytohormones. For this reason, we investigated the effect of gamma irradiation on the content of endogenous phytohormones (IAA, IBA, zeatin and ABA). After irradiation with stimulating doses, the balance of phytohormones shifted due to the increased content of growth stimulators and the decreased content of growth inhibitors (Fig. 43.4). In contrast,

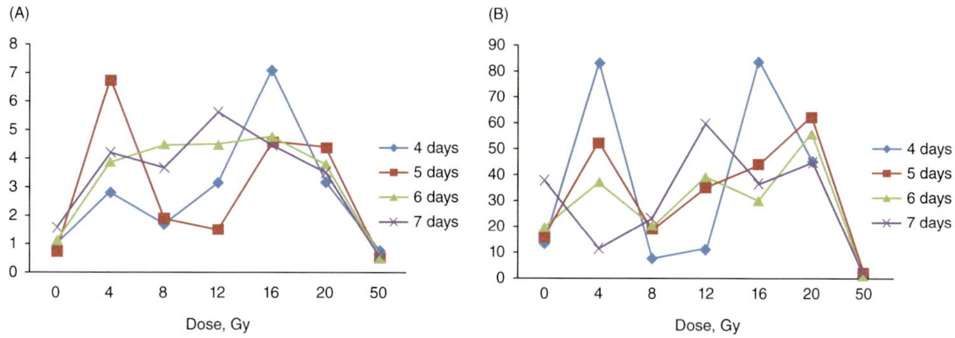

Fig. 43.4. Ratio (IAA + IBA + zeatin)/ABA, as represented by x-axis, from 4th to 7th days of germination in **(A)** shoots and **(B)** roots, depending on dose and time after germination.

the inhibitory dose of 50 Gy led primarily to a decrease in the content of growth stimulators and an increase in ABA content. Therefore, significant changes in the phytohormonal balance may be involved in the acceleration of plant growth.

3.2 Field studies

The early and simultaneous germination of irradiated seeds, as well as the increase in the lengths of root and shoot, give a significant advantage to plants in the field. Identification of the regimes of irradiation that create significant improvement in economically important traits might be used to enhance crop production. However, maintenance of the benefits obtained at the initial stages of ontogenesis until the end of the growing season depends on many factors. In our field experiments, it was shown that irradiation of seeds significantly influenced the development of plants throughout the vegetative period. As a result of the rapid passing of certain phases of ontogenesis in plants growing from irradiated seeds with stimulating doses, the phase of full ripeness for them is 5–7 days earlier than in the control (Fig. 43.5).

Moreover, barley plants growing from seeds irradiated with stimulating doses are characterized by enhanced values of economically valuable traits, including yield (Fig. 43.6). These findings indicate the positive influence of seed irradiation on the growth and development of agricultural plants under the conditions encountered in the field.

However, the effect of seed irradiation is constrained by field conditions that affect the general plant growth patterns. In drought conditions, plants growing from irradiated seeds have an advantage in obtaining nutrients and water from the soil due to a more developed root system in the early stages of ontogenesis; this will eventually affect the subsequent stages of development. From our field experiments it was shown (Churyukin and Geras'kin, 2017) that weather conditions can influence the results of pre-sowing irradiation in a different way. Indeed, in the dry 2014 season the enhanced yield was due to the increase in the number of productive stems, while under the optimal conditions of 2015, this was due to the increase in the number of grains per ear (Table 43.1).

4 Discussion

Germination of a seed is a complex process, during which external factors may activate development of the embryo. The study of growth stimulation after seed treatment with the genotoxic and reactive oxygen species (ROS)-producing agent (such as gamma-rays) will help to reveal new patterns that can be used for improvement of yield characteristics and stress resistance of important crops. The ability of seeds to germinate seems to be linked to the accumulation of a critical level of ROS, since seed imbibition entails a large increase in ROS content (Bailly *et al.*, 2008). The energy of ionizing radiation absorbed by seeds is converted mainly into ROS that can

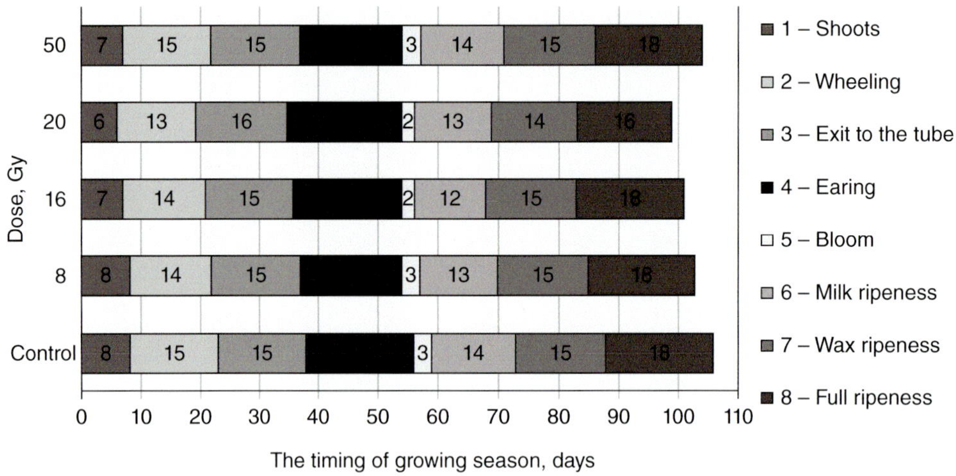

Fig. 43.5. Timing of growing season depending on dose.

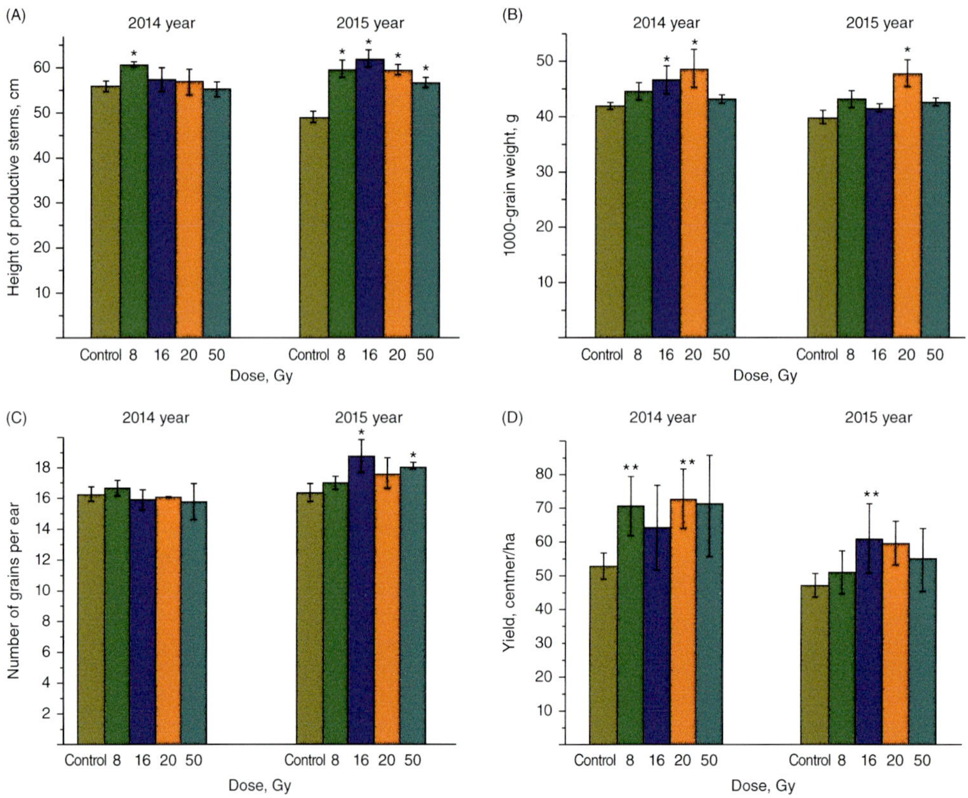

Fig. 43.6. **(A)** Height of productive stems, **(B)** 1000-grain weight, **(C)** number of grains per ear and **(D)** yield, depending on dose of gamma irradiation of barley seeds.
(* - significant difference ($p < 0.05$) from the control level, t-test; ** - significant difference ($p < 0.01$) from the control level, t-test).

Table 43.1. The effect of barley seed irradiation on yield structure under contrasting weather conditions.

	2014				2015			
	8 Gy	16 Gy	20 Gy	50 Gy	8 Gy	16 Gy	20 Gy	50 Gy
Length of stem	+8%				+18%	+20%	+18%	+15%
Weight of 1000 grains		+9%				+18%	+19%	
Weight of straw	+17%							
Weight of ears			+25%			+17%		
Number of stems		+17%	+17%	+35%				
Number of stems with ear			+12%	+25%				
Number of grains in the ear						+11%		+9%
Yield	+37%*		+38%*			+26%*		

* - significant difference from the control, p < 0.05%

exist for some time after irradiation in air-dried seeds; however, when water enters the seed, ROS become mobile and react rapidly with biologically important molecules. Reactive oxygen species can loosen the cell walls and are involved in endosperm weakening. This process is essential for overcoming the mechanical restraint provided by the seed-covering layers (Kumar *et al.*, 2015). These data provide evidence that ROS play an important role in seed dormancy release. However, overproduction of ROS can surmount the antioxidative capacity of cells and lead to oxidative stress, which may cause substantial damage to the embryo. This damage can manifest itself in the retardation of plant growth that we observed during the germination of seeds after exposure to the high dose rate (350 Gy/h) (Fig. 43.2). Therefore, the study of mechanisms of radiation-induced hormesis and its modifications by conditions of irradiation is an important step towards the creation of environmentally friendly and cost-effective technology in the pre-sowing treatment of seed crops.

The basis for the regulation of physiological processes in a plant is the balance of phytohormones, with particular importance of the ratio of growth hormone stimulators to inhibitors (IAA + IBA + zeatin)/ABA. The radiation doses of 16 and 20 Gy, which stimulated the growth of barley seedlings (Fig. 43.1), caused the maximal increase of the ratio of phytohormone growth stimulators to inhibitors (Fig. 43.4). When exposed to an inhibitory dose of 50 Gy, the balance was shifted towards ABA, because of an increase in its content and a decrease in the content of growth-stimulating hormones. Therefore, we can suggest that phytohormones are actively involved in the formation of adaptive reactions of barley plants in the early stages of ontogenesis. Significant changes in the phytohormonal balance create the necessary conditions for acceleration of plant growth and development. On the other hand, these differences may be a reflection of some other causative process. Overall, the effect of gamma irradiation on the growth of barley seedlings was estimated. The dose range and irradiation parameters that induced persistent stimulating effects were revealed, namely: (i) the dose range of apparent improvement, 10–20 Gy; (ii) the dose rate, 60 Gy/h; (iii) the moisture content, 13–15%; (iv) the imbibition immediately after irradiation; and (v) use of seeds of the first reproduction.

Any change in the growth pattern ultimately affects maturity and yield. Increased root growth in seedlings of seeds irradiated with stimulating doses was mainly observed during the first 4–6 days after germination. By this time, for control seedlings, the root length of pea and corn increased by 4–5 cm as compared with the control (Gudkov, 1991). Shoot growth is stimulated to the same extent. As a result, the passage of the first phases of ontogenesis is accelerated and the time of ripening is reduced, which, under favourable conditions, can lead to increased yield and improved quality. The early and concerted germination of irradiated seeds, as well as the increase in the length of root, give a significant advantage to plants in arid climates (Saleh and El-Shoney, 1974). Therefore, the exposure to gamma radiation may have stimulatory effects on plant growth and yield. However, retention of the benefits obtained in the initial stages of ontogenesis until the end of the growing season depends on many factors (Belz and Cedergreen, 2010). The insufficient predictability of the hormesis response is the main obstacle to the widespread use of this technique in agriculture (Sheppard and Hawkins, 1990; Kozmin *et al.*, 2015). Nevertheless, an example of the practical use of glyphosate hormesis in sugarcane (Belz and Duke, 2014) shows that at least some hormetic effects can occur in the field with sufficient regularity and predictability for practical use. A more complete understanding of the underlying mechanisms of hormesis is needed to exploit its potential benefits in crop production.

5 Conclusion

Our findings have shown that irradiation of seeds of agricultural crops can stimulate plant growth and yield in the field. However, the insufficient predictability of these effects is the main obstacle to the widespread use of this technique in agriculture. Future research will show which hormetic approaches can be efficiently and reasonably integrated in new crop production systems and which will remain only of academic interest.

References

Bailly, C. *et al.* (2008) From intracellular signalling networks to cell death: the dual role of reactive oxygen species in seed physiology. *Comptes Rendus Biologies* 331, 806–814.

Belz, R.G. and Cedergreen, N. (2010) Parthenin hormesis in plants depends on growth conditions. *Environmental and Experimental Botany* 69, 293–301.

Belz, R.G. and Duke, S.O. (2014) Herbicides and plant hormesis. *Pest Management Science* 70, 698–707.

Bisswanger, H. (2004) *Practical Enzymology*. Wiley-VCH Verlag GmbH & KGaA, Weinheim.

Bitarishvili, S. *et al.* (2018) γ-Irradiation of barley seeds and its effect on the phytohormonal status of seedlings. *Russian Journal of Plant Physiology* 65, 446–454.

Calabrese, E.J. and Blain, R.B. (2009) Hormesis and plant biology. *Environmental Pollution* 157, 42–48.

Cedergreen, N. *et al.* (2007) The occurrence of hormesis in plants and algae. *Dose-Response* 5, 150–162.

Cedergreen, N. *et al.* (2009) Chemical stress can increase crop yield. *Field Crops Research* 114, 54–57.

Churyukin, R.S. and Geras'kin, S.A. (2017) Hormesis in barley (*Hordeum vulgare* L.) plants derived from γ-irradiated seeds under contrasting weather conditions. *Agricultural Biology* 52, 820–829.

Geras'kin, S. *et al.* (2017) Radiation exposure of barley seeds can modify the early stages of plants' development. *Journal of Environmental Radioactivity* 177, 71–83.

Gudkov, I.N. (1991) *Fundamentals of General and Agricultural Radiobiology*. USXA, Kiev [in Russian].

Jan, S. *et al.* (2012) Effect of gamma radiation on morphological, biochemical, and physiological aspects of plants. *Environmental Reviews* 20, 17–39.

Kim, J.H. *et al.* (2004) Alterations in the photosynthetic pigments and antioxidant machineries of red pepper (*Capsicum annuum* L.) seedlings from gamma-irradiated seeds. *Journal of Plant Biology* 47, 314–321.

Kozmin, G.V. *et al.* (ed.) (2015) *Radiation technologies in the agriculture and food industry*. RIRAE, Obninsk [in Russian].

Kumar, S.P.J. *et al.* (2015) Seed birth to death: dual functions of reactive oxygen species in seed physiology. *Annals of Botany* 116, 663–668.

Melki, M. and Marouani, A. (2010) Effects of gamma rays irradiation on seed germination and growth of hard wheat. *Environmental Chemistry Letters* 8, 307–310.

Saleh, S. and El-Shoney, K. (1974) Effect of grain irradiation with gamma rays on growth, nitrogen content and yield of wheat plants. *Egypt Journal of Botany* 17, 27–33.

Sheppard, S.C. and Hawkins, J.L. (1990) Radiation hormesis of seedlings and seeds, simply elusive or an artefact? *Environmental Experimental Botany* 30, 17–25.

Velini, E.D. *et al.* (2008) Glyphosate applied at low doses can stimulate plant growth. *Pest Management Science* 64, 489–496.

44 Impact of Cross-breeding on the Metabolites of the *Low Phytic Acid* Rice Mutant *Os-lpa*-MH86-1

Chenguang Zhou, Yuanyuan Tan, Sophia Goßner, Youfa Li, Qingyao Shu, and Karl-Heinz Engel[1]

Abstract

Phytic acid (*myo*-inositol-1,2,3,4,5,6-hexa*kis*phosphate), the major storage form of phosphorus in cereals, is considered as an antinutrient in food and feed. During the past few years, various cereals have been subjected to mutation breeding for generating *low phytic acid* (*lpa*) crops. Recently, it was demonstrated that reduction of phytic acid in the rice mutant *Os-lpa*-MH86-1 obtained by gamma irradiation was due to a disruption of *OsSULTR3;3*, an orthologue of the sulfate transporter family group 3 genes. The application of a GC/MS-based metabolite profiling approach revealed that the reduction of phytic acid was accompanied by changes in concentrations of metabolites from different classes in the *Os-lpa*-MH86-1 mutant.

Lpa mutant lines often exhibit lower grain yield and seed viability compared with their wild-type parents. To improve the agronomic performance of the *Os-lpa*-MH86-1 mutant, cross-breeding with a commercial cultivar was performed. The resulting progenies were genotyped using molecular markers to identify homozygous wild-type and *lpa* mutants from generations F_4 to F_7. The objectives of this study were: (i) to observe the consistent metabolic changes in *Os-lpa*-MH86-1 *lpa* mutants by following their composition over several independent field trials; (ii) to investigate the impact of cross-breeding on the phytic acid content and the metabolic phenotype of the homozygous *lpa* mutant; and (iii) to assess the stability of the mutation-specific metabolite signature in the *lpa* progenies over several generations.

Statistical assessment of the data via multivariate and univariate approaches demonstrated that the *lpa* trait and the mutation-induced metabolite signature in the *lpa* progenies were comparable to the progenitor *Os-lpa*-MH86-1 mutant and consistently expressed over generations.

These findings extend the basis for implementing mutation breeding in the generation of *lpa* rice cultivars.

Keywords: rice (*Oryza sativa* L.) • *low phytic acid* (*lpa*) mutant • metabolite profiling • cross-breeding • *OsSULTR3;3*

1 Introduction

Phytic acid (*myo*-inositol-1,2,3,4,5,6-hexa*kis*-phosphate) (PA) constitutes the predominant storage form of phosphorus (P) in cereal grains and legume seeds (Raboy, 2003). PA is considered as an antinutrient since it limits the bioavailability of nutritionally relevant minerals for humans (Kumar *et al.*, 2010). Moreover, excreted phytate in manure from monogastric animals may lead

[1] Email: k.h.engel@wzw.tum.de

 Mutation Breeding, Genetic Diversity and Crop Adaptation to Climate Change (eds S. Sivasankar *et al.*)
DOI: 10.1079/9781789249095.0044

to eutrophication of water (Abdel-Megeed *et al.*, 2015). Therefore, during the past decade various cereals such as maize, barley, wheat, rice and the legume soybean have been subjected to mutation breeding to generate *low phytic acid* (*lpa*) crops (Raboy, 2009).

Induced mutations in *lpa* mutants have been shown to result not only in decreased contents of PA, but also in changed levels of the other metabolites related to the biosynthesis of phytic acid. In *lpa* rice seeds, increased levels of *myo*-inositol, galactose and raffinose compared with the corresponding wild-type rice seeds have been observed (Frank *et al.*, 2007). Application of metabolite profiling also revealed consistently changed levels of sucrose, stachyose and galactosyl cyclitols in *lpa* soybean mutants (Frank *et al.*, 2009). The *lpa* rice mutant *Os-lpa*-MH86-1 was generated from the wild-type cultivar MH86 by gamma-irradiation (Liu *et al.*, 2007). Later, it was demonstrated that this treatment resulted in a 1 bp deletion in the putative sulfate transporter gene in MH86. The consequences of this mutation were, on the one hand, the reduction of the content of phytic acid by approximately 44% and, on the other hand, significant changes in the concentrations of a broad spectrum of nutritionally relevant metabolites, such as a decrease of cysteine and increases of others such as amino acids, sugars, sugar alcohols, phytosterols and γ-aminobutyric acid (Zhao *et al.*, 2016).

The *lpa* mutant crops often exhibit inferior agronomic parameters such as low field emergence and seed-setting ratios, and cross- and selection breeding with elite wild-type cultivars is a practical way to improve the agronomic performance of *lpa* crops. For example, after crossing *lpa* soybean mutant lines with commercial cultivars, the generated *lpa* progenies exhibited remarkably increased levels of seedling emergence (Spear *et al.*, 2007; Boehm *et al.*, 2017). However, information regarding the impact of such cross-breeding of *lpa* mutants on the metabolite profiles of the resulting progenies is still missing.

To this end, the *lpa* rice mutant *Os-lpa*-MH86-1 was used as an example: (i) to observe the consistent metabolic changes in *lpa* mutants by following their compositions over several independent field trials; (ii) to investigate the impact of cross-breeding of the *lpa* mutant with a commercial cultivar on the metabolite profiles of the resulting *lpa* mutant progenies; and (iii) to assess the stability of the mutation-specific metabolite signature in the *lpa* progenies over several generations.

2 Materials and Methods

2.1 Sample materials

The wild-type MH86 was subjected to gamma irradiation to develop the progenitor *lpa* mutant *Os-lpa*-MH86-1 as previously described (Liu *et al.*, 2007). Both MH86 and the *lpa* mutant *Os-lpa*-MH86-1 were grown in four independent field trials in China over three consecutive years: Hangzhou in 2011, Hainan in 2012 and 2013 and Jiaxing in 2013 (Fig. 44.1A). The commercial rice cultivar JH99 was employed for cross-breeding with *Os-lpa*-MH86-1. The bulk-harvested F_2 seeds were grown into F_2 plants and high-resolution melting curve analysis was performed to genotype the F_2 plants into homozygous wild-types, homozygous *lpa* mutants and heterozygous progenies, as previously described (Tan *et al.*, 2013). The rice seeds from homozygous wild-type and homozygous *lpa* mutant F_2 plants were reserved until 2013 for cultivation, and those from heterozygous F_2 plants were grown into F_3 plants followed by genotyping procedure as conducted for F_2 plants. Similar selection and breeding steps were conducted until F_5. The reserved seeds harvested from homozygous F_2 to F_5, as well as two parental lines, were grown in Jiaxing in 2013 (Jiaxing 2013) to generate rice samples for PA content analysis and metabolite profiling (Fig. 44.1B).

Rough rice grains were hulled and ground into rice flour by a cyclone mill with a 500 μm sieve. The rice flour was freeze-dried for 48 h and then stored at –20°C until analysis.

2.2 PA content analysis

PA contents of the rice samples were determined using high-pressure ion chromatography (HPIC) as previously described for dried distiller's grains (Oates *et al.*, 2014). Three aliquots (100 mg) of

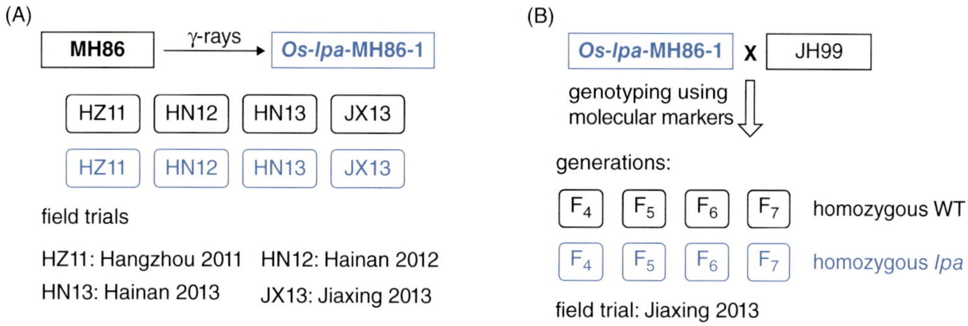

Fig. 44.1. Overview of rice sample investigations. **(A)** MH86 and the progenitor *lpa* mutant *Os-lpa*-MH86-1 harvested from four independent field trials. **(B)** The homozygous wild-type and the homozygous *lpa* mutant progenies generated via cross- and selection breeding.

rice flour were extracted with 20 ml of 0.5 M HCl in a sonication bath for 20 min. After centrifugation at 2000 ×*g* for 20 min, the supernatants were filtered using 0.22 μm polyether sulfone filters, and 100 μl were subjected to HPIC.

A Thermo Scientific Dionex (Dreieich, Germany) ICS-5000 HPIC system was utilized to determine the PA contents. The mobile phases were deionized water and 0.5 M HCl. The flow rate of the column was set to 1 ml/min. An HCl gradient elution (5% increased to 100% within 4 min and held for 10 min) was followed by post-column derivatization with ferric nitrate solution at 0.4 ml/min. Peak detection was performed at 290 nm. Quantification of phytic acid was based on an external standard calibration curve ($R^2 = 0.9997$) considering a recovery rate of 98.8%. The limit of detection and the limit of quantification were 0.7 mg/l and 2.2 mg/l, respectively.

2.3 Metabolite profiling

Extraction and fractionation of the rice flour were performed in accordance with previously described procedures (Frank *et al.*, 2007). Briefly, the rice flour was extracted with different solvents, resulting in lipophilic and polar extracts. Lipids were separated by solid-phase extraction into fraction I, containing fatty acid methyl esters (FAME) and hydrocarbons, and fraction II, containing minor lipids (free fatty acids, fatty alcohols, sterols). Fraction III (sugars and sugar alcohols) and fraction IV (acids, amino acids and amines) were separated from the polar extracts by selective hydrolysis of silylated derivatives. The four fractions were analysed by gas chromatography (GC) coupled with a flame ionization detector (FID) and a mass spectrometer (MS) under the conditions previously described (Frank *et al.*, 2007).

Rice constituents were identified by comparing retention times and mass spectra with those of reference compounds, with data from the mass spectra library NIST08 and the literature (Kamaleldin *et al.*, 1992; Xu *et al.*, 1999). The amounts of identified metabolites from fractions I–IV were expressed as relative peak intensity (i.e. (metabolite peak intensity)/(internal standard peak intensity) × 100) based on the internal standards from each fraction.

2.4 Data analysis

For each rice sample, peak heights and retention times generated from GC-FID/MS were exported to Chrom*pare* 1.1 (http://www.chrompare.com, accessed 2019) (Frenzel *et al.*, 2003) for standardization and consolidation. After data pre-treatment of transformation and scaling, principal component analysis (PCA), partial least squares–discriminant analysis (PLS-DA) and heat map analysis were performed by XLSTAT (version 19.5, France). The PLS-DA model was validated by a sevenfold cross-validation and a 1000-times permutation test. Student's *t*-test ($p < 0.05$) was performed for single metabolites that exhibited a normal distribution and homogeneity of variance; otherwise the non-parametric Mann-Whitney test ($p < 0.05$) was performed.

3 Results

3.1 PA contents in MH86 and *Os-lpa*-MH86-1 mutant

The PA contents determined in the wild-type MH86 and in the progenitor *lpa* mutant *Os-lpa*-MH86-1 harvested from four independent field trials are shown in Table 44.1. For each field trial, the PA content in the *lpa* mutant *Os-lpa*-MH86-1 exhibited a significantly lower level than the corresponding wild-type MH86. The pronounced percentage reductions of PA were comparable to those previously reported for the *lpa* mutant of this mutation type (Liu *et al.*, 2007).

3.2 PA contents in progenies of cross *Os-lpa*-MH86-1 with JH99

As shown in Table 44.2, the PA content in the homozygous F_4 *lpa* progeny (4.54 ± 0.05 mg/g) was significantly lower than that in the progenitor *Os-lpa*-MH86-1 mutant grown at the same field trial in Jiaxing 2013 (Table 44.1). However, the reduction percentage (–36.3%) was comparable to that for the *Os-lpa*-MH86-1 mutant (–35.3%), owing to the lower PA content in the homozygous F_4 wild-type progeny. For the homozygous *lpa* progenies of each generation from F_5 to F_7, the PA contents were all significantly lower than those observed for the corresponding homozygous wild-type progenies, with reduction percentages ranging from –30.2% to –37.5%. In addition, there were no consistent changes of PA contents in homozygous *lpa* progenies with progressing generations.

3.3 Multivariate analysis of rice seed metabolite profiles of MH86 and *Os-lpa*-MH86-1 mutants

The PCA score plots obtained for the combined non-polar fractions I–II and the combined polar fractions III–IV are shown in Fig. 44.2. For the non-polar fractions I–II (Fig. 44.2A), for the rice samples from the field trial Jiaxing 2013, a slight separation between the wild-types and the *lpa* mutants was observed. However, no comparable separations were observed for the other field trials. In contrast, there was a clear clustering of

Table 44.1. PA contents (mg/g) in the wild-type MH86 and in the progenitor *lpa* mutant *Os-lpa*-MH86-1.[a]

Field trial	MH86	*Os-lpa*-MH86-1	Decrease (% change)[b]
Hangzhou 2011	11.36 ± 0.13	6.61 ± 0.07	–41.8*
Hainan 2012	11.60 ± 0.06	6.72 ± 0.16	–42.1*
Hainan 2013	11.72 ± 0.07	6.57 ± 0.04	–43.9*
Jiaxing 2013	9.69 ± 0.20	6.27 ± 0.05	–35.3*

[a]Values represent means ± standard deviations resulting from the analysis of three aliquots of freeze-dried flour.
[b]Asterisks indicate statistically significant differences (Student's *t*-test, $p < 0.05$) between the *lpa* mutant and the corresponding wild-type.

Table 44.2. PA contents (mg/g dry matter) in homozygous wild-type and *lpa* mutant progenies of generations F_4–F_7.[a]

Generation	Homozygous wild-type	Homozygous *lpa* mutant	Decrease (% change)[b]
F_4	7.13 ± 0.03	4.54 ± 0.05	–36.3*
F_5	9.79 ± 0.03	6.12 ± 0.01	–37.5*
F_6	8.05 ± 0.10	5.11 ± 0.03	–30.2*
F_7	9.27 ± 0.07	6.08 ± 0.06	–34.4*

[a]Values represent means ± standard deviations resulting from the analysis of three aliquots of freeze-dried flour.
[b]Asterisks indicate statistically significant differences (Student's *t*-test, $p < 0.05$) between the *lpa* mutant and the corresponding wild-type.

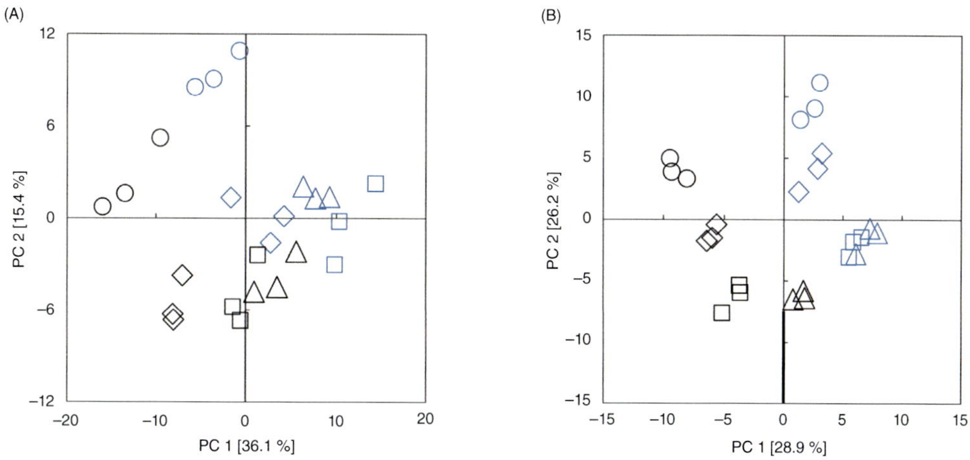

Fig. 44.2. PCA score plots of metabolite profiling data of **(A)** combined non-polar fractions I–II and **(B)** combined polar fractions III–IV of original wild-type MH86 (black) and *Os-lpa*-MH86-1 mutants (blue) from the four field trials: Hangzhou, 2011 (open diamonds); Hainan, 2012 (open triangles); Hainan, 2013 (open squares); and Jiaxing, 2013 (open circles).

the wild-types and the *lpa* mutants in the score plot of the polar fractions III–IV (Fig. 44.2B). The metabolic shift of the *Os-lpa*-MH86-1 mutant compared with the corresponding wild-type MH86 remained nearly comparable for each field trial. These results indicated that the *OsSULTR3;3* mutation resulted in more significant metabolic changes in the polar fractions than in the non-polar fractions of the *lpa* rice mutant seeds.

Individual constituents contributing to the discrimination of the metabolite profiles of the wild-types and the *lpa* mutants are summarized in the heat map shown in Fig. 44.3. The mutation resulted in changed levels of a broad spectrum of low molecular constituents, e.g. reduced level of cysteine, and increased contents of various amino acids, organic acids and other nutritionally relevant compounds, such as γ-aminobutyric acid (GABA). These changes were consistently expressed independently from the field trials.

3.4 Multivariate analysis of rice seed metabolite profiles of homozygous progenies

For the PCA score plot based on the combined non-polar fractions I–II (Fig. 44.4A), the homozygous wild-type progenies of generation F_4 clustered close to the original wild-type, and the homozygous *lpa* progenies close to the progenitor *lpa* mutant. In addition, there were metabolic shifts along PC1 and PC2 for the progenies of generations F_5 to F_7 when compared with generation F_4. However, no clear separation was found between the homozygous wild-type and the homozygous *lpa* progenies.

In the PCA plot of the combined polar fractions III–IV (Fig. 44.4B), the *Os-lpa*-MH86-1 mutant and the F_4 *lpa* progenies exhibited a clear separation from the wild-type MH86 and the F_4 wild-type progenies, which indicated that the mutation-induced metabolic changes in homozygous *lpa* mutant progenies were almost unaffected by the cross-breeding step. This signature was mainly attributed to the polar rice constituents. In addition, this separation between homozygous wild-type and homozygous *lpa* progenies was consistently expressed for the progressing generations F_5 to F_7, demonstrating that the mutation-specific metabolite signature in homozygous *lpa* progenies was stable over generations.

3.5 Metabolic differences between homozygous wild-type and homozygous *lpa* mutant progenies

The PCA loading plot was employed to identify those individual metabolites that mainly contributed to the observed clustering of wild-type

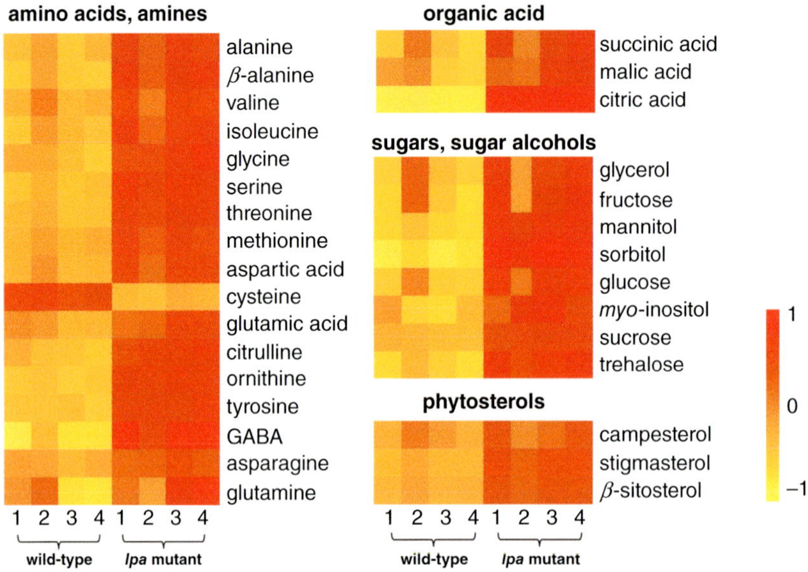

Fig. 44.3. Heat map of constituents contributing to the separation of the metabolite profiles of the wild-type MH86 and the *lpa* mutants *Os-lpa*-MH86-1 from four field trials. Lanes 1–4 represent Hangzhou 2011, Hainan 2012, Hainan 2013 and Jiaxing 2013, respectively.

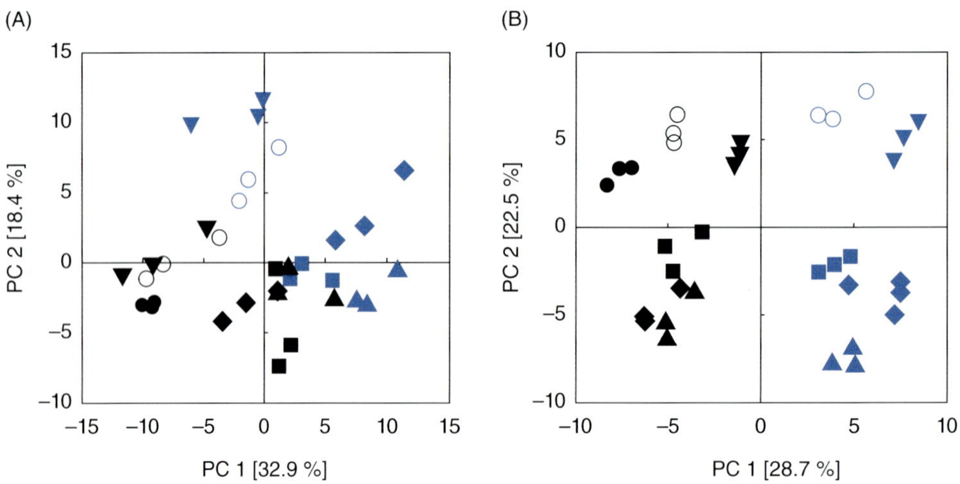

Fig. 44.4. PCA score plots of metabolite profiling data of **(A)** combined non-polar fractions I-II and **(B)** combined polar fractions III-IV of wild-type (black) and *lpa* mutant (blue) rice seeds: original wild-type MH86 (open circles) and *Os-lpa*-MH86-1 mutant (open circles); crossing parent JH99 (black solid circles); homozygous progenies of generations F_4 (inverted triangles), F_5 (diamonds), F_6 (squares) and F_7 (triangles).

and *lpa* mutants (Fig. 44.4B). These metabolites were quantitated based on their relative peak intensities and mapped in a simplified biosynthetic pathway according to the KEGG pathway database (Kanehisa *et al.*, 2000). Fig. 44.5 shows comparisons of the contents of the metabolites between: (i) the original wild-type MH86 and the progenitor *lpa* mutant *Os-lpa*-MH86-1; and

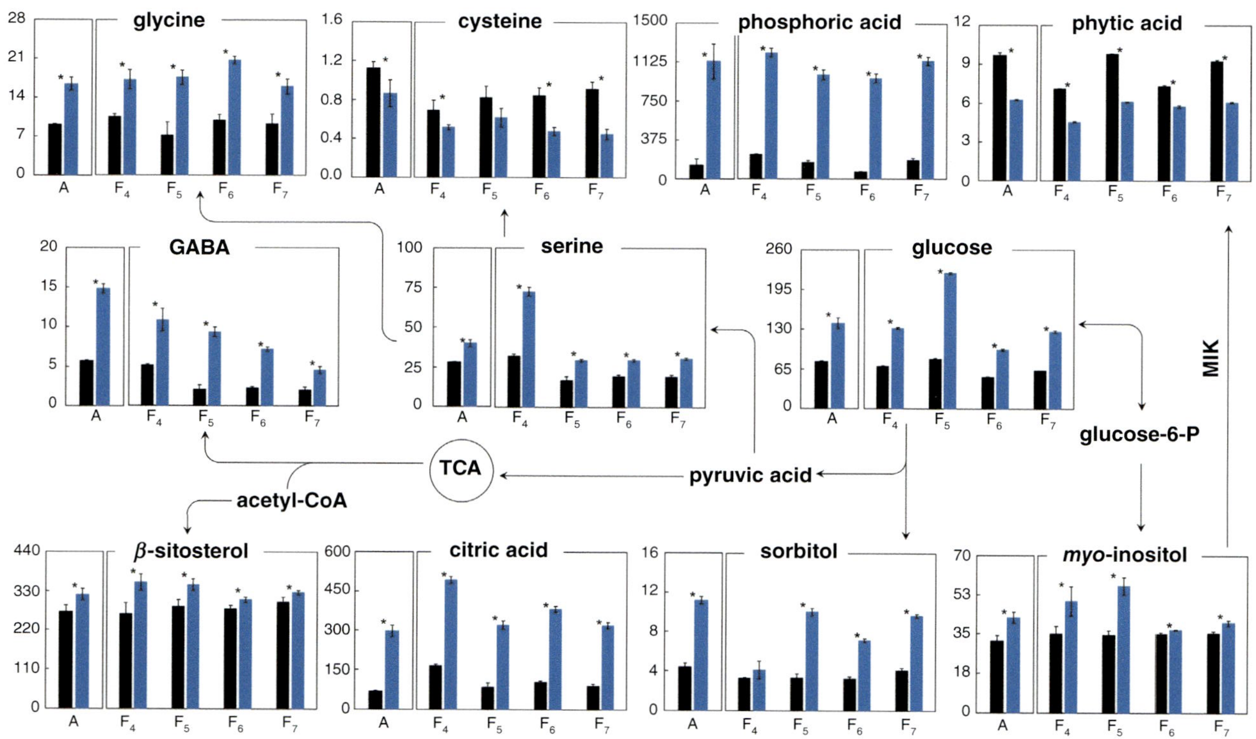

Fig. 44.5. Simplified biosynthetic pathway of selected rice seed metabolites involved in phytic acid, sugar, amino acid and phytosterol metabolism. The bars are displayed in the following order: **(A)** original wild-type MH86 (black) and progenitor *Os-lpa*-MH86-1 mutant (blue); homozygous wild-type progenies (black) and homozygous *lpa* mutant progenies (blue) of generations F_4, F_5, F_6 and F_7. The metabolites are expressed as relative peak intensities, i.e. metabolite peak intensity/internal standard peak intensity × 100. Asterisks represent statistically significant differences (Student's *t*-test or Mann-Whitney test, $p < 0.05$) between the wild-type and the corresponding *lpa* mutant. *MIK*, *myo*-inositol kinase; TCA, tricarboxylic acid cycle.

(ii) the homozygous wild-type and the homozygous *lpa* progenies of generations F_4 to F_7.

As shown in Fig. 44.5, the mutation-induced metabolite signature of the homozygous *lpa* progenies was determined not only by the remarkable reduction of the PA content but also by increases and/or decreases in the levels of the other metabolites, for example reduced contents of cysteine and increased concentrations of glucose, sorbitol, citric acid, serine and GABA. Despite variation in the absolute levels of these metabolites depending on the generations, statistically significant differences between the homozygous wild-type and the homozygous *lpa* progenies were consistently observed for the generations F_4 to F_7 (Fig. 44.5).

3.6 Supervised statistical analysis of the rice seed metabolite profiles

The results of PCA (Fig. 44.4) and univariate analyses (Fig. 44.5) strongly indicated that the homozygous *lpa* mutant progenies (F_4 to F_7) resulting from cross-breeding exhibited significantly different metabolite profiles compared with the corresponding homozygous wild-type progenies. In order to extend the validity of the conclusion that the metabolite signature of the progenitor *lpa* mutant is not changed owing to the cross-breeding step, the metabolite profiling data for MH86 and *Os-lpa*-MH86-1 from three other field trials (Fig. 44.2) were taken into account for further investigation.

The PLS-DA score plot (Fig. 44.6) displayed a clear separation between wild-type MH86 and homozygous wild-type progenies resulting from the cross-breeding on the one hand, and progenitor *lpa* mutant *Os-lpa*-MH86-1 and the homozygous *lpa* mutant progenies on the other hand. The R^2 (92.3%) and Q^2 (87.0%) values of the model parameters demonstrated the good explanatory and predictive ability, respectively. The results of permutation tests suggested a robust model without overfitting.

The metabolites contributing to the observed discrimination between wild-types and *lpa* mutants (Fig. 44.6) were determined based

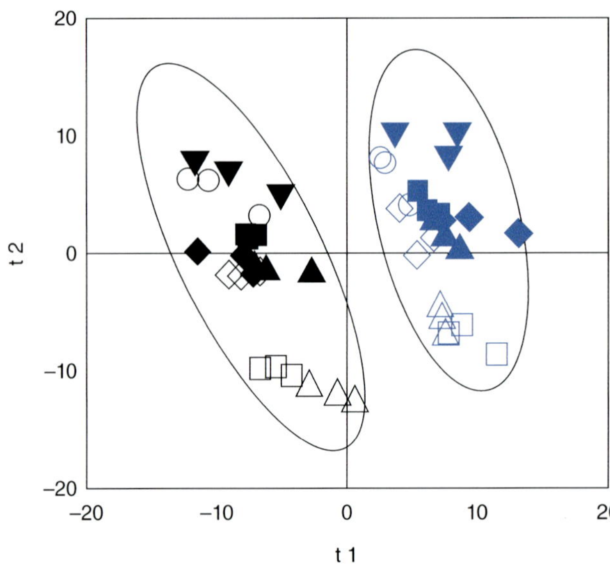

Fig. 44.6. PLS-DA score plot of metabolite profiling data. Original wild-type MH86 (black) and *Os-lpa*-MH86-1 mutants (blue) from the four field trials Hangzhou, 2011 (open diamonds); Hainan, 2012 (open triangles); Hainan, 2013 (open squares) and Jiaxing, 2013 (open circles); homozygous wild-type progenies (solid black) and homozygous *lpa* mutant progenies (solid blue) of generations F_4 (inverted triangles), F_5 (diamonds), F_6 (squares) and F_7 (triangles). The boundaries of the clusters of wild-type and *lpa* mutant seeds correspond to the 95% Hotelling's T2 ellipses.

on the variable importance in projection (VIP) values (threshold > 1) of the PLS-DA model (Mehmood *et al.*, 2012). For the main metabolites responsible for the separation of wild-type and *lpa* mutants listed in Table 44.3, univariate analyses demonstrated that, except for cysteine, all other metabolites exhibited statistically increased levels in the *lpa* mutant compared with wild-type rice seeds.

4 Discussion

After cross-breeding of the progenitor *lpa* mutant *Os-lpa*-MH86-1 with the commercial cultivar JH99, the PA contents of the resulting homozygous *lpa* progenies were significantly lower than those of the corresponding wild-type progenies as well as of the original wild-type MH86. This demonstrated that cross- and selection breeding of the *lpa* mutant *Os-lpa*-MH86-1 with a commercial cultivar did not hamper the mutation-induced *lpa* trait, and that this phenotype was stably observed in homozygous *lpa* progenies over several generations (F_4 to F_7). This is a key result demonstrating that a major requirement for the production of *lpa* rice cultivars via cross- and selection breeding is being met.

In *Os-lpa*-MH86-1, a 1 bp deletion in a putative sulfate transporter gene (*OsSULTR3;3*) was shown to be responsible for the *lpa* phenotype. The disruption of the gene *OsSULTR3;3* is thought to directly or indirectly play a role in the cross-talk between sulfate and phosphate homeostasis and/or signalling, since the expression of several genes involved in grain phosphorus and sulfur homeostasis and metabolism were up- and/or downregulated (Zhao *et al.*, 2016). As a result, the metabolite profile of *Os-lpa*-MH86-1 mutant comprised changed concentrations of a broad spectrum of low molecular grain constituents, which is very similar to that reported in plants suffering from sulfate deficiency, such as reduction of cysteine content and increased level of its precursor, serine, and increased concentrations of sugars, sugar alcohols and GABA (Nikiforova *et al.*, 2006; Hammond *et al.*, 2008).

The metabolic differences between the *lpa* mutant *Os-lpa*-MH86-1 and the wild-type MH86 were consistently observed in the different field trials over 3 years (Figs 44.2 and 44.3), indicating a stably expressed metabolite signature in *lpa* mutants independent from the environmental impact. The investigation of the metabolite profiles of the homozygous wild-type and the homozygous *lpa* mutant progenies demonstrated that the mutation-specific metabolite signature was not hampered by the cross-breeding step with the commercial cultivar and was also consistently expressed in homozygous *lpa* progenies over generations (Figs 44.4 and 44.5).

The supervised statistical analysis via PLS-DA demonstrated that it is possible to classify the homozygous progenies into the respective wild-type and *lpa* phenotypes (Fig. 44.6). The metabolite profiles of the homozygous *lpa* progenies of generations F_4 to F_7 were differentiated from that of the original wild-type MH86, on the one hand, but were still comparable to that of the progenitor *lpa* mutant *Os-lpa*-MH86-1, on the other hand. This indicated that the cross-breeding step and generation-dependent variations did not interfere with the pronounced discrimination of the metabolite profiles of wild-types and *lpa* mutants.

The metabolite signature determined in *lpa* mutant *Os-lpa*-MH86-1 and in homozygous *lpa* progenies was attributed to a number of discriminating metabolites shown in Table 44.3. Except for a reduction of the cysteine content in the *lpa* mutant, increased levels were observed for all other metabolites, among which were nutritionally relevant compounds, such as essential amino acids, phytosterols and the well-known biogenic amine GABA.

So far, the mechanism underlying the disruption of the sulfate transporter gene *OsSULTR3;3* in the *lpa* mutant has not been fully elucidated. However, the mutation-induced metabolite signature was shown to be maintained after cross- and selection breeding of *Os-lpa*-MH86-1 with a commercial cultivar. This result provides encouraging information for breeders, because it indicates that cross-breeding might be a practical strategy to obtain valuable *lpa* crops with specific traits induced by mutation.

5 Conclusion

The disruption of the *OsSULTR3;3* gene resulted not only in a pronounced reduction of the concentration of phytic acid, but also in significant

Table 44.3. Metabolites contributing to the PLS-DA separation of wild-type and *lpa* mutant rice seeds.[a]

Metabolite	Wild-type			*lpa* mutant			VIP[b]	Fold change (*lpa*/WT)	*p* value[c]
Phosphoric acid	154	±	52	1209	±	191	4.17	7.87	3.90E-26
Citric acid	104	±	31	383	±	77	3.26	3.67	1.31E-21
GABA	4.7	±	2.7	11.8	±	5.4	2.39	2.51	5.42E-7
Mannitol	16.8	±	3.0	33.8	±	8.8	2.20	2.01	1.77E-12
Sorbitol	6.1	±	3.7	15.7	±	10.4	2.10	2.59	5.72E-5
Ornithine	1.0	±	0.3	1.8	±	0.3	1.99	1.81	7.41E-9
Xylitol	16.4	±	3.7	39.8	±	14.2	1.91	2.18	7.20E-6
Tyrosine	5.4	±	1.3	9.6	±	1.8	1.90	1.77	2.99E-7
Glycine	10.2	±	2.0	16.6	±	2.2	1.85	1.63	2.45E-10
Glucose	90	±	40	154	±	43	1.82	1.71	1.04E-6
Threonine	7.9	±	0.8	12.0	±	1.9	1.69	1.51	2.60E-11
Isoleucine	10.4	±	1.7	15.7	±	2.1	1.69	1.50	2.87E-10
Citrulline	0.8	±	0.1	1.3	±	0.4	1.69	1.59	6.65E-9
Fructose	33.3	±	22	59.2	±	29.5	1.67	1.78	7.95E-4
Serine	23.1	±	5.1	37.8	±	14.7	1.63	1.64	1.13E-6
Valine	11.9	±	3.7	17.0	±	3.3	1.53	1.93	5.56E-4
Leucine	5.9	±	2.1	8.9	±	2.0	1.53	1.50	1.23E-5
Cysteine	1.1	±	0.3	0.8	±	0.3	1.27	0.70	0.0014
β-sitosterol	281	±	10.2	329	±	20.4	1.02	1.17	2.69E-9
Stigmasterol	74.2	±	6.9	92.2	±	5.9	1.21	1.24	1.70E-9

[a]For each metabolite, the relative peak intensity of the wild-type was the mean value of MH86 from four field trials and homozygous wild-type progenies of generations F_4 to F_7; the relative peak intensity of the *lpa* mutant was the mean value of *Os-lpa*-MH86-1 from four field trials and homozygous *lpa* progenies of generations F_4 to F_7.
[b]VIP values of each metabolite were obtained from the PLS-DA results.
[c]The *p* values were calculated based on the false discovery rate (FDR) correction.

other metabolic changes in the *Os-lpa*-MH86-1 mutant. The mutation-induced metabolite signature in *Os-lpa*-MH86-1 was almost unaffected by cross-breeding with a commercial cultivar and was consistently expressed in homozygous *lpa* progenies over generations. The data indicate that even for complex metabolic changes resulting from a mutation, cross-breeding can be employed as a tool to generate *lpa* rice mutant seeds stably exhibiting the mutation-induced traits.

References

Abdel-Megeed, A. *et al.* (2015) Reduction of phosphorus pollution from broilers waste through supplementation of wheat based broilers feed with phytase. *Journal of Chemistry* 3, 1–3.

Boehm, J.D. *et al.* (2017) Seed inorganic phosphorus stability and agronomic performance of two low-phytate soybean lines evaluated across six southeastern US environments. *Crop Science* 57, 2555–2563.

Frank, T. *et al.* (2007) Metabolite profiling of two low phytic acid (*lpa*) rice mutants. *Journal of Agricultural and Food Chemistry* 55, 11011–11019.

Frank, T. *et al.* (2009) Metabolite profiling of two novel low phytic acid (*lpa*) soybean mutants. *Journal of Agricultural and Food Chemistry* 57, 6408–6416.

Frenzel, T. *et al.* (2003) A methodology for automated comparative analysis of metabolite profiling data. *European Food and Research Technology* 216, 335–342.

Hammond, J.P. *et al.* (2008) Sucrose transport in the phloem: integrating root responses to phosphorus starvation. *Journal of Experimental Botany* 59, 93–109.

Kamaleldin, A. *et al.* (1992) Seed lipids of *Sesamum indicum* and related wild species in Sudan. The sterols. *Journal of the Science of Food and Agriculture* 59, 327–334.

Kanehisa, M. *et al.* (2000) KEGG: Kyoto Encyclopedia of Genes and Genomes. *Nucleic Acids Research* 28.

Kumar, V. *et al.* (2010) Dietary roles of phytate and phytase in human nutrition: a review. *Food Chemistry* 120, 945–959.
Liu, Q.L. *et al.* (2007) Generation and characterization of low phytic acid germplasm in rice (*Oryza sativa* L.). *Theoretical and Applied Genetics* 114, 803–814.
Mehmood, T. *et al.* (2012) A review of variable selection methods in Partial Least Squares Regression. *Chemometrics and Intelligent Laboratory Systems* 118, 62–69.
Nikiforova, J.V. *et al.* (2006) Effect of sulfur availability on the integrity of amino acid biosynthesis in plants. *Amino Acids* 30, 173–183.
Oates, K. *et al.* (2014) *Determination of Inositol Phosphates in Dried Distiller's Grains with Solubles.* Thermo Fisher Scientific, Sunnyvale, California.
Raboy, V. (2003) *myo*-Inositol-1,2,3,4,5,6-hexa*kis*phosphate. *Phytochemistry* 64, 1033–1043.
Raboy, V. (2009) Approaches and challenges to engineering seed phytate and total phosphorus. *Plant Science* 177, 281–296.
Spear, J.D. *et al.* (2007) Genetic improvement of seedling emergence of soybean lines with low phytate. *Crop Science* 47, 1354–1360.
Tan, Y.Y. *et al.* (2013) Functional molecular markers and high-resolution melting curve analysis of low phytic acid mutations for marker-assisted selection in rice. *Molecular Breeding* 31, 517–528.
Xu, Z.M. *et al.* (1999) Purification and identification of components of gamma-Oryzanol in rice bran oil. *Journal of Agricultural and Food Chemistry* 47, 2724–2728.
Zhao, H. *et al.* (2016) Disruption of *OsSULTR3;3* reduces phytate and phosphorus concentrations and alters the metabolite profile in rice grains. *New Phytology* 211, 926–939.

45 In Search of Mutants for Gene Discovery and Functional Genomics for Multiple Stress Tolerance in Rice

Priyanka Das, Rajeev N. Bahuguna, Rohit Joshi, Sneh Lata Singla-Pareek and Ashwani Pareek[1]

Abstract

Mutation breeding is a commanding tool, which has been adapted to generate altered genetic material to study functional genomics, including understanding the molecular basis of stress tolerance. Hitherto, several rice lines have been generated through mutagenesis and the mutated genes responsible for the 'gain of function' in terms of plant architecture, stress tolerance, disease resistance and grain quality have been characterized. *Oryza sativa* L. cv. IR64 is a high-yielding rice cultivar but sensitive to abiotic stresses such as acute temperatures, salinity and drought. In this study, a population of rice IR64 mutants was generated using gamma irradiation. The population was then subjected to a preliminary phenotypic screening under abiotic stresses such as heat and salinity at the seedling stage. On the basis of root length, shoot length, fresh weight, dry weight and chlorophyll measurements, we identified eight 'gain-of-function' mutant lines and used them for further biochemical and molecular characterization. Phenotyping results demonstrated that the identified mutant plants have gained the potential to thrive under heat and salinity conditions. This information would be of wide scientific interest and helpful for developing novel cultivars able to maintain yield in saline, hot and dry areas.

Keywords: rice-mutant • gamma irradiation • heat • drought • salinity

1 Introduction

Rice is the staple food for around 60% of the world's population. Though the yield of rice more than doubled from the 1960s to the 1990s (http://faostat.fao.org, accessed 2019), sustainable rice production for the whole population is still to be achieved. Evidence of the impact of climate change on global rice production is being documented, for example, through the frequent occurrence of different abiotic stresses such as heat, drought and flooding. In the face of these environmental constraints in the current climate, which are expected to become more frequent in the future, rice production needs to increase for feeding the growing population. Intolerance of rice to multiple interconnected stresses such as heat, drought and salinity limits its genetic yield potential (Grover and Minhas, 2000). Hence, it is an urgent requirement to find a suitable method to shape abiotic stress-tolerant rice that could give high yields even under adverse environmental situations.

Mutation breeding has made a prominent contribution towards the production of stress-tolerant crops (Cassells and Doyle, 2003; Parry *et al.*, 2009; Jankowicz-Cieslak *et al.*, 2017). Attempts

[1] Email: ashwanip@mail.jnu.ac.in

DOI: 10.1079/9781789249095.0045

have already been made through mutation breeding to generate abiotic stress-tolerant rice. Many abiotic (including salinity and extreme temperature conditions) stress-tolerant rice varieties have been generated through mutation breeding and released across the globe (Singh, 2000; Ahloowalia and Maluszynski, 2001; Baloch *et al.*, 2003; Chhun *et al.*, 2003; Ahloowalia *et al.*, 2004; Saleem *et al.*, 2005; Hayashi *et al.*, 2007; Hayashi *et al.*, 2008; Jain and Suprasanna, 2011).

Abiotic stresses have an adverse impact on plant physiology and this varies with age of the plant (Barnabás *et al.*, 2008; Sakata and Higashitani, 2008). Rice is sensitive to abiotic stresses particularly at the early seedling and flowering stages (Bahuguna *et al.*, 2015). In the recent past, several rice varieties with tolerance to one or other stresses (abiotic or biotic) have been developed using mutation breeding, as described below in the Results section. Although efforts are being focused on integrating multiple abiotic stress tolerance in rice, so far no rice genotype has been reported with tolerance to multiple abiotic stresses. In this study, we describe the screening of multiple stress-tolerant mutant rice lines at the seedling stage through their phenotypic performance.

2 Materials and Methods

2.1 Plant materials, growth conditions and stress treatments

Rice (cv. IR64) seeds were subjected to 100 Gy of ^{60}Co gamma irradiation to obtain mutant germplasm. The plant obtained from an individual mutagenized seed is referred as an individual M_1 mutant plant. The mutant population was advanced up to the M_3 generation by selfing to get the stable homozygous lines. M_3 mutants were grown hydroponically under controlled growth conditions (12 h day/night, 28°C, 10,000 Lux light and 70% humidity) for up to 7 days. Wild-type (WT) IR64 rice seedlings were grown along with the mutant seedlings as controls. Following the stress treatment and recovery, phenotypic parameters from each line were recorded in a screen for mutant lines. A schematic diagram describing screening protocol for multiple stress-tolerant mutant lines is shown in Figs 45.1 and 45.2.

For standardization of the ideal duration for heat stress and recovery, 7-day-old seedlings were subjected to high temperature (45°C for 8 h or 10 h or 12 h) in a growth chamber (Daihan Labtech Co. Ltd, India) followed by different durations of recovery (40 h or 60 h or 72 h) under controlled growth conditions. Heat shock (45°C) treatment for 12 h followed by 72 h of recovery was found to be the best to identify heat-tolerant putative mutants.

For screening purposes, 12-day-old seedlings were subjected to different stress treatments. Salinity treatment was given by transferring the plants into 200 mM NaCl solution; for heat treatment, plants were exposed to 45°C for 12 h followed by recovery for 72 h; for dehydration, seedlings were air dried at 28 ± 1°C for 6 h. Parallel control samples were prepared by keeping the plants in Yoshida nutrient media (Yoshida *et al.*, 1976).

2.2 Phenotypic analysis

All the phenotypic analyses were done at the seedling stage and parameters such as root length, shoot length, fresh weight, dry weight and total chlorophyll were considered for stress-tolerance screening.

2.3 Analysis of reports on gain-of-function mutants towards abiotic stress tolerance

Along with our experimental data, we also revisited the available reports showing identification of mutants for abiotic stress tolerance.

2.4 Statistical analysis

All the experiments were analysed as completely randomized design (CRD) with two-way Analysis of Variance (ANOVA). Means were compared using least significant difference (LSD) at $p \leq 0.05$.

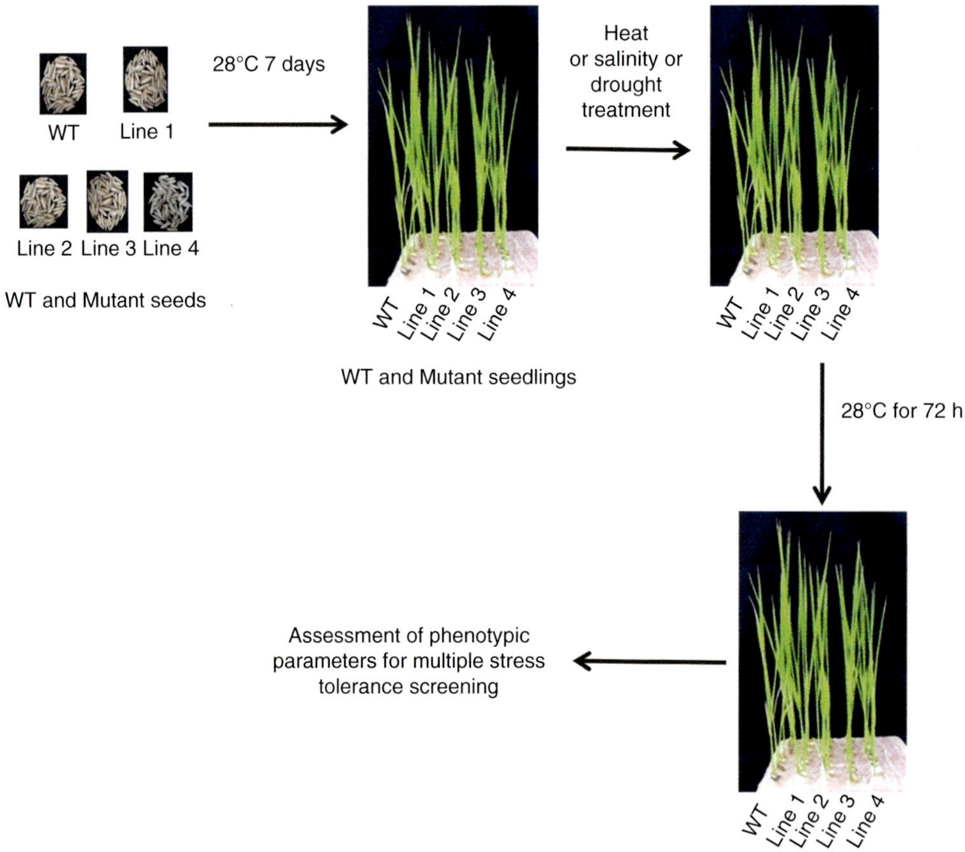

Fig. 45.1. Screening strategy for the identification of multiple stress-tolerant mutant rice lines.

3 Results

3.1 Screening for multiple stress-tolerant mutant lines

Gamma-irradiated mutant rice seedlings were first screened at 45°C for different durations of heat stress as described by Sarsu *et al.* (2018). Based on morphological appearance (leaf greenness, leaf kink and seedling height), eight mutant lines (D100/1, D100/2, D100/3, D100/4, D100/5, D100/6, D100/7, D100/8) were selected for detailed phenotyping analysis. All the eight selected mutant lines showed a healthy appearance after 12 h of heat stress (45°C) followed by a 72 h recovery period (Fig. 45.3a). These putative heat-tolerant mutant lines were then subjected to salinity and drought stress and we observed notable differences between WT and mutant lines for stress tolerance under salinity and drought conditions as well (Fig. 45.3b,c).

Imposition of heat, salinity and drought stress individually at the seedling stage affected root and shoot growth and fresh weight in WT plant as compared with the control conditions (Fig. 45.3d,e,f). Higher shoot length (ca. 11–25% under heat stress, ca. 13–20% under salinity and ca. 14–22% under drought), root length (ca. 20–33% under heat stress, ca. 19–32% under salinity and ca. 24–31% under drought) and fresh weight (ca. 10–29% under heat stress, ca. 9–23% under salinity and ca.

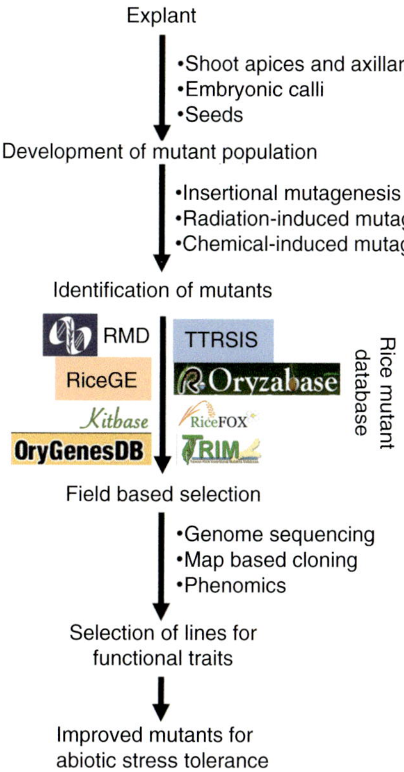

Fig. 45.2. Flow diagram showing various approaches used to develop mutants for different combinations of abiotic stress tolerance in rice. Various rice mutant databases available to the community are listed.

14–33% under drought) were recorded in selected mutant seedlings as compared with the WT (Fig. 45.3d,e,f). Similarly, the mutant lines showed higher content of chlorophyll after stress treatment as compared with the WT seedlings (Fig. 45.3g).

4 Discussion

Rice is most sensitive to heat, drought and salinity stress during the seedling and reproductive stages. The occurrence of individual stress or multiple stresses in a simultaneous or sequential manner could hamper plant growth and yield under controlled and natural field conditions (Mittler and Blumwald, 2010; Kadam *et al.*, 2014). Most previous studies on abiotic stress tolerance focused on a single stress in an isolated and controlled environment. However, crop species face more than one stress in their life cycle and plants that are tolerant to a single stress could be sensitive to another stress or a combination of more than one stress condition in the natural environment (Mittler and Blumwald, 2010). Thus, tolerance to multiple stresses is needed in crop plants to sustain agricultural production in the current and future climate change scenario. In this study, eight rice mutant lines (derived from cv. IR64) were identified from a large population and then phenotyped for multiple abiotic stress tolerance such as heat, drought and salinity at the seedling stage.

Heat stress at the seedling stage reduces plant growth and creates oxidative stress by inducing the production of reactive oxygen species (Wahid *et al.*, 2007; Shah *et al.*, 2011). Moreover, heat stress above an optimum threshold could affect photosystem-II and chlorophyll, affecting photo assimilation (Wahid *et al.*, 2007). Conversely, drought stress could hamper photosynthesis with stomatal and non-stomatal limitations (Bahuguna *et al.*, 2018). Salinity stress could induce oxidative stress and affect cellular homeostasis (Moradi and Ismail, 2007). Our previously developed mutant lines showed better tolerance

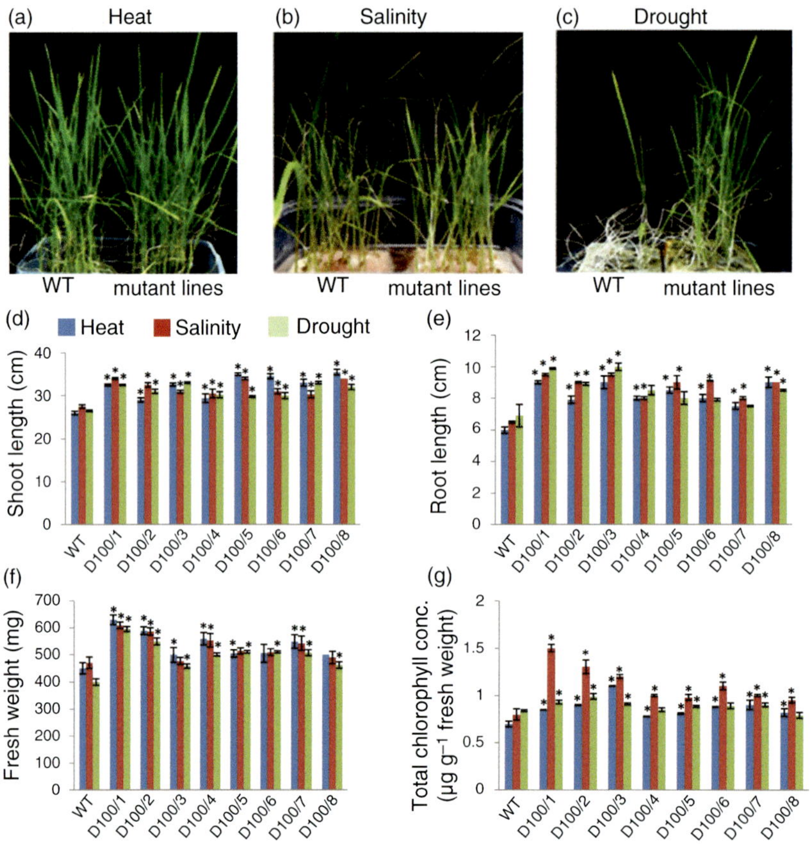

Fig. 45.3. Phenotypic analysis of selected putative stress-tolerant mutant lines. IR64 and mutant lines after **(a)** heat stress and recovery, **(b)** salt stress and recovery and **(c)** drought stress and recovery. Histograms of IR64 and selected mutant lines after heat, salinity and drought stress followed by stress recovery, showing **(d)** shoot length, **(e)** root length, **(f)** fresh weight and **(g)** chlorophyll content. Standard deviation is reflected above each bar in the figure. * indicates significance value at $p \leq 0.05$ where $n = 6$.

against high levels of salinity in comparison to WT, i.e. IR64 (Joshi *et al.*, 2016). Similarly, seedlings of these mutant rice lines were able to maintain higher growth, biomass and chlorophyll content under heat, drought or salinity. This could be attributed to novel genetic changes in the mutant lines, which could plausibly maintain a more responsive stress defence signalling providing tolerance to one or more abiotic stresses. Mittler (2006) suggested that mitochondrial respiration could play a leading role in providing energy for defence under a combination of stresses. Moreover, higher levels of antioxidants and antioxidant enzyme machinery provide an effective strategy to combat heat, drought and salinity stress (Moradi and Ismail, 2007; Farooq *et al.*, 2011; Bahuguna *et al.*, 2015).

5 Conclusion

Mutation breeding is a promising field in crop improvement. Developing multiple stress-tolerant crops is warranted to sustain agricultural production and feed the ever-growing population. We could identify the mutant rice lines that are able to sustain production under a combination of abiotic stresses. This unique feature makes them promising donors in breeding programmes for multiple abiotic stress tolerance. Also, the mechanism underlying multiple abiotic stress tolerance of these mutants needs to be explored to identify the components and pathways involved. This study has provided preliminary information on the potential of mutagenesis for developing plants that can withstand a combination of different abiotic stresses. This unique strategy could expedite crop improvement by providing novel donors to breeding programmes and a model system to study multiple abiotic stress-tolerance mechanisms in plants.

Acknowledgements

The authors would like to thank the Joint FAO/IAEA Division Plant Breeding and Genetics section for support and encouragement under the IAEA-CRP 23031 'Improving Resilience to Drought in Rice and Sorghum through Mutation Breeding' project. AP would also like to thank IUSSTF and Department of Biotechnology for financial support to his laboratory at JNU, New Delhi, India.

References

Ahloowalia, B.S. and Maluszynski, M. (2001) Induced mutations - a new paradigm in plant breeding. *Euphytica* 118, 167–173.

Ahloowalia, B.S. *et al*. (2004) Global impact of mutation-derived varieties. *Euphytica* 135, 187–204.

Bahuguna, R.N. *et al*. (2015) Physiological and biochemical characterization of NERICA-L-44: a novel source of heat tolerance at the vegetative and reproductive stages in rice. *Physiologia Plantarum* 154(4), 543–559.

Bahuguna R.N. *et al*. (2018) Mild preflowering drought priming improves stress defences, assimilation and sink strength in rice under severe terminal drought. *Functional Plant Biology* 45(8), 827–839.

Baloch A.W. *et al*. (2003) Induction of salt tolerance in rice through mutation breeding. *Asian Journal of Plant Science* 2(3), 273–276.

Barnabás, B. *et al*. (2008) The effect of drought and heat stress on reproductive processes in cereals. *Plant, Cell & Environment* 31(1), 11–38.

Cassells, A.C. and Doyle, B.M. (2003) Genetic engineering and mutation breeding for tolerance to abiotic and biotic stresses: science, technology and safety. *Bulgarian Journal of Plant Physiology Special* 2003, 52–82.

Chhun T. *et al*. (2003) Interaction between two auxin-resistant mutants and their effects on lateral root formation in rice (*Oryza sativa* L.). *Journal of Experimental Botany* 54(393), 2701–2708.

Farooq M. *et al*. (2011) Heat stress in wheat during reproductive and grain-filling phases. *Critical Reviews in Plant Sciences* 30, 491–507.

Grover, A. and Minhas, D. (2000) Towards production of abiotic stress tolerant transgenic rice plants: issues, progress and future research needs. *Proceedings Indian National Science Academy Part B* 66(1), 13–32.

Hayashi, Y. *et al*. (2007) Effects of ion beam irradiation on mutation induction in rice. In: *Proceedings of the Eighteenth International Conference on Cyclotrons and their Applications*, Giardini Naxos, Italy, 1–5 Oct 2007, pp. 237–239.

Hayashi, Y. *et al*. (2008) Characterization of salt tolerant mutants of rice induced by heavy ion irradiation. *RIKEN Accelerator Progress Report* 41, 234.

Jain, M.S. and Suprasanna, P. (2011) Induced mutations for enhancing nutrition and food production. *Gene Conserve* 40, 201–215.

Jankowicz-Cieslak, J. *et al*. (2017) Mutagenesis for crop breeding and functional genomics. In: Jankowicz-Cieslak *et al*. (eds) *Biotechnologies for Plant Mutation Breeding*. Springer, Cham, pp. 3–18.

Joshi, R. *et al*. (2016) Physiological characterization of gamma-ray induced mutant population of rice to facilitate biomass and yield improvement under salinity stress. *Indian Journal of Plant Physiology* 21(4), 545–555.

Kadam, N.N. *et al*. (2014) Agronomic and physiological responses to high temperature, drought, and elevated CO_2 interactions in cereals. *Advances in Agronomy* 127, 111–156.

Mittler, R. (2006) Abiotic stress, the field environment and stress combination. *Trends in Plant Science* 11, 15–19.

Mittler, R. and Blumwald, E. (2010) Genetic engineering for modern agriculture: challenges and perspectives. *Annual Review of Plant Biology* 61, 443–462.

Moradi, F. and Ismail, A.M. (2007) Responses of photosynthesis, chlorophyll fluorescence and ROS-scavenging systems to salt stress during seedling and reproductive stages in rice. *Annals of Botany* 99, 1161–1173.

Parry, M.A.J. *et al.* (2009) Mutation discovery for crop improvement. *Journal of Experimental Botany* 60, 2817–2825.

Sakata, T. and Higashitani, A. (2008) Male sterility accompanied with abnormal anther development in plants – genes and environmental stresses with special reference to high temperature injury. *International Journal of Plant Developmental Biology* 2, 42–51.

Saleem, M.Y. *et al.* (2005) Induced mutation and in vitro techniques as a method to induce salt tolerance in Basmati rice (*Oryza sativa* L.). *International Journal of Environmental Science and Technology* 2(2), 141–145.

Sarsu, F. *et al.* (2018) *Pre-field Screening Protocols for Heat-Tolerant Mutants in Rice*. Springer Nature Switzerland AG, Switzerland.

Shah, F. *et al.* (2011) Impact of high-temperature stress on rice plant and its traits related to tolerance. *Journal of Agricultural Science* 149, 545–556.

Singh, B.D. (2000) *Plant Breeding – Principles and Methods*. Kalyani Publishers, New Delhi.

Wahid, A. *et al.* (2007) Heat tolerance in plants: an overview. *Environmental and Experimental Botany* 61, 199–223.

Yoshida, S. *et al.* (1976) *Laboratory Manual for Plant Physiological Studies of Rice*, 3rd edn. International Rice Research Institute, Manila, Philippines.

46 Comparative Study of Mutations Induced by Carbon-Ion Beams and Gamma-ray Irradiations in *Arabidopsis thaliana* at the Genome-wide Scale

Libin Zhou[1], Yan Du, Zhuo Feng, Tao Cui, Xia Chen, Shanwei Luo, Yuze Chen, Hui Feng, Jinhu Mu, Lixia Yu and Wenjian Li

Abstract

Mutation breeding induced by irradiation with highly energetic photons and ion beams is one of the important methods to improve plant varieties, but the mutagenic effects and molecular mechanisms are often not entirely clear. Traditional research is focused on phenotype screening, chromosome aberration tests and genetic variation analysis of specific genes. The whole genome sequencing technique provides a new method to understand and undertake the comprehensive identification of mutations caused by irradiations with different linear energy transfer (LET). In this study, ten *Arabidopsis thaliana* M_3 lines induced by carbon-ion beams (CIB) and ten M_3 lines induced by gamma-rays were re-sequenced by using the Illumina HiSeq sequencing platform, and the single base substitutions (SBSs) and small insertions or deletions (indels) were analysed comparatively. It was found that the ratio of SBSs to small indels for M_3 lines induced by CIB was 2.57:1, whereas the ratio was 1.78:1 for gamma-rays. The ratios of deletions to insertions for carbon ions and gamma-rays were 4.8:1 and 2.8:1, respectively. The single-base indels were more prevalent than those equal to or greater than 2 bp in both CIB and gamma-ray induced M_3 lines. Among the detected SBSs, the ratio of transitions to transversions induced by carbon-ion irradiation was 1.01 and 1.42 for gamma-rays; these values differ greatly from the 2.41 reported for spontaneous substitutions. This study provides novel data on molecular characteristics of CIB and gamma-ray induced mutations at the genome-wide scale. It can also provide valuable clues for explaining the potential mechanism of plant mutation breeding by irradiations with different LETs.

Keywords: *Arabidopsis thaliana* • carbon-ion beams • gamma-rays • mutations • genome

1 Introduction

According to the IAEA/FAO database, as of 2018, more than 3200 new crop varieties had been obtained by mutation breeding, of which more than 80% were obtained by gamma-ray radiation mutagenesis. As an effective physical mutagen, gamma-rays can be used for mutation breeding and functional genomics research due to their characteristics. Gamma-rays are highly penetrating and can cause damage to cells and tissues when interacting with target matter (Van De Walle *et al.*, 2016). Gamma-rays can induce both DNA single-strand breaks and DNA double-strand breaks (Biermans *et al.*, 2015), as well as chromosome structural variation (SV), etc. In addition, gamma-ray irradiation has the advantage of low cost, and thus it is the most widely

[1] Email: libinzhoulz@gmail.com

 Mutation Breeding, Genetic Diversity and Crop Adaptation to Climate Change (eds S. Sivasankar *et al.*)
DOI: 10.1079/9781789249095.0046

used physical mutagenic source for mutation breeding in the world.

As an emerging, effective and unique mutagen, the heavy-ion beam has been rapidly accepted as a powerful mutagen for mutagenic breeding of plants and microorganisms. The effects of heavy-ion beams and conventional irradiations (gamma-rays, X-rays or electrons) on phenotypic mutation induction were comparatively studied in *Arabidopsis thaliana*, chrysanthemum and carnation. The results indicated that the heavy-ion beam showed high mutation frequency and a broad mutation spectrum (Tanaka *et al.*, 2010). Essentially, heavy-ion beams have unique physical and biological advantages. Compared with X- and gamma-rays, the most important physical feature of heavy-ion beams is their greater linear energy transfer (LET). Heavy-ion beams provide a higher relative biological effect (RBE) than low LET radiation (Tanaka *et al.*, 2010; Kazama *et al.*, 2011). For plant mutagenesis, various mutant populations have been obtained using heavy-ion beam radiation mutagenesis techniques: for example, *A. thaliana* (Tanaka *et al.*, 1997, 2002; Du *et al.*, 2014), *Lotus japonicus* (Oka-Kira *et al.*, 2005; Luo *et al.*, 2016), rice (*Oryza sativa* L.) (Ishikawa *et al.*, 2012; Morita *et al.*, 2017), *Petunia* hybrid (Hase *et al.*, 2010), *Tradescantia fluminensis* (He *et al.*, 2011), *Geranium* (Yu *et al.*, 2016), etc. However, the comparison between gamma-ray radiation mutagenesis and heavy-ion beam radiation mutagenesis has not been reported with respect to their mutagenic mechanism on plant genomes.

Next-generation sequencing technology (NGS) has greatly reduced the barriers to the research of plant mutation (Bolon *et al.*, 2011; Belfield *et al.*, 2012). Mutants such as single base substitutions (SBSs) and insertions and deletions (indels) of the whole genomic DNA can be detected using NGS (Hwang *et al.*, 2015). Genomic variation characteristics of *A. thaliana* induced by fast neutrons, argon-ion beams and carbon-ion beams (CIB) analysed using NGS technology have been reported (Belfield *et al.*, 2012; Hirano *et al.*, 2015; Du *et al.*, 2017, 2018; Kazama *et al.*, 2017; Hase *et al.*, 2018). However, comparisons between heavy-ion beam and gamma-ray induced *A. thaliana* genome variation have not been reported. Therefore, the use of NGS technology will complement and improve our understanding of the actual nature of the mutations caused by ionizing radiation in *A. thaliana*.

In our previous research, we obtained *A. thaliana* mutant populations induced by high-energy CIB and gamma-rays. In this study, we used NGS technology to perform genome-wide mutational spectrum analysis of the obtained M_3 *A. thaliana* mutants, such as SBSs and small indels, in order to understand the effects of high-energy CIB and gamma-rays on the *A. thaliana* genome.

2 Materials and Methods

2.1 Irradiation and mutant screening

The seeds of laboratory wild-type *A. thaliana* (Lab-WT) (Columbia genetic background) were exposed to $^{12}C^{6+}$ ions (average LET within samples was 50 keV/μm) or gamma-rays (average LET 0.2 keV/μm) at half lethal doses (LD_{50}). The LD_{50} dosage of carbon ions is 200 Gy, whereas it is 1000 Gy for gamma-rays (Du *et al.*, 2014). CIB was generated by the Heavy Ion Research Facility in Lanzhou (HIRFL) at the Institute of Modern Physics, Chinese Academy of Sciences. Methods of plant growth and mutation screening were described in a previous study (Du *et al.*, 2014). Eight CIB (C7, C116, C197, C352, C357, C541, C828, C941) and seven gamma-ray (γ200, γ240, γ266, γ287, γ320, γ431, γ692) induced mutant lines (M_3) that displayed visible and heritable traits, five mutagenesis progeny (M_3) lines without conspicuous mutant phenotypes (C1001, C1322, γ85, γ689 and γ269), along with the Lab-WT line were chosen for whole genome re-sequencing.

2.2 Whole genome re-sequencing

The genomic re-sequencing for M_3 samples derived from CIB-treated samples was performed by the Illumina HiSeq 2500 system (Illumina, Inc., San Diego, California, USA), whereas gamma-ray treatment of samples was performed by the Illumina HiSeq 4000 platform. Raw data were filtered following the standard process of Illumina, and the clean data obtained were then mapped onto the reference genome of Col-0 (TAIR10) (http://www.ncbi.nlm.nih.gov/assembly/GCF_000001735.3, accessed 2019) by using the Burrows-Wheeler Alignment tool (http://

bio-bwa.sourceforge.net/bwa.shtml, v.0.7.15, accessed 2019) (Li *et al.*, 2009a) and SAMtools (http://www.htslib.org/workflow/#mapping_to_variant, v.1.3.1, accessed 2019) (Li *et al.*, 2009b). The average read depth of the re-sequenced lines was in the 16–33-fold range.

2.3 Identification of the variants

VarScan 2 algorithms (v.3.9, http://varscan.sourceforge.net, accessed 2019) (Koboldt *et al.*, 2012) were used to read the SAMtools mpileup output, and SBSs and small indels were called as described methods and steps (Du *et al.*, 2017). Candidate mutations derived from each sequenced line were visually confirmed in the Integrative Genomics Viewer (IGV) (http://software.broadinstitute.org/software/igv/, v.2.3, accessed 2019) (Robinson *et al.*, 2011).

2.4 Mutation annotation

Variant annotation and predicted effects were performed by the SnpEff toolbox (http://snpeff.sourceforge.net/index.html, v.4.2, accessed 2019; Cingolani *et al.*, 2012). The whole analysis strategy of SBSs and small indels detection is shown in Fig. 46.1.

2.5 Verification of the mutation sites by Sanger sequencing

The specific primers corresponding to mutation sites identified by re-sequencing were designed by PRIMER3 (http://bioinfo.ut.ee/primer3-0.4.0/, accessed 2019). Genomic DNA was extracted by using the routine CTAB extraction protocol. PCR amplification was performed with an initial denaturation step at 95°C for 5 min, followed by 38 cycles at 94°C for 30 s, 58°C for 30 s and 72°C for 30 s, with a final extension step at 72°C for 10 min. The PCR products were then detected by electrophoresis on a 1.5% agarose gel; only the PCR products with one clear band of the right fragment size were used for Sanger sequencing.

2.6 Data availability

The whole-genome sequencing data reported in this study have been deposited in the Genome

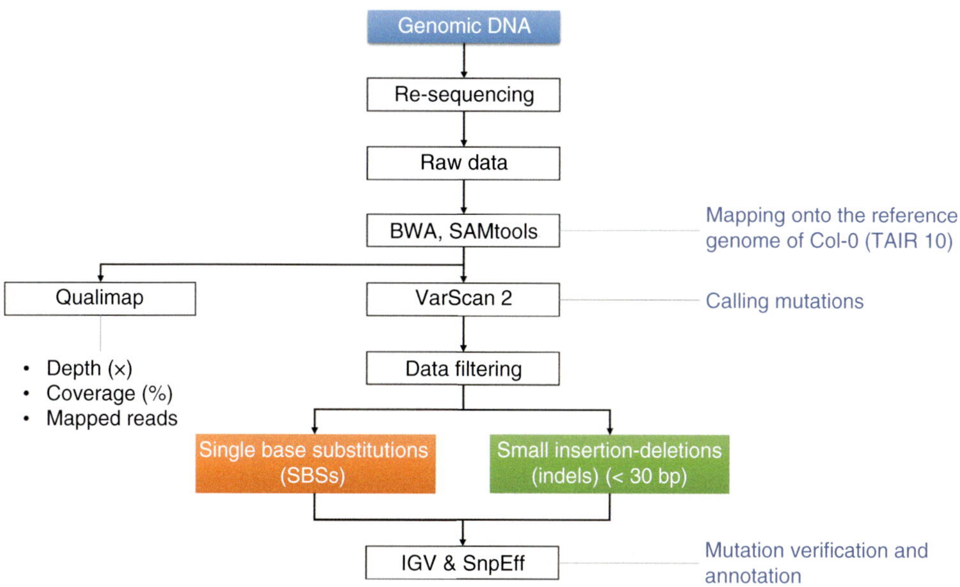

Fig. 46.1. Strategy for the detection of single base substitutions and small indels in *Arabidopsis thaliana* irradiated by carbon-ion beams and gamma-rays.

Sequence Archive (Wang *et al.*, 2017) in the BIG Data Center, Beijing Institute of Genomics (BIG), Chinese Academy of Sciences (BIG Data Center Members, 2017).

3 Results

3.1 Distribution and frequency of radiation-induced mutations

A total of 327 SBSs and 127 small indels were detected in the ten re-sequenced mutant lines that were CIB induced by combining the mutation calling by VarScan 2 and the visual confirmation by IGV. In the genome-wide range, 305 variations were in upstream and downstream regions, 113 were located in the exon, 19 were in the 3′/5′ untranslated region (UTR), 11 were in the intergenic region, three were in the splicing region (i.e. the splice donor and acceptor sites) and three were in the intron. The ratio of SBSs to small indels was 2.57:1. The number of variations per line ranged from 22 to 63. In the ten M_3 mutant lines obtained by gamma-ray mutagenesis with distinct phenotypes, we detected 215 SBSs and 121 small indels, with a total of 336 mutations. Among them, 217 occurred in the upstream and downstream regions, 82 in the exon, 20 in the 3′/5′ UTR, six in the intergenic region, six in the splicing region and five in the intron (Fig. 46.2B). The ratio of SBSs to small indels was 1.78:1. The number of variations per line ranged from 23 to 48. Significant patterns in the chromosomal distribution of variation were not found in either the CIB or the gamma-ray induced mutant lines. In addition, regarding mutations that were predicted to cause functional changes with high likelihood, the CIB induced 42 missense, 3 nonsense and 24 frameshift changes. Correspondingly, the data were 34, 3 and 24 in the mutations induced by gamma-rays, respectively (Table 46.1).

3.2 Radiation-induced single base substitutions

The types of SBSs can be divided into two categories: transitions (mutations that occur among the same type of bases; for example, purine > purine, or pyrimidine > pyrimidine) and transversions (mutations that occur between different types of bases; for example, purine > pyrimidine, or pyrimidine > purine). A total of 327 SBSs induced by CIB radiation were identified and classified, obtaining a transition to transversion (Ts/Tv) ratio of 1.01. On the other hand, the ratio of Ts/Tv induced by gamma-rays was 1.42. These results indicated that both CIB and gamma-rays induced more transitions than transversions (Fig. 46.3). In addition, the detected SBSs in both CIB and gamma-ray induced mutations showed an obvious bias for C > T transitions.

Table 46.1. Numbers of mutations predicted to cause functional changes in re-sequenced M_3 *Arabidopsis thaliana* lines induced by carbon-ion beams and gamma-ray irradiations.

	Carbon-ion beams	Gamma-rays
Missense	43	35
Nonsense	3	3
Frame shift	24	24

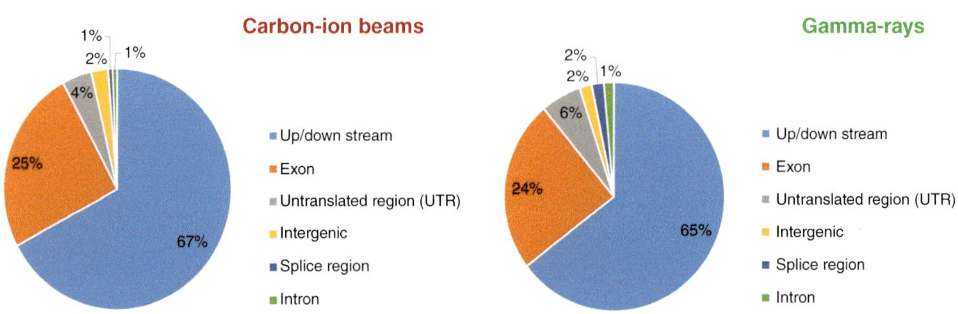

Fig. 46.2. Annotation of mutations in genomes of M_3 mutant lines of *Arabidopsis thaliana* induced by carbon-ion beams and gamma-ray irradiation.

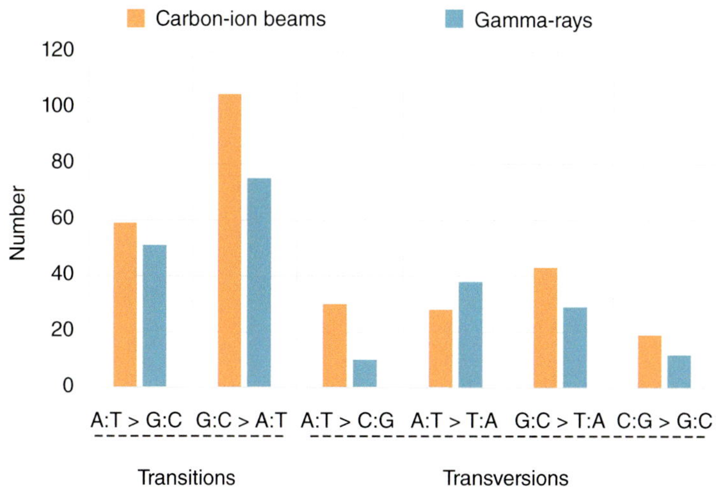

Fig. 46.3. Distribution of SBSs induced by carbon-ion beam and gamma-ray irradiation in the genomes of *Arabidopsis thaliana* M_3 lines.

3.3 Radiation-induced small indels

For a long time, deletions were thought to be the main type of mutations induced by heavy-ion beam radiation, or at least similar in frequency to SBSs, according to the sequencing results of specific genes (Shikazono *et al.*, 2005; Kazama *et al.*, 2011; Hirano *et al.*, 2015). In the *A. thaliana* mutations induced by CIB, 127 small indels were detected throughout the genome, of which 105 (82.68%) were deletion mutations and only 22 (17.32%) were insertions. Among these 105, we found 73 deletions of multiple bases (≥ 2 bp) ranging from 2 to 23 bp (Fig. 46.4). For gamma-ray induced genomic small indels, 89 deletions and 32 insertions were detected in total. Of the 89 deletions, 35 were multiple base deletions (≥ 2 bp) with a maximum size of 25 bp (Fig. 46.4).

4 Discussion

In this study, we re-sequenced ten mutagenesis progeny lines (M_3) induced by CIB and ten lines from gamma-ray radiation. Based on the collected data, we can describe the mutation effects of CIB and gamma-rays on *A. thaliana* at the whole genome level, as well as their associated molecular mutation profiles and mutation rates.

Previous studies suggested that heavy-ion beam radiation can generate mutations in the form of SBSs, small indels and structural variants (Tanaka *et al.*, 2010), but there are few reports on this characteristic induced by gamma-ray irradiation. In this study, we found that both CIB and gamma-ray radiations induced more SBSs than small indels. The ratio of SBSs to small indels was 2.6:1 for CIB and 1.78:1 for gamma-rays. For fast neutron irradiation, the ratio is 1.45:1 in *A. thaliana* and 1.26:1 in rice mutants (Belfield *et al.*, 2012; Li *et al.*, 2016). Although current results showed that CIB and gamma-ray radiations had similar effects to fast neutron radiation (inducing more SBSs), the ratio of SBSs to small indels varied with radiation quality ($\chi^2 = 9.470$, $p = 0.009$). We attempted to detect large indels by using both pINDEL (https://trac.nbic.nl/pINDEL/, accessed 2019) (Ye *et al.*, 2009) and Break Dancer Max (http://breakdancer.sourceforge.net/breakdancermax.html, accessed 2019) (Chen *et al.*, 2009) algorithms. Although hundreds of large deletions were eventually detected, most of them were confirmed to be false positives after IGV verification. In addition, most of the severe DNA damage, including radiation-induced large deletions (from kb to Mb), cannot be inherited by the offspring, as these non-hereditary mutations may be involved in genes critical for gamete development

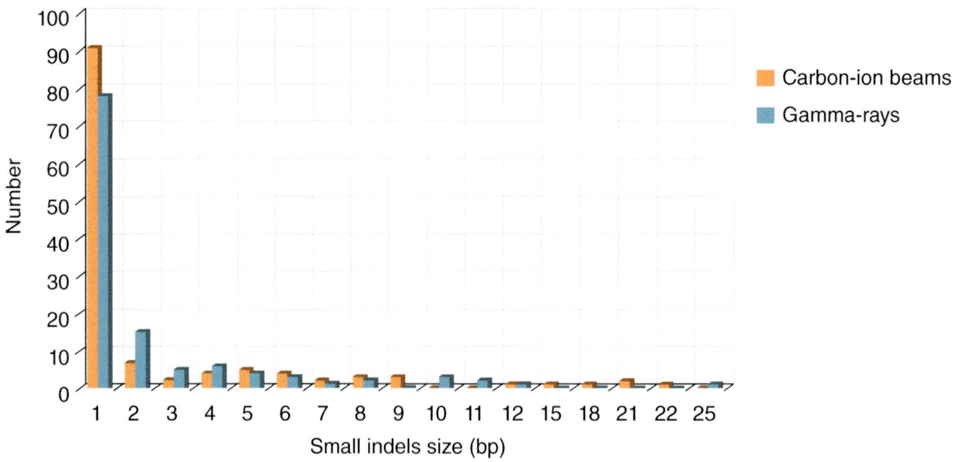

Fig. 46.4. Size distributions of small indels induced by carbon-ion beams and gamma-ray irradiation in the genomes of *Arabidopsis thaliana* M_3 lines.

or viability. Therefore, the small indels of interest in this study refer to insertions or deletion mutations of base < 30 bp. We previously thought that there might be some mutation hotspots in the genome. For example, CIB or gamma-ray radiation is likely to induce specific mutations on certain chromosomes. However, based on the rate of mutations on each chromosome in ten of the CIB and ten of gamma-ray mutagenesis lines, no so-called hotspots were observed.

In the present study, for SBSs, the ratio of transitions (Ts) to transversions (Tv) induced by CIB and gamma-ray were 1.01 and 1.42, which are quite different from the spontaneous SBSs (2.41) reported in the mutant accumulation line ($\chi^2 = 13.824$, $p = 0.001$) (Ossowski *et al.*, 2010) but are close to the SBSs (1) reported in the fast neutron-induced mutant line ($\chi^2 = 4.003$, $p = 0.135$) (Belfield *et al.*, 2012). A relatively low Ts/Tv ratio (0.92), compared with spontaneous mutations, was also characteristic of mutations observed in *in vitro* regenerant *A. thaliana* lines (Jiang *et al.*, 2011). Concerning other species, comparison of genome-wide mutations induced by ethyl methanesulfonate (EMS) and gamma irradiation in the tomato have been reported (Shirasawa *et al.*, 2016). The predominant mutations in the EMS mutants were C:G > T:A transitions, and the Ts/Tv ratio is 2.94; while for the gamma-ray mutants, C:G > T:A transitions, A:T > T:A transversions, A:T > G:C transitions and deletion mutations were equally common, with a Ts/Tv ratio of 0.87. Targeted exome sequencing of unselected heavy-ion beam irradiated rice populations showed that the total numbers of transition and transversion events were 271 and 302, respectively (ratio 0.90) (Ichida *et al.*, 2019). Therefore, for transitions, both chemical and physical mutagenesis demonstrate a bias of G:C > A:T transitions. Factors that affect the ratio of transition to transversion might be the types of mutagens, as well as the species.

Mutations involved in missense, nonsense, stop gained/lost and frame shift are predicted to affect gene function more probably. Therefore, we focused on these high-impact mutations which might disrupt the corresponding gene function. Based on genetic variation annotations and effect predictive analysis of the detected 454 mutations from ten re-sequenced lines induced by CIB, a total of 69 genes (that is, an average of approximately seven genes per genome) incurred a relatively high-impact mutation, which can disrupt the corresponding gene function. For ten mutant lines induced by gamma-ray radiation, it was found that 61 genes belong to relatively high-impact mutations. The average number of mutated genes for each line was about six per genome. A similar proportion of genes as that observed with gamma-ray radiations was affected by the CIB. Therefore, a small number of genes will be disrupted in the genomes of mutants. Based on the characteristics of heavy -ion beams and gamma-rays, both are very promising tools for plant mutation breeding and gene function study, as they can change some features of interest without interfering with other key features.

5 Conclusion

This study revealed SBSs and small indels mutations induced by high-energy CIB and gamma-ray radiations in *A. thaliana* at the whole genome level. In the current study, as an effective and unique physical mutagen, heavy-ion beam radiation can reach a mutagenesis effect similar to that with conventional photon radiations (such as gamma-rays) at a much lower dose (CIB, 200 Gy; gamma-ray, 1000 Gy). However, in this research, only the SBSs and small indels in the M_3 generation were taken into account as the limitation of the NGS technique. There is still a certain distance between the original mutations induced by irradiation and those of the advanced generation. However, we hope that our data can provide valuable clues for explaining the characteristics and potential mechanisms of heavy-ion beam and gamma-ray radiation on plant mutation breeding. We will continue to mine more information looking for insights into how radiations with different LETs induce plant mutations from multiple angles in the future.

References

Belfield, E.J. *et al.* (2012) Genome-wide analysis of mutations in mutant lineages selected following fast-neutron irradiation mutagenesis of *Arabidopsis thaliana*. *Genome Research* 22, 1306–1315.

Biermans, G. *et al.* (2015) *Arabidopsis thaliana* seedlings show an age-dependent response on growth and DNA repair after exposure to chronic gamma-radiation. *Environmental and Experimental Botany* 109, 122–130.

BIG Data Center Members (2017) The BIG Data Center: from deposition to integration to translation. *Nucleic Acids Research* 45(D1), D18–D24.

Bolon, Y.T. *et al.* (2011) Phenotypic and genomic analyses of a fast neutron mutant population resource in soybean. *Plant Physiology* 156, 240–253.

Chen, K. *et al.* (2009) BreakDancer: an algorithm for high-resolution mapping of genomic structural variation. *Nature Methods* 6, 677–676.

Cingolani, P. *et al.* (2012) A program for annotating and predicting the effects of single nucleotide polymorphisms, SnpEff: SNPs in the genome of *Drosophila melanogaster* strain w^{1118}; iso-2; iso-3. *Fly (Austin)* 6, 80–92.

Du, Y. *et al.* (2014) Mutagenic effects of carbon-ion irradiation on dry *Arabidopsis thaliana* seeds. *Mutation Research – Genetic Toxicology and Environmental Mutagenesis* 759, 28–36.

Du, Y. *et al.* (2017) Identification of substitutions and small insertion-deletions induced by carbon-ion beam irradiation in *Arabidopsis thaliana*. *Frontiers in Plant Science* 8, 1851.

Du, Y. *et al.* (2018) Strategies for identification of mutations induced by carbon-ion beam irradiation in *Arabidopsis thaliana* by whole genome re-sequencing. *Mutation Research – Fundamental and Molecular Mechanisms* 807, 21–30.

Hase, Y. *et al.* (2010) Efficient induction of flower-color mutants by ion beam irradiation in petunia seedlings treated with high sucrose concentration. *Plant Biotechnology* 27, 99–103.

Hase, Y. *et al.* (2018) Physiological status of plant tissue affects the frequency and types of mutations induced by carbon-ion irradiation in Arabidopsis. *Science Reports* 8, 1394–1404.

He, J.Y. *et al.* (2011) Pigment analysis of a color-leaf mutant in Wandering Jew (*Tradescantia fluminensis*) irradiated by carbon ions. *Nuclear Science and Techniques* 22, 77–83.

Hirano, T. *et al.* (2015) Comprehensive identification of mutations induced by heavy-ion beam irradiation in *Arabidopsis thaliana*. *Plant Journal* 82, 93–104.

Hwang, W.J. *et al.* (2015) Genome-wide analysis of mutations in a dwarf soybean mutant induced by fast neutron bombardment. *Euphytica* 203, 399–408.

Ichida, H. *et al.* (2019) Targeted exome sequencing of unselected heavy-ion beam-irradiated populations reveals less-biased mutation characteristics in the rice genome. *Plant Journal* 98, 301–314.

Ishikawa, S. *et al.* (2012) Ion-beam irradiation, gene identification, and marker-assisted breeding in the development of low-cadmium rice. *Proceedings National Academy of Sciences USA* 109, 19166–19171.

Jiang, C. *et al.* (2011) Regenerant *Arabidopsis* lineages display a distinct genome-wide spectrum of mutations conferring variant phenotypes. *Current Biology* 21, 1385–1390.

Kazama, Y. *et al.* (2011) Characterization of highly efficient heavy-ion mutagenesis in *Arabidopsis thaliana*. *BMC Plant Biology* 11, 161–171.

Kazama, Y. *et al.* (2017) Different mutational function of low- and high-linear energy transfer heavy-ion irradiation demonstrated by whole-genome resequencing of Arabidopsis mutants. *Plant Journal* 92, 1020–1030.

Koboldt, D.C. *et al.* (2012) VarScan 2: somatic mutation and copy number alteration discovery in cancer by exome sequencing. *Genome Research* 22, 568–576.

Li, G. *et al.* (2016) Genome-wide sequencing of 41 rice (*Oryza sativa* L.) mutated lines reveals diverse mutations induced by fast-neutron irradiation. *Molecular Plant* 9, 1078–1081.

Li, H. *et al.* (2009a) Fast and accurate short read alignment with Burrows-Wheeler transform. *Bioinformatics* 25, 1754–1760.

Li, H. *et al.* (2009b) The Sequence Alignment/Map format and SAMtools. *Bioinformatics* 25, 2078–2079.

Luo, S.W. *et al.* (2016) Mutagenic effects of carbon ion beam irradiations on dry *Lotus japonicus* seeds. *Nuclear Instruments and Methods in Physics Research, Section B* 383, 123–128.

Morita, R. *et al.* (2017) Heavy-ion beam mutagenesis identified an essential gene for chloroplast development under cold stress conditions during both early growth and tillering stages in rice. *Bioscience, Biotechnology and Biochemistry* 81, 271–282.

Oka-Kira, E. *et al.* (2005) *klavier* (klv), A novel hypernodulation mutant of *Lotus japonicus* affected in vascular tissue organization and floral induction. *Plant Journal* 44, 505–515.

Ossowski, S. *et al.* (2010) The rate and molecular spectrum of spontaneous mutations in *Arabidopsis thaliana*. *Science* 327, 92–94.

Robinson, J.T. *et al.* (2011) Integrative genomics viewer. *Nature Biotechnology* 29, 24–26.

Shikazono, N. *et al.* (2005) Analysis of mutations induced by carbon ions in *Arabidopsis thaliana*. *Journal of Experimental Botany* 56, 587–596.

Shirasawa, K. *et al.* (2016) Genome-wide survey of artificial mutations induced by ethyl methanesulfonate and gamma rays in tomato. *Plant Biotechnology Journal* 14, 51–60.

Tanaka, A. *et al.* (1997) A new *Arabidopsis* mutant induced by ion beams affects flavonoid synthesis with spotted pigmentation in testa. *Genes and Genetic Systems* 72, 141–148.

Tanaka, A. *et al.* (2002) An ultraviolet-B-resistant mutant with enhanced DNA repair in Arabidopsis. *Plant Physiology* 129, 64–71.

Tanaka, A. *et al.* (2010) Studies on biological effects of ion beams on lethality, molecular nature of mutation, mutation rate, and spectrum of mutation phenotype for mutation breeding in higher plants. *Journal of Radiation Research* 51, 223–233.

Van De Walle, J. *et al.* (2016) *Arabidopsis* plants exposed to gamma radiation in two successive generations show a different oxidative stress response. *Journal of Environmental Radioactivity* 165, 270–279.

Wang, Y. *et al.* (2017) GSA: Genome Sequence Archive. *Genomics, Proteomics & Bioinformatics* 15(1), 14–18.

Ye, K. *et al.* (2009) Pindel: a pattern growth approach to detect break points of large deletions and medium sized insertions from paired-end short reads. *Bioinformatics* 25, 2865–2871.

Yu, L.X. *et al.* (2016) Flower color mutants induced by carbon ion beam irradiation of geranium (*Pelargonium × hortorum*, Bailey). *Nuclear Science and Techniques* 27, 112–119.

47 The Conservation and Sustainable Use of Plant Genetic Resources for Food and Agriculture and Emerging Biotechnologies

Chikelu Mba[1] and Hans Dreyer

Abstract

The 50% increase in food production required to feed an ever-growing global population, and which must be attained under dire climate change scenarios and other constraints, will not be attained with a 'business as usual' mindset. For crops, the current cultivars will have to be replaced by ones that are more nutritious, stress tolerant and input-use efficient and that would produce higher yields with less external input. Generating such varieties requires significant efficiency enhancements to the conservation and characterization of plant genetic resources for food and agriculture and their use in plant breeding. Genome editing holds great promise in this regard. Its rapid adoption as a relatively cheap and rapid means to generate precise and predictable heritable variations and its universal applicability mirror the developments of the closely associated gene drive. Large amounts of digital sequence data are also increasingly available, while the field of synthetic biology has been expanding rapidly. This all holds great promise for improving and broadening the genetic base of crop varieties for the enhancement of crop productivity without damaging the environment. However, the pace of the scientific and technological developments for these methods has far outstripped that of the requisite policy regimes. The demonstrable potentials notwithstanding, the developments have not been universally accepted. The ongoing debates include whether the products of genome editing, with or without gene drive, should be considered living modified organisms and, if so, subject to the international framework, the Cartagena Protocol on Biosafety to the Convention on Biological Diversity. Another debate is whether digital sequence information should be subject to some access-and-benefit sharing regime, considering that, with the power of synthetic biology, products previously harnessed only from living organisms can now be produced in the laboratory once the DNA sequence is available. There are also debates about ethics. In order to avoid the mistakes of the past, a call is made for evidence-based multi-stakeholder (including especially intergovernmental) dialogues on the safety, fairness and ethics of the use of these emerging biotechnologies, as the stakes are extremely high.

Keywords: genome editing • gene drive • digital sequence data • synthetic biology • Plant Genetic Resources for Food and Agriculture

1 Introduction

The international community has, through the Sustainable Development Goals (UN General Assembly, 2015; FAO, 2017a), inter alia, committed to achieve a world free of hunger and malnutrition. Yet, barely a decade from the set target of 2030, over 821 million people are chronically

[1] Email: chikelu.mba@fao.org

 Mutation Breeding, Genetic Diversity and Crop Adaptation to Climate Change (eds S. Sivasankar *et al.*)
DOI: 10.1079/9781789249095.0047

hungry (FAO, IFAD, UNICEF, WFP and WHO, 2018) while over 2.5 billion people suffer from malnutrition globally (WHO, 2019). In fact, agricultural production must be increased sustainably, by over 50%, in order to nourish the ever-increasing human population expected to exceed 9.2 billion by 2050 (FAO, 2017a). Though crop production, for instance, has been increasing steadily for several decades, the projected demands will not be met at the current rates; a 37% rate increase is required (Tester and Langridge, 2010). Climate change – which manifests as erratic extreme weather events and new strains and biotypes of pests and diseases – exacerbates these daunting constraints (Evans, 2009; Nelson *et al.*, 2009; Hertel *et al.*, 2010; Seeberg-Elverfeldt and Tapio-Biström, 2010; Beddington *et al.*, 2011; von Grebmer *et al.*, 2011; Rosegrant, 2011; Field *et al.*, 2012; FAO, 2016a). With the increasing scarcity of arable land and water resources coupled with the prohibitive economic, health and environmental costs of excessive use of external inputs, enhanced productivity, i.e. producing more yield per unit of input, would be the most sustainable means to feed an ever-increasing population under climate change scenarios (FAO, 2011a, 2013, 2016b). In this regard, farmers need a diverse suite of the widest possible spectrum of nutritious well-adapted crops and varieties (FAO, 2011a), a goal that can only be attained through a significantly increased harnessing of the diversity that undergirds our crop production system (Mba *et al.*, 2012a).

2 Harnessing the Potential of Plant Genetic Resources for Food and Agriculture

Wild ancestors or related species of modern crops, i.e. crop wild relatives (CWR); wild plants harvested for food; unimproved and non-adapted materials; farmers' varieties/landraces and modern varieties developed by breeders – these are some examples of the diversity of plants that is critical for food security and nutrition. Generally known as plant genetic resources for food and agriculture (FAO, 2002), they may be found in genebanks, on-farm and in nature. The extremely narrow genetic base of crops and breeding materials (Martynov and Dobrotvorskaya, 2006) must be broadened in order to generate the diverse suite of well-adapted, nutritious, input-use efficient, biotic and abiotic stresses-resistant and productive crop varieties that meet the preferences of farmers and end-users. The Second Global Plan of Action (Second GPA) for Plant Genetic Resources for Food and Agriculture (PGRFA), consisting of 18 Priority Activities – which countries implement – is the internationally agreed framework for the conservation and sustainable use of PGRFA (FAO, 2011b). The Priority Activities are divided into four main areas: (i) *in situ* conservation and management; (ii) *ex situ* conservation; (iii) sustainable use; and (iv) building sustainable institutional and human capacities. The Food and Agriculture Organization (FAO) of the United Nations (UN) and partners, in providing support to countries for the implementation of the Second GPA, advocate a continuum approach to the management of PGRFA whereby institutions and their activities devoted to the conservation of PGRFA dovetail neatly with those for pre-breeding and plant breeding which, in turn, are seamlessly linked to the mechanisms for the delivery of quality seeds and planting materials of the ensuing improved crop varieties (Mba *et al.*, 2012b). The strengthening of institutional and human capacities is ideally implemented as a cross-cutting theme to ensure that adequate capacities are available for each of the three modules. Various applications of biotechnology, defined in the Convention on Biological Diversity (CBD) as 'any technology application that uses biological systems, living organisms, or derivatives thereof, to make or modify products or processes for specific use' (UN, 1992), are used in both the conservation of PGRFA and their use, especially in crop improvement.

3 Novel Biotechnologies for the Conservation and Sustainable Use of PGRFA

The continuing rapid advances in science and technology are resulting in novel biotechnological applications, procedures and tools with demonstrated potentials to increase significantly the efficiency, throughput and precision of the processes for the conservation, characterization and evaluation of PGRFA and for generating crop varieties with improved attributes. Some of the most critically important ones are discussed below.

3.1 Genome editing

The 1953 elucidation of the structure of deoxyribonucleic acid (DNA) was an early success for the discipline of molecular biology. Other subsequent milestones, each demonstrating more than the previous one an even greater power to characterize and/or otherwise manipulate heredity at previously unimaginable scopes, depths and precision, have included the polymerase chain reaction (PCR), DNA sequencing, the Human Genome Project, etc. A notable impactful outcome of these advances in the 20th century is recombinant DNA technology, whereby hereditary materials from more than one source are brought together artificially in the laboratory in order to create *de novo* DNA sequences which may lead to the expression of novel traits. Genetic engineering, genetic transformation or genetic modification (GM), used to generate genetically modified organisms (GMOs), is one such application of recombinant DNA technology.

Controversies about the release of GMOs into the environment and their use in food, livestock feeds, vaccine production, etc. are well documented (Domingo and Bordonaba, 2011). The global community, through the UN system, in recognizing the potential of GM and at the same time acknowledging concerns about the safety of GM products, agreed on the Cartagena Protocol on Biosafety to the Convention on Biological Diversity (SCBD, 2000) as a means for the exploitation of the powers of genetic engineering for the benefit of mankind without endangering the environment or human health. This has provided a unifying reference framework that relies on evidence.

Since the entry into force of the Cartagena Protocol, biotechnologies in general, and molecular biology in particular, have continued to evolve. Genome editing (or gene editing) refers to a relatively new set of technologies that are being used to make precise changes to the genetic makeup of living organisms (Gupta and Musunuru, 2014; Hsu *et al.*, 2014; Trevino and Zhang, 2014; Jiang and Marraffini, 2015; Sternberg and Doudna, 2015; Langner *et al.*, 2018). They can be used to cause heritable changes (i.e. mutations) by inserting, deleting, or substituting one or more specific nucleotides at a predetermined location in the organism's genome, such as a gene. The resulting mutant may be beneficial for agricultural production and nutrition. In fact, gene editing has been used to develop new crop varieties (Waltz, 2016; Bomgardner, 2017) and breeds of livestock (Ainsworth, 2015) and to edit a human embryo correcting a gene variant that causes a potentially life-threatening heart disease (Zeng *et al.*, 2018) and for studying fertility (Fogarty *et al.*, 2017). Genome editing is relatively inexpensive so its continuing rapid expansion, including in developing countries, should be expected.

Policies have not been developed to keep pace with the scientific and technological advances that derive from genome editing. For instance, there is no unanimity as to whether or not genome edited organisms are GMOs, and if they are, whether they are to be subjected to regulations covered by the Cartagena Protocol or whether a new framework is called for. Developed countries have been evaluating this subject and taking decisions; for example, the United States Department of Agriculture (USDA), having decided that genome edited crops 'could otherwise have been developed through traditional breeding techniques', will not regulate them (Waltz, 2016; USDA, 2018). At the other end of the spectrum, New Zealand has deemed products of genome editing as requiring the same regulations as GMOs (Fritsche, 2018). A recent ruling of the apex court of the European Union implies that genome-edited organisms should be treated in the same way as GMOs by its member countries (Callaway, 2018). However, the majority of developing countries do not appear to have commenced consultations.

3.2 Gene drive

The probability of an offspring inheriting a variant of a gene from a parent can be accurately predicted. Bias away from an expected frequency towards the prevalence of the inheritance of a particular allele or set of alleles on account of a directed mechanism, for example through genetic engineering, is known as gene drive (DiCarlo *et al.*, 2015; Gantz and Bier, 2015). With the recent demonstration that genome editing could be accompanied by gene drive, genome-edited mutant mosquito populations have been produced with a gene drive for a preponderance of males, heritable infertility in females, or a reduced ability to transmit malaria, dengue or zika diseases (Callaway, 2015, 2017; Ledford and Callaway, 2015)

While the benefits of using gene drive mosquitoes to combat these deadly diseases is undeniable, there is also a concern about the possibility of an unintended consequence: skewing permanently the population dynamics of the organism and, by extension, the overall ecosystem. Such gene drive processes have been criticized, with an international moratorium on its research and development having been unsuccessfully advocated (Callaway, 2016). Just as is the case with genome editing, there are scant dedicated policies and certainly no global mechanism for regulating this powerful technology, which has demonstrable environmental and ethical issues.

3.3 Synthetic biology

In the absence (yet) of an internationally agreed definition for synthetic biology, common distinguishing features of such applications as are categorized under this term include 'the *de novo* synthesis of genetic material and an engineering-based approach to develop components, organisms and products' (SCBD, 2015). In effect, by harnessing the advances in biology, chemistry, computer science and engineering in a concerted way, scientists are able to create DNA sequences from scratch and sequences are designed to create an organism that did not previously exist, such as 'Synthetic Yeast 2.0' and 'Mycoplasma mycoides JCVI-syn1.0' (Gibson *et al.*, 2010). Basically, scientists armed with computers and laboratory chemicals design from scratch organisms that do new things, for example producing biofuels or excreting the precursors of medical drugs. A number of products, including artemisinic acid (an anti-malarial drug), are either being commercially produced or in the pipeline.

3.4 Access and benefit sharing for digital sequence information

Amongst the principles that underpin global efforts to conserve and use genetic resources sustainably is access and benefit-sharing (ABS), which 'refers to the way in which genetic resources may be accessed, and how the benefits that result from their use are shared between the people or countries using the resources (users) and the people or countries that provide them (providers)' (SCBD, 2010). Of note, the International Treaty on Plant Genetic Resources for Food and Agriculture, adopted by the FAO Conference in 2001, became the first legally binding and operational international instrument for ABS for genetic resources. Subsequently, the Nagoya Protocol on Access to Genetic Resources and the Fair and Equitable Sharing of Benefits Arising from their Utilization to the Convention on Biological Diversity, 'an international agreement which aims at sharing the benefits arising from the utilization of genetic resources in a fair and equitable way' (SCBD, 2011) entered into force in October 2014.

Taken together, the rapid advances in molecular biology and the ancillary engineering and computer science disciplines, increasing human and institutional capacities and significantly lowered costs are resulting in huge amounts of DNA sequence data becoming available. Whole genomes can now be sequenced, and the resulting data curated at a miniscule fraction of the costs of barely 20 years ago. The ownership of the sequence data of genetic resources has recently become topical. This is mostly because of biopiracy concerns stemming from the belief that, with synthetic biology capabilities, whole organisms or just genes for commercial products could be created from digital sequence information in the laboratory (Servick, 2016). At one end of the spectrum, arguments are proffered in support of making the data available freely in support of research and development and innovation, while at the other end there are proponents for subjecting this digital information to some ABS regime. There is as yet no consensus.

3.5 Efforts to foster international agreements

Some of the factors that have constrained the widespread adoption of the products of biotechnologies, especially in food and agricultural systems – including the conservation and sustainable use of PGRFA – relate to concerns for biosafety, health, ethics, equity, fairness, intellectual property rights, public perception and acceptance and sub-optimal institutional and human capacities. The international community is certainly mindful of the pitfalls of the past and is leveraging different forums to achieve a common understanding and, hopefully, develop universal norms to guide the applications of the novel biotechnologies.

The CBD has been a leading convener of international mechanisms to devise a common understanding of, and norms for, the safe deployment of the products of biotechnologies. Initially focused on GMOs, CBD's work on biotechnologies now includes aspects of the newer techniques. For instance, sessions of the Conference of the Parties to the Cartagena Protocol on Biosafety and the Biosafety Clearing House provide platforms for international engagements on the safe use of GMOs, while National Biosafety Frameworks reflect activities in regard to countries' commitments to adhering to the Protocol. The CBD, at the 13th and 14th Sessions of its Committee of Parties in December 2016 and November 2018, respectively, decided to extend both the mandate of the current Ad Hoc Technical Expert Group on Synthetic Biology and the duration of the open-ended online forum (https://bch.cbd.int/synbio/open-ended/discussion, accessed 2019) to support its work (SCBD, 2016a). Also at the 13th session, the CBD, in noting 'the rapid advances arising from research and development in biotechnology regarding the use of digital sequence information on genetic resources and therefore recognizing the importance of addressing this matter in the framework of the Convention in a timely manner', invited the submission of views and relevant information from stakeholders and decided to establish an Ad Hoc Technical Expert Group on the theme (SCBD, 2016b, 2018a). Subsequently, at its 14th session in 2018, the CBD decided to establish a science- and policy-based process and an extended Ad Hoc Technical Expert Group on the subject (SCBD, 2018b).

On its part, in 2016, FAO organized an international symposium on 'The role of agricultural biotechnologies in sustainable food systems and nutrition' in Rome, Italy (FAO, 2016c), followed by two regional meetings – for Asia and Pacific (FAO, 2018a) and sub-Saharan Africa (FAO, 2018b) – in 2017. These underscored the critical importance of biotechnologies as part of the toolbox for supporting family farmers in their efforts to feed the world. Also, FAO's statutory bodies, especially its Commission on Genetic Resources for Food and Agriculture ('the Commission') and the sectoral Committees on Agriculture, Forestry and Fishery, all consider the applications of biotechnologies. The Commission, for instance, has consistently requested FAO both to strengthen national capacities in biotechnologies and to disseminate updated information on relevant trends and agreed that biotechnology-related issues be retained in its Multi-Year Programme of Work (FAO, 2011c, 2015a,b). The Commission had previously commenced action on the development of a Code of Conduct on Biotechnology (FAO, 2007). Additionally, the International Treaty on Plant Genetic Resources for Food and Agriculture, especially through its Ad Hoc Open-ended Working Group to Enhance the Functioning of its Multilateral System, is considering the issues relating to genetic information associated with the material accessed from the Multilateral System (FAO, 2017b).

The 'Future of Biotechnology' is one of the Global Future Councils Network of the World Economic Forum (WEF) (2019). It has the mandate to 'explore how developments in Biotechnology could impact industry (especially agriculture and health), governments and society in the future, and design innovative governance models that ensure that their benefits are maximized, and the associated risks kept under control' (https://www.weforum.org/communities/the-future-of-biotechnology, accessed 2019). The WEF has conducted a series of multi-stakeholder dialogues – with industry perspective – on this subject. Both the European Commission (2019) and the Organization for Economic Co-operation and Development (OECD, 2019) have programmes on biotechnology.

4 Discussion

A 'business as usual' mindset will be ineffective in achieving the significantly increased agricultural productivity needed to feed an ever-growing population with safe and nutritious food without damaging the environment. Deploying the same crop varieties or breeding new ones using the same parental lines or following the same methods, for instance, will not result in the desired 37% rate increase (Tester and Langridge, 2010) for crop production (Mba *et al.*, 2012a). The narrow genetic base of the parents used in breeding (Martynov and Dobrotvorskaya, 2006) must be broadened significantly (Rabinovich, 1998; Gur and Zamir, 2004; McCouch, 2004). Novel biotechnologies hold great promise for the induction, discovery and exploitation of new heritable variation needed for producing the desired nutritious and highly productive crop varieties. However, despite the demonstrable power of novel biotechnologies, especially genome editing, gene drive and various aspects of synthetic biology, it is becoming increasingly obvious that society requires a lot more than the demonstration of proofs of concept in order to accept products of new biotechnologies. These are concerns that should not be ignored.

While the activities of the multi-stakeholder platforms described above are helpful in enabling a common understanding and seeking ways to allay the various concerns that have been identified, their impacts are hard to measure currently. The danger is that the longer the studies, consultations and negotiations drag on, the more fractured societies become and the more difficult it would be to devise widely acceptable mechanisms. The diametrically opposed positions which countries have already adopted regarding genome editing, for instance, is symptomatic of this chasm – which may be too late to bridge. The gap between developed and developing countries in the applications of novel biotechnologies is also extremely wide. The characterization of genome editing as cheap, for instance, cannot be taken as universal. The developing countries that currently do not have functional facilities for routine molecular biology – because of cost and sub-optimal human capital and infrastructure – are unlikely to leapfrog to genome editing. The continuing divide certainly does not help with the passionate discourses around access and benefit-sharing.

The dilemma facing the international development community is enormous. Confronted with unprecedented challenges, it is extremely difficult to forego solutions that offer promise, such as genome editing. But it is equally difficult to ignore the risks that are being enumerated. The demonstration of the feasibility of editing the genome of a human embryo and the reported delivery of a genome-edited baby (Cyranoski and Ledford, 2018), for instance, underscore profound ethical issues that the international community must address. A plausible collateral damage for inaction or ineffectual attempts at consensus building is that the ripple effects from public outrage and fear could easily impact quite negatively on the applications of genome editing to food and agriculture. It is critically important to devise and agree on evidence-based modalities for decision making. Principles, such as those pertaining to the imperative of biosafety, might be universally applicable but the solutions must be specific with respect to the species, crop types and geographical locations. For instance, from a biosafety standpoint, it would be necessary to determine whether the deployment of a genome-edited outcrossing species in its centre of genetic diversity should be handled with the same protocol as a self-pollinating (or a clonally propagated) one being considered for release in a geographical location where its CWR are not available.

Scientists must also engage with the wider society. While it may be impractical to expect that a critical mass of scientists would begin to attend sessions of the CBD or FAO's governing bodies, they should none the less convey frankly to their nations' representatives at these normative bodies the aims of their research; the methods they adopt – and why; alternative options to the same goal; cost–benefit analyses; importance of work to society's well-being in terms of the economy, nutrition, health, the environment, etc.; and, importantly, the associated risks. There are also supposedly knotty issues that only scientists can help untangle. For instance, is genome editing synonymous with gene drive as sometimes implied? Or is gene drive necessarily associated with genome editing? Governments have as much responsibility to facilitate such dialogues as to legislate on or otherwise regulate novel biotechnologies. Information dissemination is also important as it enables people to make reasoned judgements about what is or is not desirable.

5 Conclusion

In summary, therefore, the rapid pace of the development of the scientific and technological methodologies and infrastructure for novel biotechnologies have far outstripped those of the requisite enabling policy regimes. This is clearly unsustainable. Every member of society is a stakeholder and therefore has a role to play –

through communicating effectively, facilitating engagement and listening actively to dissenting opinions with an open mind. The stakes are too high to continue to entertain the risk of the baby being ultimately jettisoned with the dirty bathwater. We advocate strengthened multilateralism, which has worked well in the past, as a means to ensuring that the benefits are enjoyed as fully as the risks are mitigated and issues of fairness, equity and ethics are addressed satisfactorily. It is essential that evidence-based benefits to society are not subjugated to commercial or even jingoistic considerations. In the same vein, every concern must be addressed and acceptable means for mitigating every risk devised. Innovations are as important as the responsibility that must accompany their deployment.

References

Ainsworth, C. (2015) A new breed of edits. *Nature* 528, S15–S16.

Beddington, J. *et al.* (2011) *Achieving food security in the face of climate change: summary for policy makers from the Commission on Sustainable Agriculture and Climate Change*. CGIAR Research Program on Climate Change, Agriculture and Food Security (CCAFS), Copenhagen, Denmark.

Bomgardner, M.M. (2017) CRISPR: A new toolbox for better crops. *Chemical & Engineering News* 95 (24), 30–34.

Callaway, E. (2015) Mosquitoes engineered to pass down genes that would wipe out their species. *Nature.* doi: 10.1038/nature.2015.18974

Callaway, E. (2016) 'Gene drive' moratorium shot down at UN biodiversity meeting. *Nature.* doi: 10.1038/nature.2016.21216

Callaway, E. (2017) Gene drives thwarted by emergence of resistant organisms. *Nature* 542, 15. doi: 10.1038/542015a.

Callaway, E. (2018) EU law deals blow to CRISPR crops. *Nature* 560, 16.

Cyranoski, D. and Ledford, H. (2018) International outcry over genome-edited baby claim. *Nature* 563, 607–608.

DiCarlo, J.E. *et al.* (2015) Safeguarding CRISPR-Cas9 gene drives in yeast. *Nature Biotechnology* 33(12), 1250–1257.

Domingo, J.L. and Bordonaba, J.G. (2011) A literature review on the safety assessment of genetically modified plants. *Environment International* 37, 734–742.

European Commission (2019) Biotechnology. Available at: http://ec.europa.eu/growth/sectors/biotechnology_en (accessed 20 January 2019)

Evans, A. (2009) *The Feeding of the Nine Billion: Global Food Security for the 21st Century*. Chatham House, London.

FAO (2002) *International Treaty on Plant Genetic Resources for Food and Agriculture*. Food and Agriculture Organization of the United Nations, Rome. Available at: http://www.fao.org/plant-treaty/overview/texts-treaty/en/ (accessed 9 April 2019)

FAO (2007) *Progress on the Draft Code of Conduct on Biotechnology as It Relates to Genetic Resources for Food and Agriculture: Policy Issues, Gaps and Duplications*. Eleventh Regular Session Commission on Genetic Resources for Food and Agriculture Rome, Italy, 11–15 June 2007. Food and Agriculture Organization of the United Nations, Rome (CGRFA-11/07/13).

FAO (2011a) *Save and Grow. A policymaker's guide to the sustainable intensification of smallholder crop production*. Food and Agriculture Organization of the United Nations, Rome.

FAO (2011b) *Second Global Plan of Action on Plant Genetic Resources for Food and Agriculture*. Food and Agriculture Organization of the United Nations, Rome.

FAO (2011c) *Report of the Thirteenth Regular Session Commission on Genetic Resources for Food and Agriculture*, Rome, Italy, 18–22 July 2011. Food and Agriculture Organization of the United Nations, Rome (CGRFA-13/11/Report).

FAO (2013) *Save and Grow: Cassava. A guide to sustainable production intensification*. Food and Agriculture Organization of the United Nations, Rome.

FAO (2015a) *Draft Elements to Facilitate Domestic Implementation of Access and Benefit-Sharing for Different Subsectors of Genetic Resources for Food and Agriculture*. Fifteenth Regular Session Commission on Genetic Resources for Food and Agriculture Rome, Italy, 19–23 January 2015. Food and Agriculture Organization of the United Nations, Rome (CGRFA-15/15/5).

FAO (2015b) *Report of the Fifteenth Regular Session Commission on Genetic Resources for Food and Agriculture*, Rome, Italy, 19–23 January 2015. Food and Agriculture Organization of the United Nations, Rome (CGRFA-15/15/Report).

FAO (2016a) *The State of Food and Agriculture: Climate Change, Agriculture and Food Security*. Food and Agriculture Organization of the United Nations, Rome.

FAO (2016b) *Save and Grow in practice: maize, rice, wheat. A guide to sustainable cereal production*. Food and Agriculture Organization of the United Nations, Rome.

FAO (2016c) *Proceedings of the FAO International Symposium on the Role of Agricultural Biotechnologies in Sustainable Food Systems and Nutrition*. Food and Agriculture Organization of the United Nations, Rome.

FAO (2017a) *The future of food and agriculture – trends and challenges*. Food and Agriculture Organization of the United Nations, Rome.

FAO (2017b) *Consideration of Issues Regarding Genetic Information Associated with Material Accessed from the Multilateral System: Note by the Co-Chairs*. Sixth Meeting of the Ad-Hoc Open-Ended Working Group to Enhance the Functioning of the Multilateral System. International Treaty on Plant Genetic Resources for Food and Agriculture, Rome, Italy, 14–17 March 2017. Food and Agriculture Organization of the United Nations, Rome (IT/OWG-EFMLS-6/17/Inf.8).

FAO (2018a) *Report on the Outcome of the FAO Regional Meeting on Agricultural Biotechnologies in Sustainable Food Systems and Nutrition in Asia-Pacific*. Thirty-fourth Session FAO Regional Conference for Asia and the Pacific, Nadi, Fiji, 9–13 April 2018. Food and Agriculture Organization of the United Nations, Rome (APRC/18/INF/9). 4 pages.

FAO (2018b) *Outcomes of the FAO Regional Meeting on Agricultural Biotechnologies in Sustainable Food Systems and Nutrition in Sub-Saharan Africa*. Thirtieth Session FAO Regional Conference for Africa, Khartoum, the Sudan, 19–23 February 2018. Food and Agriculture Organization of the United Nations, Rome (ARC/18/INF/10). 8 pages.

FAO, IFAD, UNICEF, WFP and WHO (2018) *The State of Food Security and Nutrition in the World 2018. Building climate resilience for food security and nutrition*. Food and Agriculture Organization of the United Nations, Rome.

Field, C.B. *et al*. (eds) (2012) *Managing the Risks of Extreme Events and Disasters to Advance Climate Change Adaptation: Summary for Policymakers*. Intergovernmental Panel on Climate Change (IPCC). Cambridge University Press, Cambridge, UK.

Fogarty, N.M.E. *et al*. (2017) Genome editing reveals a role for OCT4 in human embryogenesis. *Nature* 550, 67–73.

Fritsche, S. (2018) A New Zealand Perspective on the Application and Regulation of Gene Editing. *Frontiers in Plant Science* 9, 1323.

Gantz, V.M. and Bier, E. (2015) The mutagenic chain reaction: A method for converting heterozygous to homozygous mutations. *Science* 348(6233), 442–444.

Gibson, D.G. *et al*. (2010) Creation of a bacterial cell controlled by a chemically synthesized genome. *Science* 329(5987), 52–56.

Gupta, R.M. and Musunuru, K. (2014) Expanding the genetic editing tool kit: ZFNs, TALENs, and CRISPR-Cas9. *Journal of Clinical Investigation* 124(10), 4154–4161.

Gur, A. and Zamir, D. (2004) Unused natural variation can lift yield barriers in plant breeding. *PLoS Biology* 2, 1610–1615.

Hertel, T.W. *et al*. (2010) The poverty implications of climate-induced crop yield changes by 2030. *Global Environmental Change* 20, 577–585.

Hsu, P.D. *et al*. (2014) Development and Applications of CRISPR-Cas9 for Genome Engineering. *Cell* 157(6), 1262–1278.

Jiang, W. and Marraffini, L.A. (2015) CRISPR-Cas: new tools for genetic manipulations from bacterial immunity systems. *Annual Review of Microbiology* 69, 209–228.

Langner, T. *et al*. (2018) CRISPR crops: plant genome editing toward disease resistance. *Annual Review of Phytopathology* 56, 479–512.

Ledford, H. and Callaway, E. (2015) 'Gene drive' mosquitoes engineered to fight malaria. *Nature*. doi: 10.1038/nature.2015.18858.

Martynov, S.P. and Dobrotvorskaya, T.V. (2006) Genealogical analysis of diversity of Russian winter wheat cultivars (*Triticum aestivum* L.). *Genetic Resources and Crop Evolution* 53, 379–386.

Mba, C. *et al*. (2012a) Re-orienting crop improvement for the changing climatic conditions of the 21st century. *Agriculture & Food Security* 1, 7.

Mba, C. *et al.* (2012b) Mainstreaming the continuum approach to the management of plant genetic resources for food and agriculture through national strategy. *Plant Genetic Resources: Characterization and Utilization* 10(1), 24–37.

McCouch, S. (2004) Diversifying selection in plant breeding. *PLoS Biology* 2(10), e347.

Nelson, G.C. *et al.* (2009) *Climate Change: Impact on Agriculture and Costs of Adaptation. Food Policy Report No. 21*. International Food Policy Research Institute, Washington, DC.

OECD (2019) *Emerging Technologies. Organization for Economic Co-operation and Development.* Available at: http://www.oecd.org/sti/emerging-tech/ (accessed 20 January 2019)

Rabinovich, S.V. (1998) Importance of wheat-rye translocations for breeding modern cultivars of *Triticum aestivum* L. *Euphytica* 100, 323–340.

Rosegrant, M.W. (2011) Impacts of climate change on food security and livelihoods. In: Solh, M. and Saxena, M.C. (eds) *Food Security and Climate Change in Dry Areas*. Proceedings of an International Conference, 1–4 February 2010, Amman, Jordan. International Center for Agricultural Research in the Dry Areas (ICARDA), Aleppo, Syria, pp. 24–26.

SCBD (2000) *Cartagena Protocol on Biosafety to the Convention on Biological Diversity: text and annexes*. Montreal, Canada. Available at: http://bch.cbd.int/protocol/text/ (accessed 9 April 2019)

SCBD (2010) *Introduction to access and benefit-sharing.* Secretariat of the Convention on Biological Diversity, Montreal, Canada.

SCBD (2011) *Nagoya Protocol on Access to Genetic Resources and the Fair and Equitable Sharing of Benefits Arising from their Utilization to the Convention on Biological Diversity*. Secretariat of the Convention on Biological Diversity, Montreal, Canada. Available at: https://www.cbd.int/abs/about/ (accessed 9 April 2019)

SCBD (2015) *Synthetic biology*. Technical Series No. 82. Secretariat of the Convention on Biological Diversity, Montreal, Canada. 118 pages.

SCBD (2016a) *Decision Adopted by the Conference of the Parties to the Convention on Biological Diversity. Synthetic biology*. Conference of the Parties to the Convention on Biological Diversity Thirteenth meeting Cancun, Mexico, 4–17 December 2016 (CBD/COP/DEC/XIII/17). Secretariat of the Convention on Biological Diversity, Montreal, Canada.

SCBD (2016b) *Decision Adopted by the Conference of the Parties to the Convention on Biological Diversity. Digital sequence information on genetic resources*. Conference of the Parties to the Convention on Biological Diversity Thirteenth meeting Cancun, Mexico, 4–17 December 2016 (CBD/COP/DEC/XIII/16). Secretariat of the Convention on Biological Diversity, Montreal, Canada.

SCBD (2018a) *Decision Adopted by the Conference of the Parties to the Convention on Biological Diversity. Synthetic biology*. Conference of the Parties to the Convention on Biological Diversity Fourteenth meeting, Sharm El-Sheikh, Egypt, 17–29 November 2018 (CBD/COP/DEC/14/19). Secretariat of the Convention on Biological Diversity, Montreal, Canada.

SCBD (2018b) *Decision Adopted by the Conference of the Parties to the Convention on Biological Diversity. Digital sequence information on genetic resources*. Conference of the Parties to the Convention on Biological Diversity Fourteenth meeting, Sharm El-Sheikh, Egypt, 17–29 November 2018 (CBD/COP/DEC/14/20). Secretariat of the Convention on Biological Diversity, Montreal, Canada.

Seeberg-Elverfeldt, C. and Tapio-Biström, M.-L. (2010) *Global Survey of Agricultural Mitigation Projects*. Mitigation of Climate Change in Agriculture Series No. 1. Food and Agriculture Organization of the United Nations, Rome.

Servick, K. (2016) Rise of digital DNA raises biopiracy fears. *Science.* doi: 10.1126/science.aal0395.

Sternberg, S.H. and Doudna, J.A. (2015) Expanding the Biologist's Toolkit with CRISPR-Cas9. *Molecular Cell* 58(4), 568–574.

Tester, M. and Langridge, P. (2010) Breeding technologies to increase crop production in a changing world. *Science* 327, 818–822.

Trevino, A.E. and Zhang, F. (2014) Genome editing using Cas9 nickases. *Methods in Enzymology* 546, 161–174.

UN (1992) *Convention on Biological Diversity*. United Nations Organization, New York.

UN General Assembly (2015) *Transforming our world: the 2030 Agenda for Sustainable Development*, 21 October 2015, A/RES/70/1. Available at: https://www.refworld.org/docid/57b6e3e44.html (accessed 20 January 2019)

USDA (2018) Press Release No. 0070.18. Available at: https://www.usda.gov/media/press-releases/2018/03/28/secretary-perdue-issues-usda-statement-plant-breeding-innovation (accessed 19 January 2019) von Grebmer, K. *et al.* (2011) *2011 Global Hunger Index: The Challenge of Hunger: Taming Price Spikes and Excessive Food Price Volatility*. Welthungerhilfe, International Food Policy Research Institute (IFPRI), and Concern Worldwide. Bonn, Washington, DC, and Dublin.

Waltz, E. (2016) Gene-edited CRISPR mushroom escapes US regulation. *Nature* 532, 293.

WHO (2019) Malnutrition. Available at: https://www.who.int/news-room/fact-sheets/detail/malnutrition (accessed 8 April 2019)

World Economic Forum (2019) *Global Future Council on Biotechnology*. Available at: https://www.weforum.org/communities/the-future-of-biotechnology (accessed 20 January 2019)

Zeng, Y. *et al.* (2018) Correction of the Marfan syndrome pathogenic FBN1 mutation by base editing in human cells and heterozygous embryos. *Molecular Therapy* 26(11), 2631–2637.

48 Gamma-ray Induced Pedigreed Mutant Population of Tossa Jute (*Corchorus olitorius* L.): a Key Resource for Forward and Reverse Genetics

Shashi Bhushan Choudhary[1], Sanjay Kumar Jambhulkar, Hariom Kumar Sharma, A. Anil Kumar, Neetu Kumari and Dileep Kumar

Abstract

Narrow genetic diversity in available germplasm is a serious limiting factor for academic progress and agronomic improvement of crops like *Corchorus olitorius*, an economically important bast fibre crop. Mutation breeding, with its proven ability to improve qualitative as well as quantitative traits, can be employed to augment germplasm diversity. In the present study, gamma-rays were used to treat the seeds of two promising varieties, JRO 204 and JRO 8432; LD_{50} doses for gamma-rays were 200 Gy and 300 Gy for JRO 204 and JRO 8432, respectively. Irradiation of two varieties has resulted in the development of a large number of macro-mutants, such as twisted bark, extreme dwarf, non-abscission leaf, soft stem, hard stem and round pod mutants. Morphological and anatomical studies of these mutants gave new light on secondary growth in the species. In addition to the academic utility, these mutants will prove of immense importance to plant breeders aiming to improve fibre quality. Moreover, novel mutants will help to develop new plant architecture suitable for diversified applications of the genus.

Keywords: *Corchorus* • mutation • gamma-ray • pedigreed mutant population

1 Introduction

Mutation breeding is the quickest tool available to plant breeders for the improvement of crops in terms of productivity, quality or stability (Bolbhat and Dhumal, 2009). Over the years it has helped to improve crop yield and quality across species (Das *et al.*, 2000; Javed *et al.*, 2000; Abdullah *et al.*, 2004; Khatri *et al.*, 2005; Gaur *et al.*, 2007; Barve *et al.*, 2009; Naeem-ud-Din *et al.*, 2009). Mutation as a phenomenon also facilitated basic research in structural as well as functional plant genomics. To realize their vast applications, several mutagens have been employed across the plant kingdom, among which irradiation has been widely and successfully employed (Song and Kang, 2003), with gamma-rays alone accounting for the development of 60% of mutant varieties (Kharkwal *et al.*, 2004).

Tossa jute (*Corchorus olitorius*; Malvaceae) is an economically important phloem fibre crop of India and Bangladesh. Africa appears to be the primary centre of diversity of the species (Kundu, 1951; Singh, 1976; Edmonds, 1990) while India or the Indo-Myanmar region is a secondary centre of diversity (Kundu, 1951). As a result, the diversity of Indian tossa jute germplasm is narrow and deficient for supporting

[1] Email: shashigen@gmail.com

 Mutation Breeding, Genetic Diversity and Crop Adaptation to Climate Change (eds S. Sivasankar *et al.*)
DOI: 10.1079/9781789249095.0048

basic, strategic and applied research in the crop. Further, fibre quality improvement has diversified the industrial applications of the crop, which has been critical for the sustainability of jute-based industries. High lignin is the principal cause of coarseness in jute fibre, so a cultivar with low lignin is highly desirable. Similarly, cultivars that do not flower early are highly recommended to tap ecologically sustainable and economically remunerative jute–legume intercropping (Sarkar *et al.*, 2013). Moreover, tossa jute is one of the cultivated species where a classical linkage map is not available, largely due to the limited number of polymorphic traits in the germplasm. In order to address these challenges, the present study was undertaken with the objectives of inducing mutants by gamma-ray irradiation and studying the genotype-specific effects of gamma-rays on the frequency and spectrum of morphological mutants.

2 Materials and Methods

2.1 Induction of mutation and phenotyping

True-to-type seeds of tossa jute of cvs JRO 204 and JRO 8432 were used for inducing mutation by gamma-rays. A total of 200 uniform and dry seeds of each cultivar were irradiated by exposing them to 100, 200, 300 and 400 Gy of gamma-rays from a ^{60}Co source in the gamma cell at the National Botanical Research Institute (NBRI), Lucknow, India. The treated seeds, along with their respective controls, were sown immediately to grow the M_1 generation. Different biological parameters such as germination and survival of plants were recorded and expressed as a percentage of the control. Randomly marked plants for each treatment in the M_1 generation were harvested separately.

To develop the M_2 generation, seeds of randomly selected plants of the M_1 generation were space-planted in the field in three replications. The M_2 generation was screened for lethal chlorophyll mutations during the first 4 weeks after germination. Viable chlorophyll and other macro-mutants were scored throughout the crop duration according to the procedure given by Gustafsson (1947), with suitable modifications.

2.2 Microscopic anatomy of fibre stem

The bark of the promising tossa jute mutants was peeled off from the central woody core of the basal stem. Transverse sections of these bark samples were cut (free-hand) with a sharp blade (7 o'clock, Super Platinum). The sections were mounted in water, covered by a cover slip and sealed with wax. Staining was done with Maule reagent that specifically stained syringyl lignin monomers (Dashek, 1997). Observations were made under a Zeiss Axioskop 40 (Carl Zeiss, Jena, Germany) microscope with bright field illumination and images were recorded with a Canon Power Shot A80 camera system.

3 Results and Discussion

Determining the dose for mutagen treatment is critical for the success of a mutation breeding programme. Using too high a mutagen dose results in higher frequency of deleterious mutations that negatively influence agronomic values of selected mutants. On the other hand, a very low dose of mutagen decreases the frequency of mutations and consequently a large population of mutants in the M_2 is necessary to find desirable mutants. Hence, the key for success of mutation breeding lies in establishing a balance between the frequency of desired and deleterious mutants. This brings forth the concept of LD_{50} to determine the optimal dose for a breeding programme.

3.1 Sensitivity of tossa jute cvs JRO 204 and JRO 8432

The M_1 seed was considered to have germinated if the seedling achieved the first vegetative leaf stage (attainment of autotrophic status of the embryo). Control seeds of both the cultivars showed 100% germination and attainment of the autotrophic stage. This measure was used because irradiated seeds of tossa jute have shown germination comparable to the control but deviated during attainment of the first vegetative leaf stage. The finding is in conformity with Ghosh and Sen (1978), Brock (1965) and Dubinin (1961), who identified the attainment of autotrophic status as the best indicator for radiosensitivity assessment in this species. The cause for this mutagenic effect has not been explained but might be due to the injurious effect of gamma radiation on the metabolic makeup regulating the development and differentiation of the quiescent embryo in jute seeds.

Shoot growth rate was adversely affected in gamma-ray irradiated seeds in comparison with controls (Table 48.1 and Fig. 48.1). In parallel to rice (Talebi and Talebi, 2012) and finger millet (Ambavane *et al.*, 2015), the deleterious effect of the mutagen on plant growth rate was negatively associated in an almost linear fashion. In the M_1 generation, the control plants had small glossy and folded leaves which subsequently expanded into typical ovate-lanceolate leaves with fine serrations, appendages and long petiole. Among the irradiated plants a few had deformed leaves, slowly giving rise to irregularly lanceolate, oblique, thick, dark leaves with a wavy margin without a regular shape and outline. The percentage of

Table 48.1. Frequency of plants with deformed leaves and chimeras in the M_1 generation.

Treatment (kR)	Number of seeds sown	Survival as % of control	M_1 plants with deformed leaves (%)	Plants with chimeras (%)
JRO 8432				
10	200	71	1.39	0
20	200	57	1.89	0.21
30	200	49	2.56	0.98
40	200	40	3.11	1.4
JRO 204				
10	200	64	0	0
20	200	55	1.54	0.49
30	200	41	3.15	1.01
40	200	29	4.65	2.89

(1 kR = 10 Gy)

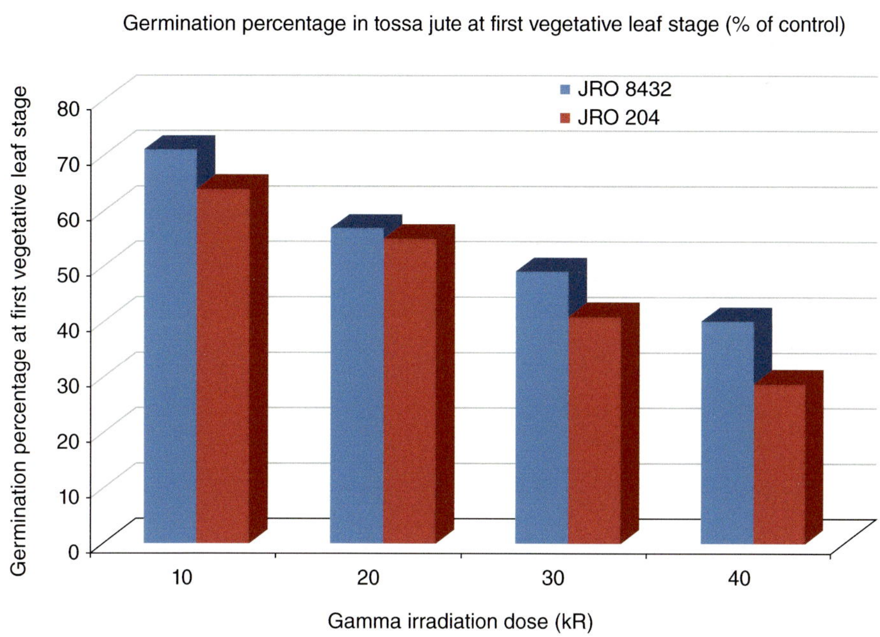

Fig. 48.1. Germination rates as a function of irradiation intensity.

seedlings with deformed leaves increased with gamma-ray doses in both the cultivars.

3.2 Critical dose of gamma-ray for jute cvs JRO 204 and JRO 8432

The mutagenic effect depends on genetic background of the treated material. Therefore, a comparative study of somatic and genetic effects induced by a range of gamma-ray doses in the two genotypes is required. For this purpose, the frequency of chlorophyll mutants among M_2 generation seedlings was recorded (Table 48.2), as suggested by Åke Gustafsson (1940) for barley. This is mainly because the mutation frequency in genes controlling chlorophyll gives a convenient number for counting. Lethal chlorophyll mutants in jute seldom grow beyond the cotyledonary stage. Hence, they were scored immediately after germination. Chlorophyll mutants were absent from the control plant. Their frequency increased with increasing doses of gamma-rays in both the varieties. This is in agreement with earlier reports in French bean (Blixt, 1961), black gram (Kumar *et al.*, 2007), cowpea (Dhanavel *et al.*, 2008) and finger millet (Ambavane *et al.*, 2015). The spectrum of chlorophyll mutants induced by gamma-rays in both varieties included *albino*, *xantha*, *alboviridis* and *viridis* types. Among these four mutant types, *xantha* was predominant in both cultivars, suggesting that genes for xanthophyll development are readily available for mutagenic action. Similar observations were made by Reddy and Annadurai (1991) in lentil. The highest frequency of chlorophyll mutants in the M_2 generation was 4.29 with a gamma-ray dose of 300 Gy in JRO 204, while the maximum number of chlorophyll mutants was recorded in JRO 8432 at a dose of 400 Gy. Hence the critical doses of gamma radiation for JRO 204 and JRO 8432 for large-scale mutation breeding programmes are estimated to be 300 Gy and 400 Gy, respectively.

3.3 Spectrum and frequency of macro-mutants

In both cultivars, gamma-rays induced a wide range of macro-mutants (Table 48.3) that were selected in the M_2 or more advanced generations. Most of these mutants were fertile, and viable seeds were harvested after selfing to maintain their genetic purity. The maximum mutation spectrum in both the cultivars was recorded at 300 Gy. Some of the outstanding mutants identified from the pedigreed mutant population of JRO 204 and JRO 8432 are presented in Fig. 48.2.

Most of the mutants exhibited additional phenotypic changes; for example, the 'epileaf' and 'nonabs' mutants had slower plant growth and reduced height besides having epinastic and non-abscising leaf, respectively. Normal flowering and pod development were recorded in these mutants. These phenotypic changes may be attributed to pleotropy or cryptic mutation

Table 48.2. Frequency and spectrum of chlorophyll mutations in the M_2 generation of jute cvs JRO 204 and JRO 8432.

Dose (kR)	Number of plants in M_2 generation	Number of mutants in M_2 generation	Number of different types of chlorophyll mutants				Chlorophyll mutation frequency (%)
			Albina	*Xantha*	*Alboviridis*	*Viridis*	
JRO 204							
10	6000	68	0	100	0	0	1.13
20	5150	145	0	37	0	73	2.82
30	4100	176	68	32	0	0	4.29
40	2800	103	32	41	13	14	3.68
JRO 8432							
10	6700	39	0	92	0	18	0.58
20	6000	44	60	0	40	0	0.73
30	4500	56	0	100	0	0	1.24
40	3250	101	23	51	12	14	3.11

(1 kR = 10 Gy)

Table 48.3. Spectrum and frequency of novel mutants in the M_2 plants of tossa jute.

Progenitor variety	Dose	No. of plants in M_2	Number of morphological mutants: Dwarf	Soft stem	Hard stem	Twisted bark	Non-abscission leaves	Palmate leaves	Round pod	Smooth pod	Total frequency
JRO 204	10	6000	3	–	–	1	–	11	1	–	1.01
	20	5150	7	5	–	5	–	16	13	–	1.15
	30	4100	16	10	9	3	–	19	6	–	1.23
	40	2800	2	14	2	–	–	–	2	–	1.14
JRO 8432	10	6700	0	–	–	–	–	–	–	–	1.04
	20	6000	6	3	6	4	–	9	8	–	2.1
	30	4500	2	1	11	8	6	12	2	–	2.09
	40	3250	–	–	4	7	1	11	–	3	2.18

Fig. 48.2. Phenotypic mutants screened from JRO 8432 and JRO 204: **(a)** chlorophyll deficient (chld), **(b)** serrated leaf, **(c)** smooth pod (smopod), **(d)** round pod (ropod) and **(e)** non-abscission (nonabs).

(Thakare *et al.*, 1973). Earlier, in tossa jute, an induced mutant with an enhanced abscission rate had been reported by Sen (1968). Since there has been no detailed study of the abscission process in the genus, the development and subsequent study of the delayed abscission mutant may be of use, particularly under stress conditions.

Further, a super dwarf mutant (*sdf*) with stunted growth was identified. The mean height of the mutant was 0.3 m in contrast to 3.2 m for the wild-type. Internodes of the mutant were highly condensed and leaves had a prominent pulvinus. The dwarf mutant flowered early with profuse flowers but the number of viable seeds per pod was low. Paria and Basak (1994) reported an induced mutant with mean height of 0.8 m. A stem mutant called low lignin phloem (*llpf*) was characterized by undulated growth of stem. The mean plant height of the *llpf* mutant is less than for the wild-type and its retting duration, under identical conditions, is one-third of that of the wild-type. Interestingly, the mutant was found to be deficient in phloem cell-wall lignification, as revealed from Maule reagent staining (Fig. 48.3).

Considerably high lignin in jute fibre limits its applications in low-value industrial applications such as sack preparation. Development of a low lignin fibre producing mutant (Choudhary *et al.*, 2017) will serve as an important source parent in fibre quality improvement programmes.

Similarly, we screened for early-flowering resistant mutants (*pfr*) under short-day conditions (sown in February) and the performance of superior mutants is presented in Fig. 48.4. In contrast to the wild-type, these mutants are significantly delayed in flowering even under early sowing conditions and so have a prolonged vegetative phase. These mutants can be directly useful for the jute–legume intercropping system.

4 Conclusion

Overall, gamma-rays induced a large number of morphological and developmental mutants

Fig. 48.3. Phenotype of the *low lignin phloem fibre* (*llpf*) mutant. **(a)** Wild-type. **(b)** JRO 204, at 65 days after sowing. Transverse section of the basal portion of the stem of *llpf* mutant **(c)** and JRO 204 **(d)** stained with Maule reagent ($KMnO_4$ - HCl) to reveal lignin deposition. PFBs, phloem fibre bundles; Ca, cambium.

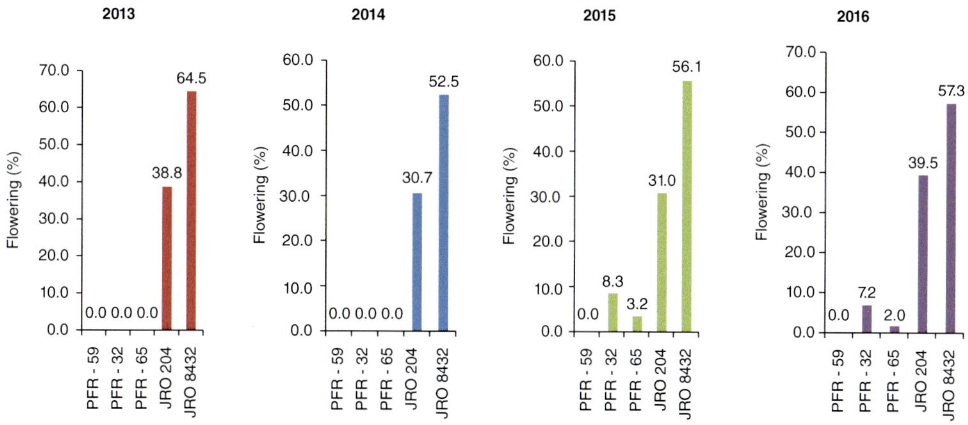

Fig. 48.4. Flowering percentage in selected *pfr* mutants compared with wild-type(s) under short-day conditions (February sowing).

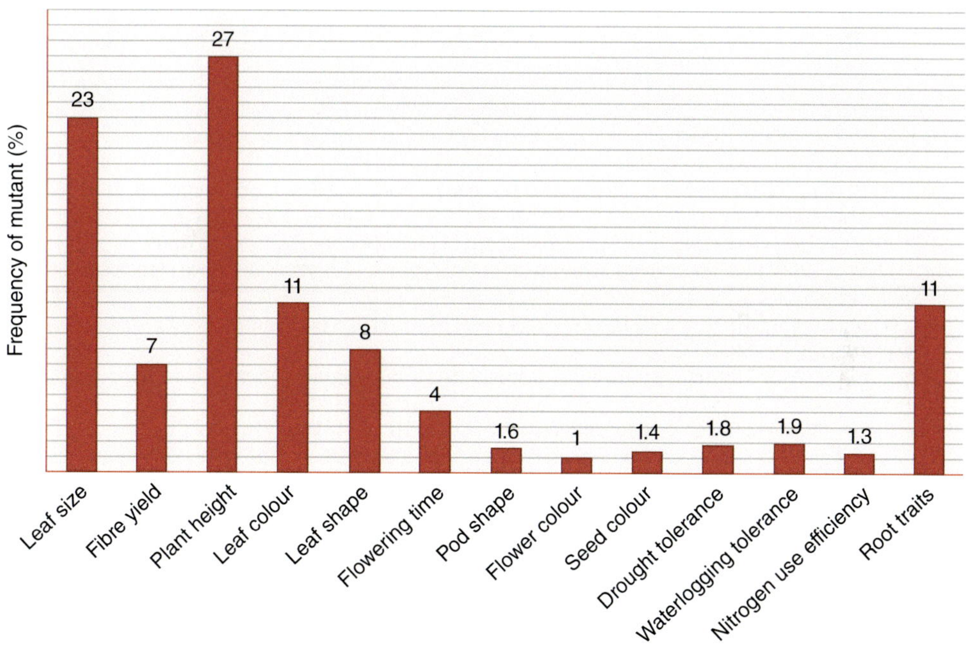

Fig. 48.5. Trait-specific frequency distribution of gamma-ray induced mutants.

in both cultivars (Fig. 48.5) and these will be of academic as well as agronomic importance. These mutants will help plant breeders aiming at the improvement of fibre yield and quality. Genomic tools can be used to exploit these resources for enhanced understanding of developmental biology of this crop.

Acknowledgement

All support rendered by the Director, ICAR-CRIJAF, Barrackpore, during the research is highly acknowledged. Financial assistance for the work was received in the form of a research grant (35/14/01/2014-BRNS) from BRNS, Department of Atomic Energy, Govt of India.

References

Abdullah, K. *et al.* (2004) Evaluation of high yielding mutants of *Brassica juncea* cv. S-9 developed through gamma rays and EMS. *Pakistan Journal of Botany* 37, 279–284.

Ambavane, A.R. *et al.* (2015) Studies on mutagenic effectiveness and efficiency of gamma rays and its effect on quantitative traits in finger millet (*Eleusine coracana* L. Gaertn). *Journal of Radiation Research and Applied Sciences* 8(1), 120–125.

Barve, Y.Y. *et al.* (2009) Induced mutations for development of *B. juncea* canola quality varieties suitable for Indian agro-climatic conditions. In: Shu, Q.Y. (ed.) *Induced Plant Mutations in the Genomics Era.* Joint FAO/IAEA Division of Nuclear Techniques in Food and Agriculture, International Atomic Energy Agency, Vienna, pp. 373–375.

Blixt, S. (1961) Quantitative studies of induced mutation in peas V, chlorophyll mutations. *Agriculture, Horticulture and Genetics* 5(3), 503–515.

Bolbhat, S.N. and Dhumal, K.N. (2009) Induced macromutations in horsegram [*Macrotyloma uniflorum* (Lam.) Verdc]. *Legume Research* 32(4), 278–281.

Brock, R.D. (1965) Response of *Trifolium subterraneum* to X-rays and thermal neutrons. *Radiation Botany* 5, 543–555.

Choudhary, S.B. *et al.* (2017) Morphological, histobiochemical and molecular characterisation of low lignin phloem fibre (llpf) mutant of dark jute (*Corchorus olitorius* L.). *Applied Biochemistry and Biotechnology* 83(3), 980–992.

Das, A. *et al.* (2000) Induction of mutations for heat tolerance in potato by using *in vitro* culture and radiation. *Euphytica* 114, 205–209.

Dashek, W.V. (1997) *Methods in Plant Biochemistry and Molecular Biology*. CRC Press, Boca Raton, Florida.

Dhanavel D. *et al.* (2008) Effectiveness and efficacy of chemical mutagens in cowpea (*Vigna unguiculata* (L.) Walp). *African Journal of Biotechnology* 7(22), 4116–4117.

Dubinin, N.P. (ed.) (1961) *Problems of Radiation Genetics*. Oliver and Boyd, Edinburgh, pp. 163–187.

Edmonds, J.M. (1990) *Herbarium survey of African Corchorus L. species*. Systematic and Ecogeographic Studies on Crop Genepools 4. International Board of Plant Genetic Resources, Rome.

Gaur, P.M. *et al.* (2007) An induced brachytic mutant of chickpea and its possible use in ideotype breeding. *Euphytica* 159(1–2), 35–41.

Ghosh, N. and Sen, S. (1978) Assessment of radio sensitivity from the germination of Jute seeds. *Biologia Plantarum* 20(4), 268–273.

Gustafsson, Å. (1940) The mutation system of the chlorophyll apparatus. *Acta Universitatis Lundensis* 36(11), 1–40.

Gustafsson, Å. (1947) Mutation in agricultural plants. *Hereditas* 33, 1–100.

Javed, M.A. *et al.* (2000) Utilization of gamma irradiation for the genetic improvement of oriental mustard (*Brassica juncea* Coss.). *Pakistan Journal of Botany* 32, 77–83.

Kharkwal M.C. *et al.* (2004) Mutation breeding for crop improvement. In: Jain, H.K. and Kharkwal, M.C. (eds) *Plant Breeding – Mendelian to Molecular Approaches*. Narosa Publishing House, New Delhi, India, pp. 601–645.

Khatri, A. *et al.* (2005) Evaluation of high yielding mutant of *Brassica juncea* cv. S-9 developed through gamma rays and EMS. *Pakistan Journal of Botany* 37(2), 279–284.

Kumar, A. *et al.* (2007) Induced mutagenesis in blackgram [*Vigna mungo* (L.) Hepper.]. *Indian Journal of Genetics and Plant Breeding* 67(1), 41–46.

Kundu, B.C. (1951) Origin of jute. *Indian Journal of Genetics and Plant Breeding* 11, 95–99.

Naeem-ud-Din *et al.* (2009) High yielding groundnut (*Arachis hypogea* L.) variety ‘Golden’. *Pakistan Journal of Botany* 41(5), 2217–2222.

Paria, P. and Basak, S.L. (1994) Inheritance of some induced mutant characters in Jute (*Corchorus olitorius* L.). *Indian Journal of Genetics and Plant Breeding* 54(4), 347–350.

Reddy, V.R.K. and Annadurai, M. (1991) Induction of mutations in pigeonpea. *Journal of Cytological Genetics* 26, 31–37.

Sarkar, S. *et al.* (2013) Strip-cropping of legumes with jute (*Corchorus olitorius*) in jute-paddy-lentil cropping system. *Journal of Crop and Weed* 9(1), 207–209.

Sen, S. (1968) An induced mutant of *Corchorus olitorius* L. with enhanced abscission rate. *Annals of Botany* 32, 863–866.

Singh, D.P. (1976) Jute. In: Simmonds, N.W. (ed.) *Evolution of Crop Plants*. Longman, London and New York, pp. 290–291.

Song, H.S. and Kang, S.Y. (2003) Application of natural variation and induced mutation in Breeding and Functional Genomics: papers for International Symposium; Current Status and Future of Plant Mutation Breeding. *Korean Journal of Breeding Science* 35(1), 24–34.

Talebi, A.B. and Talebi, A.B. (2012) Radiosensitivity study for identifying the lethal dose in MR219 (*Oryza sativa* L. spp. Indica Cv. MR219). *International Journal of Agricultural Science, Research and Technology* 2(2), 64–67.

Thakare, R.G. *et al.* (1973) Induced viable mutations in *Corchorus olitorius*. *Indian Journal of Genetics* 33, 204–228.

49 Effects of Light and UV-C Radiation on the Transcriptional Activity of *COP1* and *HY5* Gene Homologues in Barley

Irina Boycheva, Ralitsa Georgieva, Lubomir Stoilov and Vasilissa Manova[1]

Abstract

Photomorphogenic regulators COP1 (Constitutive Photomorphogenic 1) and HY5 (Elongated Hypocotyl 5) play a key role in plant development by guiding the transition from dark to light growth. In *Arabidopsis* they are also implicated in the transcriptional control of photolyase genes. Here we characterize the transcript abundance of *COP1* and *HY5* gene homologues in barley in relation to light-grown conditions and UV-damage response. Etiolated and green 6-day-old seedlings were UV-C irradiated and exposed to light or kept in darkness. The abundance of barley *COP1* and *HY5* transcripts was assessed by real-time RT-PCR. In etiolated leaves we found several-fold lower levels of *COP1* transcripts which reached the levels of the green ones after 1 h of light exposure. Barley *HY5* transcripts were very low in the dark-grown seedlings and after 1 h of illumination they increased drastically to levels significantly exceeding those measured in the green leaves. Both genes were upregulated by light in the irradiated plants as well, but to a lesser extent compared with their controls, probably due to the presence of non-repaired DNA damage in the etiolated leaves soon after irradiation. The enhanced transcription of barley *COP1* under light is unexpected in view of the well-known function of COP1 as a negative regulator of plant photomorphogenesis but conforms to the positive role reported for AtCOP1 in UV-B signalling. HY5 is recognized as a stimulator of light-inducible genes and our data support such a role for the barley *HY5* homologue as well. Our study shows that, in barley seedlings, the regulation of *COP1* and *HY5* gene expression is achieved through light-positive transcriptional modulation, suggesting that both genes contribute to the de-etiolation phase in barley. According to our knowledge, this is the first quantitation of the *COP1* and *HY5* mRNAs in barley that also regards the UV-damage response of this crop.

Keywords: gene expression • *COP1* • *HY5* • barley • ultraviolet radiation

1 Introduction

The UV-B portion of the solar spectrum is an important environmental signal that is able to modify the growth and development of plants (Jenkins, 2017). On the other hand, it is damaging for all cellular components and DNA is a primary target of short-wave UV radiation. Cyclobutane pyrimidine dimers (CPDs; covalently linked adjacent pyrimidine residues) may account for up to 90% of all lesions induced by UV-B (Dany *et al.*, 2001). Even one CPD can completely block the transcription of any gene. This might affect the functionality not only of the cell but also of the whole plant, decreasing its growth and productivity. Plants, as other

[1] Email: vmanova@bas.bg

DOI: 10.1079/9781789249095.0049

living organisms, have evolved DNA repair pathways to remove these lesions. The classical dark repair mechanisms function through replacement of the damaged nucleotides by new ones, making the repair prone to mistakes. Plants are also 'happy' to have a light-dependent error-free repairing mechanism, termed photoreactivation. As regards CPD, photoreactivation is executed by a CPD photolyase enzyme that binds to the dimer, absorbs photon from light and uses its energy to split the dimer in an error-free manner (Bray and West, 2005; Manova and Gruszka, 2015).

In our previous work we have shown that the stage of photomorphogenic development influences the proficiency of young barley seedlings to repair the UV damage and therefore may impact their ability to survive and grow under UV-stress conditions. In agreement with the DNA repair results, the transcript levels of the barley *CPD* photolyase gene were several-fold lower in etiolated leaves. However, the transcripts increased rapidly to levels significantly exceeding those contained in the green plants within 1 h of light exposure (Manova *et al.*, 2009; Manova *et al.*, 2016). To better understand the molecular basis of UV tolerance in this crop, we have further focused on the genes that are presumably engaged in the regulation of barley's UV damage response.

In plants, the photomorphogenic regulators COP1 (Constitutive Photomorphogenic 1) and HY5 (Elongated Hypocotyl 5) have been found to play important but contrasting roles in development during the transition from skotomorphogenic (dark) to photomorphogenic (light) growth. In *Arabidopsis*, COP1 is a key negative regulator of photomorphogenesis, and as such has been extensively characterized. COP1 protein functions as a RING finger E3 ubiquitin ligase in highly sophisticated multimeric protein complexes. AtCOP1 is located in the nucleus in the dark and as a result the levels of many light-responsive proteins (including AtHY5) are kept low by its ubiquitin ligase activity. Under white light, conversely, COP1 is transferred to the cytoplasm and is thus inactivated with respect to its targets in the nucleus. The lack of AtCOP1 in the nucleus allows accumulation of the bZIP transcription factor AtHY5, which is a stimulator of plant de-etiolation and activates many photomorphogenesis-promoting genes under light-growth conditions (Osterlund *et al.*, 1999; Lau and Deng, 2012; Gangappa and Botto, 2016).

Data show, however, that both genes are positively engaged in plant response to short-wave UV radiation. Under low UV-B levels, AtCOP1 functions as a crucial transcriptional stimulator of *AtHY5* which activates UV-B responsive and pro-photomorphogenesis acting genes (Oravecz *et al.*, 2006). In addition, COP1 interaction with UVR8 (the only plant UV-B photoreceptor known so far) supports the nuclear localization of the latter and is important for mediating the UV-B signalling (Favory *et al.*, 2009; Yin *et al.*, 2016). *COP1* and *HY5* are implicated not only in the UV-B driven plant growth regulation but also in the UV-B and UV-C damage response in *Arabidopsis*, as well as in the transcriptional regulation of plant photolyase genes under dark/light conditions (Oravecz *et al.*, 2006; Castells *et al.*, 2010; Li *et al.*, 2015). That is why we were interested to study the transcriptional activity of barley *COP1* and *HY5* gene homologues during seedling de-etiolation under normal conditions as well as after exposure to strong UV stress. Our results reveal that the expression levels of both *COP1* and *HY5* genes in the leaves of young barley seedlings are obviously modulated by the light-grown conditions. On the other hand, the UV-C radiation has a reducing influence on the light-promoted transcription of the two barley genes in the etiolated leaves, possibly associated with the time needed for repair of the DNA lesions during the early recovery periods.

2 Materials and Methods

2.1 Experimental material, UV-C irradiation and recovery

Spring barley (*Hordeum vulgare* L.) variety 'Freya' was used as experimental material. Seedlings were grown on moist filter papers and vermiculite at 25°C for 6 days in complete darkness (etiolated) or exposed to visible daylight (natural day/night cycle) for 1 day before irradiation (green). Plants were irradiated with UV-C light (5 kJ/m^2) in BLX 254 UV crosslinker (Life Technologies). The seedlings were incubated for recovery under high-intensity solar light or under dark conditions. The intensity of photosynthetically active radiation (PAR) was ca. 1000 μmol/m^2/s, the UV-B irradiance (280–315 nm) was glass filtered to ca. 0.05 W/m^2. Light measurements were performed by Photo-radiometer HD.2302.0 (Delta Ohm). Air-conditioning was used to ensure constant temperatures of 24–26°C for the light-recovered seedlings. Post-radiation dark incubation was performed in a thermostat at 25°C. The unfolded part of the first leaf of control and irradiated seedlings was excised immediately (0 h) or after the recovery period as specified in the text, up to 25 h later. Material was wrapped in aluminium foil and frozen in liquid nitrogen for subsequent RNA isolation.

2.2 RNA extraction, reverse transcription, primer design and real-time PCR analysis

Frozen material (ca. 100 mg) from the 6-day-old UV-C irradiated and control barley seedlings was ground in liquid nitrogen and total RNA was extracted with the GeneMATRIX Universal RNA Purification Kit (EURx, Poland) and with RNeasy plant mini kit (Qiagen). The RNA elution step was performed twice for each sample; the integrity of RNA was analysed by neutral agarose gel-electrophoresis; afterwards the eluates were combined. RNA concentrations were determined spectrophotometrically (BioSpec-nano, Shimadzu) and 1 μg of each DNA-free RNA sample was used as template for reverse transcription reactions carried out with random hexamer primers to produce cDNA with the First Strand cDNA Synthesis Kit (Thermo Scientific), following the manufacturer's instructions. The products of the reverse transcription were diluted 1:3 with ultrapure water and 3 μl of the diluted cDNA was used for the qPCR. The PCR reaction mixture was prepared according to the manufacturer's instructions (Maxima SYBR Green qPCR Mater Mix (x2), Thermo Scientific) for 10 μl of each reaction. The RT-qPCR was carried out in a PikoReal Real-Time PCR System (Thermo Scientific). The amplification conditions were: initial denaturation for 10 min at 95°C, 45 cycles of denaturation for 15 s at 95°C, annealing for 30 s at 59°/60°C, and extension for 1 min at 72°C followed by melting curve analysis. All the data were analysed using the PikoReal Software version 2.2 (Thermo Scientific). The fold changes in the expression levels of barley *COP1* and *HY5* genes were estimated applying the Pfaffl method (Pfaffl, 2001). The following primers were constructed to match the 3′ end of the barley *COP1* gene to amplify a short fragment of 138 bp: HvCOP1_F-GGACAATTGCCCACTAAGGA and HvCOP1_R-GGTTTTGAGATAGCCTTGTGGTA. Primers HvHY5_F-CCTCAAGAGGTTGCTGAGGA and HvHY5_R-CAGGATCTGTCGGAGCATCT were utilized for the amplification of 172 bp fragment of barley *HY5* sequences. Barley 18S ribosomal RNA fragment (118 bp) was used as internal control; the primers (HV_18Ssh1F CCTGCGGCTTAATTTGACTCA and HV_18SR AACTAAGAACGGCCATGCAC) were described previously (Manova *et al.*, 2016). All PCR primers were designed with the online-based program 'Primer3: WWW primer tool' (http://biotools.umassmed.edu/primer3/primer3web_input.htm, accessed 2019) (Rozen and Skaletsky, 1999).

3 Results

3.1 *In silico* identification of barley *COP1* and *HY5* coding sequences

Barley homologous sequences were identified through BLAST analysis utilizing the coding sequences of *COP1* and *HY5* genes from *Oryza sativa* and *Brachypodium distachyon* available in the National Center for Biotechnology Information (NCBI) database. The *Hordeum vulgare* full-length mRNA (2443 bp), accession no. AK374819 from cultivar 'Haruna Nijo', was found to be 84% and 88% identical to the rice and brachypodium *COP1* mRNAs at the nucleotide sequence level, respectively. The predicted polypeptide of 663 aa (accession no. BAK06015.1) showed 100% coverage and high level of identity to the E3 ubiquitin-protein ligase COP1 in *O. sativa* (87%) and *Aegilops tauschii* (96%). Two conserved domains, the RING-HC_COP1 and the WD40, were identified in the predicted barley COP1 protein. Based on that, we concluded that accession no. AK374819.1 represented the coding sequence of the barley *COP1* gene. It was further utilized to search the databases on the IPK barley BLAST Server to identify the fragment (barke_contig_59913 CAJV010043233 carma= 6HL) containing the respective barley *COP1* genomic sequence. The same approach was applied to identify the barley *HY5* gene homologue. The complete 876 bp coding region (CDS) of clone NIASHv2034M24 (accession no. AK365526) showed high similarity to the predicted *HY5* sequences in *A. tauschii* (94%) and *B. distachyon* (82%). The predicted protein of 156 aa (accession no. BAJ96729) had a high level of identity to the HY5 transcription factor in *A. tauschii* (98%), *B. distachyon* (92%) and *O. sativa* (85%). The conserved basic leucine zipper (bZIP) domain characteristic for the *Arabidopsis thaliana* transcription factor HY5 was identified in the predicted barley HY5 protein. The two accessions (AK374819 and AK365526) identified *in silico* in the current study as the putative *COP1* and *HY5* homologues in *H. vulgare* had been previously released in the NCBI as part of Barley Full-Length cDNA libraries by Matsumoto *et al.* (2011). The alignment of the *in silico* identified coding and genomic sequences allowed localization of the exon-intron junctions necessary for designing primers suitable for gene expression analysis.

3.2 Expression analysis of barley *COP1* gene in etiolated and green seedlings, effect of light and UV-C radiation.

The influence of light-grown conditions on the transcriptional activity of the barley *COP1* gene, the changes in the transcript amounts upon the first light exposure of the etiolated seedlings and the effect of UV-C radiation on the accumulation of *COP1* mRNA were monitored under high-intensity light (up to 3 h) or after 24 h of dark incubation followed by an additional 1 h of light exposure. The expression pattern of the barley *COP1* gene in the control and UV-C exposed barley seedlings is presented in Fig. 49.1. Data show about tenfold less *COP1* mRNA in the dark-grown than in the light-grown 6-day-old plants. *COP1* transcript levels almost double after 30 min, and soon after 1 h of light exposure of the etiolated leaves they reach the steady-state levels of the green ones. UV-C treatment has an inhibiting effect on *COP1* transcript accumulation in the etiolated plants at the first hour after irradiation. In the green seedlings, a tendency for increase of *COP1* transcripts is observed in the UV-C irradiated leaves in comparison with the non-treated material at all time-points examined.

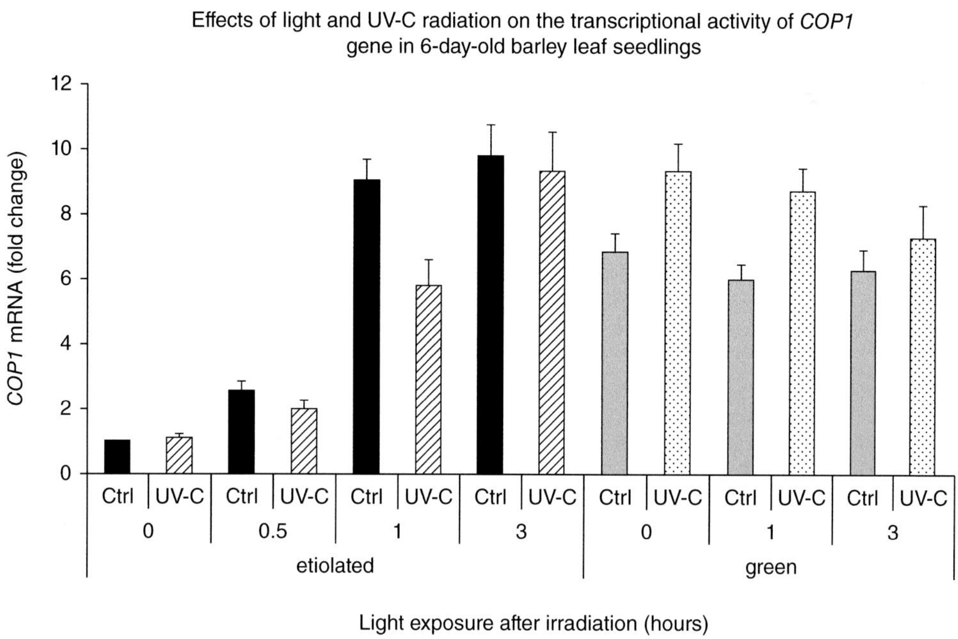

Fig. 49.1. Transcriptional profile of barley *COP1* homologue in control and UV-C treated etiolated and green seedlings exposed to high-intensity solar light. Error bars represent SE; etiolated (n = 4), green (n = 3).

The transcriptional pattern of *COP1* in the dark-recovered etiolated plants is shown in Fig. 49.2. It is evident that *COP1* levels do not change significantly during the full 24 h period both in the control and in the UV-C irradiated plants. Additional exposure for 1 h to visible light, however, immediately increases *COP1* transcripts and more strongly in the non-irradiated plants, confirming the differences between control and irradiated material at this time-point obtained in the light experiments (Fig. 49.1).

3.3 Transcriptional profile of *HY5* gene homologue in barley seedlings exposed to light and UV-C radiation

The transcript abundance of the barley *HY5* gene was investigated in the etiolated and green control and UV-C treated plants exposed to light for up to 3 h post-irradiation (Fig. 49.3).

The *HY5* mRNA amounts were very low in the 6-day-old plants, grown in complete darkness. In fact, the green seedlings, exposed to light for 24 h before the experiment, contained more than 50-fold higher amounts of the *HY5* transcript. The *HY5* transcripts accumulated very rapidly after exposure of etiolated leaves to light, especially in the control plants. More than a 20-fold change could be detected within the first 30 min. However, a drastic increase in *HY5* expression (ca. 400-fold change) was found at the 1 h time-point. As a result, the *HY5* gene transcript levels in the etiolated seedlings quickly exceeded, by several-fold, those measured in the green plants. After the peak at 1 h, *HY5* gene expression decreased 3 h later. Again, there was a significant reduction (ca. 50%) of the newly synthesized *HY5* transcripts in the UV-C treated etiolated plants. Nevertheless, with the exception of the first 30 min of light exposure after the irradiation, *HY5* transcript levels measured in the irradiated etiolated leaves were comparable or even higher than those found in the irradiated green leaves. An increase in *HY5* expression was measured in the green leaves subjected to UV-C irradiation immediately (0 h) after the treatment; however, such an effect was not evident at the other time-points.

4 Discussion

In *Arabidopsis* the DNA photorepair mechanisms and the photomorphogenic stage of plant development are tightly interconnected, ensuring plant survival after UV-C stress. More interestingly, data suggest that the molecular changes

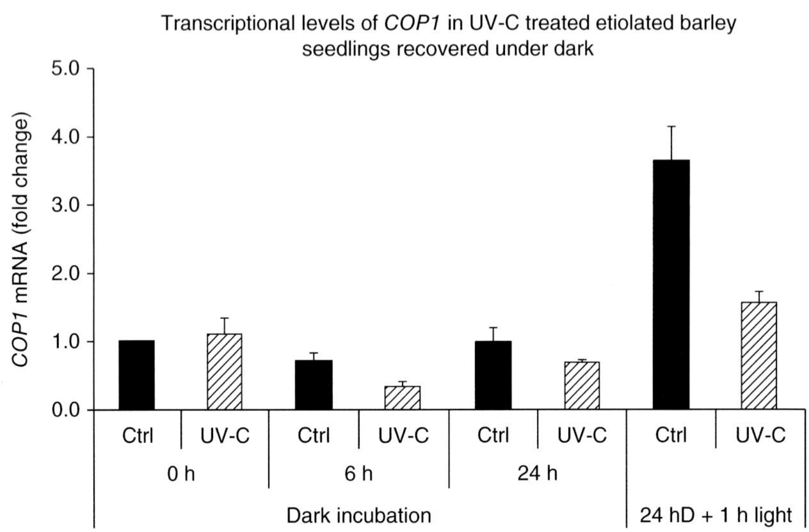

Fig. 49.2. Transcript levels of barley *COP1* gene in control and UV-C treated etiolated seedlings incubated under dark for 24 h followed by 1 h exposure to solar light. Error bars represent SE ($n = 3$).

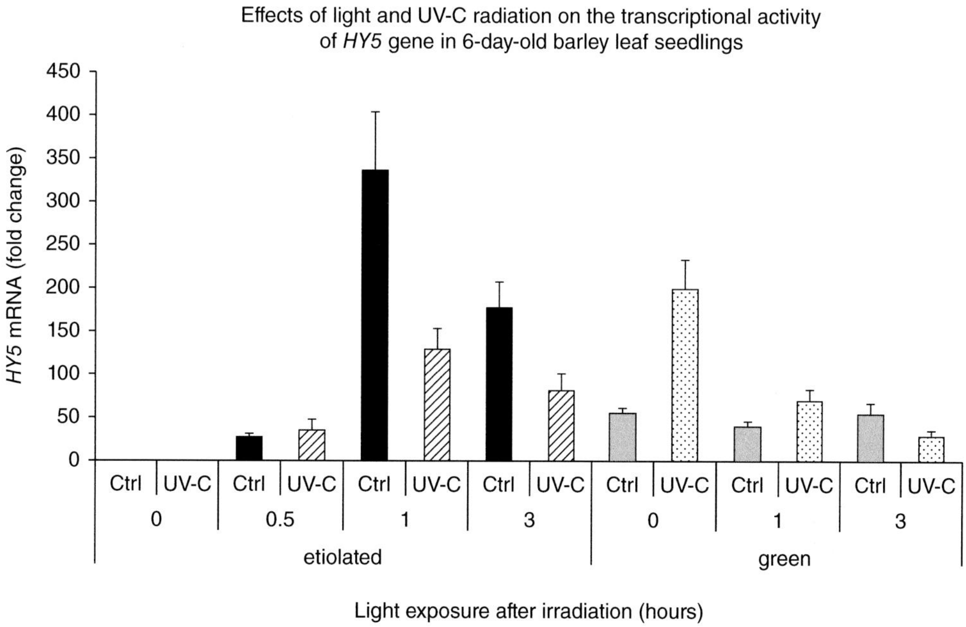

Fig. 49.3. Transcript abundance of barley *HY5* homologue in control and UV-C irradiated etiolated and green seedlings exposed to high-intensity solar light. Error bars represent SE; etiolated (n = 4), green (n = 3).

occurring during the transition stage from dark- to light-guided growth, associated with the action of photomorphogenic regulators such as COP1 and HY5, might actually stimulate the plant's DNA repair capacity (Castells *et al.*, 2010). The delayed CPD elimination from etiolated barley leaves found in our previous work suggests that photomorphogenesis influences DNA photorepair potential in barley and hence may impact its UV tolerance, especially at the stage of early seedling development (Manova *et al.*, 2016). In this respect, it is of particular interest to explore the genetic control of barley DNA repair mechanisms during photomorphogenesis and its connection with the activity of photomorphogenic regulators such as COP1 and HY5 in particular. As a first step we aimed to characterize the transcriptional profiles of *COP1* and *HY5* genes in light- and dark-grown barley seedlings and to study the modulation of their expression that may occur in plants after exposure to light. In addition, we were interested to determine the transcriptional pattern of both genes after UV-C stress, a condition when the requirements of plants for photorepair activities increase.

As there was no information about *COP1* and *HY5* genes in barley, it was necessary to search the available databases and find barley sequences homologous to *COP1* and *HY5* mRNAs in closely related species. The already released full-length barley cDNA sequences (Matsumoto *et al.*, 2011) as well as the sequence assembly of barley genome (International Barley Genome Sequencing Consortium, 2012) allowed *in silico* identification of *COP1* and *HY5* gene homologues in this crop. Our BLAST data suggest that barley *COP1* and *HY5* sequences are highly similar at the DNA and protein levels to their respective homologues in other cereals. In addition, *COP1* and *HY5* seem to be present as single copy genes in the barley genome, unlike, for example, the UV-B photoreceptor *UVR8*, which might exist in many copies in some species, including barley (Fernández *et al.*, 2016; under investigation).

The results presented here show that *COP1* mRNA amount in dark-grown barley seedlings is several-fold lower than in the green leaves; and more remarkably, it is positively controlled by light. In addition, it seems that *COP1* transcript abundance in the etiolated plants kept in darkness after irradiation is stable during the 24 h follow-up under dark and rapidly increases only upon their exposure to light. Thus our data suggest that, if there is no post-transcriptional regulation of *COP1* mRNA abundance, the regulation of *COP1* expression in barley seedlings is achieved mainly through light-positive transcriptional modulation. This is in contrast with the mechanism of *COP1* regulation in *Arabidopsis* and pea, where the *AtCOP1* transcript levels and the amounts of COP1 protein in pea do not differ between dark- and light-grown seedlings (Deng *et al.*, 1992; Zhao *et al.*, 1998). As such results were absolutely unexpected, we performed a more detailed search of the literature and found two early studies strongly supporting our data. Indeed, in rice, *COP1* activity was found to be phytochrome mediated, dependent on the tissue type and stage of plant development as well as being positively influenced by light, all of which are indicative of an important role of COP1 in the development of monocots (Raghuvanshi *et al.*, 2001; Tsuge *et al.*, 2001). Barley is a field crop that grows under high intensities of PAR and UV irradiance of solar light; in accordance it was shown that high PAR was particularly important for improvement of overall photoprotective capacity in barley (Klem *et al.*, 2015) as well as for the efficient removal of CPD from its genome (Manova *et al.*, 2016). Thus, it might be anticipated also that UV signalling would have a more substantial role in the development of plants adjusted to high-intensity solar light. In this respect, although our results appear contradictory at first glance, they may in fact reflect key specificities in the regulation of photomorphogenic development that depend on environmental light conditions or that might exist between species with different light requirements.

AtHY5 is a transcription factor, which associates with the promoters of light-inducible genes and thus directly controls the growth of plants under light conditions (Heijde and Ulm, 2012). Our data show that, similarly to COP1, the expression of HY5 in barley seedlings is also regulated by light at the transcript abundance level. The data for the barley *HY5* homologue reveals that the mRNA levels of the gene are high in green seedlings but are very low in etiolated ones. This pattern changes after exposure to high intensity light. The transcriptional activation of the barley *HY5* homologue observed in our study involves a drastic increase in amounts of *HY5* mRNA within a very short time (ca. 1 h) after the first light perception by the etiolated seedlings. Such significant (several hundred-fold) upregulation of the barley *HY5* gene homologue in etiolated seedlings is in complete accordance with the high demand for HY5 during de-etiolation and hence with its very important role as an initiator of photomorphogenesis in barley.

The impact of UV-C irradiation on the transcript abundance of barley *COP1* and *HY5* genes seems to be complicated. First, our data showed that UV-C treatment affected transcript accumulation negatively in the etiolated leaves, mainly at the time-points when maximum upregulation was detected in the respective non-treated controls, especially in the case of the *HY5* gene. The UV-C dose utilized here induced on average about 1 CPD per 3 kb of DNA in the barley genome, and such a high damage frequency affects potentially any gene (Manova *et al.*, 2016). Etiolated barley leaves repair CPDs more slowly than the green ones, which are able to remove half of the damage shortly after irradiation (Manova *et al.*, 2016). This means that, as the etiolated leaves contain a large number of DNA lesions within the first hour after irradiation, they are more sensitive to UV-C induced inhibition of transcription. Accordingly, our result is a good, although indirect, example of how a damaged template may inhibit gene expression, and more importantly shows that the genes that are in a high transcriptionally active state would be affected to a larger extent than the less active ones. Indeed, green barley seedlings, having steady-state levels of *COP1* and *HY5* expression, seem to be unaffected by the damaging influence of UV-C. Moreover, there is a tendency for an increase of their transcript abundance, which requires more detailed characterization.

5 Conclusion

This study shows that in barley seedlings undergoing de-etiolation the expression of photomorphogenic regulators COP1 and HY5 is positively controlled by light at the transcript level. The effect of UV-C radiation on the expression of barley *HY5* and *COP1* genes in the etiolated leaves is rather negative and correlates with the lower CPD repair potential of dark-grown barley seedlings.

The transcriptional responses of barley *COP1* and *HY5* genes to light presented in this study, supported also by previously published data in rice, suggest that our understanding of the mechanisms underlying seedling photomorphogenesis in these important crops is still far from clear. Based on the knowledge obtained so far in other species, this developmental phenomenon definitely deserves further investigation.

Acknowledgements

This work was supported by the IAEA national TC Project Bul5/014 'Evaluation of wheat and barley germplasm stress response by nuclear, omics and physiological approaches' and the IAEA RC BUL15481 'Isolation and characterization of genes for radiation induced DNA damage repair in barley'.

References

Bray, C.M. and West, C.E. (2005) DNA repair mechanisms in plants: crucial sensors and effectors for the maintenance of genome integrity. *New Phytologist* 168, 511–528.

Castells, E. *et al.* (2010) *det1-1*-induced UV-C hyposensitivity through *UVR3* and *PHR1* photolyase gene over-expression. *Plant Journal* 63, 392–404.

Dany, A.-L. *et al.* (2001) Repair of the main UV-induced thymine dimeric lesions within *Arabidopsis thaliana* DNA: evidence for the major involvement of photoreactivation pathways. *Journal of Photochemistry and Photobiology B: Biology* 65, 127–135.

Deng, X.W. *et al.* (1992) *COP1*, an *Arabidopsis* regulatory gene, encodes a protein with both a zinc-binding motif and a Gβ homologous domain. *Cell* 71, 791–801.

Favory, J.J. *et al.* (2009) Interaction of COP1 and UVR8 regulates UV-B-induced photomorphogenesis and stress acclimation in *Arabidopsis*. *EMBO Journal* 28, 591–601.

Fernández, M.B. *et al.* (2016) A comprehensive phylogeny reveals functional conservation of the UV-B photoreceptor UVR8 from green algae to higher plants. *Frontiers in Plant Science* 7, 1698.

Gangappa, S.N. and Botto, J.F. (2016) The multifaceted roles of HY5 in plant growth and development. *Molecular Plant* 9, 1353–1365.

Heijde, M. and Ulm, R. (2012) UV-B photoreceptor-mediated signalling in plants. *Trends in Plant Science* 17, 230–237.

International Barley Genome Sequencing Consortium (2012) A physical, genetic and functional sequence assembly of the barley genome. *Nature* 491, 711–717.

Jenkins, G.I. (2017) Photomorphogenic responses to ultraviolet-B light. *Plant, Cell & Environment* 40, 2544–2557.

Klem, K. *et al.* (2015) Ultraviolet and photosynthetically active radiation can both induce photoprotective capacity allowing barley to overcome high radiation stress. *Plant Physiology and Biochemistry* 93, 74–83.

Lau, O.S. and Deng, X.W. (2012) The photomorphogenic repressors COP1 and DET1: 20 years later. *Trends in Plant Science* 17, 584–593.

Li, N. *et al.* (2015) UV-B-induced CPD photolyase gene expression is regulated by UVR8-dependent and -independent pathways in Arabidopsis. *Plant and Cell Physiology* 56, 2014–2023.

Manova, V. and Gruszka, D. (2015) DNA damage and repair in plants – from models to crops. *Frontiers in Plant Science* 6, 885.

Manova, V. *et al.* (2009) Genomic and gene-specific induction and repair of DNA damage in barley. In: Shu, Q.Y. (ed.) *Induced Plant Mutations in the Genomics Era*. Joint FAO/IAEA Division of Nuclear Techniques in Food and Agriculture, International Atomic Energy Agency, Vienna, pp. 133–136.

Manova, V. *et al.* (2016) Efficient removal of cyclobutane pyrimidine dimers in barley: differential contribution of light-dependent and dark DNA repair pathways. *Physiologia Plantarum* 158, 236–253.

Matsumoto, T. *et al.* (2011) Comprehensive sequence analysis of 24,783 barley full-length cDNAs derived from 12 clone libraries. *Plant Physiology* 156, 20–28.

Oravecz, A. *et al.* (2006) Constitutively photomorphogenic1 is required for the UV-B response in *Arabidopsis*. *Plant Cell* 18, 1975–1990.

Osterlund, M.T. *et al.* (1999) The role of COP1 in repression of *Arabidopsis* photomorphogenic development. *Trends in Cell Biology* 9, 113–118.

Pfaffl, M.W. (2001) A new mathematical model for relative quantification in real-time RT-PCR. *Nucleic Acids Research* 29, e45.

Raghuvanshi, S. *et al.* (2001) Isolation and molecular characterization of the *COP1* gene homolog from rice, *Oryza sativa* L. subsp. *Indica* var. Pusa Basmati 1. *DNA Research* 8, 73–79.

Rozen, S. and Skaletsky, H. (1999) Primer3 on the WWW for General Users and for Biologist Programmers. In: Misener, S. and Krawetz, S. (eds) *Methods in Molecular Biology, Vol. 132 Bioinformatics Methods and Protocols*. Humana Press, Totowa, New Jersey, pp. 365–386.

Tsuge, T. *et al.* (2001) Phytochrome-mediated control of *COP1* gene expression in rice plants. *Molecular Genetics and Genomics* 265, 43–50.

Yin, R. *et al.* (2016) COP1 is required for UV-B–induced nuclear accumulation of the UVR8 photoreceptor. *Proceedings National Academy of Sciences* 113, E4415–E4422.

Zhao, L. *et al.* (1998) Molecular cloning and sequencing of the cDNA of *cop1* gene from *Pisum sativum*. *Biochimica et Biophysica Acta – Gene Structure and Expression* 1395, 326–328.

Index

Note: Page numbers in **bold** type refer to **figures**
Page numbers in *italics* type refer to *tables*